Color Atlas of Cancer Cytology

Color Atlas of Cancer Cytology

Third Edition

Masayoshi Takahashi, M.D., Ph.D., F.I.A.C.

Formerly Professor and Chairman
Department of Pathology
Gifu University School of Medicine
Gifu, Japan

Chief Pathologist
St. John's Sakuramachi Hospital
Koganei, Tokyo, Japan

Consultant Pathologist
Nerima-ku Medical Association
Tokyo, Japan

IGAKU-SHOIN Tokyo·New York

Published by

IGAKU-SHOIN Ltd.
5-24-3 Hongo, Bunkyo-ku, Tokyo 113-8719

IGAKU-SHOIN Medical Publishers, Inc.
60 Madison Avenue, New York, N.Y. 10010

ISBN 4-260-14348-4

Printed and bound in Japan

Preface to the Third Edition

Nearly twenty years have passed since the second edition was published. In these years, extraordinary development of medical electronics, computer science, molecular biology and information technology has expanded the field of clinical cytology.

Routine utilization of imaging diagnosis by computerized tomography, especially with thin-section tomograms and magnetic resonance has enabled fine needle aspiration to obtain specimens from minute lesions in deep organs with great accuracy. The thyroid, thymus, salivary glands, pancreas, liver and prostate are organs which can be reached by fine needle aspiration. Newly chosen chapters on these organs were targeted on illustrative interpretations with their histological bases. The lung, breast, and endometrium in the previous edition were described for the most part on exfoliative cytology. I attempted, in this volume, to add diagnostic information for cytology specimens taken either by needle aspiration or with a scraping device.

Basic aspects of general cytology, such as preparation, staining, traditional classification, and criteria of malignancy were omitted. The author greatly appreciates the contribution of the most important applications of immunocytochemistry by Dr. Robert M. Nakamura and Dr. Gerald M. Bordin, DNA quantitation and cell cycle analysis by Dr. Richard J. Clatch, and biological prognostic interpretation of lymphoma and allied diseases by Professor Thomas M. Grogan.

I am sure that molecular biology will play an important future role for cells obtained by needle aspiration. In this volume, genetic analysis of cytology specimens has been focused only on interphase cells which can be directly examined without specialized installations or sophisticated techniques. Fluorescence in situ hybridization (FISH) is applicable for interphase cells and may provide profound biological observations, such as chromosome and gene translocations and oncogene overexpressions. X chromatins and Y chromatins were left out, because these are easily studied in interphase cytogenetics.

The female genital tract, a major part of this volume, was extensively revised. Interpretations were based on the Bethesda System in comparison with traditional classification or terms. Endometrial cytology discussions were expanded, because the uterine corpus has become as important as the uterine cervix in the diagnosis of malignancies.

Urinary cytology is a unique subject, because subsequent specimens can be obtained. My special interest is in the cytological analysis of urinary sediments with use of supravital stain and with differential interference microscopy. The validation of routine urinalysis can be done to obtain information on renal glomerular and intrinsic tubular disorders.

I have also been involved with automated cytology for more than twenty years at the Medical School of Gifu University. In addition, I have participated in the U.S.-Japan Cytology Automation Project. I am very grateful to Professor George L. Wied, Drs. Yoshio Tenjin, Noboru Tanaka and Shoichi Inoue for their advice and assistance. It is regretful that automated cytology has not been discussed in this book. This area of automated cytology will advance in the future.

The author has discussed briefly the practice of telecytology/telepathology transmitting accurate digital images of cytology and histology, regardless of distance and time difference. The hybrid system, in particular, will be available in real-time intraoperative diagnosis which may be worthy for guidance of surgical procedures. Offline teleconsultation and teleconference may come to be of routine practice.

Finally, I gratefully express acknowledgement to Professor Leopold G. Koss for his hearty encouragement from the first edition up to the present. Many thanks to Professors Kunio Mizuguchi, Toshiro Kawai, and Masahiko Fujii for participation with the third edition. Special thanks are owed to the editorial and production staff of the publisher, Igaku-Shoin Ltd., Tokyo for their patience, help, and cooperation with complex revisions.

July 2000, Tokyo

Masayoshi Takahashi, M.D.

Preface to the First Edition

Papanicolaou's technique has come to be widely accepted as a diagnostic tool for early cancers in various organs of the body. Despite the facts that roent-genological and endoscopic diagnoses are limited to tumors of macroscopic size and that exfoliative cytology enables us to detect intraepithelial carcinoma which is absolutely curable, the first two methods still prevail clinically.

Professor Dr. Rudolf Virchow, the father of "Cellular Pathology" has left the famous saying, "Omnis cellula e cellula." Even a single cancer cell will be biologically malignant as a single cancer cell transplantation is experimentally successful, and this smallest unit may possess the distinctive features that fulfill the criteria of malignancy. Precise recognition of cellular alterations is the basis of exfoliative cytology.

Generally speaking, pathologists who are accustomed to observing structural atypism such as abnormality in the arrangement of cells, invasive growth and/or vascular permeation may hesitate to diagnose the malignancy of single cells without histologic features. Physicians, on the other hand, cannot afford the time for microscopic studies in the laboratory. Ordinarily, the task of cytology is carried out through at least three persons: the physician who must collect sufficient specimens; the cytotechnologist who must prepare smears and screen them; and the cytopathologist who is responsible for the conclusion of the diagnosis. In other words, cytodiagnosis is like a relay race where none of the runners is allowed to swerve from his course. There should be a textbook that is acceptable by these three experts. At first, this book was intended to be an atlas illustrated with real color figures as a manual for cytotechnologists. Meanwhile, the author was influenced by the excellent textbook, "Diagnostic Cytology and Its Histopathologic Bases" by Dr. Leopold G. Koss, and zealously resolved to extend the use of the book not only for technologists but for medical students and physicians.

Detailed laboratory procedures, their diagnostic values and their pathologic interpretations have been added to the atlas. Therefore the original Japanese edition in 1965 entitled "Color Atlas of Cancer Cytology" has been completely renewed in this English edition. In spite of the textbook-like content, the original name, "Color Atlas of Cancer Cytology" has remained unchanged. This book consists of two parts:

PART ONE deals with general cancer cytology; it comprises fundamental knowledge of cytology and the detailed laboratory techniques which are necessary for routine practice and occasionally available for clinical research work. As a result, some portions of this Part may include those facets unrelated to cancer cytology.

PART TWO concerns practical cytology that is applied to various organs and tissues. Each chapter is organized like a fascicle, beginning from the technique of preparing smears, and containing normal and atypical cytology with particular reference to histopathology. Consequently, any chapter on a specific organ can be understood with ease by those who are interested in that specific organ.

I am deeply indebted to Dr. B. Cornelis Hopman for his initial guidance to exfoliative cytology at the Jackson Memorial Hospital, University of Miami. I gratefully acknowledge the advice of Dr. Y. Chiba, Medical Director of the Central Hospital of the Japanese National Railways, and that of Prof. K. Hashimoto and Prof. N. Kosakai of Juntendo University. I wish to thank Prof. H. Katsuki, Prof. Y. Hayata, Dr. M. Tajima and Dr. E. Tsuboi for kindly presenting cases; Dr. Y. Tenjin, Dr. S. Noda, Dr. S. Kurita, Dr. N. Fukushima, Dr. Y. Sakai and Mr. A. Sato for the contributions of color figures; and the entire medical and laboratory staffs of the J. N. R. Hospital for their hearty support. Finally, I thank the staffs of Igaku-Shoin Ltd. for the efforts and courtesy with which they handled many problems that I asked them to solve.

April 1971, Tokyo

Masayoshi Takahashi, M.D.

Foreword to the Second Edition

It almost seems like yesterday that shortly after the publication of the first edition of this work I passed a large book store in front of the University Hospital in Vienna and saw it displayed prominently in the window, almost as if the owner of the shop was as proud of it as if he had written it himself.

During the process of preparing this revised and enlarged edition, Dr. Masayoshi Takahashi was kind enough to let me see the galley proofs. On the occasion of the first edition being awarded a prize for its contribution to the medical publishing in Japan I was called upon to deliver a speech, therefore I cannot help feeling a certain personal involvement. As I read the galley proofs, I began to be aware of a sensation of intense excitement. The reason was the growing realization that this is a book without peer in the world on the subject of cancer cytology.

The title of this book, Color Atlas of Cancer Cytology is one towards which I harbor a strong feeling of discontent. In other words I believe that I am among the many readers of the first edition who believe that this work should be renamed Textbook and Color Atlas of Cancer Cytology. By only employing the title Color Atlas, the impression that one receives is of a simple atlas and explanation, but this volume actually is based on an extremely meticulously written text and the fruits of Dr. Takahashi's famed research, which in turn allow the introduction of the atlas. Therefore the title should indicate that this book is essentially different from other books with similar titles.

The photography of the materials shown in the atlas section of the work, except for those with credits that indicate otherwise, was performed by Dr. Takahashi himself, which explains why the materials appear so alive. In this sense also this makes this volume not a mere atlas in the usual sense of the word, the specimens themselves almost appear to be contained within its pages.

This work represents a significant contribution to vital cytology by emphasizing the morphology and function of cancer cells and by thorough analysis of the appearance of cancer cells on the basis of observation by scanning electron microscopy and differential interference contrast microscopy. His evaluation of the diagnostic applicability of measuring cytochemical changes such as the deviation of isoenzymes and the appearance of tumor-associated antigens is also a remarkable achievement. Furthermore, he has included a new chapter using a fluorescence technique for the staining of Y-bodies in cancer of males. He has also added a new chapter on bone and bone marrow and expanded largely his first edition text chapters on skin diseases, lymphomas and brain tumors. Dr. Takahashi clarifies all the points of his text by color illustrations and by comparison with the histological background whenever necessary.

At the same time that I take deep pleasure in writing this introduction for a close friend and respected colleague, Dr. Takahashi, I would also like to express the hope that it will help those involved in the study of cytology to make further progress in the field.

August 31, 1981

Kazumasa Masubuchi, M.D., F.I.A.C. (HON.)
President, International Academy of Cytology
Vice-President, Cancer Institute Hospital, The
Japanese Foundation for Cancer Research, Tokyo

Contributors

Chapter 1

Robert M. Nakamura, M.D.

Chairman Emeritus, Department of Pathology
Scripps Clinic
La Jolla, California, U.S.A.

Gerald M. Bordin, M.D.

Department of Pathology
Scripps Clinic
La Jolla, California, U.S.A.

Chapter 2

Richard J. Clatch, M.D., Ph.D.

Department of Pathology
Lake Forest Hospital
Lake Forest, Illinois, U.S.A.

Chapter 7

Masahiko Fujii, M.D.

Professor, Department of Pathology
Kyorin University School of Health Sciences
Hachioji, Tokyo, Japan

Chapters 10 and 11

Toshiro Kawai, M.D.

Associate Professor, Department of Pathology
Jichi Medical School
Minami-kawachi, Tochigi, Japan

Chapter 13

Kunio Mizuguchi, M.D.

Professor, Department of Clinical Pathology
Mizonokuchi Hospital
Teikyo University School of Medicine
Kawasaki, Kanagawa, Japan

Chapter 18

Thomas M. Grogan, M.D.

Professor, Department of Pathology
University of Arizona School of Medicine
Tucson, Arizona
Chairman, Southwest Oncology Group (SWOG)
Lymphoma Biology Central Laboratory
Tucson, Arizona, U.S.A.

Contents

Part 1. General Cancer Cytology

Chapter 1
Principles of Diagnostic Immunocytochemistry................................3
Robert M. Nakamura and Gerald M. Bordin

1. Background and General Principles 3
   Introduction 3
   Glossary 3
   The avidin-biotin immunoenzyme method 4
2. Important Tissue Antigen in Diagnostic Pathology .5
   Intermediate filament proteins 5
   Actin 7
   CEA (Carcinoembryonic antigen) 7
   S100 protein 8
   Leukocyte common antigen (LCA) or T200 antigen 8
   Anti-neuroendocrine antibodies: anti-chromogranins .8
   Tumor-specific markers: anti-melanoma antibodies 9
   Antibodies to prostate acid phosphatase and prostate
     specific antigen 9
   Immunohistochemical localization of estrogen and
     progesterone receptors 9
   Antibodies to human chorionic gonadotropin 9
3. Immunohistochemical Profiles of Various
   Tumors 10
   Anaplastic tumors 10
   Lymphoid tumors 10
   Epithelial tumors 10
   Soft tissue tumors 10
   Germ cell tumors 10
4. Summary 10

Chapter 2
Cytologic DNA Content and Cell Cycle Analysis 13
Richard J. Clatch

1. Cell Cycle Basics 13
2. Laboratory Methodologies 14
   Flow cytometry 14
   Image analysis 15
   Laser scanning cytometric analysis 17
   Single parameter DNA content and cell cycle
     analysis 17
   Multiparameter DNA content and cell cycle analysis ..19

Chapter 3
X Chromatin 24

1. Configuration of X Chromatin 24
2. Significance of the X Chromatin Test 24
3. Hermaphroditism 24
4. Endocrine Intersex 25
5. Sex Reversals 26
   Testicular feminization 26
   Turner's syndrome 26
   Klinefelter's syndrome 26
6. Superfemale 27
7. The Technique of Sex Chromatin Determination .27
   Preparation of smears 27
   Staining procedure 28
   Results of the X chromatin test 29
8. X Chromatin in Cancer Cells 29
   Significance of X chromatin in prognosis 31
   Other nuclear appendages 31

Chapter 4
Y Chromatin 34

1. Configuration of Y chromatin 34
2. Evaluation of Y Chromatin Test 34
3. Technique of Y Chromatin Determination 35
   Preparation of smears 35
   Staining procedure 35
4. Y Chromatin in Cancer Cells 35

Chapter 5
Fluorescence In Situ Hybridization (FISH) 39

1. In Situ Hybridization on Human Chromosomes .40
2. Immunodetection of Probes 41
3. Procedure of In Situ Hybridization on Biopsy
   and Needle Aspirates 42
   Biopsy 42
   Needle aspirates 42
4. Selection of Probes 42
   HER-2/neu 42
   bcr/abl chimera gene 43
   PML/RARA 43
   AML1 43

Part 2. Practical Cytology of Organs

Chapter 6
Female Genital Tract 47
  1. Overview of Carcinoma of the Uterus 47
  2. Preparation of Smears 47
     Vaginal pool smear 47
     Cervical scraping smear 47
     VCE smear 48
     Endometrial aspiration 48
     Endometrial scraping smear 49
  3. Histology of the Uterus and Vagina 49
     Histology of the uterus 49
     Histology of the vagina 50
  4. Normal Cells in the Female Genital Tract 50
     Squamous cells 50
     Endocervical cells 52
     Endometrial cells 53
     Histiocytes 54
     Leukocytes 55
  5. Cyclic Hormonal Cytology in the Reproductive
     Period 55
     Cyclic hormonal secretion 55
     Cyclic cytologic changes 56
     Cytologic indices 58
  6. Cytology in Menopause 58
     Crowded menopause 58
     Advanced menopause 58
  7. Cytology before Puberty 59
     Pattern in newborn infant 59
     Pattern in childhood before puberty 60
  8. Pathological Conditions Affecting Cytohormonal
     Patterns 60
     Inflammation 60
     Nonhormonal medication 60
  9. Cytology in Hormonal Dysfunction 60
     Estrogen hyperactivity 60
     Estrogen hypoactivity 61
     Progesterone hyperactivity 61
     Androgen hyperactivity 62
 10. Cytology of Pregnancy and Its Allied Changes 62
     Cytohormonal pattern in pregnancy 62
     Cytology of pregnancy at term 63
     Cytology of the postpartum period 63
     Arias-Stella phenomenon 64
     Vernix caseosa cells 64
     Effects of contraceptives on cytology 64
 11. Benign Epithelial Cell Changes 66
     Tissue repair 66
     Squamous metaplasia 67
     Inflammatory cell changes 71
     Atrophic cervicitis with reactive changes 72
     IUD associated changes 72
     Folic acid deficiency with dysplasia-mimic changes ...73
     Vaginal flora and microorganism affections 73
     Sexually transmitted diseases (STD) 75
     Specific infections 78
     Nonorganic substances 78
 12. Squamous Intraepithelial Lesions (Borderline
     Lesions and Related Changes) 80
     Atypical squamous cells of undetermined significance
        (ASCUS) 82
     Historical review on classification of dysplasia ..82
     Koilocytotic atypia 82

     Low-grade squamous intraepithelial lesion
        (Mild dysplasia) 84
     High-grade squamous intraepithelial lesion (Moderate
        dysplasia–severe dysplasia–carcinoma in situ) ..85
     Dysplasia in pregnancy 86
     Condyloma 88
     Effect of immunosuppressive agent 88
 13. Carcinoma In Situ 88
     Concept of carcinoma in situ–With its historical
        review 89
     Management of detection of early carcinoma of
        the uterine cervix 90
     Histology of carcinoma in situ 91
     Cytology of carcinoma in situ 91
 14. Carcinoma of the Cervix 96
     Squamous cell carcinoma 96
     Verrucous carcinoma 96
     Glassy cell carcinoma 96
     Large cell carcinoma (Giant cell carcinoma) 100
     Small cell carcinoma (Apudoma) 100
 15. Adenocarcinoma and Precancerous Glandular
     Lesions 101
     Endocervical glandular atypia 101
     Adenocarcinoma in situ 101
     Invasive adenocarcinoma 102
     Combined carcinoma 102
     Extrauterine carcinoma 107
 16. Clear Cell Carcinoma 108
 17. Acute Radiation Cellular Changes 110
 18. Hyperplasia and Carcinoma of
     the Endometrium 111
     Key points of endometrial microscopy 112
     Endometrial hyperplasia 113
     Endometrial carcinoma 114
     Endometrioid adenocarcinoma 114
     Adenoacanthoma 115
     Squamous cell carcinoma 116
     Clear cell carcinoma 117
     Iatrogenic effect 117
 19. Chorionic Tumors 118
 20. Malignant Melanoma 120
 21. Lymphoid Hyperplasia and Lymphoma 121
     Chronic lymphocytic cervicitis 121
     Malignant lymphoma 122
 22. Uterine Sarcoma 122
     Müllerian mixed tumor 122
 23. Endometrial Stromal Sarcoma 124

Chapter 7
Breast 136
Masayoshi Takahashi and Masahiko Fujii
  1. Collection of Specimens and Preparation of
     Smears 136
     Nipple discharge 136
     Needle aspiration 136
     Imprint smears of biopsy specimens 137
  2. Histology of the Mammary Gland 137
  3. Benign Cellular Components in Breast Secretions
     and Aspirates 138
     Foamy cells 138
     Duct epithelial cells 139
     Myoepithelial cells 143

Squamous cells 144
4. Benign Diseases 144
   Acute mastitis 144
   Plasma cell mastitis (Mammary duct ectasia) 144
   Fibrocystic disease 144
   Intraduct papilloma (Intracystic papilloma) 149
   Fibroadenoma 150
   Adenoma 150
   Phyllodes tumor 151
   Foreign body granuloma 153
   Gynecomastia 153
5. Malignant Cells 153
   Key points of cytodiagnosis: How to approach
     the goal in fine needle aspiration cytology 153
   Duct carcinoma (Adenocarcinoma) 154
   Lobular carcinoma 154
   Intracystic papillary carcinoma 156
   Mucinous carcinoma (Colloid carcinoma) 156
   Tubular carcinoma 157
   Comedocarcinoma 157
   Medullary carcinoma 157
   Paget's disease 157
   Apocrine carcinoma 161
   Squamous cell carcinoma 161
   Lymphoma 162
6. Prognostic View 162
   Vaginal hormonal cytology management of
     advanced breast carcinoma 162
   Cytomorphological aspect 162
   Biochemical aspect 163

Chapter 8
Respiratory Tract 165

1. Steps of Pulmonary Cytology 165
2. Examination of Sputum 165
   Macroscopic observation 165
   Collection of sputum 165
   Cell concentration method 166
   Aerosol induced sputum 166
3. Selective Bronchial Scrapings 168
   Bronchoscopic cytology 168
   Selective bronchial curetting and brushing 168
4. Fine Needle Aspiration 169
   Cytodiagnosis 170
5. Histology and Normal Cell Components 171
   Histology of the respiratory tract 171
   Normal cells in the respiratory tract 172
   Differences in cytologic features between sputum
     and bronchial scrapings 175
6. Benign Epithelial Cell Changes 178
   Hyperplastic basal cells 178
   Squamous metaplastic cells 178
   Atypical bronchial and bronchiolar hyperplastic
     cells 180
7. Retrogressive Epithelial Cell Changes 184
   Ciliocytophthoria 184
   Multinucleation 184
   Pigmentation 184
   Epithelial changes in viral infection 186
8. Anormal Changes of Histiocytic Cells 186
   Lipophagocytes 186
   Hemosiderin-laden macrophages 186
   Melanin-laden macrophages 186
   Bile pigment-laden macrophages 186
   Multinucleated giant cells 186
   Epithelioid cells and Langhans giant cells 186
9. Non-Cellular Elements 187

Curschmann's spirals 187
Fiber-like substance 187
Calcific concretions 187
Corpora amylacea 188
Asbestos body 188
Proteinaceous material 189
Charcot-Leyden crystals 189
Elastic fiber 190
Other non-cellular elements and infection 190
10. Cytological Typing of Lung Cancer 194
   Squamous cell carcinoma (Epidermoid carcinoma) ..194
   Adenocarcinoma 194
   Bronchiolo-alveolar carcinoma 197
   Small cell carcinoma 199
   Large cell carcinoma 203
   Clear cell tumor 207
   Malignant lymphoma 207
   Combined carcinoma and dual carcinoma 210
   Pulmonary blastoma and carcinosarcoma 210
   Carcinosarcoma 211
   Pulmonary hamartoma 212
11. Cytopathologic Changes in Heavy Smokers 212
12. Early Carcinoma and Carcinoma In Situ 213
   Goal of cytodiagnosis 213
   Cytologic pictures of early carcinoma 213
13. Bronchial Carcinoid and Bronchial Gland
     Carcinoma 215
   General aspect 215
   Cytopathology 216
14. Primary Sarcoma 218
15. Metastatic Carcinoma 218
   Renal cell carcinoma 219
   Hepatoma (Hepatocellular carcinoma) 220
   Breast carcinoma 220
   Thyroid carcinoma 220
   Other malignant tumors 220
16. Radiation Changes 222

Chapter 9
Thyroid 229

1. Introduction 229
2. Anatomy and Histology 229
3. Procedure of Thyroid Needle Aspiration 230
4. Adenomatous Goiter (Nodular Goiter) 230
5. Chronic Thyroiditis (Hashimoto's Disease) 230
6. Subacute Thyroiditis
     (Granulomatous Thyroiditis) 231
7. Graves' Disease (Basedow's Disease) 231
8. Follicular Adenoma 231
9. Thyroid Carcinoma 233
10. Follicular Carcinoma 233
11. Papillary Carcinoma 234
12. Anaplastic Carcinoma 236
13. Medullary Carcinoma 236

Chapter 10
Thymus and Mediastinum 238
Masayoshi Takahashi and Toshiro Kawai

1. Thymocyte Maturation 238
2. Tumors of the Thymus and Mediastinum 238
   Thymoma 239
3. Carcinoid Tumor 242

Contents

Chapter 11
Salivary Glands 243
Masayoshi Takahashi and Toshiro Kawai

1. Procedure of Aspiration and Preparation 243
2. Pleomorphic Adenoma 243
3. Warthin's Tumor (Adenolymphoma)
   (Papillary Cystadenoma Lymphomatosum) 243
4. Myoepithelioma 246
5. Acinic Cell Carcinoma 246
6. Adenoid Cystic Carcinoma 247
7. Basal Cell Adenoma 247
8. Mucoepidermoid Carcinoma 247

Chapter 12
Esophagus and Intestines 249
ESOPHAGUS 249
1. Collection of Specimens and Preparation of
   Smears 249
   Esophageal washings 249
   Esophageal brushings 249
   Capsulated sponge method 249
2. Histology of the Esophagus 250
3. Benign Cellular Components 250
4. Benign Lesions 251
   Esophagitis (Reflux esophagitis) 251
   Hyperplasia (Prickle cell hyperplasia) 251
   Herpetic esophagitis 252
   Cytomegalovirus infection 252
   Candidiasis 252
5. Precancerous Lesions of the Esophagus 253
6. Malignant Cells 253
   Squamous cell type (Epidermoid type) 254
   Glandular type 254
   Adenoacanthoma 255
   Small cell carcinoma 256
7. Malignant Melanoma 256
SMALL AND LARGE INTESTINES 256
1. Collection of Specimens and Preparation of
   Smears 256
   Choledocho-pancreatic duct cytology 256
   Colon cytology 257
2. Histology of the Intestine and Its Benign Cellular
   Components 257
   Histology of the intestine 257
   Benign cellular components 257
   Noncellular components 257
3. Benign Atypical Cells 260
   Chronic ulcerative colitis 260
   Adenomatous polyp 260
4. Malignant Cells 260
   Carcinoma of biliary tract and pancreatic duct 260
   Carcinoma of intestines 262

Chapter 13
Liver, Bile Ducts and Pancreas 264
Kunio Mizuguchi

LIVER 264
1. Collection of the Specimen and Preparation of
   the Smears 264
   Procedure of liver aspiration 264
   Preparation of the aspirated material 264
2. Normal Cell Components 264
3. Non-neoplastic Lesions 265
   Abscess including amebic abscess 265
   Cyst 265
   Hydatid cyst 265
4. Benign Tumors and Tumorous Lesions 266
   Regenerative nodule 266
   Focal nodular hyperplasia 266
   Liver cell adenoma 266
   Hemangioma 266
   Inflammatory pseudotumor 266
   Angiomyolipoma 266
5. Malignant Tumors 266
   Hepatocellular carcinoma 266
   Cholangiocellular carcinoma (CCC) 267
   Hepatoblastoma 268
   Other rare tumors 270
   Metastatic tumors 270
GALLBLADDER AND
EXTRAHEPATIC BILE DUCTS 271
   Cells and material 271
PANCREAS 272
1. Diagnostic Procedures 272
2. Pancreatic Tumors 272
   Duct cell carcinoma 272
   Acinar cell carcinoma 274
   Islet cell carcinoma 274
   Undifferentiated carcinoma 274
   Miscellaneous 274

Chapter 14
Effusions in Body Cavities 277
1. Preparation of Smears 277
   Centrifugation method 277
   Cell block method 277
   Membrane filter method 277
   Thin layer preparation method 278
2. Macroscopic Examination 279
   Transudate 279
   Exudate 279
3. Histology of Mesothelium and Benign Cellular
   Components from Serous Cavities 279
   Histology of mesothelium 279
   Benign cellular components 280
4. Effusions in Benign Pathological Conditions 292
   Eosinophilia in effusions 292
   Acute suppurative inflammatory effusions 292
   Acute non-suppurative inflammatory effusions 292
   Tuberculous effusions 292
   Effusions secondary to liver cirrhosis 293
   Effusions secondary to congestive heart failure 293
5. Malignant Effusions 293
   General consideration for malignant effusions 293
   Features of malignant cells 295
6. General Cellular Pattern in Carcinomas of
   Various Organs 297

Carcinoma of the stomach 297
Carcinoma of the colon 297
Carcinoma of the ovary 297
Specific tumors of the ovary 298
Carcinoma of the lung 305
Carcinoma of the breast 308
Carcinoma of the liver 308
Leukemia and malignant lymphoma 310
Multiple myeloma 314
Malignant mesothelial cells 314
Rare tumors 317

Chapter 15
Urinary Tract 326

1. General Aspect of Urinary Cytology 326
   Routine urinalysis and cytology 326
   Gross appearance of urine 327
   Sensitivity dependent on cell typing and efficacy of
     serial urinary cytology 327
2. Collection of Specimen 327
3. Preparation of Smears 328
   Routine centrifugation method 328
   Centrifugation after pretreatment 329
   Irrigation method 329
   Modified method with proteolytic enzyme saline
     solution 329
4. Staining Technique 329
   Sternheimer-Malbin stain 329
   Sternheimer stain 329
   Iodine stain 330
   Acridine orange stain 330
   Instant Giemsa stain 330
5. Differential Interference Microscopy for
     Urinalysis 330
6. Polarized Light Illumination for Urinalysis 330
7. Standardization of Urinalysis 330
   NCCLS recommendation 331
8. Nephron and Its Derivatives 331
   Histology and function of the nephron 331
   Tubular epithelial cells 331
   Cast formation and pathognomonic evaluation 332
9. Urinary Tract and Its Benign Cellular
     Components 337
   Histology of the urinary tract 337
   Benign cellular components in urine 338
10. Benign Atypical Cells 342
    Urolithiasis 342
    Benign tumor cells 342
    Tuberculosis 343
11. Malignant Cells 345
    Staging and grading 345
    Transitional cell carcinoma 346
    Squamous cell carcinoma 353
    Adenocarcinoma 353
    Renal cell carcinoma 354
    Nephroblastoma (Wilms' tumor) 356
    Malignant lymphoma 356
    Sarcoma 358
12. Cytomegalic Inclusion Disease 358
13. Polyomavirus Infection 358
14. *Schistosoma Haematobium* and Carcinogenesis 358
15. Cytoplasmic Inclusion Bodies 360
16. Malakoplakia 360
17. Comet Cells 361
18. Glitter Cells in Urine 361

Clinical significance of glitter cells 362
19. Allograft Rejection Cytology 362
    Rejection of renal transplantation 362
    Cytologic profile of urine after renal
      transplantation 363
20. Radiation Cystitis 364

Chapter 16
Prostate 368

1. Structure of Prostatic Glands 368
2. Collection of Specimens 368
   Prostatic massage 368
   Prostatic needle aspiration 369
3. Prostatic Cytology 369
   Value of prostatic cytology 369
   Prostatitis 369
   Nodular hyperplasia 370
   Carcinoma 371
   Therapeutic effect 373
   Stilbestrol therapy 373

Chapter 17
Lymph Node 374

1. Collection of Specimens and Preparation of
     Smears 374
   Technique 374
2. Histology of Lymph Nodes 374
3. Normal Cellular Components 375
   Lymphocytes 376
   Prolymphocytes (Medium-sized lymphocytes,
     Small cleaved cells) 377
   Lymphoblasts 377
   Large lymphoid cells 377
   Reticulum cells 377
   Plasma cells 378
4. T Cell and B Cell Population of Lymphocytes 378
   Identification of human B cells 378
   Identification of human T cells 379
   Cytochemical property as a T cell marker 380
   Clinical application 381
5. Blastic Transformation of Lymphocytes by
     Mitogens 381
   Clinical application 382
   Procedure 382
6. Malignant Lymphoma 383
   General concept on classification 383
   Classification of lymphomas based on
     immunological cell typing by Lukes and Collins ...383
   Follicle center cell-derived follicular lymphoma 387
   Stem cell lymphoma (Undifferentiated malignant
     lymphoma) 389
   Burkitt's lymphoma 389
   Lymphocytic malignant lymphoma 390
   Histiocytic malignant lymphoma 391
   Hodgkin's disease
     (paragranuloma–granuloma–sarcoma)
     (lymphocytic predominance–nodular sclerosis–
     mixed cellularity–lymphocytic depletion) 391
   Histiocytic medullary reticulosis 392
7. Extranodal Lymphoma 393
   MALT lymphoma 393
8. Key Points in the Cytodiagnosis of Lymph Node
     Smears 393
   Detection of abnormal cells specific to the disease ..393
   Detection of abnormal patterns in differential cell
     counts (Lymphcytograms) 397

9. Leukemia as an Allied Disease of Lymphoma 397
    Acute lymphatic leukemia (ALL) 397
    Adult T cell leukemia-lymphoma (ATL) 397
    Chronic lymphatic leukemia (CLL) 397
    Sézary syndrome (Cutaneous T cell lymphoma) 397
    Hairy cell leukemia 400
    Immunoblastic lymphadenopathy 400
    Myeloma (Plasmacytoma) 403
10. Histiocytosis X 404

Chapter 18
Prognostic Markers in Malignant
Lymphoma 410
Thomas M. Grogan

1. Introduction 410
2. Refined Diagnosis 410
3. Phenotypic Markers of Prognosis 411
    Single institution, retrospective study 412
    Multi-institutional, prospective studies 413
    Malignant phenotypes suggest biologic principle 414
    Conclusion 414

Chapter 19
Bone and Bone Marrow 417

1. Osteogenic Sarcoma 417
2. Chondrosarcoma 417
3. Chondroblastoma 417
4. Giant Cell Tumor 421
5. Rhabdomyosarcoma 421
6. Bone Marrow Tumor 421
7. Histiocytosis X 422

Chapter 20
Central Nervous System 425

1. Collection of Specimens and Preparation of
    Smears 425
    Collection of cerebrospinal fluid 425
    Centrifugation method 425
    Membrane filtration method 426
    Sedimentation technique 426
    Aspiration cytology 427
    Imprint smear 427
    Squash method 427
2. Gross Appearance of Cerebrospinal Fluid 428
3. Benign Cellular Components in Cerebrospinal
    Fluid 428
    Non-tumorous pleocytosis 428
    Cell counting in cerebrospinal fluid 428
4. Cellular Features of Benign Cells 429
    Ependymal cells and choroidal cells 429
    Pia-arachnoid cells 430
    Astrocytes 430
    Oligodendrocytes 430
    Microglia 430
    Nerve cells 430
    Macrophages 430
5. Abnormal Cytology 430
    Structure profile diagnostic of brain tumors 431
6. Metastatic Carcinoma 432
7. Glioma 432
    Astrocytoma and glioblastoma multiforme 435
    Medulloblastoma 438
    Ependymoma 438
    Oligodendroglioma 438
8. Pinealoma 440
9. Pituitary Adenoma 443
10. Meningioma 443
    Classification 445
11. Neurinoma (Schwannoma) 445
12. Ganglioglioma and Ganglioneuroblastoma 445
13. Chordoma 445
14. Dermoid Cyst and Craniopharyngioma 445
15. Neurocutaneous Melanosis and Meningeal
    Melanoma 446
16. Cerebral Paragonimiasis Westermani 446
17. Retinoblastoma 446

Chapter 21
Skin and Its Related Mucous
Membrane 449

1. Collection of Specimens and Preparation of
    Smears 449
2. Immunofluorescence Dermatologic Cytology 451
3. General Aspect of Cutaneous Cytology 451
4. Histology of the Skin and Normal Cellular
    Components 451
5. Abnormal Cytology 451
    Viral infection 452
    Pemphigus vulgaris 453
    Disorders of horny and granular layers 454
    Carcinoma 456
    Melanoma 458
    Cutaneous Lymphoma 459
    Dermal lesions 459
    Tumors of the epidermal appendages 462

Chapter 22
Telepathology and Telecytology 464

1. Telepathology/Telecytology Systems 464
2. Telepathology 464
3. Telecytology 464

Index 467

Part 1
General Cancer Cytology

1. Principles of Diagnostic Immunocytochemistry
2. Cytologic DNA Content and Cell Cycle Analysis
3. X Chromatin
4. Y Chromatin
5. Fluorescence In Situ Hybridization (FISH)

1 Principles of Diagnostic Immunocytochemistry

Robert M. Nakamura and Gerald M. Bordin

1 Background and General Principles

Introduction

Immunochemistry was initiated by Dr. Albert Coons et al in 1941.[8] He demonstrated the presence of pneumococcal antigen in tissues by the use of fluorescent antibody. Since that time, there has been considerable progress. The fluorescent antibody method was adapted for localization of immunoglobulins, complement, and other antigens in immunologically mediated diseases such as immune glomerulonephritis. More recently, the use of newer non-fluorescent immunohistochemical methods has led to significant advances in unravelling the pathogenesis of many diseases.

The application of immunohistochemistry has also resulted in the development of new diagnostic criteria for more specific and meaningful analysis of tissue specimens. The clinical use of immunohistochemistry has clearly revolutionized the daily practice of surgical pathology and cytology.[13,14,48,68,71,77]

Today, many specific monoclonal antibodies are available for use in various cellular, tumor tissue and infectious agent assays for classification. Immunohistochemical methods are especially popular with pathologists when studies can be performed on fixed, paraffin-embedded tumor tissues. However, immunohistochemical assays are also applicable to unstained, fixed cell smears, cytocentrifuge prepared slides, and cell block material obtained for cytologic evaluation. In many laboratories, immunohistochemical assays have become routine and even automated for evaluation and differential diagnosis of human tumors.[24] Various immunohistochemical and immunocytochemical assays are currently used commonly for assays as varied as estrogen and progesterone receptors for breast cancer[69] in excisional and fine needle aspiration biopsies of breast, aberrant p53 expression in human tumors,[59,72,80] rapid analysis of acute leukemias,[26] and DNA ploidy and study of proliferative cell nuclear antigen (PCNA).[26,28,32,37,42]

This review will discuss some of the important principles of immunochemical methods with examples of applications for diagnostic surgical and cytologic pathology.

For most laboratories, the immunoenzyme methods have largely replaced the immunofluorescence methods.[16,30,64] The immunoenzyme methods have certain advantages over the immunofluorescence methods as follows:

1) No need for special fluorescence microscope.
2) Can develop permanent slides with immunoenzyme method employing non-diffusible reaction product.
3) Immunoenzyme amplification methods such as avidin-biotin immunoenzyme (ABC) are very sensitive.

Glossary

Definitions of terms used in describing immunocytochemical procedures.

Antibody: a protein immunoglobulin molecule with specific amino-acid-sequence with specific interaction with a particular epitope of an antigen.

Polyclonal Antibody: produced by immunization across animal species. The antibody may vary in class, affinity, specificity, and reactivity. Polyclonal antibody which has a single defined specificity (monospecific antibody) is often obtained by adsorption of the polyclonal antiserum with undesired antigens.

Monoclonal Antibody: an antibody preparation produced in the laboratory which is a uniform homogeneous antibody directed at a single epitope or antigenic determinant.

The belief that a monoclonal antibody is "monospecific" is not entirely correct, since a given monoclonal antibody may interact with a set of related epitopes with different affinities such as a panel of related viral proteins.[18,50,77,81]

Antigen: any compound or substance which is capable under appropriate circumstances of inducing production of specific antibodies.

Epitope: a specific antigenic determinant which is the actual site of a specific antibody binding.

Monospecific Antibody: a polyclonal or monoclonal antibody which has a single defined specificity.

Sensitivity: the ability of a test or procedure to detect a desired result.

Antibody Sensitivity for a given antigen or cell type is defined as the percentage of "true positive" cases which are identified with the particular antibody.

Method Sensitivity: defined as the minimal detectable antigen concentration with use of a particular procedure or method.

Specificity: the ability of the antibody to detect the desired antigen from a panel of "true positives" and "true negatives." However, the specificity of a particular antibody may vary according to other conditions such as type of fixation of the tissue preparation, etc.

Digestion: the treatment of slides, after deparaffinization prior to the application of antibodies in the immunoenzyme procedure. The slide is usually treated with certain proteins such as trypsin, pronase, or pepsin to "unmask" certain epitopes which may be blocked during the fixation and embedding processes.[3,6] The treatment with proteolytic enzymes breaks methylene cross-links in antigenic molecules and reconstitutes their immuno-reactivity.

Controls: cells or tissue in a given slide test which can serve as positive controls for the antibody and/or method.

■ The avidin-biotin immunoenzyme method

The method was developed by Hsu et al in 1981.[30] The enzyme used was peroxidase, but glucose oxidase can also be used with similar results.

The indirect antibody procedure is called the avidin-biotin enzyme complex (ABC) method.[30,47] The method employs a secondary antibody bridge to which biotin is synthetically complexed and relies on the high affinity of avidin for biotin.

Many pathology laboratories use the avidin-biotin immunoperoxidase ABC method as the state-of-the-art immunocytochemical procedure. The ABC technique is the method of choice in most laboratories. Commercial reagents are readily available either as individual reagents or packaged "kits" for certain assays.

1) Fixation

There are four major types of fixatives utilized today.[17,45,57,77] There is no ideal fixative which can be used for all tissue-based procedures. However, it is important to understand the differences between various fixatives and how they vary in the retention of immunogenicity and antigenicity of the tissue antigens.

a) Coagulant fixatives: Examples of this type are ethanol, methanol, acetone, and Carnoy's solution. These fixatives initiate denaturation by displacing water from the protein antigens and break the hydrogen bonds. The soluble proteins in the cytoplasm are coagulated, organelles are destroyed and nucleic acids are not precipitated. Certain secretory products such as hormones remain soluble in water even after significant fixation in alcoholic mixtures at room temperature.

These fixatives extract lipids from tissues; however, carbohydrate containing components are largely unaffected. Glycogen is retained by these fixatives.

When tissue is fixed in methanol or acetone alone for a long period of time, there may be considerable shrinking and hardening of the tissue. When Carnoy's solution (70% ethanol, 20% chloroform, 10% glacial acetic acid) is used, the hardening of the tissue is minimized. Tissue can be kept in Carnoy's fixative for several days without excessive hardening or shrinkage.

The alcohol-based fixatives have the advantage of increased retention of antigenicity of a wide range of antigens relative to tissue fixed in standard aldehyde fixatives. Antigens which are preserved by alcohol fixatives are filament proteins, extracellular matrix proteins, Factor VIII, and other antigens.

b) Formaldehyde-based fixatives: Formalin is a 37% aqueous solution of formaldehyde. The formalin fixative fixes tissue by cross-linking of proteins to form methylene bridges.[57]

Unlike alcohol and acetone-based fixatives, formaldehyde preserves most lipids and does not react significantly with carbohydrates. Long periods of storage in formaldehyde results in excessive tissue hardening and loss of antigen stainability. With acidic solutions, formalin will form brown hematin pigment from degraded hemoglobin. The acid change in pH will also cause adverse effects on the antigenicity of tissue proteins.

Formalin fixative is widely used in many laboratories. Everyone hopes to find specific antibodies which can identify antigens in formalin-fixed paraffin-embedded material. However, one should be aware that there is great variation in formalin-fixed preparations. Those variables are affected by pH, tonicity, actual formaldehyde concentration, composition of buffer, and duration of fixation.

c) Glutaraldehyde-bond fixatives: Glutaraldehyde is a bifunctional strongly reactive cross-linking fixative. Tissue which is well fixed in glutaraldehyde *is not* satisfactory for immunohistochemical studies. The fixation with glutaraldehyde renders many protein antigens non-reactive with specific antibodies. As a general rule, tissue fixed in electron microscopy fixatives containing glutaraldehyde should not and cannot be used for immunohistochemical studies.

d) Mercuric-based fixatives and miscellaneous fixatives: B5 is a mercuric and formalin-based fixative which is often used for lymphoid histology preservation.[77] Because B5 is a formalin-based fixative, it is useful in selective immunocytochemical procedures.

B5 fixation time (usually less than 8 hours) is usually controlled. However, the use of B5 fixative requires treatment of the tissue to remove the mercury pigment which can interfere with the immunoenzyme procedure.

Bouin's fixative (a picric acid based fixative) is useful for immunolocalization of hormones such as neuropeptides and biogenic amines.

Choice of Fixatives

There is no single best fixative for immunohistochemistry. Certain fixatives are preferable for various antigens as follows:

1. Ethanol or Carnoy's solution is excellent for intermediate filament proteins.
2. Bouin's fixative (a picric acid based fixative) is the best for preservation of neuropeptides and biogenic amines.
3. B5 fixatives have been shown to help preserve expression of lymphoid determinants.

e) General recommendations for tissue fixation: All tissue should be routinely fixed in a standard fixative such as neutral-buffered formalin. Also, whenever possible, a portion of the tissue should be placed in an alternative fixative alcohol-based fixative. Tissue in an alcohol-based fixative can be kept without further processing, i.e., delayed embedding until the formalin-based fixed tissue is studied.

If the alcohol-based fixed tissue is processed, it should not be placed in the same tissue processor used for formalin-based fixatives.

f) Workup of newly acquired antibodies[77]: A systematic workup of all new antibodies is recommended as follows:

1. A bank of tissue should be accumulated and used as positive controls. The banks should include poorly differentiated neoplasms in which the diagnosis has been confirmed.
2. Each new antibody should be assessed for specificity and sensitivity against known controls and test tissues.
3. Working dilutions of primary antibodies should be chosen to provide positive staining in poorly differentiated tissues with maintenance of adequate specificity.
4. New lots of antibody must be retitered and rechecked for specificity against test tissues.
5. Difference in antibody reactivity that is produced by proteolytic treatment of test tissue should be evaluated.
6. Ideally, more than one immunohistochemical system should be tested with the antibody to determine the preferred antibody-enzyme system.

g) General approach to immunocytochemical studies[19]: Panels of antibodies are recommended for immunocytochemical studies. There are several reasons why several different antibodies are used.

1. It is rare that a single antibody study can give an unequivocal result.
2. A negative result with a single antibody may be misleading because of fixation problems and lack of antigenicity of the tissue or sensitivity of the specific antibody reaction.
3. The use of a panel of antibodies will demonstrate the varying range of sensitivity of the specific antibodies in different tumors.

h) Rules in interpretation of diagnostic immunohistochemical techniques for use in cytology and tissue sections: Swanson[65] emphasized the following rules for the use and interpretation of diagnostic immunohistochemical techniques.

1. Immunohistochemical analyses can be interpreted only in the context of an informed, carefully considered clinical and histologic differential diagnosis.
2. Single immunostains are unlikely to provide a specific diagnosis even within a limited differential diagnosis.
3. Panels of immunostains are not specific for a given diagnosis; rather, reproducible immunoprofiles have distinct relative predictive values for different diagnostic alternatives.

Immunohistochemical methods have helped the pathologist to determine the histogenesis of various tumors. The classification of neoplasms is usually based on the staining reactivity of a given lesion with a panel of antibodies for a variety of different antigens along with careful assessment of tissue or cell morphology and correlation with clinical information. The determination of the histogenesis of a tumor is very important for patient management and therapeutic decisions.[63]

i) Reporting immunohistochemical stain results: Banks[2] recommends, for report clarity and the exchange of information between laboratories, the following:

1. Immunostaining results should always be reported, regardless of perceived significance.

2. A differential diagnosis justifying immunostaining methods should be provided in the report.
3. The portion of the report dealing with immunostains should include: (a) the nature of the sample (cytology, frozen section, cellular imprints, etc.); (b) the immunoreagents used; (c) the results for each antibody with detail sufficient to justify the interpretation.

2 Important Tissue Antigens in Diagnostic Pathology

■Intermediate filament proteins

There are three major filamentous or cytoskeletal systems in all mammalian cells.[20,21]

1) There are microfilaments (5-6 nm) which are clearly identified with the protein actin and prominent in muscle cells.

2) Tubulin—protein found in most cells and in high concentration in nerve cells. Tubulin is associated with the thick cytoskeletal 25 nm microtubules.

3) Intermediate-Sized Filaments—these are 10 nm filaments present in all cells. The filaments are insoluble under most physiological conditions and tend to associate with other cellular structures such as membranes, mitochondria, and polyribosomes.

There are five major classes of intermediate filaments which compare to five major tissue types.[20,21,51,63]

1) Vimentin[15,44,52]

Vimentin is a 58,000 M.W. protein present in all *mesenchymal cells*. The vimentin filaments are generally inconspicuous in fibroblasts, endothelial cells, lymphocytes, and other mesenchymal cells. The use of labeled specific antibody such as fluorescent conjugated antibody is necessary to demonstrate expression of vimentin by a particular cell type of tumor cell.

Vimentin is expressed as the "primary" intermediate filament protein in all sarcomas, i.e., malignant tumors of mesenchymal cells. This includes angiosarcomas, epithelioid sarcomas, fibrosarcomas, malignant fibrous histiocytomas, leiomyosarcomas, liposarcomas, rhabdomyosarcomas, synovial sarcomas, neurogenic sarcomas, etc.

However, many non-mesenchymal tumors do express vimentin as a "secondary" intermediate filament protein along with their primary intermediate filament protein. Normal non-mesenchymal cells *do not* express vimentin.

Malignant gliomas (astrocytomas) which express glial fibrillary acidic protein (GFAP) as the primary intermediate protein generally co-express vimentin. Leiomyosarcomas and rhabdomyosarcomas generally express *desmin* as the primary intermediate filament protein, and almost always co-express vimentin. Also, a selected subset of spindle cell carcinomas which are distinguished by expression of cytokeratin proteins may co-express vimentin. Thus, it is clear that expression of vimentin by itself does not justify the classification of a given tumor as a sarcoma. This illustrates the importance of using *vimentin* as a part of a panel of antibodies.

There is only a single vimentin species; however, different antibody preparations to vimentin react differentially with vimentin of different cells. Lymphocytes contain vimentin. However, many commercially available antibodies do not react with vimentin in most lymphocytes and lymphomas.

Alcohol-fixed or Carnoy's fixed tissue is preferred for localization studies of vimentin and all intermediate proteins. Certain commercially available vimentin monoclonal antibody may react consistently with formalin-fixed, paraffin-embedded tissues. However, the reactivity of formalin-fixed, paraffin-embedded tissue for vimentin is variable, dependent on fixation time and other variables. The vimentin expression in this situation is usually most prominent in endothelial cells. Therefore, the endothelial cells can serve as "controls" or markers for the preservation of antigenicity of vimentin in processed tissue specimens.

2) Cytokeratins[9,12,46,58]

The importance of cytokeratin is that they are highly sensitive markers for carcinomas. Cytokeratin protein expression is characteristic of epithelial differentiation. Cytokeratin antibodies are used to distinguish carcinomas from other poorly differentiated malignancies such as lymphomas, sarcomas, melanomas, etc.

There have been reported 19 cytokeratin proteins based upon molecular weight and isoelectric point. The various cytokeratins are expressed in vivo as pairs of acidic and basic cytokeratins. Further, the cytokeratin compositions and profile is characteristic for a particular epithelial cell or carcinoma. Each of the cell types express a reproducible subset of the 19 cytokeratin proteins.

For practical reasons, the surgical pathologists have referred to two subclasses of cytokeratins.
   a) Low molecular weight cytokeratins or cytokeratins of simple epithelium.
   b) High molecular weight cytokeratins or cytokeratins of complex epithelium.

Many investigators have demonstrated how the use of the two types of cytokeratin antibodies allow division of carcinomas on the basis of their histogenetic origin. In addition, the "cytokeratin profile" of specific tumor types remains constant in primary tumors and their metastases.

The various commercially available anti-cytokeratin antibodies can be divided into three groups:
   a) Antibodies reactive with virtually all cytokeratins and simple epithelia.
   b) Antibodies reactive with "low molecular weight cytokeratins"(LMW-CK) corresponding to simple or ductal epithelium. Examples of simple epithelium are hepatocytes, acinar cells of pancreas, colonic

mucosa, etc.
   c) Antibodies reactive with "high molecular weight cytokeratins"(HMW-CK) corresponds to complex epithelia such as ductal and squamous epithelium.

It should be emphasized that the various cytokeratin antibodies should be evaluated in your own laboratory. The reactivity of different lots may vary according to methods and variations in tissue fixation and processing.

Cytokeratins are expressed in poorly differentiated carcinomas. The cytokeratin panel of antibodies is useful to distinguish carcinomas from non-epithelial neoplasms such as lymphomas, melanomas, sarcomas except synovial sarcoma and epithelioid sarcomas, and central nervous system tumors.

The co-expression of other antigens such as CEA and vimentin along with differential reactivity of the individual cytokeratin antibodies with help in identification of the primary site of carcinomas (see Table 14-3).

Both vimentin and cytokeratin are co-expressed in a restricted subgroup of carcinomas which include endometrial carcinomas, renal cell carcinomas and salivary gland tumors.

Certain cell types typically express a particular intermediate filament protein. However, many of these intermediate filaments are co-expressed with cytokeratin in different tumors in a regular and predictable manner. This is summarized in Table 1-1 which is an algorithm based upon the cytokeratin profile and co-expression of other intermediate filament proteins in specific tumors.

3) Neurofilaments and glial fibrillary acid protein (GFAP)

Central and peripheral nervous system cells contain two different intermediate filament proteins, glial fibrillary acid protein (GFAP) and the neurofilament triplet protein.

GFAP is the major protein of glial filament and has a M.W. of 52,000. The GFAP is distinct and different from the neurotubular and neurofilament protein.[70]

GFAP is a useful marker for astrocytes although it has also been demonstrated in ependymal cells. The GFAP antisera identify the same cells as the Wright silver stain. The phosphotungstic acid hematoxylin (PTAH) special stain identifies glial filaments; however, GFAP immunochemistry is a superior marker for glial differentiation since unlike the PTAH, the GFAP also identified the non-filamentous form of the protein. GFAP will identify reactive astrocytes, gemistocytic astrocytes and astrocyte giant cells (Fig. 1-1).

Non-astrocytic tumors of the central nervous system such as oligodendrogliomas, meningiomas, lymphomas,

Table1-1 Classes of intermediate filament proteins

Name	Molecular weight	Normal tissues where found	Tissues where found
Vimentin	58 kd	Mesenchymal cells	Sarcomas
Cytokeratins	40-70 kd (19 different)	Epithelial cells	Carcinomas
Neurofilaments	70, 150, 210 kd	Neurons	Ganglioneuromas
Glial Fibrillary Acidic Protein	52 kd	Glial cells Ependyma	Astrocytomas Ependymomas
Desmin	55 kd	Muscle cells	Rhabdo-, leio-myosarcomas

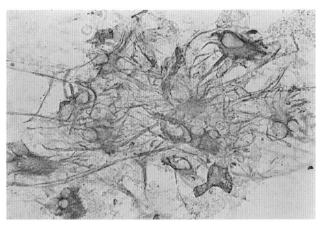

Fig. 1-1 Gemistocytic astrocytoma showing strong expression of glial fibers for GFAP.

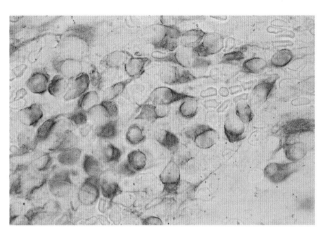

Fig. 1-2 Oligodendroglioma negative for GFAP.

sarcomas, and metastatic carcinomas are *negative* for GFAP (Fig. 1-2).[25,36,41,53,67] In the non-astrocytic tumors of the nervous system, the "reactive gliosis" may show positive staining and be confusing. Medulloblastomas may have glial differentiation and show focal GFAP positivity.[41] Besides glial cells, the ependymal cell may show GFAP positivity.[7] The reactive ependymal cells show enhanced GFAP expression. The ependymomas and subependymomas are GFAP positive.

A small fraction of peripheral nerve tumors such as schwannomas may co-express GFAP. Pleomorphic adenoma of the salivary gland may co-express GFAP along with cytokeratin. GFAP expression has not been seen in normal or reactive salivary gland tissue.

4) Neurofilament proteins
Neurofilament proteins are present in the intermediate filaments of neurons and traditionally identified by the Bodian silver method. The neurofilament proteins with M.W. of 70,000, 150,000, and 210,000 have been identified. Antibodies have been produced to each of these proteins as well as antibody which cross-reacts with each of the above neurofilament proteins. These particular antibodies do not cross-react with other classes of intermediate filaments. The antibody of choice for immunocytochemical studies is one which cross-reacts with all three neurofilament proteins.

Neurofilament proteins are expressed in ganglioneuroblastomas and pheochromocytomas and esthesioneuroblastomas. The undifferentiated neural tumors such as neuroblastomas are usually nonreactive with antibodies to neurofilament proteins.

Tumors of neuroendocrine origin will express neurofilament proteins with other intermediate filament proteins as cytokeratins.[36] The neurofilament co-expression is usually limited to more differentiated tumors such as carcinoids. Occasionally, atypical carcinoid tumors and small cell carcinomas may co-express cytokeratin and neurofilaments.

5) Desmin
Desmin is a class of intermediate filament proteins expressed in a subset of mesenchymal cells showing muscle cell differentiation. The desmin expression is seen in all forms of muscle differentiation including skeletal, smooth, and cardiac muscles.[1,43] Therefore, desmin expression is noted in muscle tumors, such as rhabdomyosarcomas, and leiomyosarcomas.[1,43]

■Actin
There are three classes of cytoskeletal proteins:
1) Microtubules
2) Intermediate filaments
3) Microfilaments
The intermediate filaments are the proteins which show tissue specific heterogeneity. The microtubules display a protein structure which is similar from tissue to tissue.

Actins can be divided into muscle and non-muscle actins.[22] Recently, specific antibodies to isotypes of actin have been produced. Thus the specific antibodies can differentiate smooth muscle actin from skeletal muscle actin. The monoclonal antibody HHF-35 is specific for a muscle actin and does not react with non-muscle actin. The HHF-35 antibody reacts with all types of muscle cells and myoepithelium. The myoepithelium is a unique epithelium which reacts with anti-cytokeratin and also co-expresses muscle specific actins. The anti-muscle actin antibody is important as a marker in rhabdomyosarcomas and leiomyosarcomas. This antibody is reactive with both alcohol and formalin-fixed tissues.[4]

■CEA (Carcinoembryonic antigen)
CEA is a tumor-associated antigen originally described in the human digestive system. CEA expression is seen in epithelial differentiation and is not seen in non-epithelial tumors. The CEA expression is directly proportional to the degree of cellular differentiation in epithelial cells.[54,73]

The interpretation of CEA antibody reactivity has been confused by a wide range of available antibodies to so-called CEA and the differential reactivity in different tissue fixatives. Cytoplasmic reactivity to CEA is more prominent in the tissue if it is alcohol-fixed.

Most CEA antisera contain antibodies cross-reacting with CEA and NCA (non-specific cross-reacting antigen). CEA antisera which react with polymorphonuclear leukocytes are not specific for CEA and contain antibody which cross-react with NCA. These serve as useful positive controls in the analysis of tissues.

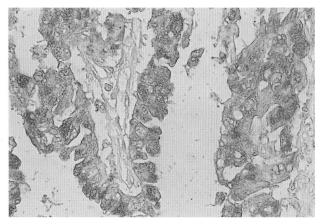

Fig. 1-3   CEA expressed diffusely in colonic adenocarcinoma. The reactivity belongs to that of group 1

The carcinomas are divided into three groups as follows[19]:

Group 1—These tumors show expression of CEA in
more than 90% of the cases.
Adenocarcinomas of the GI tract
Adenocarcinomas of the lung
Adenocarcinomas of the breast

Group 2—CEA Negative
Renal cell carcinoma
Prostate adenocarcinoma
Liver cell carcinoma
Mesotheliomas

Group 3—Variable CEA Expression
Endometrial adenocarcinoma (generally CEA negative)
Adenocarcinoma of the ovary
Transitional cell carcinoma of urothelium

CEA expression similar to cytokeratin expression is a characteristic marker for epithelial neoplasm (Fig. 1-3). CEA analysis when combined with intermediate filament and vimentin analysis are helpful in determining the site of origin and tissue histogenesis.[19]

■S100 protein

S100 protein is an acidic protein which is expressed by glial cells and gliomas, Schwann cells and schwannomas, a subset of neurons and neuroblastomas, melanomas, sweat and salivary gland tumors (e.g., pleomorphic adenomas), chondrocytes, and chondrosarcomas, Langerhans' cells and the cells of histiocytosis X, granular cell myoblastomas, chordomas, a subset of breast carcinomas, and adipose tissue and a subset of liposarcomas.[27,34,76]

The S100 protein has not been completely characterized. However, there are two components known as S100A and S100B. Most of the available antisera cross-react with both components.

The S100 expression is nuclear and is noted in both mesenchymal and epithelial cells, cells within the central nervous system, melanocytes, and a subset of histocytes.

The S100 is primarily useful in the subcategorization of vimentin positive, cytokeratin negative tumors, i.e., sarcomas.[76] Today, S100 is used to help confirm the diagnosis of melanomas. However, the use of S100 antibodies alone cannot confirm the diagnosis of melanoma. S100 and HMB 45 are used together to identify melanoma cells. A significant number of breast, salivary and sweat gland carcinomas are *S100 positive.*[10]

If a tumor is S100 positive, vimentin positive and cytokeratin negative, it is not likely to be a carcinoma.

■Leukocyte common antigen (LCA) or T200 antigen

LCA is a 200,000 M.W. cell surface associated protein found in cells of bone marrow origin and includes lymphocytes and monocytes. One hundred percent of lymphomas are positive for the LCA antigen (CD45).

Recently, specific antibodies to LCA which can react with tissue fixed in formalin or alcohol, and paraffin-embedded have been made available.[35,74,75]

One should be careful with interpretation of tissues stained with LCA antibody. Non-neoplastic infiltrating lymphocytes admixed with the tumor cells may be mistaken as positivity of the tumor itself.

The LCA is used in combination with the LEU M1 marker. The LEU M1 marker is a granulocyte antibody which reacts with "Hodgkin's cells" in fixed, paraffin-embedded tissue sections in "perinuclear dot plus membrane" pattern.[11,55,66]

Antibodies to LCA are positive on atypical cells of large cell non-Hodgkin's lymphoma. In Hodgkin's lymphoma, the Reed-Sternberg or Hodgkin's cells will be negative for LCA and positive with LEU M1 granulocyte specific antibody.[29,31]

The exception to the above rule is seen in a single subtype of Hodgkin's disease which is the nodular variant of lymphocyte-predominant Hodgkin's disease. In this later subtype, the "Hodgkin's cells" will demonstrate positivity with LCA and also with specific antibodies to B cells.

The LCA and LEU M1 antibodies are extremely useful for the evaluation of tissue sections of lymphomas which have been previously fixed in formalin and paraffin embedded.

■Anti-neuroendocrine antibodies: anti-chromogranins

Neurone-specific enolase (NSE) has been used to detect neuroendocrine tumors, however, NSE has low specificity. The currently available antibodies to NSE are not very useful for differentiating between neuroendocrine and non-neuroendocrine tumors.[19]

The anti-chromogranin markers are now available and are more useful markers for neuroendocrine differentiation. Chromogranins are acidic, phosphorylated proteins in chromaffin granules. The chromogranins are found in the adrenal medulla, anterior pituitary cells, parathyroid gland, C cells of the thyroid gland, islet cells of the pancreas, and enterochromaffin cells of the gastrointestinal tract.[60,78]

Antibodies to chromogranins are specific for tumor cells containing neuroendocrine granules.[60] Tumors which contain many granules such as pheochromocy-

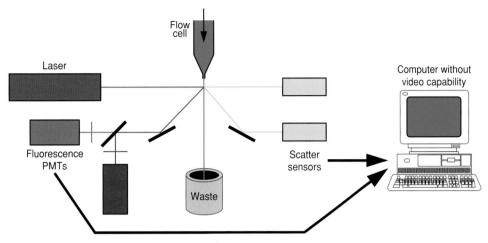

Fig. 2-5 Flow cytometer.

a series of fluorescence photomultiplier tubes. A schematic diagram of a flow cytometer showing the flow cell, laser beam, forward and side scatter light detectors, and fluorescence photomultiplier tubes (PMT) is shown in Fig. 2-5.

Analog signals simultaneously collected from each scatter detector and fluorescence PMT are digitized and stored within the memory of a computer for subsequent analysis. Each cell or particle passing through the flow cell has a single corresponding value for the light scatter and fluorescence input channels. Flow cytometers are capable of analyzing as many as several thousands of cells or particles per second, with the total number of cell analyzed for any given specimen typically ranging between 100,000 and 1,000,000. Acquisition of data is rapid and entirely automated. Data analysis is performed using computer software that interrelates the digitized data from any of the input channels. For example, a scattergram showing the forward angle and side angle light scatter values from an analysis of peripheral blood leukocytes is shown in Fig. 2-6. Because forward angle light scatter as determined via FCM roughly correlates with cell size and side angle light scatter roughly correlates with nuclear granularity, the different populations of peripheral blood leukocytes can be seen as separate populations within this scattergram. Had the specimen been reacted with propidium iodide with RNase prior to flow cytometric analysis, a DNA content histogram could be easily displayed through the software simply by displaying cell number on the y-axis and the appropriate red fluorescence PMT on the x-axis. FCM software also has built in functionality to allow sophisticated gating of the data based on results from any of the input channels. Relevant to DNA content analysis, such techniques are often employed in multi-parameter analyses using antibodies specific for cytokeratin or other lineage-specific antigens as a means of positively or negatively gating desired or unwanted populations of cells or debris. Advantages of FCM as a laboratory methodology include extremely rapid and automated data acquisition, excellent sensitivity in detecting weak fluorescent signals, and powerful data analysis subsequent and separate from data acquisition. Drawbacks include the requirement that specimen be a suspension of preferably single

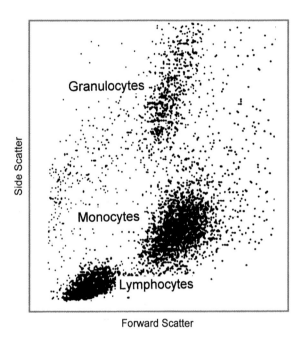

Fig. 2-6   Scattergram of peripheral blood leukocytes by FCM.

cells or particles which, for the most part, cannot be microscopically viewed during or after analysis.

◼ Image analysis

Although often considered to be fundamentally different, image analysis (IA) and FCM systems share scientific roots and have a similar fundamental purpose. Specifically, that purpose is to utilize objective measurements of cells and their compartments to generate medically useful diagnostic, prognostic, and therapeutic information. A diagram of a typical IA system is shown in Fig. 2-7. Most systems consist of a conventional light microscope to which is added a high resolution video camera, a computer, and sophisticated software programs necessary for data analysis. By definition, specimens are located on a glass microscope slide placed on the microscope stage. Depending on the purpose of the study, the specimens may be paraffin-embedded tissue sections or cytologic preparations of intact cells, and

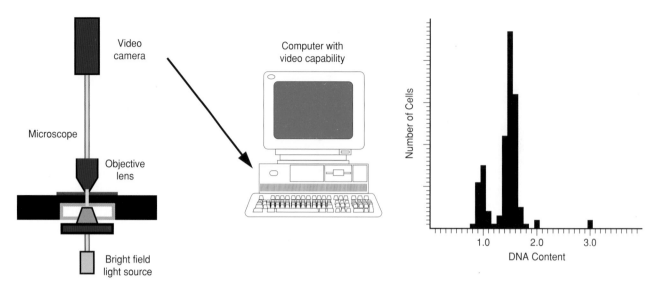

Fig. 2-7 Image analysis system.

Fig. 2-8 DNA histogram of an aneuploid tumor as determined by image analysis.

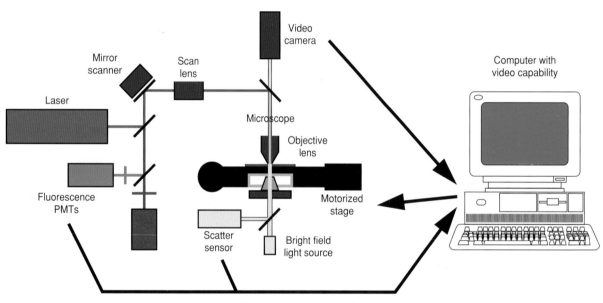

Fig. 2-9 Laser scanning cytometer.

may have been stained by any number of protocols. To make use of fluorescent reagents, the system may include an ion arc lamp or other excitation light source and appropriate filters for viewing, photographing, and capturing emitted fluorescent light.[58-67]

Using IA systems, data acquisition is anything but automated. Individual cells are chosen for analysis by the investigator, and digitized images of these cells are stored for subsequent analysis. With the recent proliferation of computer video accessories such as imaging boards, IA systems have flourished. However, data acquisition remains time consuming and limited in the number of cells that can be analyzed. Diametrically opposed to flow cytometric analysis, IA systems usually collect information from only up to several hundred cells within reasonable time frames. On the other hand, IA has the great advantage that the specimen can be microscopically viewed at any time—before, during, or after analysis. The stains utilized are for the most part optimized for light microscopy. For example, DNA content analysis by image cytometry is typically performed using a Feulgen stain, which stoichiometrically binds DNA and absorbs light with a wavelength of 546 nm. As with FCM, DNA content analysis results are displayed as a histogram relating cell number to DNA content. An example DNA histogram generated from an IA system is shown in Fig. 2-8. The relatively low number of cells analyzed results in a very coarse histogram as compared with that of FCM. Interestingly, the sensitivity of IA at detecting minor populations of aneuploid tumor cells is similar to or exceeds that of FCM, due to the data acquisition process wherein a user can actively select microscopically atypical cells only for analysis. On the other hand, accurate determinations of S-phase fraction are difficult if not impossible to achieve using IA systems, simply because data from an insufficient numbers of cells is collected.

■ Laser scanning cytometric analysis

Laser scanning cytometry (LSC) is a newly developed technology highly suited to DNA content analysis of cancer and eliminates many of the drawbacks of FCM and IA.[68-77] LSC shares the fundamental principles of FCM, but the specimen is located on a glass microscope slide mounted on a microscope stage. A block diagram of a laser scanning cytometer is shown in Fig. 2-9. An interrogating laser beam is directed in retrograde fashion through the optics of a conventional microscope to interact with the specimen that moves past the laser beam via a motorized stage. Similar to FCM, forward scattered light is collected via a light sensor beneath the stage. In addition, emitted fluorescent light transmitted back through objective lens is passed through appropriate filters to the fluorescence photomultiplier tubes. Signals from all photodetectors are digitized and the information is passed to a computer. Based on user defined threshold values for any of the scatter or fluorescence channels, the computer software automatically contours cells or particles within the specimen. A list of properties created within the software for each cell includes the location of the cell on the slide, the area of the cell, the perimeter of the cell, the peak intensity for each of the scatter and fluorescence channels within the cellular contour, and the integrated value of total scatter or fluorescence attributable to the cellular contour. Unlike FCM, the list of properties from each cell or contour is generated based on approximately 100 individual data points (pixels) for that cell for each of the scatter detector and fluorescent PMTs. In other words, the spatial resolution of LSC (while still less than IA) far surpasses that of FCM. This gives LSC some useful advantages over FCM, such as the ability to analyze fluorescence in situ hybridization (FISH) specimens in an automated fashion. Similar to FCM software, any of the aforementioned properties can be displayed relative to any other, and populations of cells can be positively or negatively gated during data analysis. Data acquisition is entirely automated with up to 5,000 cells counted per minute. With respect to DNA content analysis, this is manifest as good coefficients of variation, making possible accurate S- phase fraction determinations (Fig. 2-10). In addition, there are numerous advantages offered by the fact that the specimen is located on glass microscope slide, rather than within a fluid stream. For example, because the specimen is located on a glass slide and because the LSC is equipped with a video camera, specimens can be microscopically viewed at any time—before or after analysis. To facilitate microscopic viewing after analysis, the specimen can be removed from the laser scanning cytometer, stained with conventional dyes such as hematoxylin and eosin, and replaced on the laser scanning cytometer stage. Because the location of the individual cells remains fixed on the slide and that location is part of the permanent data generated from the automated analysis, it is a simple matter to relocalize individual cells meeting any user-definable criteria rele-

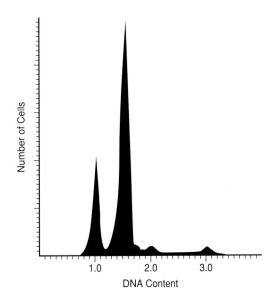

Fig. 2-10 DNA histogram of an aneuploid tumor as determined by laser scanning cytometry.

vant to the property list generated. This capability is particularly advantageous for cytologic DNA content analysis in cases for which there might be some question as to which peak of the histogram corresponds to the tumor cells, and which peak to the benign population of cells.

■ Single parameter DNA content and cell cycle analysis

Regardless of the instrumentation utilized, all methods of cytologic DNA content (DC) and cell cycle analysis (CCA) are based on stoichiometric binding of a marker dye or fluorescent molecule to DNA. For IA systems, this dye is usually the Feulgen stain, which specifically and stoichiometrically binds DNA a brown color. For FCM and LSC systems, the fluorescent reagent propidium iodide (PI) is commonly used. Because PI itself actually binds any double stranded nucleic acid (including tRNA), staining methods must utilize the enzyme RNase to destroy such double-stranded RNA molecules so that the PI binding occurs only with the double-stranded DNA that is of interest. Other fluorescent reagents such as acridine orange, are also sometimes used, but these are mostly for research rather than clinical applications.

In its simplest form, cytologic DC and CCA uses only the single observed parameter of light absorbance or emitted fluorescence specific for that dye or fluorescent reagent utilized to stoichiometrically bind nuclear DNA. As described in the "Cell Cycle Basics" section above, histograms are generated that display DNA content (as measured by light absorbance or fluorescent light emission) of individual cells on the x-axis and the number of cells with a particular DNA content on the y-axis. The actual analysis then takes the form of an interpretation of such a histogram. This interpretation is usually performed by human observation with varying degrees of assistance by statistical analysis and computer software algorithms.

For simple single parameter DNA content analysis, the goal is to detect an aneuploid population of tumor cells

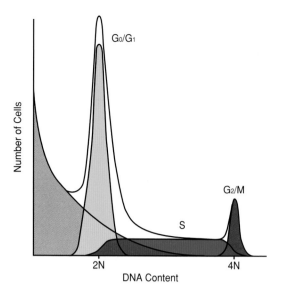

Fig. 2-11   The effect of debris on DNA histograms.

within the specimen, if one is present. This can be relatively simple for cases that contain significant numbers of aneuploid tumor cells, and in which the aneuploid DNA content of the tumor cells is significantly different than the DNA content of normal cells within the specimen (i.e. with G_0/G_1 and G_2/M peaks of DI=1.0 and 2.0, respectively). Difficulty with single parameter DNA content analysis can be encountered for any of three reasons.

First, if the population of aneuploid tumor cells constitutes a very small percentage of the specimen relative to the number of normal cells, the aneuploid "peak" within the DNA content histogram produced may be obscured.

Second, if the DNA content of the aneuploid tumor cells falls near to DI=1.0 or DI=2.0, that portion of the DNA histogram attributable to the normal cells within the specimen may again obscure the aneuploid peak. For near-diploid tumors, the DNA histogram may only show a broadening of the apparent diploid peak. For tetraploid tumors, the DNA histogram may only show an unusually large peak at DI=2.0.

The third cause of difficulty in interpreting single parameter DNA histograms for the determination of tumor cell DNA content is the presence of debris within the specimen. Debris of any external source or from the cells themselves (especially as from nuclear fragmentation) will artifactually raise the background of the DNA histogram and may obscure aneuploid peaks. Of course, all of these factors can co-exist depending on the particular specimen, making interpretation of single parameter DNA content histograms potentially very simple or very difficult.

Before moving on to single parameter cytologic cell cycle analysis, it is useful to describe in more detail the methods that have been devised to overcome the potential problems described above for DNA content analysis to make interpretation of DNA histograms reproducible and less subjective. Single-parameter FCM and LSC can typically detect minor populations of aneuploid tumor cells constituting approximately 5% of the total speci-

men *if* the aneuploid DI is not near 1.0 or 2.0 and *if* the specimen is not significantly contaminated by debris. Enhancement in this detection limit can be achieved most easily by resorting to multiparameter analyses as will be discussed in detail below.

For near-diploid tumors which result in a broadening of the "diploid" peak, many interpreters/algorithms use the coefficient of variation (CV) of that peak as an objective means to indicate the presence (or at least possible presence) of such a near-diploid tumor cell population. This process is admittedly arbitrary in that some cutoff value must be used to define low, moderate, and high CV's corresponding to the absence, suspicion, or presence of a near-diploid peak. Similarly, tetraploid or near-tetraploid tumors may result in a broadening or enlargement of the "tetraploid" peak that can be confusing to the interpretation of the histogram. Similar to the coefficient of variation cutoff values for near-diploid tumors, for tetraploid and near-tetraploid tumors, a mathematical cutoff value is again used. In this case, some value ranging from 15% to 25% is used as the highest allowable normal G_2/M population relative to the entire population of cells. In cases where the G_2/M peak constitutes more than the cutoff percentage of the entire population of cells, a tetraploid peak is inferred.

Debris such as fragmented nuclei often contaminate specimens for DNA content analysis, artifactually raising the background level and complicating the interpretation. Obviously, debris can be particularly troublesome for cases with minor populations of aneuploid tumor cells or those with diploid or tetraploid tumor cells. Many mathematical methods of debris subtraction have been developed, most based on models in which the amount of debris is inversely and exponentially related to the size of the debris particles. Some of the best debris subtraction algorithms have been developed by Rabinovitch and others, who have taken into account that debris generation is a result of degradation or fragmentation of nuclei and thus not strictly exponentially related to size. Fig. 2-11 graphically illustrates the effects of debris contamination (assuming a simple exponential model) on the DNA histogram observed. It is easy to see how excess debris could obscure small aneuploid peaks and/or affect mathematical calculations such as CVs and population percentages.

Having discussed these intricacies of cytologic DNA content analysis, we are better prepared to move on to cytological analysis of the cell cycle (or S-phase fraction analysis). This clinically relevant but poorly reproducible laboratory determination directly parallels the proliferative activity of a tumor. However, even for perfect histograms (those without debris, with a majority of the specimen being comprised of tumor cells, and without overlapping populations), S-phase fraction determination is difficult at best. In most instances, the cutoff value used to define the right hand border of the diploid peak and the left hand border of the tetraploid peak (thereby defining the S-phase fraction) are defined manually and subjectively. Needless to say, such practices result in very poorly reproducible data, especially between different interpreters/laboratories. Polynomial-based, Fourier transform-based, and other mathemati-

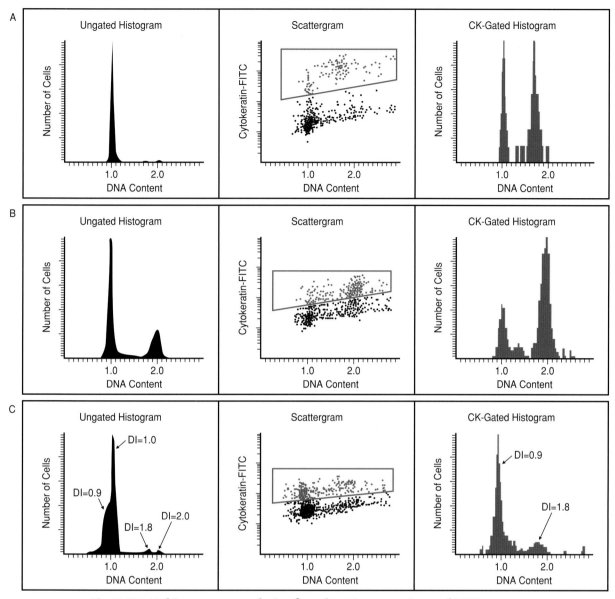

Fig. 2-12   Multiparameter analysis of cytokeratin expression and DNA content.

cal models to divide DNA histograms into the relevant phases of the cell cycle have been developed, but none satisfactorily accounts for all possible artifacts to allow truly reproducible data between laboratories.

▦Multiparameter DNA content and cell cycle analysis

Many of the problems inherent in single parameter DNA content and cell analysis can be somewhat overcome with additional information regarding the cells being analyzed.[78-95] For FCM and LSC in particular, it is very simple to collect multiparameter data such that information regarding some other cellular constituent (for example, cytokeratin expression or expression of proliferating cell nuclear antigen) is collected simultaneously with the DNA content and other physical parameters measured. In addition to staining for DNA with propidium iodide or other reagents as previously described, a specimen may be reactive simultaneously or beforehand with a fluorochrome labelled monoclonal antibody specific for some other cellular constituent. Specimens reacted with FITC labelled anti-cytokeratin antibodies provide a good example. Because of their epithelial origin, carcinoma cells will universally react with appropriate anti-cytokeratin antibodies. Therefore, cytokeratin positivity (as evidenced by green fluorescence if FITC is the fluorochrome) can be used as a means to positively select epithelial cells only (including carcinoma cells) for analysis. Even specimens taken from central portions of a tumor often contain non-tumor cells of stromal or lymphoid origin as well as debris. So, for carcinomas, multiparameter analysis using FITC labelled anti-cytokeratin antibodies is particularly useful to gate out these unwanted stromal cells, lymphoid cells, and debris. Although cytokeratin positivity in and of itself does not necessarily imply malignancy, the practice of routinely gating cytokeratin positive cells only for analysis of epithelial neoplasms is very useful.[78-95] Without such gating, numerical cutoff values to define the presence or absence of near-diploid tumor cell populations, tetraploid tumor cell populations, and S-phase fraction determinations are poor indeed. As shown in Fig. 2-12, the addition of anti-cytokeratin reagents for the purpose of multiparameter DNA content analysis markedly increases the lower limit of detectability for minor populations of aneuploid tumor cells. Because of the technical limitations described above, single parameter DNA content analysis can at best detect minor populations of tumor cells constituting approximately 4% of a given specimen. Multiparameter DNA content analysis with cytokeratin gating of epithelial cells can lower this limit to approximately 0.5% for epithelial neoplasms (carcinomas).

Cytologic cell cycle analysis (S-phase fraction analysis) also stands to benefit greatly from multiparameter analysis with the addition of cell cycle specific reagents.[96-107] Bromodeoxyuridine (BrDU) has been utilized to label the DNA of cells in the process of active DNA synthesis. Other similar but methodologically simpler techniques such as monoclonal antibody binding of antigens such as cyclins, proliferating cell nuclear antigen, Ki-67 and others are also promising. Because such reagents can specifically label S-phase cells, these methods promise to provide a means of determining S-phase fraction percentages in an objective and reproducible manner. It is certainly useful to take advantage of the large number of available fluorochromes and numerous fluorescence photomultiplier channels available in flow cytometers and laser scanning cytometers. Only in this way can we simultaneously combine information regarding cell cycle specific antigens, lineage specific antigens, and cytologic DNA content and cell cycle data to truly achieve objective and accurate results to translate into useful clinical information.

References

A: Clinical Relevance of DNA Content Analysis

1. Batsalos JG, Sneige N, el-Naggar AK: Flow cytometric (DNA content and S-phase fraction) analysis of breast cancer. Cancer 71: 2151, 1993.
2. Bauer KD, Bagwell CB, Giaretti W, et al: Consensus review of the clinical utility of DNA flow cytometry in colorectal cancer. Cytometry 14: 486, 1993.
3. Bauer KD, Duque RE, Shanke TV: Clinical Flow Cytometry. Baltimore: Williams and Wilkins, 1993.
4. Coon JS, Weinstein RS: Diagnostic Flow Cytometry. Baltimore: Williams and Wilkins, 1991.
5. Duque RE, Andreeff M, Braylan RC, et al: Consensus review of the clinical utility of DNA flow cytometry in neoplastic hematopathology. Cytometry 14: 492, 1993.
6. Hedley DW, Clard GM, Cornelisse CJ, et al: DNA Cytometry Consensus Conference: Consensus review of the clinical utility of DNA cytometry in carcinoma of the breast. Breast Cancer Res Treat 28: 55, 1993.
7. Hedley DW, Clark GM, Cornelisse CJ, et al: Consensus review of the clinical utility of DNA cytometry in carcinoma of the breast: Report of the DNA Cytometry Consensus Conference. Cytometry 14: 482, 1993.
8. Hedley DW: DNA Cytometry Consensus Conference: DNA flow cytometry and breast cancer. Breast Cancer Res Treat 28: 51, 1993.
9. Keren DF, Hanson CA, Hurtubise PE: Flow cytometry and clinical diagnosis. Chicago: ASCP Press, 1993.
10. Schroder F, Tribukait B, Bocking A, et al: Clinical utility of cellular DNA measurements in prostate carcinoma: Consensus Conference on Diagnosis and Prognostic Parameters in Localized Prostate Cancer. Stokholm, Sweden, Ma. Scand J Urol Nephrol 162(Suppl): 51; discussion 15, 1994.
11. Shankey TV, Kallioniemi OP, Koslowski JM, et al: Consensus review of the clinical utility of DNA content cytometry in prostate cancer. Cytometry 14: 497, 1993.
12. Spyratos F: DNA content and cell cycle analysis by flow cytometry in clinical samples: Application in breast cancer. Biol Cell 78: 69, 1993.
13. Wheeless LL, Badalament RA, de Vere White RW, et al: Consensus review of the clinical utility of DNA cytometry in bladder cancer: Report of the DNA Cytometry Consensus Conference. Cytometry 14: 478, 1993.

B: Cell Cycle Basics

14. Bauer KD, Duque RE, Shankey TV: Clinical Flow Cytometry. Baltimore: Williams and Wilkins, 1993.
15. Coon JS, Weinstein RS: Diagnostic Flow Cytometry. Baltimore: Williams and Wilkins, 1991.
16. Keren DF, Hanson CA, Hurtubise PE: Flow Cytometry and Clinical Diagnosis. Chicago: ASCP Press, 1993.

C: DNA Staining

17. Bauer KD, Duque RE, Shankey TV: Clinical Flow Cytometry. Baltimore: Williams and Wilkins, 1993.
18. Coon JS, Weinstein RS: Diagnostic Flow Cytometry. Baltimore: Williams and Wilkins, 1991.
19. Keren DF, Hanson CA, Hurtubise PE: Flow Cytometry and Clinical Diagnosis. Chicago: ASCP Press, 1993.
20. Schulte EK: Standardization of the Feulgen reaction for absorption DNA image cytometry: A review. Anal Cell Pathol 167, 1991.
21. Vindelov LL, Christensen IJ, Nissen NI: A detergent-trypsin method for the preparation of nuclei for flow cytometric DNA content analysis. Cytometry 3: 323, 1983.

D: Interpretation of DNA Histograms

22. Bauer KD, Duque RE, Shankey TV: Clinical Flow Cytometry. Baltimore: Williams and Wilkins, 1993.
23. Coon JS, Weinstein RS: Diagnostic Flow Cytometry. Baltimore: Williams and Wilkins, 1991.
24. Keren DF, Hanson CA, Hurtubise PE: Flow Cytometry and Clinical Diagnosis. Chicago: ASCP Press, 1993.
25. Wersto RP, Liblit RL, Koss LG: Flow cytometric DNA analysis of human solid tumors: A review of the interpretation of DNA histograms. Hum Pathol 22: 1085, 1991.

E: Original Flow Cytometry Articles

26. Kamentsky LA, Melamed MR: Instrumentation for automated examinations of cellular specimens. Proc IEEE 57: 2007, 1969.
27. Kamentsky LA: Cytology automation. Adv Biol Med Phys 14: 93, 1973.

F: Flow Cytometric Immunophenotyping

28. Ball ED, Davis RB, Griffin JD, et al: Prognostic value of lymphocyte surface markers in acute myeloid leukemia. Blood 77: 2242, 1991.
29. Bennett JM, Catovsky D, Daniel MT, et al: French-American-British (FAB) Cooperative Group; Proposals for the classification of acute leukemias. Br J Haematol 33:451, 1976.
30. Bennett JM, Catovsky DD, Daniel MT, et al: Criteria for the diagnosis of acute leukemia of megakaryocytic lineage (M7). Ann Intern Med 103: 460, 1985.
31. Bennett JM, Catovsky DD, Flandrin G, et al: Proposed revised criteria for the classification of acute myeloid leukemia. Ann Intern Med 103: 626, 1985.
32. Bowman GP, Neame PB, Soambonsrup P: The contri-

bution of cytochemistry and immunophenotyping to the reproducibility of the FAB classification in acute leukemia. Blood 68: 900, 1986.

33. Braylan RC, Benson NA: Flow cytometric analysis of lymphomas. Arch Pathol Lab Med 113: 627, 1989.

34. Braylan RC: Lymphomas. In: Bauer KD, Duque RE, Shankey TV (eds): Clinical Flow Cytometry. pp203-234. Baltimore: Williams and Wilkins, 1993.

35. Campana D, Pui CH: Detection of minimal residual disease in acute leukemia: Methodologic advances and clinical significance. Blood 85: 1416, 1995.

36. Chernoff WG, Lampe HB, Cramer H, et al: The potential clinical impact of the fine needle aspiration/ flow cytometric diagnosis of malignant lymphoma. J Otolaryngol Suppl 1: 1, 1992.

37. Diamond LW, Nathwani BN, Rappaport H: Flow cytometry in the diagnosis and classification of malignant lymphoma and leukemia. Cancer 50: 1122, 1982.

38. Duque RE: Acute leukemias. In: Bauer KD, Duque RE, Shankey TV (eds): Clinical Flow Cytometry. pp235-245. Baltimore: Williams and Wilkins, 1993.

39. Duque RE, Braylan RC: Applications of flow cytometry to diagnostic hematopathology. In: Coon JS, Weinstein RS (eds): Diagnostic Flow Cytometry. pp89-102. Baltimore: Williams and Wilkins, 1991.

40. First MIC Cooperative Study Group: Morphologic, immunologic, and cytogenetic (MIC) working classification of acute lymphoblastic leukemias. Cancer Genet Cytogenet 23: 189, 1986.

41. Foon KA, Todd RF III: Immunologic classification of leukemia and lymphoma. Blood 68: 1, 1986.

42. Geisler CH, Larsen JK, Hansen NE, et al: Prognostic importance of flow cytometric immunophenotyping of 540 consecutive patients with B-cell chronic lymphocytic leukemia. Blood 78: 1795, 1991.

43. Hanson CA, Schnitzer B: Flow cytometric analysis of cytologic specimens in hematologic disease. J Clin Lab Anal 3: 2, 1989.

44. Hanson CA: Fine-needle aspiration and immunophenotyping: A role in diagnostic hematopathology? Am J Clin Pathol 101: 555, 1994.

45. Harris NL, Jaffe ES, Stein H, et al: A revised European-American classification of lymphoid neoplasms: A proposal from the international Lymphoma Study Group. Blood 84: 1361, 1994.

46. Katz RL, Raval P, Manning JT, et al: A morphologic, immunologic, and cytometric approach to the classification of acute leukemia: The Southwest Oncology group experience. Am J Hematol 18: 47, 1985.

47. Knowles DM, Chadburn A, Inghirami G: Immunophenotypic markers useful in the diagnosis and classification of hematopoietic neoplasms. In: Knowles DM (ed): Neoplastic Hematopathology. pp73-167. Baltimore: Williams and Wilkins, 1992.

48. Neame PB, Soamboonsrup P, Browman GP, et al: Classifying acute leukemia by immunophenotyping: A combined FAB-immunologic classification of AML. Blood 68: 1355, 1986.

49. Picker LJ, Weiss LM, Medeiros LJ, et al: Immunophenotypic criteria for the diagnosis of non-Hodgkin's lymphoma. Am J Pathol 128: 181, 1987.

50. Pitts WC, Weiss LM: The role of fine needle aspiration biopsy in diagnosis and management of hematopoietic neoplasms. In: Knowles DM (ed): Neoplastic Hematopathology. pp385-405. Baltimore: Williams and Wilkins, 1992.

51. Pui CH, Behm FG, Crist WM: Clinical and biologic relevance of immunologic marker studies in childhood acute lymphoblastic leukemia. Blood 82: 343, 1993.

52. Robbins BA, Ellison DJ, Spinosa JC, et al: Diagnostic application of two-color flow cytometry in 161 cases of hair cell leukemia. Blood 82: 1277, 1993.

53. Robbins BA, Ellison DJ, Spinosa JC, et al: Diagnostic application of two-color flow cytometry in 161 cases of hairy cell leukemia. Blood 82: 1277, 1993.

54. Second MIC Cooperative Study Group: Morphologic, immunologic, and cytogenetic (MIC) working classification of acute myeloid leukemias. Br J Haematol 68: 487, 1988.

55. Sun T: Color Atlas-Text of Flow Cytometric Analysis of Hematologic Neoplasms. New York: lgaku-Shoin, 1993, pp3-8 (Instrumentation), 18-25 (Classification of hematologic neoplasms), and 206-211 (Sample preparation).

56. Willman CL: Flow cytometric analysis of hematologic specimens. In: Knowles DM (ed): Neoplastic Hematopathology. pp169-195. Baltimore: Williams and Wilkins, 1992.

57. Witzig TE, Banks PM, Stenson MJ, et al: Rapid immunotyping of B-cell non-Hodgkin's lymphomas by flow cytometry: A comparison with the standard frozen-section method. Am J Clin Pathol 94: 280, 1990.

G: Image Analysis

58. Bacus JW, Bacus JV: A method of correcting DNA ploidy measurements in tissue sections. Mod Pathol 7: 652, 1994.

59. Bacus JW: Cervical cell recognition and morphometric grading by image analysis. J Cell Biochem Suppl 23: 33, 1995.

60. Bocking A, Giroud F, Reith A: Consensus report of the ESACP task force on standardization of diagnostic DNA image cytometry. European Society for Analytical Cellular Pathology. Anal Cell Pathol 8: 67, 1995.

61. Bocking A, Giroud F, Reith A: Consensus report of the European Society for Analytical Cellular Pathology task force on standardization of diagnostic DNA image cytometry. Anal Quant Cytol Histol 17: 1, 1995.

62. Cajulis RS, Haines GK 3rd, Frias-Hidvegi D, et al: Cytology, flow cytometry, image analysis, and interphase cytogenetics by fluorescence in situ hybridization in the diagnosis of transitional cell carcinoma in bladder washes. Diagn Cytopathol 13: 214; discussion 224, 1995.

63. Gurley AM, Hidvegi DF, Bacus JW, et al: Comparison of the Papanicolaou and Feulgen staining methods for DNA quantification by image analysis. Cytometry 11: 468, 1990.

64. Montironi R, Diamanti L, Santinelli A, et al: Computed cell cycle and DNA histogram analyses in image cytometry in breast cancer. J Clin Pathol 46: 795, 1993.

65. Ross JS: DNA ploidy and cell cycle analysis in cancer diagnosis and prognosis. Oncology (Huntingt) 10: 867; discussion 887, 1996.

66. Thunnissen FB, Ellis IO, Jutting U: Interlaboratory comparison of DNA image analysis. Anal Cell Pathol 12: 13, 1996.

67. Thunnissen FB, Ellis IO, Jutting U: Quality assurance in DNA image analysis on diploid cells. Cytometry 27: 21, 1997.

H: Laser Scanning Cytometry

68. Clatch RJ, Walloch JL, Foreman JR, et al: Multiparameter analysis of DNA content and cytokeratin expression

in breast carcinoma by laser cytometry. Arch Pathol Lab Med 121: 585, 1997.

69. Clatch RJ, Walloch JL, Zutter MM, et al: Immunophenotypic analysis of hematologic malignancy by laser scanning cytometry. Am J Clin Pathol 105: 744, 1996.

70. Clatch RF, Walloch JL: Multiparameter immunophenotypic analysis of fine needle aspiration biopsies and other hematologic specimens by laser scanning cytometry. Acta Cytol 41: 109, 1997.

71. Gorczyca W, Darzynkiewicz Z, Melamed M: Laser scanning cytometry in pathology of solid tumors: A review. Acta Cytol 41: 98, 1997.

72. Gorczyca W, Melamed MR, Darzynkiewicz Z: Laser scanning cytometer (LSC) analysis of fraction of labelled mitoses (FLM). Cell Prolif 29: 539, 1996.

73. Gorczyca W, Sarode V, Juan G, et al: Laser scanning cytometric analysis of cyclin B1 in primary human malignancies. Mod Pathol 10: 457, 1997.

74. Kamentsky LA, Burger DE, Gershman RJ, et al: Slide-based laser scanning cytometry. Acta Cytol 41: 123, 1997.

75. Kamentsky LA, Kamentsky LD: Microscope-based multiparameter laser scanning cytometer yielding data comparable to flow cytometry data. Cytometry 12: 381, 1991.

76. Martin-Reay DG, Kamentsky LA, Weinberg DS, et al: Evaluation of a new slide-based laser scanning cytometer for DNA analysis of tumors: Comparison with flow cytometry and image analysis. Am J Clin Pathol 102: 432, 1994.

77. Sasaki K, Kurose A, Miura Y, et al: DNA ploidy analysis by laser scanning cytometry (LSC) in colorectal cancers and comparison with flow cytometry. Cytometry 23: 106, 1996.

I: Multiparameter DNA Content Analysis

78. Begg AC, Hofland I: Cell kinetic analysis of mixed populations using three-color fluorescence flow cytometry. Cytometry 12: 445, 1991.

79. Brown RD, Zarbo RJ, Linden MD, et al: Two-color multiparametric method for flow cytometric DNA analysis: Standardization of spectral compensation. Am J Clin Pathol 101: 630, 1994.

80. Clatch RJ, Walloch JL, Foreman JR, et al: Multiparameter analysis of DNA content and cytokeratin expression in breast carcinoma by laser scanning cytometry. Arch Pathol Lab Med 121: 585, 1997.

81. Frei JV, Martinez VJ: DNA flow cytometry of fresh and paraffin-embedded tissue using cytokeratin staining. Mod Pathol 6: 599, 1993.

82. Freo JV, Rizkalla K, Martinez VJ: Proliferative cell indices measured by DNA flow cytometry in node-negative adenocarcinomas of breast: Accuracy and significance in cytokeratin-stained archival specimens. Mod Pathol 7: 925, 1994.

83. Huffman JL, Garin-Chesa P, Gay H, et al: Flow cytometric identification of human bladder cells using a cytokeratin monoclonal antibody. Ann NY Acad Sci 468: 302, 1986.

84. Kenyon NS, Schmittling RJ, Siiman O, et al: Enhanced assessment of DNA/proliferative index by depletion of tumor infiltrating leukocytes prior to monoclonal antibody gated analysis of tumor cell DNA. Cytometry 16: 175, 1994.

85. Kimmig R, Kapsner T, Spelsberg H, et al: DNA cell-cycle analysis of cervical cancer by flow cytometry using simultaneous cytokeratin labelling for identification of tumour cells. J Cancer Res Clin Oncol 121: 107. 1995.

86. Park CH, Kimler BF: Tumor cell-selective flow cytometric analysis for DNA content and cytokeratin expression of clinical tumor specimens by "cross-gating." Anticancer Res 14: 29, 1994.

87. Savarese A, Nuti M, Botti C, et al: Sample preparation for bivariate flow cytometric analysis of breast tumor specimens. Anal Quant Cytol Histol 15: 39, 1993.

88. Visscher DW, Wykes S, Zarbo RJ, et al: Multiparametric evaluation of flow cytometric synthesis phase fraction determination in dual-labelled breast carcinomas. Anal Quant Cytol Histol 13: 246, 1991.

89. Visscher DW, Zarbo RJ, Jacobsen G, et al: Multiparametric deoxyribonucleic acid and cell cycle analysis of breast carcinomas by flow cytometry: Clinicopathologic correlations. Lab Invest 62: 370, 1990.

90. Visscher DW, Zarbo RJ, Sakr WA, et al: Flow cytometric DNA and cell cycle analysis of cytokeratin-labeled breast carcinomas: Correlations with prognostic factors. Lab Invest 60: 102A, 1989.

91. Wingren S, Stal O, Carstensen J, et al: S-phase determination of immunoselected cytokeratin-containing breast cancer cells improves the prediction of recurrence. Breast Cancer Res Treat 29: 179, 1994.

92. Wingren S, Stal O, Nordenskjold B: Flow cytometric analysis of S-phase fraction in breast carcinomas using gating on cells containing cytokeratin. Br J Cancer 69: 546, 1994.

93. Wingren S, Stal O, Sullivan S, et al: S-phase fraction after gating on epithelial cells predicts recurrence in node-negative breast cancer. Int J Cancer 59: 7, 1994.

94. Zarbo RJ, Brown RD, Linden MD, et al: Rapid (one-shot) staining method for two-color multiparametric DNA flow cytometric analysis of carcinomas using staining for cytokeratin and leukocyte common antigen. Am J Clin Pathol 101: 638, 1994.

95. Zarbo RJ, Visscher DW, Crissman JD: Two-color multiparametric method for flow cytometric DNA analysis of carcinomas using staining for cytokeratin and leukocyte-common antigen. Anal Quant Cytol Histol 11: 391, 1989.

J: Multiparameter Cell Cycle Analysis

96. Beppu T, Ishida Y, Arai H, et al: Identification of S-phase cells with PC10 antibody to proliferating cell nuclear antigen (PCNA) by flow cytometric analysis. J Histochem Cytochem 42: 1177, 1994.

97. Darzynkiewicz Z, Bruno S, Del Bino G, et al: The cell cycle effects of camptothecin. Ann NY Acad Sci 803: 93, 1996.

98. Darzynkiewicz Z, Gong J, Juan G, et al: Cytometry of cyclin proteins. Cytometry 25: 1, 1996.

99. Darzynkiewicz Z, Gong J, Traganos F: Analysis of DNA content and cyclin protein expression in studies of DNA ploidy, growth fraction, lymphocyte stimulation and the cell cycle. Methods Cell Biol 41: 421, 1994.

100. Gong J, Bhatia U, Traganos F, et al: Expression of cyclins A, D2 and D3 in individual normal mitogen stimulated lymphocytes and in MOLT-4 leukemic cells analyzed by multiparameter flow cytometry. Leukemia 9: 893, 1995.

101. Gong J, Traganos F, Darzynkiewicz Z: Discrimination of G_2 and mitotic cells by flow cytometry based on different expression of cyclins A and B1. Exp Cell Res 220: 226, 1995.

102. Gorczyca W, Sarode V, Juan G, et al: Laser scanning

cytometric analysis of cyclin B1 in primary human malignancies. Mod Pathol 10: 457, 1997.

103. Sasaki K, Kurose A, Ishida Y: Flow cytometric analysis of the expression of PCNA during the cell cycle in HeLa cells and effects of the inhibition of DNA synthesis on it. Cytometry14: 876-882. 1993.

104. Sasaki K, Kurose A, Shibata Y, Matsuta M: Varying detection of PCNA in solid tumor cells: Effects of fixation and detergent. J Histotechnol 18: 31, 1995.

105. Sasaki K, Kurose A, Uesugi N, et al: Effects of cyclohex-imide on chromatin-bound and unbound PCNA in HeLa cells. Oncology 52: 419, 1995.

106. Traganos F, Seiter K, Feldman E, et al: Induction of apoptosis by camptothecin and topotecan. Ann NY Acad Sci 803: 101, 1996.

107. Visscher DW, Wykes S, Kubus J, et al: Comparison of PCNA/cyclin immunohistochemistry with flow cytometric S-phase fraction in breast cancer. Breast Cancer Res Treat 22: 111, 1992.

3 X Chromatin

Barr and Bertram (1949)[5,6] found a large chromocenter in ganglion cells of female cats. Although it was first called a nucleolar satellite related to the nucleolus, later investigations confirmed that this so-called sex chromatin was closely related to the inner aspect of the nuclear envelope and recognizable in human females. Thus the buccal smear technique was introduced as a useful tool for determination of the genetic sex (Moore and Barr, 1955).[47] Since 1969 Y chromosomes have been found to be heteropyknotic in male interphase nuclei and recognizable as bright spots after quinacrine staining. Therefore, genetic nuclear sexing can be performed with use of X chromatin and Y chromatin stains. It is advisable to use the term X chromatin or X body instead of sex chromatin for female somatic cells.

1 Configuration of X Chromatin

The X chromatin (sex chromatin, Barr body), measuring about 0.7×1.2 μm, is planoconvex or triangular in shape with the broad base on the nuclear rim (Fig. 3-1). It usually reveals a solid configuration, and yet detailed observation may demonstrate a bipartite structure[36,43] that is composed of two hemispherical masses attached closely to each other. Using an advanced technique of microscopy with polarized dark-field oblique-illumination, Sato[65] observed a detailed sex chromatin structure forming double rings with one side open or semicircular figures. Although he has not described in detail the nature of this structure, original illustrations are suggestive of a folded X chromosome configuration (Fig. 3-3).

Regarding the origin of X chromatin there have been some speculations that a single X chromosome in males is too small to be visualized or is inhibited in its heteropyknotic condensation by the presence of the Y chromosome. The postulation[73] that the one X chromosome of the female is metabolically inactive and visible during the intermitotic phase has become universally accepted.[13,44,46] Ohno and his associates[55,56,57] confirmed the single X chromosome nature of the Barr body and noted that one of two X chromosomes in the female or the single X chromosome in the male behaved functionally in the same way as autosome, i.e., isopyknotic or euchromatic, while the second X chromosome in the female was heavily condensed to form the X chromatin.

In a practical count of X chromatin, it should be read as positive when it is located at the nuclear rim without space and measures about 1 μm as normal size. Well preserved intermediate or parabasal squamous cells should be selected. Basophilic stainability that is exactly the same as that of nuclear membrane is strictly important for identification. In case of appreciating smears with atypia or malignancy, pyknotic and degenerative nuclei with marked condensation of chromatin at the nuclear rim should be avoided for evaluation.

2 Significance of the X Chromatin Test

The technique of chromosomal analysis of marrow culture has improved so much that it becomes a routine method for estimating the number of X or Y chromosomes as well as autosomes. However, it is still a complex and time-consuming method. Genetic sex is expressed by the presence or absence of X and Y chromatins in somatic cell nuclei. Nuclear sexing or X chromatin test is simple and readily available to the laboratory of a physician's office. The evidence that one X chromosome per diploid chromosome set is euchromatic and all remaining X chromosomes are heteropyknotic is valid in detecting anomalous X chromosome numbers. The number of X chromatins is one less than the number of X chromosomes, while the number of Y chromatins corresponds directly to the number of Y chromosomes in the somatic cell. The X chromatin positive frequency in normal women reported is variable from 25% to 45%; a normal control value should be set in one's own laboratory using the same procedure of fixation and staining.

3 Hermaphroditism

Intersex means the evidence that phenotypic sex is different from true genetic sex. It is classified into true hermaphroditism showing mixed gonads like ovo-testis, female phenotype of male pseudohermaphroditism with testes only, and male phenotype of female pseudohermaphroditism with ovaries only. Most of the cases of true hermaphroditism are sex chromatin positive (XX true hermaphroditism). Female pseudohermaphroditism without obvious cause, i.e., excluding hormonal virilization is very rare and sex chromatin positive.[30] Male pseudohermaphroditism, on the other hand, has a higher incidence than female pseudohermaphroditism. Marked differences of the incidence can be conjectured from the hormonal background of development of the genitalia. Generally speaking, the sex differentiation is divided into three stages: (1) genetic and gonadal differentiation which depends on the genetic constitution, (2) intra-uterine somatic differentiation and (3) maturation at puberty, i.e., establishment of legal sex. The intra-

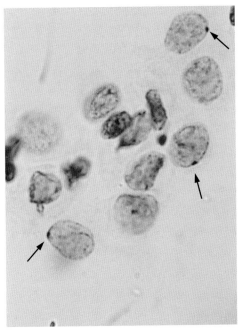

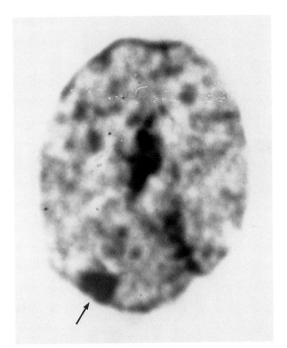

Fig. 3-1　X chromatins in normal somatic cells. Note uniform triangular X chromatins along the nuclear rim (arrows). Thionin stain. Higher magnification of endocervical cells.

Fig. 3-2　Nucleus of a normal cyanophilic superficial cell from vaginal wall. Note a condensed X chromatin seen under a bright-field illumination (arrow). Papanicolaou stain. (original magnification ×6,000)

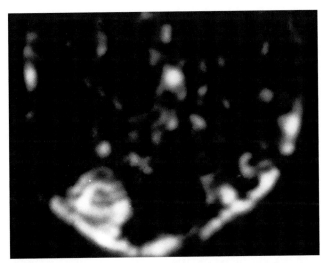

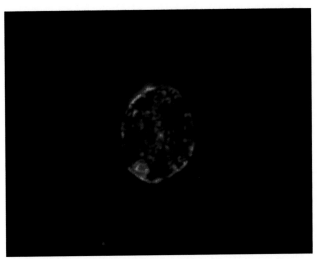

Fig. 3-3　Enlarged part of the X chromatin in the same nucleus. Observation by the polarized, dark-field, oblique illumination method. Note a semicircular double ring-like structure, a part of which appears to lie upon the nuclear membrane. (original magnification ×10,000)

Fig. 3-4　Color figure of X chromatin under observation by the polarized, dark-field, oblique illumination. (By courtesy of Mr. A. Sato, Technical Research Department of Nichimen Co., Osaka, Japan.)

uterine somatic differentiation is well explained by Jost's effect (1947). Müllerian atrophy and Wolffian development, responsible for producing male anatomy, are induced by the stimulus of embryonal testes. On the other hand, the absence of any gonad may give rise to atrophy of the Wolffian ducts and Müllerian development, i.e., female anatomy. It is probable that male pseudohermaphroditism results from a partial failure of the function of the embryonal testes.

Male pseudohermaphrodites (XY females) have 46 chromosomes with an XY sex chromosome constitution and show negative X chromatin. They produce a more or less male hormonal state and belong to a form of incomplete testicular feminization.[52,59]

4　Endocrine Intersex

Endocrine intersex is derived from various endocrine tumors or other hormonal disorders which cause virilization of the female or feminization of the male. Congenital adrenal virilization is by far the most important cause of clinically recognizable intersex. Nuclear sex

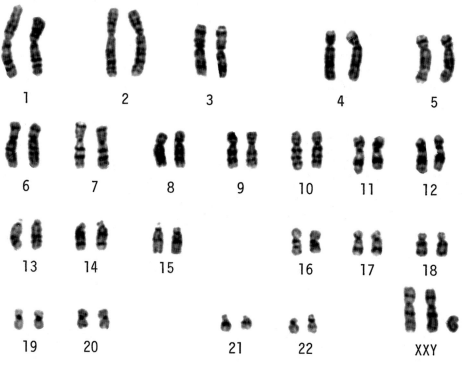

Fig. 3-5   Karyotype of Klinefelter's syndrome by G stain.

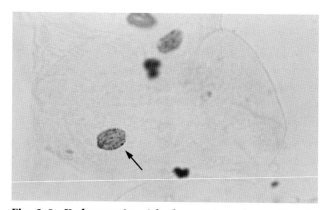

Fig. 3-6   X chromatin with sharp triangular outline in an intermediate squamous cell. From a case with Klinefelter's syndrome that is also positive for Y chromatin in Q stain.

determination in this case is no doubt positive for X chromatin.

5 | Sex Reversals

There is occasionally discrepancy between the externally apparent sex and the nuclear sex. There are three representative conditions that belong to this category.

■Testicular feminization

Testicular feminization (Morris, 1953)[51] is the extreme form of male pseudohermaphroditism with a female phenotype in spite of the presence of testis. The manifestation is not noticed until the age of puberty. The female hormonal state suggests that a metabolic disorder of the Leydig cells produces the inappropriate hormones. The sex chromatin test is negative and chromo-

some study reveals an XY chromosome constitution.[31] The intelligence is normal.

■Turner's syndrome

Turner's syndrome (pure gonadal dysgenesis) first described by Turner (1938)[76] in sex-reversed females is characterized by dwarfism, sexual infantilism, congenital webbing of the neck and cubitus valgus.[63] In addition to this syndrome, aortic coarctation and peripheral lymphedema are also occasionally present. The majority of these patients are sex chromatin-negative[27,79] and have an XO chromosome pattern with 45 modal number as the prototype.[21]

There may be some exceptions showing mosaics,[37] i.e., 45, XO/46, XY[62] or 45, XO/46, XY/47, XYY.[8] In X chromatin-negative patients of this category, mosaicism with a Y chromosome is said to be found in a proportion of about one of twenty cases of the subjects with dysgenetic ovaries.[41]

Chromatin-positive gonadal dysgenesis. Chromatin-positive ovarian dysgenesis derived from mosaics is characterized by short stature, breast atrophy, and amenorrhea, which are similar to symptoms of true Turner's syndrome. However, it is distinguished by the absence of neck webbing, peripheral lymphedema and coarctation of the aorta. The chief mosaic patterns possible are 45, XO/46, XX,[17,69] 45, XO/47, XXX,[16] and 45, XO/46, XX/46, XXr.[29] The latter is the mosaicism involving the ring X (Xr) with characteristic late replication.

■Klinefelter's syndrome

Klinefelter and associates described the syndrome of testicular dysgenesis, which is characterized by primary spermatogenic failure with atrophy of the seminiferous tubules and prominence of the Leydig cells, gynecomastia, and varying degrees of eunuchoidism. Hormonal investigation may show high concentrations of gonadotropin and low concentrations of 17-ketosteroids in urine. This syndrome represents a type of gonadal dysgenesis associated with chromosomal aberration. It is now believed that the syndrome results from a defect in the genetic sex differentiation, i.e., meiotic non-disjunction in one of the gametes. Most of these patients are positive for X chromatin as well as Y chromatin,[19,61] and have 47 chromosomes with an XXY sex chromosome pattern.

Concerning the frequency, the Klinefelter's syndrome occurs in about 1 per 800 to 1,000 liveborn phenotypic males.[25]

According to the studies by Makino and his associates on chromosomal abnormalities in the subjects with Klinefelter's syndrome, 53 cases were chromosomally abnormal out of 56 patients with the syndrome. Among them 43 cases were of the prototype with 47, XXY and the other 10 cases showed X chromosome polysomy and/or mosaicism: 2 cases with 48, XXXY, 1 case with 49, XXXXY, 3 cases with XXYY, 2 cases with 46, XX/47, XXY mosaicism and 2 cases with 46, XY/47, XXY.

The presence of an extra X chromosome increases the severity of testicular dysgenesis and mental retardation. Some other variants, polysomic (XXXY,[10,20] XXYY,[11] XXXXY[23,24]) or mosaic complex (XXY/XY)[38] can be recognized.

6 Superfemale

Superfemale[32] is a phenotype with multiple X chromosomes; this condition which was reported to show mental retardation, amenorrhea and hypoplasia of genital organs in association with polysomic X chromosome constitution. The XXX constitution having double X bodies in nuclei is by far the most common sex chromosome abnormality in newborn females.

According to the statistical study of Maclean[40] on the incidence of sex chromosome abnormalities, XXY constitution is the most common in male newborn babies (1.1/1,000), while XXYY polysomic variant (0.09/1,000) and mosaic complex XXY/XY (0.28/1,000) are very rare. Concerning sex chromosome anomalies in female babies, superfemale (XXX) is more common (0.9/1,000) than XO constitution (0.4/1,000).

7 The Technique of Sex Chromatin Determination

Although the X chromatin test cannot be a substitute for chromosomal karyotyping, it is of practical value in laboratory medicine. The X chromatin is well demonstrated in smears from the skin,[42,50] mucous membrane of the mouth,[47] vaginal and urethral mucosa,[9] etc. Histological sections from parenchymal organs in females also reveal the X chromatin with the acetocarmine squash technique. A strict evaluation of the X chromatin configuration is necessary[54] (Fig.3-12). Nuclei of granulocytes in female individuals reveal accessory nuclear appendages which were first named "drumsticks" by Davidson and Smith (1954).[15] The normal drumstick

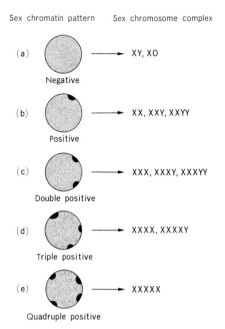

Fig. 3-7 X chromatin and X chromosome complexes. (From Barr ML: Sex chromatin techniques. In: Yunis JJ (ed): Human Chromosome Techniques. New York: Academic Press, 1965.)

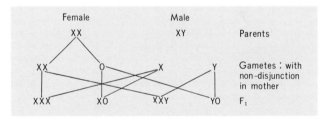

Fig. 3-8 Occurrence of sex chromosome abnormalities through maternal X chromosome non-disjunction.

measures approximately 1.5 μm in diameter and appears to be suspended by a fine thread from the nuclear lobe. The neutrophil method has the same diagnostic value as counting the X chromatin, although large numbers of granulocytes must be counted to be certain of eliminating errors because the difference between male and female percentages is not very high. Drumsticks are found in about 3% of the granulocytes in female, but seldom in males (below 0.2%).

A higher frequency of drumsticks is present in smears of newborn infants and patients with a shift to the right of differential counts; the lower frequency in the patients with Klinefelter's and Down syndrome is explained from the correlation between the frequency of drumsticks and maturity or segmentation of polymorphonuclear leukocytes.[45,75]

■Preparation of smears
Buccal smears:
1) Scrape the buccal mucosa with a wooden spatula and discard the first smear.
2) Carefully scrape the same area again and spread the smear on a clean glass slide.
3) Promptly fix the smear in equal parts of 95% ethyl alcohol and ethyl ether for about 30 min.

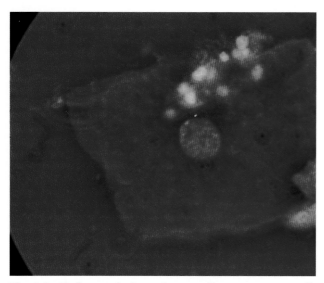

Fig. 3-9　X chromatin in an intermediate squamous cell. Differential interference phase contrast microscopy.

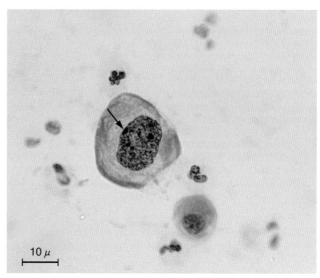

Fig. 3-10　Large X chromatin in a severe dysplastic cell. Note a large triangular X chromatin at the nuclear rim.

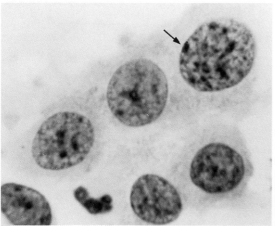
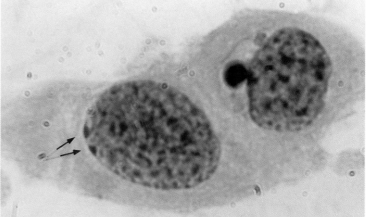

Fig. 3-11　Large X chromatin and double X chromatins in cancer cells of the uterine cervix. Although these are hardly distinguishable from heterochromatins, the characteristic location and the shape are the key points of identification (arrows).

Vaginal smears:

Scrape the vaginal wall with a wooden spatula. Care must be taken to obtain cells from deeper layers.

Schwarzacher recommends as a good fixative a mixture of 1 part acetic acid and 3 parts ethyl alcohol (96%) instead of 96% ethyl alcohol or ether-alcohol. Fixation should be carried out prior to air drying.[68]

■Staining procedure

[*Guard's stain*[28]]

1) Remove slides from the fixative and pass through 90% alcohol to 70% alcohol.

2) Stain the smear with Riebrich scarlet-Fast Green for 2 minutes.

3) Rinse in 50% alcohol for 5 min.

4) Differentiate in fast green FCF for 1 to 4 hours observing the step of differentiation until the cytoplasm and nucleus stain green, while pyknotic nuclei remain bright red.

5) Rinse in 50% alcohol for 5 min.

6) Dehydrate and mount in Permount.

　Biebrich scarlet

　　Biebrich scarlet water soluble (Harleco)···1.0 g

　　Phosphotungstic acid, C. P.················0.3 g
　　Glacial acetic acid ····················5.0 ml
　　50% ethyl alcohol ····················100.0 ml

Fast Green FCF

　　Fast Green FCF·······················0.5 g
　　Phosphomolybdic acid···················0.3 g
　　Phosphotungstic acid···················0.3 g
　　Glacial acetic acid ····················5.0 ml
　　50% ethyl alcohol ····················100.0 ml

[*Thionin stain*]

1) Remove slides from the fixative and pass through 90% alcohol, 70% alcohol to distilled water.

2) Hydrolyze in 1 N HCl for not more than 5 min at 56°C. This will remove bacteria.

3) Place in two changes of distilled water for 5 min each.

4) Stain with thionin solution for 5 min.

5) Differentiate and dehydrate in 70% alcohol and mount in Permount.

　Thionin stock solution

　　Thionin ·····························1.0 g
　　50% ethyl alcohol ····················100.0 ml

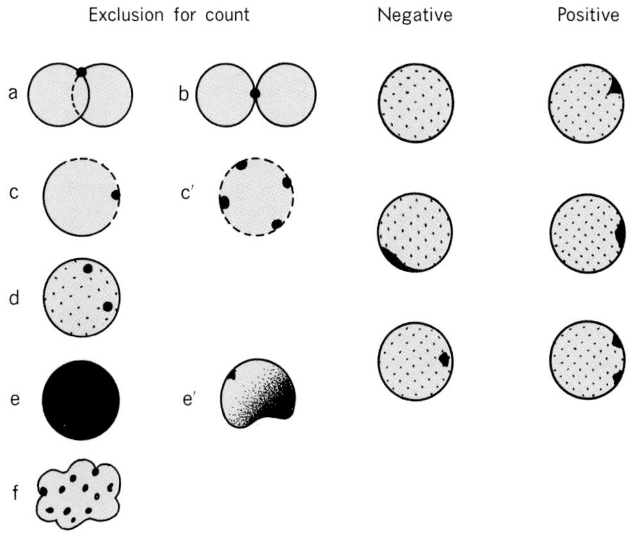

Fig. 3-12　Criteria for X chromatin count. In order to avoid overestimation of X chromatins, it is advisable to exclude overlapped cells (a) or closed cells face to face (b), degenerative cells that have lost nuclear membranes (c), cells having X body-like heterochromatin being separated from nuclear membrane (d), pyknotic cells (e), and shrunk cells (e′) from counting. (Slightly modified from Ohashi H, Takahashi M: J Jpn Soc Clin Cytol 17: 19, 1978.)

Thionin working solution

    0.1 N HCl ·· 32 ml
    Stock buffer: 9.7g of sodium acetate,
       14.7 g of sodium barbiturate, 500 ml
      of distilled water ······················· 28 ml
    Stock thionin solution ······················ 40 ml

Other stains such as hematoxylin eosin, aceto-orcein carbolfuchsin[12] and Feulgen reaction are also available.

■Results of the X chromatin test

One hundred single normal cells are examined with oil immersion. Males show less than 10% X chromatin and females more than 25% to 45%. The correlation between X chromatin patterns and sex chromosome complexes can be summarized schematically (Fig. 3-7).[4]

Generally speaking, X chromatin-positive cells in smears and skin biopsy speciments may vary from laboratory to laboratory; according to Eggen,[18] smears from females show 60% to 65% and skin biopsy specimens do about 80%. Therefore, the technique for obtaining specimens and preparing smears should be established in each laboratory. It must be kept in mind for practical use that a single count is not sufficient for the determination of X chromatin value because of variation of counts in the same individual from day to day and under various medical conditions; according to the study by Platt and Kailin[60] the counts were low in glomerulonephritis, chemical reaction syndrome, atopy and neonatus, while quite elevated in rheumatoid arthritis, gastrointestinal hemorrhage, etc. For cancer patients different results have been reported; either decreased,[72,77] normal or rather elevated counts.[60] Platt's interpretation for this discrepancy is that the stage of the cancer is responsible for the variation.

8　X Chromatin in Cancer Cells

Concerning the frequency of X chromatins in intraepithelial neoplasia of cervical lesions, positive X chromatins in severe dysplasia and carcinoma in situ are decreased in number, while no significant difference is detectable between mild dysplasia and normal epitheli-

Table 3-1　Frequency of X chromatin-positive nuclei

Lesion	Cases	Percentage (meagn± SD)	Significance
Normal	30	25.1 ± 6.4	nonsignificant
Dysplasia, mild	13	22.7 ± 4.7	p<0.05
Dysplasia, marked	11	18.1 ± 4.5	nonsignificant
Carcinoma in situ	30	15.9 ± 6.1	p<0.01
Invasive carcinoma	27	11.3 ± 4.5	
Stage 1	12	12.9 ± 4.5	
Stage 2	9	11.9 ± 2.8	
Stage 3	6	8.8 ± 4.0	

um. Overall frequency of positive X chromatins of cancer cells decreases as cervical intraepithelial neoplasia (squamous intraepithelial lesion) advances; in our study among the patients with carcinoma in situ and invasive carcinoma of the uterine cervix positive X chromatins are apparently decreased ($p<0.01$) (Table 3-1), although no significant correlation to the stage of cancer is present. The occurrence of double X chromatins or abnormally large ones in cancer cells of the uterine cervix has been pointed out (Table 3-2),[22,54] although a large proportion of cancer cells, particularly anaplastic cancer cells, are X chromatin-negative.[48,70] The X chromatin pattern studied by Atkin on cancer cells of the uterine cervix showed double X chromatins in about 10%, triple X chromatins in about 0.6% and negative chromatin in 40%.[3] Naujoks[53] investigated the distribution limited to carcinoma in situ and noted nearly the same incidence of multiple sex chromatin bodies. The study by Atkin,[1,2] on the correlation between the X chromatin and its modal DNA value proved that the cell with single X chromatin belongs to near-diploidy and the cell with multiple X chromatin belongs to near-tetraploidy which is a common feature of cancer cells.[39,71]

Enlargement of the X chromatin is progressive during DNA synthesis in the interphase, and it is most promi-

Table 3-2 Frequency of multiplication of X chromatins

	Cases	X chromatin positive cells	Double X chromatin positive cells	Triple X chromatin positive cells
Dysplasia	24	219	6/219 (2.7%)	0
Carcinoma in situ	30	221	27/221 (12.2%)	0
Invasive carcinoma	27	179	29/179 (16.2%)	1/179 (0.5%)

(From Ohashi H, Takahashi M: J Jpn Soc Clin Cytol 17: 19, 1978.)

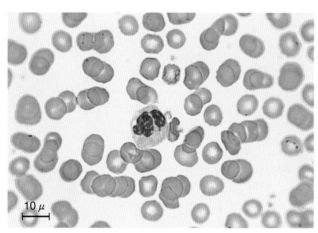

Fig. 3-13 Needle-like process referable to as malignancy associated change. From a female patient with advanced carcinoma of the stomach.

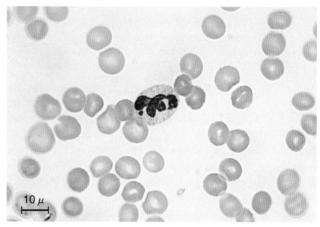

Fig. 3-14 Drumstick in a neutrophil leukocyte. A drumstick is attached to a lobe with a filament.

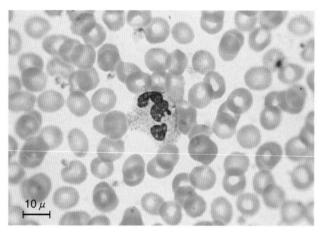

Fig. 3-15 Sessile nodule in a neutrophil leukocyte. It is considered to have the same significance as a drumstick.

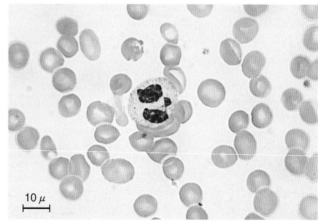

Fig. 3-16 Pseudo-lobule of a neutrophil leukocyte.

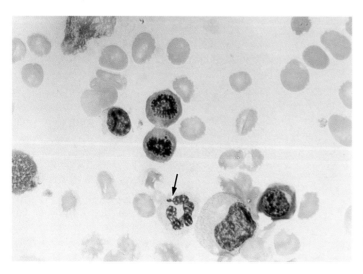

Fig. 3-17 Technically formed artificial nuclear appendage in a leukocyte. A nuclear appendage is shown with arrow. Erythroblasts in the telophase of mitosis are seen in the middle of this figure.

nent before mitosis (G_2). It is understandable that an enlargement of X chromatin would be frequently encountered in actively multiplying cancer cells compared with benign cells with variably long interval of G_1. An interesting description is the absence of sex chromatin that is suggestive of poor differentiation of cancer and unfavorable prognosis. The behavior of X chromatin in malignancy is variable and still open to discussion.

The mechanism of X-inactivation and/or X-heterochromatization remains obscure; the simple explanation that loss of X chromosome results in disappearance of X body is not appropriate for malignant tumors showing no loss of C+X group chromosomes. Straub and associates[74] suggested a reversion of cytogenetic state in some tumors to the stage of embryo, in which X-linked G-6-phosphate dehydrogenase in females reveals two-fold activity compared with that in males.

■ Significance of X chromatin in prognosis

The prognosis value of X chromatin incidence in neoplasm has been studied in cases of breast cancer since Moore and Barr[49] referred to variation of sex chromatin counts in cancer cells of the breast: It is said that a group with high X chromatin body incidence showed a good survival time or therapeutic response regardless of histologic differentiation.[7,34,35,78] To evaluate the prognosis, cases with X chromatin counts above 20% are significant as an X chromatin-positive group showing a favorable prognosis.[67]

The prognostic or biological significance of X chromatins for malignant neoplasms has not much debated since the 1980s until a new interphase cytogenetics was introduced. The fluorescence in situ hybridization is at present available as a routine interphase cytogenetics (see Chapter 5). Certain chromosome-specific probes bound with multi-color fluorescent dyes can be counted as signals. Loss of X signal was described by Powell et al.[64] There appeared a note-worthy report by Sauter et al who studied on male bladder cancers using centromeric probes for X, Y, 7, 9, and 17 chromosomes. Polysomies of X signals were shown in 30 of 65 tumors (34%) that took a rapid tumor progression.[66]

■ Other nuclear appendages

Drumsticks must be discriminated from several other nuclear projections or nuclear appendages. Small clubs which are encountered equally in males and females are small, less than 1 μm in diameter, occasionally multiple, weakly basophilic and have thin stalks. Sessile nodules, however, are regarded as a variant of the drumsticks. Accessory or minor lobes can be distinguished from drumsticks by the presence of two stalks uniting with main lobes. There are also thread-like nuclear excrescences or spikes from the lobes occurring singly or in groups (Fig. 3-17).

Johnson and his associates[33] emphasized that atypical thread-like excrescences are often associated with malignant disease and producible experimentally in tumor transplanted animals. They regarded the cases with more than 20% frequency of these excrescences as positive. This feature may belong to a category which was defined as malignancy-associated changes (MAC) by Nieburgs.

Clausen and Von Haam[14] confirmed electron microscopically the nuclear projections bound with membranes of about 400 Å thickness as a peculiar feature to be related to malignancy. Attention should be paid to the fact that air drying preparation of blood smears results in irregular shrinkage, nonspecific bodies and irregular contours of nuclei which may be mistaken for malignancy-associated changes.[58]

References

1. Atkin NB: Die chromosomale Basis von Sexchromatinabweichungen in menschlichen Tumoren. Wien Klin Wschr 76: 859, 1964.
2. Atkin NB: Sex chromatin abnormalities in carcinoma of the cervix uteri. Acta Cytol 10: 392, 1966.
3. Atkin NB: Variant nuclear types in gynecologic tumors: Observations on squashes and smears. Acta Cytol 13: 569, 1969.
4. Barr ML: Sex chromatin techniques. In: Yunis JJ (ed): Human Chromosome Methodology. New York: Academic Press, 1965.
5. Barr ML, Bertram EG: A morphological distinction between neurones of the male and female, and the behaviour of the nucleolar satellite during accelerated nucleoprotein synthesis. Nature 163: 676, 1949.
6. Barr ML, Bertram LF, Lindsay HA: The morphology of the nerve cell nucleus according to sex. Anat Rec 107: 283, 1950.
7. Blumel G, Turcic G, Regele H, et al: Zur Frage des Geschlechtsdimorphismus der Tumorzellen beim Mammakarzinom. II. Teil, Oertliches Rezidiv und zellkernmorphologisches Tumorgeschlecht. Wien Klin Wschr 75: 41, 1963.
8. Buhler EM, Muller H, Muller J, et al: Variant of the fluorescence pattern in an abnormal human Y chromosome. Nature 234: 348, 1971.
9. Carpentier PJ, Stolte LAM, Visschers GP: Sexing nuclei. Lancet 2: 874, 1955.
10. Carr DH, Barr ML, Plunkett ER, et al: An XXXY sex chromosome complex in Klinefelter subjects with duplicated sex chromatin. J Clin Endocrinol 21: 491, 1961.
11. Carr DH, Barr ML, Plunkett ER: A probable XXYY sex determining mechanism in a mentally defective male with Klinefelter's syndrome. Canad Med Assoc J 84: 873, 1961.
12. Carr DH, Walker JE: Carbol fuchsin as a stain for human chromosomes. Stain Technol 36: 233, 1961.
13. Childs B: Genetic origin of some sex differences among human beings. Pediatrics 35: 798, 1965.
14. Clausen KP, Von Haam E: Fine structure of malignancy-associated changes in peripheral human leukocytes. Acta Cytol 13: 435, 1969.
15. Davidson WM, Smith DR: A morphological sex difference in the polymorphonuclear neutrophil leucocytes. Br Med J 2: 6, 1954.
16. Day RW, Larson W, Wright SW: Clinical and cytogenetic studies on a group of females with XXX sex chromosome complements. J Pediat 64: 24, 1964.
17. De Grouchy J, Lamy M, Frezal J, et al: XX/XO mosaics in Turner's syndrome: Two further cases. Lancet 1: 1369, 1961.
18. Eggen RR: Chromosome Diagnostic in Clinical Medicine. Springfield, Ill: Charles C Thomas Publisher, 1965.
19. Ferguson-Smith MA: Chromatin positive Klinefelter's syndrome. Primary microorchidism in a mental-defi-

ciency hospital. Lancet 1: 928, 1958.

20. Ferguson-Smith MA, Johnston AW, Handmakers SD: Primary amentia and microorchidism associated with an XXXY sex-chromosome constitution. Lancet 2: 184, 1960.

21. Ford CE, Jones KW, Polani PE, et al: A sex-chromosome anomaly in a case of gonadal dysgenesis (Turner's syndrome). Lancet 1: 711, 1959.

22. Forni A, Miles CP: Sex chromatin abnormalities in carcinoma of the cervix uteri. Acta Cytol 10: 200, 1966.

23. Fraccaro M, Klinger HP, Schult W: A male with XXXXY sex chromosomes. Cytogenetics 1: 52, 1962.

24. Fraser JH, Boyd E, Lennox B, et al: A case of XXXXY Klinefelter's syndrome. Lancet 2: 1064, 1961.

25. Gelehrter TD, Collins FS: Cytogenetics. In: Principles of Medical Genetics. pp159-189. Baltimore: Williams and Wilkins, 1990.

26. German J: Sex chromosomal abnormalities. South Med J 64: 73, 1971.

27. Grumbach MM, Van Wyk JJ, Wilkins L: Chromosomal sex in gonadal dysgenesis (ovarian agenesis): Relationship to male pseudohermaphroditism and theories of human sex differentiation. J Clin Endocrinol 15: 1161, 1955.

28. Guard HR: A new technic for differential staining of the sex chromatin, and the determination of its incidence in exfoliated vaginal epithelial cells. Am J Clin Pathol 32: 145, 1959.

29. Higurashi M: Application of autoradiography technique on the chromosome analysis. Acta Paediat Jpn 72: 1291, 1968.

30. Iffy L, Ansell JS, Bryant JS: Nonadrenal female pseudohermaphroditism: An unusual case of fetal masculinization. Obstet Gynecol 26: 59, 1965.

31. Jacobs PA, Baikie AG, Court Brown WM, et al: Chromosomal sex in the syndrome of testicular feminization. Lancet 2: 591, 1959.

32. Jacobs PA, Baikie AG, Court Brown WM, et al: Evidence for the existence of the human "super female." Lancet 2: 423, 1959.

33. Johnson B, Brady JM: Malignancy related changes in the peripheral blood of animals following transplant of tumors. Acta Cytol 13: 443, 1969.

34. Kallenberger A: Geschlechtschromatin bei Mammakarzinomen. Schweiz Med Wschr 94: 1450, 1964.

35. Kallenberger A, Hagmann A, Descoeudres C: The interpretation of abnormal sex chromatin incidence in human breast tumors on the basis of DNA measurements. Eur J Cancer 3: 439, 1968.

36. Klinger HP: The fine structure of the sex chromatin body. Exp Cell Res 14: 207, 1958.

37. Lennox B: Nuclear sexing and the intersexes. In: Harrison CV (ed): Recent Advance in Pathology. London: J & A Churchill, 1960.

38. Lubs HA Jr: Testicular size in Klinefelter's syndrome in men over fifty. Report of a case with XXY/XY mosaicism. N Engl J Med 267: 326, 1962.

39. Lubs HA Jr, Clark R: The chromosome complement of human solid tumors. I. Gastrointestinal tumors and technic. N Engl J Med 268: 907, 1963.

40. Maclean N, Harnden DG, Court Brown WM, et al: Sex-chromosome abnormalities in newborn babies. Lancet 1: 286, 1964.

41. Makino S: Human Chromosomes. Tokyo: Igaku-Shoin, 1975.

42. Marberger E, Nelson W: Geschlechtsbestimmung in der menschlichen Haut. Bruns' Beitr Klin Chir 190: 103, 1955.

43. Miles CP: Sex chromatin in cultured normal and cancerous human tissues. Cancer 12: 299, 1959.

44. Mittwoch U: Sex chromatin. J Med Genet 1: 50, 1964.

45. Mittwoch U: Frequency of drumsticks in normal women and in patients with chromosomal abnormalities. Nature 201: 317, 1964.

46. Mittwoch U: Sex chromosomes and sex chromatin. Nature 204: 1032, 1964.

47. Moore KL, Barr ML: Smears from the oral mucosa in the detection of chromosomal sex. Lancet 2: 57, 1955.

48. Moore KL, Barr ML: The sex chromatin in benign tumors and related conditions in man. Br J Cancer 9: 246, 1955.

49. Moore KL, Barr ML: The sex chromatin in human malignant tissues. Br J Cancer 11: 384, 1957.

50. Moore KL, Graham MA, Barr ML: The detection of chromosomal sex in hermaphrodites from a skin biopsy. Surg Gynecol Obstet 96: 641, 1953.

51. Morris JM: The syndrome of testicular feminization in male pseudohermaphrodites. Am J Obstet Gynecol 65: 1192, 1953.

52. Morris JM, Mahesh VB: Further observations on the syndrome "testicular feminization." Am J Obstet Gynecol 87: 731, 1963.

53. Naujoks H: Sex chromatin in exfoliated cells of cervical carcinoma in situ. Acta Cytol 13: 634, 1969.

54. Ohashi H, Takahashi M: The behavior of sex chromatin in the uterine cervical lesions. J Jpn Soc Clin Cytol 17: 19, 1978.

55. Ohno S: The origin of the sex chromatin body. Acta Cytol 7: 147, 1963.

56. Ohno S, Hauschka TS: Allocycly of the X-chromosome in tumors and normal tissues. Cancer Res 20: 541, 1960.

57. Ohno S, Makino S: The single-X nature of the sex chromatin in man. Lancet 1: 78, 1961.

58. Paegle RD: Ultrastructural alterations induced by air drying, Wright's and PAS staining and their relationship to malignancy associated changes (MAC) in leukocytes. Acta Cytol 16: 466, 1972.

59. Philip J, Trolle D: Familiar male hermaphroditism with delayed and partial masculinization. Am J Obstet Gynecol 93: 1076, 1965.

60. Platt LI, Kailin EW: Buccal X-chromatin frequency in numerous diseases. Acta Cytol 13: 700, 1969.

61. Plunkett ER, Barr ML: Testicular dysgenesis affecting the seminiferous tubules principally, with chromatin-positive nuclei. Lancet 2: 853, 1956.

62. Polani PE: Chromosome phenotypes— Sex chromosome. In: Frazer FC, McKusick VA (eds): Congenital Malformations. pp233-250. Amsterdam: Excerpta Medica, 1970.

63. Polani PE, Hunter WF, Lennox B: Chromosomal sex in Turner's syndrome with coarctation of the aorta. Lancet 2: 120, 1954.

64. Powell I, Tyrkus M, Kleer E: Apparent correlation of sex chromosome loss and disease course in urothelial cancer. Cancer Genet Cytogenet 50: 97, 1990.

65. Sato A: Fine structure of human nuclear chromatin in interphase as observed by polarized dark-field oblique-illumination microscope. Acta Cytol 13: 218, 1969.

66. Sauter G, Moch H, Wagner U, et al: Y chromosome loss detected by FISH in bladder cancer. Cancer Genet Cytogenet 82: 163, 1995.

67. Savino A, Koss LG: The evaluation of sex chromatin as a prognostic factor in carcinoma of the breast: A prelimi-

nary report. Acta Cytol 15: 372, 1971.

68. Schwarzacher HG: Analysis of interphase nuclei. In: Schwarzacher HG, Wolf U (eds): Methods in Human Cytogenetics. pp207-234. Berlin: Springer-Verlag, 1974.

69. Simpson JL, Allen FH, German J: Abnormalities of human sex chromosomes. II. Turner's syndrome associated with the mosaicism 45, X/46, XXp—. Ann Génét 14: 105, 1971.

70. Siracky J: Sex chromatin in cancer of the uterine cervix. Acta Cytol 11: 486, 1967.

71. Spriggs AI, Boddington MM, Clarke CM: Carcinoma-in-situ of the cervix uteri. Some cytogenetic observations. Lancet 1: 1383, 1962.

72. Stanley MA, Bigham DA, Cox RI, et al: Sex-chromatin anomalies in female patients with breast carcinoma. Lancet 1: 690, 1966.

73. Stewart JSS: Genetic mechanisms in human intersexes. Lancet 1: 825, 1960.

74. Straub DG, Lucas LA, McMahon NJ, et al: Apparent reversal of X-condensation mechanism in tumors of the female. Cancer Res 29: 1233, 1969.

75. Tolksdorf M: The diagnosis of X-chromatin by the leukocyte test. In: Achwarzacher HG, Wolf U (eds): Methods in Human Cytogenetics. Berlin: Springer-Verlag, 1974.

76. Turner HH: A syndrome of infantilism, congenital webbed neck, and cubitus valgus. Endocrinology 23: 566, 1938.

77. Von Yamana K: Geschlechtschromatin unter Verschiedenen Endokrinen Funktionen: Klinische und Tierexperimentelle Untersuchungen. Endokrinologie 47: 63, 1964.

78. Wacker B, Miles CP: Sex chromatin incidence and prognosis in breast cancer. Cancer 19: 1651, 1966.

79. Wilkins L, Grumbach MM, Van Wyk JJ: Chromosomal sex in "ovarian agenesis." J Clin Endocrinol 14: 1270, 1954.

4 Y Chromatin

The application of quinacrine fluorochromes to cytogenetic studies has become a routine technique to demonstrate reproducible chromosomal bandings. A brilliant fluorescence at the distal segment of the long arms of Y chromosome is recognized as a bright spot in the interphase nucleus.

The advent of fluorescence in situ hybridization method (FISH) has taken a major place of interphase cytogenetics. As far as Y chromatin is concerned, the quinacrine stain is no doubt simple and inexpensive method for numerical study of Y chromosomes. However, FISH technique with use of other chromosomal and/or oncogene-specific probes has broadened its application.

1 Configuration of Y Chromatin

The Y chromatin or Y body is a bright fluorescent dot measuring about 0.3-1.0 μm in diameter (Fig. 4-1). In some nuclei it takes a double structure of twin bodies having two halves separated (Fig. 4-2).[11,19]

This structure is said to be observed in G_1 as well as G_2 phase.[21] A large nucleus in active growth reveals some dispersion of the Y chromatin in the intermitotic phase (Schwarzacher). While X chromatin locates at the nuclear rim, Y chromatin is spatially related to the nucleolus; Bobrow et al described a para-nucleolar position of the Y chromatin in about 59% of normal somatic cells and Litton et al in about 37% of malignant cells.[3,12] An interesting interpretation has been given on the position of the Y body by Therkelsen and Petersen,[25] who consider that the frequency of centrally locating Y bodies increases during growth phase. It is assumed that the distal segment of the long arm of the Y chromosome contains a specialized form of genetic material concerned with nucleolar organization.[3]

2 Evaluation of Y Chromatin Test

The number of Y chromatin corresponds directly to the number of Y chromosomes. A Y chromatin count is useful for screening of numerical abnormalities of the Y chromosomes; YY syndrome, Klinefelter syndrome, male pseudohermaphroditism (XY female) and their mosaics can be suggested with a combination of X chromatin test. Exceptional is the case when the karyotype with an extremely small, weakly fluorescent Y chromosome lacks Y body in the interphase nucleus.[4,20] As drumstick count is applied to nuclear sexing, quinacrine fluorescence of polymorphonuclear leukocytes has the same diagnostic value. Hale et al[7] have described that the Y body is positive in 79 ± 5.2% of male polymorphonuclear leukocytes and located mainly at the nuclear rim. Because of low frequency of the positive fluorescent body in female

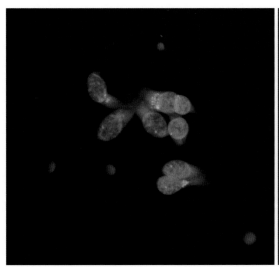

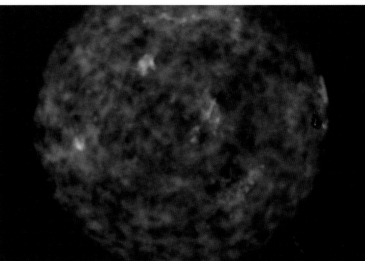

Fig. 4-1 Normal ciliated columnar cells showing single bright fluorescent Y bodies in their nuclei. Yellowish particles are consistent with lipofuscin granules.

Fig. 4-2 Pulmonary adenocarcinoma cell showing double Y bodies with a structure that is composed of twin bodies. Minute fluorescent particles are not Y bodies.

leukocytes below 9%, the screening of a small number of leukocytes is sufficient for nuclear sexing.

An important point to avoid misinterpretation is the presence of tiny fluorescent dots similar to Y bodies; these are considered to be derived from intense fluorescent segments of other autosomes such as the centromere of chromosome 3, the satellite of chromosome 13 and others (acrocentric D and G group chromosomes). Therefore a tiny fluorescent dot should not be read as a true Y body. On the other hand, absence of Y body does not necessarily imply the lack of the Y chromosome. The X chromatin is occasionally visible with Q-staining and should not be interpreted as Y body from its larger size, weaker fluorescence and location at the nuclear rim.[9]

3 Technique of Y Chromatin Determination

As a fluorochrome for Y chromatin study, quinacrine mustard and quinacrine dihydrochloride are used. The quinacrine mustard has a 10 fold fluorescence intensity compared with quinacrine dihydrochloride, so that the former is suitable for a quantitative study.[5]

■Preparation of smears
After gargling, scrape the buccal mucosa firmly several times with a spatula and spread the specimen on clean glass slides. Fix the slides immediately with 96% ethyl alcohol or acetic acid-alcohol for one hour.

For a specimen scraped from the lesion or the cut surface of tumors, suspend the specimen in a physiologic saline to be centrifuged. Spread the cell pellet evenly on clean glass slides and fix them prior to drying.[22]

Addendum: In order to avoid miscounting heterochromatin of autosomes, suspension of the scraped specimen from the tumor into 0.9% sodium citrate solution for 15 minutes is recommended (Takahashi). Mild hydrolysis (5N HCl for 3 minutes at room temperature) is also advised to prevent misinterpretation.[13]

■Staining procedure
For smear:
1) Remove slides from the fixative and pass through graded alcohol to water.
2) Stain with 0.005% aqueous quinacrine mustard solution (or 0.5% quinacrine dihydrochloride) for 20 min at 20°C.
3) Place the slides for 5 min in phosphate buffer solution (pH 6.0).
4) Mount with buffer solution.
For tissue section (Kovács et al)[11]:
1) Deparaffinize tissue sections cut at about 4 μm thick.
2) Pass through graded alcohol solutions.
3) Place the sections in 0.25% trypsin (Difco) balanced salt solution for 30 min at 37°C, then wash with distilled water.
4) Dip in 2% acetic acid.
5) Stain with 0.5% quinacrine dihydrochloride 2% acetic solution for 5 min.
6) Wash the sections with 2 changes of distilled water.
7) Mount with distilled water.

4 Y Chromatin in Cancer Cells

X chromatin abnormalities in cervical neoplasia have been much discussed and detecting missing, duplicated and/or triplicated X chromatins has been applied to screening of early carcinoma of the uterine cervix,[6] urinary bladder,[2] etc. Concerning the behavior of Y chromatins in malignancy of the male, contrastive reports have appeared (Tables 4-1 and 4-2).

Atkin described that Y chromatins were present in the cells of most tumors of males.[1] This evidence was accepted by Litton et al[12] who noticed very high frequencies of fluorescent bodies ranging from 78.1% in a case of ganglioneuroblastoma to 100% in a case of reticulum cell sarcoma (Figs. 4-3 and 4-5). On the other hand, Y-negative neoplasms have been encountered not infrequently; the case with Y chromatin frequency of 10% or less referable to be Y body-negative, was found in about 21%[23] to 27%[24] of all malignancies (Fig. 4-6).

Table 4-1 Y bodies in cancer cells of lung

Case	Sex	Age	Histodiagnosis	Cells counted	Y bodies 1+	2+	3+	4+	%
Mi.G.	M	53	Epidermoid ca.	78	16	0	0	0	20.5
Ta.K.	M	68	Epidermoid ca.	120	37	16	1	0	45.0
Ig.R.	M	45	Large cell ca.	151	18	2	0	0	13.2
Oh.K.	M	45	Small cell ca.	156	5	1	0	0	3.8
Ok.T.	M	38	Small cell ca.	280	12	1	0	0	4.6
Sh.I.	M	48	Small cell ca.	474	175	6	0	0	38.2
Ki.J.*	M	57	Small cell ca.	211	5	0	0	0	2.4
			Adenocarcinoma	153	34	3	0	0	24.2
Sa.T.	M	53	Adenocarcinoma	321	105	15	19	0	43.3
Ta.T.	M	48	Adenocarcinoma	204	58	44	12	2	56.9
Ta.N.	M	38	Adenocarcinoma	300	115	130	20	0	88.3
Og.S.	M	45	Choriocarcinoma	101	53	25	13	0	90.1
Control		n=20	Lymphocytes	2,000					77.6 ± 5.4
			Bronchial cells	159	130				81.8

*This case is double cancer showing small cell carcinoma in the right lung and adenocarcinoma in the left lung. (From Takahashi M: Acta Cytol 21: 132, 1977. ©1977 International Academy of Cytology.)

Table 4-2 Y bodies in cancer cells of gastrointestinal tract

Case	Sex	Age	Histodiagnosis	Cells counted	1+	2+	3+	4+	%
Ka.S.	M	59	Stomach: adenoca.	65	15	0	0	0	23.1
Na.T.	M	47	Stomach: adenoca.	68	10	2	0	0	17.6
Ya.R.	M	43	Stomach: adenoca.	109	11	3	0	0	12.8
Ta.M.	M	51	Stomach: adenoca.	199	7	1	0	0	4.0
Oh.K.	M	49	Stomach: adenoca.	300	26	0	0	0	8.7
Ya.T.	M	52	Stomach: adenoca.	88	41	4	1	0	52.3
Ko.M.	M	70	Stomach: early ca.	281	115	75	1	0	68.0
Ts.H.	M	49	Stomach: scirrhous ca.	35	23	1	0	0	68.6
			Benign mucosa	104	90	1	0	0	87.5
Ha.B.	M	76	Stomach: adenoca.	71	5	0	0	0	7.0
			Benign mucosa	70	52	0	0	0	74.3
Ta.Z.	M	48	Rectum: adenoca.	351	40	21	16	2	22.5
En.R.	M	48	Rectum: adenoca.	289	105	65	2	0	59.5
Ho.R.	M	39	Sigmoid: adenoca.	120	5	1	0	0	5.0
			Benign mucosa	235	190	0	0	0	80.9
Ta.N.	M	47	Rectum: adenoca.	67	2	0	0	0	3.0
			Benign mucosa	230	172	0	0	0	74.8
Se.T.	M	49	Rectum: adenoca.	104	21	1	0	0	21.2
Ki.S.	M	49	Pancreas: adenoca.	105	5	2	0	0	6.7
Og.W.	M	47	Stomach: ulcer	87	70	0	0	0	80.5
Ok.T.	M	40	Chronic gastritis	25	20	0	0	0	80.0
Na.M.	M	48	Chronic gastritis	38	30	0	0	0	78.9

(From Takahashi M: Acta Cytol 21: 132, 1977. ©1977 International Academy of Cytology.)

Table 4-3 Y bodies and Y chromosomes in cultured cancer cells

Cell Line	Sex	Primary site and histodiagnosis	Subculture passages	Y body or Y chromosome	Numbers counted	1+	2+	3+	4+	%
PC4	M	Lung: Adenocarcinoma	45	Y body	526	44	13	0	0	10.8
PC5	M	Lung: Epidermoid carcinoma	47	Y body	497	22	1	0	0	4.6
			73	Y body	163	7	1	0	0	4.9
PC6	M	Lung: Oat cell carcinoma	25	Y body	520	330	78	16	3	82.1
			25	Y chromosome	112	54	14	1	0	61.6
PC7	M	Lung: Adenocarcinoma	72	Y body	606	53	10	1	0	10.6
			72	Y chromosome	63	4	0	0	0	6.3
PC8	M	Lung: Adenocarcinoma	62	Y body	200	28	1	0	0	14.5
			115	Y body	263	43	5	0	0	18.3
			115	Y chromosome	12	1	0	0	0	8.3
RH1	M	Kidney: Renal cell carcinoma	25	Y body	307	16	3	0	0	6.2
PC3	F	Lung: Adenocarcinoma	92	Y body	540	0	0	0	0	0

(From Takahashi M: Acta Cytol 21: 132, 1977. ©1977 International Academy of Cytology.)

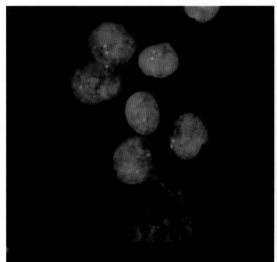

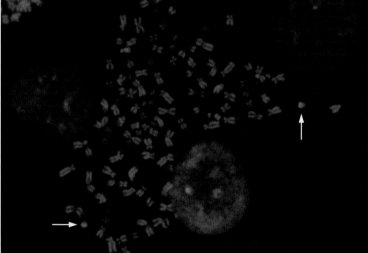

Fig. 4-3 Small adenocarcinoma cells showing many double Y bodies. From a case with pulmonary adenocarcinoma.

Fig. 4-4 Metaphase chromosomes of oat cell carcinoma in culture. Note double Y chromosomes yielding brilliant fluorescence at the distal segment of long arms (arrows).

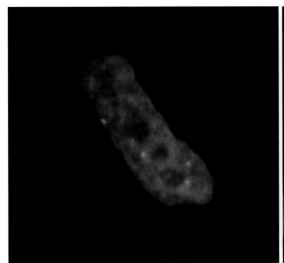

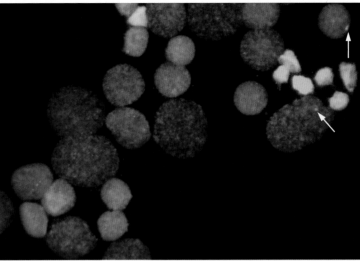

Fig. 4-5 Adenocarcinoma cell having quadruple Y bodies. From a male with adenocarcinoma of the rectum.

Fig. 4-6 Oat cell carcinoma with low frequency of Y bodies. Tiny weak fluorescent dots indicate Y chromatins (arrows). The majority of cancer cells lack in the fluorescent Y bodies.

Sandberg and Sakurai[14] noted missing Y chromosomes in 4 of 103 patients with chronic myelogenous leukemia; although no definite evidence suggestive of a prognostic significance could be verified, the fact that these patients never entered the blastic phase was mentioned.

The difference of Y chromatin behavior between benign and malignant cells consists not only in the frequency but also in multiplicity. According to the studies by Litton et al, duplication and triplication have occurred with high frequency in addition to the high positive rates in total cell counts in malignant neoplasms of the male.

It is noteworthy that the appearance of multiple Y chromatins has been encountered even in the case of relatively low Y chromatin frequency. Clearly understandable explanations on numerical variations of Y chromatins have not appeared. Several questions may occur; (1) although the loss of the Y bodies has been mentioned to occur in an early or precancerous state of neoplasms,[23] it should be clarified whether or not the evidence is related to initiation or promotion of the tumor, (2) as a practical point, benign epithelial and stromal cells may be misread within the tumor cells in dealing with surgical specimens even if the tumor is solid and localized, (3) it is uncertain whether increasing and missing Y bodies really mean additions and losses of Y chromosomes. A comparative study on metaphase and interphase nuclei of cultured cancer cells has given some proof for these questions (Fig. 4-4, Table 4-3). As Peterson et al[14] have emphasized a decrease in Y chromosomes after about 30 passages, the Y body frequency was considerably low in our culture studies, less than 18% in five of six cell lines, and remained fairly constant in continuous subcultures. One cell line which was derived from pulmonary oat cell carcinoma possessed multiple Y bodies as well as double and triple Y chromosomes. In spite of frequent, multiple Y chromatins more than three, rarity of triple and quadruple Y chromosomes cannot be simply explained from numerical increase of Y chromosomes. It is important to count carefully Y bodies with constant configuration and brilliant fluorescence from well preserved and well stained smears, because some autosomes may emit heteropyknotic fluorescence with or without alteration of A-T to G-C content ratio.

As a reliable method to identify sex chromosomes for interphase and metaphase cells. FISH technology is very valuable. A loss of Y chromosomes was ever suggested to be related with poor prognosis.[8,15,17] However, the result of the study by Sauter et al[18] was controversial; the result that used simultaneously multiple probes specific to chromosomes 7, 9 and 17 showed no association between chromosome Y and grade of malignancy or stage but correlation to aging of the patient.

Further clinical studies on the significance of Y signal nullisomy and polysomy in various neoplasms are required. From the aspect of practice of interphase cytogenetics each laboratory must have cut-off value with many normal controls.[10]

References

1. Atkin NB: Y chromosomes and quinacrine fluorescence technique. Br Med J 4: 118, 1970.
2. Atkin NB, Petkovic I: Variable sex chromatin pattern in an early carcinoma of the bladder. J Clin Pathol 26: 126, 1973.
3. Bobrouw M, Pearson PL, Collacott HEAC: Para-nucleolar position of the human Y chromosome in interphase nuclei. Nature 232: 556, 1971.
4. Borgaonkar DS, Hollander DH: Quinacrine fluorescence of the human Y chromosome. Nature 230: 52, 1971.
5. Caspersson T, Lomakka G, Zech L: The 24 fluorescence patterns of human metaphase chromosomes: Distinguishing characters and variability. Hereditas 67: 89, 1971.
6. Goldstein AI, Kent D, Ketchum M: A new method of screening for malignancy of the cervix. Lancet 1: 146, 1973.

7. Hale MT, Vaughn WK, Engel E: Fluorescent male sex chromatin in white blood cells. South Med J 66: 340, 1973.

8. Holmes RI, Keating MJ, Cork A, et al: Loss of the Y chromosome in acute myelogenous leukemia: A report of 13 patients. Cancer Genet Cytogenet 17: 269, 1985.

9. Iinuma K, Nakagome Y: Fluorescence of Barr body in human amniotic-fluid cells. Lancet 1: 436, 1973.

10. Jenkins RB, LeBeau MM, Kraker WJ, et al: Fluorescence in situ hybridization: A sensitive method for trisomy 8 detection in bone marrow specimens. Blood 79: 3307, 1992.

11. Kovacs M, Vass L, Sellyei M: Detection of Y chromatin and Barr bodies in histological sections by quinacrine fluorescence. Stain Technol 48: 94, 1973.

12. Litton LE, Hollander DH, Borgaonkar DS, et al: Y-chromatin of interphase cancer cells: A preliminary study. Acta Cytol 16: 404, 1972.

13. Pearson PL, Bobrow M, Vosa CG: Technique identifying Y chromosomes in human interphase nuclei. Nature 226: 78, 1970.

14. Peterson WD Jr, Simpson WF, Ecklund PS, et al: Diploid and heteroploid human cell lines surveyed for Y chromosome fluorescence. Nature New Biol 242: 22, 1973.

15. Powell I, Tyrkus M, Kleer E: Apparent correlation of sex chromosome loss and disease course in urothelial can-

cer. Cancer Genet Cytogenet 50: 97-101, 1990.

16. Sandberg AA, Sakurai M: The missing Y chromosome and human leukaemia. Lancet 1: 375, 1973.

17. Sandberg AA: Chromosome changes in bladder cancer: Clinical and other correlations. Cancer Genet Cytogenet 19: 163, 1986.

18. Sauter G, Moch H, Wagner U, et al: Y chromosome loss detected by FISH in bladder cancer. Cancer Genet Cytogenet 82: 163, 1995.

19. Schwarzacher HG: Analysis of interphase nuclei. In: Schwarzacher HG, Wolf U (eds): Methods in Human Cytogenetics. pp207-234. Berlin: Springer-Verlag, 1974.

20. Schwinger E: Non-fluorescent Y chromosome. Lancet 1: 437, 1973.

21. Schwinger E, Pera F: On the splitting of the Y-fluorescent body in man. Humangenetik 14: 107, 1972.

22. Sellyei M, Vass L: Absence of fluorescent Y bodies in some malignant epithelial tumours in human males. Eur J Cancer 8: 557, 1972.

23. Sellyei M, Vass L: Sex-chromosome loss in human tumors. Lancet 1: 1041, 1975.

24. Takahashi M: A behavior of Y chromatin in cancer cells of males. Acta Cytol 21: 132, 1977.

25. Therkelsen AJ, Peterson GB: Frequency of "Y chromatin body" in human skin fibroblasts in tissue culture and its relation to growth phase. Exp Cell Res 65: 473, 1971.

5 Fluorescence In Situ Hybridization (FISH)

According as molecular genetic analyses undergo a rapid progress, human cancers with genetic instability are characterized by activation of oncogenes, loss of tumor suppressor genes, excessive production of oncogene or antioncogene products, etc. which are associated with chromosomal abnormalities such as translocation, deletion or loss and insertion (Tables 5-1 and 5-2). Karyotypic changes in leukemia and associated diseases with monoclonality have been extensively studied because adequate sample is obtainable. For solid tumors, however, metaphase chromosome analyses after short-term culture are tedious and not always successful.

Advent of fluorescence in situ hybridization (FISH) of interphase nuclei may supplant conventional cytogenetics in some part, because FISH is the simplest, the most facile and the most reliable method for detecting chromosomal aberrations. FISH technology still has a shortage of probes that cover the whole genome. As the second shortcoming of FISH, most of FISH probes require targets on chromosomes of the order of 10,000 or more base pairs.[7]

A large scale clinical study on progression from intraepithelial neoplasia to invasive carcinoma and further to metastasis of prostatic carcinoma by Qian et al have found a gain of chromosome 8 was the most common numerical alteration and became more pronounced with increasing the stage and the grade. The trisomy 8 was followed by gains of chromosome 7 and 10.[14] The presence of chromosome 7 trisomy has significantly poor survivorship[8] (Figs. 5-1 and 5-2).

FISH technology for detecting chromosomal aberrations of human tumors in reference to clinical behaviors has been reported in various organs, the urinary bladder,[8] breast,[3,5,9] ovary,[12] colon,[3] and desmoid tumor.[6]

In addition to confirmation of chromosomal aberrations such as polysomy, monosomy, nullisomy and chimera formation, oncogene amplification analysis using the targeted gene and its related chromosome is a valuable method; increased signals (extrachromosomal amplification) and aggregation of signals (intrachromosomal amplification) can be easily examined for interphase nuclei of tumor cells obtained by fine needle aspiration[16] (Fig. 5-3)

A metaphase FISH is also available to get localization of the target probe; the c-myc amplification can be observed as aggregated signals (Fig. 5-4).

A multi-target FISH method using various probes bound with different fluorochromes may become valuable for interphase cytogenetics, because pleural chromosomal aberrations occur secondarily in malignant neoplasia. An example of multi color FISH for a normal cell is illustrated in Fig. 5-5.

Table 5-1 Chromosomal changes diagnostic of the tumor involved

Tumor	Chromosomal abnormality
	Translocations
Myxoid liposarcoma	t(12;16)(q13;p11)
Rhabdomyosarcoma	t(2;13)(q37;q14)or t(1;13)(p13;q14)
Chondrosarcoma	t(9;22)(q31;q25)
Leiomyoma	t(12;14)(q14-15;q23-24)
Lipoma	t(12;?)(q14;?)
Ewing's sarcoma	t(11;22)(q24;q12)
Synovial sarcoma	t(x;18)(p11;q11)
Pleomorphic adenoma	t(3;8)(p21;q12)or t(9;12)(p13;q13)
	Deletions and other changes
Germ cell tumors	i(12p)
Meningioma	-22/22q-
Retinoblastoma	13q-
Wilms' tumor	11p-
Kidney tumors	3p-
Small-cell cancer of lung	3p-

(From Sandberg AA: The cytogenetics of solid tumors. Adv Oncol 8: 3-9, 1992.)

Table 5-2 Some of the more recurrent but not necessarily diagnostic chromosomal changes in tumors

Bladder cancer Structural changes of chromosomes 1, 3, and 11 i(5p)/5q-,+7,del(8p),-9/9q-	**Lipoma** del(13)(q12q22), der(6)(p21-23), der(11)(q13)	**Retinoblastoma** Structural change of chromosome 1 i(6p)
Breast cancer Structural changes of chromosomes 1(1p35) and 7 t/del(16q)	**Liposarcoma** **(well differentiated and pleomorphic)** Telomere association, rings Many abnormal markers	**Small-cell lung cancer** del(3)(p14p23),17p-
Colon cancer Structural changes of chromosomes 1,17, and 18 +7,+12,del(5q),del(10q)	**Malignant melanoma** t/del(1)(p12-22), t(1;19)(q12;p13), t/del(6q)/i(6p), +7	**Testicular tumors** Structural and numerical changes of chromosome 1
Glioma Double minute chromosomes +7,–10	**Neuroblastoma** del(1)(p31-32) Double minute chromsomes Homogeneously staining regions	**Thyroid tumors** 19q13 (follicular adenoma) **Uterine cancer** Structural and numerical changes of chromsome 1
Kidney cancer t(5;14)(q13;q22)	**Ovarian cancer** Structural changes of chromosome 1 t(6;14)(q21;q24), del(6q)	**Wilms' tumor** Structural changes of chromosome 1
Leiomyoma +12,del(7)(q21,2q31,2),del/inv(1q),–22	**Prostate cancer** del(7)(q22), del(10)(q24)	

(From Sandberg AA: The cytogenetics of solid tumors. Adv Oncol 8: 3-9, 1992.)

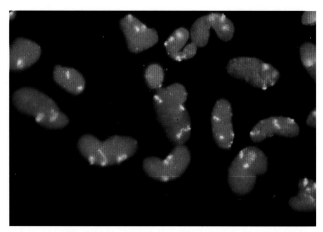

Fig. 5-1 FISH for metaphase nuclei of urinary bladder cancer using the chromosome 7 probe. Many cells contain two normal signals and cancer cells show three signals as a result of trisomy 7. (By courtesy of Dr. Avery A. Sandberg, The Genetics Center and The Cancer Center, Scottsdale, Arizona, U.S.A.)

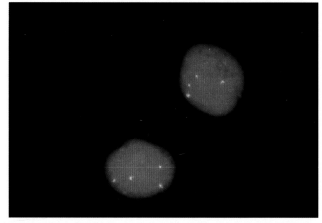

Fig. 5-2 Fish for metaphase nuclei of urinary bladder cancer using the chromosome 8 probe. Note two extrachromosomes resulting in tetrasomy (+8,+8). (By courtesy of Dr. Avery A. Sandberg, The Genetics Center and The Cancer Center, Scottsdale, Arizona, U.S.A.)

1 In Situ Hybridization on Human Chromosomes

1) Prepare 200 ml of RNase solution of final concentration of 100 μg/ml. Preheat the solution at 37°C for 20 min. Immerse the slides for 1 hour at 37°C and rinse three times in 2 × standard saline citrate (SSC) for 10 min each at room temperature.

2) Dehydrate the slides in 50%, 75% and 100% ethanol. Air dry the slides at room temperature.

3) Prepare 100 ml of 70% formamide, 2 × SSC solution to be separated into two baths of 50 ml. Preheat at 70°C for 20 min before use.

4) Dilute DNA probes in the hybridization buffer (pH 7) consisting of 50% deionized formamide, 2 × SSC pH 7, 40 mM phosphate buffer, Denhardt's solution, 0.1% SDS and 10% dextran sulfate (Pharmacia LKB). Keep on ice. Hybridization is carried out with two or three different concentrations of biotinylated DNA. (The optimal concentrations of DNA varies according to the probe size. For large probes over 4 kb long 5, 7, 10 ng/μl in a final volume of 15 μl are sufficient. For 1-3 kb probes 7, 10, 15 ng/μl are used. For small probes below 1 kb 15 and 20 ng/μl are recommended.)

5) Denature the slides for 2 min in 70% formamide, 2 × SSC at 70°C. Then rinse the slides for 2 min in 200 ml of 2 × SSC (pH 7) placed on ice.

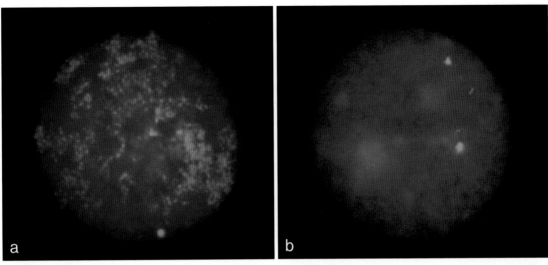

Fig. 5-3 C-myc amplification of colon cancer derived cell (COLO 320-DM) with control of peripheral blood mono-cyte.
a. Marked increased signals of LS1 c-myc spectrum orange (Vysis) and double green chromosome 8 alpha-satellite signals (Vysis).
b. Each two signals using the same probes.
(Figures 5-3 to 5-5 by courtesy of Mr. M. Maekawa, Medical Supplies & Systems, Fujisawa Pharmaceutical Co., Ltd., Osaka, Japan.)

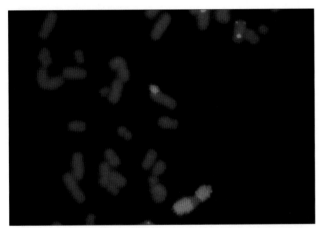

Fig. 5-4 Metaphase FISH of a COLO 320 cell derived from colon cancer. The c-myc amplification is identified on an abnormal homogeneously staining region as aggregated LS1 c-myc (8q24) spectrum orange signals. Note CEP8 spectrum green signals (Vysis).

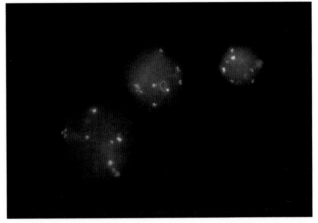

Fig. 5-5 Multi-target FISH of a normal metaphase nucleus from peripheral blood. A useful FISH technique to detect genetic abnormalities of targeted chromosomes using combining probes provided by Vysis (Downers Grove, Illinois, U.S.A.). This figure is showing chromosome 13 (spectrum green), chromosome 18 (spectrum green + spectrum orange), chromosome 21 (spectrum orange), chromosome X (spectrum orange + spectrum aqua) and chromosome Y (spectrum aqua).

6) Dehydrate in 50%, 75%, and 100% ethanol placed on ice for 2 min each. Air dry the slides.

7) After addition of the denatured probes on to the chromosome spreads, cover with a small piece of high temperature-resistant plastic film (15 mm × 15 mm). Incubate the slides overnight at 37 °C.

8) Perform posthybridization washes with 50% formamide, 2×SSC, and 2×SSC (pH 7).

2 Immunodetection of Probes

Probes are detected by two-step immunoreactions using the primary antibody goat antibiotin (Vector Laboratories Inc., Burlingame, California, U.S.A.) and the second antibody rabbit anti-goat IgG conjugated to FITC (Nordic Laboratories, France) are preferable to a direct reaction using avidin or streptavidin conjugates.[19] A

counterstain with propidium iodide is better than with rhodamine or Texas red.[19]

1) After posthybridization washes, immerse the slides in 200 ml modified PBS (PBT) that contains 0.4% BSA bovine serum albumin (IBF Laboratories, France) and 0.1% Tween 20 (Sigma Chemical Co., St. Louis, Missouri, U.S.A.) for 5 min.

2) Place the slides without dry into 200 ml fresh PBT. Remove from PBT solution and add 100 μl of anti-biotin antibody diluted in PBT (final concentration of 5 μg/ml) to each slide.

3) Put a glass coverslip and incubate in a moist chamber for 45 min at 37 °C. Thereafter rinse for 5 min in 200 ml of PBT solution at room temperature.

4) Place the slides in 200 ml of fresh PBT. Remove from

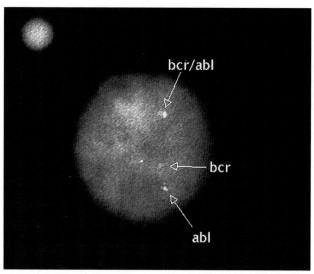

Fig. 5-6a An interphase nucleus from a patient with CML. M-bcr/abl chimera showing fused fluorescent signals. Note an isolated red bcr signal and a green abl signal.

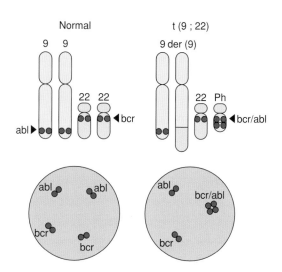

Fig. 5-6b Schematic illustration of t(9;22) by FISH.

PBT and add 100 μl of anti-goat IgG FITC conjugate(30:500 diluted in PBT to a final concentration of 60 μg/ml) to each slide. Cover with a glass coverslip (22 mm × 32 mm) and incubate in a dark moist chamber at 37℃ for 45 min.

5) For counterstain add 100 μl of propidium iodide of final concentration of 1 to 5 μg/ml for 4 min at room temperature without removing the fluorescein conjugate.

6) Remove the coverslip and rinse twice with PBS using a Pasteur pipette. Put a few drops of PBS and cover with a glass coverslip (24 mm × 60 mm) to avoid dry. Then store the slides at 4℃ in the dark.

7) For observation remove the coverslip, add 10 to 20 μl of the antifading agent PPD (p-phenylenediamine) and cover with a glass coverslip (24 mm × 60 mm).

3 Procedure of In Situ Hybridization on Biopsy and Needle Aspirates

■Biopsy

1) Place biopsy material into a sterile beaker containing 10 ml of RPMI 1640 10 ml. When tissue specimens are submitted, cut it into small pieces with sterile scissors.

2) Add 10 mg collagenase (0.1% collagenase/PBS) onto tissue pieces and incubate for 15 to 30 min at 37℃.

3) Remove 1.5 to 2.0 ml of cell suspension to a homogenizer, and gently homogenize.

4) Filter the cell suspension through a nylon mesh placed on another beaker (pore size 100 μm).

5) Remove the dispersed cell suspension to a microtube and centrifuge at 1,000 rpm for 5 min.

6) Decant the supernatant and resuspend in 10 ml 0.075 M KCl hypotonic solution and stir by tapping. Leave the tube at room temperature for 30 min.

7) Centrifuge at 1500 rpm for 5 min and decant leaving about 2 ml of supernatant.

8) Stir using a tip of the pipette and make the cell suspension.

9) Prepare 10 ml of Carnoy solution (methanol:acetic acid, 3:1), on which drop the cell suspension one by one.

10) Cap the microtube and shake several times and centrifuge at 1,500 rpm for 5 min. Repeat this step 2 times

changing the Carnoy solution.

11) Place a drop of cell suspension onto a clean glass slide. Fix the glass slide with vapor on boiling water and air dry overnight.

■Needle aspirates

1) After fine needle aspiration, aspirate 500 μl of 0.05 M KCl using the syringe which was used for needle aspiration biopsy.

2) Mix well by tapping with an index finger. Replace the specimen to another microtube that contains fresh 500 μl Carnoy solution. Leave for 10 min.

3) Centrifuge at 1,500 rpm for 5 min and decant the supernatant, leaving about 2 ml of supernatant.

Following steps are same as the step 8.

4 Selection of Probes

■HER-2/neu

Targets of selecting probes for FISH have been placed on chromosomes which are frequently aberrated in malignancies and or related to activation of oncogenes by amplification and translocation.

The HER-2/neu (c-erb-B2) amplification is shown in about 30% of mammary carcinomas. An overexpression HER-2/neu gene product P185[HER-2] that is highly predictive of a poorer outcome was used to be measured by immunohistochemistry and Western blotting.[18] A recent report by Sauter et al has proved the gene amplification applying a dual-labeling technique for the erb-B2 cosmid probe (17q21) and pericentromeric probe of chromosome 17 that include p53 (17p13.1) nm 23 and the postulated familial breast cancer gene, BRCAL.[16] Heterogeneity of the mammary carcinoma is evidenced by increase and variation in number of erb-B2 signals. In addition, as an advantage of FISH the targeted gene amplification is demonstrable not only by increased signals but also by the appearance of clustered signals.[12,17]

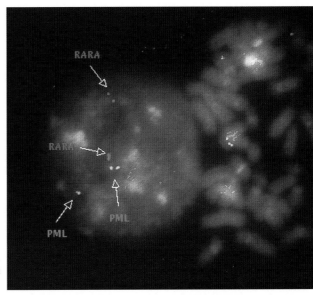

Fig. 5-7 An interphase nucleus from a patient with APL.
Negative for PML/RARA chimera gene. Both green PML gene and red RAR gene are isolated.

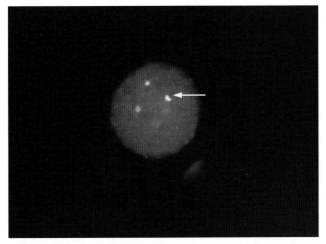

Fig. 5-8a An interphase nucleus from a patient with APL.
Note a fused PML/RARA chimera gene.

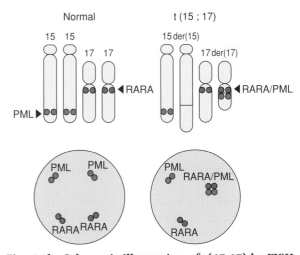

Fig. 5-8b Schematic illustration of t(15;17) by FISH.

■bcr/abl chimera gene

The Philadelphia chromosome (Ph[1]) occurs in more than 90% of chronic myelogenous leukemia (CML): it results from a translocation t(9;22)(q34;q11) that fuses c-able oncogene on chromosome 9 to bcr (breakpoint cluster region) on chromosome 22. The majority of bcr breakpoints in CML occur within the narrow 5.8 kb major breakpoint cluster region (M-bcr). The Ph[1] is also observed in about 20% to 30% of acute lymphocytic leukemia (ALL) in adults. In children about 2% to 5% of ALL are associated with Ph[1]. The patient with ALL positive for Ph[1] take a downward prognosis with frequent involvement of the brain and resistance to chemotherapy.[2,13] The bcr breakpoint occurs in the more centromeric locus as called minor breakpoint cluster region (m-bcr). The fused bcr/abl chimera gene can be detected by FISH using a dual color probe (Fig. 5-6). Fusion of dual color signals results from M-bcr or m-bcr.

For distinction between M-bcr and m-bcr, a reverse transcription polymerase chain reaction (RT-PCR) method to measure mRNA encoded by Mbcr/abl and mbcr/abl genes is required.

■PML/RARA

The majority of promyelocytic leukemia (APLM3) is associated with t(15;17)(q22;q11.2) which results in chimera gene formation consisting of PML(myl) on chromosome 15q22 and RARα (retinoic acid α receptor) on chromosome 17q11.2-q12 (Figs. 5-7 and 5-8). The evidence that the locus of breakpoint 17q is in close proximity to G-CSF, NGFR(nerve growth factor receptor), MPO (myeloperoxidase) and HER-2/neu genes was thought to be concerned with onset of APL.[4,20]

The FISH method is extremely important to detect residual leukemic cells of the patient in a clinical remission. Zhao et al have proved that the FISH results correlated well with those of a two-round nested RT-PCR assay.[21]

■AML1

In about 20% of acute myelogenous leukemia (AML, M2), translocation between chromosome 8 and chromosome 21 occurs and it indicates a good convalescence. As a result of translocation the MTG8 gene locating on the 8q22 and the AML1 gene locating on the 21q22[11] form the chimera AML1/MTG8 gene. The FISH method detects splitting AML1 signals using AML1 cosmid clone. In account of numerical abnormalities of fluorescent signals, distribution of signals must be compared with those of healthy persons.

References

1. Abe R, Umezu H, Uchida T: Myeloblastoma with 8;21 chromosome translocation in acute myeloblastic leukemia. Cancer 58: 1260, 1986.

2. Berger R, Chen SJ, Chen Z: Philadelphia-positive acute leukemia: Cytogenetic and molecular aspects. Cancer Genet Cytogenet 44: 143, 1990.

3. Cajulis R, Frias-Hidvegi D: Detection of numerical chromosomal abnormalities in malignant cells in fine needle aspirates by fluorescence in situ hybridization of interphase cell nuclear with chromosome-specific probes. Acta Cytol 37: 391, 1993.

4. Chang KS, Schroeder W, Siciliano MJ: Localization of the human myeloperoxidase gene is in close proximity to the translocation breakpoint in acute promyelocytic leukemia. Leukemia 1: 458, 1987.

5. Devilee P, Van Vliet M, Bardel A, et al: Frequent somatic imbalance of marker alleles for chromosome 1 in human primary breast carcinoma. Cancer Res 51: 1020, 1991.

6. Fletcher JA, Naeem R, Xiao S, et al: Chromosome aberrations in desmoid tumors. Cancer Genet Cytogenet 79: 139, 1995.

7. Fox JL, Hsu P-H, Legator MS, et al: Fluorescence in situ hybridization: Powerful molecular tool for cancer prognosis. Clin Chem 41: 1554, 1995.

8. Hopman AHN, Poddighe PJ, Smeets AW, et al: Detection of numerical chromosome aberrations in bladder cancer by in situ hybridization. Am J Pathol 135: 1105, 1989.

9. Ichikawa D, Hashimoto N, Hoshima M, et al: Analysis of numerical aberrations in specific chromosomes by fluorescent in situ hybridization as a diagnostic tool in breast cancer. Cancer 77: 2064, 1996.

10. Kallioniemi O, Kallioniemi A, Kurisu W, et al: C-erbB-2 oncogene amplification in breast cancer analyzed by fluorescence in situ hybridization. Proc Natl Acad Sci USA 89: 5321, 1992.

11. Miyoshi H, Shimizu K, Kozu T: t(8;21) breakpoint on chromosome 21 in acute myeloid leukemia are clustered within a limited region of a single gene AML1. Proc Natl Acad Sci 88: 10434, 1991.

12. Nederlof PM, Robinson R, Abuknesha J, et al: Three-color fluorescence in situ hybridization for the simultaneous detection of multiple nucleic acid sequences. Cytometry 10: 20, 1989.

13. Pui C-H, Crist WM, Look AT: Biology and clinical significance of cytogenetic abnormalities in childhood acute lymphoblastic leukemia. Blood 76: 1449, 1990.

14. Qian J, Bostwick DG, Takahashi S, et al: Chromosomal anomalies in prostatic intraepithelial neoplasia and carcinoma detected by fluorescence in situ hybridization. Cancer Res 55: 5408, 1995.

15. Sandberg AA: The cytogenetics of solid tumors. Adv Oncol 8: 3, 1992.

16. Sauter G, Feichter G, Torhorst J, et al: Fluorescence in situ hybridization for detecting ErbB-2 amplification in breast tumor fine needle aspiration biopsies. Acta Cytol 40: 164, 1996.

17. Sauter G, Moch H, Moore D, et al: Heterogeneity of erbB-2 gene amplification in bladder cancer. Cancer Res 53: 2199, 1993.

18. Szöllösi J, Balázs M, Feuerstein BG, et al: ERBB-2 (HER2/neu) gene copy number, p185 overexpression, and intratumor heterogeneity in human breast cancer. Cancer Res 55: 5400, 1995.

19. Viegas-Pequignot E: In situ hybridization to chromosomes with biotinylated probes. In: Wilkinson DG (ed): In Situ Hybridization: A Practical Approach. pp137-158. Oxford: IRL Press, 1992.

20. Weil SC, Rosner GL, Reid MS: Translocation and rearrangement of myeloperoxidase gene in acute promyelocytic leukemia. Science 240: 790,1988.

21. Zhao L, Chang K-S, Estey EH, et al: Detection of residual leukemic cells in patients with acute promyelocytic leukemia by the fluorescence in situ hybridization method: Potential for predicting for relapse. Blood 85: 495, 1995.

Part 2
Practical Cytology of Organs

6. Female Genital Tract

7. Breast

8. Respiratory Tract

9. Thyroid

10. Thymus and Mediastinum

11. Salivary Glands

12. Esophagus and Intestines

13. Liver, Bile Ducts and Pancreas

14. Effusions in Body Cavities

15. Urinary Tract

16. Prostate

17. Lymph Node

18. Prognostic Markers in Malignant Lymphoma

19. Bone and Bone Marrow

20. Central Nervous System

21. Skin and Its Related Mucous Membrane

22. Telepathology and Telecytology

6 Female Genital Tract

1 Overview of Carcinoma of the Uterus

Although mortality of the uterine cervix has been decreased, the incidence differs between developed countries and developing countries; according to estimates of the incidence of common cancers in 1985 by Parkin et al,[175] carcinoma of the cervix is third in developing countries, while it is 10th in developed countries. Cytology is the most reliable method for detecting early cancer as well as precancerous lesions. An appropriate management of cytology screening is proceeded as follows; sampling smears, slide preparation, interpretation of smears and reporting system. Since the National Cancer Institute Workshop in 1988, the Bethesda System has been widely adopted instead of Papanicolaou classification. As cytology of human papillomavirus (HPV) infection is included in the category of squamous intraepithelial lesions, current molecular techniques implicated HPV as the cause of cervical carcinoma; subtypes HPV 16, 18, 30, 31, 33, 35, and 39 seemed to be associated with oncogenesis.

Full genome of HPV 16 or 18 inserted into Syrian hamster embryonic cells which were immortalized by HSV-2 gave rise to morphologic transformation.[92] According to postulate of zur Hausen, synergism of HPV and HSV-2 on oncogenesis of cervical carcinoma referred to HSV-2 infection as initiating function.[286]

The incidence of endometrial carcinoma is 12th in developed countries, while it is 16th in developing countries.[175] The ratio of cervical cancer to endometrial cancer is nearly 1.02: 1.0 in affluent caucasians. The disease occurs at the age of menopause and older. Cytoscreening is recommended for women after menopause with high risks; as physiologic factors obesity, diabetes, hypertension, infertility and anovulation, nulliparity and Stein-Leventhal syndrome, and as exogenous or environmental factors high dietary fat and hormone replacement therapy with unopposed estrogen are pointed out.[256]

2 Preparation of Smears

There are various methods for the preparation of smears. Each has specific value for its own purpose in screening and diagnosis of cancers or hormonal evaluation.

■ Vaginal pool smear

Method: Aspirate secretions from the posterior fornix using a pipette and spread thinly on a clean glass slide with the pipette. Vaginal secretions can also be taken with a wooden spatula or cotton swab.

Cellular components: These consist mainly of superficial and intermediate squamous cells, a small number of parabasal and endocervical cells, and very occasionally a few endometrial cells. Histiocytic cells are often seen. The Döderlein bacilli are frequently found in normal vaginal flora and cause cytoplasmic degeneration of the intermediate and superficial cells.

Application: This smear can be applied to hormonal evaluation; however, accurate hormonal cytology is obtained only from the scraping smear of the upper lateral vaginal wall. It is also used in routine examinations for cancer detection, but may fail to detect cervical carcinoma unless a cervical scraping smear is studied simultaneously.

■ Cervical scraping smear

Method: Insert a cotton swab or an Ayre's wooden cervical-scraper and scrape the portio vaginalis around the external os including a portion of the cervical canal. Make smears by rolling the pre-moistened, nonabsorbent cotton swab on clean glass slides. Do not rub. A tight cotton swab is suitable to obtain the specimen from the endocervix (*endocervical smear*). An ordinary wooden tongue blade or Ayre's wooden spatula is commonly used to collect the specimen from the ectocervix especially around the external os (Fig. 6-1).

A brushing device like cytobrush (Fig. 6-2) is commercially available and most effective to get sufficient cells from the cervical canal and cervical os.

Cellular components: The number of shed cells and the size of cell clusters differ by the procedure taking specimens. Endocervical cells exfoliate in large numbers, especially forming cell clusters by use of a wooden spatula. A substantial number of superficial, intermediate and parabasal squamous cell are present. Histiocytic cells are found occasionally.

Application: This method is reliable for the detection of cervical carcinoma, either adenocarcinoma or squamous

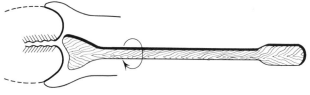

Fig. 6-1 Wooden spatula used by Ayre for cervical scrapings.

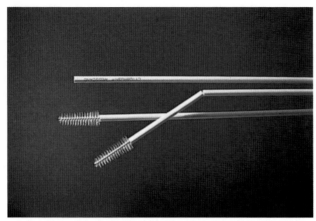

Fig. 6-2 **Cytobrush for endocervical brushing.** After gently brushing the endocervix, make a direct smear or place a tip into preservation fluid.

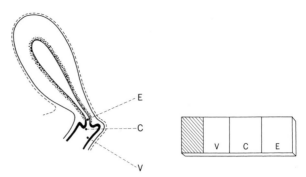

Fig. 6-3 Preparation of VCE smear.

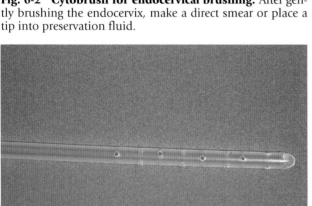

Fig. 6-4 Tip of endometrial aspiration device by Masubuchi. A polyethylene-made disposable cannula 3 mm in diameter is designed with small holes.

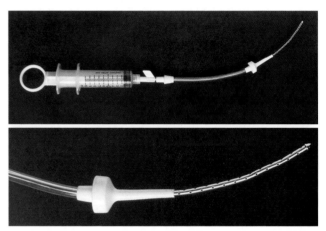

Fig. 6-5 **Endometrial cell sampler of Curity-Isaacs.** Note the tip with many bilateral holes.

cell carcinoma including in situ carcinoma. In routine cytology it is desirable to use the vaginal pool smear in addition to the cervical smear for detection of endometrial carcinoma.

■VCE smear

Preparation of VCE smears was designed by Wied[276] to smear specimens obtained from three parts on the same glass slide (Fig. 6-3). Examination of VCE smears is valuable for routine cytological screening of adult females regardless of clinical signs or symptoms. A slide is divided into three parts, V for the specimen from the upper lateral wall of the vagina obtained with a wooden blade, C for the specimen from the ectocervix around the external os obtained with another wooden blade, and E for the specimen from the endocervical canal by means of a pre-moistened cotton swab. Practically, soon after spreading the endocervical specimen on the E portion, spreading the specimens from C and V should be promptly done in order to avoid air drying during preparation.

Addendum: **Self sampling device**

A proportion of women who participate in cervical cancer screening project is large in western countries. However, there are still large numbers of women who are not aware of the significance of cancer screening or not will-

ing to be examined at clinic in Asian countries.

Application: This method is suitable for mass screening survey in districts where medical facilities are insufficient, and for women who hesitate to visit gynecologists. Mailability and preservability of cells for a week are advantageous to mass screening. Application is limited and only useful to promote unconcerned people.[39] Negative results do not assure the absence of cervical cancer. Positive and suspicious cases detected by this technique indicate the need for further investigation by gynecologists.

■Endometrial aspiration

Method: Without dilatation, insert a cannula 3-4 mm in diameter with four or five small holes at the end into the cervical canal. Aspirate secretions or rinse with a few milliliters of physiological saline. Place a drop of the aspirated secretions on a clean glass slide and smear (Figs. 6-4 and 6-5).

In order to prevent spreading cancer cells by vigorous washing through the fallopian tube into the peritoneal cavity, Dowling and Gravlee designed a disposable double-lumened polyethylene tube with holes in each of two tubes along the distal end that is able to rinse the endometrial cavity under negative pressure; since the fluid bathing the endometrial cavity is aspirated by suc-

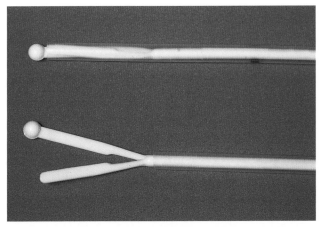

Fig. 6-6 Endocyte for endometrial scraping. A transparent outer tube has three marks at distance of 4, 7 and 10 cm to guide insertion of the device. Rotate a V-shaped device 1 or 2 time after insertion.

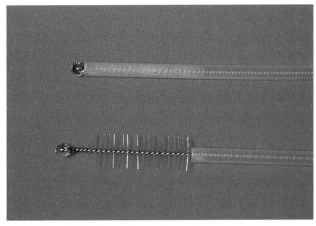

Fig. 6-7 Endobrush for endometrial brushing. Swedish endometrial sampler is placed in a plastic tube to avoid cell contamination through a cervical canal.

tion, there is no limit to the amount of irrigation fluid. Usually 40 to 50 ml sterile isotonic saline is used for cytology.[47] For simple endometrial aspiration we use the Masubuchi's cannula having small holes (Fig. 6-4). The cannula is designed to have two marks at distance of 7 and 10 cm in order to guide insertion into the uterine cavity. The endometrial cell sampler designed by Curity-Isaacs (Fig. 6-5)[2,219] was reported by several experts to be useful with satisfactory accuracies of endometrial cytology at the International Congress of Cytology in Munich (1980).

■Endometrial scraping smear
Recently scrapers of the endometrium are adopted in order to obtain sufficient numbers of endometrial cells instead of endometrial aspiration.

Endocyte is one device designed by Cohen to scrape endometrium with Y-shaped bars which is placed inside of the mantle to prevent contamination of cervical and vaginal cells until the device is inserted into the cavity of uterine corpus. Endobrush is another device to scrape the endometrium by fine brushes (Figs. 6-6 and 6-7).

Cellular components: Endometrial cells are sufficiently obtained as tubules or sheets. Stromal cells are contained in groups or singly. Mixture of histiocytes and leukocytes depends on the sexual cycle of reproductive females.

Contraindication: In doubt about acute endometrial inflammation and pregnancy, endometrial cytology should not be performed. But postmenopausal women with high risks have no such a problem.

Application: Endometrial aspiration or scraping is a tool for the detection of endometrial carcinoma and is an adjunct to endometrial curettage. If atypical cells with a glandular pattern or necrotic cell debris are present in cervical smears without any manifestations of cervical lesions, endometrial cytology is advisable. Even if benign-looking, when endometrial cells appear in cervical or vaginal pool smear after the 10-12th day of the menstrual cycle and in the postmenopausal phase, further studies for endometrial carcinoma should be performed.

3 Histology of the Uterus and Vagina

■Histology of the uterus
The uterus is composed of two main parts: the cervix and the corpus. The cervix is anatomically divided into (1) the endocervix and (2) the portio vaginalis. The endocervix is 2-3 mm thick and lined with a single layer of tall columnar epithelium; there are ciliated columnar cells and mucus secreting cells with basally located nuclei. The compound racemose glands of the cervix secrete a viscid mucus during the reproductive years. It is influenced by cyclic changes of ovarian hormones. The alkaline secretion is increased in amount and decreased in viscosity during the estrogen phase. After ovulation the mucus becomes thick under the influence of progesterone. The portio vaginalis lying external to the primary squamo-columnar junction (external os) is covered by the stratified squamous epithelium.

The uterine corpus consists of the endometrium and myometrium. The former is important for exfoliative cytology and varies in structure according to the menstrual cycle. In the estrogen or follicular phase the endometrial glands grow and its lining epithelium becomes taller and produces alkaline phosphatase activity which may participate in protein synthesis and glycogenesis. Intense stain of alkaline phosphatase is noticeable along the cellular surface facing to lumens (see Fig. 6-32). Glucose-6-phosphatase and β-glucuronidase show also elevated activities in the proliferative glandular epithelium.

In the progesterone or corpus luteum phase the endometrial glands change from ovoid to sawtooth-like structures and begin to secrete. Glycogen formation is remarkable. The functional layer of the endometrium will be denuded by menstruation leaving the thin basal layer. The surface epithelium of the endometrium is lined with a single layer of low columnar cells which are occasionally ciliated. Alkaline phosphatase activity is hormone dependent and becomes negative towards the late secretory phase (see Fig. 6-33). On the other hand, there occur elevations in activities of the following

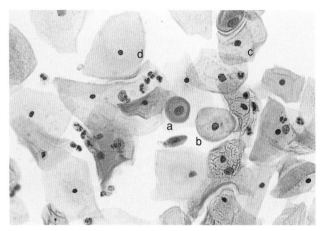

Fig. 6-8　Variable types of squamous epithelial cells.
a. Inner parabasal cell having round sharply bordered cytoplasm and an oval vesicular nucleus.
b. Outer parabasal cell having slightly larger, spherical cytoplasm and a small vesicular nucleus.
c. Intermediate cell having transparent broad cytoplasm and a small vesicular nucleus.
d. Superficial cell having broad eosinophilic cytoplasm and a tiny pyknotic nucleus.

enzymes such as acid phosphatase, lactate dehydrogenase, DPNH, etc.[42]

■Histology of the vagina

The vagina with prominent rugae is covered with stratified squamous epithelium which is responsive to estrogen. Epithelial proliferation is stimulated by estrogen during the reproductive years. Progesterone effects on glycogen deposition are remarkable during the corpus

luteum phase by the premenstrual period when glycogenesis considerably drops. Since the cyclic changes of the squamous epithelium reflect the hormonal status, vaginal cytology is applied to hormonal evaluation unless the patient has vaginitis (see page 74).

4 Normal Cells in the Female Genital Tract

■Squamous cells

1) Superficial cells

Superficial cells originate from the superficial layer of the nonkeratinizing stratified squamous epithelium. These are the most common epithelial cells at a preovulatory phase of reproductive women.

Cytoplasm: The cytoplasm is thin broad (45-50 μm in diameter) and polyhedral. Cytoplasmic borders are irregular but clearly defined, and have little curlings. Staining reaction is usually orangeophilic or eosinophilic and occasionally cyanophilic. Small dot-like granules of keratohyaline are occasionally found around the nucleus.

Nucleus: The nucleus is small (5-7 μm), shrunken, round or oval in shape, and reveals condensed chromatins. Most of the superficial cells contain pyknotic nuclei. The decisive criterion for the superficial cell is pyknosis of the nuclei regardless of the staining reaction of the cytoplasm (Figs. 6-8 to 6-10).

2) Intermediate cells

Intermediate cells originate from the middle, prickle cell layer of the stratified squamous epithelium. These are the most common epithelial cells at a postovulatory progesterone phase of reproductive women.

Cytoplasm: The cytoplasm is adequate (40-50 μm in

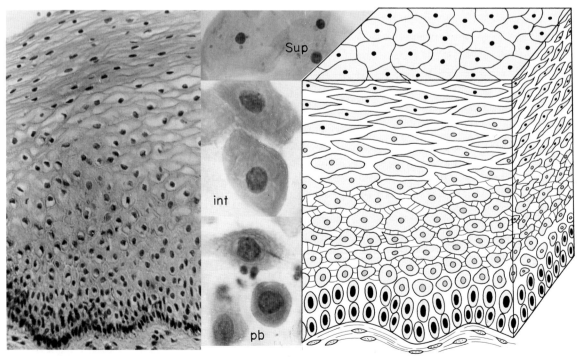

Fig. 6-9　Histology of the portio vaginalis and its corresponding cytologic features. Superficial cells (Sup), intermediate cells (int), and parabasal cells (pb).

50

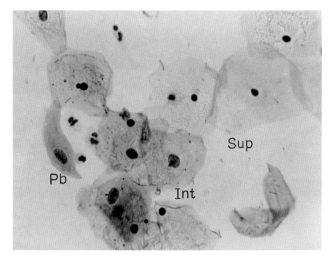

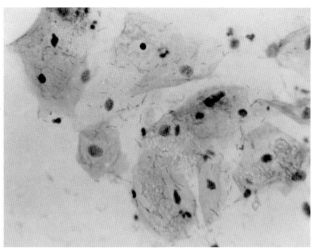

Fig. 6-10 Normal vaginal smear showing superficial cells (Sup), intermediate cells (Int) and an outer parabasal cell (Pb).

Fig. 6-11 Cytoplasmic degeneration of superficial and intermediate cells by Döderlein bacilli.

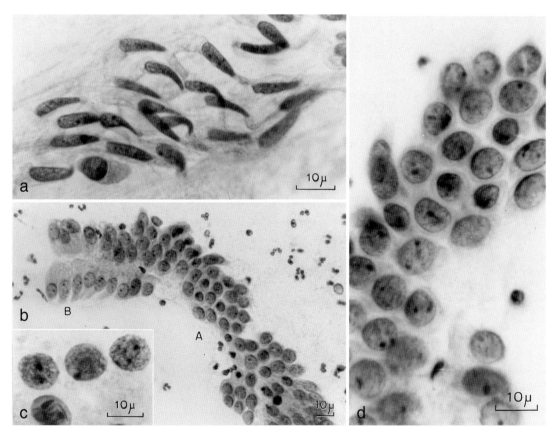

Fig. 6-12 Endocervical cells.
a. Lateral view of endocervical cells. The columnar cytoplasm is degenerative but visible. The nuclei are elongated and their chromatins are smoothly distributed and finely granular.
b. Large sheet of well preserved endocervical cells. A group of endocervical cells seen on end reveals a honeycomb appearance (A), while another group of the cells placed laterally shows a definite columnar configuration (B).
c. Denuded endocervical nuclei. Hypertrophy of the nuclei and a somewhat coarsely stippled chromatin pattern should not be regarded as malignant criteria. Although some karyosomes are shown, uniform shape and size of the nuclei and the same chromatin content should be noted.
d. Normal endocervical cells seen in clusters. Note a uniform shape of the nuclei and evenly distributed chromatin patterns. Higher magnification from a part of Fig. 6-12b.

diameter), thin, transparent and polygonal. The staining reaction is pale, bluish green. The cytoplasmic border is folded.

Nucleus: The nucleus is round or oval, larger (9-11 μm) than that of superficial cells, and has a vesicular appearance. It is characterized by a delicate and distinct nuclear rim, finely granular chromatin and some noticeable karyosomes. Sex chromatins can be distinguishable at nuclear borders (Figs. 6-8 and 6-10).

Generally speaking, intermediate cells tend to exhibit

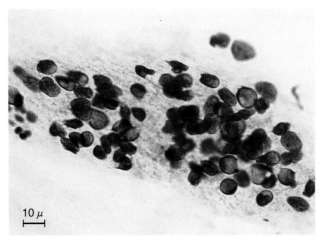

Fig. 6-13 Denuded endocervical cells in advanced degeneration. See denuded nuclei showing condensation of chromatin material towards the nuclear rim. There is moderate variation in nuclear size.

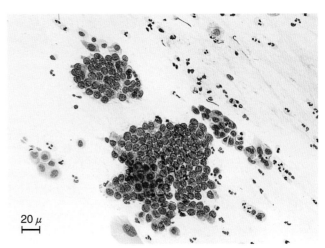

Fig. 6-14 Clusters of endocervical cells with a characteristic honeycomb appearance. The nuclei are round to oval and uniform in size. The columnar cytoplasm is recognized at the border of cell clusters.

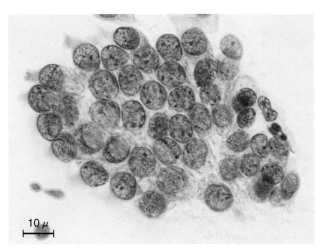

Fig. 6-15 Larger magnification of endocervical cells of a secretory type. The cytoplasm is transparent, columnar but ill-defined. The nuclear outline is distinct. The chromatin is finely granular and evenly distributed.

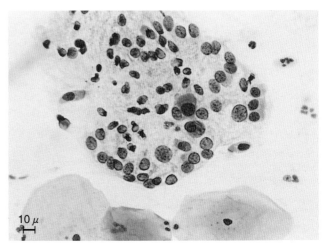

Fig. 6-16 Free endocervical cells including well preserved ciliated cells. Well preserved cells possess round nuclei with fine chromatin network, while degenerative cells are denuded and show pyknotic or liquefied nuclei.

cytolysis in the progesterone phase by the presence of Döderlein bacilli (Fig. 6-11).

Navicular cells. These are variants of intermediate cells with boat-shaped cytoplasm of yellowish hue resulting from increased glycogen content. The cellular border appears to be heavy because of wrinklings (see Fig. 6-44). Exfoliation of this specific type of cell, particularly in groups, occurs in certain conditions such as pregnancy and crowded type of menopause.

3) Parabasal cells

Shedding of parabasal cells originating from the deep layer is infrequent in normal reproductive women. They appear physiologically in prepuberty, postmenopause and lactational period (see Figs. 6-42 and 6-43).

Cytoplasm: These are smaller than superficial or intermediate cells (15-30 μm) and exhibit definite cellular borders. The outer parabasal cells are transparent, lightly bluish green with Papanicolaou stain, and polyhedral in shape. The inner parabasal cells are elliptical in shape, dense, and stain deep bluish green (see Fig. 6-8).

Nucleus: Round or oval (8-12 μm) and slightly hyper-

chromatic. The chromatin is finely granular and evenly distributed.

4) Basal cells

These are from a single basal layer. They do not appear in ordinary scrapings unless basal cell hyperplasia occurs. The cytoplasm is scant and deeply basophilic. The nucleus is relatively large and hyperchromatic.

■**Endocervical cells** (Figs. 6-12 to 6-17)

The endocervical cells tend to degenerate and appear often as single stripped nuclei. Chromatin condensation at the nuclear rim forms knot-like or nipple-like protrusions (Figs. 6-13 and 6-17). If well preserved, they can be distinguished into ciliated and secretory types.

Cytoplasm: The shape and amount vary greatly depending on different conditions and the state of degeneration. They are better preserved in the cervical scraping smears than in vaginal smears. Well preserved endocervical cells display columnar cytoplasm with slender tails and the nuclei are eccentric in position. If endocervical cells are viewed from above, a definite honeycomb

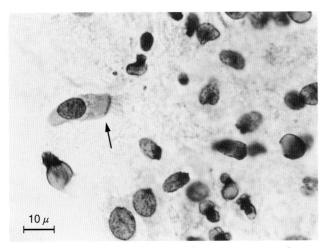

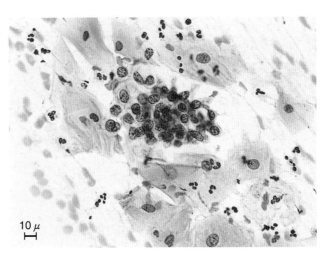

Fig. 6-17 Larger magnification of degenerative endocervical cells. Nuclear chromatins are liquefied and show condensation towards the nuclear rim. See chromatin nipples projecting out of the nuclei. There is a well preserved ciliated cell showing cilia and distinct terminal plate (arrow).

Fig. 6-18 A cluster of endometrial glandular cells shed on the 10th menstrual cycle. The nuclei are uniform, usually round, and tend to overlap.

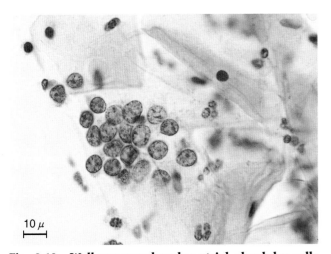

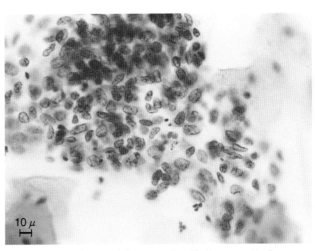

Fig. 6-19 Well preserved endometrial glandular cells shed in a cluster. The endometrial cells found in cervical smears are usually stripped of the cytoplasm. The nuclei are smaller than those of endocervical cells, and their chromatins are arranged in clumps that may give an active appearance.

Fig. 6-20 Endometrial stromal cells in a cervical smear. Note characteristic shedding of stromal cells in loosely aggregated sheet. The stripped nuclei are oval or elliptic in form and delicately outlined.

appearance with clear cytoplasm is apparent. By focusing up and down one may see distinct cytoplasmic borders. Cilia are occasionally seen at the broad cellular borders of ciliated endocervical cells. Even though delicate cilia are invisible, lavender terminal plates can be distinguishable in well preserved cells (ciliated form)(Figs. 6-16 and 6-17). Nonciliated endocervical cells have semitranslucent, finely vacuolated or lacy cytoplasm. In the state of hypersecretion, the cytoplasm is distended by mucus secretion (secretory form)(Fig. 6-15).

Nucleus: Round or oval in shape and eccentric in position. Well preserved cells reveal a delicate but well defined nuclear rim. The chromatin is finely and evenly distributed. As cells degenerate, the cytoplasm is lost and stripped nuclei retaining fairly intact chromatin may be observed. At the end of degeneration, the chromatin liquefies in association with local condensation at the nuclear rim(Fig. 6-17).

■Endometrial cells (Figs. 6-18 to 6-22)
1) Endometrial cells in cervico-vaginal smear
The appearance of endometrial cells in vaginal or cervical smears occurs during or immediately after menstruation.[188] Generally speaking, the presence of endometrial cells after the first ten-day period in the cycle may inform some pathologic condition of the endometrium.[128] An important exception is the case of use of intrauterine contraceptive devices when endometrial cells may be shed at midcycle.[111] Although the endometrial cells should be distinguished into glandular and stromal types, identification of these two types is usually difficult in vaginal or cervical smears. Well preserved cells originating from endometrial glands are in a tight cluster with nuclear overlapping. The endometrial cells derived from the stroma are arranged in flat sheets or single files. The nuclei of stromal cells are somewhat elongated.

Epithelial cells. Well preserved cells are aggregated

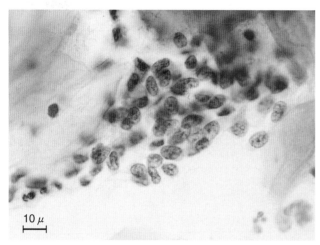

Fig. 6-21 Higher magnification of endometrial stromal cells in a cervical smear. The stromal cells are distinguishable from glandular cells by (1) cell arrangement with loose syncytial sheet, (2) elongated oval shape of the nuclei, (3) delicate nuclear outlines, and (4) fine lacy chromatin network.

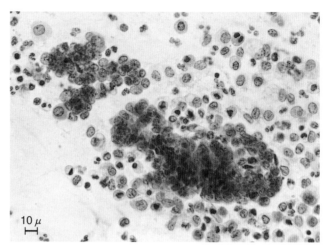

Fig. 6-22 "Exodus" of endometrial cells in a cervical smear. Tight clusters of endometrial glandular cells which are surrounded by histiocytes with pale chromatin pattern are called "exodus."

in tight clusters or balls. Differentiation from the endocervical cells is based on the regularity of the nuclear size (smaller than that of endocervical cells) and scantiness and indistinctiveness of the cytoplasm (Figs. 6-18 and 6-19).

Cytoplasm: In ordinary vaginal smears the cytoplasm is lost as a result of rapid degeneration. When the cytoplasm is well preserved, it is seen as a faint background around the nucleus. The staining reaction is feebly basophilic.

Nucleus: As the cells appear in clusters, stripped nuclei have a tendency to clump and overlap. Well preserved endometrial cells disclose round or oval nuclei with delicate nuclear borders and evenly dispersed but somewhat condensed chromatin particles. Thus the endometrial cells display a darker and more active appearance than endocervical cells (Fig. 6-19).

On the other hand, degenerating cells show variations in nuclear shape rather than in size. The chromatin may be condensed towards the nuclear border.

Stromal cells. The endometrial stromal cells shed in irregular sheets with less overlapping than glandular epithelial cells. Indistinctiveness of the cytoplasm and elongated nuclear shape are characteristic of stromal cells (Figs. 6-20 and 6-21).

Addendum: Exodus. Exodus means exfoliation of endometrial cells in cluster at the late menstrual or postmenstrual phase accompanied by a number of histiocytic cells (Fig. 6-22).[136]

2) Endometrial cells in scraping smear
Endometrial epithelial cells shed in large sheets which are folded or tubes with openings. Their nuclei which are regularly and evenly arranged disclose a honeycomb appearance. The cells at periphery of clusters appear to stand side by side. Stromal cells are isolated or in loose, flat sheets. Distinction of stromal cells from epithelial cells consists in spindly or elliptic nuclei, delicate nuclear rim, bland chromatin and loose arrangement of cluster.

Table 6-1 Endometrial exfoliation in different phases

Phase	Number of women	Endometrial cells in smears	
		definite	Equivocal
Menstrual	258	90 (35%)	5
Proliferative	990	54 (5%)	11
Secretory	726	9 (1%)	7
Premenstrual	449	5 (1%)	4
	2,423		27

(From Liu W, et al: Acta Cytol 7: 212, 1963. © 1963 International Academy of Cytology.)

In proliferative phase: Epithelial cells display compact clusters; the cytoplasm is small in amount, the nuclei are fairly small in size and densely stained, and chromatins are condensed (see page 111, Fig. 6-172a).

In secretory phase: The cytoplasm is broadened and cell to cell interrelation is loosened; lightly stained cytoplasm is ill-defined. The nuclei are round to oval in shape and larger in size than in a proliferative phase. Vesicular nuclei are also characteristic (Fig. 6-172b).

■**Histiocytes** (Figs. 6-23 to 6-26)
Histiocytes do not shed normally after the 10th day of the menstrual cycle in cervical or vaginal smears unless the mucosal epithelium lining the female genital tract is eroded or inflamed. There are two types of histiocytes having single or multiple nuclei.
1) Common mononuclear histiocytes
Cytoplasm: Commonly round or oval in shape, but varies markedly in size. The appearance is foamy and finely vacuolated. Ordinarily it stains faint greyish green. Phagocytosed foreign particles, blood pigments and fat droplets can be recognized (Fig. 6-24). Degenerated histiocytes have fuzzy cytoplasm and indistinct cellular borders. The cytoplasm may be lost in advanced degeneration. Leukocytes are not infrequently superimposed.

Nucleus: Oval, round or bean-shaped nuclei are located eccentrically. The nuclear rim is delicate but sharply

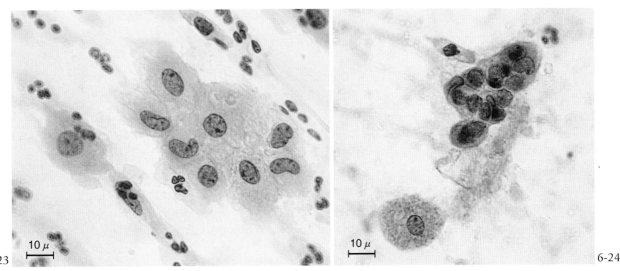

6-23

6-24

Fig. 6-23　Histiocytes appeared in a loose aggregate from cervical scrapings. These are ill-defined from each other. The nuclei are eccentric in position, oval or elliptic, and delicately outlined.
Fig. 6-24　Histiocyte in aspiration smear from a vaginal cyst. Diagnostic points: (1) vesicular cytoplasm with fuzzy border, (2) phagocytosis full of pigmented material, and (3) paracentral, oval nucleus. Clustered cuboidal cells are indicative of Gartner's duct cyst.

Fig. 6-25　Histiocyte with double nuclei, which are bean-shaped and eccentric in position. The chromatin is smoothly distributed except for a few karyosomes. The cytoplasm is foamy in appearance, and its border is indistinct.
Fig. 6-26　Multinucleated giant histiocyte of foreign body type. A number of oval or elliptic nuclei are clumped centrally in abundant, lacy cytoplasm, the borders of which are fuzzy in appearance. Compare this with Langhans giant cell having peripherally locating nuclei.

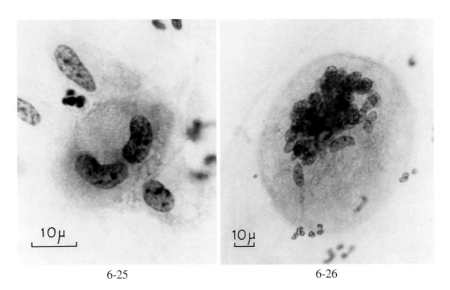

6-25

6-26

defined. The chromatin is finely granular.

2) Multinucleated histiocytes

Cytoplasm: The cytoplasm is huge and varies in shape. Cellular borders are obscure. There is occasional superimposition of leukocytes.

Nucleus: Each of the multiple nuclei reveals similar findings characterized by an oval shape with slight variation in size, finely granular chromatin and peripheral location with overlapping (Figs. 6-25 and 6-26).

■Leukocytes

The presence of polymorphonuclear leukocytes is a common finding in vaginal smears. The number of leukocytes is not always correlated with inflammation but is greatly reflective of the menstrual cycle. A smear indicating a high estrogen level appears clean; however, a smear after ovulation is accompanied by a large number of leukocytes. The occurrence of plasma cells having peripherally located oval nuclei with eccentric, coarsely clumped chromatin is indicative of chronic cervicitis.

Light perinuclear halo is a distinguishable feature of plasma cells.

5　Cyclic Hormonal Cytology in the Reproductive Period

■Cyclic hormonal secretion

The reproductive function of females is regulated by the nerve cells in the hypothalamus. Secretion of gonadotropin, follicle stimulating hormone (FSH) and luteinizing hormone (LH) liberated from the anterior lobe of the pituitary gland is under the control of hypothalamic-pituitary portal system. The female sexual or menstrual cycle is controlled by interplay between ovarian and hypothalamic-pituitary hormones. Rhythmic changes of interplay among the related organs are summarized as follows[76](Fig. 6-27):

1) Postovulatory secretion of the ovarian hormones and depression of gonadotropin occur between ovulation and menstruation. Production of large amounts of

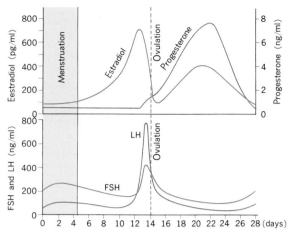

Fig. 6-27 **Cyclic hormonal change in menstruation.** (From Guyton AC: Basic Human Physiology, 2nd ed. Philadelphia: WB Saunders, 1977.)

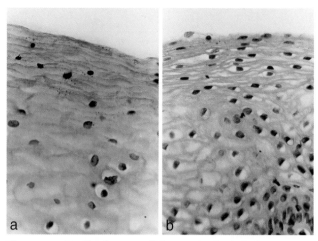

Fig. 6-28 **Cyclic change of the vaginal epithelium.**
a. Estrogen phase. Thickening and maturation of the mucosal epithelium are evident as a response to estrogen stimulation. See the abundance of cytoplasm of the superficial cells.
b. Progesterone phase. Predominance of intermediate cells of the mucosal epithelium is shown as crowded nuclei.

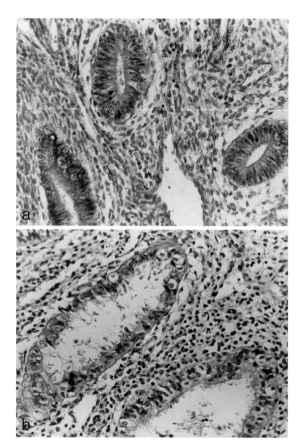

Fig. 6-29 **Histological changes of the endometrium in menstrual cycle.**
a. Proliferative or estrogen phase showing small ovoid glands with frequent mitoses. The stroma is compact.
b. Secretory or progesterone phase showing apparent secretion into the lumens of the elongated endometrial glands. The stroma is slightly edematous.

progesterone and estrogen gives rise to negative feedback decrease in secretion of gonadotropin, LH and FSH.

2) Follicular growth is induced according to involution of the corpus luteum. Depression of secretion of both estrogen and progesterone may release the hypothalamus from the feedback effect of the estrogen and may stimulate secretion of both FSH and LH. A gradual increase in estrogen secretion reveals the first peak at about the 13th day.

3) Ovulation follows abrupt surge of production of LH and FSH on about the 11th day, which is considered to be caused by a positive feedback effect of high estrogen level.

■**Cyclic cytologic changes** (Figs. 6-28 to 6-34)
Cyclic changes of the ovarian hormones, estrogen and progesterone, are reflected on maturation of the stratified squamous epithelium of the portio vaginalis and the upper portion of the vagina. Pool smears can be used for hormonal cytology, however, they are often influenced by degeneration and inflammation, i.e., cytolysis, pyknosis and/or cytoplasmic eosinophilia. If these findings are encountered, the pool smear will be incorrectly evaluated in hormonal cytology. Therefore, smears gently scraped from the outer wall of the upper third of the vagina are preferable to pool smears. In order to avoid contamination of cells from the lower part of the vagina showing cornification, smears should be collected with a wooden spatula[193,275] or a cotton-tipped applicator[77] after insertion of a dry speculum. Estrogen produced from the granulosa cells and theca cells of the Graafian follicles may stimulate continuous proliferation of the endometrium until ovulation. The upper part of the vaginal epithelium also shows a response to estrogen, a kind of growth hormone, and increases in thickness. Superficial cells are the main component of exfoliated cells in the vaginal flora. After ovulation the follicle undergoes metamorphosis into the corpus luteum, which produces progesterone in addi-

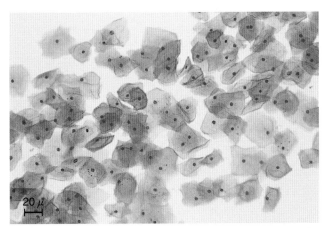

Fig. 6-30 Estrogen effect of a vaginal smear pattern. The estrogen pattern is characterized by a preponderance of superficial cells with eosinophilia and karyopyknosis, and clear background with scant mucus and few leukocytes.

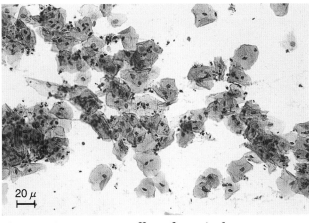

Fig. 6-31 Progesterone effect of a vaginal smear pattern. The progesterone pattern is characterized by a preponderance of intermediate cells with crowding and curling.

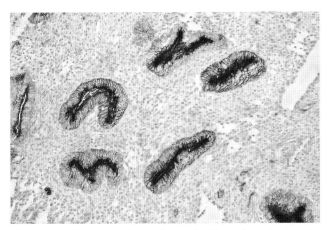

Fig. 6-32 Hormone-dependent alkaline phosphatase in the estrogen phase. This is a heat-labile (65°C for 10 min) isoenzyme. Normal proliferative endometrium is positive towards the ovulation.

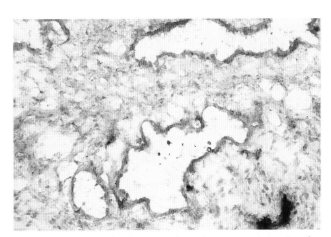

Fig. 6-33 Loss of alkaline phosphatase activity in the progesterone phase. Alkaline phosphatase becomes negative after the 20th day of menstrual cycle through the mosaic pattern being composed of positive and negative glands after ovulation.

tion to estrogen. A rapid secretion of progesterone gives rise to a secretory response of the endometrial glands and exfoliation of intermediate cells from the portio vaginalis.[197] Towards the late secretory phase, the intermediate cells become crowded and folded, accompanied by leukocytic infiltration and mucus secretion. In physiological conditions acting with estrogen, progesterone shows an antiestrogenic effect; so that cessation of maturation of the vaginal epithelium and massive desquamation of intermediate cells will be seen in the secretory phase.

Early proliferative phase: Intermediate cells with broad, folded cytoplasm are predominant. In addition, outer parabasal cells, endometrial cells and histiocytes are occasionally present.

Late proliferative phase: Superficial cells with broad, eosinophilic cytoplasm and pyknotic nuclei increase in number and are most numerous at ovulation. These cells are mostly flat and present singly. A high karyopyknotic index indicates the peak of the estrogen effect. Keratohyaline granules appear around the nuclei of the superficial cells. There are few leukocytes and little mucus; therefore, the background of the smear appears clean (Fig. 6-30).

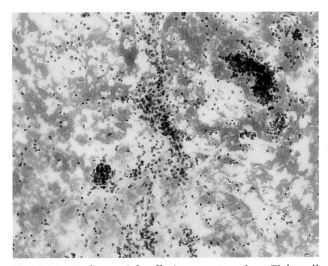

Fig. 6-34 Endometrial cells in menstruation. Tight cell clusters from endometrial glands and loose syncytial sheets from stroma are dispersed in bloody discharge.

Table 6-2   Cytologic pattern affected by ovarian hormones

Phase	Main cell components in vaginal smear	Other cell components in vaginal smear and its background
A) Before puberty		
Newborn	Intermediate cells	
Childhood	Parabasal cells	Scant cellularity
B) Reproductive period		
Menstrual phase	Erythrocytes	Degenerate endometrial cells in cluster and leukocytes
Early proliferative phase	Intermediate cells	Occasionally endometrial cells and/or histiocytes
Late proliferative phase	Superficial cells lying flat and singly	Clean
Early secretory phase	Superficial cells with foldings	Some intermediate cells and leukocytes
Late secretory phase	Intermediate cells with foldings and crowding	Abundant mucus and many leukocytes, and with cytoplasmic degeneration
Pregnancy	Intermediate cells of "navicular type" in cluster	
Postpartum period	Parabasal cells with glycogen, "postpartum cells"	
C) Menopause		
Crowded type	Intermediate cells and outer parabasal cells	Responsive to estrogen
Atrophic type	Parabasal cells in a small number	Senile colpitis often associated

Early secretory phase: As the corpus luteum develops, intermediate cells with folded cytoplasm become prominent in association with leukocytic infiltration and mucus production. The cells tend to aggregate in clusters (Fig. 6-31).

Late secretory phase: Intermediate cells with marked foldings and curlings of the cytoplasm are predominant. Döderlein bacilli and leukocytes markedly increase in number and may bring about cytoplasmic degeneration (cytolysis).[231,275]

Menstrual phase: In addition to numerous red cells, leukocytes and endometrial cells with stripped nuclei shed in large numbers (Fig. 6-34).

■Cytologic indices

Several indices for the description of cytohormonal patterns were introduced by the Panel on Hormonal Cytology of the Second International Congress of Cytology.[278] These included the following indices (Fig. 6-35):
- a) Karyopyknotic index[280]
- b) Eosinophilic index
- c) Maturation index[60]
- d) Folded index[280]
- e) Crowded index[280]

Meisels[139] introduced a numerical value to estimate the level of cell maturation. Differential cell counts were made on a minimum of 200 well-preserved squamous cells which were assigned specific values according to their five cell types. These cell types which were represented as percentages were multiplied by their own values and then added: superficial eosinophilic cells 1.0, superficial cyanophilic cells 0.8, large intermediate cells 0.6, small intermediate cells 0.5, and parabasal cells 0. For a simpler calculation Meisels[140] assigned the following values: Superficial cells 1.0, intermediate cells 0.5, and parabasal cells 0. For an accurate assessment of the estrogen effect the karyopyknotic index is a more reli-

able method than the eosinophilic index, which is variable by fixing and staining techniques.

6 Cytology in Menopause

Estrogenic activity is reflected in proliferation and maturation of the stratified squamous epithelium of the vagina. In menopause insufficiency of estrogen secretion may inhibit growth of the epithelium and retard maturation at the intermediate cell layer. Exfoliated cells consist of parabasal cells and intermediate cells. The onset of the menopause is gradual and indefinite; diminished estrogenic activity begins with decrease in the number of eosinophilic superficial cells and predominance of intermediate cells. In perimenopausal period, the stage showing proliferative, protective thickness of the epithelium with intermediate cell maturation is called *estatrophy* (lack of estrogenic effect; Frost). Thereafter, gradual development of atrophy of the vaginal epithelium is cytohormonally characterized by the shift to the parabasal cell predominance; *teleatrophy* (Frost) is the extreme atrophic stage as the result of withdrawal of maturing substances. Cytohormonal patterns are represented as follows.

■Crowded menopause

Moderate depression of estrogenic activity brings crowding of deep intermediate cells and/or outer parabasal cells. As a result glycogen deposition these cells impart a yellowish hue to the cytoplasm.

■Advanced menopause

With atrophy of the portio vaginalis in the advanced menopause, exfoliated cells are scanty and consist mainly of parabasal cells (Fig. 6-36). The cervical epithelium also atrophies and is lined with low columnar or cuboidal epithelium. Secretion of mucus from the cervi-

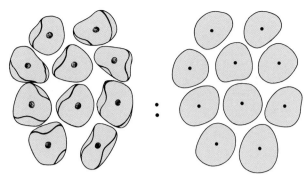

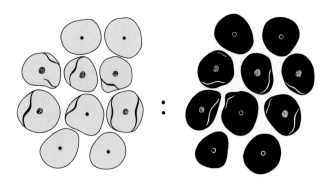

a. Karyopyknotic index. Relation of mature superficial cells to mature intermediate cells regardless of staining reaction.

b. Eosinophilic index. Relation of eosinophilic mature squamous cells to cyanophilic mature squamous cells regardless of nuclear appearance.

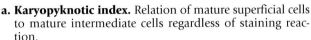

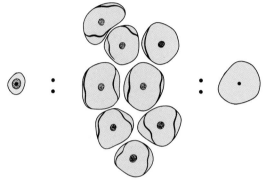

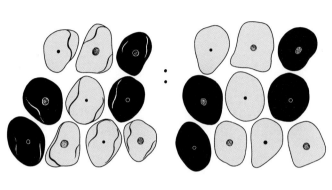

c. Maturation index. Relation of superficial cells to intermediate cells to parabasal cells expressed in percentages, and written as follows; e. g. 10/80/10.

d. Folded cell index. Relation of folded mature squamous cells to flat mature squamous cells, regardless of staining reaction and regardless of nuclear appearance.

e. Crowded cell index. Relation of mature squamous cells in clusters of cells of four or more cells to cells lying either singly or in clusters of less than four cells, regardless of staining reaction, nuclear appearance and cellular folding or flatness.

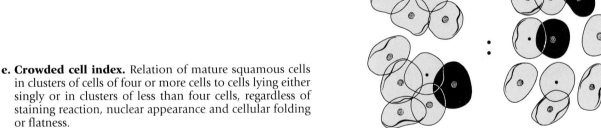

Fig. 6-35 Cytologic indices for the cytohormonal evaluation. (From Wied GL, Bibbo, M : Evaluation of the endocrinologic condition of the patient by means of vaginal cytology. In: Compendium on Diagnostic Cytology, 4th ed. Chicago: Tutorials of Cytology, 1979.)

cal glands decreases. This condition is susceptible to inflammation, i.e., senile colpitis and cervicitis (Fig. 6-37). Strong eosinophilia of parabasal cells with karyopyknosis, which is frequently encountered as inflammatory cell changes, should not be mistaken for dyskeratosis. A dyskeratotic cell can be distinguished by the presence of a normal-looking nucleus. In the case of an inflammatory process, polymorphonuclear leukocytes are numerous and some histiocytes, occasionally multinucleated, are present in smears. The above cytologic pattern is characteristic and estrogenic activity diminishes gradually with increasing age to the stage of epithelial atrophy. Some aged females in the postmenopause maintain moderate levels of estrogenic activity and may show shedding of superficial cells. Stone and his associates reported a high

maturation value in 8.9% of women as late as 20 years postmenopause.[240] The evidence that oophorectomy does not result in the atrophic pattern supports the concept that the adrenal cortex could be responsible for estrogenic activity[46,60,137,192,240]

7 Cytology before Puberty

■Pattern in newborn infant

Newborn females are under progesterone effect which is brought about by her mother in delivery. Thus, principal cells of vaginal smears are composed of intermediate cells without mixture of leukocytes. Döderlein bacilli are absent. This pattern is maintained up to about the 4th week.

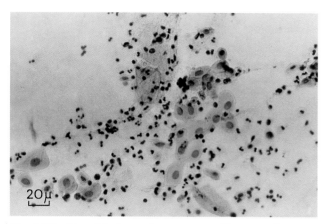

Fig. 6-36 Atrophic type of menopause. The smear is scanty in cells. Inner and outer parabasal cells are the main cells constituents of atrophic smear.

Fig. 6-37 Atrophic epithelium of the portio vaginalis of an aged woman associated with senile vaginitis. The stratified squamous epithelium is lined by only several layers of cells. The subepithelial stroma reveals heavy inflammation with foreign body giant cell reaction.

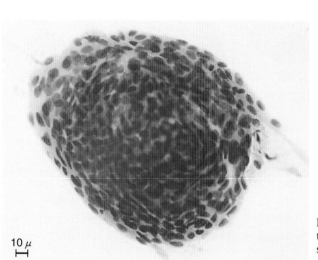

Fig. 6-38 Clustered parabasal cell in a cervical smear of teleatrophy. Such a cell ball consisting of parabasal cells should not be confused with cell clumping of cancer cells or endometrial cells in exodus.

■Pattern in childhood before puberty

(Figs. 6-39 and 6-40)

After release from the progesterone effect by her mother, the cytological pattern maintains a parabasal cell predominance until puberty when ovarian function begins to act on vaginal smears. Thus, infants before *menarche* reveal a cytohormonal pattern similar to the postmenopausal atrophic pattern; a predominance of parabasal cells in such a grade as maturation index from 70/30/0 to 100/0/0.

8 Pathological Conditions Affecting Cytohormonal Patterns

■Inflammation

Inflammatory changes, trichomonas infestation in particular, are apt to make a mistake for dyskeratosis because of intense eosinophilic stain of the affected cells. Early menopausal females in the healing process of true erosion may show maturation of squamous cells. Reproductive females having actual erosion show exfoliation of parabasal cells. This should not be interpreted as depressed estrogenic effect (see Figs. 6-74 to 6-76).

■Nonhormonal medication

Several investigators reported a high maturation index in aged females who have been receiving digitalis for two years or longer.[24,158] The similar molecular structure of digitalis glycoside to that of estrogen may explain the effect on the stratified squamous epithelium of the portio vaginalis. Vitamin A deficiency may cause a shift to the right of the maturation index by generalized hyperkeratosis (Naib).

9 Cytology in Hormonal Dysfunction

Exogenous medication of estrogenic substances must be excluded for cytohormonal evaluation.

■Estrogen hyperactivity

Feminizing tumors producing estrogen such as granulosa cell tumor and thecoma exert persistently a high estrogenic effect on cytology in association with various clinical features. Prepuberal girls show precocious puberty and irregular or excessive menstruation. Postmenopausal patients exhibit a return of irregular menstruation, while reproductive women manifest amenorrhea of irregular menstruation.

Certain tumors are not productive of estrogen but are

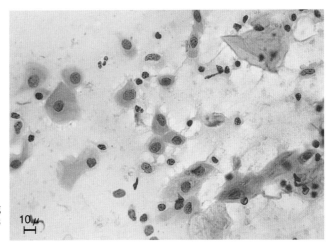

Fig. 6-39 Vaginal cytology of a 3-year-old girl showing an atrophic smear pattern. Note scarcity of squamous epithelial cells and predominance of parabasal cells.

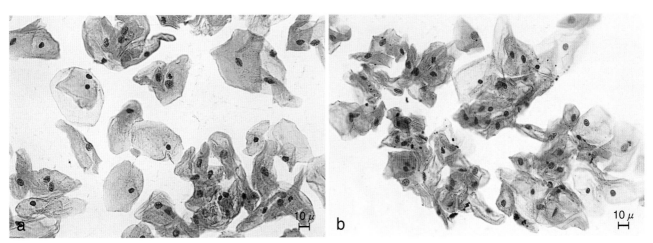

Fig. 6-40a,b Vaginal cytology of precocious puberty. This 8-month-old girl visited our clinic because of swelling of the breast and genital bleeding. Vaginal smears showed high cellularity and predominance of superficial and intermediate cells. Some alteration of cellular pattern can be noted by serial cytologic studies.

occasionally associated with an abnormal estrogen level. Endometrial carcinoma is a typical example with this condition and other tumors, such as carcinoma of the fallopian tube, carcinoma of the ovary, carcinoma of the breast and glandular hyperplasia of the endometrium may elevate estrogenic activity.

Cirrhosis of the liver or other liver cell damage may cause insufficiency of the function that inactivates estrogenic activity and may result in a relative increase in the estrogenic effect.

■Estrogen hypoactivity

Hypoestrogenism is primarily caused by ovarian failure, agenesis, or atrophy. The clinical manifestations are amenorrhea, sterility and absence of sexual maturity. The cytohormonal pattern is absolutely atrophic. Primary ovarian insufficiency, Turner's syndrome or ovarian dysgenesis reveals complete absence of maturation, atrophy of endometrium and negative X chromatin body. Secondary ovarian insufficiency or pituitary hypogonadism that is characterized by amenorrhea or delayed menarche and infantile sexual organs shows also atrophic vaginal smear and atrophic endometrium. Other endocrine disorders showing insufficient sexual maturation and loss of estrogenic effect on vaginal smears are Sheehan's syndrome and Simmonds' disease; the matu-

ration index reveals progressive shift to the left. Polycystic ovarian disease (Stein-Leventhal syndrome) is a type of ovarian insufficiency and is manifested by bilateral polycystic enlargement of the ovaries, amenorrhea or oligomenorrhea and sterility. However, it has an inconstant cytological pattern,[12] i. e., low estrogenic or progesterone-like effects in general and occasionally even moderate estrogenic effects. The absence of normal cyclic changes by repeated smear tests is a noteworthy finding for the diagnosis of Stein-Leventhal syndrome.

■Progesterone hyperactivity

Endogenous progesterone is physiologically produced by the corpus luteum or placenta. Ovarian luteoma and lutein cysts may produce biologically a high progesterone level and clinically a delay of menstruation. While hyperestrogenism can readily be detected by cytohormonal tests, hyperactivity of progesterone is less easily detected by cytological investigation. The progesterone effect is interpreted by the following features[280]:

1) Crowding of intermediate cells with broad and cyanophilic cytoplasm
2) Folding of the cytoplasm of individual cells
3) Disappearance of discrete cell boundaries

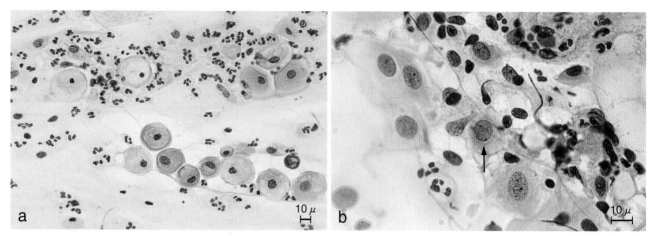

Fig. 6-41a,b Atrophic smear in gonadal dysgenesis of a young female aged 20 having mosaicism of 45, X/46, XXqi in karyotype. Six of 200 parabasal cells are X chromatin-positive. Occasionally a large X body is seen; the large X body is considered to be derived from the iso-X chromosome.
a. Atrophic smear showing scanty cellularity; **b.** X body-positive cell is rarely included (arrow).

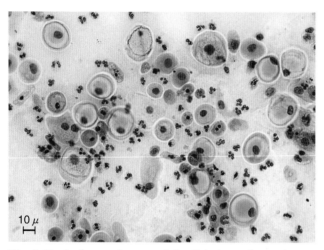

Fig. 6-42 Androgen hyperactivity. Relative androgen hyperactivity in a vaginal smear caused by intense treatment of testosterone after mastectomy because of mammary carcinoma. Parabasal cells with cytoplasmic yellowish hue are predominant. Medication of androgen leads to proliferation of parabasal cells and small intermediate cells with thickened borders.

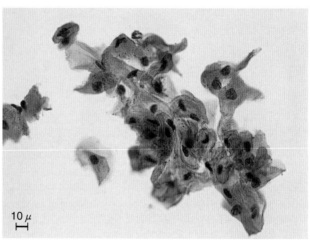

Fig. 6-43 Cytohormonal pattern in a case with arrhenoblastoma showing predominant intermediate and parabasal cells. Preponderance of testosterone secretion over estrogen is reflected in cytology without atrophy. From a 36-year-old female who showed amenorrhea, deepening of voice and flattening of the breast. Unilateral tumor was composed of imperfect testicular tubules and immature mesenchymal cells. Before operation: urinary 17 KS 9.57 mg/day; serum testosterone 2.67 ng/ml; urinary testosterone 61.6 ng/liter.

■Androgen hyperactivity

Excessive androgen is seen in cases of masculinizing tumors, i.e., arrhenoblastoma, virilizing hilus cell tumor, and adrenal rest tumor (masculinovoblastoma), which are characterized by amenorrhea, defeminization, vaginal atrophy and raised 17-ketosteroid excretion. The effect of androgen on vaginal smears[20,70,187] is the presence of a number of small intermediate and/or parabasal cells with centrally located small nuclei and condensed cytoplasmic borders. These cells have a yellowish hue around the nucleus as a result of glycogen deposition. This pattern is not clearly distinguished from depressed estrogenic activity barring the absence of cyclic changes (Fig. 6-43).

10 Cytology of Pregnancy and Its Allied Changes

Female sex hormones in the first trimester of pregnancy are secreted by the corpus luteum and later by the placenta (trophoblasts). Moderate estrogenic activity which has been maintained during the first trimester decreases gradually and is replaced by the progesterone effect produced by the placenta.

■Cytohormonal pattern in pregnancy

Ordinary navicular type: This cytological pattern[89,138,142,259] is characterized by deep intermediate cells crowded in clusters at the end of the first trimester. The cell of this type is called a "navicular cell" because of its boat-

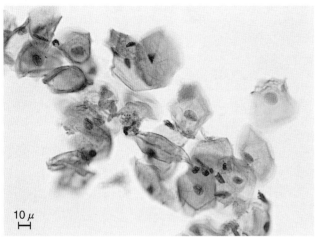

Fig. 6-44 Ordinary navicular type in pregnancy. The parabasal and intermediate cells transform to so-called navicular cells which are characterized by heavy cellular border with curling and intracellular glycogen deposition.

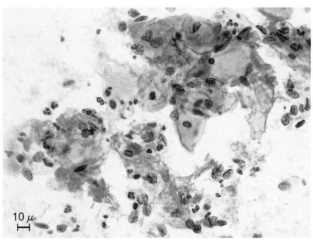

Fig. 6-45 Cytolytic type in pregnancy. Cytoplasmic borders of intermediate and parabasal cells become indistinct because of cytolysis. Several navicular cells with yellowish brown glycogen deposition remain.

like shape with thickened cellular border and an eccentrically locating nucleus (*oyster shell cell*).[174] Occasionally the cytoplasm of these intermediate cells has a yellowish hue caused by glycogen deposition (Fig. 6-44). The presence of navicular cells is not always diagnostic of pregnancy but is encountered in conditions of depressed estrogenic activity or increased progesterone activity including pregnancy.[264]

Cytolytic type: This is another cytohormonal pattern of the ordinary type. Cytolysis and appearance of free nuclei are frequently observed in association with marked predominance of Döderlein bacilli,[274,279] even though no inflammatory signs are present (Fig. 6-45).

Inflammatory type: Pregnant women are more susceptible to infection than non-pregnant women. A certain proportion of smears shows an inflammatory pattern associated with numerous leukocytes, abundant mucus, and bacterial flora. It is believed that local application of antibiotics returns it to the usual navicular pattern.[231]

Estrogenic type: This pattern is characterized by a predominance of eosinophilic superficial cells. This type of smear occurs during the first six weeks of gestation. However, its persistence reflects hormonal disturbance. For instance, Nesbitt et al[159] reported 27.5% abortions in patients with a karyopyknotic index over 20, whereas only 0.3% abortions with a normal pregnancy smear pattern. Koller and Artner[105] divided the estrogenic type into the precornification type and the cornification type according to the kind of predominant cells, whether eosinophilic superficial cells or cyanophilic superficial cells. Their cytohormonal studies on 725 pregnant women represented by the percentage of several characteristic cell type as mentioned above are as follows:

1. Navicular type	63.4%
2. Cytolytic type	15.6%
3. Inflammatory type	5.0%
4. Precornification type	13.1%
5. Cornification type	2.9%

The cytologic patterns are not absolutely diagnostic of pregnancy.[64,191] The diagnostic rates by cytology which

have been reported are variable: 72% by Pundel,[191] 68% by Meyer and von Haam,[146] and 96% to 98% by Hopman.[89]

■Cytology of pregnancy at term

Cytology of the pregnancy at term[126,231] is characterized by a decrease in the number of navicular cells and an increase in the eosinophilic or karyopyknotic index. In addition, the navicular cells as well as the superficial cells tend to be isolated and single. A large proportion of antepartum women[150] may have the inflammatory smear pattern abundant in leukocytes and mucus; therefore, local administration of antibiotics is desirable for the precise cytohormonal evaluation.

■Cytology of the postpartum period

Cytology of the immediate postpartum period: For about 10 days after delivery the cytological picture is characterized by the appearance of polymorphonuclear leukocytes and histiocytes, followed by the predominance of round or oval parabasal cells which are celled "postpartum cells", they have much glycogen, thickened cytoplasmic borders and pyknotic nuclei. After the second week, these characteristic postpartum cells return to normal parabasal cells (Fig. 6-46).

Cytology of the prolonged postpartum period: During lactation and for about four months before the return of menstruation the cytological picture is dominated by parabasal cells because of the suppression of maturation of squamous cells.

Cytology of threatened abortion: This pattern is characterized by an increase of more than 10% in the eosinophilic or karyopyknotic index in association with a disappearance of navicular cells.[41,237] If hormonal therapy has adjusted the cytohormonal pattern to the normal, gestation would continue and delivery would be expected. On the other hand, the cytology as such being followed by the replacement of increased superficial cells by parabasal cells and so-called postpartum cells[53] without response to the therapy may be indicative of fetal death.[234]

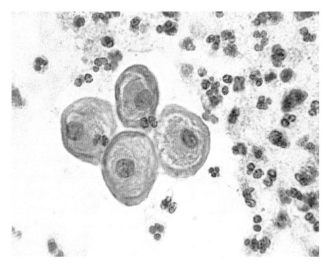

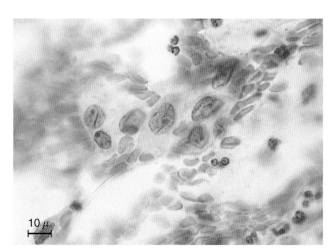

Fig. 6-46 Postpartum cells in a cervical smear soon after delivery. The cervical smear is characterized by a predominance of round parabasal cells which are called "postpartum cells." Note round or oval parabasal cells with thickened cytoplasmic borders and glycogen deposition.

Fig. 6-47 Decidual cells in a cervical smear after incomplete abortion. Note loose sheet of ill-defined cells with pale, moderate amount of cytoplasm and polyhedral or oval nuclei; the chromatin is finely granular and smoothly distributed. The nuclear membrane is delicate but wrinkled.

Table 6-3 Differential characteristics of single trophoblastic cells from the third type (differentiated parabasal type) and from undifferentiated cancer cells in a vaginal smear

Criteria	Cell		
	Single trophoblastic cell	Third type cancer cell	Undifferentiated cancer cell
Size of cell	Large (variable)	Small	Marked variation
Shape of cell	Round	Round	Marked irregularity
Nuclear/cytoplasmic ratio	Increased	Increased	Increased
Cellular border	Indistinct	Sharp	Variable
Nuclear membrane	Thick, sharp, regular	Thin, sharp, slightly irregular	Marked variation, irregular
Chromatin pattern	Bland	Granular, irregular clumping	In clumps with dotty distribution
Number of nuclei	1-3, piled up	Usually single	Usually single

(From Naib ZM: Cancer 14: 1183, 1961.)

The appearance of multinucleated syncytiotrophoblasts and/or cytotrophoblasts in addition to numerous red cells and neutrophil leukocytes indicates incomplete abortion.[231] They do not appear into vaginal smears in normal deliveries. Although cytotrophoblasts are rarely found, they must be distinguished from cancer cells (Table 6-3).[156]

Syncytial cells have pyknotic nuclei often overlapping (see Fig. 6-195); these differ from herpes-infected cells having moulded nuclei with a ground glass appearance. Immunocytochemical stain using anti-gonadotropin and placental lactogen anti-serum is useful for identification of trophoblasts.[55]

Decidual cells can be obtained[109] by endometrial aspiration or curettage in some cases of incomplete abortion. They are obtained as single cells or cell clusters in a flat sheet. As a rule decidual tissue from incomplete abortion has been more or less degenerated, thus the borders of fairly abundant, homogeneous cytoplasm are indistinct and the nucleus is hazy in appearance.

■Arias-Stella phenomenon

Arias-Stella reaction with endometrial atypia was originally noted in ectopic pregnancy.[4] The progesterone effect is thought to be responsible for adenocarcinoma-like hyperplasia. Cytologists must keep in mind that the Arias-Stella phenomenon of endometrial glands can be found in abortion; the nuclei are enlarged and hyperchromatic (Figs. 6-48 and 6-49).[104,152] Dense nuclear stain and vacuolation of the cytoplasm may lead to misinterpretation as clear cell carcinoma.

■Vernix caseosa cells

In case rupture of the fetal membrane one sees characteristic vernix caseosa cells desquamated from the degenerated squamous cells of the fetal skin in the vaginal fluid. They are anucleated, polygonal in shape, translucent and stain lightly yellow (Fig. 6-50). As a practical technique to examine the vernix caseosa cells, the smear should be taken as near the cervical os as possible (Hopman). According to Hopman et al,[90] they are of diagnostic value after the 28th week of gestation.

■Effects of contraceptives on cytology

A cytohormonal pattern of reproductive women who have been taking oral contraceptives is likely of early stage of pregnancy with predominance of folded intermediate cells resembling navicular cells.

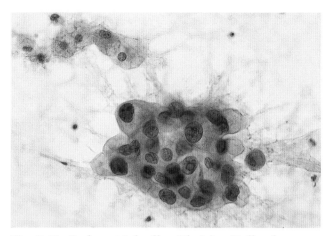

Fig. 6-48 Endometrial cells with Arias-Stella phenomenon. Endometrial aspiration from a 50-year-old woman who complained of vaginal bleeding because of incomplete abortion. Enlarged nuclei are densely stained and the cytoplasm is vacuolated.

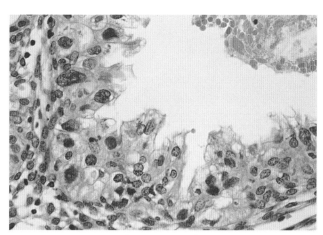

Fig. 6-49   Arias-Stella pattern of endometrial glands. Note enlarged and hyperchromatic nuclei of the lining epithelial cells. They may lead to a pitfall of overdiagnosis.

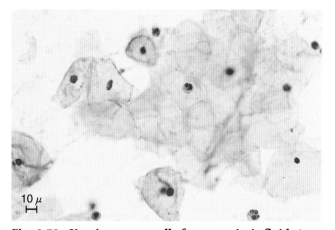

Fig. 6-50 Vernix caseosa cells from amniotic fluid. Anucleated scaly cells with yellowish, transparent cytoplasm originate from the fetal vernix caseosa. If these are found in a cervical smear, rupture of fetal membrane is suggested.

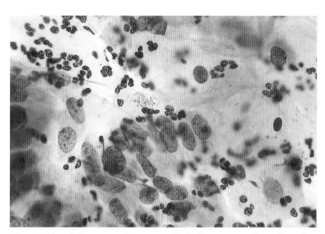

Fig. 6-51 Adenocarcinoma-mimic hypertrophy of endocervical cells. From a woman with so-called "pill cervix" who had a long-term use of oral contraceptives. Enlarged columnar cells show bland chromatins which are finely and evenly distributed.

Table 6-4 Histological types of cervical carcinoma in relation to OCH intake

Patients	Squamous cell carcinoma		Mixed carcinoma		Adenocarcinoma	
	No.	Mean age	No.	Mean age	No.	Mean age
New Zealander						
OCH user	140 (65.1%) *		59 (27.4%) **		16 (7.4%)	
White	101 (63.1%)	39.3	46 (28.8%)	37.9	13 (8.1%)	38.5
Nonwhite	39 (82.5%)	38.1	13 (23.6%)	34.8	3 (5.5%)	39.3
OCH nonuser	376 (82.5%) *		38 (8.3%) **		42 (9.2%)	
White	278 (82.2%)	59.0	27 (8.0%)	52.7	33 (9.8%)	57.2
Nonwhite	98 (83.1%)	51.0	11 (9.3%)	42.3	9 (7.6%)	55.9
Japanese						
OCH nonuser	397 (84.5%)	54.1	33 (7.0%)	52.7	40 (8.5%)	51.1

* Incidence significantly decreased by OCH intake ($p<0.01$)
** Incidence significantly by OCH intake ($p<0.01$)

Cytological importance is an occurrence of adenocarcinoma-mimic lesion of the cervix in long-term use of oral contraceptives (OCH)(Fig. 6-51). Epidemiological study in New Zealand by Takahashi et al[245] found significantly increased proportion of mixed carcinoma regardless of the race compared with squamous cell carcinoma (Table 6-4). Glassy cell carcinoma and mucoepidermoid carcinoma which are subtypes of mixed carcinoma are rather frequent in OCH users and thought to be induced by long lasting hormonal influence.

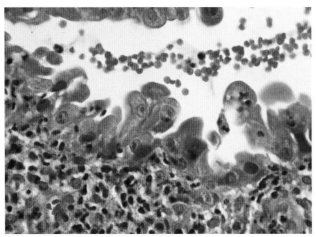

Fig. 6-52 Hypertrophy of endocervical cells in association with tissue repair. Note nuclear enlargement, occasional binucleation and prominence of nucleolus with eosino-philia.

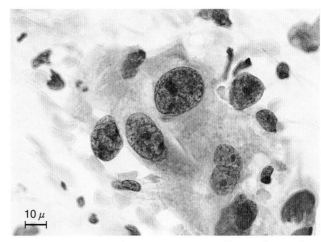

Fig. 6-53 Hypertrophic tissue repair cells in a loose cluster. The nuclei are enormously enlarged and show large nucleoli. The cytoplasm, ill-defined, is also increased in size. The chromatin pattern is bland.

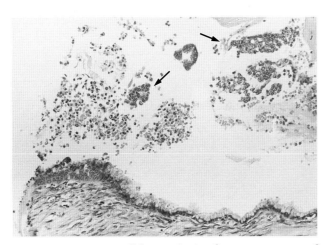

Fig. 6-54 Reserve cell hyperplasia after cryosurgery of the uterine cervix. Desquamation of clustered reserve cells is seen in exudate (arrows).

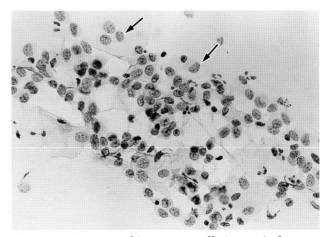

Fig. 6-55 Hypertrophic reserve cells in cervical scrapings after cryosurgery. The nuclei appearing almost denuded are densely stained but uniform in size and shape. Note the presence of X chromatin (arrows).

11 Benign Epithelial Cell Changes

There are several pathological conditions which may produce suspicious cells to be distinguished from cancer cells.

Tissue repair

Regenerative or repair cells in cervical lesions are encountered in a tissue reparatory process following the defect of mucous membranes; there are various causative conditions such as chronic cervicitis, cervical polyp, cryosurgery, electrocauterization, and irradiation. By different causative factors general morphological alteration of tissue repair will be modified. Although both epithelial and stromal tissues are concerned in reparatory process, epithelial cells should not be overestimated to be malignant.

General aspect: Reparatory cells exfoliate as large cell clusters in sheet by mechanical scraping. Cellular boundaries are obscure in cell clusters. The nuclei is regular in arrangement and nuclear polarity is observed in the cell cluster. It is difficult to determine the origin of repair cells from the endocervix or ectocervix, because endocervical repair cells become metaplastic in the process.

Cellular features: There is a moderate degree of nuclear enlargement ranging from 10 to 15 μm in diameter. Although slight anisonucleosis is associated, the nuclei are uniform in shape, round or ovoid, and possess delicate nuclear membranes. The chromatin is finely granular and evenly distributed. Eosinophilic macronucleoli, usually single but occasionally multiple, are of characteristic features (Fig. 6-56).

Other peculiarities: Tissue repair following cryosurgery is characterized by occurrence of marked reserve cell and basal cell hyperplasia after cryonecrosis that is followed by squamous metaplastic change (Figs. 6-54 and 6-55). Such reparatory proliferation occurs about three weeks after cryosurgery. Since these hyperplastic reserve cells are accompanied by hyperchromatism, irregularity of nuclear membrane and coarse chromatins, a diagnostic pitfall to carcinoma in situ should be avoided (Figs. 6-55 and 6-57). An attention for sampling cervical smears using the cytobrush or similar device is advocated,

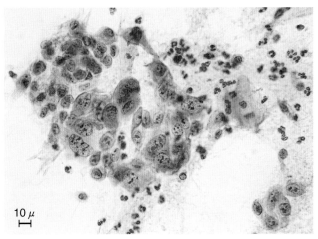

Fig. 6-56   Tissue reparatory hyperplasia of epithelium of reserve cell origin. The nuclei are enlarged with prominence of nucleoli, but uniform in shape. Beginning of squamous metaplastic change is associated. Note indistinct cytoplasm of fair amount.

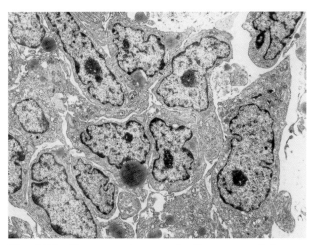

Fig. 6-57   Electron micrograph of reserve cell hyperplasia in tissue repair. From the same case as Figs. 6-54 and 6-55. Note a high N/C ratio, condensation of chromatin at nuclear envelopes and prominence of nucleolus. See the difference from cancer cells in chromatin pattern (see Fig. 6-136).

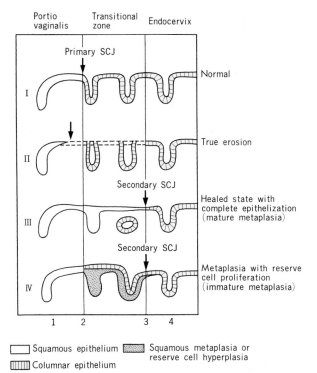

Squamous epithelium   Squamous metaplasia or reserve cell hyperplasia

Columnar epithelium

Fig. 6-58   Squamo-columnar junction (SCJ). Concept of primary and secondary SCJ. The cervix is histologically divided into three zones: (1) the portio vaginalis covered with stratified squamous epithelium, (2) the endocervix covered with columnar surface and glandular epithelia, and (3) the transitional (transformation) zone. This transitional zone which was originally covered with columnar epithelium is apt to develop to reserve cell hyperplasia, squamous metaplasia and dysplasia. Therefore, the squamo-columnar junction may move inside as a result of epidermization. (Slightly modified from Johnson LD, et al: Cancer 17: 231, 1964. The concept of the secondary SCJ was introduced by K. Masubuchi.)

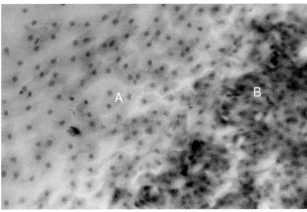

Fig. 6-59   Colpomicroscopic findings showing the squamo-columnar junction. Squamous zone (A) and columnar zone (B). Since superficial squamous cells have a broad scaly cytoplasm, the squamous zone appears to be sparsely dotted with round, small nuclei, and is distinguishable from the crowded columnar zone.

Colpomicroscopy is a valuable technique for defining the distribution of intraepithelial neoplasia in vivo. The cervix should be cleaned with 2% acetic acid before observation. One percent toluidine blue is applied with use of a cotton swab.

because the squamo-colmunar junction is lifted into the cervical canal after repair in the patient with a history of cryotherapy.[87]

Tissue repair occurring in association with chronic cervicitis or polyp is variable from case to case, because inflammatory and metaplastic changes are superimposed. In case of postirradiation, some persistent radiation effect (postradiation dysplasia)[179] can be noticed (see page 85).

■Squamous metaplasia

Metaplasia means production of one type of adult tissue by cells which normally differentiate to another type of tissue. Chronic inflammation and persistent stimulation are the main causes of squamous metaplasia in colum-

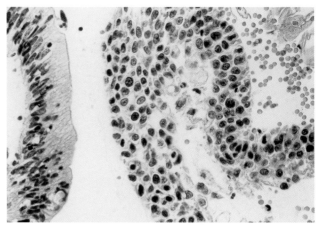

Fig. 6-60 Reserve cell hyperplasia of the uterine cervix. The endocervical mucosa is covered by multilayered reserve cells with spindly nuclei. These nuclei are basally situated retaining cell polarity.

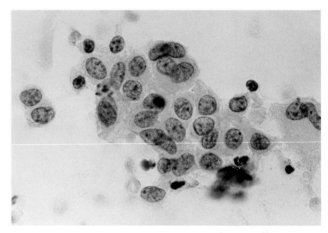

Fig. 6-61 Higher magnification of histology of reserve cell hyperplasia. Strips of exfoliated reserve cells tend to appear in sheet or cluster. Refer to exfoliative cytology (Figs. 6-62 and 6-63). Note a horizontal cut section at right showing pronounced anisonucleosis.

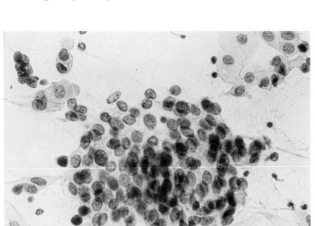

Fig. 6-62 Cluster of reserve cells in a cervical smear from a case with reserve cell hyperplasia. These cells appear to be denuded because of scant cytoplasm. Diagnostic points to be distinguished from carcinoma in situ: (1) uniformity of nuclear size and shape, (2) hyperchromatic but even compact chromatin particles, (3) invisibleness of nucleoli, and (4) regular nuclear arrangement leaving polarity.

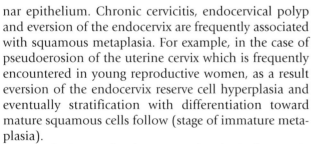

Fig. 6-63 Immature squamous metaplastic cells in cervical smears. Note a loose cell cluster with a cobblestone pavement or tiled arrangement interspaced with slits, that is referable to as epidermization. Note slight to moderate amount of immature cytoplasm.

nar epithelium. Chronic cervicitis, endocervical polyp and eversion of the endocervix are frequently associated with squamous metaplasia. For example, in the case of pseudoerosion of the uterine cervix which is frequently encountered in young reproductive women, as a result eversion of the endocervix reserve cell hyperplasia and eventually stratification with differentiation toward mature squamous cells follow (stage of immature metaplasia).

A single layer of columnar cells which frequently remains lying on the layer of proliferated reserve cells may slough off after differentiation to stratified squamous epithelium (stage of mature metaplasia). Metaplasia involves the epithelium of endocervical glands as well as the surface epithelium (Figs. 6-64 and 6-67).

Cytology: There is no definite criterion of mature metaplastic cells to be distinguished from original prickle cells. However in immature stage, metaplastic cells usually exfoliate as cell clusters or sheets arranged flatly in a cobblestone pattern; metaplastic cells are polygonal and have a fairly abundant cytoplasm. Cellular borders are indistinct and polyhedral or irregular in shape. Squamous metaplasia is frequently accompanied by chronic cervicitis, which makes the cellular border hazier and produces vesicular vacuolation or perinuclear halo of the cytoplasm.

The staining reaction of the cytoplasm, cyanophilic and pale in the immature stage (Fig. 6-63), tends to be intense with maturation (Figs. 6-67 to 6-70).

The nuclei are ovoid, somewhat enlarged and more or less hyperchromatic. The chromatin, however, is smoothly distributed except for a few prominent karyosomes. Smooth and delicate nuclear outline is one of the most important features of squamous metaplasia to be distinguished from dysplasia.

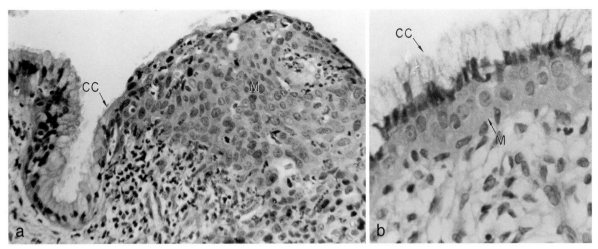

Fig. 6-64 Squamous metaplasia of the endocervix next to the columnar epithelium.
a. Solid sheets of epithelial cells showing squamous differentiation (M) are considered to arise from indifferent reserve cells beneath the columnar cells (CC).
b. At the beginning of squamous metaplasia, multiplication of reserve cells (M) is seen beneath the preexistent columnar cells (CC).

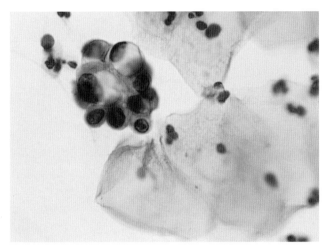

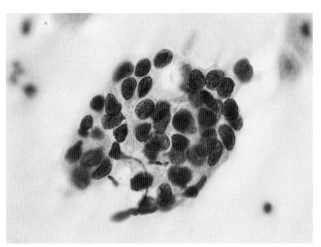

Fig. 6-65 Endocervical dysplasia. These are endocervical cells grouping in a small tubular fashion. The nuclei are small but vary in size and hyperchromatic. The cytoplasm is scant in comparison with endocervical cells and vacuolated. These findings are those referred to as AGUS category.

Fig. 6-66 Reserve cell hyperplasia in a cervical smear. Cluster formation and stratification of denuded cells with hyperchromatic, oval or elliptic nuclei are indicative of reserve cell hyperplasia.

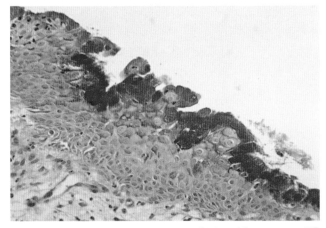

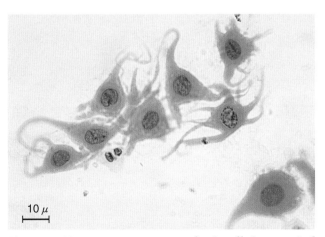

Fig. 6-67 Mature squamous metaplasia with extreme differentiation and well development of stratification. Well differentiated stratified squamous epithelium is shown as though it were developed creeping beneath the preexistent columnar epithelium that is strongly positive for PAS reaction.

Fig. 6-68 Mature squamous metaplastic cells in a cervical smear. The nuclei are oval, slightly enlarged and their chromatins are slightly roughened. The cytoplasm is thickened, well-stained and shows cytoplasmic processes like a star fish.

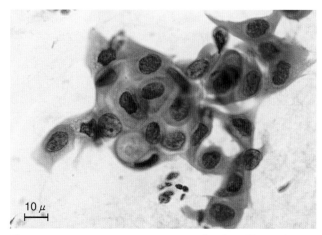

Fig. 6-69 Mature squamous metaplastic cells in sheet. A tiled arrangement of parabasal-like cells with abundant cytoplasm is indicative of mature squamous metaplasia. Intercellular bridges are discernible. Refer to the histology in Fig. 6-67.

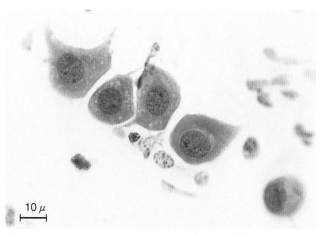

Fig. 6-70 Mature squamous metaplastic cells in a cervical smear. The cytoplasm is adequate in amount, opaque and thickened. Diagnostic points: (1) angulated or tailed cellular border, (2) finely vesicular cytoplasm though appears thickened, and (3) flat, stone-pavement arrangement.

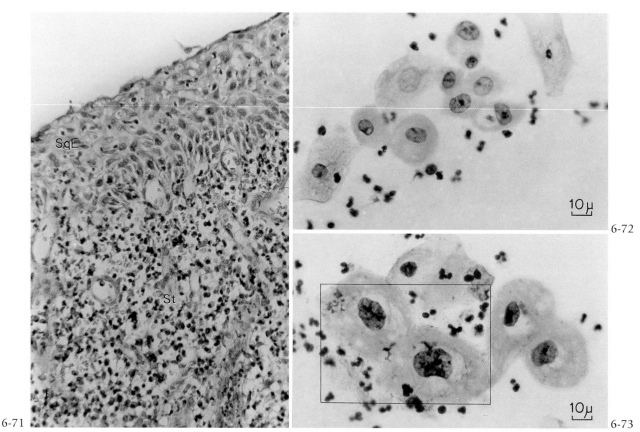

Fig. 6-71   Severe cervicitis in association with trichomoniasis and inflammatory hyperplastic epithelial response. Note a striking inflammatory cell infiltration in the mucosal stratified squamous epithelium (SqE) and its underlying stroma (St). Intraepithelial edema and leukocytic infiltration distort the arrangement of epithelial cells.

Figs. 6-72 and 6-73   Dyskaryosis in response to inflammation. Chronic and subacute inflammatory processes not infrequently give rise to a hyperplastic reactive response such as nuclear enlargement, hyperchromatism and prominence of karyosomes. Note cytoplasmic vacuolation as a result of intraepithelial edema.

General criteria for interpretation of inflammatory cell changes are as follows:

(1) Dense stain of nuclei, though occasionally coarsely granular in appearance, is due to condensation of chromatins. (2) The nuclear-cytoplasmic ratio is not much distorted. (3) Cytoplasmic vacuolization and perinuclear halos are frequently observed (frame). (4) Phagocytic infiltration of polymorphonuclear leukocytes is often seen within the cytoplasm. (5) Perinuclear halos or clearing should not be misinterpreted as HPV infection. Note indistinctly outlined vacuolization.

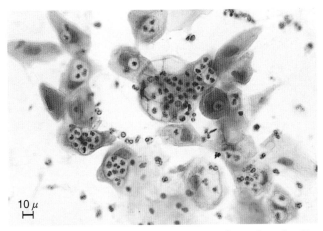

Fig. 6-74 Inflammatory cell changes of parabasal cells. Engulfment of many leukocytes, cytoplasmic vacuolation and eosinophilic staining reaction (pseudoeosinophilia) are the result of inflammation.

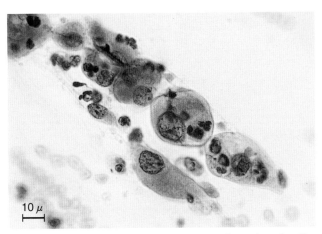

Fig. 6-75 Inflammatory cell changes of parabasal cells. Intracellular leukocyte infiltration and slight nuclear hypertrophy are seen as a reactive cell change.

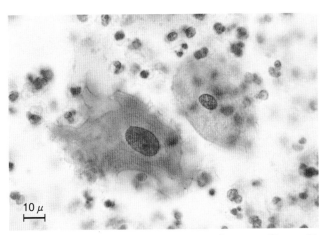

Fig. 6-76 Inflammatory cell changes of intermediate cell. Intense eosinophilic stain reaction, worm-eaten cytolysis, nuclear enlargement without chromatin abnormality, and perinuclear halo are characteristic features associated with inflammation.

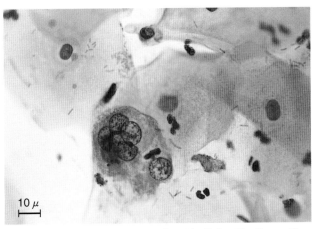

Fig. 6-77 Multinucleation of epithelial cells. See uniformity of nuclear size and shape. The chromatin appears somewhat coarsely granular. Note smooth nuclear membranes of tightly overlapped nuclei.

■Inflammatory cell changes

Inflammation may play a causative role in exfoliation of various abnormal cells. Insufficient maturation of the stratified squamous epithelium in prepuberal and postmenopausal periods or in deficiency of estrogen hormones may be susceptible to bacterial infection. It may be caused by many bacteria, such as streptococci, staphylococci, gonococci and *Neisseria catarrhalis*. Histologically the mucosa is edematous and true erosion may occur.

Chronic productive inflammation that may bring about proliferation of epithelial cells is more important than acute inflammation in cytological interpretation. It is most likely due to complexity of chronic inflammatory change in association with epithelial growth-stimulating (regenerative), desmoplasia-producing and host resistant immunocellular reactions.

Cytology:

1) Acute and subacute inflammation: The smear is smudgy in appearance because of the preponderance of neutrophilic leukocytes, degenerative or necrotic cells, cellular debris, and causative microorganism or protozoa. In the subacute stage, a number of histiocytes with single and occasionally multiple nuclei are often pre-

sent. Although superficial and intermediate cells are usually involved, parabasal cells are often found in the presence of true erosion. Inflammation should not be mistaken for postmenopausal smears. Dependable cell changes in inflammation are characterized by cytolysis as shown by vesiculation, vacuolation, feeble and irregular cell borders, perinuclear halos, leukocytic migration into the cytoplasm, and eosinophilic staining reaction. Nuclear changes are also remarkable. Pyknosis and enlargement of nuclei with bland (normochromatic) and hazy chromatin may be frequently found. Concerning cytolysis, it should be remembered that bacterial cytolysis may also occur without any evidence of inflammation. A good example is glycolysis of squamous cells by Döderlein bacilli composing a normal vaginal flora.

2) Chronic inflammation: There are no peculiar cytologic features of chronic inflammation. Nuclear hypertrophy, dense stain, remarkable karyosomes and occasional multinucleation that may mimic malignancy are often noticeable (Figs. 6-75 to 6-78 and 6-80).

Benignancy of these changes can be determined with ease by N/C ratio within benign limits and a smoothly distributed chromatin pattern regardless of presence of

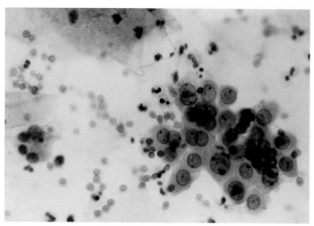

Fig. 6-78 Endocervical cells with multinucleation. The nuclei are round and clearly outlined. Multinucleation that is composed of many small nuclei occurs as a response to chronic inflammation. Absence of molding of the nuclei and/or perinuclear halo differ from multinucleation in herpes infection.

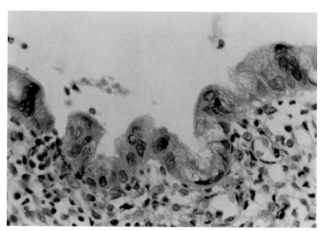

Fig. 6-79 Endocervix showing multinucleation in association with chronic cervicitis. Note oval nuclei clumping and overlapping on the surface mucosal epithelium. The subepithelial stroma is infiltrated by inflammatory cells and rich in blood capillaries.

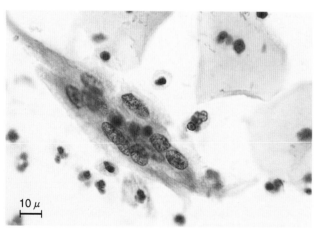

Fig. 6-80 Multinucleation of epithelial cells. Multinucleation with formation of syncytium of the cytoplasm in the endocervix in association with chronic cervicitis and repair. Uniformity of nuclear size and broad cytoplasm are the points of distinction from malignant multinucleation. Rigid nuclear rim is due to chromatin condensation at periphery by degeneration.

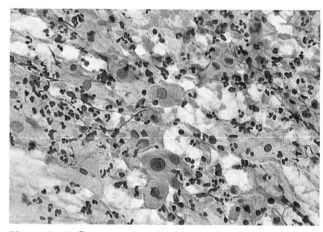

Fig. 6-81 Inflammatory cell changes in atrophic cervicitis. Note predominance of intermediate and parabasal cells with abundance of polymorph nuclear leukocytes. Nuclei are hypertrophic and the cytoplasm tends to be acidophilic. Chromatins are evenly distributed and bland.

some karyosomes. It should be remembered that metaplastic cells are the most common accompaniment of chronic inflammation. We must note that nuclear stain, though normochromatic in inflammation, may vary according to association with degenerative or reparatory process. Vooijs mentions that chromatins are more often hypochromatic rather than hyperchromatic in inflammation-associated cell changes.[261]

■Atrophic cervicitis with reactive changes

As the most common benign cellular change atrophic cervicitis (senile colpitis) may reveal generalized nuclear hypertrophy of parabasal and intermediate cells. When dyskeratosis or eosinophilia (orangeophilia) is associated with nuclear enlargement and moderate homogeneous hyperchromasia, careful attention should be paid for differentiation from cancer of the cervix. The accompaniment of many polymorph nuclear leukocytes, macrophages, multinucleated giant cells and basophilic

bodies that are called "blue blods" may become a clue of diagnosis of atrophic cervicitis.[120] Chapman mentions difficulty of differentiation of blue blods from stripped nuclei of cancer cell.[38]

■IUD associated changes

Nonspecific chronic endometritis is a frequent occurrence in about 10% of women with use of intrauterine device (IUD). Chronic inflammatory cell infiltration localized at IUD lodgment is associated with foreign body giant cell reaction. Thus the cervicovaginal smear of IUD users often displays peculiar findings such as degenerative endometrial cell balls, multinucleated macrophages and calcific bodies (Fig. 6-82). The calcific bodies are often surrounded by macrophages and leukocytes. Exfoliation of degenerated endometrial cells mixed with leukocytes, macrophages and stromal cells is suggestive of "exodus."

We must note that women wearing intrauterine con-

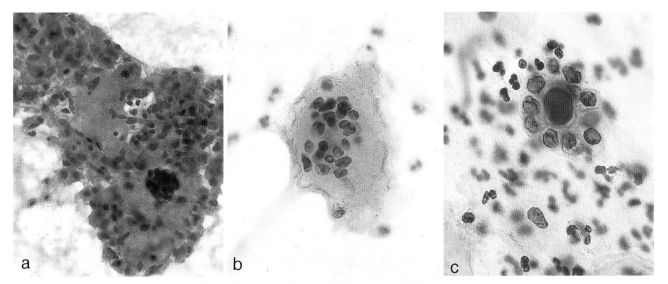

Fig. 6-82 Variable cell changes in cervical smear of IUD user.
a. Degenerated glandular cell ball (exodus-like) surrounded by macrophages and leukocytes.
b. Multinucleated foreign body giant cell.
c. Calcific body surrounded by macrophages.

traceptive users are liable to be affected by amoebic and *Actinomyces* infection.[6] As the cause of simultaneous infection of *Entamoeba gingivalis* and *Actinomyces* series, symbiotic correlation is suggested.[44] Differentiation between *Entamoeba gingivalis* and *Entamoeba histolytica* is difficult except for grave symptoms in *E. histolytica* infection. Cervical smears of IUD users occasionally shows calcific bodies to be surrounded by macrophages and leukocytes.[14]

■ Folic acid deficiency with dysplasia-mimic changes

Cellular morphological abnormalities in folate deficiency are not only confined to hematopoietic cells such as macrocytic red cells and hypersegmentation of neutrophils but to various epithelial cells. Cervical epithelial changes[102,103,258] are characterized by binucleation, increased nuclear cytoplasmic ratio and cytoplasmic vacuolation (Fig. 6-83). Vitamin B_{12} deficiency produces the similar cytology. Folate deficiency causes impaired DNA synthesis, which is reflected in the cell nucleus. Asynchronous maturation of cells as shown in retarded nuclear DNA synthesis and normally maturing cytoplasm is apparently encountered in red cell series. Pregnant women who have stopped the pill (combination type oral contraceptive) before pregnancy have a risk to folate deficiency. It has been reported that about one fifth of women who took oral contraceptives showed similar cytologic changes to those with folate or vitamin B_{12} deficiency because of impairment of folate metabolism, although less severely involved quantitatively. It has been known that the group taking the pill produced lower serum and red cell folate and higher urinary formiminoglutamic acid excretion than the normal control group.[224] It is noteworthy that the cellular abnormalities show a good response to the folate therapy.[273] The evidence suggests that the dysplastic change that occurs in the patient who has taken long term oral contraceptives[7,29] may be the reaction identical with mega-

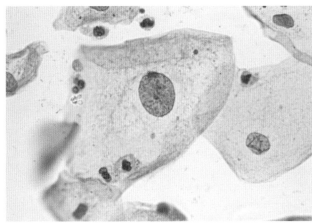

Fig. 6-83 Vaginal cytology reflecting the change by folic acid deficiency. Note enlargement of the nuclei and of the whole cell that is one of the most important features in the case with folic acid deficiency. (By courtesy of Prof. W.A. van Niekerck, The University of Stellenbosch, Stellenbosch, South Africa.)

loblastic anemia or the results of complication of megaloblastic anemia.[232,241]

■ Vaginal flora and microorganism affections

Renkonen et al[201] reported bacteriological studies on vaginal discharge from 200 patients, in which trichomoniasis were 18.5%, candidiasis 32.5%, *Haemophilus vaginalis* infection 30.5%, and enterococcus infection 52.5%. *Haemophilus vaginalis* is never found to be the only organism to cause vaginitis but found in association with infection of cocci. Bibbo and her associates[16] described a general vaginal flora studied on vaginal Papanicolaou smears taken from 14,212 women: Döderlein bacilli, 18.3%; bacteria 36.1%; cocci or coccoid bacteria 11.1%; *Haemophilus vaginalis* 5.2%; *Leptothrix* 2.4%; *Trichomonas vaginalis* 16.7%; fungi 6.3%, and none 3.9%, respectively.

Table 6-5 Organisms isolated in abundance from the female genital tract

Microorganism	Vaginal swab		Cervical swab	
	No./70	%	No./70	%
Acid tolerant	54	77	43	61
(typical lactobacilli)	(38)	(44)	(29)	(41)
Staphylococcus aureus	6	9	4	6
Staphylococcus sp.	24	34	13	19
Faecal streptococci	15	21	12	17
Anaerobic streptococci	24	34	16	23
Other streptococci	6	9	6	9
Diphtheroids	33	49	29	41
Haemophilus vaginalis	11	16	11	16
Bacteroides sp.	13	19	11	16
Coliforms	12	17	8	11
Proteus sp.	2	3	1	1
Other enterobacteriaceae	4	6	2	3
Candida albicans	4	6	3	4

Note: Used media (1) 5% horse blood agar, (2) MacConley's bile salt agar, (3) Sabouraud's glucose-peptone agar, (4) 0.0075% Neomycin blood agar, (5) tomato juice agar (Oxoid) (From Corbishley CM: Clin Pathol 30: 745, 1977.)

It should be kept in mind that the sensitivity of detecting microbiological infection is not equally evaluated with Papanicolaou stain; trichomoniasis can be detected as well as with a wet film, but candidiasis must be studied by a culture technique.[252]

The microbial flora of the vagina and cervix investigated by means of culture media[35] was more variable in organism and more frequent in incidence than that studied by the smear method (Table 6-5). No noticeable variation in the pattern of microbial flora has been described in relation to different phases of sexual cycle or to the use of contraceptives.[35]

1) *Haemophilus* vaginitis

Characteristic of *Haemophilus* vaginitis is the appearance of so-called "clue cells" which are closely packed and marginated by small, gram-negative bacilli; this characteristic feature is called "grainy" by Bibbo[16] (Fig. 6-85) The background of vaginal smears is also diagnostic; (1) lactobacilli are absent, (2) leukocytes are scant, and (3) admixture with small rods or cocci is commonly found.

2) *Leptothrix* infestation

Leptothrix is not a pathognomonic inhabitant and identified microscopically as hair-like or filamentous rods with occasional branching.

It should be remembered that the bacteria called *Leptothrix* include *Lactobacillus* and actinomycete. The *Lactobacillus* cannot be distinguished from Döderlein bacillus in the Papanicolaou stain (Fig. 6-84). According to Bibbo's report,[15] the frequency of actinomycete was 2.2 % in patients from the University of Chicago Clinics and symbiotic association with other organisms was frequent; 74.9% of the patients with occurrence of *Leptothrix* were associated with infestation of trichomonads and Döderlein bacilli and 21.7% of the patients were associated with infestation of trichomonads and coccoid bacteria.

3) Moniliasis (Candidiasis)

Fungus vaginitis is commonly due to infection of *Candida albicans*. Candidiasis is a low grade inflammatory response accompanied by thin vaginal discharge. *Candida albicans*, which belong to bacterial flora of the vagina, is conditioned by a depression of normal acidity to invade the tissues through the epithelium. The invasion is favored by diabetes mellitus, pregnancy, and the administration of broad-spectrum antibiotics.[215] Morphologically it consists of pseudomycelium with constrictions to which oval, yeast-like spores are attached (Fig. 6-87). Squamous cell changes in smears with candidiasis are characterized by radially clumped arrangement, vacuous cytoplasm, perinuclear halos and nuclear

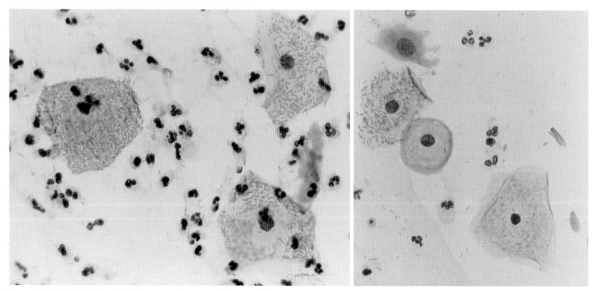

Fig. 6-84 Döderlein bacilli lying on intermediate cells giving rise to cytolysis. Döderlein bacilli that belong to a heterogeneous group of lactobacilli predominate in the second half of the menstrual cycle, pregnancy or premenarche when intermediate cells are predominant.

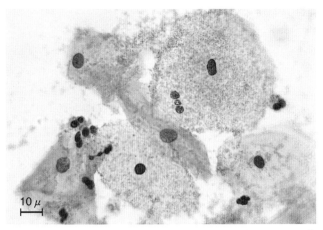

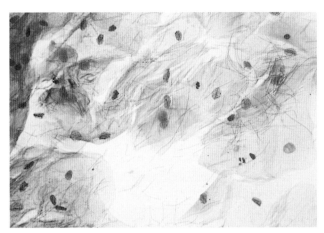

Fig. 6-85 *Haemophilus vaginalis* **in a cervical smear.** Rod-like coccobacilli overshadow squamous epithelial cells. The presence of such "clue cells" with a grainy appearance and lack of lactobacilli in the background are diagnostic of *Haemophilus* vaginitis.

Fig. 6-86 *Leptothrix vaginalis* **in a vaginal smear.** These organisms are pinkish grey, poorly stained, filamentous or hair-like structures.

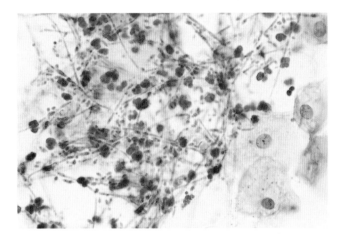

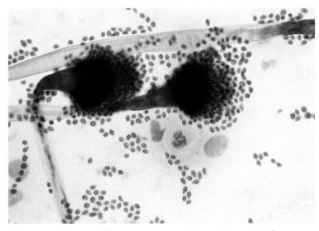

Fig. 6-87 **Candidiasis in a vaginal smear.** Slightly eosinophilic, septate filaments (hyphae) and conidia are seen.

Fig. 6-88 *Aspergillus* **as a contaminant in vaginal smear.** This is not a pathognomonic organism.

abnormalities such as irregular contours, hyperchromasia and swelling. These changes are particularly pronounced in the case including hyphae and blastospores .[82)]

4) *Aspergillus*

Aspergillus species are often found in vaginal smears but they are considered to be one of contaminants like alternaria (Fig. 6-88).[16)]

5) *Alternaria*

Alternaria sporidium is not pathognomonic but often encountered as one of laboratory contaminants. The dark brown conidia are muriform with transverse and longitudinal cross walls (Fig. 6-89).

■Sexually transmitted diseases (STD)

There are many pathogenic organisms which are transmitted by sexual contact. Diseases by classic definition are limited to venereal diseases but a current concept of sexually transmitted diseases (Center for Disease Control, 1985) includes a large variety of infectious agents because of a great importance of major public health problems such as AIDS, HPV and hepatitis B infections. According to the current epidemiologic study on cervical infectious diseases in 49,219 reproductive women by

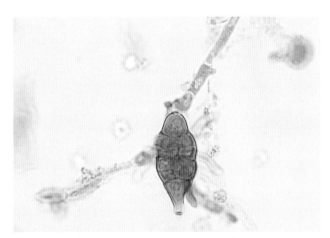

Fig. 6-89 *Alternaria* **as a contaminant.** *Alternaria* is a common contaminant in cytology. Note a characteristic racquet-shaped conidium that colors deeply brown in Papanicolaou stain.

Table 6-6 Sexually transmitted diseases (STD) and pathologic agents

Organism		Disease
Bacteria	*Treponema pallidum*	Syphilis
	Neisseria gonorrhoeas	Gonorrhea
	Haemophilus ducreyi	Chancroid
	Calymmatobacterium donovani	Granuloma inguinale
Virus	Herpes simplex virus	Herpes genitalis
	Human papillomavirus	Condyloma
	Molluscum contagiosum virus	Molluscum contagiosum
	Hepatitis B virus	Viral B hepatitis
	Cytomegalovirus	Cytomegalovirus infection
	Epstein-Barr virus	Infectious mononucleosis
	Human immunodeficiency virus	AIDS
Chlamydia	*Chlamydia trachomatis* D-K	Urethritis/cervicitis
	Chlamydia trachomatis L1-L3	Lymphogranuloma venereus
Mycoplasma	*Ureaplasma urealyticum*	Urethritis
Fungus	*Candida albicans*	Candidiasis
Protozoa	*Trichomonas vaginalis*	*Trichomonas* vaginitis
	Entamoeba histolytica	Ameba cervicitis

Sardana et al,[214] *Trichomonas vaginalis* comprised 5.1%, HPV 0.82%, *Candida albicans* 0.41%, herpes simplex virus 0.06%, *Chlamydia trachomatis* 0.06% and others, although cytology based incidence of infection may differ from microbiological detection in sensitivity. Cytologically detectable are described in this chapter.

1) *Chlamydia* infection

Chlamydia is classified into three species, *Chlamydia trachomatis*, *Chlamydia psittaci* and *Chlamydia pneumoniae*. *Chlamydia trachomatis* is one of the most common STD in recent years. The cervix shows shallow mucosal erosions followed by squamous metaplasia. Nearly 50% of nongonococcal urethritis is due to *Chlamydia* infection, which is manifested by mucopurulent exudate from the urethra. This organism is transmitted from pregnant women to infants. Thus screening tests of *Chlamydia trachomatis* IgG and IgA enzyme immunoassays are used for pregnant women.

Cytomorphological detection of chlamydiae is low in sensitivity in routine Papanicolaou smears; the sensitivity and specificity have been reported 23% and 91% by Shiina.[223] Chlamydiae take a unique developmental cycle; 1) attachment and penetration of elementary body, 2) transition from initial inert elementary body to active reticulate body, 3) division (the organization into smaller infectious form), 4) maturation to infectious elementary bodies, and 5) release from the host cell.

The affected cell is identified by characteristic intracytoplasmic inclusions having minute elementary bodies and by "moth-eaten appearance"[74] or "nebular inclusion" (Fig. 6-90).[223] In order to detect the antigen of *Chlamydia trachomatis*, either enzyme-linked immunosorbent assay (ELISA) or direct fluorescent antibody method (FA) are sensitive and specific; the sensitivity of FA is high ranging from 70% to 100%.[31,125]

2) Trichomoniasis

The *Trichomonas* is a relatively large organism, measuring 10 to 30 μm in length, with pear-shaped cytoplasm. There are four anterior flagella, one posterior flagellum and an undulating membrane. An oval eccentric nucleus and eosinophilic granules will be found in a faint greyish or greyish-green cytoplasm.

As a cause of inflammatory cell changes, trichomoniasis most commonly accompanies cellular atypia by inflammation. *Trichomonas vaginalis*, a flagellated protozoa tends to grow during depression of the normal acidity of the vagina. It is the most common cause of leukorrhea in women of reproductive ages. An incidence of the infection varies from 13% to 50%[21]; in reproductive women aged 30 to 35, overall incidence of trichomoniasis is about 14%.[153] Grossly spotty hyperemic lesions are scattered on the congested mucosa and superficial erosions may occur in severe cases. The squamous epithelium of the lesion reveals edematous thickening, cytolysis and various nuclear atypia. In the stage of florid infection nonspecific inflammation predominates, while in the latent infection atypical epithelial hyperplasia leads to bizarre nuclear changes.

Cytology: The cytology in trichomonas infestation is an extreme form of inflammatory change. The nuclei show various alterations mixing hyperplasia and degeneration, i.e., bizarre hypertrophy associated with occasional hyperchromasia, pyknosis and disintegration. As a result of cytolysis, the cytoplasmic border shows haziness and a worm-eaten appearance. The staining reaction of the cytoplasm is markedly acidophilic (Figs. 6-91 and 6-93).

3) Herpes virus infection

Herpes genitalis is most commonly caused by the type II virus of the pox virus group through sexual intercourse, but type I virus that is known to affect the oropharyngeal mucous membrane can be causative occasionally.[155] The incidence of herpes virus type II infection of the female genital tract is relatively low ranging from 0.6 to 4.3 per 1,000. Comparative studies by Jordan et al[96,157,282] on

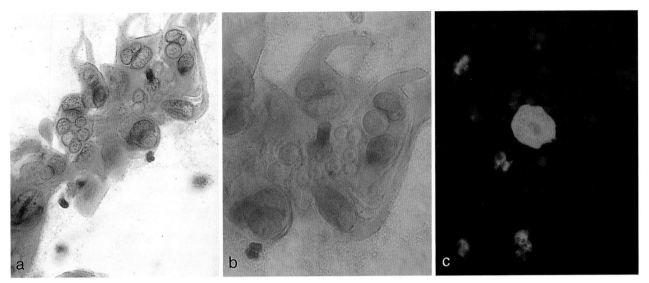

Fig. 6-90   *Chlamydia* infection in cervical smear.
a. *Chlamydia* infected metaplastic cells. There are minute dots (coccoid bodies) insides of inclusions. Papanicolaou stain.
b. Intracytoplasmic inclusion containing minute dots under differential interference microscopy. Papanicolaou stain.
c. Fluorescein-conjugated monoclonal antibody for *Chlamydia* depicts minute elementary bodies within the cytoplasm.

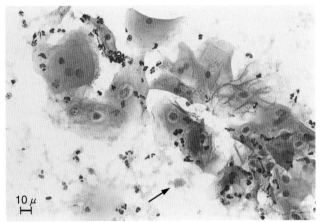

Fig. 6-91   Inflammatory cell change associated with *Tri-chomonas* infestation. Note intense eosinophilia of cyto-plasm, worm-eaten appearance of cellular borders due to cytolysis and perinuclear halos. Greyish green trichomonas are seen (arrow).

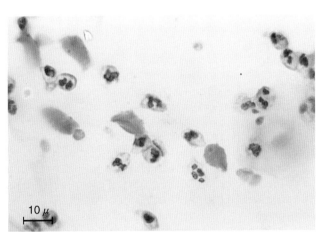

Fig. 6-92   Greyish green *Trichomonas* organisms in a vaginal smear. The cytoplasm is hazy and poorly defined. A small nucleus is recognizable, but no flagella can be seen in a Papanicolaou's smear.

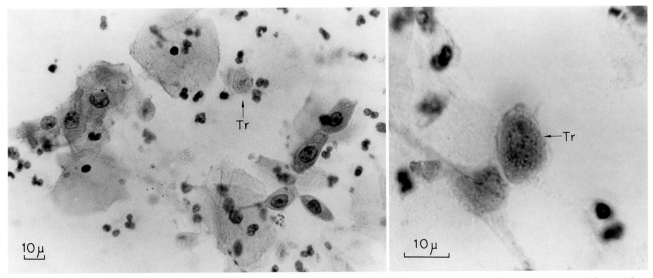

Fig. 6-93   *Trichomonas vaginalis* identified in vaginal smears. They appear as grey-green, hazy structures with ovoid or pear-like shape (Tr).

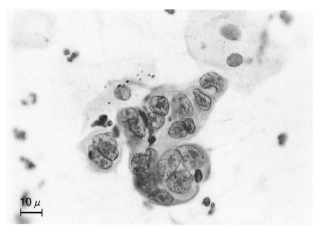

Fig. 6-94 Herpes simplex infection. In addition to multinucleation with molding, single acidophilic inclusion bodies are seen within the nuclei. Intranuclear inclusion could not be misinterpreted as nucleolus.

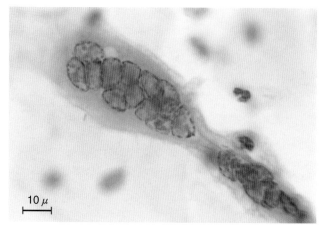

Fig. 6-95 Herpes simplex infection. The affected cell is characterized by (1) multinucleation with molding, (2) loss of granularity of chromatin with a ground glass appearance, (3) margination, and (4) thickening of the cytoplasm.

cytologically detected herpes virus infection by ethnic groups showed the similar frequency of the infection between rural American Indians and Caucasian, 0.9 and 0.6 per 1,000, respectively. On the other hand, a much higher rate (10.9 cases per 1,000)[8,154,198] was seen in Negro although statistically small in number. There have appeared several reports referring to a higher positive rates with antibodies to herpes virus type II among the patients with cervical carcinoma compared with control subjects from the same socioeconomic group.[65,157,184,198] The relationship of herpes virus infection to cervical carcinoma is also suggested epidemiologically; statistical difference of the affected age between herpes virus primary infection (10 to 30) and cervical carcinoma (30 to 50) may suggest the initiating factor of viral infection in carcinogenesis. Herpetic vesicles followed by shallow ulceration occur after incubation for 2 to 7 days more commonly in the external genitalia than in the cervix. The minimal symptomatic lesions are often overlooked in spite of the most common affliction of the external genitalia. Cytological and immunoserological studies are important to rule out syphilitic and granulomatous diseases.

Cytology: The nuclei of the affected cells are enlarged and multinucleated; each marginated nucleus is face to face with another as called nuclear molding (Fig. 6-94). Margination of the nucleus is due to condensation of chromatin at the nuclear envelope. The nuclear chromatin pattern is lost and characterized by a smudgy, greysh ground glass appearance (Fig. 6-95). Eosinophilic or cyanophilic, intranuclear inclusion body with halo is an important diagnostic feature. The cytoplasm is densely stained by Light Green and its border is indistinct. The most frequent findings occurring in all affected epithelial cells out of these cytologic changes are ground glass appearance of the nuclei (in 75%) and intranuclear inclusions (in 20%).[108] Multinucleation and nuclear margination are also indispensable alterations in herpes virus infection. Multinucleation is considered to be the result of cell adhesion[281] rather than amitotic cell division (see Fig. 6-95).

■ Specific infections

1) Tuberculosis

Tuberculous cervicitis is usually secondary to extragenital tuberculosis and manifests as extension from tuberculous salpingitis and endometritis. It affects rather young women and forms pseudotumorous lesions. The patient with cervical tuberculosis visits gynecologists with complaints of contact bleeding.[263] Characteristic epithelioid cells and Langhans giant cells from specific granulomatous tubercles are of diagnostic importance; however, they cannot be observed cytologically unless the lesion is externally eroded. The epithelioid cells have elliptical nuclei with light, vesicular chromatin network. The Langhans giant cells possess peripherally locating multiple nuclei with the same features as those of epithelioid cells (Figs. 6-96 and 6-97). Foreign body type giant cells having multiple, centrally locating, enrounded nuclei to be distinguished from the Langhans giant cells often appear in cervico-vaginal smears with senile colpitis.

2) *Enterobius vermicularis* infestation

This rarely occurs in association with leukemia and cervicitis.[109,196] The vulva and vagina are inflamed with pruritus. The embryonatal eggs are occasionally submitted to cytology. This seatworm was formerly called *Oxyuris vermicularis* or *Ascaris vermicularis* (Fig. 6-98).

3) Amebiasis

Amebiasis involving the vulva and vagina occurs rarely as extension from intestinal amebiasis in children. More ulceronecrotic amebiasis of the cervix was reported by Cohen.[34] An encysted form of the protozoa having an eccentric nucleus should be distinguished from a macrophage (Fig. 6-99).

■ Nonorganic substances

1) Pollen

Pollens are common contaminants in cytology. Various shapes of pollens look like parasite ova. Refractive, thickened cellular wall and appearance of the interior devoid of miracidium will be recognizable as plant cell (Fig. 6-100).

2) Starch powder

This lavender-colored, structureless corpuscle with occa-

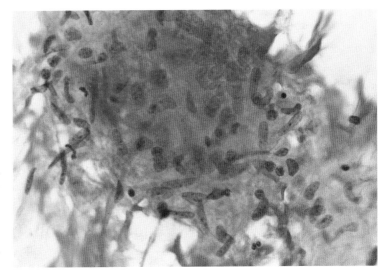

Fig. 6-96 Langhans giant cell multinucleated at periphery. A cervical scraping smear. Note that elongated nuclei are arranged in a periphery of the merged cytoplasm. The cellular border is ill-defined. Diagnostic points: (1) elongated and occasionally curved nuclei locating peripherally, (2) very fine lacy chromatins, and (3) thin but distinct nuclear rim.

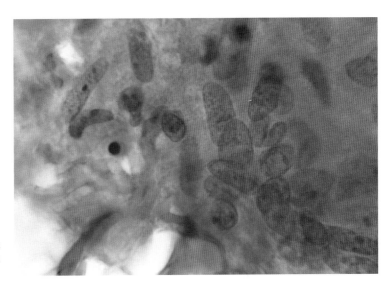

Fig. 6-97 High magnification of Langhans giant cell in a cervical smear from a patient with tuberculosis.

sional crackled clefts is one of the common contaminants which are derived from a gynecologist's glove used for internal examination (Fig. 6-101).

3) Psammoma bodies
Laminated calcific concretions called psammoma bodies in cervico-vaginal smears can be derived from various sites of the female genital tract and caused by various pathologic conditions. Papillary adenocarcinoma of the ovary is of the most probable origins, but endometrial papillary adenocarcinoma is also productive of calcific bodies.[50,62,81] Psammoma bodies that occur passing through the fallopian tube are found in a fairly clean background. Observation on cells accompanying with psammoma bodies gives a clue to the nature whether malignant or benign (see Fig. 6-164); the psammoma bodies are rarely found in chronic salpingitis or in use of the intrauterine contraceptive device (Fig. 6-82).

The majority of psammoma bodies are associated with malignancy and derived from ovarian serous cystadeno-carcinoma.[101] Endometrial adenocarcinoma should be also considered as the cause of psammoma bodies appearing in a cervicovaginal smear.[61,101] Although less than half of prevalence of psammoma bodies, the occurrence in benign conditions such as IUD use, endosalpingiosis and tuberculosis is important.[101]

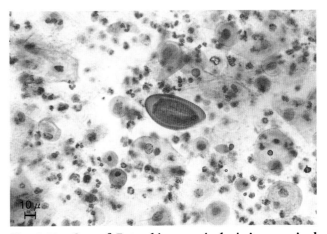

Fig. 6-98 Ova of *Enterobius vermicularis* in a vaginal smear. An embryonated egg containing folded larva is flat on the ventral side and convex on the other side. It measures approximately 50 to 60 by 20 to 30 microns.

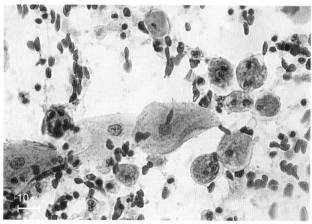

Fig. 6-99 Vaginal amebiasis from a cervico-vaginal smear. Note histiocyte-like *Entamoeba histolytica* which can be distinguished by a small eccentric nucleus. (By courtesy of Dr. C.R. Vieirafe Silva, São Paulo, Brasil.)

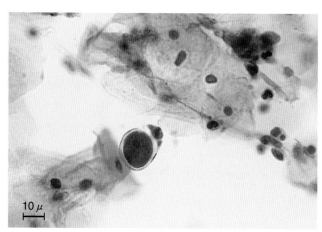

Fig. 6-100 Parasite egg-like pollen in a vaginal smear. A pollen in a vaginal smear. A pollen is characterized by strikingly refractive wall and amorphous structure inside that is different from miracidium.

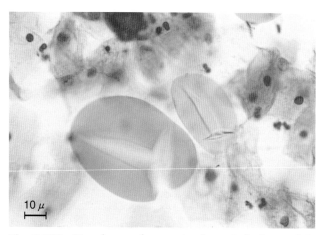

Fig. 6-101 Starch powder contaminated during pelvic examination from a gum glove.

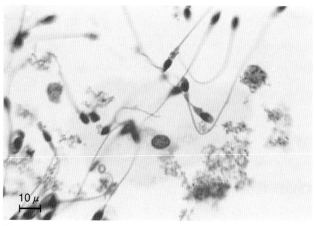

Fig. 6-102 Spermatozoa in a vaginal smear. Each has a long tail and pear-shaped head.

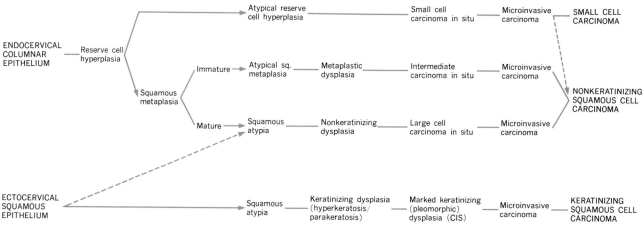

Fig. 6-103 Concept of morphogenesis of carcinoma of uterine cervix. (Modified from Patten SF Jr: Diagnostic Cytopathology of the Uterine Cervix. Monographs in Clinical Cytology, Vol. 3, 2nd revised edition. Basel: S Karger, 1978.)

12 Squamous Intraepithelial Lesions (Borderline Lesions and Related Changes)

Histological abnormalities or atypia, which used to be called borderline lesions or precancerous lesions, are frequently observed in the adjacent epithelium of carcinoma in situ[135,200] or microinvasive carcinoma. Although they do not always progress to carcinoma, these lesions are considered to be a state that may precede to overt carcinoma.[182,200] Various nomenclatures have been given to this state (Table 6-7), but the term "dysplasia" (Palmer and de Brux), as the histological term for borderline or precancerous lesions of the uterine cervix (at Vienna, 1961), has been used in the nineteen-seventies and eighties.

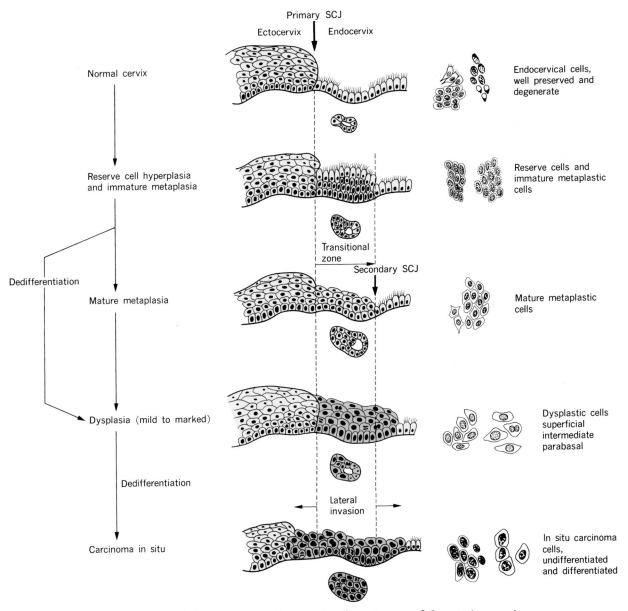

Fig. 6-104   Schema on morphogenesis of carcinoma of the uterine cervix.

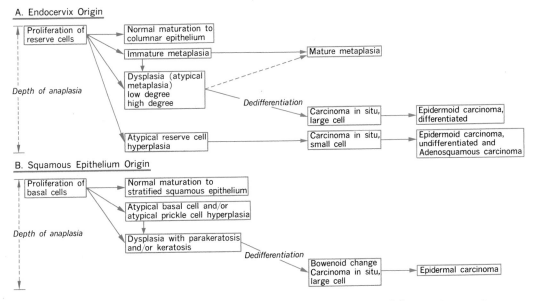

Fig. 6-105   Hypothetical schema on morphogenesis of carcinoma of the uterine cervix.

Table 6-7 Terminology of pathological epithelium of the uterine cervix

Reagan et al	Slight atypical hyperplasia	Atypical hyperplasia (moderate and marked)	Carcinoma in situ
Palmer and de Brux	Regular dysplasia	Irregular dysplasia	Intraepithelial carcinoma
Borst	Atypical epithelium	Strong atypical epithelium	
Held	Abnormal epithelium	Irregular (unquiet) epithelium	Noninvasive, atypical squamous epithelium
Fluhmann	Dysplasia (stage I)	Dysplasia (stage II and III)	Carcinoma in situ
Nieburgs	Benign dysplasia	Atypical dysplasia	Incipient carcinoma
Friedell et al	Prickle cell hyperplasia	Anaplasia (slight and marked)	Epidermoid carcinoma in situ
International Agreement on Histological Terminology for Lesions of the Uterine Cervix (Vienna, 1961)	Dysplasia, low degree	Dysplasia, high degree	Carcinoma in situ
Richart	CIN grade 1	CIN grade 2	CIN grade 3
Patten	Dysplasia, mild; metaplastic nonkeratinizing keratinizing	Dysplasia, moderate and marked; metaplastic nonkeratinizing keratinizing	Carcinoma in situ; small cell intermediate large cell
Koss	Early borderline lesion	Advanced borderline lesion	Carcinoma in situ; small cell anaplastic keratinizing moderately well-differentiated
The Bethesda System	Low-grade SIL*	High-grade SIL	

*SIL : squamous intraepithelial lesion

The nomenclature of cervical intraepithelial neoplasia (CIN) proposed by Richart and Barron has been also adopted universally as a simple terminology. Four grades initially defined have come to be simplified into three grades; CIN grade 3 equates severe dysplasia/carcinoma in situ. From the side of clinicians, carcinoma in situ was apt to be treated as more grave than severe dysplasia, whereas severe dysplasia would be underestimated with a conservative clinical approach.[48] As a result of semantic discussions on borderline or precancerous cervical lesions, the National Institute of Health provocated a comprehensive term "squamous intraepithelial lesion (SIL)"; the term "lesion" is preferred to "neoplasia" because biological potential of cervical abnormalities that encompass mild to severe dysplasia/carcinoma in situ is uncertain whether regress or progress.[120]

■Atypical squamous cells of undetermined significance (ASCUS)[120]

Cellular abnormalities of squamous cells within benign limits that are characterized by simple nuclear hypertrophy are occasionally encountered without evidence of reactive, inflammatory or dysplastic changes. Cytological findings as a whole fall short of diagnostic criteria of SIL or dysplasia. Instead of an equivocal term, the use of other descriptions such as "borderline nuclear changes" in Britain[26] and "nuclear enlargement of squamous cells not reaching the level of dysplasia" in Germany is described.[216] The ASCUS/AGUS Task Force members have confirmed the necessity of the remain of categories reflecting diagnostic uncertainty.[49]

■Historical review on classification of dysplasia

Concerning morphological subclassification of dysplasia or preinvasive state most histopathologists or cytopathologists have used the classification according to the stage of development or graveness. Patten proposed a unique cytomorphological subclassification including the concept of morphogenesis of intraepithelial neoplasia[176] such as keratinizing, nonkeratinizing and metaplastic dysplasia (Fig. 6-103). Keratinizing dysplasia occurs in the ectocervical site with hyperkeratosis and parakeratosis as a consequence of prosoplasia with atypia. Metaplastic and nonkeratinizing dysplasia are referred to as the dysplastic reaction at the endocervical site following or in the process of immature and mature squamous metaplasia respectively (Fig. 6-105). The author considers that it is probable that keratinizing dysplasia is preceded by atypical prickle cell hyperplasia of the ectocervical mucosa (Fig. 6-105); prickle cell hyperplasia is compatible with acanthosis in association with nuclear hypertrophy that is characterized by thickening of the prickle cell layer irrespective of nuclear atypia. As to the morphogenesis of metaplastic and nonkeratinizing dysplasia of Patten's concept, these seem to belong to the same pathologic entity and to be different in the stage of development; metaplastic dysplasia may develop to nonkeratinizing dysplasia as an advanced state of atypical squamous metaplasia at the transitional zone of the endocervix.

■Koilocytotic atypia

Koilocytotic atypia (KA) was defined by Koss for a borderline atypia having nuclear abnormalities in association with curious perinuclear vacuolization of the cyto-

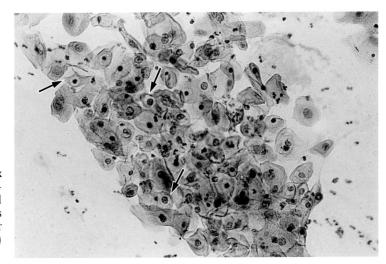

Fig. 6-106 Prickle cell hyperplasia of ectocervix (koilocytotic atypia). Crowding of mature intermediate or outer parabasal cells having enlarged nuclei, occasionally binucleated, is referable to as prickle cell hyperplasia. The presence of perinuclear halos reflects koilocytotic atypia (HPV infection) (arrows).

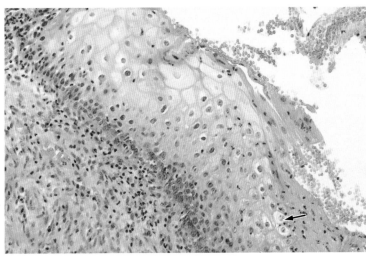

Fig. 6-107 Histology of LSIL with HPV infection. The stratified squamous epithelium of the cervix designated flat condyloma accompanies koilocytosis (arrow) in the upper half of the epithelium. LSIL that corresponds to mild dysplasia is characterized by basal and parabasal cell proliferation, partial interruption of stratification and nuclear abnormality in addition to koilocytosis.

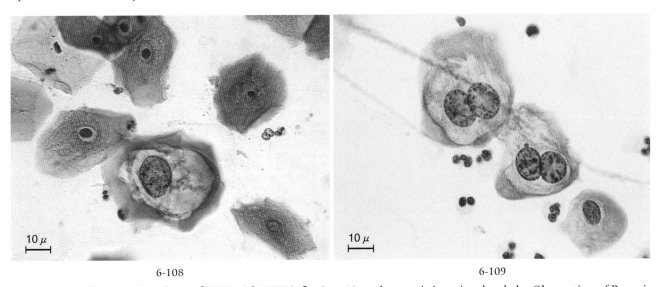

6-108   6-109

Figs. 6-108 and 6-109 Cytology of LSIL with HPV infection. Note characteristic perinuclear halo. Observation of Papanicolaou-stained smear under differential interference microscopy (left) discloses particle-like precipitates in the cleared cytoplasm. Nuclear enlargement, hyperchromatism and binucleation disproportionate to cytoplasmic maturation are noteworthy findings in LSIL with HPV infection.

plasm.[113] Since KA includes a spectrum of squamous epithelial changes ranging from condyloma without overt nuclear atypia to dysplasia with marked nuclear atypia, the cases of KA are classified into three grades depending on nuclear atypism. According to the study by Komorowski and Clowry[106] on 217 cases of KA from 858 cervical biopsies, grade 1 KA is much more frequent (17.0%) than grade 2 KA (6.2%) and grade 3 KA (2.0%). A noteworthy finding in KA is the presence of nuclear atypism (dyskaryosis) beyond the one third of

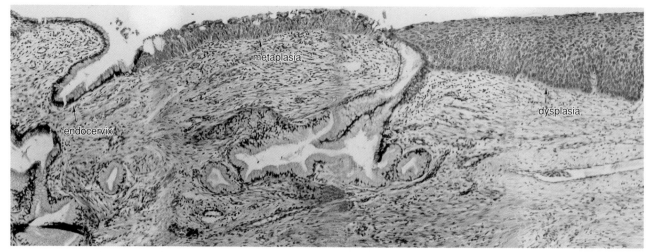

Fig. 6-110   Development of squamous metaplasia and dysplasia in the transitional (transformation) zone of the endocervix. Metaplastic change is known to begin with reserve cell proliferation as seen in the center. A single layer of columnar cells still remains lying on the metaplastic epithelium. Adjacent to metaplasia is a development of dysplasia of moderate degree (HSIL), which is characterized by aglycogenous state, distortion of stratification and proliferation of hyperchromatic parabasal cells. The arrangement of these dysplastic cells is fairly regular. Surface differentiation is retained. Refer to a diagram on the squamo-columnar junction in Fig. 6-58.

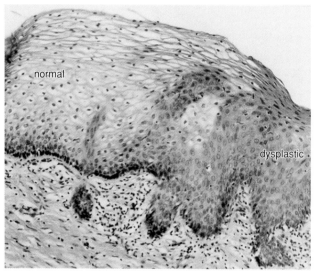

Fig. 6-111   LSIL (mild dysplasia) showing proliferation of deep epithelial cells with spindly or oval nuclei. Dysplastic zone in aglycogenous state is darkly stained and crowded with hyperchromatic cells. Note the tendency to downgrowth of the epithelium in dysplastic zone that is referable to the beginning of bulky outgrowth.

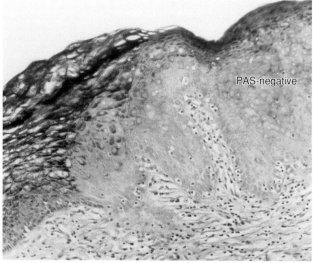

Fig. 6-112   Aglycogenous state in LSIL (mild dysplasia). This condition is represented by loss of PAS-stainability. The PAS-negative dysplastic zone corresponds to the darkly stained (right side) area in hematoxylin eosin stain as illustrated in Fig. 6-111.

the stratified squamous epithelium. The deep parabasal and basal layers are free from the change. Occasional vesicle formation[106] and recognition of virus particles[85,124] are suggestive of causative relation to viral infection.

■Low-grade squamous intraepithelial lesion (Mild dysplasia)

Low-grade squamous intraepithelial lesion (LSIL) encompasses abnormalities of squamous epithelium which have been defined as mild dysplasia, CIN 1 and flat condyloma with koilocytotic atypia.

Histologically LSIL is characterized by nuclear atypia in mature prickle cells rather than in deep basal cells, aglycogenous state, disorderly maturation and partial

interruption of stratification by abnormal cell growth. Although there may be some proliferation of basal and pasabasal cells, they are regularly arranged and retain the polarity. The basement membrane and epithelial polarity remain intact. The PAS stain is useful as the simplest and most reliable technique to recognize the degree of epithelial maturation. When maturation proceeds normally, PAS-stainability will increase towards the surface of the epithelium.[88,161] The PAS-stainable material may be lost in SIL (Fig. 6-112).

Cytology: General smear pattern. LSIL is characterized by a predominance of superficial and intermediate dyskaryotic cells and a clear background regardless of coexistence with erythrocytes and neurophils. The dyskaryotic index, i.e., immature dyskaryotic/mature dyskaryotic ratio, is low. Exfoliation of the dyskaryotic cells in number is less marked in LSIL than in HSIL. When dyskaryotic cells shed as cluster by vigorous scraping, the nuclear arrangement is regular in sheets.

Cellular feature: The nuclei are moderately enlarged although the cytoplasm is mature and well preserved. They are round or ovoid in shape. The nuclear rim is delicate and smooth; however fissure-like wrinklings of nuclear membranes are remarkable. The chromatin is slightly increased in amount, but the distribution is finely granular and evenly stained. Several karyosomes are visible, but a ground glass appearance is a distinctive feature of the chromatin texture. Nucleoli are not prominent. If present it is inconspicuous. X bodies along the nuclear membrane are normally preserved; the incidence of X bodies in LSIL is about the same as in normal epithelial cells.

LSIL is often associated with HPV infection; perinuclear halos, binucleation or multinucleation and dark, smudged chromatin pattern will be displayed.

■High-grade squamous intraepithelial lesion (Moderate dysplasia – severe dysplasia – carcinoma in situ)

This category defined by the Bethesda System (TBS) encompasses wide morphologic varieties which have been designated moderate dysplasia (CIN 2), severe dysplasia (CIN 3) and carcinoma in situ (CIN 3). There are some variations in histologic features according to the range of abnormalities.

Basic histologic changes comprise (1) abnormal basal and parabasal cell proliferation with distortion of cellular arrangement (distortion of polarity/loss of polarity), (2) disorganized epithelial maturation with distortion of stratification, (3) presence of mitotic figures above the middle of thickness of the epithelium with mixture of abnormal mitoses such as two group or three group metaphase, tripolar or multipolar anaphase and asymmetric chromosome separation, (4) nuclear atypia as depicted hyperchromatism, densely granular chromatins, anisonucleosis and high N/C ratio, and (5) indistinctiveness of X chromatins.

Cytology: The major cell components that define HSIL are parabasal type abnormal cells shedding singly, in loose sheets or in lines used to be called "Indian file appearance." The nuclei are hyperchromatic, densely granular and have distinctive chromocenters; chromatins are rather evenly distributed than there of cancer cells. Nucleoli are indistinctive. According to the grade of immaturity of the epithelium, the cytoplasm becomes small in amount, lacy in appearance and weakly stained (Figs. 6-114 and 6-115).

Addendum 1: Dyskaryosis and dyskeratosis
Dyskaryosis defined by Papanicolaou implies abnormal hypertrophy of nucleus disproportionate to normally differentiated cytoplasm.

Dyskeratosis means overmaturation of the cytoplasm of squamous cells in spite of normal nuclear maturation; cytoplasmic overmaturation is characterized by intense orangeophilia and/or appearance of kerato-hyaline granules.

Addendum 2: Metaplastic dysplasia
Although metaplastic dysplasia, nonkeratinizing dysplasia and keratinizing dysplasia based on cytoplasmic differentiation (Patten) are not or TBS used in the nomenclature published by the WHO or TBS, these terms are unique to understand the morphogenesis of a broad spectrum of dysplasia. Nonkeratinizing dysplasia is the most common ordinary type. Metaplastic dysplasia defined as metaplastic cellular disorders with certain nuclear abnormalities may develop in the transitional zone of the uterine endocervix. Main cellular constituents are round dyskaryotic cells, isolated or in sheets, chromatins of which are uniformly distributed and finely granular.

Addendum 3: Keratinizing dysplasia
Keratinizing dysplasia is a variety of dysplasia showing parakeratosis and/or hyperkeratosis and occurs in the ectocervix.[177] This type of dysplasia differs from that arising from the transitional zone of the endocervix by way of squamous metaplasia in association with varying degrees of nuclear abnormalities. Exfoliation of keratinizing superficial dyskaryotic cells becomes a pitfall of misinterpretation of early keratinizing squamous cell carcinoma. (1) Hyperkeratotic dyskaryotic cells should be distinguished from differentiated cancer cells showing a similar fiber-like feature; dysplastic cells are smaller in size and more uniform in shape than cancer cells. (2) Furthermore, attention should be paid to the absence of tumor diathesis from a whole area of the smear. (3) While in keratinizing dysplasia well preserved dyskaryotic cells derived from the deep layer are admixed with small hyperkeratotic cells, evidently malignant cells of the parabasal type can be detected in early invasive squamous cell carcinoma.

Keratinizing dysplasia that equates HSIL with dyskeratosis can be distinguished by following findings; 1) dyskeratotic atypical cells are less polymorphic and smaller in size than squamous cancer cells, and 2) occurrence of dysplastic parabasal cells mixed with above described small dyskeratotic cells (Figs. 6-116 and 6-117).

Addendum 4: Postirradiation dysplasia
Postirradiation dysplasia which was defined by Patten et al[180] as abnormal epithelial change following a long latent period of postirradiation is characterized by enlarged nuclei, increased nuclear-cytoplasmic ratio, marked hyperchromatism with finely granular or densely uniform texture and presence of surface differentiation (hyperkeratosis and parakeratosis); its morphology is similar to naturally arising dysplasia. These cellular changes differ from acute radiation effects such as marked increase in cell size with bizarre shape, cytoplasmic vacuolization, giant cell formation, perinuclear halo and phagocytosis. The biological behavior of postirradiation dysplasia is obscure, but possesses a biological potential of progression[178] particularly in the case when it occurs within three years after irradiation therapy.[270] In order to distinguish postirradiation dysplasia from coexistent classical dysplasia, Patten[178] emphasized several cytologic points: (1) appearance of abnormal cells in sheets rather than as single isolated cells, (2) prominence of cytoplasmic eosinophilia, and (3) prominence of oval

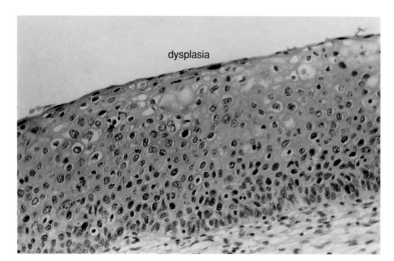

Fig. 6-113 HSIL (moderate dysplasia) of the uterine cervix. There is distortion of epithelial cells in stratification and polarity. The nuclei vary in size and shape. Surface differentiation and cellular polarity remain in the basal cell layer. Note koilocytosis in the superficial and intermediate cell layer.

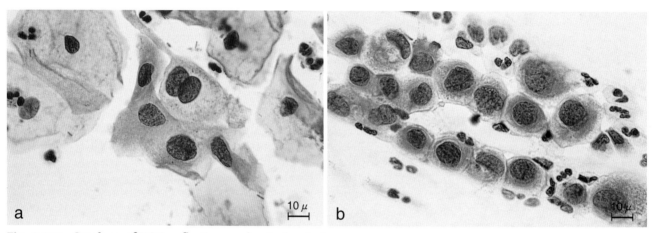

Fig. 6-114 Cytology of HSIL reflecting moderate dysplasia.
a. Dysplastic intermediate and outer parabasal cells. Note moderate nuclear atypia that is disproportionate to adequate mature cytoplasm of their normal counterparts. Chromatins are increased in amount but finely and evenly distributed.
b. Dysplastic parabasal cells are arranged in lines. The nuclei are enlarged and hyperchromatic but finely and evenly distributed. Nuclear rims are delicate and show convolution-like indentations. The cytoplasm is immature and stains lightly green.

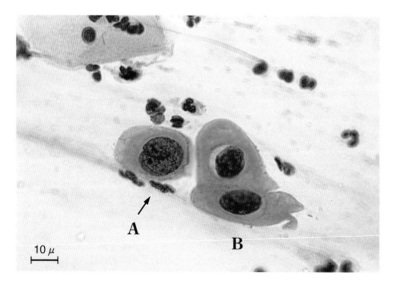

Fig. 6-115 Cytology illustrative of HSIL. A diagnostically important cell (A) exhibits high N/C ratio and marked hyperchromatism. Chromatins are increased in amount but finely and evenly distributed. Indentation of nuclear rim is in favor of SIL (arrow). Two mature dysplastic cells are seen (B).

and otherwise "irregular" cell configurations (Figs. 6-119 and 6-120).

■Dysplasia in pregnancy
Cytology under the hormonal environment of pregnancy is referred to as mucosal proliferation and eversion of

the endocervical epithelium that is followed by squamous metaplasia. Whereas epidemiologist's insistence of carcinogenic potency of pregnancy on cervical carcinoma[19] or the evidence of regression of dysplasia as well as carcinoma in situ after parturition[227] seems to be concerned with carcinogenesis, the incidence of carcinoma

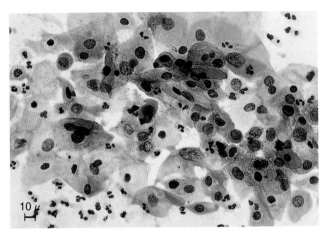

Fig. 6-116 Cytologic overview of HSIL associated with dyskeratosis. Preponderance of dyskeratotic superficial and intermediate cells; orangeophilia of the cytoplasm is reflection of hyperkeratosis.

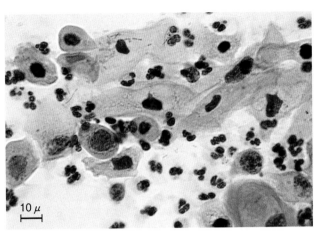

Fig. 6-117 Higher magnification of HSIL with marked dyskeratosis. Nuclear pyknosis and marked orangeophilia of superficial and intermediate cells are suggestive of keratinization. Note the occurrence of dysplastic parabasal cells.

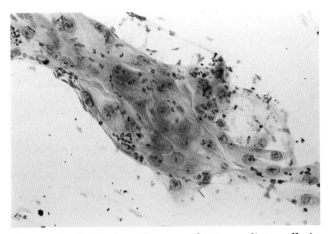

Fig. 6-118 Radiation changes of intermediate cells in vaginal smear. These intermediate cell clusters showing increase in nuclear and cytoplasmic size are likely to be from dysplasia. Differentiation from malignancy is made by (1) uniformity of nuclear size and shape, (2) lack of chromatin abnormality, and (3) benign limits of N/C ratio.

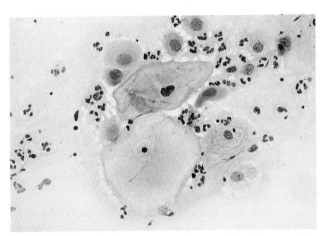

Fig. 6-119 Persistent radiation changes after four months of cobalt-60 teletherapy. Note enormous cytoplasmic enlargement, bizarre cellular elongation, and polychromasia of the cytoplasm.

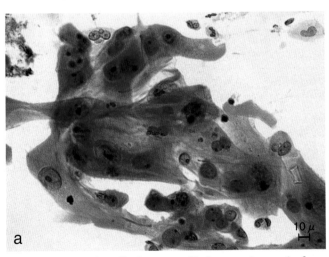
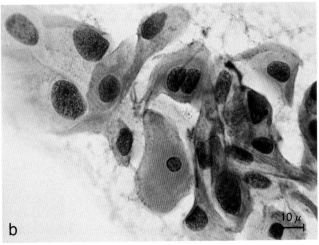

Fig. 6-120 Postirradiation on cell changes in cervical smears (postirradiation dysplasia).
a. From a 60-year-old female 50 days after ^{60}Co irradiation (tumor dosis 8,100 rad). Note nuclear enlargement and nucleolar prominence in association with bizarre shape of cells and fibrillar structure in the cytoplasm.
b. From a 51-year-old female 100 days after ^{60}Co irradiation (tumor dosis 9,280 rad). The nuclei are enlarged and occasionally binucleated. The cytoplasm is fibrillar and swollen in bizarre form. Abrupt eosinophilia is also notable.

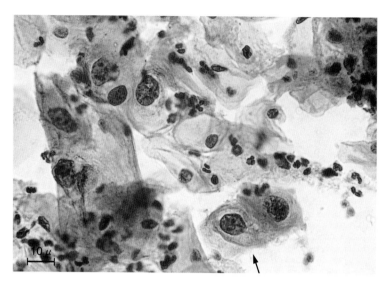

Fig. 6-121 HSIL (moderate dysplasia) in pregnancy. Dyskaryosis occurs in parabasal cells with a navicular feature (arrow). Although pregnancy shows no direct relationship to progression of dysplasia, cytology is particularly important for management of the pregnant patient. This smear is from a 34-year-old woman in the 30th week of pregnancy.

in situ and invasive carcinoma in pregnancy is about the same as in non-pregnant women.[18,230] The importance of cytology is to pursue squamous epithelial alteration with the use of colposcopy, since surgical assessment by multiple punch biopsies or cold knife conization is accompanied by grave complications such as infection, hemorrhage and abortion. Cytology is not specific but retains the pregnancy pattern as represented by characteristic navicular configuration. Their nuclei are dyskaryotic within a fair amount of cytoplasm showing thickened or folded borders (Fig. 6-121).

Currently Meisels described HPV-infection cytology was found twice as frequent during the second half of pregnancy as in the first half.[141] Its rate in late weeks of pregnancy (7.37% after 21 pregnancy week) is also more often than that in non-pregnant young women. The case shown in Fig. 6-121 is from a 34-year-old pregnant woman associated with koilocytosis.

■Condyloma

General aspect: Condyloma acuminatum is a warty growth in the vulva, vagina and occasionally cervix showing marked papillomatosis, acanthosis with elongation of rete ridges. It is most likely to be induced by viral infection[124,176] and transmittable by sexual intercourse, but squamous cell papilloma that occurs as s solitary benign tumor cannot be distinguished from condyloma only by histologic features. Meisels et al[143,144] include a flat, acanthotic change as flat condyloma and also endophytic inverted condyloma into the broad concept of "condylomatous lesion" with classical condyloma acuminatum. They referred to the frequent association of the cytological pattern with mild dysplasia in young women and emphasized a potential relation to intraepithelial neoplasia (CIN). Hyperkeratosis may accompany in the site of external genitals. Cytoplasmic vacuolization occurs frequently in the stratum malpighii.

Concerning the prevalence of HPV infection in cancer screening program according to ages based on cytomorphologic criteria,[112] the most frequent infection was seen in ages under 24; 1.86% in 191,475 women between ages 15 and 19 and 1.85% in 233, 419 between ages 20 and 24, respectively.[207] Molecular laboratory testing using dot blot hybridization and in situ hybridization with type-specific probes may get more sensitive diagnostic rates.[45,265] Since subclinical flat condyloma due to HPV infection was noticed as koilocytosis or koilocytotic atypia, attention has been paid to make reference to precancerous state.[207] Irrespective of nuclear changes, HPV-associated changes are classified into LSIL by the Bethesda System.

Cytology: Dyskeratotic spindly cells may shed from the surface with parakeratosis and hyperkeratosis. The prickle cells, particularly rounded deep intermediate or outer parabasal cells are of a principal cell constituent of acanthosis. Therefore, condyloma is characterized by a predominance of slightly dyskaryotic, deep intermediate cells, which are called "condylomatous intermediate cells" by Purola and Savia.[194] Their nuclei devoid of degeneration are only slightly enlarged and bland in chromatin pattern. Many nuclei are degenerative and tend to be pyknotic. Binucleation and occasional multinucleation are also characteristically seen. In addition to these intermediate and dyskeratotic superficial cells, the appearance of intermediate cells with distinctive perinuclear halo, so-called koilocytotic cells is emphasized to be typical of condyloma acuminatum (Figs. 6-122 and 6-123).[194]

■Effect of immunosuppressive agent

There is a marked increase in the incidence of intraepithelial carcinoma of the cervix in renal homograft recipients.[186] Immunosuppressive therapy such as busulfan, cyclophosphamide and azathioprine also brings about dysplastic change, in which bizarre configuration and cellular enlargement are characteristically seen.[110]

13 Carcinoma In Situ

As far as the stage classification of Tis by TNM and stage 0 by Federation Internationale de Gynecologie et d'Obstetrique is universally used for an appropriate therapy, the traditional concept of carcinoma in situ in the category of HSIL should be understated.

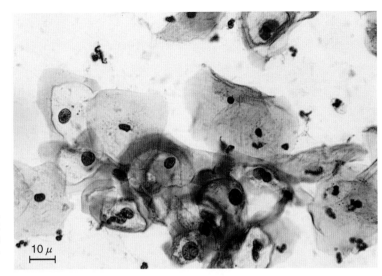

Fig. 6-122 Vaginal smear from condyloma acuminatum. Increased number of hypertrophic prickle cells (outer parabasal and intermediate cells) are derived from papillary fronds. Note characteristic koilocytotic perinuclear vacuolation.

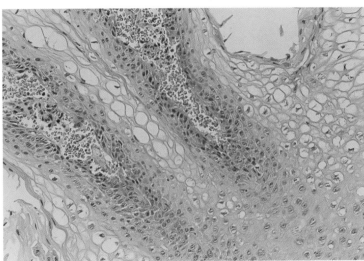

Fig. 6-123 Histology of classical condyloma acuminatum. Note papillomatosis and thickening of the epithelium by increased prickle cells.

■Concept of carcinoma in situ —With its historical review

Carcinoma in situ or intraepithelial carcinoma is a state of cytomorphological malignant change (confined to the mucosal epithelium) regardless of the involvement of cervical glands. However, there is no unanimous agreement on the entity of carcinoma in situ. There are some conservative opinions[17,23,78,80,107] regarding its morphogenesis as a transitory stage of cancer, namely a form of severe dysplastic change, because of the following facts: (1) cure or regression may occasionally occur after simple removal of tissue for biopsy or after delivery[80,82,160,182,217]; (2) not all cases of carcinoma in situ develop to invasive carcinoma[116,183,283]; (3) the difference in the average age between patients with carcinoma in situ and invasive carcinoma seems to be too great; (4) classical carcinoma in situ is rarely seen at margins of microinvasive squamous cell carcinoma.

However, there have been many other reports indicating that carcinoma in situ is real cancer from clinical and biological standpoints: (1) the average age of patients with carcinoma in situ is statistically always several years younger than that of patients with early invasive carcinoma[32,54,99,272]; (2) both carcinoma in situ and invasive carcinoma are less common in Jewish than in non Jewish white women; (3) according to a study by

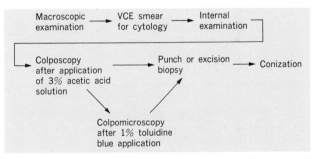

Fig. 6-124 Procedure of practical diagnosis for SIL and carcinoma of the cervix.

Michalkiewicz et al,[147] a recurrence rate of carcinoma in situ after several therapeutic procedures is low (3,8%) but it is nearly eight times as high as that of dysplasia (0.5%); and (4) a long-term observation of carcinoma in situ by Koss et al,[115] confirms it to be a precursor of invasive carcinoma without evidence of spontaneous involution.

It is evident that carcinoma in situ which originates from a wide field of abnormal epithelium is followed by prong-like stromal invasions.

Bangle and his associates[13] have emphasized that well differentiated squamous carcinoma probably bypasses dysplasia or carcinoma in situ because invasive kera-

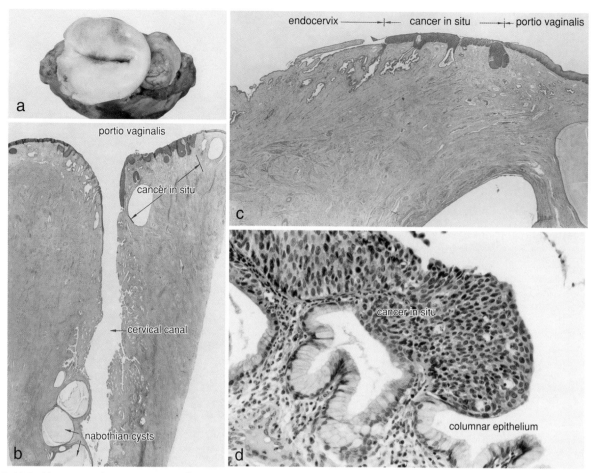

Fig. 6-125 **Macroscopic appearance of mucous membrane in carcinoma in situ showing no abnormality (a) and histology of carcinoma in situ of the uterine cervix (b to d).** Cancerization is limited to the epithelium regardless of the width of superficial spread at the squamo-columnar junction area. Field cancerization due to progressive neoplastic conversion of the epithelium, to some extent, is more probable than simple replacing growth arising from a unicentric origin. Possibility to detect carcinoma in situ by exfoliative cytology is understandable from this intraepithelial spread if a proper technique of sampling is applied.

tinizing large cell carcinoma occurs next to the surface epithelium appearing almost normal or only slightly dysplastic. There may be occasions, however, when the normal epithelium is sharply bordered by carcinoma because of the replacing growth of carcinoma (Fig.6-125).[79,172] Therefore, it is not always right to discuss the morphogenesis of carcinoma only from the intraepithelial changes contiguous to stromal invasion. Furthermore, many basic studies, i.e., the nuclear DNA content by the microspectrophotometric method,[91,211,212,260] behavior in tissue culture,[285] abnormality in chromosome analysis[235,266] and the value of anaerobic glycolysis indicate malignant characteristics of carcinoma in situ.

Concerning the terminology of precancerous and cancerous lesions of the uterine cervix, the nomenclature adopted by the Editorial of *Acta Cytologica*, i.e., dysplasia graded into two classes, carcinoma in situ and invasive carcinoma (1962) has been prevailing until the Bethesda System introduced the term "squamous intraepithelial lesion" and replaced dysplasia and carcinoma in situ. Kurman and Solomon have stated that the term "lesion" instead of neoplasia is used to convey the uncertain biological potential. Richart[202,203] proposed a histogenetic term, intraepithelial neoplasia (CIN) based on the concept that the disease entity of dysplasia and carcinoma in

situ is referred to as the continuous abnormalities; "CIN is designated as a spectrum of intraepithelial abnormalities that begins as a well differentiated intraepithelial neoplasm which has been traditionally classified as mild dysplasia and leads to invasive carcinoma through carcinoma in situ." CIN is subclassified according to the degree of differentiation; CIN grade 1 corresponds to mild dysplasia, grade 2 to moderate dysplasia, grade 3 to severe dysplasia, and carcinoma in situ respectively.

The continuous progression from CIN 1 to CIN 3 and further to early invasive carcinoma is proved by significant correlation between CIN grades and intensity of expression of p53 protein[185] and *ras* oncogene p21.[208] The more grave of epithelial abnormalities, the more pronounced in PCNA immunoreactivity and in aberrant location of PCNA-positive nuclei.[225]

■Management of detection of early carcinoma of the uterine cervix

The procedure of detection and decision of early carcinoma of the uterine cervix should be carried out in accurate steps at a gynecologist's office (Fig. 6-124). Colposcopy has come to be widely applied to screen cervical abnormalities and to localize the lesion to be biopsied. For the patients with a broad spectrum of CIN, aiming at

Table 6-8 Comparison between histology and cytology of squamous intraepithelial lesion

Classification Bethesda	Traditional	Histology	Cytology
LSIL	CIN 1 Mild dysplasia	Abnormal parabasal cell proliferation (in middle one third of epithelium) Disorganized maturation of epithelium (surface differentiation remaining) Partial interruption of stratification Frequent association of koilocytosis Occasional mitoses (in lower half of epithelium) Aglycogenous state of epithelium	Predominance of dysplastic superficial/ intermediate cells with mild nuclear atypia: slight increase in N/C ratio moderate hyperchromatism with finely evenly dispersed chromatins enough amount of mature cytoplasm Koilocytotic atypia
HSIL	CIN 2 Severe dysplasia Carcinoma in situ (large cell type)	Abnormal basal/parabasal cell proliferation (in lower two thirds of epithelium) Disorganized maturation of epithelium (tendency to surface differentiation) Distortion of stratification Distortion of polarity Frequent mitoses with occasional abnormal forms (in two thirds of epithelium)	Predominance of dysplastic parabasal cells with marked nuclear atypia: marked increase in N/C ratio marked hyperchromatism densely granular or smudged chromatins scant and immature cytoplasm
	CIN 3 Carcinoma in situ (small cell type)	Replacement of epithelium by small cancer cells (immature basal/parabasal type) Loss of stratification Loss of polarity Frequent mitoses of abnormal forms (in and over two thirds of epithelium)	Predominance of basal/parabasal type cancer cells: marked increase in N/C ratio marked hyperchromatism densely or coarsely granular chromatins scant and immature cytoplasm

the points of biopsy or excision is important to determine whether or not conization is performed. Although colposcopy is not competitive to cytology, preclinical cervical neoplasia that has been missed by cytology is occasionally detected by colposcopy.

1) Colposcopy
Colposcopy allowing a stereoscopic magnification of the mucosal membrane is as useful as cytology; whereas cytology is diagnosable of squamous intraepithelial lesions, their location and extension cannot be determined. Schiller's iodine stain was a procedure to recognize aglycogenous abnormal epithelium. Colposcopy is able to appreciate the abnormal area in the transformation* zone and to indicate the proper point of punch biopsy. Abnormal upward growth of vascular stromal papillae in borderline lesions is responsible for a characteristic mosaic pattern. If the stromal papillae are compressed by bulky outgrowth of the epithelium, superficial vasculatures will be recognized as punctuations. Thickening of the epithelium in the transformation zone with dyskeratosis and parakeratosis produces the white epithelium. Cytology and colposcopy are not competitive as a diagnostic procedure of cervical pathology but complementary each other.[27]

2) Colpomicroscopy
Colpomicroscopy that is able to visualize surface epithelial alterations such as variation of the nuclei in size and shape, dense cellularity and increased N/C ratio is a worthy technique for defining the distribution of intraep-

ithelial lesion in vivo.[202] The cervix should be cleaned with 2% acetic acid before observation. One percent toluidine blue is applied with use of a cotton swab (see Fig. 6-59).

■ Histology of carcinoma in situ
(1) The entire thickness of the mucosal epithelium is replaced by irregularly crowded small round or spindly cancer cells showing marked nuclear hyperchromatism and high N/C ratio (undifferentiated or small cell type)(Fig. 6-126). Carcinoma in situ includes cases which tend to undergo surface differentiation. Although the cytoplasm is somewhat broad and polygonal, the nuclei within the surface layers are hyperchromatic, irregular in shape and obviously malignant (differentiated or large cell type)(Fig. 6-133). (2) Polarity and stratification are lost. Disappearance of polarity in the basal layer is a reliable criterion to differentiate carcinoma in situ from classic dysplasia. (3) Mitoses are frequent not only in the basal layer but also in the intermediate and superficial layers over the lower two-thirds of the entire thickness of the epithelium. The axial direction of the spindles is not always perpendicular towards the basement membrane. (4) Abnormal mitoses are frequently observed. Multipolar mitoses, especially tripolar ones are notable findings. Asymmetric chromosomes in anaphase and polar chromosomes in metaphase known as two group and three group metaphase are also of frequent occurrence of abnormal mitoses (Fig. 6-127).

■ Cytology of carcinoma in situ
The cytology of carcinoma in situ, particularly small cell type, is characterized by a predominance of cancer cells

*The transitional zone of a histopathological term at the squamocolumnar junction that is liable to be transformed is referred to as the transformation zone by colposcopists.

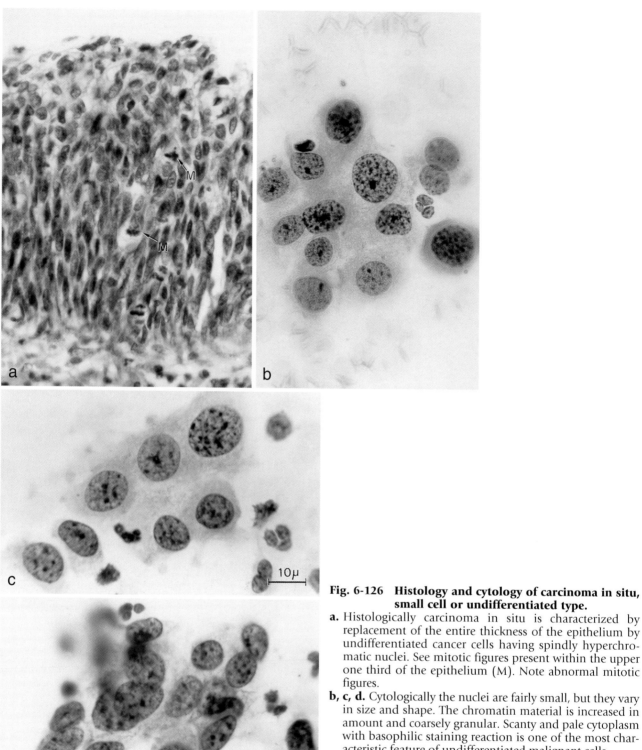

Fig. 6-126 Histology and cytology of carcinoma in situ, small cell or undifferentiated type.

a. Histologically carcinoma in situ is characterized by replacement of the entire thickness of the epithelium by undifferentiated cancer cells having spindly hyperchromatic nuclei. See mitotic figures present within the upper one third of the epithelium (M). Note abnormal mitotic figures.

b, c, d. Cytologically the nuclei are fairly small, but they vary in size and shape. The chromatin material is increased in amount and coarsely granular. Scanty and pale cytoplasm with basophilic staining reaction is one of the most characteristic feature of undifferentiated malignant cells.

Note the monotonous cellular pattern in differential cell counts or cytogram of the smear from carcinoma in situ; the cells are composed almost entirely of small undifferentiated malignant cells without evidence of necrosis.

of the basal type having scanty cytoplasm with pale cyanophilic staining reaction. The nuclei are hyperchromatic and densely granular in chromatin distribution. This compact chromatin pattern filled with fine to medium-sized granules and indistinctiveness of nucleolus are important features of in situ cancer cells; the coarsely granular chromatin is considered to appear in a degenerative process. Okagaki et al[171] have mentioned that

cervical smears showing over 30% basal type of cancer cells in differential counts of malignant cells are most likely of carcinoma in situ. As a general feature these cancer cells are isolated or tend to loosely aggregate; Cancer cells arranged in line along the smearing direction present a so-called "Indian file appearance" (Fig. 6-128). The smear shows no tumor diathesis. The presence of erythrocytes and leukocytes depends on the

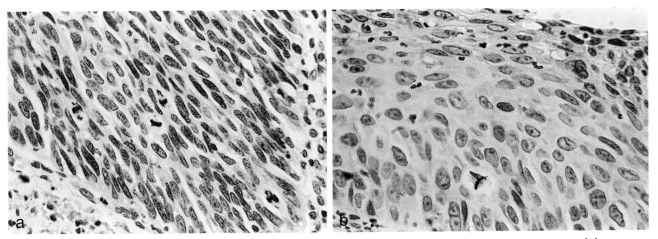

Figs. 6-127a,b Abnormal mitosis in carcinoma in situ. Polar chromosomes which are known as two group and three group metaphase are common in carcinoma in situ. A tripolar mitosis as shown in **b** is more rare than polar chromosomes. The presence of abnormal mitotic figures is one of important histologic findings of carcinoma in situ.

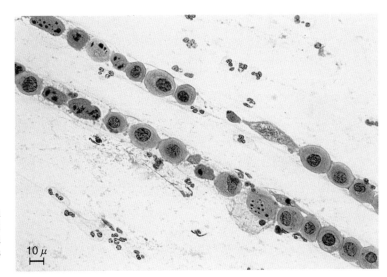

Fig. 6-128 General appearance of carcinoma in situ. An arrangement of cancer cells of parabasal type in line is called "Indian file appearance" as a characteristic feature of carcinoma in situ. There is no tumor diathesis.

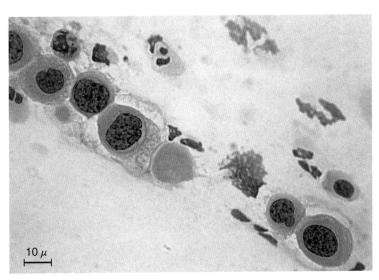

Fig. 6-129 High magnification of cytology. Carcinoma in situ of undifferentiated type. The nuclei are round or oval and hyperchromatic with a dense granular chromatin pattern. The cytoplasm is scant and occasionally poorly outlined.

coexistence of true erosion.

Large cell type carcinoma in situ shows a predominance of parabasal type of cancer cells having more abundant cytoplasm than basal type cancer cells in small cell carcinoma in situ (Figs. 6-130 and 6-131). In addition, a small number of malignant cells with pyknotic nuclei and dyskeratotic cytoplasm may occasionally be present; shedding of these cells reflects surface differentiation of carcinoma in situ. These dyskeratotic cells differ from tadpole or fiber cells shedding from invasive carcinoma in size and polymorphism. An occasional appearance of dysplastic cells of the parabasal and/or

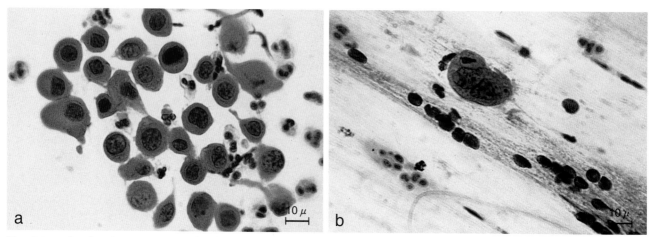

Fig. 6-130a,b   Carcinoma in situ of intermediate type. Cytology is occupied by parabasal type cancer cells showing small to moderate amount of cytoplasm. The cytoplasm is well stained and preserved because of some degrees of differentiation. This cytologic pattern differs from that of classical, small celled (basal cell type) carcinoma in situ (see Fig. 6-127) type. The chromatin pattern is coarsely granular and shows distinctive karyosomes.

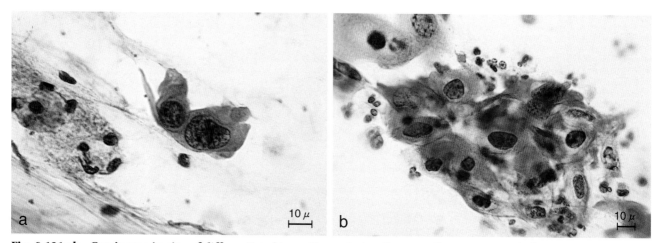

Fig. 6-131a,b   Carcinoma in situ of differentiated type. The majority of cancer cells are composed of polyhedral cancer cells having a varying amount of cytoplasm. The cancer cells derived from a deeper layer reveals cytoplasm well stained by Light Green, but keratinizing cancer cells from the upper layer are orangeophilic or eosinophilic with tendency to pyknosis. This cytologic pattern is referable to keratinizing squamous cell carcinoma in situ. Well preserved chromatins are densely granular with distinctive clear interchromatin spaces. There is no tumor diathesis.

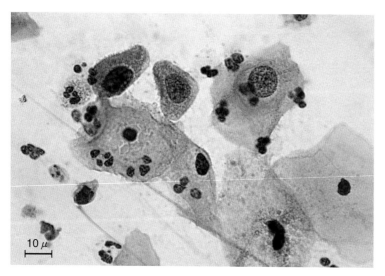

Fig. 6-132   High magnification of parakeratotic in situ cancer cells. Note a parabasal type cancer cell with large densely stained nucleus and intensely orangeophilic cytoplasm in the left upper side. There is no tumor diathesis.

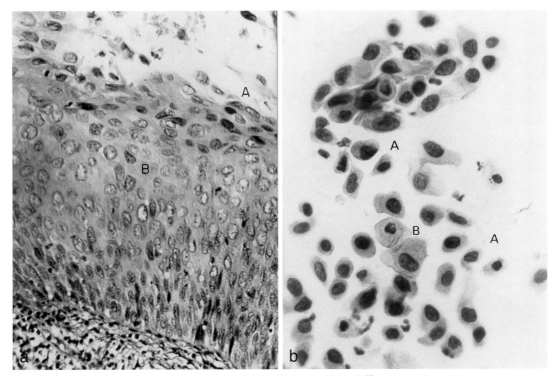

Figs. 6-133 Histology and cytology of carcinoma in situ, large cell or differentiated type.
a. There is a tendency to undergo epidermoid differentiation towards the surface. The superficial layer is parakeratotic with cytoplasmic eosinophilia and nuclear pyknosis (A), and the intermediate layer appears similar to the prickle cell layer with polygonal cytoplasm (B). The lower half of the epithelium is composed of undifferentiated small cells with spindly or elliptic nuclei with marked hyperchromatosis.
b. Cervical smear showing differentiated parabasal cancer cells with polygonal cytoplasm (B). Some cancer cells reveal "India ink" droplet-like pyknosis and dyskeratosis with cytoplasmic orangeophilia (A). The smear is generally clean in appearance and shows no tumor diathesis.

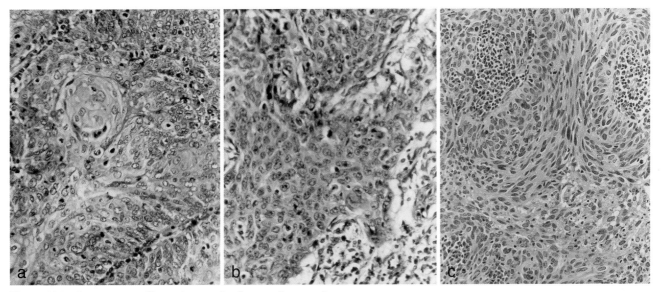

Fig. 6-134 Squamous cell carcinoma in the uterine cervix.
a. Spinal cell type. Although pearl formation is present within cancer cell nests, their outer zones consist of less differentiated cancer cells with scanty cytoplasm; thus the histology may reflect the frequent occurrence of round, parabasal type cancer cells in cervical smears.
b. Transitional cell type. Actively growing outer zones of cancer cell nests are mostly composed of basal and parabasal types of cancer cells. It is understandable from the histology that a certain number of less differentiated parabasal malignant cells exfoliate in cervical scraping smears.
c. Spindle cell type. The most undifferentiated type is composed of bands of basal or spindle cell type cancer cells. Compare with Fig. 6-145.

Table 6-9 Frequency of histologic types of cervical carcinoma
in reference to oral contraceptive hormones intake

Type of carcinoma of uterine cervix	Nonuser				User	
	Japanese		New Zealander		New Zealander	
Squamous carcinoma	397	84.5%	376	82.5%*	140	65.1%*
Adenocarcinoma	40	8.5%	42	9.2%	16	7.4%
Mixed carcinoma	33	7.0%	38	8.3%**	59	27.4%**
mature			6	(1.3%)	3	(1.4%)
immature			24	(5.3%)	28	(13.0%)
mucoepidermoid			4	(0.9%)	14	(6.5%)
glassy cell			4(5)°	(0.9%)	14	(6.5%)
Total	470		456		215	

Note : * significant decrease by OCH intake ($p < 0.01$)
    ** significant increase by OCH intake($p < 0.01$)
    ° one patient pregnant
(From Takahashi M, Aoki K, Noda S, et al: In: Kurihara S, et al (eds): Cervical Pathology and Colposcopy. pp21-27. Amsterdam: Elsevier Science, 1985.)

intermediate types may give rise to confusion in differentiation between keratinizing dysplasia and large cell carcinoma combined with dysplasia.*

Classification of carcinoma in situ and its allied changes is based upon abnormal cell constituents and the predominance of cancer cells either of the basal or parabasal type. In addition, the general appearance of the smear is also important. In dysplasia encompassing carcinoma in situ the smear is clean because of the absence of tumor diathesis (see Figs. 6-116 and 6-128).

14 Carcinoma of the Cervix

■Squamous cell carcinoma

Squamous cell carcinoma, comprises most of carcinomas of the uterine cervix, about 95%. Cancer cells are arranged in anastomosing sheets and columns with a tile-like pattern. Histologic classification is determined by cancer cell differentiation. Martzloff [132] has suggested three representative grades, i.e., spinal cell type, transitional cell type and spindle cell type. The common subclassification of worldwide use divides squamous cell carcinoma into large cell keratinizing, large cell nonkeratinizing and small cell types.[199] Large cell keratinizing or spinal cell type is the most differentiated one and is composed of cancer cells having an abundant eosinophilic cytoplasm. Epithelial pearl formation is often seen inside cancer cell nests. The most common type, large cell nonkeratinizing or transitional cell type, consists of intermediate cancer cells with a polyhedral cytoplasm. Spindle cell or basal cell type is anaplastic and lacks cellular differentiation. The nomenclature of small cell carcinoma defined by Reagan [199] as the most poorly differentiated type of squamous cell carcinoma is obscure and leaves confusion with oat cell-like carcinoma that belongs to APUD series.

Cytology: Smear in the advanced stage are smudgy

because of mixture with large number of red cells, leukocytes and degenerated or necrotic cancer cells (tumor diathesis). There are pyknotic cells with India ink droplet-like nuclei and eosinophilic cytoplasm or liquefied cells having hazy nuclei and indistinct cytoplasm. The spinal cel type in particular is characterized by various kinds of malignant cells such as tadpole cells, snake cells and fiber cells (Figs. 6-138 to 6-141). The transitional cell type carcinoma exhibits a preponderance of the intermediate type cancer cells having a small to moderate amount of cyanophilic or orangeophilic cytoplasm with polyhedral shapes (Figs. 6-143 and 6-144).

Has microinvasive carcinoma distinctive cytologic features? The cellular pattern in very early stromal invasion resembles that of in situ carcinoma, but shows often larger, more pleomorphic cancer cells. Microinvasive carcinoma tends to shed cancer cells with larger clumps or sheets by scraping. Necrotic cell debris are absent or scarcely present (Fig. 6-137a).

■Verrucous carcinoma

A rare variant of well differentiated squamous cell carcinoma occurs in the cervix as well as the vulva as a slow-growing and locally invading lethal tumor. This exophytic tumor was ever called giant condyloma.[129]

Histology is characterized by frond-like, papillary and bulky downward growths without fibrovascular stalks. The tumor cells lack distinctive features of malignancy, however, tend to show keratinization (Fig. 6-138a). Punch biopsy taken from the exophytic outgrowth often fails to demonstrate cancer cells.

Cytological diagnosis is very difficult unless gross findings are not informed to a cytology laboratory at sample receipt, because of lack of morphologic malignancy; only several isolated cells having enlarged, hyperchromatic nuclei with broad mature cytoplasm are shed. No tumor diathesis is associated (Fig. 6-146b,c).

■Glassy cell carcinoma

Glassy cell carcinoma which was first described by Glücksmann and Cherry [69] in 1956 as a poorly differentiated adenoacanthoma of the cervix is an extremely

---

*It is not rare to see smears in which severe dysplastic cells and in situ cancer cells are mixed together. Histological studies on conization specimens also reveal coexistence or transition of severe dysplasia to in situ carcinoma. This evidence supports the importance of concept of CIN as well as SIL (see Table 6-7).

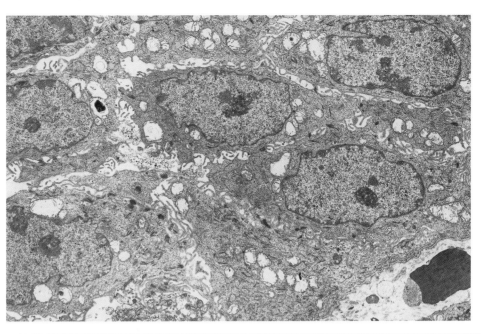

Fig. 6-135 Electron micrograph of differentiated carcinoma of the cervix. Moderately differentiated cancer cells show well development of desmosomes and tonofilaments that is the parameter of differentiation of squamous cell carcinoma. Intercellular digitations are also increased in number. The cytoplasm of a moderate amount contains an increased number of mitochondria and endoplastic reticulum. Nucleoli are visible. (original magnification ×4,000)

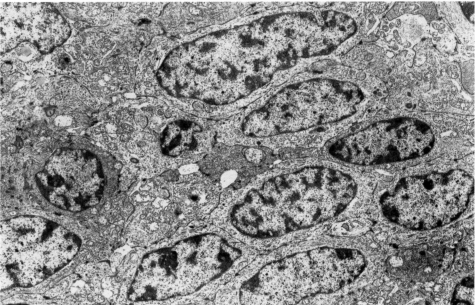

Fig. 6-136 Electron micrograph of undifferentiated cancer cells showing more simple interrelation among cells except for the desmosomes. Cytoplasmic organella are poorly developed. Note a speckle-like condensed distribution of heterochromatins, simple straight nuclear rim and indistinctive nucleoli. (original magnification ×4,000)

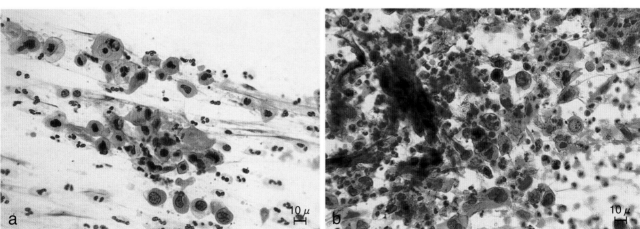

Fig. 6-137 Invasive squamous cell carcinoma of the uterine cervix.
a. Early invasive carcinoma. Cytology is characterized by well preserved squamous cancer cells varying in shape. Cellular polymorphism and nuclear atypism are more marked than in keratinizing dysplasia. Note an apparent nuclear malignancy. "Tumor diathesis" is not obvious.
b. Advanced invasive carcinoma. Characteristics of cytology are (1) exfoliation of innumerable cancer cells as single cells and in sheets, (2) marked cellular and nuclear polymorphism and (3) presence of "tumor diathesis" or malignant background.

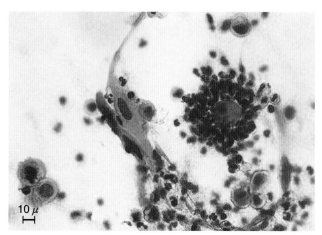

Fig. 6-138 Cannonball and fiber cells in a smear from invasive squamous cell carcinoma. Note a heavy clump of polymorphonuclear leukocytes around degenerate cancer cell. Differentiated fiber cells with intense orangeophilia are seen.

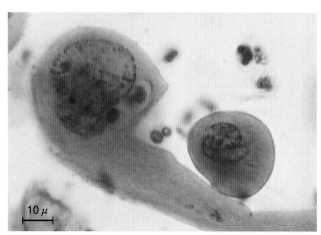

Fig. 6-139 Higher magnification of tadpole cell and third type cell from differentiated squamous cell carcinoma. A large degenerate nucleus is included in a head and tail-like cytoplasm is finely fibrillated. A parabasal cell-like, differentiated cancer cell at right is called third type cell by Graham. Note intense orangeophilia as a result of keratinization.

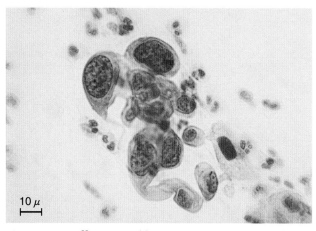

Fig. 6-140 Differentiated keratinizing squamous cell carcinoma cells in early invasion. Note polymorphism of the cytoplasm with cyanophilic or acidophilic staining reaction, high N/C ratio, and coarse clumping of chromatin. The background is clean in early invasive carcinoma.

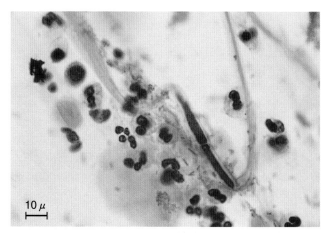

Fig. 6-141 Fiber cells from differentiated squamous cell carcinoma. The cytoplasm is elongated like a fiber and well defined. Orangeophilia of the cytoplasm reflects keratinization of squamous cell carcinoma.

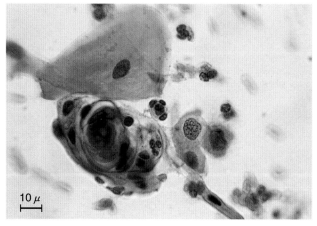

Fig. 6-142 Malignant pearl formation from well differentiated squamous cell carcinoma. Note a characteristic whorl formation by pyknotic malignant nuclei. Aggregated dense nuclei are in the center of the pearl.

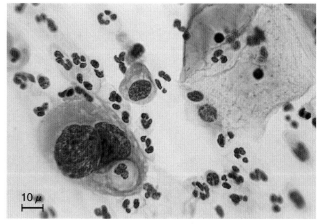

Fig. 6-143 Nonkeratinizing squamous carcinoma cells with binucleation. Such enormously large cells with dense, coarsely granular chromatins are from invasive carcinoma.

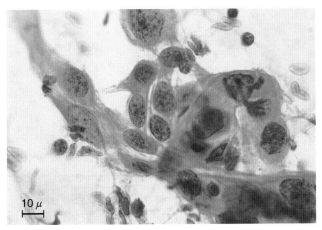

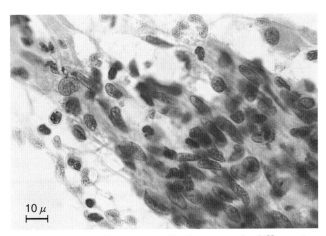

Fig. 6-144   Nonkeratinizing epidermoid carcinoma of transitional type. This common type of squamous cell carcinoma is defined as transitional or intermediate type. The cytology is characterized by (1) a moderate amount of cytoplasm, (2) cell cluster formation with a cobblestone pattern, and (3) the presence of slit-like intercellular space.

Fig. 6-145   Spindly cancer cells from poorly differentiated squamous cell carcinoma. These poorly differentiated cancer cells should be distinguished from the fiber or snake cells of differentiated epidermoid carcinoma and from sarcoma cells. Diagnostic points: (1)poor preservation of the cytoplasm, (2) thickening of nuclear rims by chromatin condensation that is distinguishable from sarcoma, and (3) exfoliation of malignant cells in a large cluster that does not occur in carcinoma in situ.

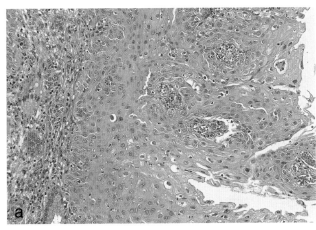

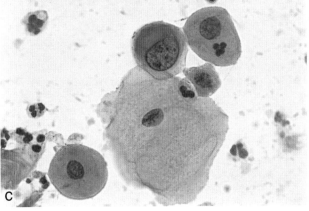

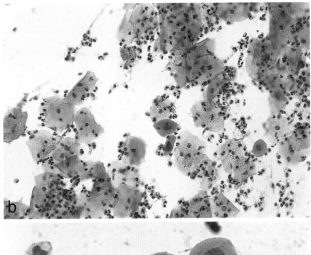

Fig. 6-146   Verrucous carcinoma. Histology and cytology.
a. Verrucous carcinoma showing exophytic papillary growth of neoplastic epithelium. The cytoplasm is mature, broad and well stained. The nuclei are occasionally enlarged and hyperchromatic.
b. General view of cytologic smear from verrucous carcinoma. Cellularity is normal except for a few abnormal cells present in the middle of the field; isolated, hyperchromatic cells with high N/C ratio are seen.
c. A single cancer cell having a large, hyperchromatic nucleus. A high N/C ratio and coarse chromocenters are notable. Note cytoplasm indicative of spotty individual cell keratinization.

malignant carcinoma that develops a progressive downhill course with extensive infiltration and extrapelvic metastases. This tumor is resistant to radiotherapy. It is of interest that the tumor is frequently related to recent pregnancy.[221] The frequency of the tumor was estimated about 1.2%[127] or 1.3%[221] of cervical carcinomas, and the mean age was 36 lower than the average age of patients with epidermoid carcinoma.[221] When Glücks-

mann and Cherry first described, increased incidence of this tumor was noticed in women in pregnancy or after delivery. A significant correlation between increased incidence of glassy cell carcinoma and mucoepidermoid carcinoma and long term use of oral contraceptive hormones (OCH) is reported (Table 6-9); intake of OCH for 12 months or longer is mentioned to be particularly significant.[245]

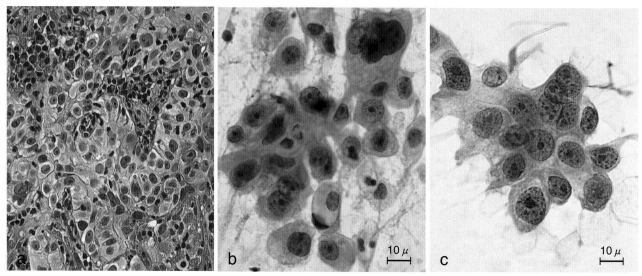

Fig. 6-147   Glassy cell carcinoma of uterine cervix from a 31-year-old woman.
a. Histology showing anastomosing solid sheets of polyhedral cancer cells. A moderate amount of cytoplasm is lacy and gives a ground glass appearance. Note thin stroma that is heavily infiltrated by lymphocytes and plasma cells.
b. Sheet of cancer cells with distinct nuclear membrane and central, macronucleoli.
c. Well preserved cancer cells in imprints from biopsy specimen. An arrangement of flat sheet is likely of squamous cell origin, but faintly blue, lacy cytoplasm and round nuclei with central macronucleoli reveal the property of adenocarcinoma.

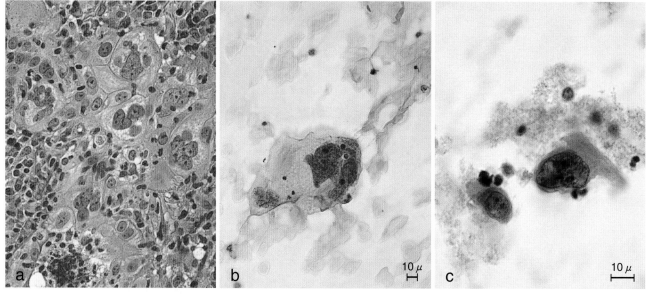

Fig. 6-148   Undifferentiated giant cell carcinoma of the uterine cervix.
a. Histology shows solid nests being composed of polymorphic cancer cells that possess bizarre nuclei.
b. A giant cell with bizarre nucleus.
c. A multinucleated giant cell with indistinctive cytoplasm.

Histologically the tumor is composed of solid sheets of cancer cells separated by thin fibrous septa in which plasma cells, eosinophils and other inflammatory cells heavily infiltrate. Cytologically the cancer cells are large, enrounded cells with moderate amount of lacy cytoplasm that gives a ground glass appearance. The nuclei show prominent nucleoli and thick nuclear membrane. Exfoliation of the tumor cells in a flat sheet differs from the cluster of clear cell carcinoma (Fig. 6-147a, b).

■**Large cell carcinoma (Giant cell carcinoma)**
This type is seen as a variety of undifferentiated squamous cell carcinoma. Cytology shows characteristically

bizarre cancer cells with polymorphic and often multi-nucleated giant nuclei. These cells must be differentiated from similar-looking mesenchymal cells of mesodermal mixed cell tumor (Fig. 6-148).

■**Small cell carcinoma (Apudoma)**
This rare tumor of the cervix resembling oat cell carcinoma of the lung is considered to belong to the APUD system[131,249] and arise from argyrophil cells[57] of neural crest origin. This is an aggressive tumor that gives rise to widespread metastases.[257]

Cytology: Cytologically the tumor cells have scanty cytoplasm and appear to be almost denuded. They scat-

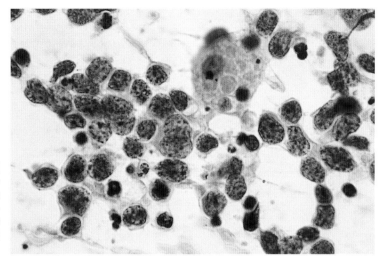

Fig. 6-149 Small cell carcinoma in a cervical smear. From a 31-year-old female. Note similarity to oat cell carcinoma of the lung. The tumor cells differ from poorly differentiated squamous cell carcinoma in nuclear feature and arrangement; they are markedly hyperchromatic with coarsely stippled chromatins and liable to disintegrate.

ter as single cells, but epithelial binding can be recognized; cell inclusion and "side by side" arrangement are noticeable. The nuclei are markedly hyperchromatic with coarsely stippled chromatins. The nuclear rims are irregularly contoured but not thickened (Fig. 6-149).

15 Adenocarcinoma and Precancerous Glandular Lesions

Adenocarcinoma of the cervix constitutes about 5% to 7% of cancers of the cervix.[148,248] While the incidence of carcinoma of the cervix is decreasing in recent years, the proportion of adenocarcinoma in cervical carcinomas is increasing. There is significant difference of survival rate between adenocarcinoma and squamous cell carcinoma although the majority of adenocarcinoma are morphologically differentiated; Tasker and Colins[248] reported that the cumulative 5-year survival rate in the stages I and II was 61.0 ± 6.4% in adenocarcinoma and 73.9 ± 5.7% in squamous cell carcinoma, and in the stages III and IV was 39.1 ± 12.3% in adenocarcinoma and 46.3 ± 12.3% in squamous cell carcinoma.

In the process of cytodiagnosis of cervical adenocarcinoma following points must be cleared up; 1) distinction from endocervical atypia or endocervical dysplasia, 2) identification of highly differentiated endocervical adenocarcinoma, 3) determination of origin of adenocarcinoma, either endocervical, endometrial or extrauterine.

■Endocervical glandular atypia
Endocervical atypia that is referable to endocervical glandular dysplasia is thought to be a derivative of reserve cell hyperplasia of endocervical glandular epithelium with atypia.

Although the concept of endocervical glandular dysplasia defined as cervical intraepithelial glandular neoplasia was ever introduced with grading of atypicality by Gloor and Hurlimann, its biological behavior has not been clarified.[67] It is most probable that endocervical type adenocarcinoma originates from reserve cell hyperplasia with atypia by electron microscopic studies.[118]

Cytologically it is characterized by nuclear hypertro-

phy, simple hyperchromatism without abnormality in chromatin pattern and high N/C ratio (see Figs. 6-65 and 6-66).

Addendum: AGUS
When atypical glandular cells which are not referred to as endocervical AIS occur, the definition of "atypical glandular cells of undetermined significance (AGUS)" is applied according to the Bethesda System. It should be remembered that AGUS includes also atypical endometrial cells having characteristics; i.e., formation of small clusters, slight nuclear enlargement, scant cytoplasm and cytoplasmic vacuolation (see Fig. 6-65).

■Adenocarcinoma in situ
Studies on the entity of adenocarcinoma in situ, analogous to squamous carcinoma in situ have been sporadically reported since Helper et al (1952) and Friedell and McKay (1953) described the preinvasive endocervical adenocarcinoma. Coexistence of adenocarcinoma in situ (AIS) with squamous carcinoma in situ is frequent; Weisbrot et al[195,269] described it in 6 of 8 cases, and Qizilbash in 5 of 14 cases. Therefore the cytology may be obscured by the extent of squamous carcinoma in situ.

Histology: The glandular structures are lined with pseudostratified columnar cells which are well differentiated; the nuclei are fairly uniform in size and shape and arranged in palisade with polarity. Microinvasion is hardly determined unless structural atypism such as confluence of glands, i.e., the "back to back lesion" and irregular budding growth is demonstrated.

Cytology: Exfoliation of clusters and sheets composed of hyperchromatic columnar cells with clear background is characteristic. The tumor cells in a cluster are piled up but pseudostratified with polarity; one side of the border that is correspondent to the glandular luminary surface exhibits occasionally the distinct terminal plate devoid of the nuclei. The structural pattern of these cell clusters is rosette-like[117] (over-view) or tubular (side-view). The sheets shed from endocervical surface transformed to AIS resemble those of endocervical epithelium but are larger and irregular in shape. The tumor cells comprising the gland or sheet retain a columnar shape. The nuclei

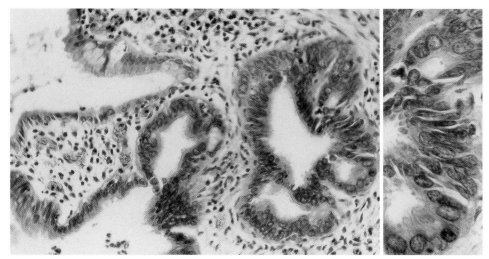

Fig. 6-150   Atypical hyperplasia of the endocervical epithelium. Endocervical dyskaryosis is distinguished from the normal columnar epithelium by multiplication of the lining epithelial cells, the nuclei of which are hyperchromatic and enlarged. However, a normal racemose pattern of the endocervical glands is not distorted and cellular polarity remains intact.

are uniformly enlarged and the nuclear size and shape variations from cell to cell are less marked than in invasive adenocarcinoma. The chromatins consisting of medium-sized granules are evenly dispersed. One or two nucleoli are discernible but not prominent (Figs. 6-151 and 6-152); prominence of nucleoli is rather suggestive of invasive carcinoma.

■Invasive adenocarcinoma
Besides stage classification, cervical adenocarcinoma is divided histologically according to the grade of differentiation. While poorly differentiated adenocarcinoma consists of solid masses of cancer cells with poor glandular structures, well differentiated adenocarcinoma shows varied growth patterns such as gyriform, papillary and ordinary cervical gland types.

Histology: The gyriform type reveals branching glandular structures with fern-like foldings which are covered by clear, tall columnar cancer cells. The ordinary cervical gland type (endocervical type) is characterized by a preexisting gland-like arrangement except for increased cellularity and atypia. An extremely differentiated type of cervical adenocarcinoma resembles mucus-rich, preexisting cervical glands is also described as adenoma malignum or minimal deviation adenocarcinoma.[52,97] Thus biopsy diagnosis is liable to overlook the neoplasia unless pathologists receive a precise information on the uterine cervix from gynecologists. The endometrioid type is composed of densely packed glands which are covered by columnar, darkly stained cytoplasm. The diagnosis of endometrioid adenocarcinoma of the cervix is possible only when the endometrium can be confirmed to be absolutely normal. In another words, this type of adenocarcinoma is apt to be interpreted as endometrial carcinoma in cervical cytology. The endometrioid type comprises about 24%[209] to 40%[251] of cervical adenocarcinomas. Adenoid cystic type is a rare growth pattern that illustrates cribriform structures consisting of densely stained basal cell-like cancer cells. The stroma is scant like in endometrioid carcinoma. This type may be developed in early stage of endometrioid cervical carcinoma (Fig. 6-153).

Cytology: In general, cervical adenocarcinoma cells are clustered with overlappings. Irregular arrangement (distorted polarity), varied distances from "nucleus to nucleus" and nuclear size variation are distinctive features. The nuclei are more well preserved and larger in size than in endometrial carcinoma. On the other hand, nuclear shape variation is inconspicuous, round or oval, and smoothly outlined. Nucleolar prominence, especially presence of acidophilic macronucleoli and variation in their number, are important findings to determine malignancy regardless of cytoplasmic differentiation. Proof of mucus production by PAS and/or alcian blue stain[40] becomes a clue to a mucus-producing variety of differentiated cervical adenocarcinoma. Even poorly differentiated form reveals often spotty PAS-positive stains in the cytoplasm.

Distinction between endocervical type and endometrioid type cervical adenocarcinoma can be possible only when cytoplasmic differentiation is evident such as tubular or solid aggregates with intensely stained cytoplasm in the endometrioid type (Fig. 6-154) and a fair amount of columnar cytoplasm with light translucent stain (Figs. 6-155 and 6-162).

■Combined carcinoma
It is not infrequent to encounter combined carcinomas with both types of adenocarcinoma and squamous cell carcinoma. An increased incidence of the combined carcinoma has been pointed out to be related to pregnancy.[68,238] Since there is different morphogenesis of the combined carcinoma, its subclassification should be precisely interpreted. According to the WHO's classification of tumors of the uterine cervix, only the term of adenosquamous carcinoma is used as a tumor in which adenocarcinomatous and squamous carcinomatous elements are intermingled. There have appeared several

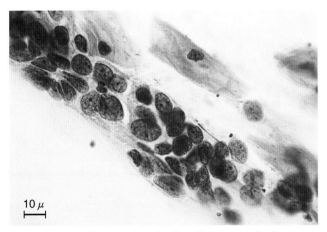 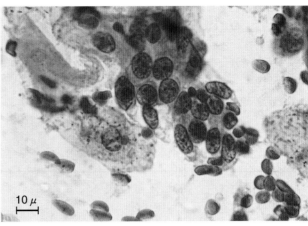

Fig. 6-151   Adenocarcinoma in situ in cervical smear. Cancer cells of endocervical origin tend to be denuded of cytoplasm. Diagnostic points: (1) irregular cluster of elliptic and spindly nuclei, (2) hyperchromatism full of compact chromatin particles, (3) presence of nucleoli, and (4) absence of tumor diathesis.

Fig. 6-152   Adenocarcinoma in situ in cervical smear. Note cluster formation of elongated cells having elliptic, hyperchromatic nuclei. Nucleoli are visible.

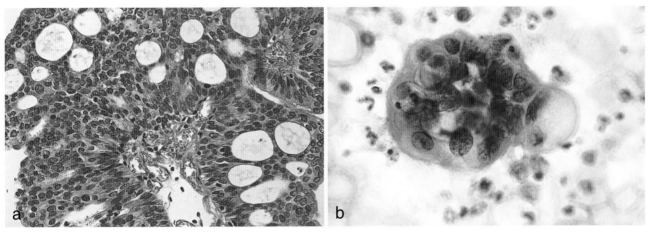

Fig. 6-153   Cytology and histology of adenoid cystic carcinoma of cervix resembling endometrioid adenocarcinoma. From a 52-year-old woman.

a. Histology. This rare tumor arising from endocervical glands is composed of hyperchromatic ovoid cancer cells that resemble basal cells. Note cribriform lumens containing fibrillar material. The stroma is scant as in endometrial carcinoma.

b. Cytology. The nuclei are uniformly oval in shape and hyperchromatic. The tumor cells are characteristically arranged in a tight cluster with a tendency to form lumen or vacuolation. Nucleolus is indistinct like in carcinoma in situ. This cancer cell ball derived from the cervix is well preserved and no smudgy background is visible.

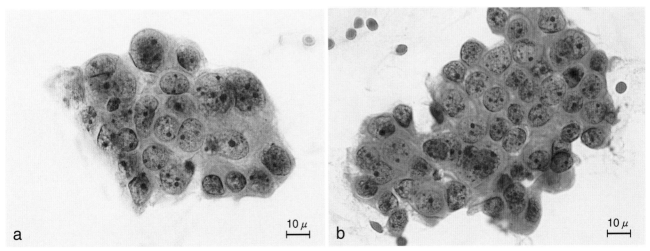

Fig. 6-154a,b   Well differentiated adenocarcinoma cells of the cervix. Endometrioid type. A large sheet having the feature distinctive of cervical adenocarcinoma. Diagnostic points: (1) large sheets varying in size, (2) eosinophilic, multiple macronucleoli, (3) delicate, smooth nuclear rims, and (4) evenly dispersed but compact chromatin particles. Well preserved cells possess a moderate amount of well stained cytoplasm but irregularly bordered.

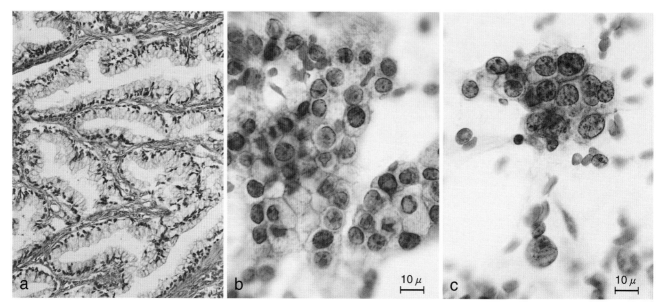

Fig. 6-155 Well differentiated adenocarcinoma cells seen in a large sheet or tight cluster. Endocervical type.

a. Histology. Gyriform type showing the structure of fern-like cervical glands: Their lining epithelium is composed of tall columnar cells with basally locating nuclei.

b, c. Cytology. The nuclei are round and uniform in size and shape. The cytoplasm is abundant and clear with much mucin content. In addition to the above cells, note a tight cell cluster having overlapping, hyperchromatic, round nuclei that is diagnostic of adenocarcinoma of endocervical type.

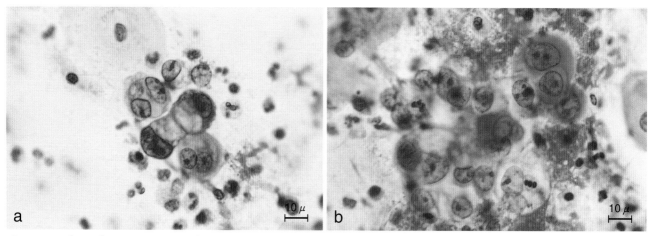

Fig. 6-156a,b Adenocarcinoma cells with mucin production and focal squamous metaplasia of the cervix. These are typical clustered cells from the cervical adenocarcinoma showing excretory vacuoles, eccentric nuclei with distinctive nuclear rims and eosinophilic macronucleoli. Note a focal squamous metaplasia that is characterized by opaque cytoplasm intensely stained by Light Green, distinct cellular borders and eosinophilic macronucleoli persistent.

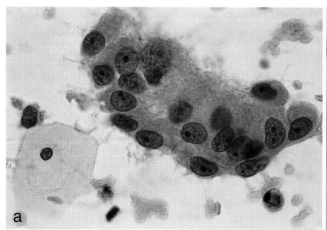
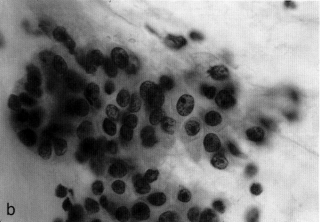

Fig. 6-157a Well differentiated adenocarcinoma of the cervix. Endocervical type. Note a characteristic feature of well preserved cytoplasm that is apparently columnar in shape and stains lightly green.

Fig. 6-157b Poorly differentiated adenocarcinoma of the cervix. Endocervical type. The cytoplasm is poorly defined and stains faintly bluish green. Very hyperchromatic nuclei retain a tubular arrangement.

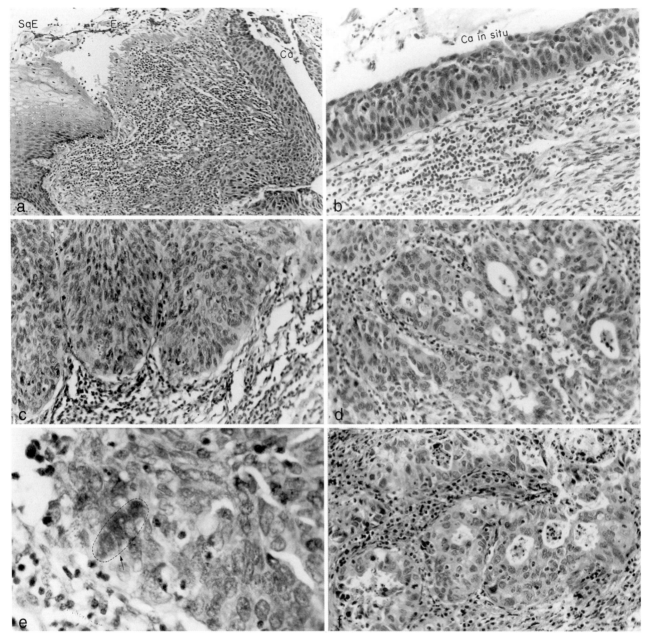

Fig. 6-158 High power view showing details of invasive adenosquamous carcinoma.
a. The superficial area showing erosion closed to the squamous epithelium. Erosion of the surface is quite commonly associated with micro-invasive carcinoma. Incorrect punch biopsy or scraping from the erosive area may lead to false negative. Er: erosion, SqE : stratified squamous epithelium, Ca : carcinoma in situ
b. Carcinoma in situ extending proximally along the endocervix.
c. The profile of the tumor with bulky outgrowth that is composed of solid nests of anaplastic spindly cancer cells.
d-f. Adenocarcinomatous component of the tumor. Note a glandular arrangement of cancer cells and PAS-positive granules in solid cell nests (arrow).

terms such as mixed carcinoma (Skinner and McDonald,[228] Glücksmann and Cherry[69]), adenosquamous carcinoma (Ng,[163] Steiner and Friedell[238]), adenoid cystic basal cell carcinoma (Moss and Collins[151]) and adenoacanthoma (Ayre,[9] Novak and Nalley[170]). Mixed carcinoma should not be confused with Müllerian mixed tumor (carcinosarcoma).

1) Double cancers, either metachronous or synchronous, are those that arise from separate sites of the endocervix and ectocervix.

2) Adenosquamous carcinoma that is composed of two distinct cancer cell components is the type showing differentiation towards both squamous and glandular structures of bipotential neoplastic cells which may originate from reserve cells beneath the columnar endocervical epithelium (Fig. 6-158).

3) Adenoacanthoma is a variety of adenocarcinoma showing mature squamous metaplasia in situ (Fig. 6-156); metaplastic squamous elements appear to be well differentiated and are present within the glands.

4) Mucoepidermoid carcinoma that is characterized by clusters of mucin-producing cells within solid epidermoid cancer cell nests. This type of mixed carcinoma is not so common in the uterine cervix as in the salivary

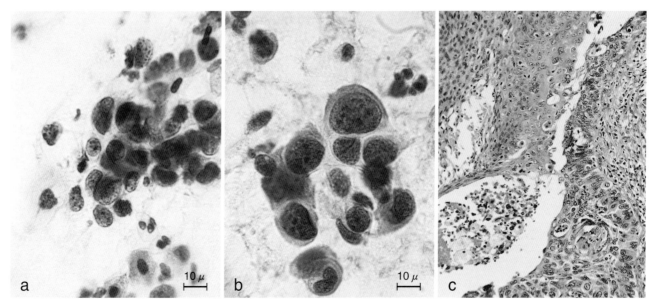

Fig. 6-159 Malignant cells from combined adenosquamous carcinoma coexistent with cancer cells showing both differentiation to glandular and epidermoid types.
a. Endocervical adenocarcinoma cells seen in a clump which are considered to be derived from early adenocarcinoma.
b. Squamous carcinoma cells of the parabasal type. Note scant but thickened cytoplasm.
c. Histology showing differentiation both to adenocarcinoma (early stage) and squamous cell carcinoma.

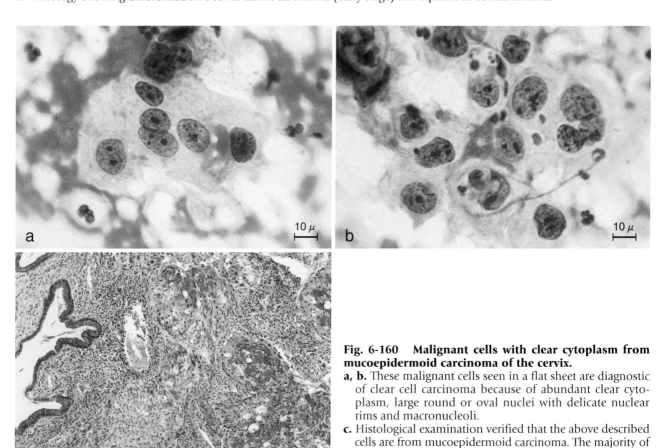

Fig. 6-160 Malignant cells with clear cytoplasm from mucoepidermoid carcinoma of the cervix.
a, b. These malignant cells seen in a flat sheet are diagnostic of clear cell carcinoma because of abundant clear cytoplasm, large round or oval nuclei with delicate nuclear rims and macronucleoli.
c. Histological examination verified that the above described cells are from mucoepidermoid carcinoma. The majority of cancer are composed of solid sheets of squamous cancer cells, in which alcian green-positive cells are found.

gland or the bronchus. "Signet-ring cell variant of mixed carcinoma" described by Glücksmann and Cherry (1956)[69] may belong to this group.

Cytology: The cytological pattern in mixed adenosquamous carcinoma is characterized by mixture with malignant cells of both epidermoid and glandular types; they are distinct from each other. The number of either kind of cancer cells depends upon the dominance in histology. Adenoacanthoma is suspected only when squamous metaplastic cancer cells appear in continuity with cell clusters of adenocarcinoma origin. Squamous metaplastic cancer cells possess the nuclei leaving features of

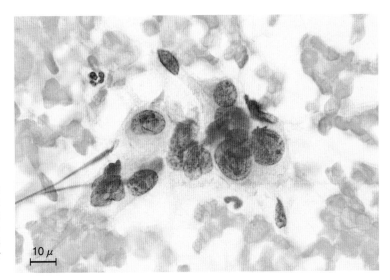

Fig. 6-161 Extrauterine adenocarcinoma cells by direct extension to the cervix from rectum carcinoma. Well preserved cancer cells shed in a tight cluster are found in a hemorrhagic smear. Piling of the nuclei with a palisading arrangement is suggestive of tubular adenocarcinoma. Tumor diathesis is not seen.

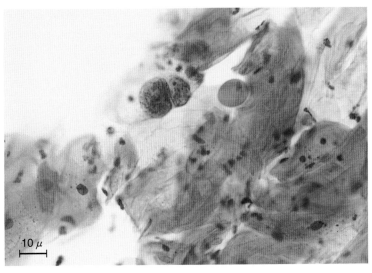

Fig. 6-162 Extrauterine carcinoma cells derived from ovarian carcinoma. High nuclear cytoplasmic ratio, marked nuclear hyperchromatism and cytoplasmic vacuolation are suggestive of cancer cells probably of glandular type. The absence of tumor diathesis is a characteristic feature of extrauterine carcinoma.

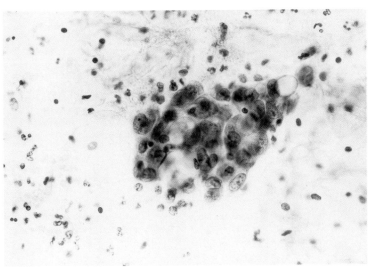

Fig. 6-163 Metastatic adenocarcinoma of the endometrium in a vaginal smear. Cancer cells of gastric origin showing overlapping cell cluster. No tumor diathesis is present in the smear from a patient with extrauterine carcinoma.

a glandular type in spite of epidermoid differentiation of the cytoplasm; nuclear rims are smooth and nucleoli are prominent with eosinophilia (Fig. 6-156).

■Extrauterine carcinoma

Extrauterine carcinoma is detected in cervical smears. Ng and Reagan[162] detected malignant cells of extrauter-ine origin in 66 cases from approximately 600,000 cytology specimens; the most common primary site was the ovary (42.1%), followed by the gastrointestinal tract (19.8%), the tube (7.5%), the pancreas (6.1%), etc. There is also a possibility that arises from the direct extension of cancer in the adjacent organs such as the rectum and the urinary bladder. It is difficult to deter-

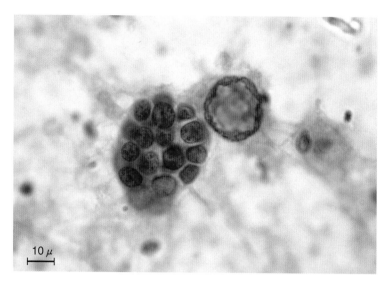

10 μ

Fig. 6-164 Extrauterine carcinoma associated with psammoma body in endometrial aspiration. A cancer cell cluster with fairly small, hyperchromatic nuclei is from metastatic mammary carcinoma. Many psammoma bodies are found to be isolated or together with cancer cells. (By courtesy of Dr. I. Fujimoto, and Dr. M. Masubuchi, Cancer Institute Hospital, Tokyo.)

mine whether the malignant cells in cervical smears are of primary or metastatic tumor only from cytologic features. The evidence suggestive of extrauterine origin is summarized as follows:

(1) Tumor diathesis is absent or inconspicuous irrespective of the presence of hemorrhage; (2) Tumor cells derived from adenocarcinoma of other sites are commonly glandular or papillary in arrangement and occasionally singly isolated; (3) There is no coexistent or related atypical alteration (dysplastic change) of cervico-endometrial epithelial cells; (4) Abnormal findings which are unusual in the uterine tumors may occur such as psammoma bodies, huge papillary clusters, and morula-like cell balls. Psammoma bodies and huge papillary cell clumps are peculiar to ovarian carcinoma, and morula-like cell balls are common in carcinoma of the breast and pancreas (Figs. 6-161 to 6-164).

16 Clear Cell Carcinoma

Clear cell carcinoma, in a broad sense, includes paramesonephric clear cell carcinoma and mesonephric carcinoma. In addition, endometrioid carcinoma and serous cystadenocarcinoma showing the similar clear cell pattern are occasionally included.[213] There is no doubt that clear cells originated from the para-mesonephric duct, Müllerian duct, are encountered in the endosalpinx, endometrium and endocervix. The mesonephric Wolffian duct which develops to the epididymis, the vas deferens and common ejaculatory duct in males undergoes regression in females without development to a functional reproductive system.

Paramesonephric clear cell carcinoma: This is a solid or cystic tumor which is composed of sheets or cords of large clear cells with round, hyperchromatic nuclei. Histochemical distinction is the positive reaction for neutral mucopolysaccharides and mucicarmine.[93]

Mesonephroma: The malignant tumor arising from the mesonephric nests as called mesonephroma (Schiller, 1939) is characterized by glandular (glomerular) and papillary growth of clear, columnar cells showing pear-shaped vacuoles secreted into the lumina (hobnailed pattern)(Fig. 6-165). The tumor cells are rich in glyco-

gen, occasionally positive for fat stain,[168] but negative for mucicarmine stain.

Adenosis and clear cell adenocarcinoma in young girls: In recent years a prenatal exposure to maternal use of diethylstilbestrol and other nonsteroid synthetic estrogens has come to be the cause of adenosis and clear cell adenocarcinoma of the cervix and vagina in young females after menarche and it is implicated in the evolution of clear cell carcinoma.[254,262] The ratio of frequency of the tumor in the vagina and cervix is 1.6 to 1.0 in the registry of clear cell adenocarcinoma of the genital tract in young females.[243] According to the study by Sherman et al[222] who examined 487 young girls aged 13 to 25 with a clinical history of in utero exposure to synthetic estrogens (5 to 25 mg daily dose), 71% disclosed abnormal nonglycogenated vaginal epithelium in iodine staining. Although the association of vaginal adenosis with prenatal use of diethylstilbestrol is very frequent, the risk of adenocarcinoma is said to be low, varying from 0.1% (Ulfelder)[255] to 5% (Stafl).[236] The state of adenosis is characterized by a single layer of mucinous, columnar cells of endocervical type as a result of Müllerian dysgenesis. This mucinous columnar cells do not only replace the vaginal squamous epithelium but enter into the underlying stroma as a glandular island (Fig. 6-166).

Cytology of clear cell carcinoma: Identification of clear cell carcinoma is often difficult because of the resemblance of the cancer cells to benign endocervical cells. The features of the cancer cells to be distinguished from the endocervical cells are larger and more polymorphic nuclei. One or more, prominent, eosinophilic nucleoli and clear cytoplasm containing glycogen are diagnostic of the tumor (Fig. 6-168). In case of unexplained abnormal bleeding and discharge in young girls, confirmation of maternal history of uptake of stilbestrol in pregnancy and cytological examination are important.

Addendum: Endodermal sinus tumor
Most germ tumors arise in the gonads, but some occur in the anterior mediastinum, sacrococcygeal region, pineal gland, etc. Endodermal sinus tumor is defined by Teilum[250] as an extra-embryonic derivative of yolk sac endoderm. This is a highly malignant germ cell tumor functioning to produce α-fetoprotein. Historically, this

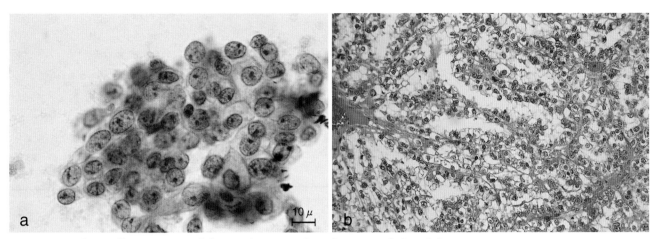

Fig. 6-165   Clear cell carcinoma of the cervix suggestive of mesonephric origin.
a. Cytology is characterized by clear, translucent cytoplasm and fairly small round nuclei with one or two prominent nucleoli. Cytologically there is no positive finding to point out a mesonephric origin.
b. Histology shows branching columns or tubules of clear cells with round uniform nuclei. Note a characteristic hobnail appearance.

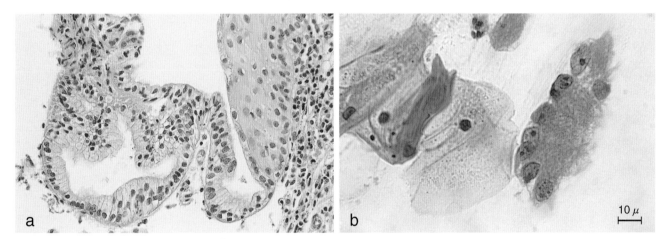

Fig. 6-166   Histology and cytology of vaginal adenosis.
a. Histology. This 18-year old girl reveals the lining of the vagina by tall columnar epithelial cells. The history of prenatal intake of diethylstilbestrol is obscure in this case.
b. Cytology. Cytology of the vagina shows mixture with tall columnar cells with basally locating hypertrophic nuclei and squamous cells.

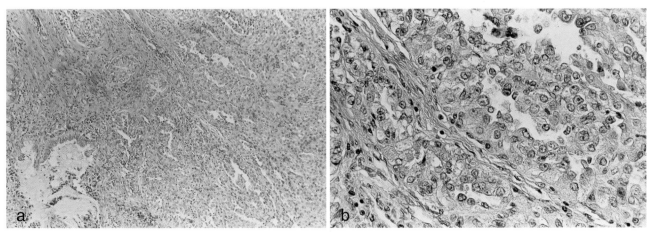

Fig. 6-167   Clear cell carcinoma arising from vaginal adenosis in a young girl. The tumor is proliferating as solid sheets or glandular structures. The nuclei are round or polygonal, and not markedly enlarged. The cytoplasm is abundant and lightly stained. (By courtesy of Dr. R. Ohyasu, North Western University School of Medicine, Chicago, U.S.A.)

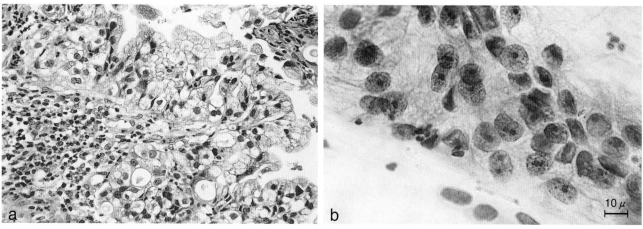

Fig. 6-168 Histology and cytology of clear celled adenocarcinoma of the cervix.
a. Mucin-secreting adenocarcinoma of the cervix is often watery clear in appearance. PAS stain with resistant diastase digestion reveals mucin content of the cytoplasm.
b. Cervical smear shows abundant clear cytoplasm which is ill-defined. Note brownish hue due to mucin content in translucent cytoplasm.

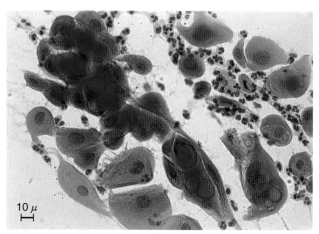

Fig. 6-169 Acute radiation change of benign epithelial cells in cervical smear. Postoperative irradiation with 5,730 rad after hysterectomy. Note more marked changes of the cytoplasm such as bizarre shape, vacuolation and intense eosinophilia than nuclei.

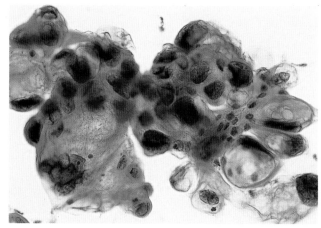

Fig. 6-170 Acute radiation change of squamous cancer cells after ⁶⁰Co radiotherapy. From a woman with stage IIIa cervical carcinoma. Note bizarre cells with enormously enlarged cells with vacuolation, condensed chromatin of enlarged nuclei and eosinophilia.

tumor has been erroneously classified in the group of mesonephroma (Schiller). A loose vacuolated meshwork with honeycomb-like microcysts of the tumor is referred to *magma reticulare* of exocoelom or extraembryonic mesoderm. The tumor cells, small anaplastic cells with round hyperchromatic nuclei, may line the sinus-like cystic spaces showing a papillary or papillotubular pattern. There may be found eosinophilic hyaline globules which are positive for diastase digestion-resistant PAS stain. The surrounding stroma is loose and reticular. Embryologically interesting is the coexistence of the tumor with other embryonal carcinoma or malignant teratoma, which is associated with production of α-fetoprotein. This evidence strongly sustains the germ cell-derived nature of the tumor.[56,167,247]

17 Acute Radiation Cellular Changes

Radiotherapy of ionizing radiation is performed for women in the II B or higher stages of cervical cancer not amenable to radical hysterectomy. Although the biological phenomenon of common 60 cobalt irradiation that

kills cancer cells more selectively than benign cells is not fully understood, dividing cells are sensitive to irradiation. Radiation cellular changes are divided into acute and chronic; the latter that occurs for a long occult period even several years after radiotherapy is encountered as postirradiation dysplasia (see page 85).

Acute radiation changes differ from tissue to tissue. The most radiosensitive cells killed or seriously injured by such as 2,500 rad or less are lymphoid cells, bone marrow cells of myeloblastoid and erythroblastoid series, and germ cells of the ovary and testis. The uterine cervix covered with stratified squamous epithelium may belong to radioresponsive tissues that are killed or severely injured by such as 2,000 to 5,000 rad (Fig. 6-169). Squamous cell carcinoma of the cervix is one of moderately radioresponsive tumors and shows various acute morphological changes as follows[233,242,267]:

(1) marked nuclear and cytoplasmic swelling with often bizarre configurations, (2) enormous cytoplasmic vacuolation, (3) pyknosis and karyorrhexis, and (4) eosinophilic stain of the cytoplasm. Although cellular changes of cancer cells are greater than those of benign

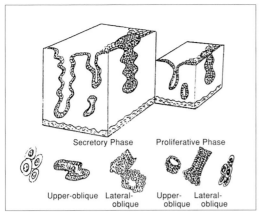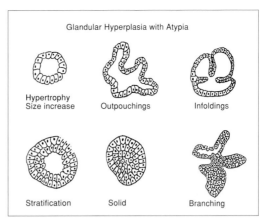

Fig. 6-171 Schematic drawing of endometrial cell changes.

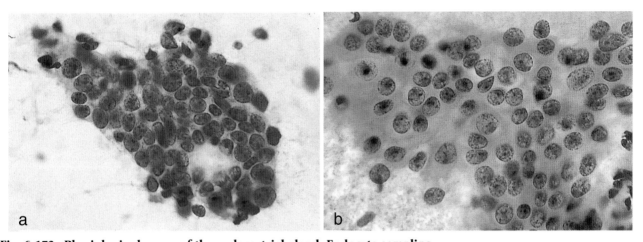

Fig. 6-172 Physiologic changes of the endometrial gland. Endocyte sampling.
a. Endometrial gland of proliferative phase. Closely packed glandular cells arranged in a tubular fashion. The nuclei are round to oval and uniform in size. Chromatins are condensed but bland.
b. Endometrial gland of secretory phase. The cytoplasm is broadened, vesicular and ill-defined. The nuclei are ovoid, uniform in size and vesicular. Note a honeycomb appearance.

cells, separation of cancer cells from benign cells is often difficult[267] (Figs. 6-169 and 6-170).

18 Hyperplasia and Carcinoma of the Endometrium

The incidence rate of endometrial carcinoma differs geographically and accounts for about 15 in 100,000 in all western countries whereas the incidence of cervical carcinoma ranges from about 10 to 20. Endometrial carcinoma is getting more common than previously. According to the report of Cancer Institute Hospital of Tokyo, the proportion of endometrial carcinoma in all uterine malignancies has steadily increased; it was about 4.0% in 1950s, from 10% to 18% in 1970s and over 30% after 1986. Thus endometrial cytology has enhanced the importance of application for detection of endometrial carcinoma. It should be applied to women of menopausal ages who showed following signs or findings;

1) Shedding of endometrial cells in cervico-vaginal smears after the 12th day of menstrual cycle.

2) Right shift of maturation index as a result of estrogen effect.

3) Irregular vaginal bleeding most likely due to cystic glandular hyperplasia.

4) Irregular vaginal bleeding derived from various pathologic lesions such as hyperplasia, neoplasia and incomplete abortion.

For asymptomatic women after menopause endometrial cytology can find incipient carcinomas. Koss and his associates[114] strongly suggested its significance and described the prevalence and incidence rates of 6.96 and 1.71 per 1,000, respectively. The target of endometrial cytology should be placed on menopausal women with significant risk factors[256]; physiologically obesity, diabetes, hypertension, infertility, nulliparity and Stein-Leventhal syndrome are considered. Exogenous or environmental factors are intake of high caloric, high fat diet and administration of hormone replacement therapy with unopposed estrogen.

For symptomatic patients we need also clinical informations in details which are concerned with above factors and sexual cycles. Status of vaginal bleeding, spotty, irregular or persistent bleeding must be clarified. Iatrogenic conditions such as hormone treatment, pill intake and use of IUD are also noteworthy.

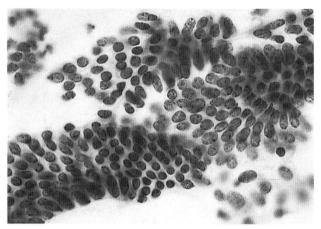

Fig. 6-173 Atrophic endometrial gland from a postmenopausal woman. Endocyte sampling. Atrophic glands appear to be tightly clustered in tubules or in sheets. The nuclei are smaller than those of proliferative glands. The chromatin pattern is bland.

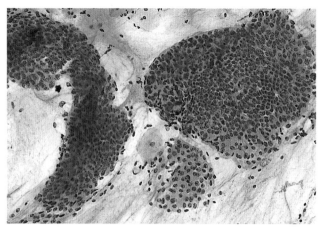

Fig. 6-174 Mixed endometrium showing an admixture of proliferative glands with secretory glands. Endocyte sampling from a menopausal patient complaining of dysfunctional hemorrhage.

Table 6-10 Alteration of hormone-dependent endometrium

| Physiology | Histology | | Cytology |
	Menopause	Postmenopause	
Depression of ovarian hormones	Resting endometrium	Atrophic endometrium	Scarcity of cells having small, enrounded nuclei with dense, homogeneous chromatin
Dysfunction of ovarian hormones	Partly proliferative, partly secretory, phase side by side	Cystic, atrophic endometrium	Overlapping cell clumps and sheets with honeycomb pattern Decrease in cellularity
Hyperestrogenism	Glandular cystic hyperplasia with Swiss-cheese appearance	Regression/ adenomatous hyperplasia (simple/complex)	Increased cellularity; flat cell sheets with scant cytoplasm/outpouchings, branchings, infoldings

Table 6-11 Differentiation between endocervical and endometrial adenocarcinoma cells

Feature	Endometrial carcinoma	Endocervical carcinoma
General appearance	1) Small in number 2) Cell ball or grape-like cluster 3) Dirty, bloody background	1) Large in number 2) Irregular sheet with "side by side" pattern 3) Fresh bloody background may occur, but relatively clean
Cytoplasm	1) Small amount 2) Translucent, lacy or vesicular, and often vacuolated	1) Moderate amount 2) Granular, well stained and often mucin-productive
Nuclei	1) Fairly small 2) Markedly hyperchromatic	1) Large 2) Slightly hyperchromatic
Nucleoli	1) Fairly small 2) Varied number 3) Either basophilic or eosinophilic	1) Prominent 2) One or two, occasionally multiple 3) Eosinophilic

■ Key points of endometrial microscopy

For cytological interpretation of the endometrium we need to reconstruct three dimensional endometrial glands that alter the form according to a menstrual cycle. Pathological changes in hyperplasia and cancer can be schematically illustrated (Fig. 6-171).

1) Sampling procedures

Cellularity, components of cells and shape of glandular cell clusters are the target of cytological observation. However these features vary by sampling procedures. Direct scraping smears are higher in cellularity than aspiration smears. Glandular cells appear to be torn out of endometrium accompanying some stromal cells. Aspiration smears acquire a small number of glandular cells which tend to clump as three dimensional cell balls. Other cell components, histiocytes, leukocytes, endocervical cells and squamous cells are admixed in aspirates.

2) Menstrual cycles

Endometrial glandular cells vary in shape according to menstrual cycle in reproductive women.

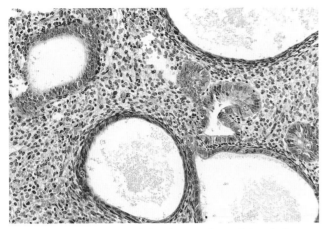

Fig. 6-175 Cystic glandular hyperplasia (simple hyperplasia). Cystic endometrial glands are lined with flat epithelium. Smaller glands disclose one or two layers of somewhat hypertrophic columnar epithelium.

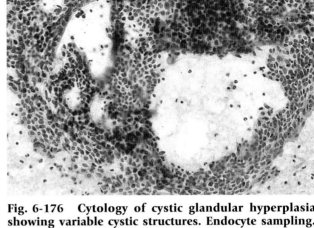

Fig. 6-176 Cytology of cystic glandular hyperplasia showing variable cystic structures. Endocyte sampling. The lining epithelium consists of low columnar epithelium.

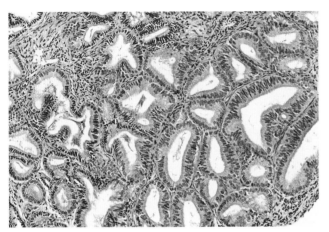

Fig. 6-177 Adenomatous hyperplasia (complex hyperplasia). Note crowded endometrial glands varying in size. Closely packed glands with lining of hyperchromatic elliptic nuclei.

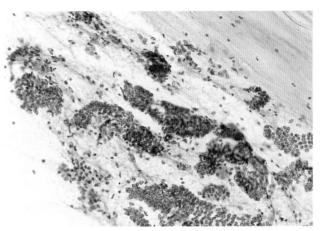

Fig. 6-178 Cytology of adenomatous hyperplasia (complex). Endocyte sampling. Note high cellularity consisting of tubules or solid sheets of hyperchromatic, columnar cells.

Proliferative phase is characterized by compact cell clusters which are overlapping and arranged in a cylinder or trabecular fashion. Their nuclei are round or oval and uniform in size and shape. They stain densely with compact, granular chromatins and have distinctive chromocenters. The cytoplasm is small in amount. Thereby the N/C ratio is increased (Fig. 6-172a).

Ovulation phase situating between proliferative and secretory phase demonstrates the tendency to secretory phase in places.

Secretory phase is characterized by cell sheets varying in size. The cytoplasm is broadened, lightly stained lacy in appearance and often vacuolated. The nuclei are enlarged, enrounded and vesicular. Since vesicular nuclei locate centrally, a honeycomb appearance is characteristically observed (Fig. 6-172b).

Postmenopausal atrophy represents small cell sheets consisting of small, round, densely stained cells with scant cytoplasm. Low cellularity with clean background is representative of atrophic phase, unless abnormality of the endometrium is present (Fig. 6-173).

◼ Endometrial hyperplasia

Endometrial hyperplasia is in the new International Society of Gynecological Pathologists classification separated into endometrial hyperplasia and atypical endometrial hyperplasia.[25,226] Further subclassification into simple and complex types is applied to each of these entities. In complex endometrial hyperplasia, an increase of endometrial glands in number and volume is associated with some architectural irregularity; cytologically the epithelial cells appear to be overlapped and crowded but lack nuclear atypia. The nuclei are uniform in size and shape. The polarity is retained in cell clusters.

Endometrial carcinoma of well differentiated type should be distinguished from endometrial hyperplasia with atypia; endometrial hyperplasia displays compact clusters having basally locating, hypertrophic, uniformly round or oval nuclei measuring about 10 μm in diameter. Staining reaction is about the same. Presence of a single micronucleolus is not infrequent. Atypical endometrial hyperplasia of borderline lesion of the endometrium is characterized by multilayered papillary pattern with alteration of polarity. Nuclei are hypertrophic and possess a few micronucleoli.

113

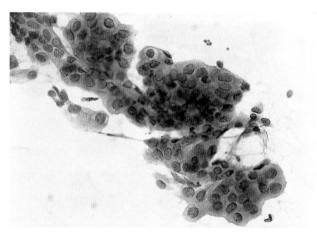

Fig. 6-179 Cytology of adenomatous hyperplasia (complex). Endocyte sampling. Higher magnification shows tightly clustered uniform, ovoid nuclei with fair amount of cytoplasm.

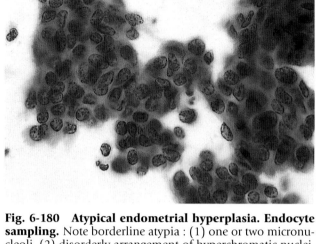

Fig. 6-180 Atypical endometrial hyperplasia. Endocyte sampling. Note borderline atypia : (1) one or two micronucleoli, (2) disorderly arrangement of hyperchromatic nuclei, and (3) slight to moderate nuclear variation in size and shape.

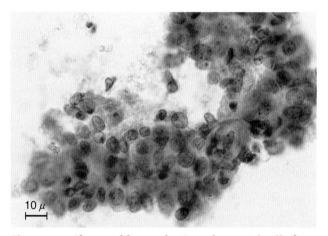

Fig. 6-181 Clustered hyperplastic endometrial cells from a menopausal woman with metropathia. The nuclei are small and uniform in size and shape. These are basally locating and orderly arranged. Their chromatin pattern is bland.

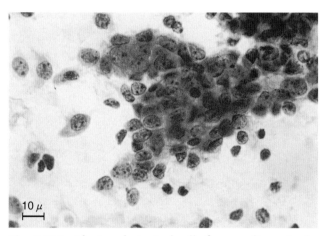

Fig. 6-182 Clustered atypical cells from malignant adenoma by endometrial scraping. The nuclei are fairly uniform in size but contain a few micronucleoli. Note papillary pattern with disorderly arrangement of the nuclei.

■Endometrial carcinoma

Endometrial carcinoma is less frequent than cervical carcinoma, and the ratio is 1 to 7 or 8. In recent years this proportion has gradually increased. Age incidence for carcinoma of the corpus is older than that for carcinoma of the cervix, and the peak is seen in ages after 55 years.

A high risk of endometrial carcinoma is noted in women who have taken conjugated estrogen. [73,229,284] The effect of estrogen on Müllerian tissues is known as an initiator or promotor of carcinoma. It is noteworthy that vaginal cytology often demonstrates a pattern of high estrogen effect.

Since the majority of endometrial carcinoma are seen in the aged group after the 5th decade, a cytological pattern of high estrogen effect such as a high karyopyknotic index and right shift of maturation index is indicative of the necessity of searching for feminizing tumors and endometrial carcinoma.

■Endometrioid adenocarcinoma

Well differentiated adenocarcinoma or grade 1 carcinoma is also called adenoma malignum. The atypical glands, which are lined with a few layers of tall columnar cells, are placed closely each other and have a "back to back" appearance. Undifferentiated carcinoma or grade 3 carcinoma is characterized by solid, alveolar masses of anaplastic columnar cells. Dealing with cervico-vaginal smears with endometrial carcinoma, one may find necrotic debris to be seized with many neutrophils and fibrinated blood. A cytohormonal pattern of high maturation index, although not specific, is not infrequently encountered in postmenopausal women with endometrial carcinoma.

Cytology: Cancer cells of well differentiated type appear as tight cell clumps forming nuclear overlappings. The nuclei are eccentrically located and fairly small compared with those of cervical adenocarcinoma. Appearance of plural micronucleoli (less than 2.5 μm in diameter) 1 to 3 is notable. Cellular borders are distinct because of cell ball formation and the cytoplasm is often

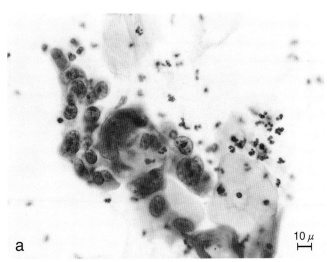
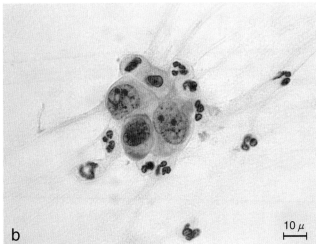

Fig. 6-183 Adenocarcinoma cells of the endometrium shed in a cervical smear showing a papillary pattern.
a. A papillary column of columnar cancer cells, the nuclei of which are hyperchromatic and pilling. Note disorderly budding of the nuclei in a cluster and prominence of nucleolus.
b. A higher magnification reveals well preserved cancer cells showing high nuclear-cytoplasmic ratio and two or three prominent nucleoli.

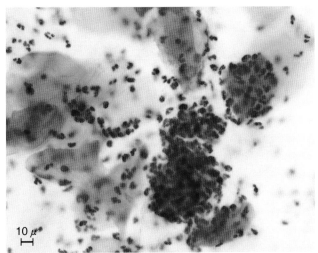
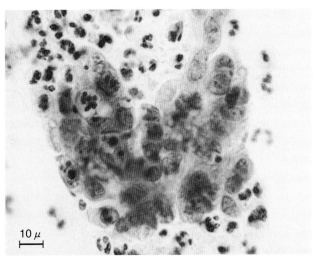

Fig. 6-184 Canon balls in a cervical smear from endometrial carcinoma. Heavy clumps of polymorphonuclear leukocytes around the degenerate cells are often encountered in advanced carcinoma. Chemotaxis of leukocytes is suggested.

Fig. 6-185 Cluster of degenerate endometrial adenocarcinoma cells exfoliated in cervical smear. The endometrial adenocarcinoma cells in a cervical smear are more or less degenerative; the nuclei are poorly preserved. Distortion of nuclear arrangement and pilling are indicative of adenocarcinoma.

vacuolated; infiltration of neutrophil leukocytes into the vacuolated cytoplasm may be observed.

Cancer cells of undifferentiated type are loosely clustered with irregular nuclear piling. The cytoplasmic border is poorly defined (Fig. 6-187). A single macronucleolus is a remarkable finding.

Secretory carcinoma is a variety of endometrioid adenocarcinoma that is composed of well differentiated adenocarcinoma with vacuolated cytoplasm suggestive of a secretory pattern. A favorable prognosis is pointed out by Tobon and Watkins[253](Fig. 6-188a).

Serous carcinoma which was first described by Lauchlan[123] is found in postmenopausal women over 60 of age. As an important clinical behavior this tumor tends to show deep myometrial invasion and serosal spread. Because of unfavorable prognosis[43,206] the tumor is referred to as grade 3 tumor irrespective of nuclear grad-

ing of atypicality. It is not infrequent that serous carcinoma is found. Cytologically the tumor is characterized by papillary growth of fairly low columnar cancer cells (Fig. 6-189). A high positivity of p53 is reported by Moll et al[149] in advanced stage with peritoneal surface spread.[66]

■Adenoacanthoma

Adenoacanthoma is not a specific type of endometrial carcinoma but squamous cell differentiation in adenocarcinoma cells. In case of mixture of squamous cell component composing over 10% of total tumor areas, it is defined as mixed carcinoma. Pure squamous cell carcinoma is very rare and can be defined in the case that the tumor is not associated with adenocarcinoma of the endometrium and not connected with squamous cell carcinoma of the cervix.

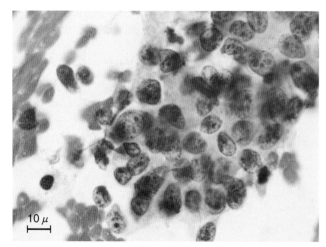

Fig. 6-186 Endometrioid adenocarcinoma cells seen in a cluster by endometrial aspiration. Well preserved nuclei are medium-sized and possess a few micronucleoli. Note abnormalities of the nuclei in size, chromatin pattern, and arrangement comparing with benign endometrial cells in Figs. 6-181 and 6-182.

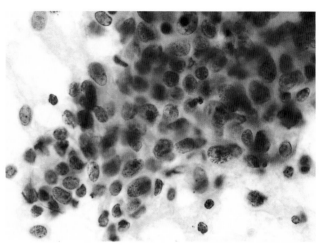

Fig. 6-187 Poorly differentiated adenocarcinoma in scraping smear by Endocyte. The grade 3 carcinoma demonstrates sheets of ill-defined cancer cells showing high N/C ratio and densely stained nuclei. The anisokaryotic nuclei are disorderly arranged and the cytoplasm is scant and ill-defined. No tubular pattern can be recognized.

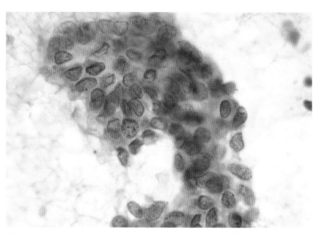

Fig. 6-188a Secretory carcinoma resembling glandular cells in a secretory phase. Endocyte sampling. The nuclei are fairly small but vary in shape and disorderly arranged. Note characteristic cytoplasmic vacuolization.

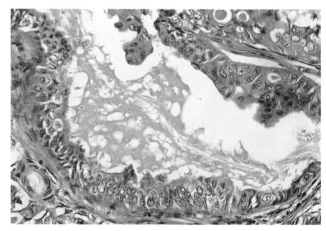

Fig. 6-188b Histology of secretory carcinoma of the endometrium. Note eosinophilic amorphous secretes into the lumen that is referable to the secretory background around a cancer cell cluster in Fig. 6-188a.

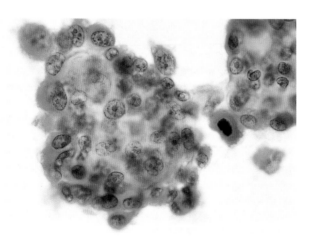

Fig. 6-189 Serous carcinoma of endometrium showing papillary cancer cell clusters. Endocyte scraping smear. Note typically large, vesicular nuclei and one or two prominent nucleoli. No psammoma body is associated in these cells.

In adenocarcinoma with higher degrees of differentiation, squamous metaplasia occasionally occurs. Solid masses of cancer cells with squamous metaplasia may be found inside the glandular structure (Figs. 6-191). Although highly differentiated cells are benign-looking, these squamous components are malignant.

Cytology: Squamous metaplastic cancer cells possess hybrid properties with epidermoid as well as glandular differentiation. Broadening and thickening of the cytoplasm with intense stainability for Light Green reflects squamous metaplasia leaving glandular characteristics in their nuclei such as round or oval configuration and prominence of nucleoli (Fig. 6-190).

■Squamous cell carcinoma

Pure squamous cell carcinoma is extremely rare.[145] Most of squamous cell carcinoma of the endometrium are those extended from cervical carcinoma or accompanied by ichthyosis uteri of rare occurrence in the endometrium.[119,220] Primary squamous cell carcinoma can be acceptable only when endometrial adenosquamous car-

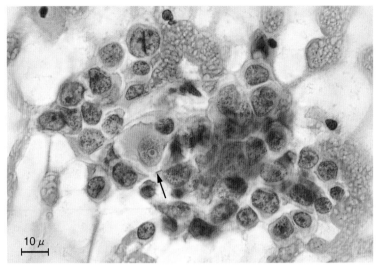

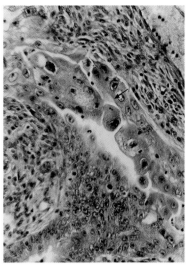

Fig. 6-190 Well preserved adenoacanthoma of the endometrium appeared in a cervical scraping smear. This 52-year-old woman showed extension of endometrial carcinoma down to the cervix. The nuclei are not as large as those of cervical adenocarcinoma. Adenoacanthoma is diagnosable from a cancer cell with broad, opaque cytoplasm and eosinophilic prominent nucleolus (arrow).

Fig. 6-191 Histology of adenoacanthoma, disclosing well differentiated glandular structures and direct metaplastic alteration to squamous carcinoma (arrow).

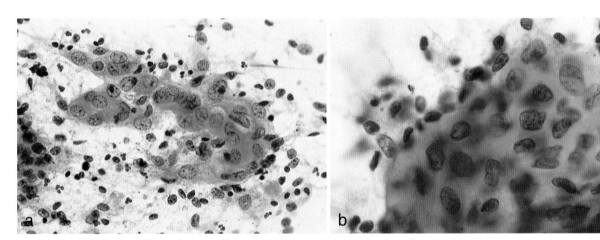

Fig. 6-192 Squamous cell carcinoma of the endometrium. Endocyte sampling from a 55-year-old woman. Coexistence of squamous carcinoma of the cervix is only clinically excluded.
a. Squamous cancer cells in sheet showing a cobblestone appearance.
b. Larger magnification clustered cancer cells showing spindly or polyhedral nuclei with rigid rims.

cinoma and cervical squamous carcinoma are histologically ruled out (Fig. 6-192).

Clear cell carcinoma

This rare type about 5% of entire endometrial carcinoma develops in postmenopausal women over 60 of age. Clear-celled change is not referred to as mesonephroid carcinoma but reflective of excess of glycogen. The relation of clear cell carcinoma to tamoxifen treatment[38] has not been clarified. Survival of five years for clear cell carcinoma is worse than for endometrioid adenocarcinoma.[1]

Cytology is characterized by large, vesicular nuclei and prominent nucleoli. The cytoplasm is clear and lightly stained (Fig. 6-193).

Iatrogenic effect

Hormonal therapy administrating medroxyprogesterone acetate either intramuscularly or orally is mostly respon-

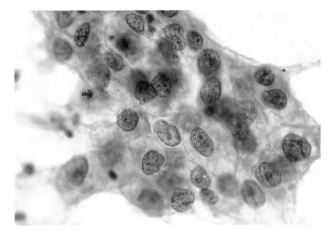

Fig. 6-193 Clear cell carcinoma of endometrium. Endocyte scraping smear. Enlarged ovoid or elliptic nuclei have prominent, eosinophilic nucleoli. Note characteristically cleared, fairly abundant cytoplasm.

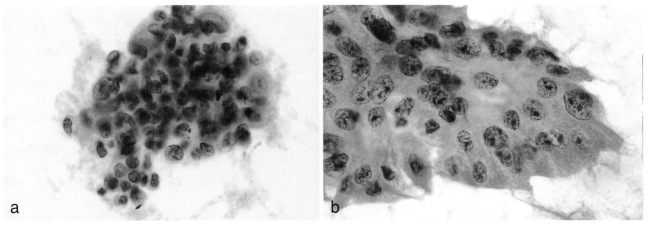

Fig. 6-194 Iatrogenic effect of endometrial adenocarcinoma. Endocyte scraping.
a. Before therapy. A three-dimensional tight cluster showing high N/C ratio, densely stained nuclei and scant cytoplasm.
b. After administration of medroxyprogesterone. A remarkable response is evidenced by increase of cytoplasm in amount, altered cytoplasmic stain, and distinctiveness of nucleoli.

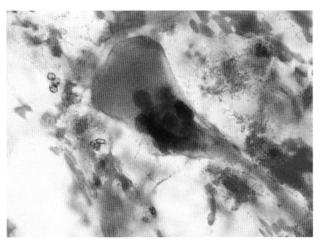

Fig. 6-195 **Syncytiotrophoblast from imprint of endometrial curettage.** From a case with hydatidiform mole at the 16th week of gestation. Multiple pyknotic nuclei with overlapping are included in strongly eosinophilic cytoplasm.

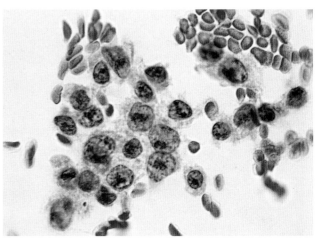

Fig. 6-196 **Cytotrophoblasts in a loose cluster from hydatidiform mole.** The same smear as Fig. 6-195. The nuclei are round, vesicular and contain prominent nucleoli. The cytoplasm is scant and poorly defined.

sive in well differentiated endometrioid carcinoma because they have steroid receptors. In about 33% of endometrioid adenocarcinoma with distant metastasis, progestin therapy leads to tumor regression and disappearance of metastasis.[256]

Morphological response to progestational agents is characterized by increase in volume of nuclei and cytoplasm, prominence of nucleoli, acidophilic stain of cytoplasm, etc. (Fig. 6-194)

A close linkage of long-term use of oral contraceptives to alteration of histologic types of cervical carcinomas has been pointed out by Takahashi and his associates.[245] Epidemiologically the Centers for Disease Control reported decreased risk for endometrial carcinoma in women who have used oral contraceptives; the relative risk of developing endometrial carcinoma was 0.5 in women of pill intake compared with nonusers[28]. Epidemiological and histological studies at National Women's Hospital of Auckland proved a significant shift of cervical carcinoma to immature mixed, mucoepidermoid and glassy cell carcinomas in women who took contraceptive pills for longer than 12 months. The pro-

portion of glassy cell carcinoma that is a peculiar type of mixed carcinoma comprised 6.5% of 215 pill users, whereas 1.1% of 456 nonusers.[256]

19 Chorionic Tumors

Pathologically chorionic tumors consist of hydatidiform mole, invasive mole (chorioadenoma destruens). In addition, the concept of placental site trophoblastic tumor (PSTT) was added to chorionic tumors by Scully and Young[218] as a tumor consisting of mononuclear intermediate trophoblastic tumor cells invading among smooth muscle fibers of the myometrium.

From a clinical standpoint the diagnosis of grave choriocarcinoma should not rely on cytologic findings but on whole laboratory examinations such as echography, HCG measurement, histology, etc.

Choriocarcinoma is most malignant in the group of chorionic tumors that tend to metastasize rapidly to the lung. Thus it is occasionally diagnosed from sputum. Needle aspirates from a pulmonary mass rarely present enlarged tumor cells which are misinterpreted as large

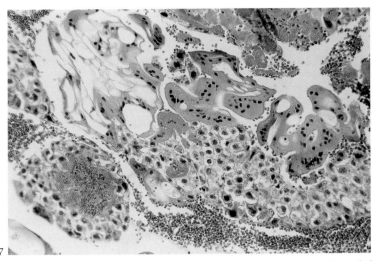
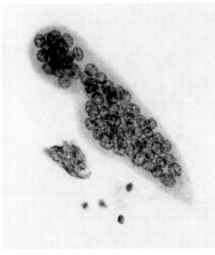

6-197

6-198

Fig. 6-197 Histology of destructive mole that is characterized by solid sheets of cytotrophoblasts and syncytiotrophoblasts. Presence or absence of chorionic villi is an important point to differentiate invasive mole from choriocarcinoma. Note capping of syncytiotrophoblasts on solid nest of cytotrophoblasts.

Fig. 6-198 Well preserved syncytiotrophoblast containing overlapped multiple nuclei. The nuclei are uniform in size and shape. The chromatin is coarsely clumped, but evenly distributed. The arrangement with nuclear overlapping differs from molding in herpes infection.

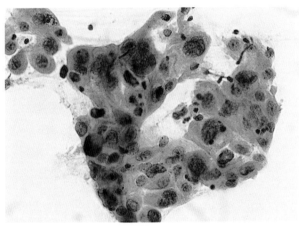
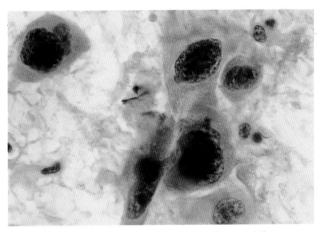

Fig. 6-199 Choriocarcinoma in endometrial aspirates. Choriocarcinoma is characteristically composed of polymorphic cytotrophoblasts (Langhans type) arranged in large sheets. Note marked variation in size of nucleus and densely clumped chromatins. For differentiation from destructive mole absence of syncytiotrophoblasts must be carefully examined.(From Kuramoto H: Color Atlas Screening of Cancer of Uterine Corpus. Tokyo: Ishiyaku Shuppan, 1987.)

Fig. 6-200 Choriocarcinoma in endometrial scrapes showing bizarre cytotrophoblastic cells. Huge trophoblastic cancer cells appear in part to be multilobulated and possess eosinophilic macronucleolus.

cell carcinoma of the lung.[37]

Diagnostic clues to choriocarcinoma are the appearance of malignant trophoblasts in imprint smears from endometrial curettings or products of conception and absence of chorionic villi in macroscopic examination of the specimen. While benign syncytiotrophoblasts which are found in incomplete abortion possess multiple, uniform nuclei and enough amount of cytoplasm, malignant syncytiotrophoblasts disclose markedly enlarged, polymorphic nuclei and scant cytoplasm. Malignant cytotrophoblasts appearing singly or in a loose cluster have large round nuclei with coarse chromatins and prominent nucleoli. The nucleoli are usually single and centrally positioned. The accurate diagnosis of chorionic tumors based on cellular components, cellular atypia, invasion of trophoblasts into the normal

tissues, and presence or absence of chorionic villi can be established only by histological examination. The cytodiagnosis is limited to cellular atypism, but a certain number of cases shedding abnormal trophoblasts are suggestive of the presence of chorionic tumors; differentiation between destructive mole and choriocarcinoma from a cytologic base is questionable.[246] In summary, in choriocarcinoma Langhans type trophoblastic cancer cells predominate (Fig. 6-200) and syncytial trophoblastic cancer cells are seldom found. Binucleated or trinucleated cells seem to be of the Langhans type. Immunocytochemical or biochemical examination for human chorionic gonadotropin (HCG) is also valuable to reach a conclusive diagnosis. A positive immunocytochemistry against human placental lactogen (HPL) instead of negative reactivity against HCG is strongly emphasized to be a clue to identification of PSTT of rare occurrence.[98,121]

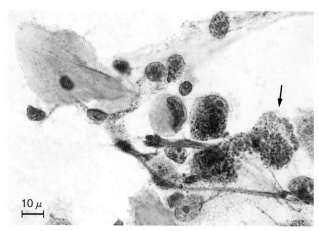

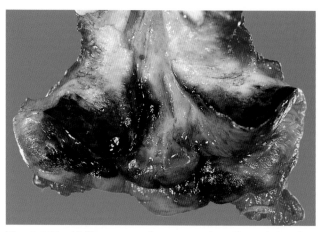

Fig. 6-201 Malignant melanoma cells in a cervical scraping smear. Large round or oval nuclei possess characteristic prominent nuclei. Note variation of melanin pigments in amount from cell to cell. A phagocyte filled with the pigment is also seen (arrow).

Fig. 6-202 Malignant melanoma primarily originating from endocervical mucosa.

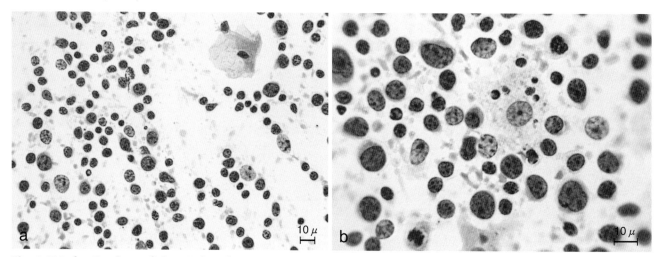

Fig. 6-203a,b Cytology of chronic lymphocytic cervicitis. Diagnostic points: (1) Mixture of mature lymphocytes and immature, large lymphocytes, (2) presence of histiocytic cells phagocytosing cell debris, and (3) absence of malignant background are the point to differentiate from malignant lymphoma.

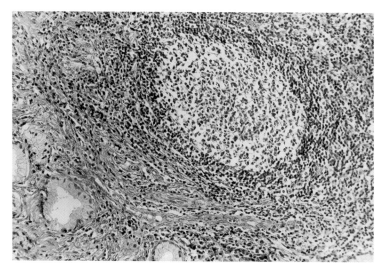

Fig. 6-204 Histology of chronic lymphocytic cervicitis. Note heavy lymphocytic infiltration and hypertrophy of a germinal center, from which immature lymphocytes (lymphogonia) and phagocytosing histiocytic cells are derived.

20 Malignant Melanoma

General aspect: Primary melanoma of the female genital tract is rare except for that arising from the vulva.[33] The juxtacutaneous mucous membranes such as the vagina, oral mucosa, nasopharynx, and anorectal mucosa are occasionally affected.

Gupta[75] reported 26 patients with malignant melanoma of the vulva and vagina who were treated at the Memorial Center for Cancer and Allied Diseases; 23 cases of vulval melanoma constituted 3.1% of all cuta-

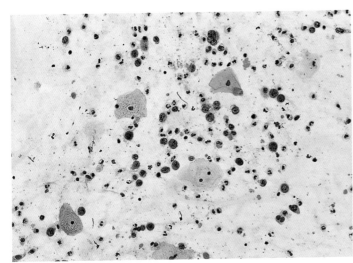

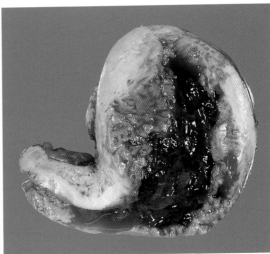

Fig. 6-205　Low power view of malignant lymphoma in cervical smear. Note a characteristic pattern consisting of isolated malignant cells without epithelial arrangement.

Fig. 6-206　Macroscopic figure of primary lymphoma of uterus. Greyish-white homogeneous tumor locating at the cervix anteriorly is extending to the uterine corpus. The endocervix is erosive and may shed lymphoma cells by cervical scraping.

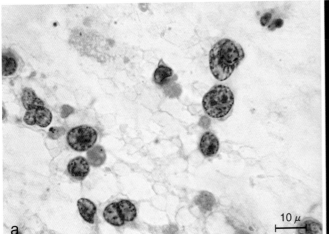

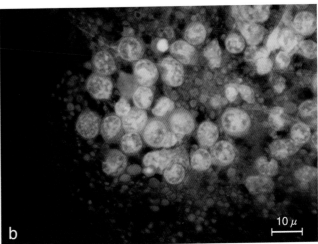

Fig. 6-207　Malignant lymphoma cells in cervical smear. This case is that of rare malignant lymphoma affecting the cervix primarily. Cytology is characterized by innumerable lymphosarcoma cells with necrotic cell debris. Diagnostic points: (1) large round lymphoblastoid cells with coarsely stippled chromatin pattern, (2) almost naked or very scant cytoplasm, (3) occasional prominent nucleoli, and (4) tumor diathesis. Papanicolaou stain (a) and acridine orange fluorescence (b).

neous melanoma. Melanoma of the cervix is the least common form and only nine cases have been reviewed by Jones.[95] Morphogenesis of malignant melanoma is considered to be related to the activity of melanocytes that exist in the mucosal epithelium[166] regardless of their neurocrest or epithelial origin.

Cytology: Since some of malignant melanoma are amelanotic or scant in amount of melanin pigment, they are apt to be misinterpreted as carcinoma or sarcoma; amelanotic melanoma was found in one of four cases of vaginal melanoma by Masubuchi et al,[133] and in two of 13 cases by Norris and Taylor.[169]

The amelanotic epithelioid type is similar to adenocarcinoma and the spindle cell type is like fibrosarcoma or leiomyosarcoma. A variety of the epithelioid type which is comprised of small cells is often misjudged as anaplastic carcinoma or malignant lymphoma. Scant melanin pigment can be detected more easily by Giemsa method than by ordinary Papanicolaou method. The

specific Dopa reaction that converts tyrosine to 3, 4-dihydroxyphenylalanine is suitable for identification of amelanotic melanoma cells. In addition, positive reaction of Fontana-Masson silver impregnation and complete bleaching of brownish pigment by oxidation with 0.25% potassium permanganate solution (for over 30 minutes) are indicative of melanin.

21　Lymphoid Hyperplasia and Lymphoma

■Chronic lymphocytic cervicitis
This is a variety of chronic cervicitis that is characterized by severe infiltration of subepithelial stroma by lymphoid cells either of mature or of immature form.

Since follicle formation is often associated, this is also called follicular cervicitis. According to the study by Roberts and NG lymphoid follicles are observed in about 37% of chronic lymphocytic cervicitis (follicular

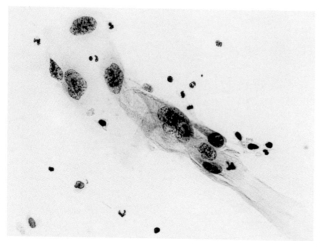

Fig. 6-208 Sarcomatous component showing oval or elliptic nuclei with delicate nuclear rims. The chromatin is increased in amount but more finely granular than that of cancer cells, although pyknotic nuclei resemble spindly cancer cells. The cellular arrangement is apparently loose compared with that of the carcinomatous component.

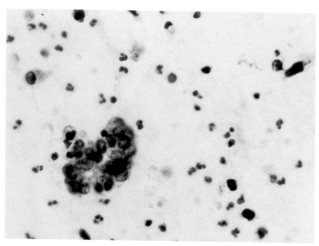

Fig. 6-209 Carcinomatous component composed of tight cell cluster showing a glandular pattern. Their nuclei are fairly small and uniform in shape, thus the cellular atypism is not sufficient for the criteria of malignancy. However such a structure is not encountered in cytology from the benign lesion of the endocervix.

cervicitis). Lymphocytic cervicitis is of fairly rare occurrence and found in women of post menopausal age. The overall frequency of chronic lymphocytic cervicitis studied with cellular samples is said to be 4.7 per 10,000 cases.[205]

Cytologically it is characterized by a predominance of mature and immature lymphocytes mixed with reticular histiocytic cells and plasma cells. Histiocytic cells showing phagocytosis in particular are derived from activated germinal centers. Immature lymphocytes can be identified by large round nuclei often with indentation or folding of the nuclear rim and very scant, pale cytoplasm. The chromatin shows a stippled pattern with a few karyosomes.

■Malignant lymphoma

The majority of extranodal lymphomas are instances of secondary involvement. However the genital organs are involved infrequently and rarely associated with gynecologic symptoms. Primary lymphoma of the female genital organs is very rare. In a large series of 1,467 extranodal lymphoma cases, only 6 cases involving the uterus and 2 cases involving the ovary were described by Freeman et al.[41] A case operated at the Tokyo Railway Hospital disclosed a hen egg-sized tumor of the uterine cervix without dissemination to other organs. The tumor was elastic, firm in consistency, pinkish grey-white in color and homogeneous because of massive infiltration of the entire wall of the cervix by lymphosarcoma cells. Cytological features in cervical scraping smears were essentially the same as those of malignant lymphoma arising from lymph nodes (Figs. 6-205 and 6-207).

For discrimination of lymphoma cells from undifferentiated carcinoma, immunocytochemical stains against leukocyte common antigen (LCA), epithelial membrane antigen (EMA) and cytokeratin are worthy of application.[134]

Table 6-12 Histopathologic classification of uterine sarcoma

I. Pure Sarcoma
  A. Pure Homologous
    1. Leiomyosarcoma
    2. Stromal sarcoma
    3. Endolymphatic stromal myosis
    4. Angiosarcoma
  B. Pure Heterologous
    1. Rhabdomyosarcoma (including sarcoma botryoides)
    2. Chondrosarcoma
    3. Osteosarcoma
    4. Liposarcoma
II. Mixed Sarcoma
  A. Mixed Homologous
  B. Mixed Heterologous (with or without homologous elements)
	III. Malignant Mixed Mullerian Tumors (mixed mesodermal tumors)
	  A. Homologous Type (carcinoma plus leiomyosarcoma, stromal sarcoma or fibrosarcoma, or mixture of these sarcomas)
	  B. Heterologous Type (carcinoma plus heterologous sarcoma with or without homologous sarcoma)
	IV. Sarcoma, Unclassified
	V. Malignant Lymphoma

(From Kempson RL, Bari W: Hum Pathol 1 : 373, 1970.)

22 Uterine Sarcoma

■Müllerian mixed tumor

General concept of uterine sarcoma: Cytopathology of uterine sarcoma is not fully understood because of rare incidence of sarcoma and rare exfoliation of mesenchymal tumor cells into the cervical smear; although leiomyosarcoma is of the next frequent tumor of the uterus, its exfoliative cytology is not of practical importance. Mixed mesodermal tumor and endometrial sarcoma are the subject of interest. Patten and his associates[181] described the frequency of uterine sarcoma in their 26 cases; malignant mixed mesodermal tumor in 65.4%, leiomyosarcoma in 15.4%, endometrial stromal sarcoma in 7.7%, rhabdomyosarcoma in 7.7% and

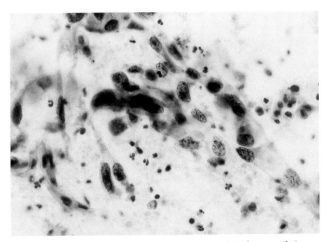

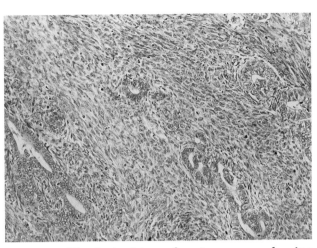

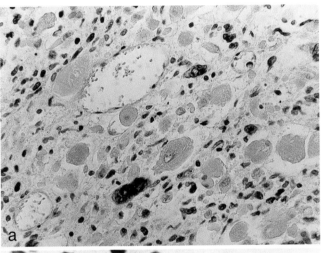

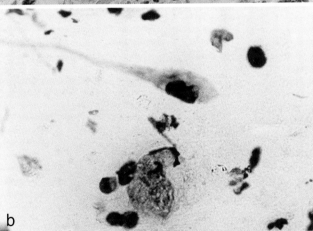

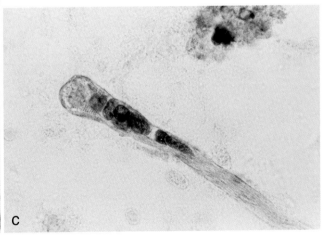

Fig. 6-210   Sarcomatous component. The nuclei are spindly or elliptic in shape. Well preserved cells are characterized by delicate nuclear membrane and fine chromatin distribution, although the pyknotic cells are indistinguishable from spindle cancer cells.

Fig. 6-211   Histologic picture of carcinosarcoma showing collision of adenocarcinoma. Note a striking resemblance of this histology to pulmonary blastoma being composed of slit-like cuboidal adenocarcinoma and dense sarcoma.

Fig. 6-212   Mesodermal mixed tumor of the cervix.
a. Histology showing diffuse infiltration of degenerative anucleated cells and polymorphic sarcoma cells with pyknotic, bizarre nuclei.
b, c. Cytology. Fiber-like or snake-like polymorphic cells are identified singly in the cervical smear. Pyknotic cells are likely to be tadpole or fiber cells of epidermoid carcinoma, but detection of cross striation in the cytoplasm is indicative of rhabdomyosarcoma element.

fibroxanthosarcoma in 3.8% according to the classification by Kempson and Bari[100] (Table 6-12).

1) Carcinosarcoma
Carcinosarcoma is composed of two components, carcinoma and sarcoma of homologous nature (Figs. 6-208 to 211).

Cytology: Cancer component is usually of a glandular type detected forming cell aggregates. Mesodermal components shed singly or in a loose cluster without any

organoid pattern. The nuclei of stromal sarcoma are elongated or spindly in shape with smooth, delicate nuclear rims, and vary in size. The chromatin is not much increased in amount but coarsely distributed.

2) Mesodermal mixed tumor
Histology of this tumor varies from case to case; in addition to homologous sarcomatous components such as stromal sarcoma, leiomyosarcoma and fibrosarcoma, heterologous rhabdomyosarcoma and chondrosarcoma

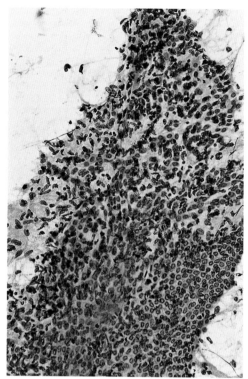

Fig. 6-213 Low power view of endometrial stromal sarcoma. Endocyte scraping. Note an admixture of endometrial epithelial cells in sheet (lower part) with loose cell cluster having elliptic or spindly nuclei. Compare with very polymorphic tumor cells of mesodermal mixed tumor (see Fig. 6-212).

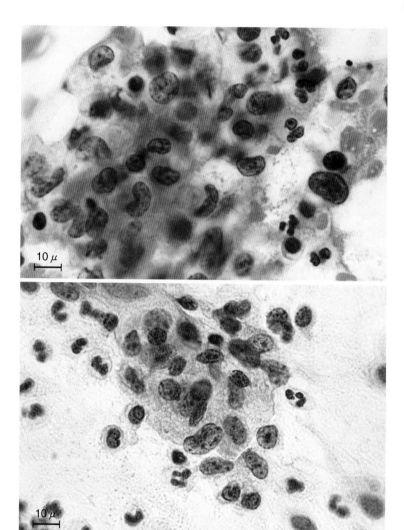

Fig. 6-214 Endometrial sarcoma cells. Pure stromal sarcoma. The nuclei are elongated and curved. The distinctive features of sarcoma are (1) delicate nuclear membrane, (2) finely granular chromatins, and (3) marked variation in size and shape of nuclei.

are frequently admixed. It is emphasized that the presence and malignancy of epithelial components most likely concern with prognosis of the tumor[210]; two of 20 cases of mixed mesodermal tumors that were absent in epithelial elements and four cases that contained benign glandular structures were favorable in their survival.

Nielsen et al mention, however, prognostic significance of surgical stage and depth of invasion irrespective of malignancy grade of cancer as well as histologic type of sarcoma in their study on 60 patients with mixed mesodermal tumors.[165]

Cytology: Specific differentiation to rhabdomyoblasts with cross striations and to chondroblasts or osteoblasts in addition to the presence of carcinoma cells is indicative of this tumor. Sarcoma botryoides, a variant of Müllerian mesodermal mixed tumor, is a malignant grape-like polypoid tumor that occurs principally from the vagina of infants and less commonly from the uterine cervix or corpus in adults. The tumor consists of various mesodermal tissue components such as striated muscle cells, cartilaginous or bony tissues in addition to glandular epithelial elements; fibromyxomatous tissues and

striated muscle cells particularly predominate (Fig. 6-212).

23 Endometrial Stromal Sarcoma

Endometrial stromal sarcoma is a rare tumor that belongs to the pure homologous sarcoma. It consists of endolymphatic stromal meiosis and endometrial stromal sarcoma. The former occurs in younger women and involves rarely the myometrium, vagina and pelvis. On the other hand, endometrial stromal sarcoma bulges into the endometrial cavity and invades the myometrium. Although stromal sarcoma is classified into low-grade sarcoma and high-grade sarcoma based on atypicality and frequency of mitotic figures, the former having smaller nuclei is hardly distinguished in cytologic smears.[71,98] Thereby, endometrial cytology may encounter sheets of malignant stromal cells of high-grade variety. Cytologically this tumor is characterized by spindly cells arising from the stromal cells of the endometrium (Fig. 6-214). The nuclei are elongated and hardly distinguishable from leiomyosarcoma cells having somewhat blunt poles.

Appendix 1. The Bethesda System for Reporting Cervical/Vaginal Cytological Diagnoses

Adequacy of the Specimen
  Satisfactory for evaluation
  Satisfactory for evaluation but limited by... (specify reason)
  Unsatisfactory for evaluation... (specify reason)
General Categorization (optional)
  Within normal limits
  Benign cellular changes: See descriptive diagnosis
  Epithelial cell abnormality: See descriptive diagnosis
Descriptive Diagnoses
BENIGN CELLULAR CHANGES
INFECTION
  Trichomonas vaginalis
  Fungal organisms morphologically consistent with *Candida* spp.
  Predominance of coccobacilli consistent with shift in vaginal flora
  Bacteria morphologically consistent with *Actinomyces* spp.
  Cellular changes associated with herpes simplex virus
  Other*
REACTIVE CHANGES
  Reactive cellular changes associated with:
  Inflammation (includes typical repair)
  Atrophy with inflammation ("atrophic vaginitis")
  Radiation
  Intrauterine contraceptive device (IUD)
  Other
EPITHELIAL CELL ABNORMALITIES
SQUAMOUS CELL
  Atypical squamous cells of undetermined significance (ASCUS): qualify**
  Low-grade squamous intraepithelial lesion (LSIL) encompassing: HPV*/mild dysplasia/CIN1
  High-grade squamous intraepithelial lesion (HSIL) encompassing: moderate dysplasia and severe dysplasia, and CIS/CIN2 and CIN3
  Squamous cell carcinoma
GLANDULAR CELL
  Endometrial cells, cytologically benign in a postmenopausal women
  Atypical glandular cells of undetermined significance: qualify**
  Endocervical adenocarcinoma
  Endometrial adenocarcinoma
  Extrauterine adenocarcinoma
  Adenocarcinoma, NOS
  Moderate dysplasia/CIN2
  Severe dysplasia/CIN3
OTHER MALIGNANT NEOPLASMS: SPECIFY
HORMONAL EVALUATION (APPLIES TO VAGINAL SMEARS ONLY)
  Hormonal pattern compatible with age and history
  Hormonal pattern incompatible with age and history: specify
  Hormonal evaluation not possible due to: specify

  *Cellular changes of human papillomavirus (HPV) — previously termed *koilocytosis, koilocytotic atypia* and condylomatous atypia — are included in the category of LSIL.

  **Atypical squamous or glandular cells of undetermined significance should be further qualified, if possible, as to whether a reactive or premalignant/malignant process is favored.
    Acta Cytologica 37: 115-124, 1993.
    The Bethesda System. Berlin: Springer-Verlag, 1994.

Appendix 2. Histological Classification of Tumours of the Female Genital Tract (WHO 1994)

■ Uterine Corpus

1	**Epithelial Tumours and Related Lesions**
1.1	*Endometrial hyperplasia*
1.1.1	Simple
1.1.2	Complex (adenomatous)
1.2	*Atypical endometrial hyperplasia*
1.2.1	Simple
1.2.2	Complex (adenomatous with atypia)
1.3	*Endometrial polyp*
1.4	*Endometrial carcinoma*
1.4.1	Endometrioid
1.4.1.1	Adenocarcinoma
1.4.1.1.1	Secretory (variant)
1.4.1.1.2	Ciliated cell (variant)
1.4.1.2	Adenocarcinoma with squamous differentiation (adenoacanthoma; adenosquamous carcinoma)
1.4.2	Serous adenocarcinoma
1.4.3	Clear cell adenocarcinoma
1.4.4	Mucinous adenocarcinoma
1.4.5	Squamous cell carcinoma
1.4.6	Mixed carcinoma
1.4.7	Undifferentiated carcinoma
2	**Mesenchymal Tumours and Related Lesions**
2.1	*Endometrial stromal tumours*
2.1.1	Endometrial stromal nodule
2.1.2	Endometrial stromal sarcoma, low grade (endolymphatic stromal myosis)
2.1.3	Endometrial stromal sarcoma, high grade
2.2	*Smooth muscle tumours*
2.2.1	Leiomyoma
2.2.1.1	Cellular (variant)
2.2.1.2	Epithelioid (variant)
2.2.1.3	Bizarre (variant)
2.2.1.4	Lipoleiomyoma (variant)
2.2.2	Smooth muscle tumour of uncertain malignant potential
2.2.3	Leiomyosarcoma
2.2.3.1	Epithelioid (variant)
2.2.3.2	Myxoid (variant)
2.2.4	Other smooth muscle tumours
2.2.4.1	Metastasizing leiomyoma
2.2.4.2	Intravenous leiomyomatosis
2.2.4.3	Diffuse leiomyomatosis
2.3	*Mixed endometrial stromal and smooth muscle tumours*
2.4	*Adenomatoid tumour*
2.5	*Other mesenchymal tumours*
2.5.1	Homologous
2.5.2	Heterologous

3 Mixed Epithelial and Mesenchymal Tumours
3.1   *Benign*
3.1.1   Adenofibroma
3.1.2   Adenomyoma
3.1.2.1   Atypical polypoid adenomyoma (variant)
3.2   *Malignant*
3.2.1   Adenosarcoma
3.2.1.1   Homologous
3.2.1.2   Heterologous
3.2.2   Carcinofibroma
3.2.3   Carcinosarcoma (malignant mesodermal mixed tumour; malignant müllerian mixed tumour)
3.2.3.1   Homologous
3.2.3.2   Heterologous

4 Miscellaneous Tumours
4.1   Sex cord-like tumours
4.2   Tumours of germ cell type
4.3   Neuroectodermal tumours
4.4   Lymphoma and leukemia
4.5   Others

5 Secondary Tumours

6 Tumour-like Lesions
6.1   *Epithelial metaplastic and related changes*
6.1.1   Squamous metaplasia, including morula formation
6.1.2   Mucinous metaplasia, including intestinal type
6.1.3   Ciliated cell metaplasia
6.1.4   Hobnail cell metaplasia
6.1.5   Clear cell change
6.1.6   Eosinophilic cell metaplasia, including oncocytic
6.1.7   Surface syncytial change
6.1.8   Papillary change
6.1.9   Arias-Stella change
6.2   *Stromal metaplastic and related changes*
6.2.1   Smooth muscle metaplasia
6.2.2   Osseous metaplasia and retention of fetal bone
6.2.3   Cartilaginous metaplasia and retention of fetal cartilage
6.2.4   Adipocyte metaplasia
6.2.5   Retention of fetal glial tissue
6.2.6   Foam cell change
6.3   Adenomyosis
6.4   Epithelial cysts of myometrium
6.5   Chronic endometritis
6.6   Lymphoma-like lesions
6.7   Inflammatory pseudotumour
6.8   Others

Gestational Trophoblastic Disease

1 Hydatidiform Mole
1.1   Complete
1.2   Partial

2 Invasive Hydatidiform Mole
(chorioadenoma destruens)

3 Choriocarcinoma

4 Placental Site Trophoblastic Tumour

5 Miscellaneous Trophoblastic Lesions
5.1   Exaggerated placental site
5.2   Placental site nodule and plaque

6 Unclassified Trophoblastic Lesions

Uterine Cervix

1 Epithelial Tumours and Related Lesions
1.1   *Squamous lesions*
1.1.1   Squamous papilloma
1.1.2   Condyloma acuminatum
1.1.3   Squamous metaplasia
1.1.4   Transitional metaplasia
1.1.5   Squamous atypia
1.1.6   Squamous intraepithelial lesions (dysplasia-carcinoma in situ; cervical intraepithelial neoplasia (CIN))
1.1.6.1   Mid dysplasia (CIN 1)
1.1.6.2   Moderate dysplasia (CIN 2)
1.1.6.3   Severe dysplasia (CIN 3)
1.1.6.4   Carcinoma in situ (CIN 3)
1.1.7   Squamous cell carcinoma
1.1.7.1   Keratinizing
1.1.7.2   Nonkeratinizing
1.1.7.3   Verrucous
1.1.7.4   Warty (condylomatous)
1.1.7.5   Papillary
1.1.7.6   Lymphoepithelioma-like carcinoma
1.2   *Glandular lesions*
1.2.1   Endocervical polyp
1.2.2   Müllerian papilloma
1.2.3   Glandular atypia
1.2.4   Glandular dysplasia
1.2.5   Adenocarcinoma in situ
1.2.6   Adenocarcinoma
1.2.6.1   Mucinous adenocarcinoma
1.2.6.1.1   Endocervical type
1.2.6.1.2   Intestinal type
1.2.6.2   Endometrioid adenocarcinoma
1.2.6.3   Clear cell adenocarcinoma
1.2.6.4   Serous adenocarcinoma
1.2.6.5   Mesonephric adenocarcinoma
1.3   *Other epithelial tumours*
1.3.1   Adenosquamous carcinoma
1.3.2   Glassy cell carcinoma
1.3.3   Adenoid cystic carcinoma
1.3.4   Adenoid basal carcinoma
1.3.5   Carcinoid tumour
1.3.6   Small cell carcinoma
1.3.7   Undifferentiated carcinoma

2 Mesenchymal Tumours
2.1   Leiomyoma
2.2   Leiomyosarcoma
2.3   Endocervical stromal sarcoma
2.4   Sarcoma botryoides (embryonal rhabdomyosarcoma)
2.5   Endometrioid stromal sarcoma
2.6   Alveolar soft-part sarcoma
2.7   Others

3 Mixed Epithelial and Mesenchymal Tumours
3.1   Adenofibroma
3.2   Adenomyoma
3.2.1   Atypical polypoid adenomyoma (variant)
3.3   Adenosarcoma
3.4   Malignant mesodermal mixed tumour (malignant

müllerian mixed tumour; carcinosarcoma)
3.5 Wilms tumour

4 **Miscellaneous Tumours**
4.1 Melanocytic naevus
4.2 Blue naevus
4.3 Malignant melanoma
4.4 Lymphoma and leukemia
4.5 *Tumours of germ cell type*
4.5.1 Yolk sac tumour (endodermal sinus tumour)
4.5.2 Dermoid cyst (mature cystic teratoma)

5 **Secondary Tumours**

6 **Tumour-like Lesions**
6.1 Endocervical glandular hyperplasia
6.2 Cysts
6.3 Tunnel cluster
6.4 Microglandular hyperplasia
6.5 Arias-Stella change
6.6 Mesonephric remnants
6.7 Mesonephric hyperplasia
6.8 Ciliated cell metaplasia
6.9 Intestinal metaplasia
6.10 Epidermal metaplasia
6.11 Endometriosis
6.12 Ectopic decidua
6.13 Placental site nodule and plaque
6.14 Stromal polyp (pseudosarcoma botryoides)
6.15 Postoperative spindle cell nodule
6.16 Traumatic (amputation) neuroma
6.17 Retention of fetal glial tissue
6.18 Lymphoma-like lesions

■Vagina

1 **Epithelial Tumours and Related Lesions**
1.1 *Squamous lesions*
1.1.1 Squamous papilloma
1.1.2 Condyloma acuminatum
1.1.3 Transitional metaplasia
1.1.4 Squamous atypia
1.1.5 Squamous intraepithelial lesions (dysplasia-carcinoma in situ; vaginal intraepithelial neoplasia (VAIN))
1.1.5.1 Mid dysplasia (VAIN 1)
1.1.5.2 Moderate dysplasia (VAIN 2)
1.1.5.3 Severe dysplasia (VAIN 3)
1.1.5.4 Carcinoma in situ (VAIN 3)
1.1.6 Squamous cell carcinoma
1.1.6.1 Keratinizing
1.1.6.2 Nonkeratinizing
1.1.6.3 Verrucous
1.1.6.4 Warty (condylomatous)
1.2 *Glandular lesions*
1.2.1 Müllerian papilloma
1.2.2 Adenosis
1.2.3 Atypical adenosis
1.2.4 Adenocarcinoma
1.2.4.1 Clear cell adenocarcinoma
1.2.4.2 Endometrioid adenocarcinoma
1.2.4.3 Mucinous adenocarcinoma
1.2.4.3.1 Endocervical type
1.2.4.3.2 Intestinal type
1.2.4.4 Mesonephric adenocarcinoma
1.3 *Other epithelial tumours*

1.3.1 Adenosquamous carcinoma
1.3.2 Adenoid cystic carcinoma
1.3.3 Adenoid basal carcinoma
1.3.4 Carcinoid tumour
1.3.5 Small cell carcinoma
1.3.6 Undifferentiated carcinoma

2 **Mesenchymal Tumours**
2.1 Leiomyoma
2.2 Rhabdomyoma
2.3 Leiomyosarcoma
2.4 Sarcoma botryoides (embryonal rhabdomyosarcoma)
2.5 Endometrioid stromal sarcoma
2.6 Others

3 **Mixed Epithelial and Mesenchymal Tumours**
3.1 Mixed tumour
3.2 Adenosarcoma
3.3 Malignant mesodermal mixed tumour (malignant müllerian mixed tumour; carcinosarcoma)
3.4 Mixed tumour resembling synovial sarcoma

4 **Miscellaneous Tumours**
4.1 Melanocytic naevus
4.2 Blue naevus
4.3 Malignant melanoma
4.4 Tumours of germ cell type
4.4.1 Yolk sac tumour (endodermal sinus tumour)
4.4.2 Dermoid cyst (mature cystic teratoma)
4.5 Adenomatoid tumour
4.6 Villous adenoma
4.7 Lymphoma and leukemia

5 **Secondary Tumours**

6 **Tumour-like Lesions**
6.1 Stromal polyp (pseudosarcoma botryoides)
6.2 Postoperative spindle cell nodule
6.3 Vault granulation tissue
6.4 Prolapse of fallopian tube
6.5 Endometriosis
6.6 Ectopic decidua
6.7 *Cysts*
6.7.1 Epidermal
6.7.2 Müllerian
6.7.3 Mesonephric
6.8 Microglandular hyperplasia
6.9 Lymphoma-like lesions

■Vulva

1 **Epithelial Tumours and Related Lesions**
1.1 *Squamous lesions*
1.1.1 Epithelial papillomas and polyps
1.1.1.1 Vestibular squamous papilloma (vestibular papilloma)
1.1.1.2 Fibroepithelial polyp
1.1.2 Condyloma acuminatum
1.1.3 Seborrheic keratosis
1.1.4 Keratoacanthoma
1.1.5 Squamous intraepithelial lesions (dysplasia-carcinoma in situ, vulvar intraepithelial neoplasia (VIN))
1.1.5.1 Mild dysplasia (VIN 1)
1.1.5.2 Moderate dysplasia (VIN 2)

1.1.5.3 Severe dysplasia (VIN 3)
1.1.5.4 Carcinoma in situ (VIN 3)
1.1.6 Squamous cell carcinoma
1.1.6.1 Keratinizing
1.1.6.2 Nonkeratinizing
1.1.6.3 Basaloid
1.1.6.4 Verrucous
1.1.6.5 Warty (condylomatous)
1.1.6.6 Others
1.1.7 Basal cell carcinoma
1.2 *Glandular lesions*
1.2.1 Papillary hidradenoma
1.2.2 Clear cell hidradenoma
1.2.3 Syringoma
1.2.4 Trichoepithelioma
1.2.5 Trichilemmoma
1.2.6 Adenoma of minor vestibular glands
1.2.7 Paget disease
1.2.8 *Bartholin gland carcinomas*
1.2.8.1 Adenocarcinoma
1.2.8.2 Squamous cell carcinoma
1.2.8.3 Adenoid cystic carcinoma
1.2.8.4 Adenosquamous carcinoma
1.2.8.5 Transitional cell carcinoma
1.2.9 Tumours arising from ectopic breast tissue
1.2.10 Carcinomas of sweat-gland origin
1.2.11 Adenocarcinomas of other types

2 Soft Tissue Tumours
2.1 *Benign*
2.1.1 Lipoma and fibrolipoma
2.1.2 Haemangiomas
2.1.2.1 Capillary
2.1.2.2 Cavernous
2.1.2.3 Acquired
2.1.3 Angiokeratoma
2.1.4 Pyogenic granuloma
2.1.5 Lymphangioma
2.1.6 Fibroma
2.1.7 Leiomyoma
2.1.8 Granular cell tumour
2.1.9 Neurofibroma
2.1.10 Schwannoma (neurilemoma)
2.1.11 Glomus tumour
2.1.12 Benign fibrous histiocytoma
2.1.13 Rhabdomyoma
2.2 *Malignant*
2.2.1 Embryonal rhabdomyosarcoma (sarcoma botryoides)
2.2.2 Aggressive angiomyxoma
2.2.3 Leiomyosarcoma
2.2.4 Dermatofibrosarcoma protuberans
2.2.5 Malignant fibrous histiocytoma
2.2.6 Epithelioid sarcoma
2.2.7 Malignant rhabdoid tumour
2.2.8 Malignant nerve sheath tumours
2.2.9 Angiosarcoma
2.2.10 Kaposi sarcoma
2.2.11 Haemangiopericytoma
2.2.12 Liposarcoma
2.2.13 Alveolar soft-part sarcoma

3 Miscellaneous Tumours
3.1 *Melanocytic tumours*
3.1.1 Congenital melanocytic naevus
3.1.2 Acquired melanocytic naevus
3.1.3 Blue naevus
3.1.4 Dysplastic melanocytic naevus
3.1.5 Malignant melanoma
3.2 Lymphoma and leukemia
3.3 Yolk sac tumour (endodermal sinus tumour)
3.4 Merkel cell tumour

4 Secondary Tumours

5 Tumour-like Lesions and Nonneoplastic Disorders
5.1 Pseudoepitheliomatous hyperplasia
5.2 Endometriosis
5.3 Ectopic decidua
5.4 Langerhans cell histiocytosis (eosinophilic granuloma; histiocytosis X)
5.5 Benign xanthogranuloma
5.6 Verruciform xanthoma
5.7 Desmoid tumour (fibromatosis)
5.8 Sclerosing lipogranuloma
5.9 Nodular fasciitis
5.10 Naevus lipomatosus superficialis
5.11 Crohn disease
5.12 *Cysts*
5.12.1 Bartholin duct cyst
5.12.2 Mucinous cyst
5.12.3 Epidermal cyst
5.12.4 Mesonephric cyst (wolffian duct cyst; Gartner duct cyst)
5.12.5 Ciliated cyst
5.12.6 Mesothelial cyst (cyst of the canal of Nuck)
5.12.7 Paraurethral cyst
5.13 *Other epithelial disorders*
5.13.1 Lichen sclerosus
5.13.2 Squamous cell hyperplasia
5.13.3 Other forms of dermatosis and dermatitis

References

1. Abeler VM, Vergote IB, Kjorstad KE, et al: Clear cell carcinoma of the endometrium: Prognosis and metastatic pattern. Cancer 78: 1740, 1996.
2. An-Foraker SH, Kawada, CY, McKinney D: Endometrial aspiration studies on Isaacs cell sampler with cytohistologic correlation. Acta Cytol 23: 303, 1979.
3. Arata T, Sekiba K, Kato K: Appraisal of self-collected cervical specimens in cytologic screening of uterine cancer. Acta Cytol 22: 150, 1978.
4. Arias-Stella J: On the importance of the atypical endometrial changes associated with the diagnosis of pregnancy. Am J Pathol 34: 601, 1958.
5. Arrighi AA: Exfoliative cytology of leukoplakia. Acta Cytol 5: 108, 1961.
6. Arroyo G, Quinn JA Jr: Association of amoebae and actinomyces in an intrauterine contraceptive device user. Acta Cytol 33: 298, 1989.
7. Attwood, ME: Cytology and the contraceptive pill. J Obstet Gynaec Br Commonw 73: 662, 1966.
8. Aurelian L, Royston I, Davis HJ: Antibody to genital herpes simplex virus: Association with cervical atypia and carcinoma in situ. J Natl Cancer Inst 45: 455, 1970.
9. Ayre JE: Adenoacanthoma of the uterus. Am J Obstet Gynecol 49: 261, 1945.
10. Bajardi F: Histomorphology of leukoplakia. Acta Cytol 5: 103, 1961.
11. Bajardi F: Interrelationship: Leukoplakia and cervical carcinoma. Acta Cytol 5: 129, 1961.

12. Bamford SB, Mitchell GW Jr, Bardawil WA, et al: Vaginal cytology in polycystic ovarian disease. Acta Cytol 9: 322, 1965.
13. Bangle R Jr, Berger M, Levin M: Variations in the morphogenesis of squamous carcinoma of the cervix. Cancer 16: 1151, 1963.
14. Barter JF, Orr JW Jr, Holloway RW, et al: Psammoma bodies in a cervicovaginal smear associated with an intrauterine device. A case report. J Reprod Med 32: 147, 1987.
15. Bibbo M, Harris MJ: Leptothrix. Acta Cytol 16: 2, 1972.
16. Bibbo M, Harris MJ, Wied GL: Microbiology and inflammation of the female genital tract. In : Wied GL, Koss LG, Reagan JW (eds): Compendium on Diagnostic Cytology, 4th ed. pp 61-75. Chicago: Tutorials of Cytology, 1979.
17. Bickenbach W, Soost H-J: Nomenclature of the atypical epithelium. Acta Cytol 5: 348, 1961.
18. Boutselis JG: Intraepithelial carcinoma of the cervix associated with pregnancy. Obstet Gynecol 40: 657, 1972.
19. Boyd JT, Doll R: A study of the aetiology of carcinoma of the cervix uteri. Br J Cancer 18: 419, 1964.
20. Boschann HW: Effect of administered androgens on the vaginal epithelium of women exhibiting the atrophic menopausal cell type. Acta Cytol 1: 87, 1957.
21. Bredland R: Trichomoniasis. Tskr Norske Laegeforg 441, 1962.
22. Bret J-A, Coupez R: Symposium on premalignant cervical lesions. III. Leukoplakia. Acta Cytol 5: 117, 1961.
23. Bret J-A, Coupez R: In: Probable percentage of regression of carcinoma in situ (symposium). Acta Cytol 6: 195, 1962.
24. Brunori IL: Estrogenic effect produced by digitalis preparations on the vaginal epithelium of woman of very advanced age. Riv Ital Ginec 49: 261, 1965.
25. Buckley CH, Fox H: Biopsy Pathology of the Endometrium. London: Chapman and Hall Medical, 1989.
26. Buckley CH, Herbert A, Johnson J, et al: Borderline nuclear changes in cervical smears. Guidelines on their recognition and management. J Clin Pathol 47:481, 1994.
27. Cecchini S, Iossa A, Ciatto S, et al: Routine colposcopic survey of patients with squamous atypia. A method for identifying cases with false-negative smears. Acta Cytol 34: 778. 1990.
28. Center for Disease Control: Cancer in steroid hormone study: Oral contraceptive use and the risk of endometrial cancer. JAMA 249: 1600, 1983.
29. Chai MS, Johnson WD, Tricomi V: Five year's experience with contraceptive pills: Cervical epithelial changes. NY State J Med 70: 2663, 1970.
30. Chapman PA: Female genital tract: Infection, inflammation and repair. In: Clinical Cytotechnology. pp172-173. London: Butterworths, 1989.
31. Chernesky MA, Mahony JB, Castriciano S, et al: Detection of Chlamydia trachomatis antigens by enzyme immunoassay and immunofluorescence in genital specimens from symptomatic and asymptomatic men and women. J Infect Dis 154: 141, 1986.
32. Christopherson WM, Parker JE: Microinvasive carcinoma of the uterine cervix. Cancer 17: 1123, 1964.
33. Chung AF, Woodruff JM, Lewis JL Jr: Malignant melanoma of the vulva: A case report of 44 cases. Obstet Gynecol 45: 638, 1975.
34. Cohen C: Three cases of amoebiasis of the cervix uteri. J Obstet Gynaecol Br Commonw 80: 476, 1973.
35. Corbishley CM: Microbial flora of the vagina and cervix. J Clin Pathol 30: 745, 1977.
36. Corscaden JA: Gynecologic Cancer, 2nd ed. Baltimore: Williams and Wilkins, 1956.
37. Craig LD, Shum DT, Desrosiers P, et al: Choriocarcinoma metastatic to the lung. Acta Cytol 27: 647 1983.
38. Dallenbach-Hellweg G, Hahn U: Mucinous and clear cell adenocarcinomas of the endometrium in patients receiving antiestrogens (tamoxifen) and gestagens. Int J Gynecol Pathol 14: 7, 1995.
39. Davis HJ: The irrigation smear. A cytologic method for mass population screening by mail. Am J Obstet Gynecol 84: 1017, 1962.
40. Davis JR, Moon LB: Increased incidence of adenocarcinoma of uterine cervix. Obstet Gynecol 45: 79, 1975.
41. Dellepiane G: Vaginal cytology in abortion. Acta Cytol 3: 282, 1959.
42. Demopoulos RI: Normal endometrium. In: Blaustein A (ed): Pathology of the Female Genital Tract. pp211-242. New York: Springer-Verlag, 1977.
43. Demopoulos RI, Genega E, Vamvakas E, et al.: Papillary carcinoma of the endometrium: Morphometric predictors of survival. Int J Gynecol Pathol 15: 110, 1996
44. de Moraes-Ruehsen M, McNell RE, Frost JK, et al: Amoebae resembling Entamoeba gingivalis in the genital tracts of IUD users. Acta Cytol 24: 413, 1980.
45. De Villiers EM, Wagner D, Schneider A, et al: Human papillomavirus infections in women with and without abnormal cervical cytology. Lancet 2: 703, 1987.
46. de Waard F, Schwarz F: Weight reduction and post-menopausal estrogenic effect. Acta Cytol 8: 449, 1964.
47. Dowling EA, Gravlee LC, Hutchins KE: A new technique for the detection of adenocarcinoma of the endometrium. Acta Cytol 13: 469,1969.
48. Drake M: Nomenclature of precancerous lesions of the uterine cervix. A position paper. Acta Cytol 28: 527, 1984.
49. Editorial Acta Cytologica: Future directions in cervical cytology. Acta Cytol 42: 1, 1998.
50. Factor SM: Papillary adenocarcinoma of the endometrium with psammoma bodies. Arch Pathol 98: 201, 1974.
51. Farhood AI, Abrams J: Immunohistochemistry of endometrial stromal sarcoma. Hum Pathol 22: 224, 1991.
52. Ferenczy A: Carcinoma and other malignant tumors of the cervix. In: Blaustein A (ed): Pathology of the Female Genital Tract, 2nd ed. New York: Springer-Verlag, 1982.
53. Ferreira CA: Vaginal cytology in abortion. Acta Cytol 3: 283, 1959.
54. Fidler HK, Boyes DA: Symposium on probable percentage of invasive carcinoma passing through the stage of carcinoma in situ. Acta Cytol 6: 188, 1962.
55. Fiorella RM, Cheng J, Kragel PJ: Papanicolaou smears in pregnancy positivity of exfoliated cells for human chorionic gonadotropin and human placental lactogen. Acta Cytol 37: 451, 1993.
56. Forney JP, Disaia PJ, Morpow CP: Endodermal sinus tumor: A report of two sustained remissions treated postoperatively with a combination of actinomycin D, 5-fluorouracil, and cyclophosphamide. Obstet Gynecol 45: 186, 1975.
57. Fox H, Kazzaz B, Langley FA: Argyrophil and argentaffin cells in the female genital tract and in ovarian mucinous cysts. J Pathol Bacteriol 88: 479, 1964.

58. Freeman C, Berg JW, Cutler SJ: Occurrence and prognosis of extranodal lymphomas. Cancer 29: 252, 1972.

59. Friedell GH, Hertig AT, Younge PA: The problem of early stromal invasion in carcinoma in situ of the uterine cervix. Arch Pathol 66: 494, 1958.

60. Frost JK: Gynecologic and obstetric cytopathology. In: Novak ER, Woodruff JD(eds): Novak's Gynecologic and Obstetric Pathology, 7th ed. pp 634-728. Philadelphia: WB Saunders, 1974.

61. Fujimoto I, Masubuchi S, Miwa H, et al: Psammoma bodies found in cervicovaginal and/or endometrial smears. Acta Cytol 26: 317, 1982.

62. Fujumoto I, Miwa H, Masubuchi S, et al: Psammoma bodies found in cervico-vaginal and/or endometrial cytology. Read in 7th International Congress of Cytology, Munich, 1980.

63. Galvin GA, Telinder RW: The present-day status of non-invasive cervical carcinoma. Am J Obstet Gynecol 57: 15, 1949.

64. Gaudefroy M: Diagnosis of pregnancy by means of cytology. Acta Cytol 3: 294, 1959.

65. Gilman SC, Docherty JJ, Clarke A, et al: Reaction patterns of herpes simplex virus type 1 and type 2 proteins with sera of patients with uterine cervical carcinoma and matched controls. Cancer Res 40: 4640, 1980.

66. Gitsch G, Friedlander ML, Wain GV, et al: Uterine papillary serous carcinoma: A clinical study. Cancer 75: 2239-2243, 1995.

67. Gloor E, Hurlimann J: Cervical intraepithelial glandular neoplasia (adenocarcinoma in situ and glandular dysplasia). Cancer 58: 1272, 1986.

68. Glücksmann A: Relationships between hormonal changes in pregnancy and the development of "mixed carcinoma" of the cervix. Cancer 10: 831, 1957.

69. Glücksmann A, Cherry CP: Incidence, histology and response to radiation of mixed carcinomas (adenoacanthomas) of the uterine cervix. Cancer 9: 971, 1956.

70. Gompel C: Is there a physiological cell type which may be defined as "androgenic cell type"? Acta Cytol 1: 83, 1957.

71. Gompel C, Silverberg SG: Pathology in Gynecology and Obstetrics 4th ed. pp258-271. Philadelphia: JB Lippincott.

72. Graham RM: The Cytologic Diagnosis of Cancer, 3rd ed. Philadelphia: WB Saunders, 1972.

73. Gray LA, Christopherson WM, Hoover RN: Estrogens and endometrial carcinoma. Obstet Gynecol 49: 385, 1977.

74. Gupta PK: Microbiology, inflammation and viral infections. In: Bibbo M (ed): Comprehensive Cytopathology. pp115-152. Philadelphia: WB Saunders, 1991.

75. Gupta TD, D'Urso J: Melanoma of female genitalia. Surg Gynecol Obstet 119: 1074, 1964.

76. Guyton AC: Basic Human Physiology: Normal Function and Mechanisms of Disease, 2nd ed. pp852-867. Philadelphia: WB Saunders, 1977.

77. Hammond DO: Vaginal cytology at the end of pregnancy. Acta Cytol 10: 230, 1966.

78. Hamperl H: Definition and classification of the so-called carcinoma is situ. In: Cancer of the Cervix. Ciba Foundation Study Group, London: J & A Churchill, 1959.

79. Hamperl H: Über das infiltrierende (invasive) Tumorwachstum. Virchows Arch Path Anat 340: 185, 1966.

80. Hamperl H, Kaufmann C, Ober KG: Histologische Untersuchungen an der Cervix schwangerer Frauen; die Erosion und das Carcinoma in situ. Arch Gynäk 184: 181, 1954.

81. Hamud K, Morgan DA: Papillary adenocarcinoma of endometrium with psammoma bodies. Cancer, 29: 1326, 1972.

82. Heller C, Hogt V: Squamous cell changes associated with the presence of candida sp. in cervical-vaginal Papanicolaou smears. Acta Cytol 15: 379, 1971.

83. Henderson PH Jr, Buck CE: Cervical leukoplakia. Am J Obstet Gynecol 82: 887, 1961.

84. Herbst AL, Ulfeder H, Poskaer DC: Adenocarcinoma of the vagina. Association of maternal stilbestrol with tumor appearance in going woman. N Engl J Med 284: 878, 1971.

85. Hills E, Laverty CR: Electron microscopic detection of papilloma virus in selected koilocytotic cells in a routine cervical smear. Acta Cytol 23: 53, 1979.

86. Hinselmann H: Die Diagnose des Uterus carcinoms. Klin Wschr 9: 1507, 1930.

87. Hoffman MS, Gordy LW, Cavanagh D: Use of the cytobrush for cervical sampling after cryotherapy. Acta Cytol 35: 79, 1991.

88. Hopman BC: Histochemistry of carcinoma in situ. Acta Cytol 5: 361, 1961.

89. Hopman BC: The Cytologic Diagnosis of Pregnancy. Miami: Miami Post Publishing, 1963.

90. Hopman BC, Wargo JD, Werch SC: Cytology of vernix caseosa cells. Obstet Gynecol 10: 656, 1957.

91. Hrushovetz SB, Lauchlan SC: Comparative DNA content of cells in the intermediate and parabasal layers of cervical intraepithelial neoplasia studied by two-wavelength Feulgen cytophotometry. Acta Cytol 14: 68, 1970.

92. Iwasaka T, Yokoyama M, Hayashi Y, et al: Combined herpes simplex virus 2 and human papillomavirus type 16 or 18 deoxyribonucleic acid leads to oncogenic transformation. Am J Obstet Gynecol 159: 1251, 1988.

93. Janovski NA, Parmandhan TL: Ovarian Tumors. Stuttgart: Georg Thieme, 1973.

94. Johnson LD, Earsterday CL, Gore H, et al: The histogenesis of carcinoma in situ of the uterine cervix: A preliminary report of the origin of carcinoma in situ in subcylindrical cell anaplasia. Cancer 17: 213, 1964.

95. Jones HW, Droegemueller W, Makowski EL: A primary melanocarcinoma of the cervix. Am J Obstet Gynecol 111: 959, 1971.

96. Jordan SW, Evangel E, Smith NL: Ethnic distribution of cytologically diagnosed herpes simplex genital infections in a cervical cancer screening program. Acta Cytol 16: 363, 1972.

97. Kaku T, Enjoji M: Extremely well-differentiated adenocarcinoma (adenoma malignum) of the cervix. Int J Gynecol Pathol 2: 28, 1983.

98. Kashimura M, Kashimura Y, Oikawa K: Placental site tropho-lastic tumor: Immunohistochemical and nuclear DNA study. Gynecol Oncol 38: 262, 1990.

99. Kasper TA, Smith ESO, Cooper P, et al: An analysis of the prevalence and incidence of gynecologic cancer cytologically detected in a population of 175,767 women. Acta Cytol 14: 261, 1970.

100. Kempson RL, Bari W: Uterine sarcomas: Classification, diagnosis and prognosis. Human Pathol 1: 373, 1970.

101. Kern SB: Prevalence of psammoma bodies in Papanicolaou-stained cervicovaginal smears. Acta Cytol 35: 81, 1991.

102. Kitay DZ, Wentz WB: Cervical cytology in folic acid

deficiency of pregnancy. Am J Obstet Gynec 104: 931, 1969.

103. Klaus H: Quantitative criteria of folate deficiency in cervico-vaginal cytograms with a report of a new parameter. Acta Cytol 15: 50, 1971.

104. Kobayashi TK, Okamoto H: Arias-Stella changes in cervicovaginal specimens. Cytopathology 8: 289, 1997.

105. Koller A, Artner J: Die Zytologie der normalen Schwangerschaft. Gynaecologia 136: 137, 1953.

106. Komorowski RA, Clowry LJ Jr: Koilocytotic atypia of the cervix. Obstet Gynecol 47: 540, 1976.

107. Korte W: Discussion in symposium on probable or possible premalignant cervical lesions. Acta Cytol 5: 280, 1961.

108. Koshino M: Detection of herpes virus infection of the female genitalia by cytosmear and its incidence. J Jpn Soc Clin Cytol 16: 77, 1977.

109. Koss LG: Diagnostic Cytology and Its Histopathologic Bases. Philadelphia: JB Lippincott, 1968.

110. Koss LG: Diagnostic Cytology, 3rd ed. p529. Philadelphia: JB Lippincott, 1979.

111. Koss LG: The role of the vaginal pool smear in the diagnosis of endometrial carcinoma. In: Wied GL, Koss LG, Reagan JW(eds): Compendium on Diagnostic Cytology, 4th ed. pp176-177. Chicago: Tutorials of Cytology, 1979.

112. Koss LG: Cytologic and histologic manifestations of human papillomavirus infection of the female genital tract and their clinical significance. Cancer 60: 1942, 1987.

113. Koss LG, Durfee GR: Unusual patterns of squamous epithelium of uterine cervix: Cytologic and pathologic study of koilocytotic atypia. Ann NY Acad Sci 63: 1245, 1956.

114. Koss LG, Schreiber K, Oberlander SG, et al: Detection of endometrial carcinoma and hyperplasia in asymptomatic women. Obstet Gynecol 64: 1-11, 1984

115. Koss LG, Stewart FW, Foote FW, et al: Some histological aspects of behavior of epidermoid carcinoma in situ and related lesions of the uterine cervix. Cancer 16: 1160, 1963.

116. Kottmeier HL: Carcinoma of the Female Genitalia. Baltimore: Williams and Wilkins, 1953.

117. Krumins I, Young Q, Pacy F, et al: The cytologic diagnosis of adenocarcinoma in situ of the cervix uteri. Acta Cytol 21: 320, 1977.

118. Kudo R: Cervical adenocarcinoma. In: Sasano N (ed): Current Topics in Pathology. pp81-111. Berlin: Springer-Verlag, 1992.

119. Kurman RJ, Norris HJ: Endometrial neoplasia: Hyperplasia and carcinoma. In: Blaustein A (ed): Pathology of the Female Genital Tract. pp311-351. New York: Springer-Verlag, 1982.

120. Kurman RJ, Solomon D: The Bethesda System. New York: Springer-Verlag, 1994.

121. Kurman RJ, Young RH, Norris HJ, et al: Immunocytochemical localization of placental lactogen and chorionic gonadotropin in the normal placenta and chorionic tumors with emphasis on intermediate trophoblast and the placental site trophoblastic tumor. Int J Gynecol Pathol 3:101,1984.

122. Lange P: Clinical and histological studies on cervical carcinoma: Precancerosis, early metastases and tubular structures in the lymph nodes. Acta Pathol Microbiol Scand 50(Suppl 143): 1, 1960.

123. Lauchlan SC: Tubal (serous) carcinoma of the endometrium. Arch Pathol Lab Med 105: 615, 1981.

124. Laverty CR, Russell P, Hills E, et al: The significance of noncondylomatous wart virus infection of the transformation zone: A review with discussion of two illustrative cases. Acta Cytol 22: 195, 1978.

125. Lefedvre J, laperiere H, Rousseau H, et al: Comparison of three techniques for detection of Chlamydia trachomatis in endocervical specimens from asymptomatic women. J Clin Microbiol 26: 726, 1988.

126. Lichtfus C, Pundel JP, Gander R: Le frottis vaginal á la fin de la grossesse. Gynéc Obstét 57: 380, 1958.

127. Littman P, Clement PB, Henriksen B, et al: Glassy cell carcinoma of the cervix. Cancer 37: 2238, 1976.

128. Liu W, Barrow MJ, Spitler MF, et al: Normal exfoliation of endometrial cells in premenopausal woman. Acta Cytol 7: 211, 1963.

129. Lucas WE, Benirschke K, Lebherz TB. Verrucous carcinoma of the female genital tract. Am J Obstet Gynecol 119; 435-1974

130. Lurain JR, Gallup DG: Management of abnormal Papanicolaou smears in pregnancy. Obstet Gynecol 53: 484, 1979.

131. MacKay B, Osborne BM, Wharton JT: Small cell tumor of cervix with neuroepithelial features. Cancer 43: 1138, 1979.

132. Martzloff KH: Carcinoma of the cervix uteri. A pathologic and clinical study with particular reference to relative malignancy of the neoplastic process as indicated by the predominant type of cancer cell. Bull Johns Hopkins Hosp 34: 141, 1923.

133. Masubuchi S Jr, Nagai I, Hirata M, et al: Cytologic studies of malignant melanoma of the vagina. Acta Cytol 19: 527, 1975.

134. Matsuyama T, Tsukamoto N, Kaku T, et al: Primary malignant lymphoma of the uterine corpus and cervix report of a case with immunocytochemical analysis. Acta Cytol 33: 228-232, 1989.

135. McKay DG, Terjanian B, Poschyachinda D, et al: Clinical and pathologic significance of anaplasia (atypical hyperplasia) of the cervix uteri. Obstet Gynecol 13: 2, 1959.

136. McLennan MT, McLennan C: Significance of cervicovaginal cytology after radiation therapy for cervical carcinoma. Am J Obstet Gynecol 121: 96, 1975.

137. Meisels A: Vaginal smear in the menopause. Acta Cytol 4: 151, 1960.

138. Meisels A: Le diagnostic cyto-hormonal durant la grossesse. Laval Medical 34: 552, 1963.

139. Meisels A: The menopause: A cytohormonal study. Acta Cytol 10: 49, 1966.

140. Meisels A: The maturation value. Acta Cytol 11: 249, 1967.

141. Meisels A: Cytologic diagnosis of human papillomavirus. Influence of age and pregnancy stage. Acta Cytol 36: 480, 1992.

142. Meisels A, Dubreuil-Charrois M: Hormonal cytology during pregnancy. Acta Cytol 10: 376, 1966.

143. Meisels A, Fortin R: Condylomatous lesions of the cervix and vagina: I. Cytologic patterns. Acta Cytol 20: 505, 1976.

144. Meisels A, Fortin R, Roy M: Condylomatous lesions of the cervix and vagina: II. Cytologic, colposcopic and histopathologic study. Acta Cytol 21: 379, 1977.

145. Melin JR, Wanner L, Schulz DM, et al: Primary Squamous cell carcinoma of the endometrium. Obstet Gynecol 53: 115, 1979.

146. Meyer IO, von Haam E: cited from von Haam, The Cytology of Pregnancy. Acta Cytol 5: 320, 1961.

147. Michalkiewicz W, Przybora LA, Simm S, et al: Recurrence and therapeutic problems in cervical dysplasia and in situ cancer. Cancer 16: 1212, 1963.

148. Mikuta JJ, Celebre JA: Adenocarcinoma of the cervix. Obstet Gynecol 33: 753, 1969.

149. Moll UM, Chalas E, Auguste M, et al: Uterine papillary serous carcinoma evolves via a p53-driven pathway. Hum Pathol 27: 1295, 1996.

150. Montalvo L: Vaginal cytology during delivery. Acta Cytol 9: 337, 1965.

151. Moss LD, Collins DN: Squamous and adenoid cystic basal cell carcinoma of the cervix uteri. Am J Obstet Gynecol 88: 86, 1964.

152. Mulvany NJ, Khan A, Ostor A: Arias-Stella reaction associated with pregnancy: Report of a case with a cytologic presentation. Acta Cytol 38: 218, 1994.

153. Naguib SM, Comstock GW, Davis HJ: Epidemiologic study of trichomoniasis in normal women. Obstet Gynecol 27: 607, 1966.

154. Nahmias AJ, Josey WE, Naib Z, et al: Antibodies to herpes virus hominis types 1 and 2 in humans. Am J Epidemiol 91: 547, 1970.

155. Nahmias AJ, Roizman B: Infection with herpes simplex viruses 1 and 2. N Engl J Med 289: 667, 1973; 289: 719, 1973.

156. Naib ZM: Single trophoblastic cells as a source of error in the interpretation of routine vaginal smear. Cancer 14: 1183, 1961.

157. Naib ZM, Nahmias AJ, Josey WE, et al: Genital herpetic infection: Association with cervical dysplasia and carcinoma. Cancer 23: 940, 1969.

158. Navab A, Koss LG, Ladue JS: Estrogen-like activity of digitalis. Its effect on the squamous epithelium of the female genital tract. JAMA 194: 30, 1965.

159. Nesbitt REL Jr, Garcia R, Rome DS: The prognostic value of vaginal cytology in pregnancy. Obstet Gynecol 17: 2, 1961.

160. Nesbitt REL Jr, Hellman LM: The histopathology and cytology of the cervix in pregnancy. Surg Gynecol Obstet 94: 10, 1952.

161. Nesbitt REL Jr, Stein AA: Histochemistry of carcinoma in situ. Acta Cytol 5: 365, 1961.

162. Ng ABP, Reagan JW: Cellular manifestations of extrauterine cancer. In: Wied GL, Koss LG, Reagan JW (eds): Compendium on Diagnostic Cytology, 4th ed. pp198-203. Chicago: Tutorials of Cytology, 1979.

163. Ng ABP, Reagan JW, Storaasli JP, et al: Mixed adenosquamous carcinoma of the endometrium. Am J Clin Pathol 59: 765, 1973.

164. Nieburgs HE: Diagnostic Cell Pathology in Tissue and Smears. New York: Grune & Stratton, 1967.

165. Nielsen SN, Podratz KC, Scheithauer BW, et al: Clinico-pathologic analysis of uterine malignant mixed müllerian tumors. Gynecol Oncol 34: 372, 1989.

166. Nigogosyan G, Dela Pava S, Pickren JW: Melanoblasts in vaginal mucosa. Cancer 17: 912, 1964.

167. Nogales-Fernandez F, Silverberg SG, Bloustein PA, et al: Yolk sac carcinoma (endodermal sinus tumor): Ultrastructure and histogenesis of gonadal and extragonadal tumors in comparison with normal human yolk sac. Cancer 39: 1462, 1977.

168. Norris HJ, Robinowitz M: Ovarian adenocarcinoma of mesonephric type. Cancer 28: 1074, 1971.

169. Norris HJ, Taylor GB: Melanomas of the vagina. Am J Clin Pathol 46: 420, 1966.

170. Novak ER, Nalley WB: Uterine adenoacanthoma. Obstet Gynecol 9: 396, 1957.

171. Okagaki T, Leach V, Younge PA, et al: Diagnosis of anaplasia and carcinoma in situ by differential cell counts. Acta Cytol 6: 343, 1962.

172. Oota K, Tanaka M: On histogenesis of cervical cancers of the uterus. A histological study on in situ and early carcinomas. Gann 45: 567, 1956.

173. Oriel JD, Almeida JD: Demonstration of virus particles in human genital warts. Br J Vener Dis 46: 37, 1970.

174. Papanicolaou GN: Diagnosis of Uterine Cancer by the Vaginal Smear. New York: Commonwealth Foundation, 1943.

175. Parkin DM, Pisani P, Ferlay J: Estimates of the worldwide incidence of eighteen major cancers in 1985. Int J Cancer 54: 594, 1993.

176. Patten SF Jr: Morphologic subclassification of preinvasive cervical neoplasia. In: Wied GL, Koss LG, Reagan JW (eds): Compendium on Diagnostic Cytology, 4th ed. pp112-124. Chicago: Tutorials of Cytology, 1976.

177. Patten SF Jr: Diagnostic Cytopathology of the Uterine Cervix. Monographs in Clinical Cytology, Vol 3, 2nd revised edition. Basel: S. Karger, 1978.

178. Patten SF Jr: Postirradiation dysplasia of the uterine cervix: Cytopathology and clinical significance. In: Wied GL, Koss LG, Reagan JW (eds): Compendium on Diagnostic Cytology, 4th ed. pp268-274. Chicago: Tutorials of Cytology, 1979.

179. Patten SF Jr: Postradiation dysplasia of the uterine cervix: Cytopathology and clinical significance. In: Wied GL, et al (eds): Compendium on Diagnostic Cytology. pp83-87. Chicago: Tutorials of Cytology, 1988.

180. Patten SF Jr, Reagan JW, Obenauf M, et al: Postirradiation dysplasia of uterine cervix and vagina: An analytical study of the cells. Cancer 16: 173, 1963.

181. Patten SF Jr, Woodworth FE, Bonfiglio TA: Cytopathology of uterine sarcomas. In: Wied GL, Koss LG, Reagan JW (eds): Compendium on Diagnostic Cytology, 4th ed. pp204-209. Chicago: Tutorials of Cytology, 1979.

182. Petersen O: Precancerous changes of cervical epithelium in relation to manifest cervical carcinoma: Clinical and histological aspects. Acta Radiol Suppl 127: 1, 1955.

183. Petersen O: Spontaneous course of cervical precancerous conditions. Am J Obstet Gynecol 72: 1063, 1956.

184. Plummer G, Masterson JG: Herpes simplex virus and cancer of the cervix. Am J Obstet Gynecol 111: 81, 1971.

185. Pollanen R, Soini Y, Vahakangas K, et al: Aberrant p53 protein expression in cervical intra-epithelial neoplasia. Histopathology 23: 474, 1993.

186. Porreco R, Penn I, Droegemueller W, et al: Gynecologic malignancies in immunosuppressed organ homograft recipients. Obstet Gynecol 45: 359, 1975.

187. Pundel JP: Is there a physiological cell type which may be defined as "androgenic cell type" ? Acta Cytol 1: 82, 1957.

188. Pundel JP: Are degenerative cell changes in endometrial cells due to inadequate preparation technique of smears? Acta Cytol 2: 588, 1958.

189. Pundel JP: Incidence of cytolysis in vaginal smear during pregnancy. Acta Cytol 3: 219, 1959.

190. Pundel JP: Vaginal cytology as prognostic method in pregnancy disorders. Acta Cytol 3: 231, 1959.

191. Pundel JP: Diagnosis of pregnancy by means of cytol-

ogy. Acta Cytol 3: 295, 1959.

192. Pundel JP: Vaginal smears in the menopause. Acta Cytol 4: 152, 1960.

193. Pundel JP: Vaginal cytology at the end of pregnancy. Acta Cytol 10: 228, 1966.

194. Purola E, Savia E: Cytology of gynecologic condyloma acuminatum. Acta Cytol 21: 26, 1977.

195. Qizilbash AH: In-situ and microinvasive adenocarcinoma of the uterine cervix: A clinical, cytologic and histologic study of 14 cases. Am J Clin Pathol 64: 155, 1975.

196. Rad MJ, Jeannot A: Letters to the editors. Enterobius vermicularis in cervical smear. Acta Cytol 14: 466, 1970.

197. Rakoff AE: The vaginal cytology of gynecologic endocrinopathies. Acta Cytol 5: 153, 1961.

198. Rawls WE, Gardner HL, Kaufman RL: Antibodies to general herpesvirus in patients with carcinoma of the cervix. Am J Obstet Gynecol 107: 710, 1970.

199. Reagan JW, Fu YS: Histologic types and prognosis of cancers of the uterine cervix. Int J Radiat Oncol Biol Phys 5: 1015, 1979.

200. Reagan JW, Hicks DJ, Scott RB: Atypical hyperplasia of uterine cervix. Cancer 8: 42, 1955.

201. Renkonen OV, Widholm O, Vartianen E: Microbiological findings (Vaginal infection, I.). Acta Obstet Gynecol Scand 49 (Suppl 2): 3, 1970.

202. Richart RM: Colposcopic studies of cervical intraepithelial neoplasia. Cancer 19: 395, 1966.

203. Richart RM: Natural history of cervical intraepithelial neoplasia. Clin Obstet Gynecol 10: 748, 1967.

204. Richart RM, Barron BA: A follow-up study of patients with cervical dysplasia. Am J Obstet Gynecol 105: 386, 1969.

205. Roberts TH, Ng ABP: Chronic lymphocytic cervicitis: Cytologic and histopathologic manifestations. Acta Cytol 19: 235, 1975.

206. Rosenberg P, Blom R, Hogberg T, et al: Death rate and recurrence pattern among 841 clinical stage 1 endometrial cancer patients with special reference to uterine papillary serous carcinoma. Gynecol Oncol 51: 311, 1993.

207. Sadeghi SB, Sadeghi A, Cosby M, et al: Human papillomavirus infection. Frequency and association with cervical neoplasia in a young population. Acta Cytol 33: 319, 1989.

208. Sagae S, Kuda R, Kuzumaki N, et al: Ras oncogene expression and progression in intraepithelial neoplasia of the uterine cervix. Cancer 66; 295, 1990.

209. Saigo PE, Cain JM, Kim WS, et al: Prognostic factors in adenocarcinoma of the uterine cervix. Cancer 57: 1584, 1986.

210. Sakamoto A, Sugano H: Mixed mesodermal tumor of the uterine body: Relationship between histology and survival. Gann 67: 263, 1976.

211. Sandritter W, Carl M, Ritter W: Cytophotometric measurements of the DNA content of human malignant tumors by means of the Feulgen reaction. Acta Cytol 10: 26, 1966.

212. Sandritter W, Fischer R: Über den DNS-Gehalt des normalen Plattenepithels des Carcinoma in situ und des invasiven Karzinoms der Porito. 1. Internationaler Kongress für exfoliative Zytologie, Gynecol, Austria, 1961.

213. Saphir O, Lacker JE: Adenocarcinoma with clear cells (hypernephroid) of the ovary. Surg Gynecol Obstet 79: 539, 1944.

214. Sardana S, Sodhani P, Agarwal SS, et al: Epidemiologic analysis of trichomonas infection in inflammatory smears. Acta Cytol 38: 693, 1994.

215. Schaberg A, Hildes JA, Wilt JC: Disseminated candidiasis. AMA Arch Intern Med 95: 112, 1955.

216. Schenk U, Herbert A, Solomon D, et al: Terminology IAC task force summary. Acta Cytol 42: 5, 1998.

217. Schleifstein J: Changes in the uterine cervix associated with pregnancy and epidermoid carcinoma in situ. NY J Med 50: 2795, 1950.

218. Scully RJ, Young RH: Trophoblastic pseudotumor. A reappraisal. Am J Surg Pathol 5:75, 1981.

219. Segadal E, Iverson OE: The Isaacs cell sampler: An alternative to curettage. Br Med J 281: 364, 1980.

220. Seltzer VL, Klein M, Beckman EM: The occurrence of squamous metaplasia as a precursor of squamous cell carcinoma of the endometrium. Obstet Gynecol 49: 34, 1977.

221. Seltzer V, Sall S, Castadot M-J, et al: Glassy cell cervical carcinoma. Gynecol Oncol 8: 141, 1979.

222. Sherman AI, Goldrath FM, Berlin FA, et al: Cervical-vaginal adenosis after in utero exposure to synthetic estrogens. Obstet Gynecol 44: 531, 1974.

223. Shiina Y: Cytomorphologic and immunocytochemical studies of Chlamydia infections in cervical smears. Acta Cytol 29: 683, 1985.

224. Shojania AM, Hornady GJ, Barnes PH: The effect of oral contraceptives on folate metabolism. Am J Obstet Gynecol 111: 782, 1971.

225. Shurbaji MS, Brooks SK, Thumond TS: Proliferating cell nuclear antigen immunoreactivity in cervical intraepithelial neoplasia and benign cervical epithelium. Am J Clin Pathol 100: 22, 1993.

226. Silverberg SG, Kurman RJ: Atlas of Tumor Pathology: Tumors of the Uterine Corpus and Gestational Trophoblastic disease. Washington, DC: Armed Forces Institute of Pathology, 1992.

227. Singer A: The uterine cervix from adolescence to the menopause. Br J Obstet Gynaecol 82: 81, 1975.

228. Skinner IC, McDonald JR: Mixed adenocarcinoma and squamous cell cancer of the uterus. Am J Obstet Gynecol 40: 258, 1940.

229. Smith DC, Prentice R, Thompson DJ, et al: Association of exogenous estrogen and endometrial carcinoma. N Engl J Med 293: 1164, 1975.

230. Smith MR, Figge DC, Bennington JL: The diagnosis of cervical cancer during pregnancy. Obstet Gynecol 31: 193, 1968.

231. Smolka H, Soost H-J: Grundriss und Atlas der Gynäkologischen Zytodiagnostik. Stuttgart: Georg Thieme, 1965.

232. Snyder LM, Necheles TF: Malabsorption of folate polyglutamates associated with contraceptive therapy. Clin Res 17: 602, 1969.

233. Soost H-J, Baur S: Zellveränderungen durch ionisierende Strahlen und Zytostatika. In: Gynäkologische Zytodiagnostik. pp 265-275. Stuttgart: Georg Thieme, 1980.

234. Soszka S, Wisniewski L: Cytologic evaluation of fetal death and an attempt to determine the time of its occurrence. Acta Cytol 11: 403, 1967.

235. Spriggs AI, Boddington MM, Clarke CM: Carcinoma-insitu of the cervix uteri. Some cytogenetic observations. Lancet 1: 1383, 1962.

236. Stafl A, Mattingly RF: Vaginal adenosis: A precancerous lesion ? Am J Obstet Gynecol 120: 666, 1974.

237. Stamm O, Rawyler V, Riotton G: Vaginal cytology in abortion. Acta Cytol 3: 283, 1959.

238. Steiner G, Friedell GH: Adenosquamous carcinoma in situ of the cervix. Cancer 18: 807, 1965.

239. Stevens LC: The biology of teratomas including evidence indicating their origin from primordial germ cells. Ann Biol 1: 585, 1962.

240. Stone D F, Sedlis A, Stone ML, et al: Estrogen-like effects in the vaginal smears of postmenopausal women. Acta Cytol 11: 349, 1967.

241. Streiff R: Folate deficiency and oral contraceptives. JAMA 214: 105, 1970.

242. Sugimori H, Taki I: Radiosensitivity test for cervical cancer. Acta Cytol 16: 331, 1972.

243. Taft PD, Robboy SJ, Herbst AL, et al: Cytology of clear-cell adenocarcinoma of genital tract in young females: Review of 95 cases from registry. Acta Cytol 18: 279, 1974.

244. Takahashi M: In: Takahashi M(ed): Early Diagnosis of Uterine Cancer. Tokyo: Igaku-Shoin, 1968.

245. Takahashi M, Aoki K, Noda S, et al: Some effect of long-term use of oral contraceptives on cervical neoplasia. In: Kurihara S, et al (eds): Cervical Pathology and Colposcopy. pp21-27. Amsterdam: Elsevier Science, 1985.

246. Taki I: Cytology of hydatidiform mole, destructive mole and choriocarcinoma. In: Wied GL, Koss LG, Reagan JW (eds): Compendium on Diagnostic Cytology, 4th ed. pp210-214. Chicago: Tutorials of Cytology, 1979.

247. Talerman A, Haije WG: Alpha-fetoprotein and germ cell tumors: A possible role of yolk sac tumors in production of alpha-fetoprotein. Cancer 34: 1722, 1974.

248. Tasker JT, Collins JA: Adenocarcinoma of the uterine cervix. Am J Obstet Gynecol 118: 344, 1974.

249. Tateishi R, Wada A, Hayakawa K, et al: Argyrophil cell carcinoma (apudomas) of the uterine cervix. Virchows Arch Pathol Anat Histol 366: 257, 1975.

250. Teilum G: Classification of endodermal sinus tumor (mesoblastoma vitellinum) and so-called "embryonal carcinoma" of the ovary. Acta Pathol Microbiol Scand 64: 407, 1965.

251. Teshima S, Shimosato Y, Kishi K, et al: Early stage adenocarcinoma of the uterine cervix. Histopathologic analysis with consideration of histogenesis. Cancer 56: 167, 1985.

252. Thin RNT, Atia W, Parker JDJ, et al: Value of Papanicolaou-stained smears in the diagnosis of trichomoniasis, candidiasis, and cervical herpes simplex virus infection in women. Br J Vener Dis 51: 116, 1975.

253. Tobon H, Watkins GJ: Secretory adenocarcinoma of the endometrium. Int J Gynecol Pathol 4: 328, 1985.

254. Tsukada Y, Hewett WJ, Barlow JJ, et al: Clear cell adenocarcinoma (mesonephroma) of the vagina: Three cases associated synthetic nonsteroid estrogen therapy. Cancer 29: 1208, 1972.

255. Ulfelder H: Stilbestrol, adenosis, and adenocarcinoma. Am J Obstet Gynecol 117: 794, 1973.

256. Ulmer HU, Hossfeld DK, Love RR: Cancers of the uterine cervix and endometrium and gestational trophoblastic disease. In: Manual of Clinical Oncology. pp369-386. Geneve: UICC, 1994.

257. van Nagell JR Jr, Donaldson ES, Wood EG, et al: Small cell cancer of the uterine cervix. Cancer, 40: 2243, 1977.

258. van Niekerk WA: Cervical cytological abnormalities caused by folic acid deficiency. Acta Cytol 10: 67, 1966.

259. von Haam E: The cytology of pregnancy. Acta Cytol 5: 320, 1961.

260. von Kother L, Sandritter W: Über den DNS-Gehalt des Carcinoma in situ. Gynaecologia 157: 9, 1964.

261. Vooijs GP: Benign proliferative reactions, intraepithelial neoplasm and invasive cancer of the uterine cervix. In: Bibbo M (ed): Comprehensive Cytopathology. pp168-172. Philadelphia: WB Saunders, 1991.

262. Vooijs PQ, Ng ABP, Wentz WB: The detecting of vaginal adenosis and clear cell carcinoma. Acta Cytol 17: 59, 1973.

263. Vuong PN, Houissa-Vuong S, Bleuse B, et al: Pseudotumoral tuberculosis of the uterine cervix: Cytologic presentation. Acta Cytol 33: 308, 1989.

264. Wachtel E: In: Symposium on hormonal cytology. Acta Cytol 12: 101, 1968.

265. Wagner D, Ikenberg H, Boehm H, et al: Identification of human papillomavirus in cervical swabs by deoxyribonucleic acid in situ hybridization. Obstet Gynecol 64: 767, 1984.

266. Wakonig-Vaartaja R, Kirkland JA: A correlated chromosomal and histopathologic study of "pre-invasive" lesions of cervix. Cancer 18: 1101, 1965.

267. Walloch JL, Young Hong H, Bibb LM: Effects of therapy on cytologic specimens. In: Bibbo M (ed): Comprehensive Cytopathology. Philadelphia: WB Saunders, 1991.

268. Waltz W: Symposium on premalignant cervical lesions. Acta Cytol 5: 120, 1961.

269. Weisbrot IM, Stabinsky C, Davis AM: Adenocarcinoma in situ of the uterine cervix. Cancer 29: 1179, 1972.

270. Wentz WB, Reagan JW: Clinical significance of postirradiation dysplasia of the uterine cervix. Am J Obstet Gynecol 106: 812, 1970.

271. Wespi H: Early carcinoma of the uterine cervix. (Transl. by Schiller M). New York: Grune & Stratton, 1949.

272. Wheeler JD, Hertig AT: The pathologic anatomy of the uterus. I. Squamous carcinoma of the cervix. Am J Clin Pathol 25: 345, 1955.

273. Whitehead N, Reyner F, Lindenbaum J: Megaloblastic changes in the cervical epithelium: Association with oral contraceptive therapy and reversal with folic acid. JAMA 226: 1421, 1973.

274. Wied GL: Der zytologische Ausstrichtyp der Patientin mit klimakterschen Ausfallsbeschwerden. Zbl Gynäk 75: 1578, 1953.

275. Wied GL: Importance of the site from which vaginal cytologic smears are taken. Am J Clin Pathol 25: 742, 1955.

276. Wied GL: Definitions of "cytolysis" and "autolysis." Acta Cytol 2: 55, 1958.

277. Wied GL: Techniques for collection and preparation of cytologic specimens from the female reproductive tract. In: Wied GL, Koss LG, Reagan JW(eds): Compendium on Diagnostic Cytology, 4th ed. pp9-16. Chicago: Tutorials of Cytology, 1979.

278. Wied GL, Bibbo M: Evaluation of the endocrinologic condition of the patient by means of vaginal cytology. In: Wied GL, Koss LG, Reagan JW(eds): Compendium on Diagnostic Cytology, 3rd ed. Chicago: Tutorials of Cytology, 1975.

279. Wied GL, Christiansen W: Die Cytolyse von Epithelien des Vaginalsekretes. Geburtsh Frauenheilk 13: 986, 1953.

280. Wied GL, Del Sol JR, Dargan AM: Progestational and androgenic substances tested on the highly proliferated vaginal epithelium of surgical castrates. I. Progestational substances. Am J Obstet Gynecol 75: 98, 1958.

281. Wilbanks GD, Campbell JA, Kaufmann LA: Cellular changes of normal human cervical epithelium infected in vitro with herpesvirus hominis, type two (herpes

simplex). Acta Cytol 14: 538, 1970.

282. Wolinska W, Melamed M: Herpes genitalis in women attending planned parenthood of New York City. Acta Cytol 14: 239, 1970.

283. Younge PA: The conservative treatment of carcinoma in situ of the cervix. Proc Second Natl Cancer Conf 1: 668, 1952.

284. Ziel HK, Finkle WD: Increased risk of endometrial carcinoma among users of conjugated estrogens. N Engl J Med 293: 1167, 1975.

285. Zinser HK: Tissue culture and carcinoma in situ. Acta Cytol 6: 159, 1962.

286. zur Hausen H: Human genital cancer: Synergism between a virus infection and initiating events? Lancet 2: 1370, 1982.

7 Breast

Masayoshi Takahashi and Masahiko Fujii

1 Collection of Specimens and Preparation of Smears

Of all cancers affecting women, breast carcinoma is one of the most important targets of cancer mass screening because the mortality trends continue to increase since the 1960s (Fig. 7-1). Although the incidence of breast carcinoma is much lower in Asian countries than in Northwestern Europe and North America, the fact that its incidence of Japanese American living in the Pacific coast is about two times higher than that of the Japanese may point out the importance of dietary, environmental and endocrinological factors than the race.

■Nipple discharge

Cytology with nipple discharge for the detection of breast carcinoma is performed with ease but has a disadvantage that is a limited number of patients produce breast secretion. Generally speaking, no nipple discharge is normally recognizable and the incidence of bloody discharge in cases of cancer of the breast is said to be as few as 2% to 5% (Table 7-1). According to the description by Ringrose,[49] however, breast secretion can be obtained from as high as 27% of nulligravid premenopausal females and 40% of premenopausal multiparous females by gentle bilateral compression of the breast. Breast secretion is obtained occasionally in the presence of acute infection, duct papillomatosis, ductal cystic dilatation and malignancies. At any rate, macro-

scopic examination of breast secretion is important and its appearance, whether bloody, purulent or serous, should be described because bloody discharge shows considerably higher incidence of cancer, as frequent as 14% to 52% (Table 7-2). In addition to cancer, duct papillomatosis (mastopathy) or fibrocystic disease are also frequent causes of bloody discharge (Table 7-3).

Although aspiration biopsy cytology (ABC) has come to be common at present and nipple discharge cytology is liable to be neglected, there is an interesting report by Takeda and his associates[57] that pointed out the importance of nipple discharge cytology; among 410 women with bloody or serous discharge in the absence of palpable lesion cancer was detected in 50 (12.2%).

Procedure of massage: The patient should be positioned lying down on the examining table. The mammary gland is supported by one hand and the base of the nipple is tightly grasped between the thumb and the index finger of another hand. As the nipple is rolled by these fingers from the base towards the tip of the nipple, the acinar secretion stored in the ampulla will be squeezed out. A clean glass slide is placed on the nipple and the secretion is thinly spread. The smear must be promptly fixed in 95% ethanol or with coating fixative.

■Needle aspiration

Since nipple discharge is obtainable only in limited cases of benign as well as malignant diseases, the procedure using a fine needle aspiration measuring 0.7-0.9 mm in outer diameter is a very reliable and harmless

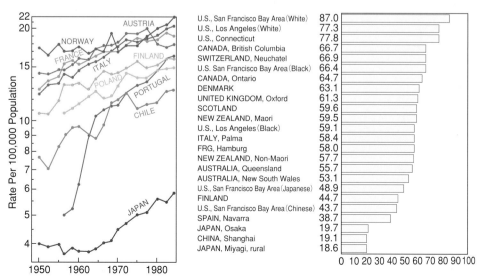

Fig. 7-1 Trends of age-adjusted mortality and rates of prevalence of breast carcinoma.
(By courtesy of Dr. Kunio Aoki, Nagoya, Japan.)

Table 7-1 Incidence of bloody discharge from the nipple in cases of cancer of the breast

Author	No. of cases of mammary cancer	Percent with bloody discharge
Geschickter (1945)	2,393	4(%)
Nohrman (1949)	1,042	5
Haagensen (1956)	546	2
Hultborn (1960)	517	2
Sunada (1966)	186	4
Fujii (1994)	273	7

Table 7-2 Incidence of cancer of the breast in cases of bloody discharge from the nipple

Author	No. of cases of bloody discharge	Percent with mammary cancer
Geschickter (1945)	287	36(%)
Nohrman (1949)	109	39
Kirgor (1953)	103	14
Sunada (1966)	32	22
Fujii (1994)	102	20

Table 7-3 Abnormal nipple discharge and its histologic bases

	Cancer	Duct papilloma	Fibrocystic disease	Fibro-adenoma	Mastitis	Total
Tumor palpable	26	11	15	10	2	64
Tumor not palpable	14	23	2	0	0	39
Total	40 (38.8%)	34 (33.0%)	17 (16.5%)	10 (9.7%)	2 (1.9%)	103

(From Fujii M: At Tokyo Metropolitan Cancer Detection Center, 1985-1989.)

technique to determine malignancy. It is not worthy of discussion to compare the accuracy of diagnosis between needle aspiration diagnosis and open frozen section diagnosis. The needle aspiration biopsy cytology (ABC), which is highly reliable, should be performed as the first step of examination in association with mammography and echography if some lump is palpated.[6,9,67] The skin over the site of needle aspiration is cleaned, sterilized and anesthetized. Then, the skin to be punctured is cut. A set of 18 gauge needles of different lengths and a 10 ml syringe are prepared. A syringe designed to be handled by a single hand for aspiration cytology is preferable (see Fig. 17-1).[20] The tumor must be fixed with another hand while the needle is inserted. Retraction of the piston is discontinued just before withdrawal of the needle. The aspirate is immediately smeared on a clean glass slide and fixed. The danger of tumor transplantation in ABC will be less frequent than in surgical biopsy, which may cut lymphatic or blood vessels. It is negligible from the statistical standpoint.[5] In order to avoid cell loss within the syringe, saline suspension and centrifugation or membrane filtration should be performed simultaneously.[21]

For non-palpable lesion that was pointed out by mammography in cancer screening, fine needle aspiration cytology is recommended with stereotactic localization[32].

■Imprint smears of biopsy specimens
Since the breast is an organ to which frozen section diagnosis is frequently applied, rapid cytodiagnosis by imprint smears is of great value as an adjunct to frozen section diagnosis. The smear is imprinted by pressing the fresh cut surface onto a clean glass slide. It must be fixed immediately to prevent drying. A rapid Diff-Quik stain is preferable to Papanicolaou stain. Although cellular features are similar to those in needle aspirations, epithelial cells tend to display sheet-like flat clusters.

Imprint smear cytology is worthy of application as an adjunct to frozen section diagnoses, because fine cellular features are well preserved compared with those in frozen sections of rash preparation. Imprint smear is indispensable for intraoperative real-time telediagnosis (refer to page 464).

2 Histology of the Mammary Gland

The female breast is composed of branched tubuloalveolar glands aggregated into 15 to 20 lobules which are surrounded by fibroadipose tissue. The secretion is discharged through the alveolar ducts and larger excretory ducts called ampullae (sinus lactiferus) to the orifice at the nipple. The alveolar ducts are lined by low columnar cells and the lactiferous sinuses at the nipple are lined with stratified squamous cells. The glands showing functional activity contain fat in the lumen and in the secreting epithelial cells (Fig. 7-2). During gestation, especially in the later stage, the proliferated epithelial cells begin to secrete colostrum, which is rich in lactoprotein (Figs. 7-3 and 7-4) Bargmann and Knoop[4] have observed electron microscopically the fact that large lipid droplets are cast off from a cell with enclosure by a cell membrane and thus secreted with a mechanism of apocrine secretion, while proteinaceous material is concentrated in the Golgi apparatus, transported to the cell surface as granules, then discharged as merocrine secretion. Even the resting glands show cyclic changes of functional activity during reproductive age. Early in the cycle, the glands are composed of solid trabeculae of epithelial cells without luminae. Late in the cycle, the glands are lined with low columnar epithelial cells forming evident luminae.

137

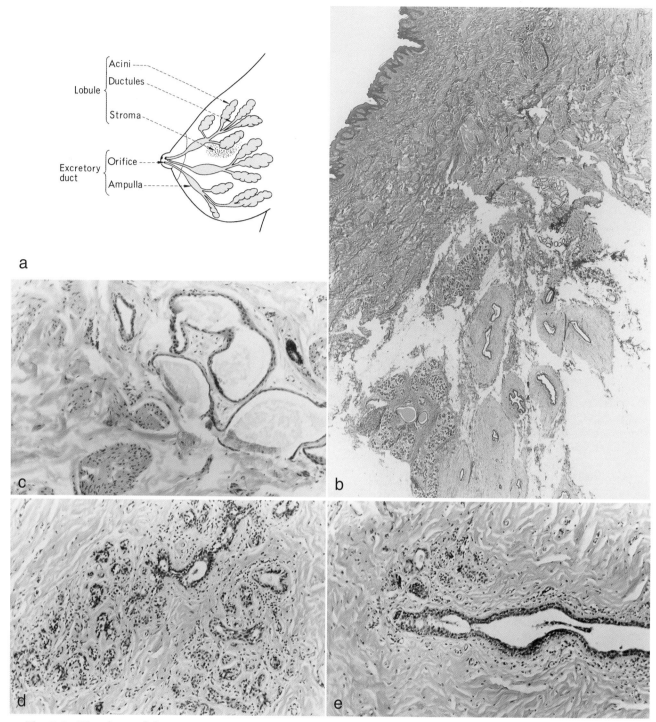

Fig. 7-2 Histology of the mammary gland.
a. Diagram of the mammary gland.
b. Low power magnification of the gland composed of lobules surrounded by fibroadipose tissue and lactiferous ducts.
c. Dilated ampulla containing thinly staining excreta.
d. Mammary lobules composed of resting alveoli and excretory ductules.
e. Larger lactiferous duct lined by a few layers of low columnar epithelial cells.

3 Benign Cellular Components in Breast Secretions and Aspirates

■ Foamy cells

Although the entity of foamy cells either of histiocytic or duct epithelial origin is not fully clarified, they are morphologically consistent with histiocytic cells and inevitably present in nipple discharge. They are frequently positive for fat stain and periodic acid-Schiff (PAS) reaction. Intense positivity of foamy cells for PAS reaction differs from ordinary histiocytes in other organs and supports the concept of secretory cell origin (Figs. 7-7 and 7-8).

Cytoplasm: The cytoplasm is abundant, finely vacuolated and foamy in appearance.

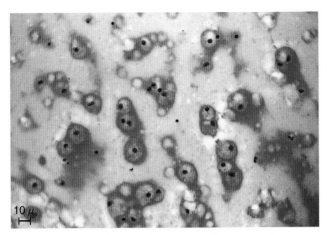

Fig. 7-3 Lipid vacuole-forming cells in a colostrum. Small nuclei are central or paracentral in position in ample vacuolated cytoplasm. Lipoproteinous material is also characteristic.

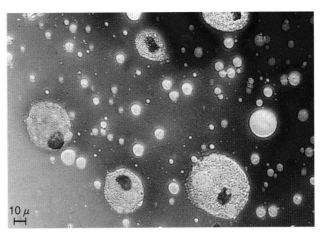

Fig. 7-4 Differential interference microscopy of colostrum. Note several foamy cells and many lipid droplets.

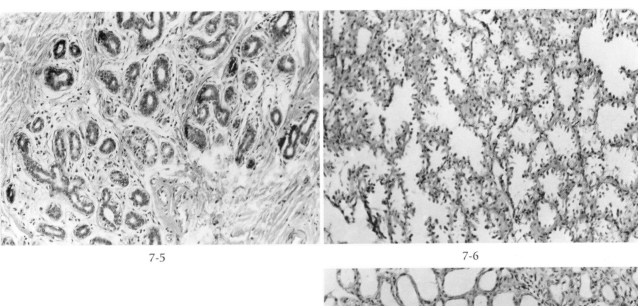

7-5

7-6

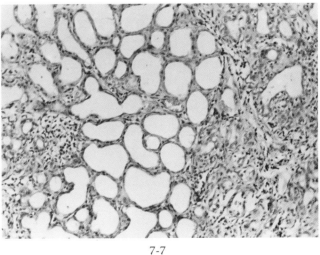

7-7

Fig. 7-5 Resting mammary gland. The lobules separated by dense connective tissue possess elongated ductules and alveoli.

Fig. 7-6 Lactating breast with active mammary gland. The alveoli are increased in number and distended. Columnar epithelial cells project into the lumen as dome-shaped protrusions.

Fig. 7-7 Galactocele of lactating mammary gland. The dilated alveoli are lined by flattened epithelium. Note a slight inflammatory reaction in the stroma.

Nucleus: The nucleus is small, oval or kidney-shaped and peripherally located. The chromatin is finely granular. In addition to large foamy cells, small histiocytic cells showing phagocytosis of blood pigments or cell debris are also seen.

■Duct epithelial cells

They appear in clusters and occasionally singly. The clusters of cells take a trabecular form and have varying lengths and sizes. The cells are low columnar or cuboidal. The nuclei are usually oval, fairly small and equal in size, but appear to be compressed in compact

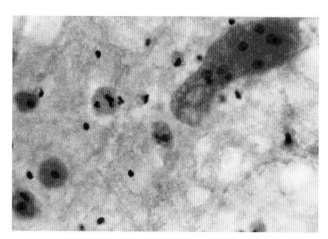

Fig. 7-8　PAS-positive foamy cells in a bloody discharge from a case of mastopathia. The strong PAS-stainability differs from the behavior of ordinary histiocytes.

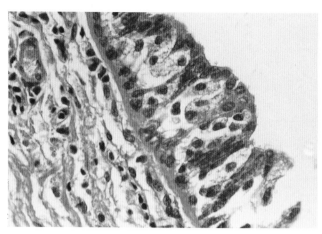

Fig. 7-9　Foamy appearance of epithelial cells of the mammary duct. Clear foamy cytoplasm is formed in the duct epithelium.

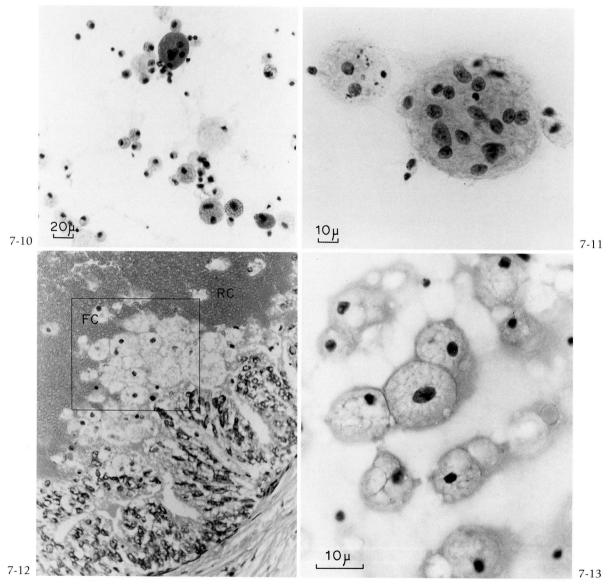

Fig. 7-10　Foamy cells with multinucleation and small histiocytes.
Fig. 7-11　Multinucleated foamy cell. Many small nuclei are scattered in the abundant cytoplasm with a foamy appearance.
Fig. 7-12　Histology from a case with bleeding nipple discharge. Note a dilated duct showing papillary epithelial proliferation. The lumen is filled with red cells (RC) and foamy cells (FC).
Fig. 7-13　Foamy cells in bloody discharge. The nuclei are small, ovoid and eccentric in position. The cytoplasm is abundant and finely vacuolated.

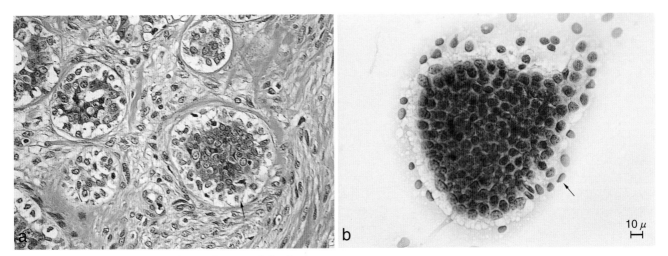

Fig. 7-14 Histology and cytology of acini of the breast.
a. Histology showing acinar epithelial cells and clear myoepithelial cells in outer side (arrow).
b. Imprint smear showing a cluster of uniform epithelial cells and loosely scattered myoepithelial cells. The cytoplasm of myoepithelial cells is clear and indistinctly outlined (arrows).

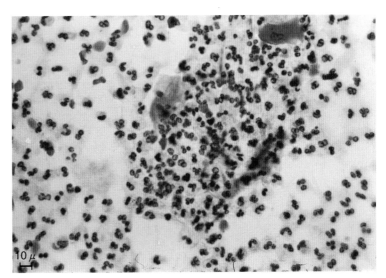

Fig. 7-15 Purulent nipple discharge in acute lactation mastitis. There is a large number of neutrophils mixed with erythrocytes and squamous epithelial cells derived from the nipple.

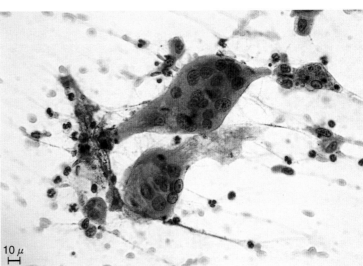

Fig. 7-16 Multinucleated lining epithelial cells, macrophages and leukocytes in a nipple discharge from a case with mammary duct ectasia. Imprinted smear of a resected lump proved many plasma cells and foamy cells.

clusters. Exfoliation of the cells with such clusters is the result of benign papillary proliferation of duct epithelial cells. Apocrine metaplasia of duct epithelial cells is characterized by an abundant eosinophilic cytoplasm with fine granules and small rather pyknotic nuclei.

Occurrence of apocrine metaplastic cells and duct epithelial cells in a papillary frond is per se pathological and suggestive of fibrocystic disease or mastopathia (Figs. 7-17 and 7-18).

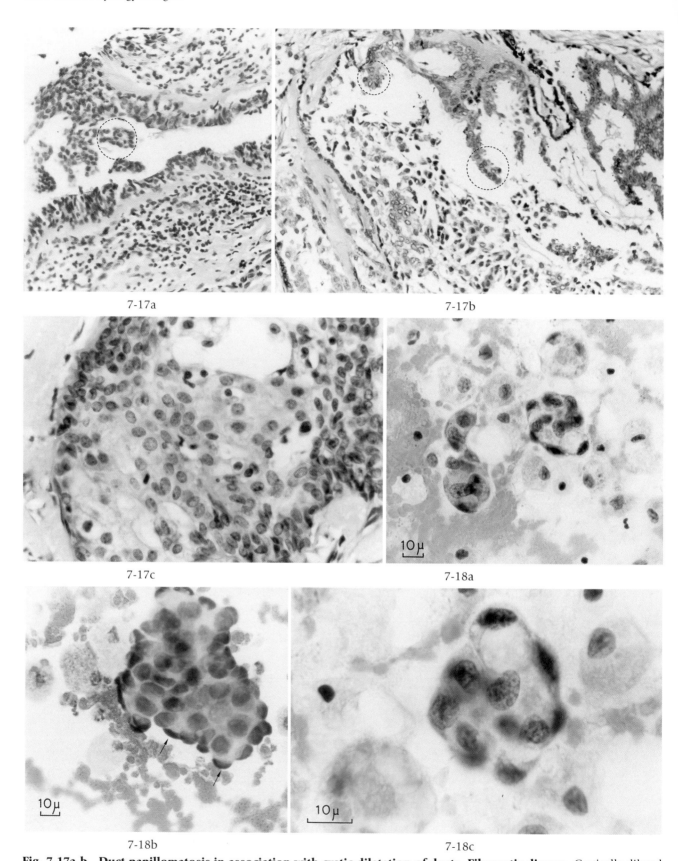

7-17a

7-17b

7-17c

7-18a

7-18b

7-18c

Fig. 7-17a,b   Duct papillomatosis in association with cystic dilatation of ducts. Fibrocytic disease. Cystically dilated ducts reveal papillary proliferation of the lining epithelial cells. These appear to be easily detachable (encircled).

Fig. 7-17c   Higher magnification of duct papillomatosis. A duct is almost filled with proliferated epithelial cells. The nuclei in solid sheets of cells are equal in size and shape.

Fig. 7-18a-c   Benign epithelial cells arranged in acinar clusters or in large sheets from smears of bloody discharge. The smears are from the same case as Fig. 7-17. The nuclei forming cell clusters are variable in number; therefore a large cell clump should not be mistaken for malignancy (b). These cell clusters are distinctly outlined and the nuclei present at periphery are compressed (nuclear capping shown with arrow). Note cytoplasmic vacuolation which is occasionally found in fibrocystic disease.

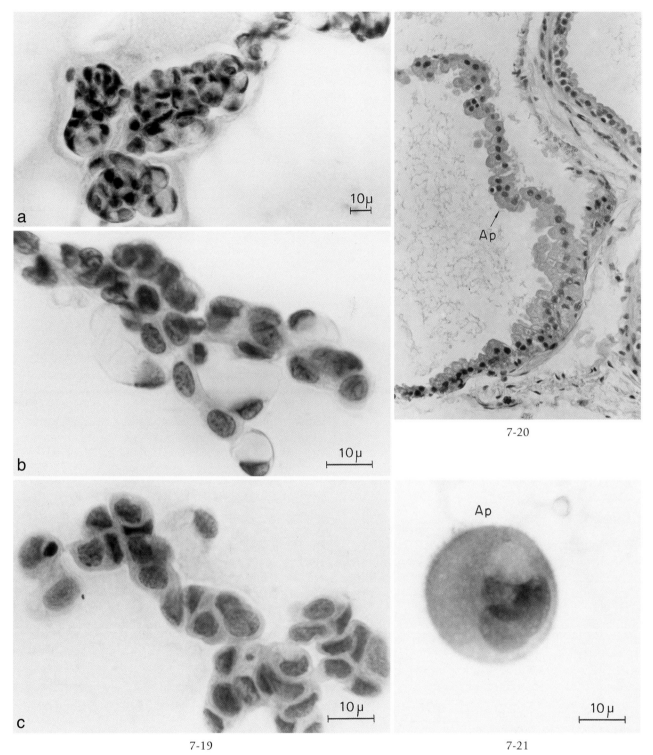

7-19

7-21

Fig. 7-19a-c Benign epithelial cells seen in large clusters. Exfoliation of large clusters of cells arranged in long columns is the result of intraductal papillomatosis (fibrocystic disease). Although the peripherally locating cells show compressed nuclei with or without cytoplasmic vacuolation ("nuclear capping"), the nuclei are generally equal in size and show finely granular distribution of chromatins.

Fig. 7-20 Cystic dilatation of ducts with apocrine metaplasia. Cystically dilated ducts are lined with apocrine cells having small pyknotic nuclei and intensely eosinophilic, thick cytoplasm.

Fig. 7-21 Apocrine cell in nipple discharge. The cytoplasm is thick, strikingly eosinophilic, opaque and distinctly outlined. The nuclei, probably binucleated, are small and peripherally located. Ap: apocrine cell.

■Myoepithelial cells

The myoepithelium locates between the basal lamina and the basal pole of the secretory epithelium. Myoepithelial cells appear to embrace the glandular cells as octopus and participate actively in secretion of the glands (Fig.7-14).

Cytologically they have centrally locating small nuclei and star-shaped, ill-defined cytoplasm. From a cyto-

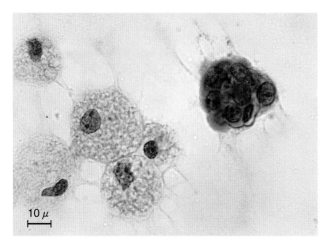

Fig. 7-22　Hyperplastic epithelial cells shed in a clump in nipple discharge. The nuclei are flattened in a free cell ball and show a bland chromatin pattern, although nuclear rims are thickened by chromatin condensation at periphery. No nuclear buddings are seen. The boundary of the cell ball is enrounded in secretion. Note exfoliation of foamy cells.

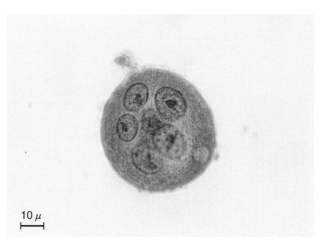

Fig. 7-23　A multinucleated giant cell of epithelial origin in nipple discharge. The same case as Fig. 7-22 with mastopathia. The nuclei having distinct nucleolus are round or oval, equally sized and show even, delicate chromatin pattern. The cytoplasm is thick and sharply outlined.

diagnostic point identification of myoepithelial cells becomes a key of differentiation between benignancy and malignancy.[40]

Squamous cells

A few squamous cells, usually keratinized form, may be occasionally present. They are derived from the nipple or orifice of lactiferous ducts (Fig. 7-15).

4 Benign Diseases

Acute mastitis

The female in postpartum and lactation is susceptible to bacterial infection. Acute inflammation with swelling, reddening and pain can be diagnosed easily from the clinical signs.

Purulent discharge from the nipple is not infrequent and proves its inflammatory evidence by numerous pus cells (polymorph nuclear leukocytes) with necrotic debris (Fig. 7-15). Although cytodiagnosis of acute and subacute mastitis is not difficult, differentiation between acute mastitis and inflammatory carcinoma is problematic from a clinical standpoint.

Addendum: **Acute inflammatory carcinoma.** Acute inflammatory carcinoma is also called erysipeloid carcinoma. This is not a specific disease entity but a clinical variant of breast carcinoma that is manifested by redness, tenderness and diffuse swelling. Histologically, permeation of subepidermal lymphatics by cancer cells is a characteristic feature.

Plasma cell mastitis (Mammary duct ectasia)

The occurrence of this peculiar, benign disorder of the breast is uncommonly found in women who have born children. The lesion is an ill-defined induration and histologically characterized by marked periductal plasma cell infiltration followed by dilatation and accumulation of inspissated secreta, in which plasma cells, lipid-laden histiocytes, degenerate epithelial cells, leukocytes and necrotic cell debris are contained. Cytologically, the preponderance of plasma cells becomes a clue to the diagnosis of plasma cell mastitis. As this neoplasia-mimic disorder is also called mammary duct ectasia or comedomastitis, cytologists may find necrotic cell debris, histiocytes and degenerate lining epithelial cells (Fig. 7-16).

Fibrocystic disease

Fibrocystic disease is the most common disorder of the breast to be distinguished from carcinoma because of the fact as follows: (1) This lesion occurs commonly in the late 4th decade of cancer ages, (2) nipple discharge, either sanguineous or turbid milky, occurs in nearly the same frequency, (3) ill-defined, firm lumps are palpated, and (4) association with carcinoma is considerably frequent. So far as the relationship of fibrocystic disease to carcinoma is concerned, the question whether precancerous or paracancerous has been argued. Coincidence of fibrocystic disease and cancer has been emphasized by many workers, but the incidence of fibrocystic disease in breast cancer-bearing patients, as reported by Davis et al (39.1%)[13] and Silverberg et al (39.4%),[53] does not differ so much from that in female autopsy cases without breast carcinoma.[19] For the purpose of differential diagnosis from malignancy for patients having a tumor mass in the breast, fine needle aspiration is inevitably important.

Fibrocystic disease, i.e., mammary dysplasia or mastopathia (mastopathy) in a broad sense, includes various histologic varieties such as fibrosclerosis, cystic hyperplasia and adenosis.

1) Fibrosclerosis

Fibrosclerosis is a variant of fibrocystic disease showing stromal overgrowth so much as to compress ductules and glands. Epithelial cells are hardly obtained by needle aspiration. There is no difficulty in interpretation of cellular features.

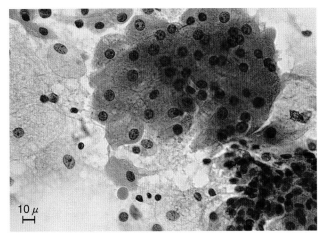

Fig. 7-24 **Apocrine metaplasia of duct epithelial cells in imprints from a patient with mastopathia.** These cells are arranged in a flat sheet with a cobblestone pavement appearance. Note a characteristic finely granular cytoplasm with acidophilic staining reaction.

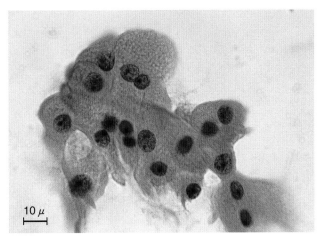

Fig. 7-25 **Higher magnification of apocrine metaplastic cells.** The identification of apocrine metaplasia is based on following features: (1) acidophilic, fine granules in the cytoplasm, (2) thick, polyhedral cytoplasm with amphoteric stainability, (3) round, equally sized nuclei, and (4) single, prominent nucleoli.

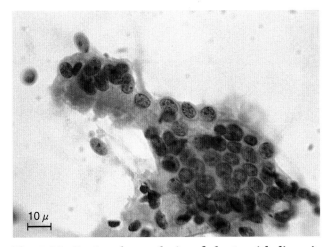

Fig. 7-26 **Benign hyperplasia of duct epithelium in imprints from a case with mastopathia.** The nuclei are hyperchromatic but equally sized and shaped. These cells with indistinctly outlined cytoplasmic protrusions are referred to as "gasoline pump cells" which are encountered in mastopathia.

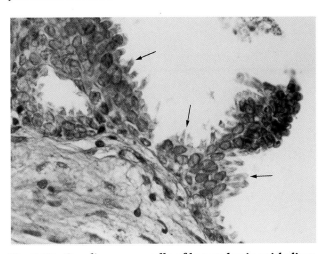

Fig. 7-27 **Gasoline pump cells of hyperplastic epithelium in mastopathia.** Note characteristic cytoplasmic protrusions (arrows).

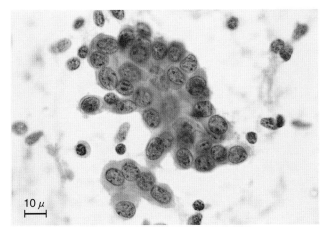

Fig. 7-28 **Benign hyperplasia in imprints from a case with mastopathia.** The nuclei are generally enlarged, hyperchromatic, but equally sized and shaped. The chromatin is finely and evenly distributed.

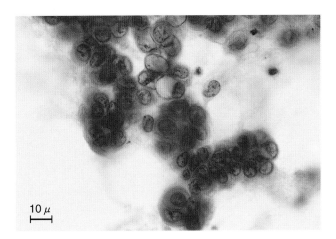

Fig. 7-29 **Intracystic papillary tumor cells seen in a tight cluster.** Aspiration smears from the cyst show tightly clustered cells with overlapped, but fairly small uniform nuclei. Cytoplasmic vacuolation is in favor of benignancy.

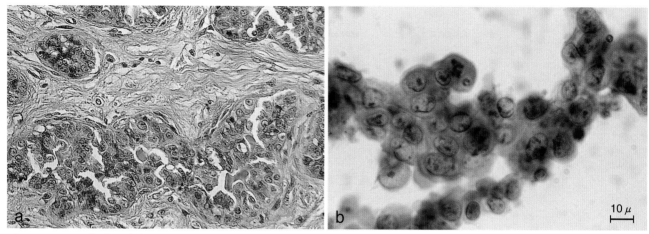

Fig. 7-30 Borderline lobular hyperplasia.
a. Histology. Hyperplastic acini and ductules vary in size and are lined by layers of hypertrophic epithelial cells.
b. Needle aspiration cytology showing atypical ductal cell cluster. This irregularly elongated ductal cluster is composed of uniformly round cells. Although one or two nucleoli are present, the nuclei are smoothly outlined and bland.

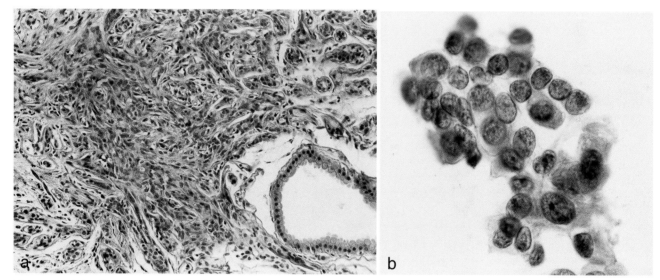

Fig. 7-31 Histology and cytology from sclerosing adenosis.
a. Histologically hyperplastic epithelial cells are irregularly arranged by fibrosis. This pattern mimics scirrhous carcinoma.
b. Cytology in a needle aspirate or imprint shows nuclear enlargement and increased chromatin amount. However, bland chromatin pattern and uniformity in nuclear size and shape are indicative of benign hyperplasia.

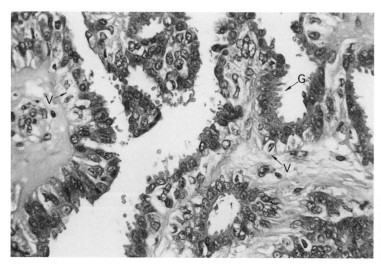

Fig. 7-32 Hyperplastic epithelium in intracystic papilloma. The nuclei are hyperchromatic and slightly variable in size. An appearance of "gasoline pump" cells (G) and intracellular vacuolation (V) are notable findings in benign papilloma.

146

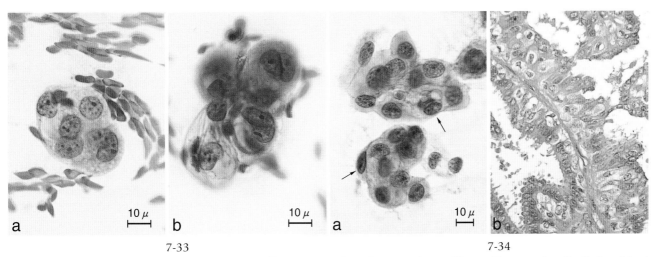

7-33   7-34

Figs. 7-33   Benign atypical hyperplasia in needle aspirates from intracystic papilloma. These are hardly distinguished from cancer cells because of nuclear enlargement and prominence of nucleoli (a). Attention should be paid to finely granular, bland chromatin and abundance of cytoplasm. Note presence of different type of cell with clear cytoplasm that may be derived from myoepithelium(b).

Fig. 7-34   Imprint cytology and histology after excision of intracystic papilloma. From the same case as Fig. 7-33.
a. Cytology shows uniform nuclei with bland chromatins. The cell with vacuolated cytoplasm and a smaller nucleus may be myoepithelial in origin (arrows). Note difference of cell cluster in shape by sampling. Tumor cells shed into the cyst tend to form cell balls.
b. Histology shows lining by darkly stained cells with secretory budding and hypertrophic, vacuolated myoepithelial cells.

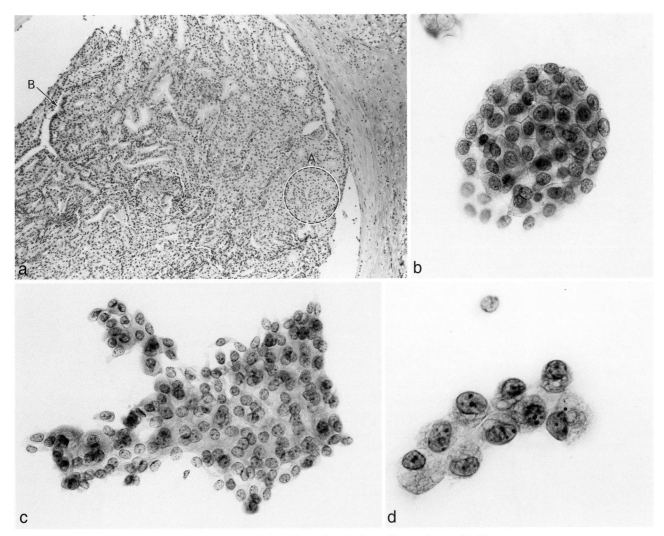

Fig. 7-35   Intracystic papilloma. Histology (a) and cytology of needle aspirates (b-d).
a. Histology of intracystic papilloma.
b. Benign cell ball with uniform round nuclei comparable to solid sheets of cells (circle A). Note distinctive cell borders.
c, d. Irregularly outlined papillary cell sheet or tubular cell cluster is derived from the histology with tubular structure (arrow B).

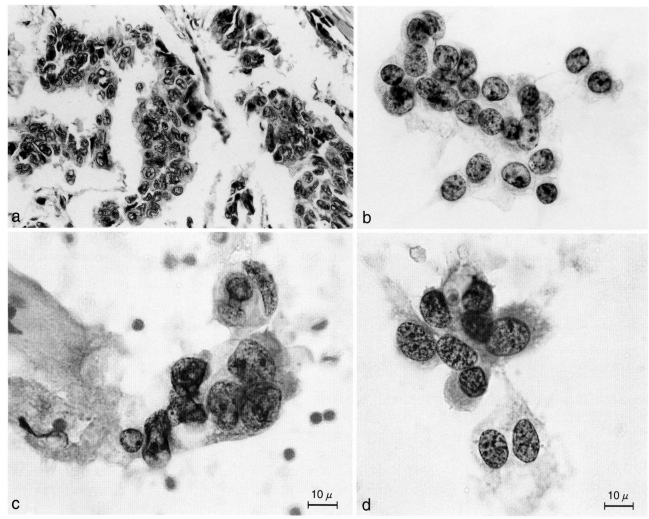

Fig. 7-36 Histology and cytology from a patient with intracystic carcinoma.
a. Cancer cell cords of papillary growth reveal nuclear size and shape variation. Nuclear stain varies in intensity.
b. The nuclei are of medium size and fairly uniform, but coarse chromatin pattern differs from benign papilloma cells.
c, d, Cancer cells in aspirated fluid from a case with intracystic carcinoma. The nuclei are of medium size, but larger and more variable in size than those of intraductal papilloma. A fairly fine granular but compact chromatin pattern is an important diagnostic finding. The cellular boundaries are irregular and not enrounded.

2) Cystic hyperplasia

This pathologic change is the most common variant of mammary dysplasia, and includes various lesions as called Schimmelbusch's disease, Semb's fibroadenomatosis cystica and regular typical epithelial proliferation in ducts and lobules. Histological features are variable from case to case and place to place: (1) cystic ectasia is frequently found in association with intracystic epithelial proliferation (duct papillomatosis), (2) epithelial hyperplasia appears with a form of lobular hyperplasia or miniature fibroadenomatosis, (3) apocrine metaplasia occurs in the lining epithelium. Cystic hyperplasia is one of the main causes of bloody nipple discharge and yields many clusters of hyperplastic epithelial cells that may be misinterpreted as malignant by beginners (Figs. 7-17, 7-18 and 7-20).

Cytology: Cell clusters (cell fronds) exfoliated from intraductal papillary proliferation display malignancy-mimic changes such as nuclear hypertrophy, hyperchromatism and presence of distinct nucleoli (Fig. 7-22). Cytologically, breast carcinoma is the problematic sub-

ject since a suspicious group in cytodiagnosis is of high frequency. Zajicek (1976)[68] reported in his huge investigation that suspected cases were as many as 11.3% of 1,068 histologically proven carcinomas, whereas false negative were 9.9%. Kline[31] described also 18.1% suspicious cases. The reason why cytodiagnosis of breast carcinoma bothers its accuracy cannot be fully understandable, but carcinomas of low grade cell atypism referable to as a suspicious group are frequent in the breast.

Benign mammary dysplasia is characterized by (1) cellular arrangement suggestive of papillary fronds (Fig.7-19), (2) round to oval, occasionally flattened nuclei at periphery with smooth delicate rim as called "nuclear capping"(Fig. 7-19a), and (3) evenly distributed, fine chromatin in spite of presence of one or two visible nucleoli. In addition, association with apocrine metaplasia and exfoliation of many foamy cells (Figs. 7-24 and 7-25) are suggestive of mammary dysplasia.

For differential diagnosis, if one or more following features are encountered, classification is favorable to cancer; (1) markedly increased nuclear-cytoplasmic

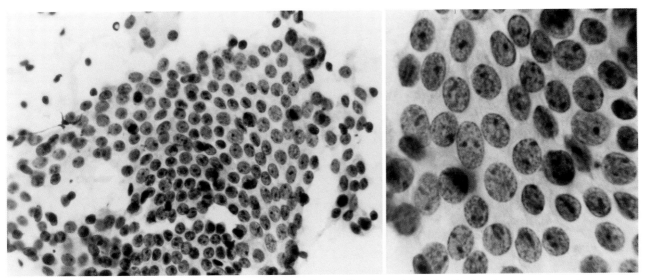

Fig. 7-37 Imprint smear from fibroadenoma. A large sheet of benign epithelial cells showing a honeycomb appearance. Note uniformity in size and shape of the nuclei, evenly spaced nuclear arrangement, smoothly distributed chromatins and delicate nuclear rims.

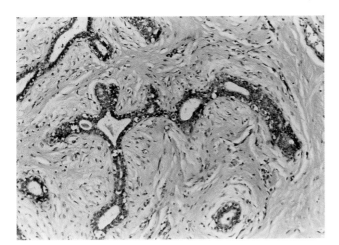

Fig. 7-38 Histology of intracanalicular fibroadenoma. Hyperplastic ductal structures are elongated, curved and branched in association with pronounced growth of stromal connective tissue.

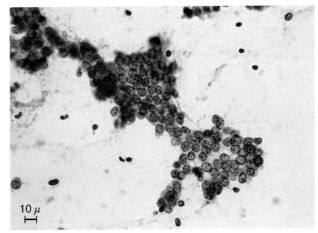

Fig. 7-39 Fibroadenoma in imprints. Compact, flat sheet formation of uniform round nuclei, i.e., a honeycomb appearance, is the characteristic feature of fibroadenoma. Small pyknotic cells are attached.

ratio, (2) marked size variation of nuclei, (3) appearance of multiple macronucleoli over two, (4) macronuclei larger than 15 microns in diameter, (5) nuclear "crescent" holding another cell (see Fig. 7-48), and (6) budding of nuclei outwards beyond the cellular outline in cluster.

3) Adenosis and sclerosing adenosis

Adenosis is defined by WHO histological classification as "increased cellularity due to regular proliferation of the breast epithelium." It is often accompanied by varying degrees of myoepithelial proliferation and fibrosis. As a result of fibrosis, proliferated tubular structures are distorted in such a degree as to mimic invasive growth.

Cytology: Clusters of cells with a tubular or acinar pattern show nuclear uniformity in size and shape, and evenly distributed chromatins. Lopes Cardozo[35] demonstrated bipolar, elongated myoepithelial cells attached to the cell cluster to be of value to confirm benignancy (Fig. 7-14). Absence of "tumor diathesis" in the smear is important as well.

■ Intraduct papilloma (Intracystic papilloma)

Elongated columns or rounded aggregates of epithelial cells with hypertrophic ovoid nuclei are found in nipple discharge, either serosanguinous or not bloody, mixed with many foamy cells. Vacuolation of the cytoplasm is favorable to duct papillomatosis. As one of the characteristic features of fibrocystic disease multiple intraduct papillomas may occur. If apocrine metaplasia (Figs. 7-24 and 7-25) and gasoline pump cells (Fig. 7-32) are detected, differentiation from intraduct carcinoma is not difficult. Papilloma with components of two type epithelial cells and carcinoma with a cribriform pattern of monomorphic cell layering can be histologically distinguished.[33,59] However, cytologists who cannot rely on such structural abnormalities are confronted with difficulty; it is not rare to see highly differentiated carcinoma that is composed of monomorphous small cells. According to the study of Feulgen cytophotometry on mammary tumors by Boquoi and associates,[8] doubtful smears from well differentiated carcinoma show the sim-

149

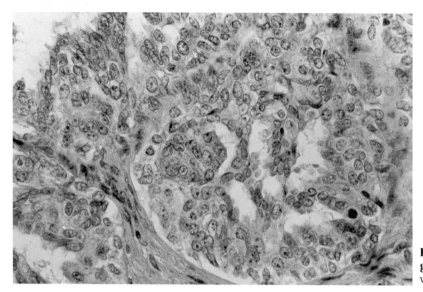

Fig. 7-40 Histology of adenoma in a young girl aged 12. Heaping-up of epithelial cells with some papillary pattern is noted.

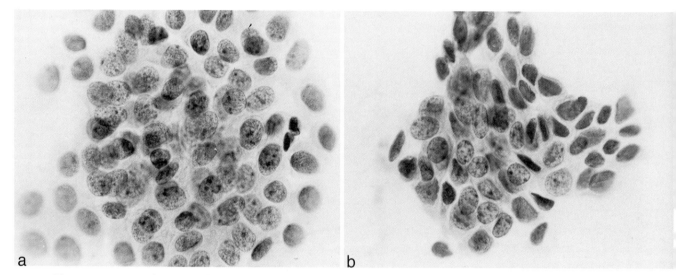

a   b

Fig. 7-41 Hyperplastic epithelial cells in imprints from adenoma. This is from the same case as Fig. 7-40.
a. The nuclei are enlarged, but uniform in size and shape. The chromatin is finely and evenly granular.
b. Elliptic myoepithelial cells are admixed with hyperplastic epithelial cells.

ilar DNA histograms to those of benign tumors with peaks at diploid set of chromosomes and DNA-synthesizing tetraploid cells. If we encounter differentiated carcinoma that is composed of fairly small, uniform nuclei without obvious atypism, we must carefully observe variation in chromatin distribution; although cancer cells appear bland histologically, well fixed cytology smears reveal densely packed chromatin particles distinguishable each in cancer cells (Fig. 7-36).

In addition, abnormal configuration of cell clusters disoriented irregularly is an important finding suggestive of malignancy. Cohesion of the tumor cells is less marked in cancer than in papilloma (see Figs. 7-34, 7-36 and 7-68). It should be kept in mind that the prognosis of well differentiated carcinoma is favorable even when axillary lymph nodes are positive for metastasis.[59]

■Fibroadenoma

A cytological approach to fibroadenoma is only available by needle aspiration before excisional biopsy. Imprint preparation is as valuable as needle aspiration as an adjunct to frozen section diagnosis.

Cytology: Flat sheets of uniform epithelial cells, orderly placed, are observed characteristically in fibroadenoma. The two cell pattern remains either in fine needle aspiration or imprint smear (Fig. 7-39). The nuclei are round to oval, equal in size and possessed of delicate, smooth nuclear membrane. The chromatin pattern is bland with evenly distributed, fine particles.

■Adenoma

Adenoma is a type of fibroadenoma of rare occurrence in a young woman. This tumor is characterized by preponderance of proliferated glands. The epithelial cells which are closely packed possess hypertrophic nuclei; uniformly enlarged nuclei show one or two nucleoli.

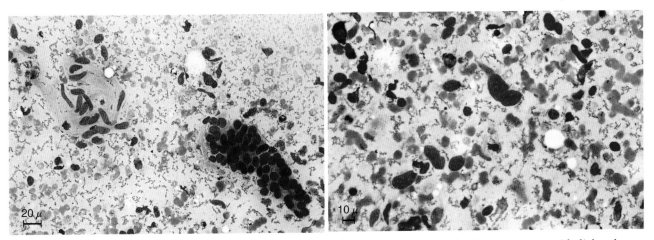

Figs. 7-42   Cystosarcoma phylloides in imprints. Giemsa stain. There are two types of cell components, epithelial and mesenchymal. The term of cystosarcoma derives from highly cellular sarcoma-like stroma; polymorphic mesenchymal cells are characterized by bizarre shape of nuclei and isolation as single cells. Duct epithelial cells with uniform, oval or round nuclei are identified by cluster formation. (By courtesy of Dr. T. Takeda, Miyagi Adult Disease Center, Sendai, Japan.)

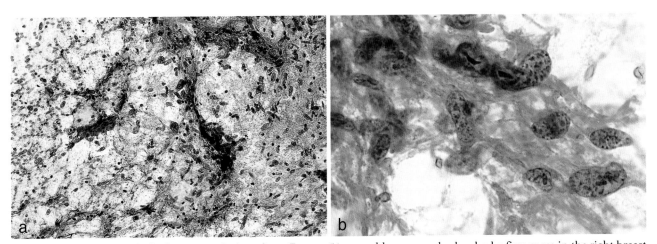

Figs.7-43   Malignant phyllodes tumor in imprints. From a 51-year-old woman who has had a firm mass in the right breast for more than 10 years and showed aggressive enlargement in late two months.
a. Epithelial cell clusters are deformed slit-like. Note many isolated fibrosarcomatous stromal cells in myxoid texture.
b. Large magnification illustrates markedly enlarged, elliptic or spindly nuclei with macronucleoli. Chromatins are increased in
   content and coarsely granular. The cytoplasm is ill-defined.
(By courtesy of Prof. R. Yatani, Mie University School of Medicine, Mie, Japan.)

The chromatins are increased in amount but finely granular. The nuclear membrane is smooth and delicate. Association with myoepithelial cells is one of diagnostic key points (Fig. 7-41).

■Phyllodes tumor

A rare variant of fibroadenoma is referred to as giant fibroadenoma, because it grows fairly rapidly up to the size measuring about 10 cm or larger. This tumor was ever defined as cystosarcoma phylloides regardless of malignancy. Although phyllodes tumor is essentially benign, there is a malignant potential in about 20%: in combination of epithelial and stromal components, the latter takes a sarcomatous transformation. The average age of patients with phyllodes tumor around 45 years is older than that of benign fibroadenoma. Whether phyllodes tumor arises de novo or develops in preexisting fibroadenoma is not clear. Histologically epithelial elements are compressed like leaves by proliferating stroma which are edematous and often highly cellular.

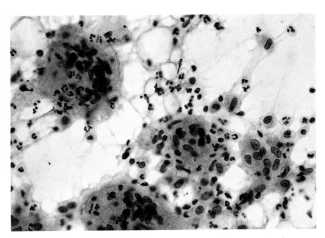

Fig. 7-44   Multinucleated giant cells of foreign body type in needle aspirates. Oval or elliptic nuclei tend to aggregate in the center of the cytoplasm in foreign body giant cells.

151

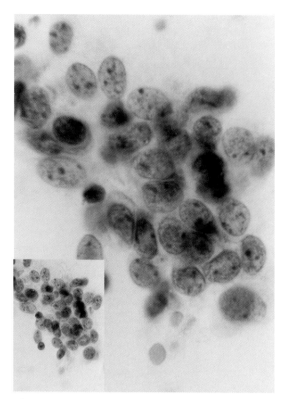

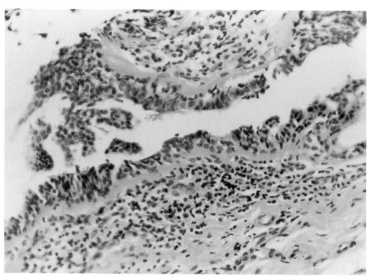

Fig. 7-46 **Gynecomastia with hyperplasia of duct epithelium.** There is proliferative change of duct epithelium; heaping out of epithelial cells and papillary projections are seen.

Fig. 7-45 **Imprint smear of gynecomastia.** The nuclear structures are essentially the same as shown in Figs. 7-37 and 7-38 whereas somewhat disorderly arrangement is often encountered because of hyperplasia of the epithelium.

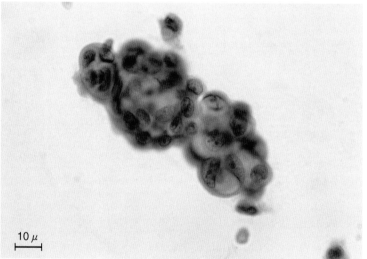

Fig. 7-47 **Hyperplastic epithelial cells in nipple discharge from an 80-year-old male with gynecomastia.** The nuclei in a cell cluster appear to be compressed from each other. Note uniformity of the size and shape of the nuclei. There is cytoplasmic vacuolation that is commonly seen in benign hyperplasia.

10 μ

Table 7-4 Cytology of nipple discharge in the case of breast carcinoma and its histology

Histology	No. of cases	Cytodiagnosis of discharge Positive*	Negative	Discharge not Productive
Noninvasive ca percent/cases	17	6 (66.7%)* (35.3%)	3 (17.6%)	8 (47.1%)
Invasive duct ca percent/cases	154	16 (66.7%)* (10.4%)	9 (5.8%)	129 (83.8%)
Scirrhous ca percent/cases	86	1 (33.3%)* (1.2%)	2 (2.3%)	83 (96.5%)
Special types percent/cases	15	0	2 (13.3%)	13 (86.7%)
Paget's disease percent/cases	1	1 (100%) (100%)	0	0
Total	273	24 (60.0%)* (8.8%)	16 (5.9%)	233 (85.3%)

*Percentage indicates positive rates in cases with discharge.
(From Fujii M: At Tokyo Metropolitan Cancer Detection Center 1985-1989.
In: Sakamoto G (ed): Atlas of Differential Diagnosis of Tumors: Breast. Tokyo: Bunkodo, 1994.)

Table 7-5 Differential diagnostic points between benignancy and malignancy in ABC

Items	Benignancy	Malignancy
Cellularity	small to moderate	large
Relation to fatty tissue	indifferent	closely admissible*
Relation to myoepithelium	adherent	not associated
Cell clusters	sheets, branching evenly arranged with polarity	papillary, tubular irregularly overlapping without polarity
Nuclear atypia : high N/C ratio anisokaryosis hyperchromasia	none to slight	moderate to marked
Nucleolus	indefinite	macronucleolus
Background	clean	necrotic debris

*In invasion to fatty tissue

Table 7-6 Sixty-eight suspicious aspirates from 3,809 benign lesions

Histologic diagnoses	Number of cases
Fibrocystic disease	37 (54.4%)
Fibroadenoma	9 (13.2%)
Granulation tissue	5 (7.4%)
Papilloma	4 (5.9%)
Gynecomastia	3 (4.4%)
Pregnancy hyperplasia	3 (4.4%)
Atypical ductal hyperplasia	7 (10.3%)
Total	68 (100%)

(From Kline TS: Handbook of Fine Feedle Aspiration Biopsy Cytology. St. Louis: CV Mosby, 1981.)

Malignant transformation is suggested by rapid growth of stromal tissue, invasion to surrounding tissues and occasional distant metastases to the lung and bones. The phyllodes tumor is classified into benign, borderline and malignant types.[46,52] Clear-cut distinction in ABC, however, is difficult because it depends on cellularity, polymorphism and atypicality of tumor cells of stromal component (Fig. 7-42). Fibroadenoma is essential in histologic pattern irrespective of ominous clinical behaviors of phyllodes tumors (Fig. 7-43).

■Foreign body granuloma
Granuloma with foreign body giant cells is formed as a result of traumatic fat necrosis and infusion of foreign substances. Traumatic injury is so light that the patient does not realize a lump. In needle aspirates one may find histiocytes, foreign body giant cells mixed with leukocytes. Foamy cells that phagocytosed fat are predominant in case of fat necrosis (Fig. 7-44).

■Gynecomastia
Mammary hyperplasia of males in childhood and old age may occur as a result of elevated estrogen levels that may be produced by estrogen or stilbestrol therapy, estrogen-producing neoplasms and depressed estrogen inactivation as in liver cirrhosis.

Cytology: Hyperplasia of duct epithelium occurs particularly in the subareolar region. Secretions accumulated in the excretory ducts will be found as nipple discharge (Figs. 7-45 to 7-47).

5 Malignant Cells

■Key points of cytodiagnosis: How to approach the goal in fine needle aspiration cytology
Needless to say, adequate specimen collection is prerequisite to refer to the goal of ABC. Image analysis using mammography and echography should be compared if possible; a stellate configuration in mammography, for example, becomes the initial sign of tubular carcinoma.

Table 7-7 Classification of breast carcinoma by WHO

1a Intraductal carcinoma
1b Lobular carcinoma in situ
2a Invasive ductal carcinoma
2b Invasive ductal carcinoma with a predominant intraductal component
2c Invasive lobular carcinoma
2d Mucinous carcinoma
2e Medullary carcinoma
2f Papillary carcinoma
2g Tubular carcinoma
2h Adenoid cystic carcinoma
2i Secretory (juvenile) carcinoma
2j Apocrine carcinoma
2k Carcinoma with metaplasia squamous type
　Carcinoma with metaplasia spindle-cell type
　Carcinoma with metaplasia cartilagenous and osseous type
2l Others
3 Paget's disease of the nipple

General information: step 1
General informations under 10×10 and 20×10 magnification observations may lead to almost diagnostic decision:
(1) cellularity of epithelial cells,
(2) features of cell clusters such as flat sheet, tubular (club-shaped), papillary (branching),
(3) relation of epithelial cells to surrounding stromal cells, i.e., fibrocytes, fatty cells, histiocytes, lymphocytes, etc.,
(4) association with myoepithelial cells,
(5) background features like tumor diathesis and mucus lake.

Detailed information: step 2
High power magnification focused on target cells makes conclusion:
(1) loss of polarity of cell axis (irregularity of cell to cell arrangement),
(2) prominence of nucleolus,
(3) anisonucleosis,
(4) variation of chromatin in content (nuclear stain).

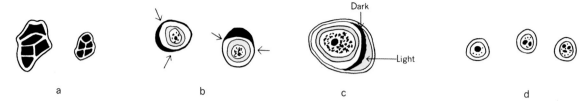

Fig. 7-48 Diagrams of distinguished cellular patterns of breast carcinoma.
a. Nuclear molding. Nuclei appear very close to one another, leaving a scanty amount of cytoplasm in between. The space may form a "Y" shape or a "+" form among three or four nuclei. Cytoplasmic border in this space is usually not seen.
b. Crescent nuclear shape. When two or more cancer cells are shed together, the outer nucleus takes the shape of a crescent. The two ends appear sharp and angular.
c. Dark inside, light outside. When the crescent nucleus is large enough for detail examination of chromatin aggregates, the nuclear border against the inner cell looks much darker and thick, and there is a lighter appearing space in the opposite direction.
d. Nucleoli. Nuclei often contain enlarged nucleolus or nucleoli, and various numbers of nucleoli are visible.
(By courtesy of Dr. T. Masukawa, Dept. of Pathology and Laboratory Medicine, University of Cincinnati, Ohio, U.S.A.)

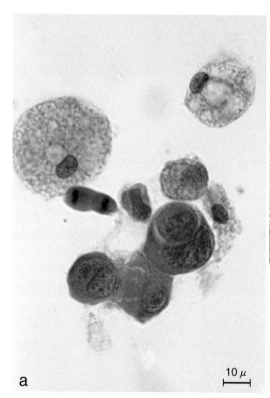

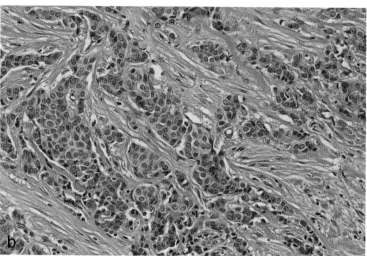

Fig. 7-49a Clustered cancer cells in non-bloody nipple discharge from a case with common infiltrating duct carcinoma. There are evidently malignant features: (1) marked hyperchromatism with coarsely granular chromatin, (2) variation in nuclear size and shape, and (3) a mitotic figure in anaphase.
Fig. 7-49b Infiltrating duct carcinoma with marked desmoplasia. The tumor is composed of irregularly shaped columns or nests of common cancer cells.

The attempt to lessen false negatives brings about high suspicious rates (10-23%).[30] Suspicious reports for benign lesions cannot be avoided; the major benign lesions are referred to as fibrocystic disease, fibroadenoma, ductal papilloma, etc.[1,30]

The WHO International Reference Center for Classification of Tumours proposed three main categories for breast carcinoma (Table 7-7).

Duct carcinoma (Adenocarcinoma)
Malignant cells of the duct origin in the nipple discharge show the same cytological features regardless of infiltration beyond the basement membrane of the ducts. According to the histological classification of infiltrating breast carcinoma by McDivitt et al,[37] infiltrating duct carcinoma comprises 78.1%, infiltrating lobular carci-

noma 8.7%, comedocarcinoma 4.6%, medullary carcinoma 4.3%, colloid carcinoma 2.6% and papillary carcinoma 1.2%. Cytologically, these cancer cells disclose similar malignant features as the glandular type, except for colloid carcinoma, comedocarcinoma, and Paget's disease.

Cytology with general standpoint: Masukawa[36] illustrated several distinguished points in cytodiagnosis of breast carcinoma (Fig. 7-48) in addition to common malignant criteria such as hyperchromatism, nuclear enlargement, irregular chromatin pattern, increased N/C ratio, macronucleolus or macronucleoli and general findings with high cellularity and tumor diathesis.

Lobular carcinoma
The incidence of lobular carcinoma varies geographically. It accounts for about 10% or over, but less common

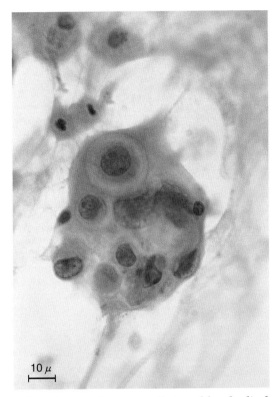

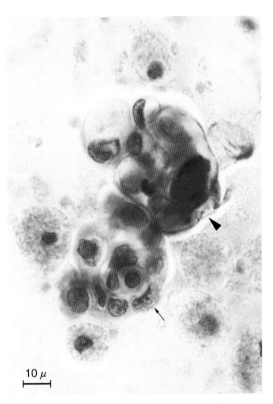

Fig. 7-50 Cluster of cancer cells in a bloody discharge from a 32-year-old female. In such a young woman, bloody discharge should be examined carefully whether or not malignant. Diagnostic points: (1) Variation in nuclear size, (2) cluster formation with enrounded large nuclei, (3) prominence of nucleoli, (4) occurrence of cannibalism, and (5) mitosis are the features diagnostic of cancer.

Fig. 7-51 Cluster of cancer cells in a bloody discharge from a 37-year-old female. These polymorphic nuclei show several characteristic findings such as (1) calcification (psammoma body, arrowhead), (2) whorled pattern with nuclear molding, and (3) crescent formation with *dark inside-light outside* pattern (arrow) (see Fig.7-48).

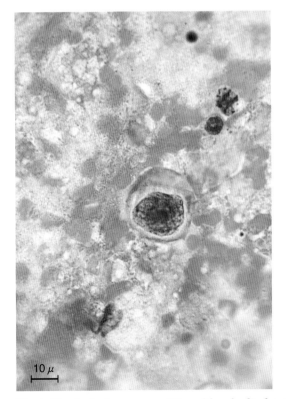

Fig. 7-52 A single cancer cell in a bloody discharge. Even a single cell fulfills the malignant criteria; high N/C ratio, hyperchromatism, coarse clumping of chromatin and enormously large nucleolus are seen. This is a sex chromatin positive cancer cell.

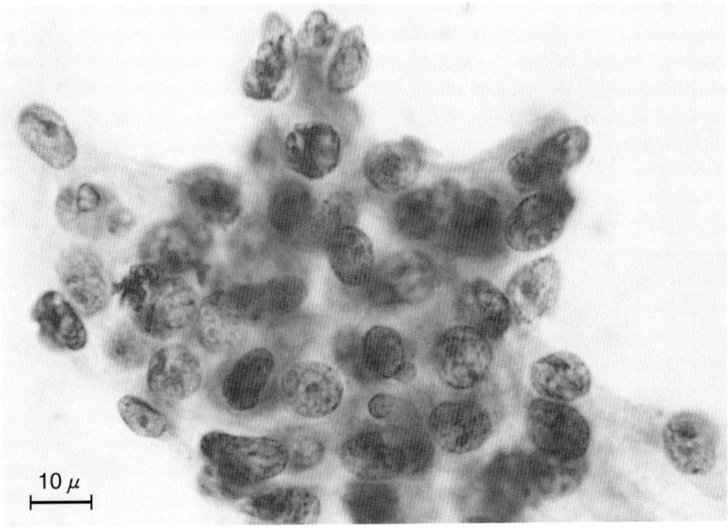

Fig. 7-53 Duct carcinoma in needle aspirates. While cancer cells in nipple discharge tend to form clumps or balls by surface tension (refer to Figs. 7-50 and 7-51), cell clusters torn by aspiration appear indistinctly outlined. Diagnostic points: (1) Distortion in arrangement, (2) poorly defined border, (3) variation in nuclear size and shape, (4) different stainability of each nucleus, and (5) macronucleolus are indicative of malignancy.

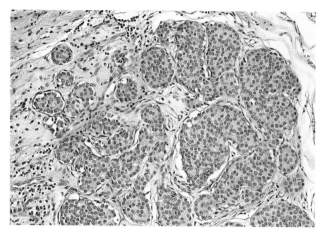

Fig. 7-54 **Lobular carcinoma in an early stage.** The tumor is fairly well localized. Each lobule packed with small hyperchromatic, ovoid cancer cells varies in size.

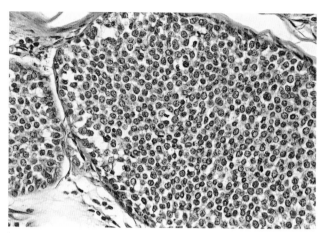

Fig. 7-55 **Higher magnification of lobular carcinoma.** A compact mass of cancer cells is filled within a lobular pattern. The nuclei are densely stained and fairly uniform in size and shape. The cytoplasm is scant in amount.

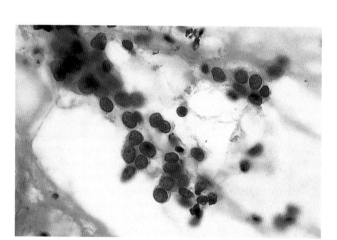

Fig. 7-56 **Lobular carcinoma cells by ABC.** The small cancer cells look like oat cells or lymphoma cells. They display a distinctive cell to cell epithelial arrangement.

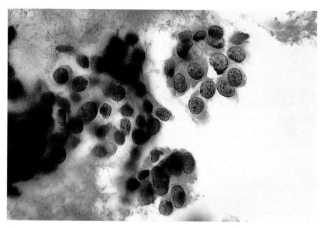

Fig. 7-57 **Lobular carcinoma cells by ABC.** The small cancer cells possess a small amount of cytoplasm that stain intensely green with Papanicolaou stain. Note disorderly arrangement and variation of nuclei in size.

in area with low incidence of breast cancer.[56]

In early stage that is called lobular carcinoma in situ, the acini of affected lobules are distended by accumulated cancer cells (Figs. 7-54 and 7-55). The pattern of infiltration of lobular carcinoma is characterized by diffuse extensive spread into desmoplastic stroma as irregular nests or trabeculae and single files.

Cytology: The cancer cells are small, hyperchromatic and have round or oval nuclei which appear to be denuded because of scanty cytoplasm. They retain an epithelial "cell to cell" arrangement (Figs. 7-56 and 7-57).

■Intracystic papillary carcinoma

Intracystic papillary carcinoma accounts for a small portion of breast carcinomas (about 0.5% by Gatchell et al[23]). Pathologically it takes a papillary growth into cystically dilated space and is hardly distinguished from intracystic papilloma.[28,54] Cytologically attention should be focused on the lack of myoepithelial accompaniment, the lack of apocrine metaplasia and the chromatin pattern with granular distribution (Fig. 7-36).

■Mucinous carcinoma (Colloid carcinoma)

This mucin-producing carcinoma is a type of duct epithelial origin and referred to as carcinoma of low incidence of metastasis to the axillary lymph node and of longer survival than in any other duct caricinoma[26]. This tumor comprises about 2% of breast cancers.

Histologically cancer cells are aggregated into groups and look like a float in mucus lake. Mucinous carcinoma is classified into a pure form and a mixed form that is associated with common ductal carcinoma. Gupta et al[26] reported 13 cases of mucinous carcinomas out of 614 cancers in which 9 were of the pure form. Regarding prognosis, 5-year-survival rates in the pure form showed favorable outcome compared with those in the mixed form.[55,62]

Cytology: A mucinous substance surrounding grouped or isolated cells is characteristic of mucinous carcinoma. Mucus is poorly stained and apt to be overlooked if we are not aware of it. Mucinous content in the cell is recognized as amphoteric stain (Figs. 7-58 to 7-60).

Addendum: **Signet ring cell carcinoma.** This type of carcinoma is common in the gastrointestinal tract as a variety of mucinous carcinoma. The possibility of metas-

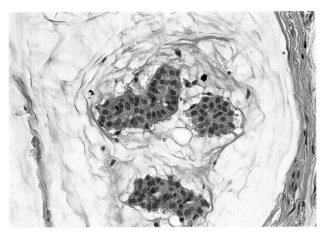

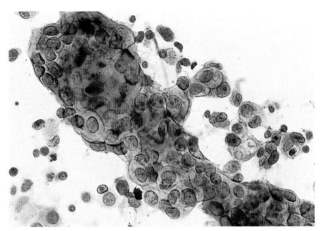

Fig. 7-58   Mucinous carcinoma. From a 41-year old woman who had an ill-defined firm tumor in the right breast. A large amount of mucinous material surrounds clusters of tumor cells as if these were on the float in mucus lake.

Fig. 7-59   A club-shaped cluster of mucinous cancer cells. Imprint cytology during operation. Brownish hue is most likely to be mucus content. Note overlapping and irregular arrangement of tumor cells.

tasis to the breast must be first of all borne in mind. In the breast, however, the evidence that signet ring cells containing mucin in the cytoplasm are noted in acinar nests of lobular carcinoma in situ (LCIS) has convinced pathologists of acinar or lobular origin of this tumor.[3,10,22,39] Since then, signet ring cell carcinoma is not discussed in mucinous carcinoma but in LCIS and invasive lobular carcinoma. A case of invasive signet ring cell carcinoma illustrated here showed remarkable intracellular and extracellular mucus production (Fig. 7-61).

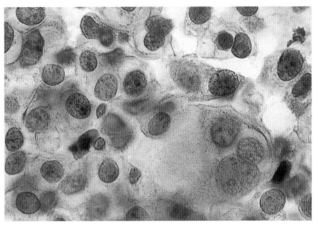

Fig. 7-60   Larger magnification of mucinous carcinoma. The fairly abundant cytoplasm illustrates a brownish hue by a fair amount of intracellular mucus content. Anisonucleosis and prominence of nucleolus are seen.

■Tubular carcinoma
Tubular carcinoma with a favorable prognosis is a variant of invasive ductal carcinoma that is well differentiated histologically and fairly small in size (6-19 mm).[14] This tumor is composed of variously sized tubules lined by a single layer of low columnar cancer cells: they are separated by dense fibrous tissue which may develop to radial scars. A characteristic mammography showing irregular stellate outlines is representable.[17]

Cytology: First of all, one must be aware of rarely occurring tubular carcinoma, because of the fact that the cancer cells show less polymorphism than other ductal cancers and often accompany myoepithelial cells leads to misinterpretation as a benign lesion.[7,25] The tubular fashion is recognized in low power magnification (Fig. 7-61).

■Comedocarcinoma
Comedo-like plugs will be expressed by massage. Cancer cells themselves are not peculiar cytologically to this type (Figs. 7-64 and 7-65), but admixture with necrotic cell debris and pyknotic malignant cells may be encountered in smears (Fig. 7-67).

■Medullary carcinoma
Medullary carcinoma forms a globose, circumscribed tumor having a low infiltrative growth. Cytologically the cancer cells are large, round or polygonal in shape and possess vesicular, centrally locating, large nuclei. Nucleoli are prominent. The cytoplasm is scant and ill-defined. The cancer cells appear as large sheets or aggre-

gates without heavy overlapping (Fig. 7-70). As one type of this tumor, an admixture with dense lymphocytes can be discernible. Longer survival after resection of the tumor and lower frequency of axillary lymph node metastasis than common ductal carcinomas are notable[34,47,65].

Medullary carcinoma with lymphoid stroma (Figs. 7-71 and 7-72) that accounts for about 3% of infiltrating carcinomas should be cytologically defined from this prognostic view.

■Paget's disease
Paget's disease shows an eczematoid appearance of the nipple (Fig. 7-73). The entity of this disease is not cancer affecting primarily the skin but slow-growing carcinoma of the large excretory duct epithelium. Extensive growth of cancer involves the epidermis of the nipple and areola (Fig. 7-74). Paget's cell is a large cancer cell having an opaque or pale-staining abundant cytoplasm and a large round or ovoid nucleus[16](Fig. 7-75). Formation of perinuclear clear zone is the result of ballooning degeneration.

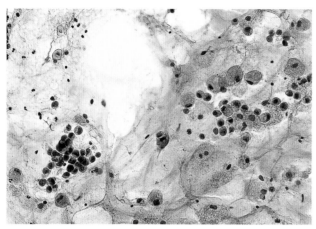

Fig. 7-61　Signet ring cell carcinoma. Isolated signet ring cells of needle aspirates. Amphoteric stain of cytoplasm is referred to as content of mucinous material. Mucinous background is informative for diagnosis. Needle aspirates.

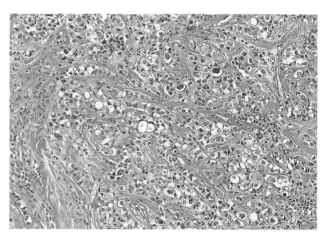

Fig. 7-62　Signet ring cell carcinoma. A variety of mucinous carcinoma. Accumulation of intracellular mucus displaces nuclei to periphery. Cancer cells invade diffusely as single cells or in loose clusters.

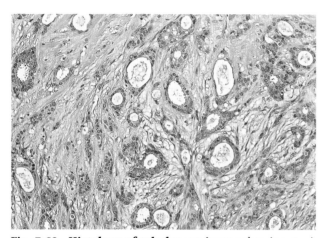

Fig. 7-63　Histology of tubular carcinoma showing atypical tubules varying in size. These tubules are surrounded by dense fibrous tissue rich in elastic fibers.

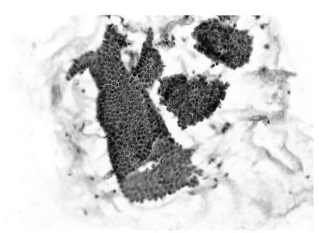

Fig. 7-64　Tubular carcinoma in needle aspirates. In low power magnification one can see a typical tubule and sheets of cancer cells.

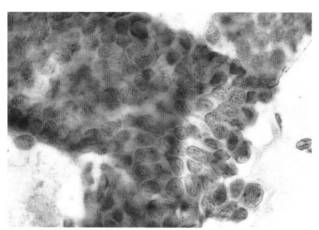

Fig. 7-65　Higher magnification of a cluster of tubular cancer cells from ABC. Note an opening of cancer tubule like a cave. The nuclei are fairly uniform in size and shape.

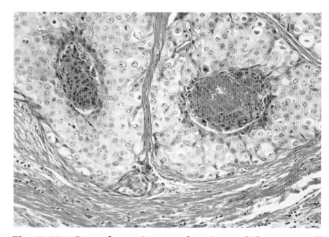

Fig. 7-66　Comedocarcinoma showing solid cancer cell nests with typical central necrosis. Cancer cells of intraductal growth possess round nuclei and a moderate amount of cytoplasm. There is a resemblance to apocrine cells.

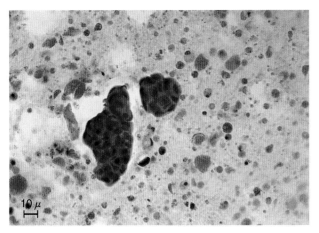

Fig. 7-67 Comedocarcinoma in nipple discharge expressed by palpation. Note a large amount of necrotic debris and pyknotic cancer cells in a cluster that appear to be derived from the central necrotic area.

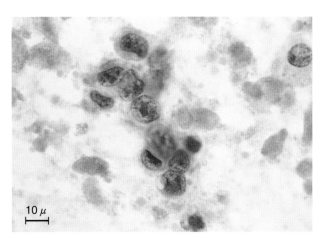

Fig. 7-68 Comedocarcinoma cells in thick nipple discharge. Comedocarcinoma has several characteristic features as follows: (1) large enrounded nuclei with vesicular appearance, (2) prominent, centrally locating nucleolus, and (3) mixture with necrotic debris.

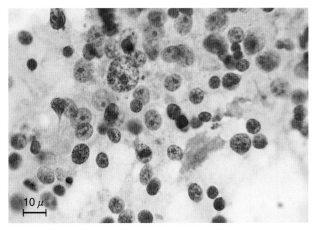

Fig. 7-69 Infiltrating medullary carcinoma in imprints. This special variant with minimal amount of stroma shows distinctive cellular features. Diagnostic points: (1) large, round or oval nuclei of varying size, (2) vesicular nuclei with delicate nuclear rim, (3) one or two, prominent nucleoli with acidophilic reaction, and (4) exfoliation as single cells or in loose cell cluster.

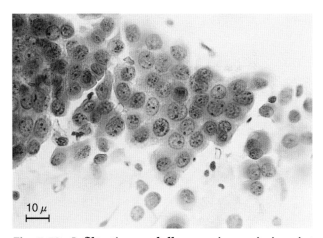

Fig. 7-70 Infiltrating medullary carcinoma in imprints showing a characteristic arrangement of flat sheet. Vesicular, enrounded nuclei are the characteristic features. There is an admixture with a small number of lymphocytes.

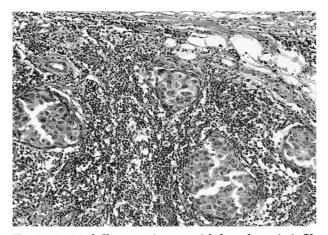

Fig. 7-71 Medullary carcinoma with lymphocytic infiltration. Several solid nests of cancer cells appear to be compressed by dense lymphocytic infiltrates. The cancer cells possess fair amount of eosinophilic cytoplasm.

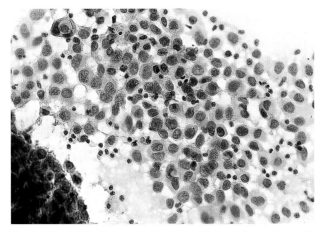

Fig. 7-72 Medullary carcinoma with lymphocytic infiltration by ABC. Massive cancer cells are arranged in a flat sheet. No three-dimensional clusters are recognizable. Epithelial cell cannibalism is seen (upper left). Note a characteristic feature of lymphocyte mixture.

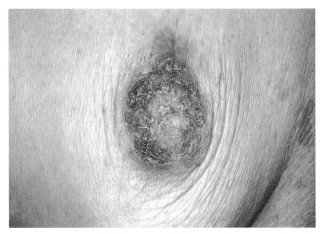

Fig. 7-73 Paget's disease of the nipple. Note chronic eczematous nipple showing thickened epidermis and covering crust.

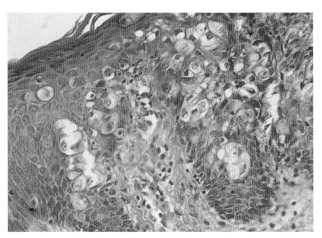

Fig. 7-74 Histology of Paget's disease showing infiltration of the epidermis by Paget's cells. These are round cancer cells with abundant, clear cytoplasm.

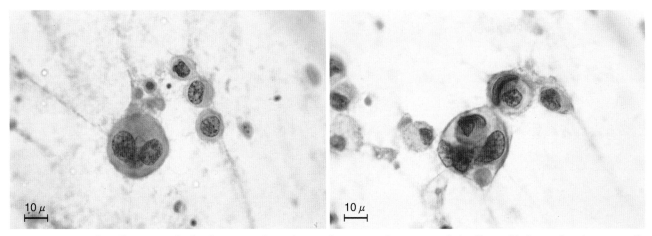

10 μ

10 μ

Fig. 7-75 Paget's cells in nipple discharge. Multinuclear and mononuclear enrounded cells are likely parabasal cancer cells in shape. The cytoplasm is of a moderate amount and shows pale stain. Note halo-like clear cytoplasm.

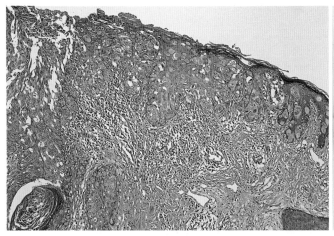

Fig. 7-76a Advanced Paget's disease with infiltrating duct carcinoma. In spite of absence of palpable tumor around the areola, there is invasion of duct carcinoma around the lactiferous duct and within the epidermis.

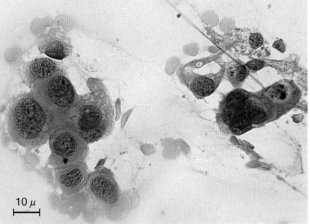

10 μ

Fig. 7-76b Adenocarcinoma cells in imprint from a resected Paget's disease. These well preserved cancer cells arising from the large duct beneath the nipple are arranged in a flat sheet. The cytoplasm is moderate in amount and densely stained.

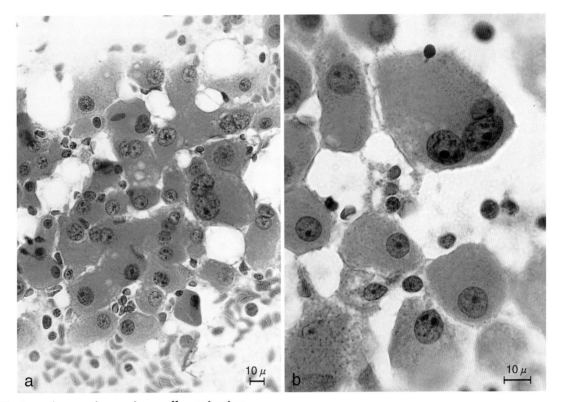

Fig. 7-77　Apocrine carcinoma in needle aspiration.
a. Apocrine carcinoma cells arranged in a flat sheet. The nuclei are round and contain one or two macronucleoli. The cytoplasm is abundant and opaque.
b. Higher magnification of apocrine carcinoma. The nuclei are round and equally shaped, but variable in size. A large amount of the cytoplasm reveals a characteristic, finely granular texture.

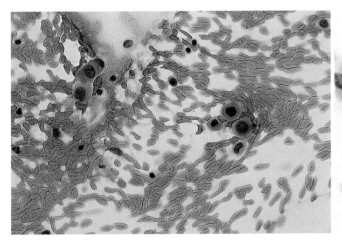

Fig. 7-78a　Squamous cell carcinoma arising from a cystic wall of the breast. Needle aspiration from a 36-year-old woman. Note isolated cells with Indian ink droplet-like nuclei and a small file of cancer cells.

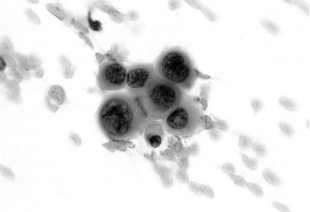

Fig. 7-78b　Higher magnification of squamous cell carcinoma. Hyperchromatic cells with high N/C ratio exhibit a cobblestone pavement arrangement. Thickened, opaque cytoplasm is also a characteristic feature of squamous cell carcinoma.

■Apocrine carcinoma

Adenocarcinoma with apocrine metaplasia (apocrine carcinoma) is also a peculiar type to breast carcinoma. This tumor is also called sweat gland carcinoma (Ewing) from the concept that the tumor originates from the aberrant sweat glands. Although there is the histologic similarity to the sweat gland, it can be produced by apocrine metaplastic change of duct carcinoma as seen in benign duct epithelium of mastopathia (see Fig. 7-25). The cytoplasm is thick, opaque and abundant; there may be a fine granular texture (Fig. 7-77). The granules

toward the apex of the cytoplasm are electron microscopically membrane bound and have homogeneous osmiophilic cores with occasional crystalloid arrays.[43]

■Squamous cell carcinoma

Squamous cell carcinoma is a rare variety (approximately 1 or 2 per 1,000 invasive carcinomas) and is thought to represent a metaplastic change of duct carcinoma. If squamous metaplasia is advanced and involves a large area, epidermoid type of malignant cells may be shed (Fig. 7-78). Pure squamous cell carcinoma is extremely

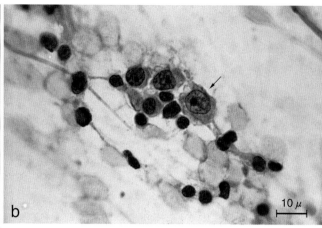

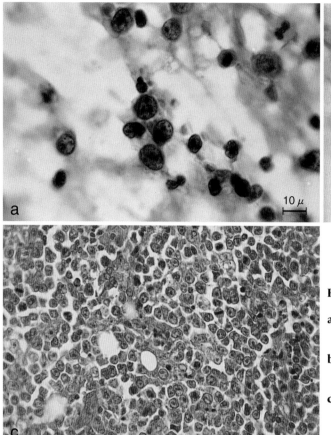

Fig. 7-79 Histology and cytology of non-Hodgkin's lymphoma.
a. Scattered lymphoma cells are almost denuded of cytoplasm. The hyperchromatic nuclei disclose prominent karyosomes.
b. The largest cell is possessed of ill-defined cytoplasm and vesicular nucleus, in which a single prominent nucleolus is found (arrow).
c. Histology is characterized by dense masses of immature lymphosarcoma cells. The cytoplasm is very scant. The nuclei are large, hyperchromatic and possess prominent nucleolus.

rare[11,44]; its incidence reported in the literature is 0.02% and 0.075%.[15,61]

The prognosis is most likely comparable to that of common ductal carcinoma.[11,44] Either primary or metaplastic origin can not be distinguished by ABC except for the case which is mixed with common type cancer cells.

■Lymphoma

Malignant lymphoma that affects primarily the breast. is rare and accounts for about 0.1% of all malignant mammary tumors.[40] It is hardly distinguished from anaplastic carcinoma histologically as well as cytologically. Lymphoma cells are identified singly without formation of cell cluster. Lymphoma cells are monomorphous and structurally the same as those in other organs (see Fig. 6-63).

6 Prognostic View

■Vaginal hormonal cytology management of advanced breast carcinoma

Evaluation of cytohormonal states in advanced breast carcinoma becomes a clue to proper hormonal therapy. If there is an extreme left shift of the maturation index, estrogenic therapy may give a good clinical response. On the other hand, in case of the right shift by estrogenic enhancement, androgen and/or estrogenic ablative therapy may be adopted expecting a good clinical response. When the cytohormonal alteration could be expectedly

obtained, the therapeutic management may be successful.

■Cytomorphological aspect

Histological grading of malignancy in reference to survival has been emphasized by Scarff and Torloni.[51] The criteria of grading based on (1) tubule formation, (2) hyperchromatism and frequency of mitosis, and (3) variation in size, shape and stainability of nuclei. Three classes for each item, i.e., mild, moderate and marked, are awarded with points from one to three. Thus, three grades of malignancy are used practically with total counts; low malignancy for points from 3 to 5, intermediate malignancy for points 6 or 7, and high malignancy for points 8 or 9. A significant correlation between the histological grades and survival rates is reported.[51] Since these features can be evaluated cytologically, the statistical meaning must be proved for further clinical investigation. A recent information from the National Institutes of Health Consensus Conference on breast carcinoma[45] has recommended an additional comment on nuclear grading based on enlargement, shape of nuclear membrane, uniformity, chromatin distribution, presence of nucleolus; grade 3, the most anaplastic type, is defined by marked hyperchromatism, irregular contour, coarsely granular chromatin and macronucleolus. Nuclear grading is related to aggressiveness of growth[12,18,42] and applicable to cytology.[42]

■Biochemical aspect

Both estrogen and progesterone receptors are bound to respective hormones. Nearly two thirds of breast cancers having either receptor or both are susceptible to hormonal therapy.

These receptor-positive cancers show less recurrence than those that do not express receptors. Levels of steroid hormone receptors are predictable for response to the therapy; breast cancers having more than 10 fmol/mg of ER or PgR protein may expect regression of the tumor.[48]

As other prognostic markers, HER2/neu or ErbB-2 that is genetically homologous to the gene for the epidermal growth factor receptor (EGF-R) has the risk of recurrence.[58,60] A linear correlation between the incidence of X body and survival rates or length has been noted by Wacker and Miles.[64]

ErbB-2 gene amplification brings about overexpression of ErbB-2 protein that is measured by immunocytochemical technique.

Depressed expression of erbB-2 gene product seems to predict favorable outcome.[2] FISH technique[29] that is referable to an increase of copy in number as gene amplification is a very worthy method of detecting not only gene amplification but deletion of chromosomes. Accumulation of analytic studies regarding gene amplification and prognosis is needed to resolve the conflicting opinions.[50]

Proliferation markers such as proportion of S-phase (flow cytometry)[38] and Ki67 expressing proliferating cells[63,66] has been currently reported.

References

1. Al-Kaisi N: The spectrum of the "gray zone" in breast cytology: A review of 186 cases of atypical and suspicious cytology 38: 898, 1994.
2. Allan SM, Fernando IN, Sandle J, et al: Expression of the c-erbB-2 gene product as detected in cytologic aspirates in breast cancer. Acta Cytol 37: 981, 1993.
3. Andersen JA, Vendelboe ML: Cytoplasmic mucous globules in lobular carcinoma in situ: Diagnosis and prognosis. Am J Surg Pathol 5: 251, 1981.
4. Bargmann W, Knoop A: Über die Morphologie der Milchsekreiton: Licht und elektronmikroskopische Studien an der Milchdruse der Ratte. Zeitschr f Zellforsch 49: 344, 1959.
5. Berg JW, Robbins GF: A late look at the safety of aspiration biopsy. Cancer 15: 826, 1962.
6. Black MM: Fine needle aspiration and breast disease. Lancet 1: 284, 1981.
7. Bondeson L, Lindholm K: Aspiration cytology of tubular breast carcinoma. Acta Cytol 34: 15, 1990.
8. Boquoi E, Krebs S, Kreuzer G: Feulgen-DNA-cytophotometry on mammary tumor cells from aspiration biopsy smears. Acta Cytol 19: 326, 1975.
9. Bothmann G, Rummel H, Kubli F: Zur Stellung der Aspirationszytologie bei der Frühdiagnostik des Mammakarzinoms. Geburtsh Frauenheik 34: 287, 1974.
10. Breslow A, Brancaccio ME: Intracellular mucin production by lobular breast carcinoma cells. Arch Pathol Lab Med 100: 620, 1976
11. Chen KTK: Fine needle aspiration cytology of squamous cell carcinoma of the breast. Acta Cytol 34: 664, 1990.
12. Dabbs DJ: Role of nuclear grading of breast carcinomas in fine needle aspiration specimens. Acta Cytol 37: 361, 1993.
13. Davis HH, Simons M, Davis JB: Cystic disease of the breast: Relationship to carcinoma. Cancer 17: 957, 1964.
14. de la Torre, M, Lindholm K, Lindgren A: Fine needle aspiration cytology of tubular breast carcinoma and radial smear. Acta Cytol 38: 884, 1994.
15. Eggers JW, Chesney TM: Squamous cell carcinoma of the breast: A clinicopathological analysis of eight cases and review of the literature. Hum Pathol 15: 526, 1984.
16. Eisen MJ, Taft RH: Cytologic diagnosis of mammary cancer associated with incipient Paget's disease of the nipple. Cancer 4: 150, 1951.
17. Fischler DF, Sneige N, Ordonez N, et al: Tubular breast carcinoma (TCB): Cytologic features in fine needle aspirations (FNA) and application of monoclonal anti-alpha smooth muscle actin (SMA) in diagnosis. Diagn Cytopathol 10: 120, 1994.
18. Fisher ER, Redmond C, Fisher B, et al: Pathologic findings from the National Surgical Adjuvant Breast and Bowel Projects. Cancer 65: 2121, 1990.
19. Frantz VK, Pickren JW, Melcher GW, et al: Incidence of chronic cystic disease in so-called "normal breast": Study based on 225 postmortem examinations. Cancer 4: 762, 1951.
20. Franzen S, Giertz G, Zajicek J: Cytological diagnosis of prostatic tumours by transrectal aspiration biopsy: A preliminary report. Br J Urol 32: 193, 1960.
21. Furnival CM, Hughes HE, Hocking MA, et al: Aspiration cytology in breast cancer: Its relevance to diagnosis. Lancet 2: 446, 1975.
22. Gad A, Azzopardi JG: Lobular carcinoma of the breast: A special variant of mucin secreting carcinoma. J Clin Pathol 28: 711, 1975.
23. Gatchell FG, Dockerty MB, Clagett OT: Intracystic carcinoma of the breast. Surg Gynecol Obstet 106: 347, 1958.
24. Geschickter CF: Diseases of the Breast. Philadelphia: JB Lippincott, 1945.
25. Gupta RK, Dowle C: Fine-needle aspiration cytology of tubular carcinoma in a young woman. Diagn Cytopathol 7: 72, 1991.
26. Gupta RK, McHutchison AGR, Simpson JS, et al: Value of fine needle aspiration cytology of the breast, with an emphasis on the cytodiagnosis of colloid carcinoma. Acta Cytol 35: 703, 1991.
27. Haagensen CD: Diseases of the Breast. Philadelphia: WB Saunders, 1956.
28. Jeffrey PB, Ljung BM: Benign and malignant papillary lesions of the breast: A cytomorphologic study. Am J Clin Pathol 101: 500, 1994.
29. Kallioniemi O, Kallioniemi A, Kurisu W, et al: C-erbB-2 oncogene amplification in breast cancer analyzed by fluorescence in situ hybridization. Proc Natl Acad Sci USA 89: 5321, 1992.
30. Kline TS: Handbook of Fine Needle Aspiration Biopsy Cytology. St Louis: CV Mosby, 1981.
31. Kline TS, Neal HS: Needle aspiration of the breast—why bother? Acta Cytol 20: 324, 1976.
32. Kopans DB, Lindfors K, McCarthy KA, et al: Spring hookwire breast lesion localiser: Use with rigid compression mammographic systems. Radiology 157: 537, 1985.
33. Kraus FT, Neubecker RD: The differential diagnosis of

papillary tumors of the breast. Cancer 15: 444, 1962.

34. Kurtz JM, Jacqemier J, Torhorst J, et al: Conservation therapy for breast cancers other than infiltrating ductal carcinoma. Cancer 63: 1630, 1989.

35. Lopes Cardozo P: Atlas of Clinical Cytology. Targa b. v.'s-Hertogenbosch, the Netherlands, 1976.

36. Masukawa T: Breast cytology. In: Wide GL, Koss LG, Reagan JW (eds): Compendium on Diagnostic Cytology, 4th ed. pp497-508. Chicago: Tutorials of Cytology, 1979.

37. McDivitt RW, Stewart FW, Berg JW: Tumors of the Breast, Fascicle 2 in Atlas of Tumor Pathology. Washington, DC: Armed Forces Institute of Pathology, 1968.

38. McGuire WL, Clark GM: Prognostic factors and treatment decisions in axillary-node-negative breast cancer. N Engl J Med 326: 1756, 1992.

39. Merino MJ, Li Vols VA: Signet ring carcinoma of the female breast: A clinicopathologic analysis of 24 cases. Cancer 48: 1830, 1981.

40. Millis RR: Papillary lesion of the breast. In: Atlas of Breast Cytopathology. pp 51-56. Lancaster: MTP, 1984.

41. Misu Y, Watanabe H, Fujita K: Abnormal nipple discharge. Fifth Breast Cancer Research Congress in Japan, Osaka, December 4, 1966.

42. Moriquand J, Gozlan-Fior M, Villemain D, et al: Value of cytoprognostic classification in breast carcinomas. J Clin Pathol 39: 489, 1986.

43. Mossler LA, Barton TK, Brinkhous AD, et al: Apocrine differentiation in human mammary carcinoma. Cancer 46:2463, 1980.

44. Nakayama K, Abe R, Tsuchiya A, et al: Squamous cell carcinoma of the breast. Report of a case diagnosed by fine needle aspiration cytology. Acta Cytol 37: 961, 1993.

45. National Institutes of Health Consensus Development Conference Statement on the Treatment of Early Stage Breast Cancer. Oncology 5: 120, 1991.

46. Pietruszka M, Barnes L: Cystosarcoma phyllodes: A clinicopathologic analysis of 42 cases. Cancer 41: 1974, 1978.

47. Rapin V, Contesso G, Mouriesse H, et al: Medullary breast carcinoma: A reevaluation of 95 cases of breast cancer with inflammatory stroma. Cancer 61: 2503, 1986.

48. Ravdin PM: Malignant diseases and their treatment. Breast cancer. In: Weiss GR (ed): Clinical Oncology. pp139-145. Norwalk, Connecticut: Appleton & Lange, 1993.

49. Ringrose CAD: The role of cytology in the early detection of breast disease. Acta Cytol 10: 373, 1966.

50. Sauter G, Feichter G, Torhorst J, et al: Fluorescence in situ hybridization for detecting ErbB-2 amplification in breast tumor fine needle aspiration biopsies. Acta Cytol 140: 164, 1996.

51. Scarff RW, Torloni H: Histological typing of breast tumours. Geneva: WHO, 1968.

52. Shimizu K, Masawa N, Yamada T, et al: Cytologic evaluation of phyllodes tumors as compared to fibroadenomas of the breast. Acta Cytol 38: 891, 1994.

53. Silverberg SG, Chitale AR, Levitt SH: Prognostic implications of fibrocystic dysplasia in breasts removed for mammary carcinoma. Cancer 29: 574, 1972.

54. Sneige N, Staerkel GA: Fine-needle aspiration cytology of ductal hyperplasia with and without atypia and ductal carcinoma in situ. Hum Pathol 25: 485, 1994.

55. Snyder M, Tobon H: Primary mucinous carcinoma of the breast. Breast 3: 17, 1977.

56. Stalsberg H, Thomas DB, Noonan EA: WHO collaborative study of neoplasia: Histological types of breast carcinoma in relation to international variation and breast cancer risk factors. Int J Cancer 44: 399, 1989.

57. Takeda T, Matsui A, Sato Y, et al: Nipple discharge cytology in mass screening for breast cancer. Acta Cytol 34: 161, 1990.

58. Tandon AK, Clark GM, Chamness GC, et al: Her-2/neu oncogene protein and prognosis in breast cancer. J Clin Oncol 7: 1120, 1989.

59. Taylor HB, Norris HJ: Well-differentiated carcinoma of the breast. Cancer 25: 687, 1970.

60. Thor AD, Schwartz LH, Koerner FC, et al: Analysis of c-erbB-2 expression in breast carcinomas with clinical follow-up. Cancer Res 49: 7147, 1989.

61. Toikkanen S: Primary squamous cell carcinoma of the breast. Cancer 48: 1629, 1981.

62. Toikkanen S, Kujari H: Pure and mixed mucinous carcinoma of the breast: A clinicopathologic analysis of 61 cases with long-term follow-up. Hum Pathol 20: 758, 1989.

63. Veronese SM, Gambacorta M, Gottardi O, et al: Proliferation index as a prognostic marker in breast cancer. Cancer 71: 3926, 1993.

64. Wacker BA, Miles CP: Sex chromatin incidence and prognosis in breast cancer. Cancer 19: 1651, 1966.

65. Wargotz ES, Silverberg SG: Medullary carcinoma of the breast: A clinicopathologic study with appraisal of current diagnostic criteria. Hum Pathol 19: 1340, 1988.

66. Weikel W, Beck T, Mitze M, et al: Immunohistochemical evaluation of growth fraction in human breast cancers using monoclonal antibody Ki-67. Breast Cancer Res Treat 18: 149, 1991.

67. Zajdela A, Ghossein NA, Pilleron JP, et al: The value of aspiration cytology in the diagnosis of breast cancer: Experience at the Fondation Curie. Cancer 35: 499, 1975.

68. Zajicek J: Aspiration biopsy in the diagnosis of palpable mammary lesions. In: Wied GL, Koss LG, Reagan JW (eds): Compendium on Diagnostic Cytology, 4th ed. p555. Chicago: Tutorials of Cytology, 1979.

Fig. 8-1   A Saccomanno's centrifuge and its spare tube. Prototype instrument.

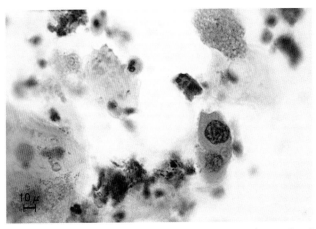

Fig. 8-2   Sputum cytology by Saccomanno's method showing a non-keratinizing squamous cancer cell. Squamous cancer cells are usually not distorted in shape.

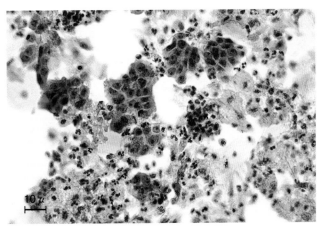

Fig. 8-3   Sputum cytology by Saccomanno's method showing many adenocarcinoma cells seen in clusters. From a case with bronchiolo-alveolar carcinoma. The cancer cells are distorted in shape and their nuclei are condensed.

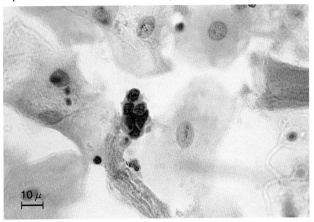

Fig. 8-4   Sputum cytology by Saccomanno's method showing oat cells appeared in clump. Loose cell clusters of oat cells appear to be shrunken and their nuclei become pyknotic. Note cluster formation hardly distinguishable from lymphocytes.

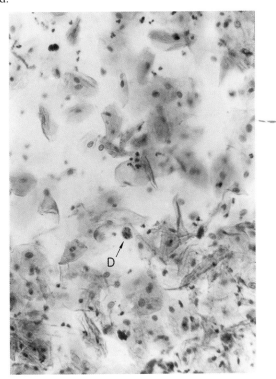

Fig. 8-5   Spontaneously expectorated sputum before the aerosol method. Insufficient specimen mainly consisting of saliva. Squamous cells are numerous and only a few dust cells (D) are recognizable.

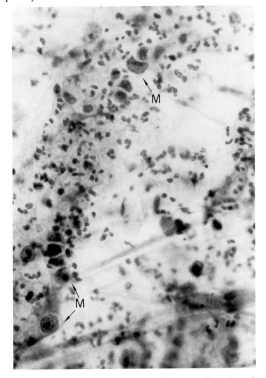

Fig. 8-6   Sputum expectorated immediately after the aerosol chymotrypsin method. Note malignant cells arranged in rows (M).

167

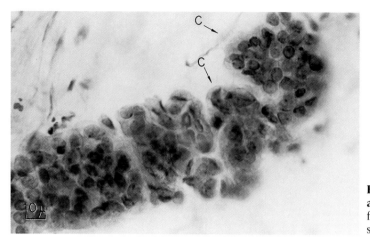

Fig. 8-7 Clusters of ciliated columnar cells obtained after the aerosol chymotrypsin method. These are freshly desquamated cell clusters showing well preserved cilia (C).

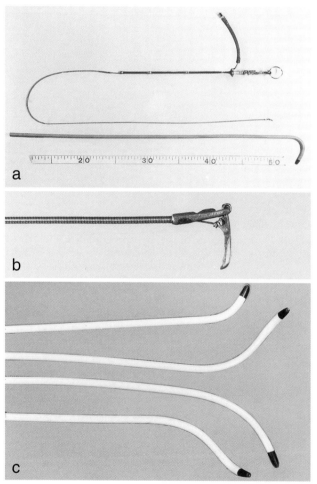

Fig. 8-8 Instrument of transbronchial biopsy (Tsuboi).
a. A whole instrument and Metras' catheter.
b. High power magnification of the tip of the instrument. The curette is made to be easily manipulated.
c. Metras' catheters. The catheter fit for the bronchus of the lesion must be used.

3 Selective Bronchial Scrapings

■Bronchoscopic cytology

Bronchoscopic cytology is recommended because the specimen is directly obtained from the lesion without contamination by saliva or nasopharyngeal secretions.[132] However, there is some disadvantage that limits considerably the field of vision during the examination.

The specimen is submitted after direct scraping, aspiration of bronchial secretes or lavage with a small amount of physiological saline solution. Introduction of a special aspirating cannula is capable of lavage of the distal bronchi. Direct scrapings can be performed with the use of a gauze pledget or nylon brush. If the returns in washing are small in amount, it is appropriate to add saline solution enough for centrifugation or membrane filtration.

A flexible bronchofiberscope has considerably increased the value of this procedure as a cytological tool. The flexible fiberoptic scope has widened visualized area down to the subsegmental bronchi of the 4th or the 5th branching order and made possible to examine the upper lobe bronchi. Insertion of biopsy forceps or brush under fluoroscopic guidance can be manipulated with ease and elevated diagnostic accuracy for peripheral carcinomas; Solomon and associates reported that 30 of 36 peripheral cancers were cytologically positive.[168] Diagnostic accuracy as high as 91.67 % for 72 cases of bronchial washing and/or bronchial brushing through the flexible fiberscope[164] also confirms its diagnostic reliability.

■Selective bronchial curetting and brushing

Sputum or bronchial aspirations are rarely available in cases of minute carcinomas, particularly in those located in the segmental or more peripheral bronchi. A device to obtain specimens directly from the lesion through a Metras' catheter has been successfully employed by Tsuboi[188] (Figs. 8-8 and 8-9) and Hattori.[66,67] Selective bronchial brushing cytology is particularly important to get accurate localization of occult lung cancer not only roentgenographically but bronchoscopically. Lung cancers may exist beyond the range of bronchoscopic visibility.[162]

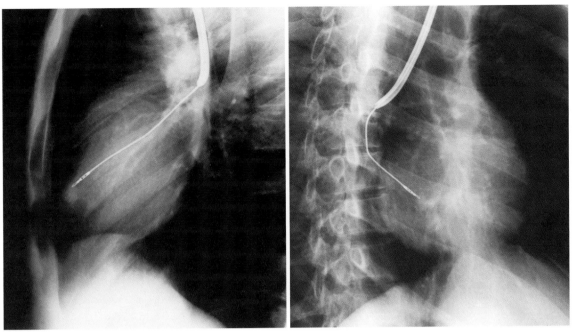

Fig. 8-9 Transbronchial biopsy by Tsuboi. The curette is placed towards a nodular opaque, coin lesion through the Metras' catheter. In order to select a proper Metras' catheter, localization of the coin lesion should have been determined beforehand.

1) Transbronchial biopsy cytology

A Metras' catheter most suitable for the tumor-bearing bronchus should be selected after bronchography (Fig. 8-8). An appropriate Metras' catheter can be introduced into the diseased segmental or subsegmental bronchus under X-ray television fluoroscopy with occasional changes in the patient's posture. The catheter is then passed peripherally until it reaches as close as possible to the lesion.

The curette placed inside the catheter must be gently advanced under fluoroscopy until it reaches the tumor (Fig. 8-8c). The lesion is scraped two or three times by manipulating the handle of the instrument. A curettage specimen is spread evenly on slides. The pull-apart method is available when enough amount of specimen is taken.

2) Transbronchial brushing method

Hattori and his associates[66] utilize two types of small nylon brushes rather than a biopsy curette, i.e., a linear brush and a manually-controlled flexible brush which can be inserted into a lesion through a Metras' catheter (Fig. 8-9). The specimen on the brush is then directly smeared on clean glass slides and immediately fixed in ethanol solution.

Fennessy[47] described a similar transbronchial brushing technique using a radiopaque, vascular catheter under fluoroscopic guidance. The advantage to this technique may be the employment of a polyethylene catheter which is inserted with the aid of a flexible coil spring guidewire.[54,120] No grave complication is considered during manipulation of the catheter and the brush except for haphazard complications such as pneumothorax and loss of brush. Catheter insertion to approach to the lesion in the upper lobe with proper guidance of fluoroscopy is less effective than that to the lesion in the middle or lower lobes.[97]

Bronchial washings or bronchoalveolar lavage fluid (BAL) are worthy of use especially for detection of microorganisms.[31] *Pneumocystis carinii* can be detected with routine Papanicolaou stain, although observation of fine structures is not so accurate as Grocott stain.[61] Introduction of fluorescent microscopy for Papanicolaou-stained smears is strongly emphasized to enhance the sensitivity detecting *Pneumocystis carinii*.[139]

Other indication of bronchial washings are for diseases of pulmonary parenchyma such as sarcoidosis, fibrosing alveolitis, allergic alveolitis, eosinophilic pneumonitis, asbestosis, etc.[180]

Addendum: **Cell block of bronchial washings**

Centrifuge bronchial washings at 2,000 rpm for 5 minutes. Drop three drops of plasma onto the sediment after decant and mix. Thereafter pour three drops of thrombin solution. The clot is fixed with 10% formalin and embedded in paraffin to be cut like tissue specimen. Cell blocks increase diagnostic accuracy and worthy as an adjunct to pulmonary cytology.[49]

4 Fine Needle Aspiration

As radiological methods progress, the localization of tumors has come to be proved with greater accuracy. Therefore, accurate needle aspiration of the tumor is becoming a more generally used procedure. Some objections to this method are the fear of complications of pneumothorax, pyothorax, hemorrhage, air embolism and implantation metastasis.[41,59] Since the incidence of these complications is rare,[119,155,182] needle aspiration is now preferable in cases of metastatic tumor or peripheral tumor which may not produce sputum.[71,86,94,127] Needle aspiration can be performed safely using image-intensified fluoroscopy with television monitoring. A fine needle designed by Nordenström is 0.6 mm in diameter (Fig. 8-10). Nasiell utilized also a hypodermic needle with a diameter of 0.6 to 0.8 mm.[127]

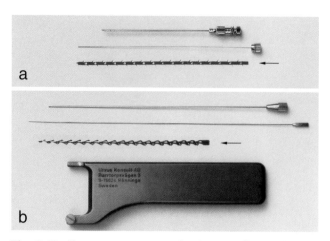

Fig. 8-10 **Transcutaneous aspiration needles with high magnification of each needle (arrows). a.** Designed by Hayata and Oho. **b.** Designed by Nordenström.

Table 8-2 Comparative result of diagnostic procedures

Method	Tumor size	
	Greater than 2.0 cm	2.0 cm or less
Sputum cytology	58/126 (46.0%)	5/23 (21.7%)
Rigid bronchoscopy	10/37 (27.0%)	0/3
Bronchofiberscopy	25/34 (73.5%)	0/7
TV-brushing cytology	34/57 (59.6%)	14/21 (66.7%)
Needle biopsy	60/67 (89.6%)	12/16 (75.0%)

(From Hayata Y, et al: Acta Cytol 17: 469, 1973. © 1973 International Academy of Cytology)

The author uses a fine jagged needle with a length of 9.5cm and a diameter of 0.8mm which was designed by Hayata and his associates at the Tokyo Medical College in Tokyo[71] (Figs. 8-10 and 8-11). The jagged needle, placed in a mantle, is manipulated in the same way as in use of the Vim-Silverman's needle. The mantle needle should be gradually advanced under X-ray television fluoroscopy until it reaches the tumor, which may often be ascertained by the slight resistance as the needle attains its goal. The advantage of this needle is in the ease and safety to obtain cytology specimens during rotating maneuver of the inner jagged needle instead of aspiration; when we use the Franzen equipment, aspiration should be carefully performed while moving the thin needle to and fro.[52] The tissue material, attached to the fine protrusions at the tip of the needle, is spread on clean glass slides.

■Cytodiagnosis

In comparative studies on accuracy of cytodiagnosis for minute peripheral cancers, needle aspiration brought about better result than bronchial brushing.[20,70] The superiority of needle aspiration over bronchial brushing (Table 8-2) is in accessibility to the apical lesion like Pancoast tumor,[193] and in obtainability of a good bacteriological specimen. However, interpretation of malignant cells or cell typing for needle aspirates is rather more difficult than for sputum specimens, because representative features of malignancy in sputum cytology,

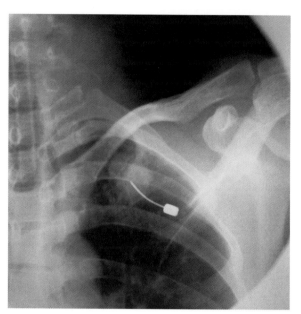

Fig. 8-11 **The needle inserted from the ventrolateral site towards a nodular opaque lesion.**

such as coarse chromatin clumping, thickening and/or irregularity of nuclear membranes appear to be exaggerated. The cellular pattern that is illustrative of the histological type of carcinoma is also hardly discriminated. Exfoliation of cancer cells in three dimensional tight cell clusters in adenocarcinoma and in isolated single cells in epidermoid carcinoma are often undetectable if they are poorly differentiated; desquamation of cells in sheet with more or less overlapping occurs in any type of pulmonary carcinoma. Small cell carcinoma appearing as scattered lymphocyte-like pyknotic cells in sputum reveals also different features; undifferentiated small cell or oat cell carcinoma may possess delicate nuclear membrane, compact chromatin granules and small amount of cytoplasm (see Table 8-6; page 195).

Fraire and associates[51] reported positives of fine needle transbronchial or transthoracic aspiration cytology in 82 of 132 patients (62.1%) who were clinically confirmed malignancy. A large scale study by Zaman et al[211] have described the diagnostic sensitivity as high as 88% of transthoracic fine needle aspirates of 1,059 patients. A noteworthy comparative study by Johnston[79a] on diagnostic rates of fine needle aspirates versus sputum or bronchial material has pointed out the great advantage of fine needle aspiration technique; in 168 patients (72.6%) of 768 patients with lung cancers fine needle aspirates were diagnostic whereas in 9 patients (5.4%) sputum cytology was positive. We must note in his study that necessity of needle aspiration was higher in large cell carcinoma (90.2%) and adenocarcinoma (79.5%) or adenosquamous carcinoma (83.3%) than in squamous cell carcinoma (58.2%) and small cell carcinoma (66.7%).

Diagnostic sensitivity depends on accurate approaches to pulmonary lesions rather than cytology interpretation.

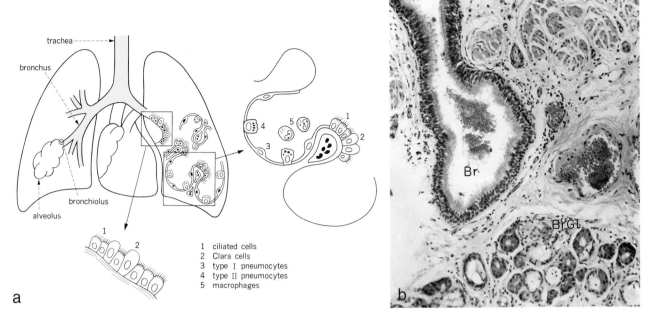

Fig. 8-12 Histology of normal bronchial tree.
a. Diagram of the lower respiratory tract. Schema depicts histology of terminal bronchiole and alveolar duct.
b. Bronchus with a lining of pseudostratified ciliated columnar epithelium. Shown beneath the mucosal epithelium of a bronchus (Br) are the seromucinous mixed glands (BrGl).

5 Histology and Normal Cell Components

■ Histology of the respiratory tract

The airways of the respiratory tract concerned with pulmonary cytology consist of the larynx, the trachea, the bronchi, the bronchioles, the alveolar ducts and the alveoli (Fig. 8-12a). The anterior surface and the upper half of the posterior surface of the epiglottis, and the vocal cords are covered with stratified squamous epithelium. From the base of the epiglottis to the bronchi the mucosa is lined by pseudostratified ciliated columnar epithelium with some intervening goblet cells (Fig. 8-13b and 8-15). Muco-serous glands are open to the epithelial surface with their ducts (Fig. 5-12b). They become more mucous and fewer in number as they branch, and they may be present until the level of the bronchioles. The undifferentiated basal cells or reserve cells lying on the basement membrane are small, darkly stained and competent for multipotential differentiation (Fig. 5-12b).

Bronchioles are airways with a diameter of 1 mm or less, where goblet cells and mucous glands are absent. Whereas prebronchioles are lined by a single layer of ciliated columnar cells, the terminal bronchioles show a low columnar epithelium with few or no cilia. The respiratory bronchioles retain the lining by cuboidal epithelium until they communicate with the alveolar ducts. It was thought that the alveolar walls were bare and the blood capillaries were in direct contact with the alveolar airways until Low[101,102] and Karrer[82] established the lining by an attenuated layer of epithelium. The alveolar surface epithelium is an endothelium-like simple squamous epithelium measuring about 0.1 μm in thick-

ness and resting on the superficial reticulin membrane[101] (Fig. 8-17).

The lining epithelial cells of alveoli are composed of two types of cells. The type I alveolar cells (Campiche, 1960[27]) or type A cells (Takagi and Yasuda, 1958) are greatly attenuated covering alveolar walls with membranous cytoplasm in which cytoplasmic organella are seldom seen. A small number of mitochondria are present beneath the nuclei (Fig. 8-17). Since these cells occupy the great areas of air-blood barrier, they were ever termed alveolar epithelial cells[123] concerned with gas exchange. The type II or type B cells are cuboidal cells provided with short microvilli on the free surface. They are rich in organelles such as Golgi bodies, rough endoplasmic reticulum and mitochondria, and characterized by a large number of osmiophilic lamellar bodies which are referred to as secretory granules[143]; phosphatidylcholine and phosphatidylethanolamine are rich in amount.[205] Products from the multilamellar bodies form surfactant that is important in lowering the surface tension and in stabilizing the alveolar sacs (Fig. 8-18).

Clara cells which have been considered to be concerned with production of surfactant (Azzopardi, Nagaishi) or mucus (Hayek) are of electron microscopic observation. Electron dense granules (0.3-0.6 μm in diameter) in the apical cytoplasm are considered to be secreted with a process of apocrine secretion (Fig. 8-16). The function of secretory Clara cells is not fully understood, but is in favor of phospholipid production.[165]

These are found to be protruding into the bronchiolar space as if they were intercalated between ciliated columnar cells. The cytoplasm is abundant in smooth surfaced endoplasmic reticulum and large enrounded mitochondria (Fig. 8-16). Recently electron microscopic investigation found the presence of bronchioloalveolar

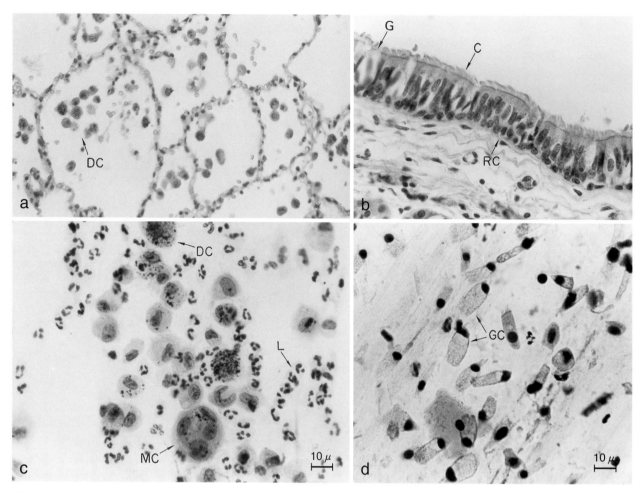

Fig. 8-13 Histology and cytology of bronchial trees.
a. Alveoli containing dust cells (DC) which are one of normal cellular components in sputum.
b. High power magnification of the bronchial epithelium consisting of ciliated cells (C) and a few goblet cells (G). Multipotent
   reserve cells (RC) are present at the bottom of the pseudostratified ciliated columnar epithelium.
c. A multinucleated histiocyte (MC), dust cells (DC) and leukocytes (L) in sputum.
d. Ciliated columnar cells and goblet cells (GC) in sputum.

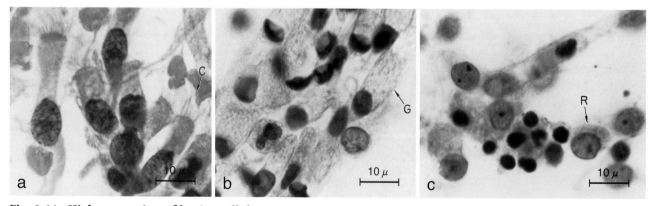

Fig. 8-14 High power view of benign cellular components in a brushing smear.
a. Ciliated columnar cells. Cilia are well preserved (C). The nuclei are oval and uniform in size and shape.
b. Cluster of goblet cells (G). The cytoplasm is distended by accumulation of mucus and the nuclei are extremely eccentric in
   position.
c. Basal or reserve cells in sputum (R). Note small ovoid nuclei and scant pale cytoplasm. The chromatin is evenly and
   smoothly distributed.

adenocarcinoma of Clara cell origin that is characterized
by protruding columnar cancer cells with small number
of apical microvilli and well developed Golgi bodies.[109]

Normal cells in the respiratory tract
1) Ciliated columnar cells
Cytoplasm: Well preserved cells reveal a columnar cytoplasm with a broad end and a short tail. Cilia are visible
and a terminal plate may be seen at the broad end (Figs.

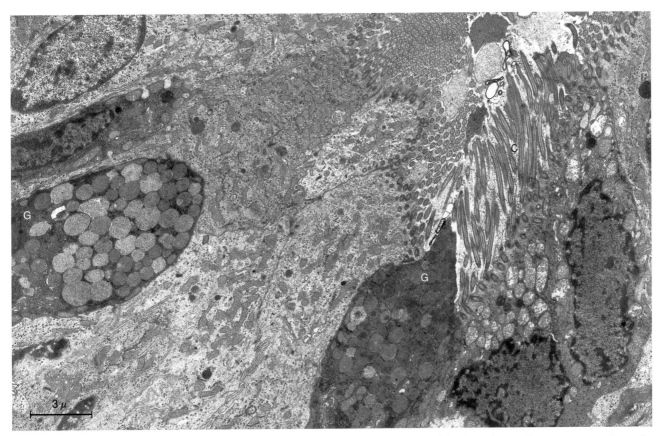

Fig. 8-15 Electron micrograph of bronchial epithelium. The epithelial lining of the bronchus. A few mucus secreting goblet cells (G) are seen among many ciliated columnar cells. In the lumen a number of cilia (C) in transverse or oblique sections are seen.

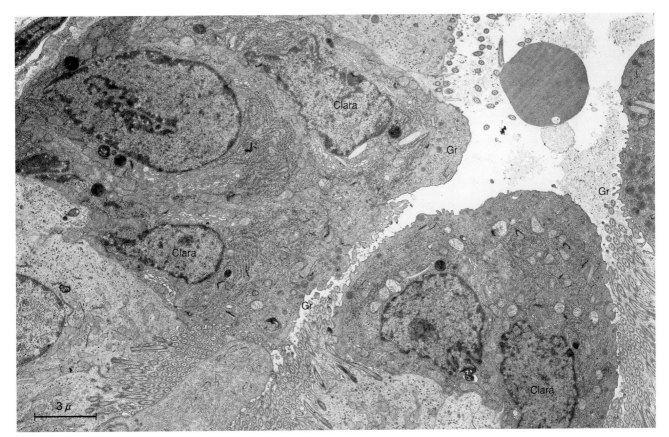

Fig. 8-16 Electron micrograph of bronchiolar epithelium. The epithelial lining of the terminal bronchiole. Non-ciliated cells (Clara cells) are protruded into the lumen and are taller than the ciliated cells. A moderate number of dense granules are located in the apex of the non-ciliated cells (Gr).

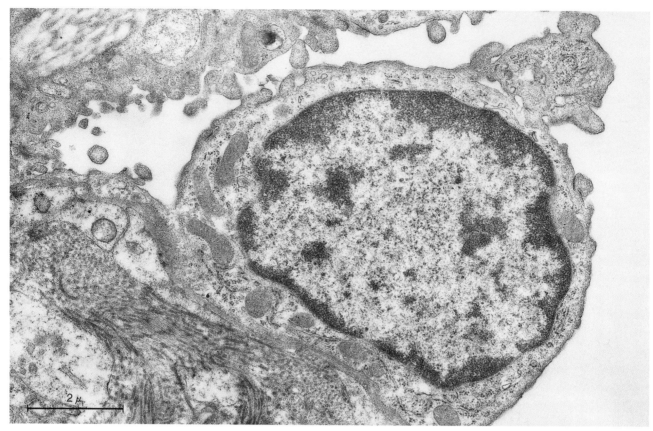

Fig. 8-17 Electron micrograph of type I cell. Elongated cytoplasm lining the alveolar wall is scant in organelles.

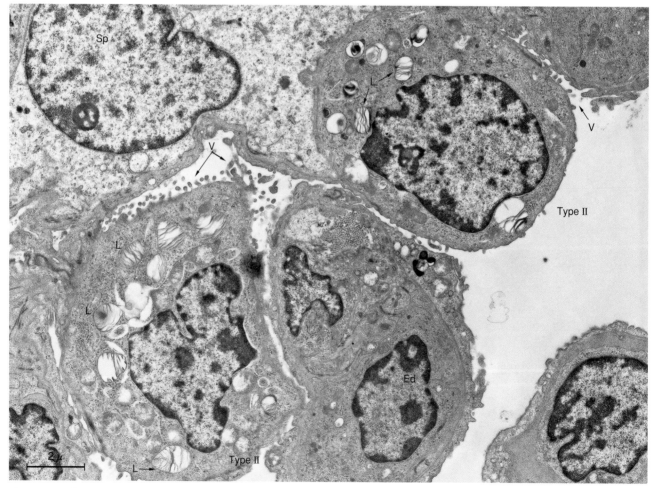

Fig. 8-18 Electron micrograph of type II cells which are characterized by osmiophilic lamellar bodies (L) and short microvilli (V). A fibroblast or septal cell is present (Sp). Ed: endothelial cell.

8-14a and 8-19). The staining reaction is bluish green or brownish green. The ciliated columnar cells are subject to degeneration and often lose their subnuclear tails.

Nucleus: Oval-shaped nuclei are eccentrically located. Well preserved cells showing a piling up of nuclei are arranged in parallel with a unipolar position. This cellular arrangement may be referable to as pseudostratification of the epithelium with polarity.

The chromatin is evenly and smoothly distributed except for a few karyosomes. Variation of nuclei is seen in size rather than in shape. Multinucleation occasionally occurs (see Figs. 8-26 and 8-43).

2) Goblet cells
Cytoplasm: The goblet type of columnar cell is stout and lantern-shaped. It is usually only lightly stained because of its mucus content (Figs. 8-13d and 8-14b). The periodic acid-Schiff reaction or Hale's colloidal iron method shows a positive stain. Cilia are absent.

Nucleus: It is often pyknotic and extremely eccentric in position.

3) Deep cells (basal cells)
Round small cells lying at the basal layer of the bronchial epithelium correspond to the reserve cells (Halper) as described in the chapter on the female genital tract (Fig. 8-14c). Desquamation of the deep cells solely into sputum is infrequent and thereby their identification is difficult. When they are present, attached to ciliated columnar cells in the cell groups which are obtained by scrapings their identification is relatively easy.

Cytoplasm: The cytoplasm is sparse and is present just along one side around the nucleus. The staining reaction is basophilic and bluish green by the Papanicolaou method (Fig. 8-21).

Nucleus: The nuclei are small and oval-shaped, and are uniform in size and shape. The chromatin is evenly distributed.

4) Squamous epithelial cells
Squamous epithelial cells derived from the mouth and the pharynx are unavoidable elements in a sputum specimen. Since deep squamous cells exfoliate in the presence of erosion or ulceration of the mouth and pharynx, they should not be mistaken for squamous metastatic columnar cells or squamous cancer cells. The squamous cells in sputum are similar to those seen in the vaginal smear. The superficial cells with broad transparent cytoplasm and a dark pyknotic or oval vesicular nucleus are by far the most frequent cells seen in sputum smears.

5) Macrophages
Macrophages are the so-called alveolar histiocytes or alveolar dust cells which are the representatives of the reticulo-endothelial system in the lung. They are phagocytic and function to clean the air passages (Fig. 8-13a). The presence of these cells in sputum usually indicates that the specimen is sufficient material for cytologic study. In other words, the specimen containing only squamous cells and no histiocytes is not worthy of pulmonary cytology.

Counts of alveolar macrophages reflect the lung reaction against inhaled dusts caused by various occupational pollution.[122] The dose-effect relationship cannot be clearly proved. The count is markedly influenced synergistically by tobacco smoking.[122]

Cytoplasm: The cytoplasm is broad, finely vacuolated, and contains phagocytosed dust particles. This is the reason why alveolar histiocytes are often referred to as "*dust cells.*" The staining reaction is bluish green to brownish green.

Nucleus: The nucleus is oval or bean-shaped and eccentrically located. The chromatin is finely and evenly distributed except for a few clumps (chromocenters). The occasional prominence of the nucleolus should not be mistaken for a malignant feature. Multinucleation of macrophages is not infrequent (Figs. 8-13c and 8-23).

Hemosiderin-laden macrophages: In case of heart failure, extravasation of erythrocytes following chronic congestion of pulmonary circulation may occur. It results in mobilization of hemosiderin-laden macrophages which are also called "*heart failure cells.*" However, these cells are not specific only to heart failure but are also derived from the various causes of hemorrhage in the lung. Since lung cancer often brings about occult hemorrhage, appearance of hemosiderin-laden macrophages is one of the indirect criteria of malignancy.

Lipophagocytes: The cytoplasm is abundant and finely vacuolated because of its lipoid content. The identification of lipophagocytes must depend on specific fat stains such as sudan III or oil red O stain. The entity of lipophagocytes will be summarized in two forms: (1) Lipophagocytes phagocytosing exogenous lipid material, i.e., olive oil used for nasal drops and lipiodol used as a contrast medium in bronchography. (2) Lipophagocytes phagocytosing endogenous lipid, mainly cholesterol originating from the blood serum or extruded from necrotic tissues. Therefore, the presence of lipophagocytes in sputum may suggest clinically not only lipid pneumonia but also necrobiosis of malignant tumor tissues or inflammatory tissues. Cytologically lipophagocytes with foamy abundant cytoplasm appear to show phagocytosis of few or no dust particles; phagocytosis of macrophages solely of lipid is referred to be as blocking with lipid (Fig. 8-24).

▮ Differences in cytologic features between sputum and bronchial scrapings
The columnar cells submitted by mechanical scrapings such as curetting, brushing and swabbing may appear in cell groups. The cytoplasm is well preserved and retains its ciliation at the broad cellular rim. The nuclei appear somewhat larger in scraping smears than in sputum smears. Furthermore, the nuclei show a finely granular or reticular chromatin pattern and delicate, smooth nuclear outlines. The cellular components are usually simple and composed of ciliated columnar cells, goblet cells and some dust cells. Numerous erythrocytes are often present because of direct scraping of the mucous membrane.

The presence of erythrocytes and/or hemosiderin-laden macrophages in sputum, however, has some pathognomonic meaning suggestive of intraparenchymal hemorrhage.

It is summarized that bronchial scrapings may represent pathologic changes of the bronchial mucosa; shift

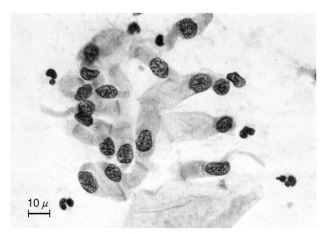

Fig. 8-19 Well preserved ciliated columnar cells and goblet cells in a bronchial scraping smear. Cilia and terminal plates are clearly seen. The nuclei are uniform in shape and possess finely granular, evenly distributed chromatins. The lantern-shaped goblet cells have broad clear cytoplasm and basally situated nuclei.

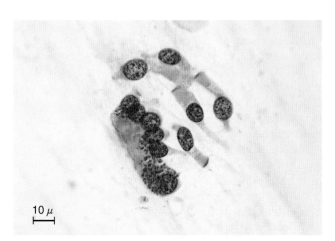

Fig. 8-20 Ciliated columnar cells containing lipofuscin granules. The nuclei are oval and regularly arranged side by side. Note deposition of brownish granules in the supranuclear portion of the cytoplasm.

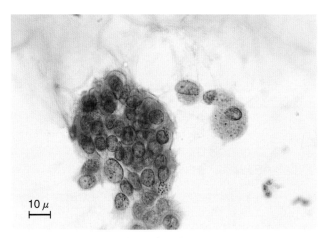

Fig. 8-21 Ciliated columnar cells seen in a cluster and a few dust cells in aerosol-induced sputum. A tight overlapping aggregate of small nuclei is referable to as hyperplasia of basal cells. Diagnostic points: In spite of heavy clumping of the nuclei, (1) uniformity of nuclear shape, (2) a bland chromatin pattern, and (3) preservation of terminal plate are indicative of benignancy.

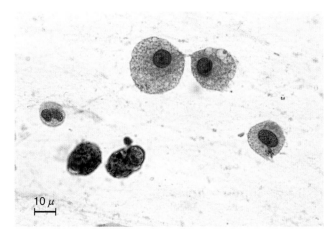

Fig. 8-22 Dust cells and lipophagocytes in sputum. A so-called dust cell contains tiny dust particles, whereas a lipophagocyte is filled with sudanophilic droplets varying in size. The presence of a large number of lipid-laden macrophages is indicative of lipid pneumonia.

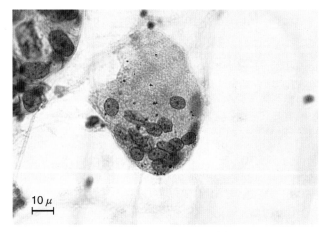

Fig. 8-23 Multinucleated histiocyte of foreign body type in sputum. Multiple oval or round nuclei are contained in abundant cytoplasm, cellular border of which is poorly defined. Several tiny dust particles are contained in the cytoplasm.

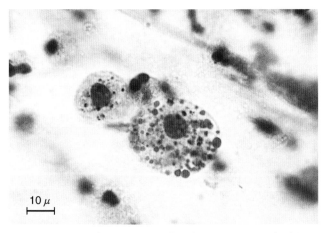

Fig. 8-24 Lipophagocytes in lipid pneumonia. The histologic section shows a fulfill of alveolar space by sudanophilic lipophagocytes.

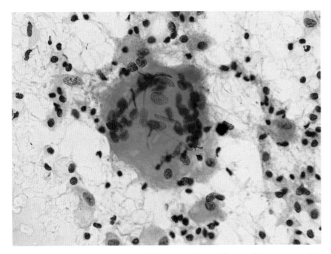

Fig. 8-25 Multinucleated histiocyte of Langhans type in bronchial brushing from a case of sarcoidosis. The elliptic nuclei arranged along the periphery of the cell may suggest the so-called Langhans type of histiocytes.

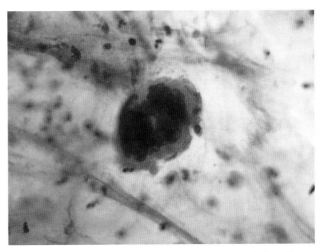

Fig. 8-26 Multinucleated columnar epithelial cells in sputum of a heavy smoker. Many nuclei appear to be fused together within the same cytoplasm like syncytial cells. Note nuclear pyknosis.

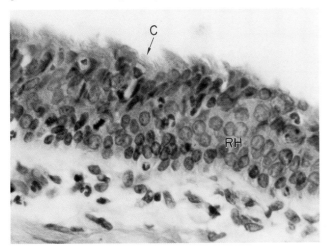

Fig. 8-27 Basal cell hyperplasia without atypia. The initial stage of metaplasia begins with proliferation of multipotential basal cells (reserve cells) with ovoid nuclei and scanty cytoplasm (RH). These nuclei are uniform in size and shape, and regularly arranged. Ciliated columnar cells remain on the surface (C).

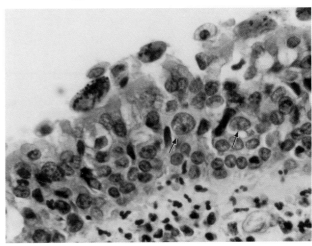

Fig. 8-28 Basal cell hyperplasia with atypia. Accompanied with proliferation of basal or reserve cells are the size variation of nuclei (arrows) and distortion of polarity.

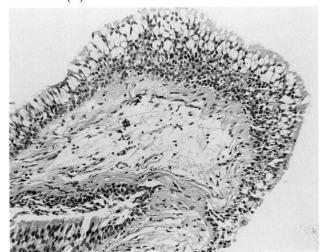

Fig. 8-29 Marked goblet cell and basal cell hyperplasia of bronchial epithelium from a heavy smoker. The ciliated mucosal epithelium is largely replaced by goblet cells in association with moderate basal cell hyperplasia. Increase in goblet cells is not specific but a frequent association in heavy smoker.

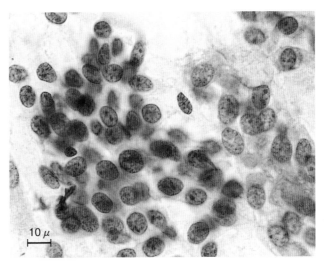

Fig. 8-30 Basal cell hyperplasia in bronchial brushing from a heavy smoker. A large cluster of uniform round nuclei with bland chromatin pattern indicates basal cell hyperplasia. There is no variation of the nuclei in size and in shape. Ciliated cells are seen in the right side.

in population of cell components and morphological alteration become the target of observation. Goblet cell hyperplasia (Fig. 8-29) and squamous cell metaplasia of the bronchial epithelium that often occur in heavy smokers will be represented by increased number of goblet cells and appearance of polyhedral epithelial cells arranged in sheet.

Non-cellular components in sputum are of diagnostic value as the products of airways with pathologic changes; Curschmann's spirals are considered to be derived from inspissated secrete in bronchioli, and a number of lipophagocytes are referred to as lipid pneumonitis. Amorphous proteinaceous mass is the product of alveolar proteinosis. The columnar cells in sputum are scattered singly and more or less degenerated. Cilia are usually obscure or lost. The nuclei appear smaller and their chromatin pattern is more roughened or pyknotic. Cellular components are variable because of contamination with saliva and nasopharyngeal secretions; there may be a large number of superficial and intermediate squamous cells, alveolar histiocytes and some tall columnar cells from the nasal cavity.

6 Benign Epithelial Cell Changes

■Hyperplastic basal cells

Basal cell hyperplasia is a state which is characterized by proliferation of subcolumnar basal cells (Fig. 8-27). Multiple layers of small cells can be recognized beneath the columnar cells. The nuclei are round to oval and fairly uniform in shape. There is a slight to moderate variation in nuclear size. The chromatin is slightly increased in amount, somewhat granular but evenly distributed except for a few karyosomes. The cytoplasm is scanty, cyanophilic and present just around the nucleus. Well preserved cells may disclose cuboidal cytoplasm with somewhat eccentric nuclei. If the clusters of these cells are present beside the ciliated columnar cells, they may be easily identified as hyperplastic basal cells (Figs. 8-21 and 8-30).

■Squamous metaplastic cells

Squamous metaplasia of the bronchial epithelium is frequently observed in various pathologic conditions[55] such as organizing pneumonitis, infarction, bronchiectasis, healing abscess, tuberculosis, heavy smoking, emphysema and carcinoma. Precancerous metaplasia, meaning a pathologic state which may precede cancerization, was noted by Lindberg (1935) not only in the mucous membrane adjacent to the carcinoma but also in the remote bronchi. Such abnormal squamous metaplasia as referred to as precancerous metaplasia or atypical metaplasia (dysplasia), shows a distortion of cellular arrangement and variation of nuclear size. Definition of precancerous metaplasia was given because of much higher frequency of atypical metaplasia in cancer cases particularly of squamous and undifferentiated histologic types than in the control cases.[126,154,191] A long-term study by Saccomanno et al[157] on sputum specimens of uranium miners who developed pulmonary carcinoma displayed the fact that supports the theory: atypia of squamous metaplastic cells became progressively more marked up to the time when the clinical malignancy developed.

As regular metaplasia advances, it may mature to the spinal type of metaplastic cells. However, most metaplastic changes in the respiratory tract belong to the transitional stage (Fig. 8-33). It is difficult to identify mature metaplastic cells in a sputum specimen which is mixed with saliva and nasopharyngeal secretions. Koss[89] emphasizes the impossibility of cytological identification of metaplastic cells except for a limited condition when the problematic cells are found to be associated with bronchial columnar cells. However, such conditions are very rarely encountered. Since metaplastic changes are histologically extensive, some of these metaplastic cells may be unquestionably coughed up. Squamous metaplasia of the bronchus can be determined when we encounter a sputum smear of deep cough containing such cellular constituents as groups of polyhedral cells arranged in sheet with a cobblestone

Table 8-3 Cytological changes in squamous metaplasia

Cytology	Histology	Regular metaplasia		Atypical metaplasia	
		Immature metaplasia	Mature metaplasia	Mild atypical metaplasia	Marked atypical metaplasia
Cytoplasm	Arrangement	Cells in sheets		Cells in sheets and occasionally single	Predominantly single
	Shape	Spherical	Polyhedral	Polygonal	Irregular
	Amount	Moderate	Abundant	Moderate	Moderate
	Staining	Basophilic to cyanophilic	Cyanophilic to acidophilic	Acidophilic	Acidophilic (orageophilic) with refractiveness
	Density	Thin and indistinct	Distinctive	Distinctive	Thickened
Nuclei	Shape	Round to oval		Polygonal	Variable
	Size	Uniform		Slightly enlarged	Markedly enlarged
	Chromatin	Fine and powdery		Finely granular and increased amount	Roughly condensed or pyknotic, and markedly increased amount

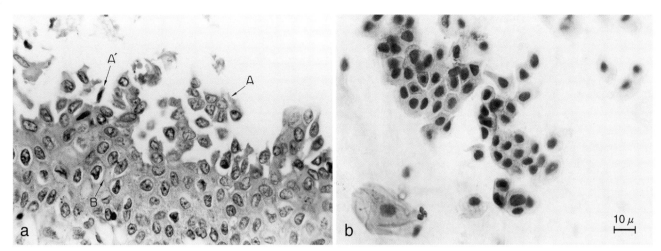

Fig. 8-31   Histology of squamous metaplasia and its exfoliative cytology.
a. Histology of typical squamous metaplasia showing the development of intercellular bridges and some parakeratosis.
b. Metaplastic cells in sputum probably desquamated from the parakeratotic surface of metaplastic epithelium (arrow A, A' in Fig. a). The cytoplasm is moderate in amount, polygonal and orangeophilic. Slit-like intercellular spaces are suggestive of intercellular bridges in squamous epithelium. Pyknosis of the nuclei may prove them to be peeled off from the parakeratotic epithelial surface.

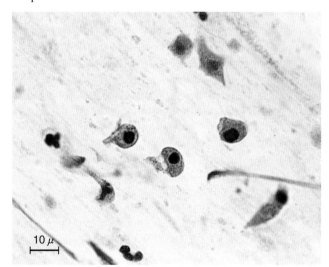

Fig. 8-32   So-called Pap cells in sputum. Note pyknotic nuclei with eosinophilic cytoplasm which were found in prolonged bronchitis.

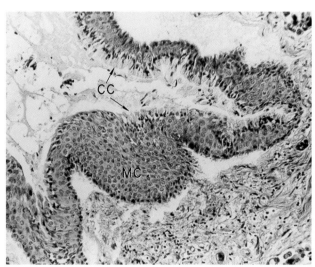

Fig. 8-33   Widely extended squamous metaplasia of transitional type. Since the ciliated columnar cells (CC) still remain lying on the surface of metaplastic cells (MC), it can be also called pretransitional epithelium as a precursor to transitional metaplasia.

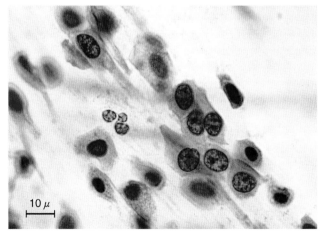

Fig. 8-34   Squamous metaplastic cells in sputum. Transitional to mature regular metaplasia. Diagnostic points: (1) polyhedral cytoplasm, (2) focal acidophilia, and (3) uniform nuclei arranged in a stone-pavement pattern. Somewhat immature (transitional) metaplastic cells retain smaller, thin cytoplasm.

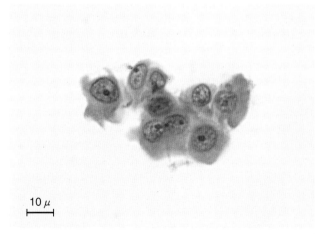

Fig. 8-35   Squamous metaplastic cells in sputum. Regular metaplastic change with moderate amount of cytoplasm. Diagnostic points: (1) dense cytoplasm with irregular polyhedral contours, (2) uniform nuclear shape though size variation is present, and (3) stone-pavement arrangement are distinctive of squamous metaplasia.

179

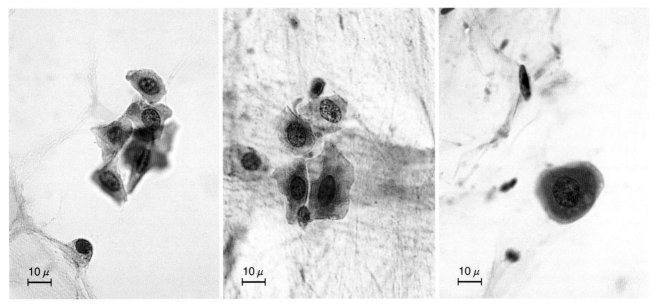

Fig. 8-36 Squamous metaplastic cells with marked atypia and dyskeratosis in sputum of a heavy smoker. The nuclei are enlarged and increased in chromatin content. However the chromatin distribution is even and fine. The N/C ratio is not much distorted. As metaplasia ripens, the cytoplasm becomes more abundant and thickened. Polychromasia of staining reaction depends on the grade of dyskeratosis.

appearance accompanying many columnar cells. Bronchial scraping smears, however, exhibit the confirmatory cellular features of squamous metaplastic cells (Fig. 8-31).

Cytology: The cytoplasm is small in amount and polyhedral, cubic or rectangular in shape. The cellular border is indistinct. The staining reaction is from cyanophilic to orangeophilic. Metaplastic cells seen in groups exhibit a characteristic cobblestone appearance (Fig. 8-31).

The nuclei are oval and fairly uniform in size. The chromatin is finely granular. Since the metaplastic cells in sputum are often degenerate, the nuclei are pyknotic and darkly blue as if they were painted. Atypical squamous metaplasia is usually classified into two or three grades, mild–marked or mild–moderate–marked.[156] Atypical metaplasia accompanies nuclear enlargement, anisokaryosis and hyperchromatosis of the nuclei, and intense orangeophilia of the cytoplasm; the grades of dyskeratosis of the cytoplasm with refractiveness and nuclear atypia associated with increased N/C ratio come to the point of differentiation between precancerous metaplasia and carcinoma in situ (Fig. 8-37) (see page 213). Dyskeratotic metaplastic cells with atypia tend to exfoliate singly rather than in sheet into sputum.

"Pap cells" are small elliptic cells with pyknotic nuclei and eosinophilic cytoplasm.[138] The name was assigned by Papanicolaou to a cell type observed in his sputum specimen when he was attacked by acute respiratory infection. The Pap cells are most likely pyknotic deep cells with a metaplastic process (Fig. 8-32).

■Atypical bronchial and bronchiolar hyperplastic cells

Atypical hyperplasia of bronchial and bronchiolar epithelial cells is not specific but is a frequently associated finding in bronchogenic carcinoma.[12,84,115,125] This term is equivalent to dysplasia that is commonly used in the cytopathology of the uterine cervix. Atypical hyperplasia of the bronchial epithelium is histologically characterized by a proliferation of the bronchial cells in association with irregular arrangement of cells such as micropapillomatosis (Figs. 8-39 and 8-40) and with moderate variation in the size of the nuclei. Cytologically the epithelial cells are in a cluster with overlapping of the nuclei. Variation of the nuclei is larger in size than in shape. There is no coarse clumping of chromatin. The nuclear rim is smoothly and distinctly outlined. The cytoplasm is fairly abundant, occasionally vacuolated and the cilia are retained at the broad border of the cell. The presence of cilia and terminal plate is one of the most reliable criteria of benignancy. Vacuolation of the cytoplasm has nothing to do with the differential diagnosis.

Atypical hyperplasia of the bronchioles and the alveolar ducts may occur extensively in pulmonary parenchyma adjacent to the peripheral carcinoma and in diffuse pulmonary fibrosis, bronchiolectasis and organizing pneumonitis, etc. Histologically the bronchiolo-alveolar ducts are lined by proliferated low columnar or cuboidal cells usually without cilia. Thus the histology is similar to bronchiolar carcinoma. When the cellular growth is solid, it is likely to be tumorlets (Whitwell).

Cytologically the hyperplastic bronchiolar cells appear in cluster with a papillary or acinar pattern. The nuclei are fairly small, round or oval, non-hyperchromatic, and generally uniform in size and shape. The cytoplasm is scanty, cuboidal in shape and mostly nonciliated (Figs. 8-44 and 8-45). Occasional cells may reveal cilia at the cytoplasmic rim. They may differ from bronchiolar carcinoma cells in the following cytologic features: (1) Sheets or clusters of cells exfoliated in sputum are larger in size and number in bronchiolar carcinoma than in

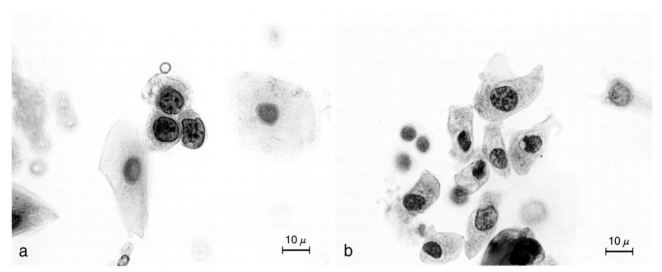

Fig. 8-37 Marked atypia (dysplasia) suspicious for in situ carcinoma in sputum of a heavy smoker. Small atypical cells with scant cytoplasm are from the deep layer (a) and polymorphic cells having polyhedral cytoplasm are from the superficial layer of dysplastic epithelium (b). Diagnostic points: (1) Although columnar shape of the cytoplasm still remains, abundant polyhedral cytoplasm is indicative of squamous metaplasia. (2) Pale cyanophilic staining reaction reflects a transitional stage of squamous metaplasia. (3) Hyperchromasia and size variation of the nuclei are referred to as marked atypia suggestive of in situ carcinoma.

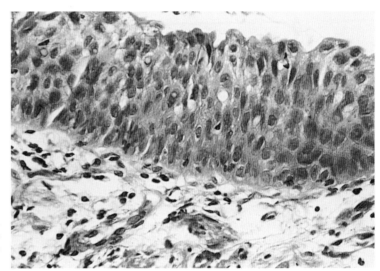

Fig. 8-38 Histology of marked atypia (dysplasia) of a heavy smoker. Note marked cellularity with loss of ciliation and surface differentiation. The deep cell layer reveals marked hyperchromatism, polymorphism and irregular arrangement.

bronchiolar hyperplasia. (2) Whereas the nuclei of bronchiolar carcinoma cells are rather small and uniform in size for cancer cells, they are apparently larger in size and possess more prominent nucleoli than the nuclei of bronchiolar hyperplastic cells. (3) The cytoplasm in bronchiolar carcinoma retains a columnar shape and is clear with transparency like goblet cells. The N/C ratio is conversely depressed compared with that of cuboidal bronchiolar cells.

Pulmonary tumorlet: Tumorlet (Whitwell[200]) is a small peripheral lesion in scarred pulmonary parenchyma that is composed of nests of hyperchromatic, spindly or round cells. The chromatin is finely and evenly distributed. There is no marked size variation of the nuclei. Although they resemble oat cells in shape, bland nuclear features differ from those of oat cells. Practically speaking, it is rare to encounter the tumorlet cells from severely scarred parenchyma: Bronchiectasis and bronchiolectasis are of the most predominant association with

tumorlets.[19,200] Tumorlet was defined as simple hyperplasia of bronchiolar epithelium accompanying chronic inflammation and fibrosis.[63,200] However, detection of argyrophil neurosecretory granules leads to consideration as a variant of peripheral carcinoid.[19,33] The concept of tumorlet should not be discarded until innocent biological behavior, lack of carcinoid syndrome and preponderance in females can be elucidated.

Cytology of tumorlet consisting of fairly uniform, small, round or oval cells with bland chromatins leads to false positive interpretation[104] (Fig. 8-45).

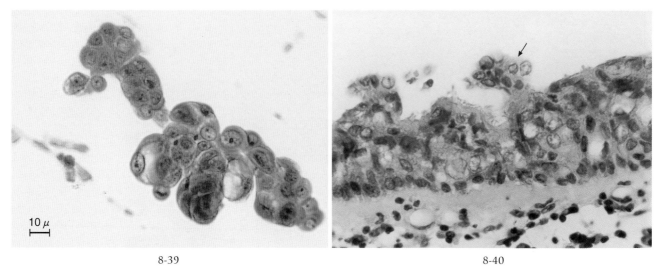

8-39   8-40

Fig. 8-39   Micropapillomatosis of bronchial epithelium. Grape-like clustered cells with a papillary or branching pattern look like cells derived from adenocarcinoma. However, cell constituents possess uniform oval nuclei with single small nucleoli.
Fig. 8-40   Atypical epithelium with micropapillomatosis. Note irregular stratification, distorted polarity, size variation of nuclei and minute papillary protrusions (arrow). Cilia are still retained.

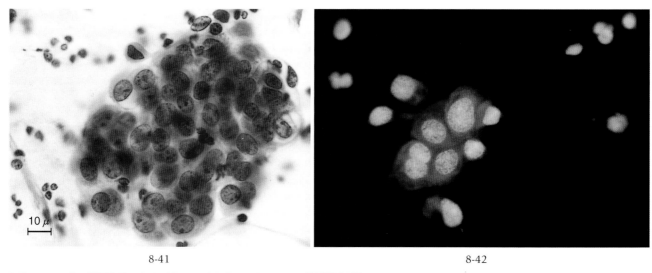

8-41   8-42

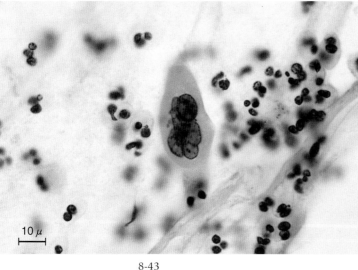

8-43

Fig. 8-41   Hyperplastic columnar cells shed in a cluster from a patient with chronic organizing pneumonitis. Such a large cell ball comprised of hyperchromatic cells may lead to misdiagnosis of adenocarcinoma. These are called "asthma cells" since often produced in asthmatics. The nuclei overlapping are uniform in size and shape. The terminal plate is recognizable by changing the focus.
Fig. 8-42   Hyperplastic columnar cells with slight atypia. Enlargement of nuclei with size variation and prominence of nucleoli are due to benign hyperplasia. Smooth delicate nuclear rims and finely granular chromatins are characteristic features of benignancy. Acridine orange fluorescence.
Fig. 8-43   Multinucleated columnar cells with squamous metaplasia in sputum from a patient with chronic organizing pneumonitis. Dense aggregation of hyperchromatic nuclei and squamous metaplastic alteration become a pitfall for malignancy. Diagnostic points: (1) The fact that aggregation of the nuclei is simply due to overlapping can be proved by changing the focus up and down. (2) The chromatin appears increased in amount but condensed at periphery. (3) Sharp cytoplasmic outline may be indicative of the terminal plate.

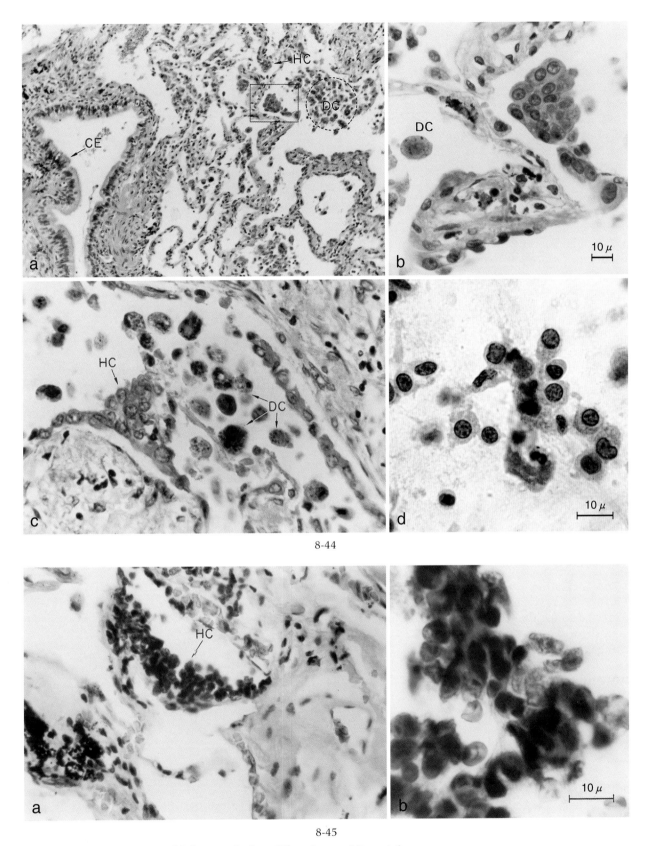

8-44

8-45

Figs. 8-44 and 8-45 Bronchiolar atypical proliferation and its cytology.
Fig. 8-44a, b Bronchus lined with pseudostratified ciliated columnar epithelium (CE) and alveolar ducts showing focally hyperplastic cuboidal cells (HC). Dust cells (DC) are present within the alveolar spaces.
Fig. 8-44c Solid cluster of hyperplastic cuboidal cells (HC) at the lining epithelium of the bronchiole.
Fig. 8-44d Group of hyperplastic bronchiolar cells in sputum. No cilia can be recognized. Note absence of coarse chromatin and a ground glass appearance of the nuclei; these features are sufficient to confirm their benignancy.
Fig. 8-45a Focal atypical proliferation of the terminal bronchiole. It is very likely "tumorlet."
Fig. 8-45b Large cluster of small hyperchromatic cells, which are quite similar to lymphoid cells except for a tight cell clumping and irregular indentation of nuclear borders.

7 Retrogressive Epithelial Cell Changes

■Ciliocytophthoria

Ciliocytophthoria designated by Papanicolaou[136,137] is a peculiar type degeneration of ciliated columnar epithelial cells. It has been emphasized that it can be observed in inflammatory disease, particularly in viral infections such as viral influenza and adenovirus infection.[137,142] In the case of viral infection, proliferation and desquamation of bronchial cells are pronounced.[172] However, it is not specific to viral infection but is observed in various diseases such as primary bronchogenic carcinoma, metastatic carcinoma, bacterial infection and bronchiectasis[153] (Table 8-4). It must be noted also that ciliocytophthoria is a much more frequent finding in sputum specimens than in bronchial scrapings or washings (Table 8-5).

Comparative cytological observations limited to curable cancer exhibit a different incidence of ciliocytophthoria in between sputum and bronchial scrapings. This evidence may support the proposition that ciliocytophthoria is the result of degeneration of bronchial epithelial cells. There are the following representative cellular changes, among which transitional forms can be recognized.

1) Pyknosis and karyorrhexis. The chromatins condense as a compact, densely stained material (pyknosis) or aggregate as coarse granules at the nuclear rim (karyorrhexis). The nuclear membrane comes to be disintegrated and lost.

2) Eosinophilic inclusions. A tail-like subnuclear portion of the cytoplasm is lost. The degenerate cytoplasm occasionally contains eosinophilic inclusion bodies which are positive for periodic acid-Schiff stain (Fig. 8-47).

3) Tuft formation. Bushes of cilia arising from a narrowed rim of degenerate cytoplasm look like tufts (Fig. 8-46). The cytoplasm remaining consists of the anucleated proximal portion that is formed as a result of the pinching off process.

■Multinucleation

Masugi and Minami (1938) first reported multinucleation of the bronchial epithelium in measles. It is thought that binucleation and multinucleation may occur as a reactive phenomenon in various diseases such as tuberculosis, bronchiectasis, pneumonia, carcinoma and also in irradiation.[63] Regarding the entity of multinucleation it is considered that the fusion of cells is more probable than an activated mitotic division or amitotic duplication of nuclei. Multinucleation of the pseudostratified columnar epithelium is present in the superficial layer where mitotic activity is usually absent. Two or three nuclei are arranged along the long axis of the cells. The nuclei are round or oval, fairly uniform in size and bordered with smooth and distinct membranes. Each of these multiple nuclei is occasionally indistinguishable because of pyknosis. The evidence may indicate a degenerating process to be responsible for multinucleation (see Fig. 8-26).

Table 8-4 Incidence of ciliocytophthoria (CCP) in pulmonary disease

Pulmonary disease	Number of cases	Number of CCP positive	Percent of CCP positive
Chronic pulmonary tuberculosis	62	6	9.7
Bronchiectasis	10	1	10.0
Chronic lung abscess	4	0	0
Chronic bronchitis and emphysema	28	2	8.0
Congestive heart failure	6	0	0
Allergic bronchial asthma	4	0	0
Bacterial pneumonia	26	3	11.5
Upper respiratory infection (viral)	52	13	25.0
Viral pneumonia	41	11	26.8
Metastatic lung cancer	8	4	50.0
Bronchogenic carcinoma	44	25	57.0
Acute bronchitis (viral)	16	12	75.0
Total	301	77	25.6

(From Rosenblatt MB, et al: Dis Chest 43: 605, 1963.)

Table 8-5 Comparison of cytology between sputum and bronchial scrapings: limited to curable cancer cases

Cytology	Cases	26 curable cancers Sputum	26 curable cancers Selective scrapings
Hyperplasia (Dyskaryosis) mild		4 (15.4%)	10 (38.5%)
Hyperplasia (Dyskaryosis) severe		0	3 (11.5%)
Metaplasia without atypia		4 (15.4%)	6 (23.1%)
Metaplasia with atypia		0	2 (7.7%)
Multinucleation		1 (3.8%)	8 (30.7%)
Ciliocytophthoria		13 (50.0%)	2 (7.7%)
Curschmann's spiral		2 (7.7%)	0
Positive cytology		9 (34.6%)	23 (88.5%)

■Pigmentation

Ciliated columnar cells occasionally contain yellowish brown granules in the supranuclear portion of the cytoplasm (see Fig. 8-20). This granules differs from hemosiderin or melanin cytochemically and is likely to be a lipofuscin that is also called waste pigment. This pigment is not soluble in acid, alkali and various lipid solvents, but is stained with basic fuchsin, methylene blue, Nile blue sulfate and various dyes for lipids. It shows a positive periodic acid-Schiff reaction and a negative iron stain. Schmorl's method and Mallory's method are both useful for demonstration of the pigment but not absolutely specific because lipofuscin is not a pure substance. These granules may be admixtures of unsaturated lipids, lipoproteins and some carbohydrates. Although the pathogenic nature of the pigments in bronchial epithelium is not clear, they seem to be remarkable in old age groups with bronchogenic carcinoma and after radiation therapy.

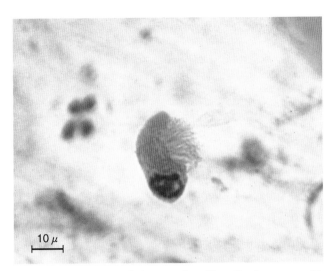

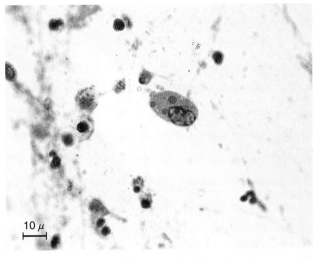

Fig. 8-46   Ciliocytophthoria of a ciliated columnar cell. Note tuft formation of cilia. Karyorrhexis with chromatin condensation at the nuclear rim and striking eosinophilia of the cytoplasm are seen. Papanicolaou stain.

Fig. 8-47   Ciliocytophthoria of a ciliated columnar cell. Intensely eosinophilic inclusions are included in the cytoplasm. Cilia are lost and chromatins are condensed at the nuclear rim. Papanicolaou stain.

Fig. 8-48   Bronchial epithelial change in herpes virus infection. The nuclei are enlarged and show a ground glass appearance. Double nucleation is seen.
Fig. 8-49   Cytomegalic viral infection of columnar epithelial cells. The affected cells are enlarged and stain darker (more darkly blue) than the noninfected epithelial cells. The enlarged nucleus containing a dark amphoteric inclusion surrounded by halo reveals a characteristic owl-eyed appearance. **a.** Bronchial aspiration. **b.** Imprint of pulmonary lesions.

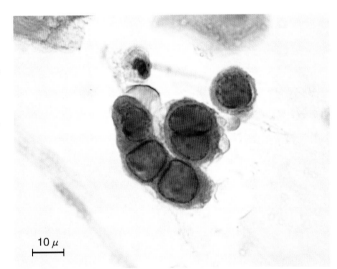

8-48

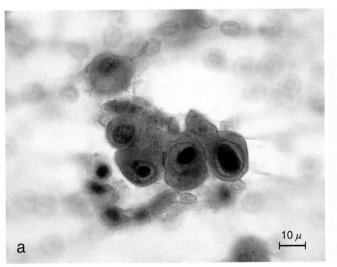

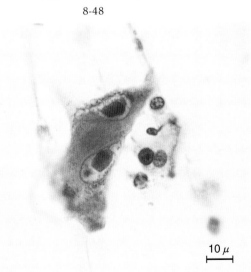

a

b

8-49

■ Epithelial changes in viral infection

Herpetic tracheobronchitis is not a rare occurrence in children as well as in adults within the immunodeficient state. Debilitated patients with advanced carcinoma are occasionally affected with other visceral involvement.[90,149] Columnar epithelial cells enlarged by herpes simplex infection should be differentiated from malignant cells. Attention should be paid for other distinctive features such as eosinophilic intranuclear inclusion, multinucleation and ground glass appearance (Fig. 8-48).

Most fetal cytomegalovirus pneumonitis found in infants is caused by intrauterine infection. The adult who has been administered immunosuppressive therapy is also affected by cytomegalovirus as an opportunistic infection (Fig. 8-49). An association with *Pneumocystis carinii* pneumonia (see Fig. 8-72) and other opportunistic infections such as systemic fungus infection may occur. The nuclei of affected bronchiolar and bronchial epithelial cells become enlarged four to six times and contain amphoteric intranuclear inclusion with halo. The inclusion body consists of morula-like, aggregated granules at first and become a large condensed body mass. There is formation of cytoplasmic inclusions which are intensively positive for PAS reaction.

In *adenovirus infection*, formation of intranuclear inclusions is somewhat different according to its types; the types 1, 2, 5 and 6 which are the cause of mild upper respiratory infection may initially produce a small inclusion with acidophilic reaction and later on a large basophilic inclusion with connection to the nuclear rim by chromatin threads. The type 3 or 4, occasionally giving rise to acute inflammation of the lower respiratory tract, produces several dispersed inclusion bodies with basophilia.

Cytologically infected cells are characterized by enlarged nuclei which stain homogeneously basophilic or amphoteric (smudge cells).[11]

8 Abnormal Changes of Histiocytic Cells

Various kinds of substances phagocytosed by macrophages may suggest the pathologic state of the respiratory tract. An example is the so-called "heart failure cell" that is a hemosiderin-laden macrophage in chronic congestive heart disease.

■ Lipophagocytes

A large number of lipid-laden macrophages with sudanophilia are expectorated in lipid pneumonia (Fig. 8-24), but they are not specific to lipid pneumonia; nonspecific inflammatory process and necrobiosis of tissues regardless of malignancy are responsible for production of lipophagocytes.

■ Hemosiderin-laden macrophages

Patients with congestive heart failure and industrial workers exposed to iron oxide dusts cough out hemosiderin-laden macrophages with brownish granular pigments. This pigment is positive for Prussian blue and frequently for PAS reaction, and is distinguishable from sudanophilic lipofuscin pigment with ease.

■ Melanin-laden macrophages

Melanophages are rarely observed in cases where melanoma involves the lung primarily or secondarily. Even in the absence of melanoma cells, the presence of melanophages becomes a clue to detect malignant melanoma affecting the respiratory tract. It is important to distinguish melanophages from neoplastic melanoma cells. This rare primary tumor has been reported to arise from the large bronchi as a polypoid tumor with dark brownish color.[127,135] For pigmented cells the bleaching technique is of practical use, whereas the Dopa reaction is appreciative for identification of amelanotic melanoma. Melanoma cells are variable in shape, varying from polygonal to spindly, and possess large droplet-like nucleoli. The melanin pigment should be confirmed by Fontana's silver impregnation method and bleaching method.

■ Bile pigment-laden macrophages

Macrophages phagocytosing bile pigments which are recognizable as greenish yellow droplets may occur in rare cases with metastatic hepatocellular carcinoma producing bile pigments. The bile pigments can be proved by Gmelin reaction.

■ Multinucleated giant cells

Two types of multinucleated histiocytes are encountered in pulmonary cytology. Langhans type discloses the same nuclear feature as epithelioid cell which is described below. Foreign body type giant cells appear in the vicinity of metabolic by-products and foreign bodies. As one good example, multinucleated histiocytes are frequently seen around necrosis of cancer tissues in good radiation response. Attention should be paid for phagocytosed material either inorganic substances or organisms in multinucleated giant cells. The important organisms giving rise to pneumonitis are *Paracoccidioides brasiliensis*, *Blastomyces dermatidis*, *Cryptococcus neoformans*, *Histoplasmosis capsulatum*, etc. Blastomycosis is particularly a point of consideration when many multinucleated giant cells are encountered in sputum in endemic areas (see Fig. 8-69).

Cytologically the cytoplasm is abundant, foamy and ill-defined. Ingested leukocytes or cellular debris are often contained. The nuclei are oval or somewhat elongated with bean-shaped contour. The nuclear border is delicate but distinct. Chromatins are finely granular or reticular. Usually single nucleolus is present in each nucleus (Fig. 8-23). Differentiation between histiocytic giant cells and epithelial giant cells is especially based on cytoplasmic features. The epithelial type giant cells possess a distinct cellular border with often terminal plate at the broad rim and a dense staining reaction of the cytoplasm. Phagocytosis is not found in the epithelial type (Fig. 8-26).

■ Epithelioid cells and Langhans giant cells

Epithelioid cells are ordinarily not expectorated in spontaneous sputum, but recognizable in needle aspirates or

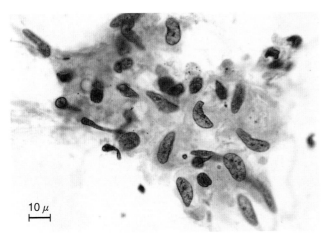

Fig. 8-50 Epithelioid cells in a needle aspirate. Transcutaneous needle aspiration from the pulmonary lesion showing a nodular X-ray shadow reveals many epithelioid cells, the nuclei of which are elongated and curved in shape. Their chromatin pattern is bland. The cellular borders are obscure. If necrotic debris is associated, it is suggestive of tuberculous caseous necrosis.

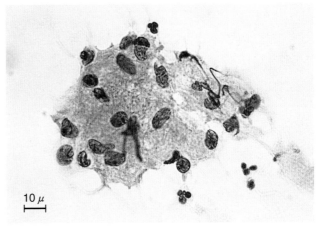

Fig. 8-51 Shorr-stained epithelioid cells in bronchial brushings from a 41-year-old woman showing bilateral hilarlymph node adenopathy. The nuclei are elongated, slightly curved and show fine chromatins. Occurrence of epithelioid cells from beneath the mucosal membrane in absence of necrosis is suggestive of sarcoidosis. Clinical investigation is necessary for diagnosis of sarcoidosis.

bronchial brushings. These are transformed macrophages and characterized by somewhat curved, elliptic nuclei with pale vesicular chromatin framework and delicate nuclear membrane (Figs. 8-50 and 8-51). Multinucleated Langhans giant cells have peripherally locating nuclei with a semicircular arrangement (Fig. 8-25). The nuclear feature is exactly the same as that of epithelioid cells. We must take into account various epithelioid cell granuloma formation such as tuberculosis, sarcoidosis and berylliosis for its pathological interpretation. Sarcoid reaction in association with malignant neoplasm is rare but may occur. The mixture of necrotic debris with scant inflammatory cell reaction reflects caseous necrosis in tuberculosis. Typical asteroid body and Schaumann body are suggestive of sarcoidosis, but their appearance is rare even in histology. A report by Aisner et al[3] on sputum cytology in patients with sarcoidosis mentions significant cytologic features in 6 cases of 16 patients as summarized below; (1) epithelioid cells with one or many nuclei, (2) clean background with little mucus and inflammation, and rarely (3) Schaumann body and asteroid body in giant cells. In our experience of histologic study on bronchoscopic punch biopsy specimens, 16 of 39 cases (41.0%) with clinical manifestation of sarcoidosis disclosed characteristic epithelioid cell granuloma with mantle of lymphocytic infiltration in the bronchial mucosa; bronchoscopic scraping smear that is characterized by epithelioid cells and lymphocytes within a clean background is favorable to sarcoidosis.

9 Non-Cellular Elements

■Curschmann's spirals
Patients suffering from chronic bronchial disorders particularly with incomplete bronchial obstruction and marked mucus production may cough out inspissated mucous casts in bronchioli which are called Cursch-

mann's spirals. This is a coil-like structure which is composed of an axis densely stained with hematoxylin and a translucent mucus mantle or sheath. Thus the latter is evidently positive for periodic acid-Schiff reaction as well as alcian blue stain. Attention must be paid to the fact that expectoration of Curschmann's spirals is a frequent accompaniment of bronchogenic carcinoma as a result of incomplete obstruction of the affected bronchi and bronchioli (Figs. 8-52 and 8-53).

■Fiber-like substance
It is not infrequent to see fiber-like structures in sputum smears of patients suffering from suppurative bronchial disorders such as bronchitis, bronchiectasis and pulmonary abscess. They are found more frequently in purulent sputum than in mucous or serous sputum. The structures are densely stained by hematoxylin and positive for Feulgen reaction (Fig. 8-54). They are liable to be spread in the same direction as that of smearing and show knot-like protrusions corresponding to the nuclei of polymorphonuclear leukocytes. There is no doubt that the structures are derived from the nucleoprotein of degenerative polymorphonuclear leukocytes.

■Calcific concretions
Calcific concretions are rarely encountered in sputum. Needle aspiration and bronchial brushings may demonstrate calcific concretions showing basophilic lamellar structures (Figs. 8-55 to 8-57). The center of the structures usually stains denser than the periphery. Papillary adenocarcinoma, tuberculosis and alveolar microlithiasis may be the cause of calcific concretions. The latter is a rare condition which was first described by Harbitz (1918); it is characterized by a filling of alveoli and bronchioles with psammoma-like bodies termed calcospherites. It is considered that alveolar microlithiasis may be associated with chronic mitral stenosis and pulmonary fibrosis.[172]

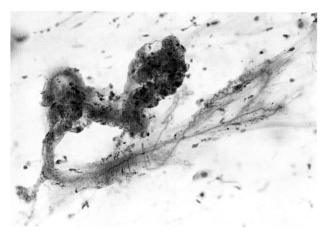

Fig. 8-52 Curschmann's spirals in sputum. Spirally twisted masses of mucus concentrate with a coiled axis that is stained darkly by hematoxylin. A semitranslucent sheath enwrapping the axis is confirmed to be composed of mucus material by PAS reaction.

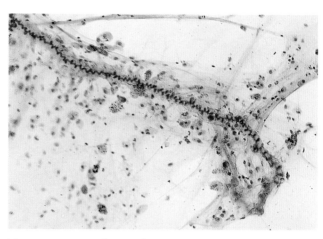

Fig. 8-53 Curschmann's spirals in sputum from a patient with pulmonary carcinoma. Note a central spiral axis that is stained intensely by hematoxylin. The mucoid sheath is translucent.

Fig. 8-54 Fiber-like structure in a muco-purulent sputum. These are often found in smears of purulent sputum as a fiber-like structure that is stained intensely by hematoxylin.

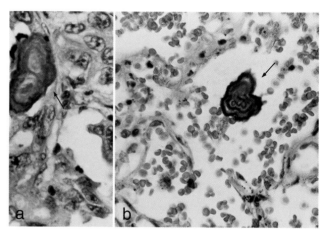

Fig. 8-55 Psammoma body in the alveolar duct from a case of bronchogenic papillary adenocarcinoma. This laminated calcific concretion was found in the alveolar duct (b) distant from the primary tumor. A part of papillary adenocarcinoma with a psammoma body is shown (a).

■Corpora amylacea

Corpora amylacea are lamellar structures measuring 60 to 100 μm in diameter. They were first found in sputum by Friedrich (1856). The concentric ringed lines are 100 to 150 Å thick by electron microscopic observation. This homogeneous substance, looking like calcific concretions, stains lightly with eosin because of the absence of calcium and is negative for periodic acid-Schiff (PAS) reaction. Histologically they are found in the alveoli with weak tissue reaction. The origin of *corpora amylacea* is still unknown.

■Asbestos body

Occupational high risk of pulmonary carcinoma and pleural mesothelioma in asbestosis[202] is related to dose, shape and size of asbestos fibers and length of duration of exposure.[58] As a cofactor to carcinogenicity, multiplicative effect of cigarette smoking is emphasized.[17,190] According to histological typing of asbestos cancers by Whitwell and associates,[201] asbestos dusts lying in distal parts of the bronchial trees are more frequently coexis-

tent with adenocarcinoma than with other cell types of carcinoma. Asbestos body appears, however, in other cell types of carcinoma.[40] Asbestos body appears in sputum as an elongated needle having a yellowish, ferruginous coat by iron salt and protein with beaded shape.

Asbestos bodies coated with ferroprotein matrix are recently called ferruginous bodies, because a variety of inhaled microscopic fibers can be nonspecifically converted to beaded bodies. Strictly speaking the term of asbestos bodies should be used only when the central fibers are composed of asbestos.[150] It has been experimentally verified that glass fibers inhaled in guinea pigs are coated with ferroprotein, fragmented, and beaded like asbestos bodies.[21] The ferruginous bodies can be quantitatively estimated from lung parenchyma by chemical digestion using domestic laundry bleaching fluid (5.25% sodium hypochlorite).[166] The residue which is suspended with equal amount of chloroform and 50% ethanol is concentrated by centrifugation. The sediment is resuspended with 95% ethanol and filtered through Nuclepore with 5 μm pore size. It is ready for

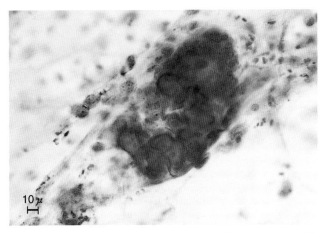

Fig. 8-56 Psammoma body in sputum from a case with bronchogenic adenocarcinoma. Note irregularly aggregated calcific bodies with thickened oyster shell-like outlines. The dark bluish brown coloration of the border is due to calcium deposition.

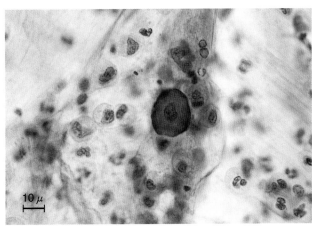

Fig. 8-57 Psammoma body in sputum from a case with tuberculosis. Hematoxylin-stained laminated corpuscle is not specific to tuberculosis. Over all observation to exclude papillary adenocarcinoma is necessary.

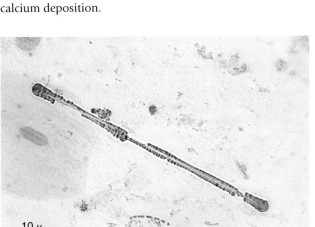

Fig. 8-58 Asbestos body in sputum. A long foil-like structure with a grip is yellow to dark brown in color; variation of color depends on the amount of coated ferroprotein.

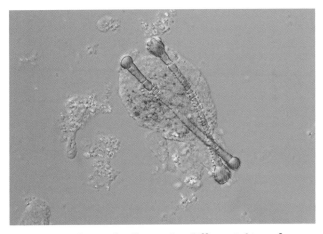

Fig. 8-59 Asbestos bodies under differential interference microscopy. The characteristic structure remains in unfixed specimen without stain.

counting by ordinary light microscopy without staining.

Cytology: According to the description on various types of the ferruginous body by Lopes Cardozo, the ferruginous thread with a few yellow deposits in an early stage becomes a foil-like body with a yellow grip at one or both ends. Lastly yellowish heme crystals deposit on the thread; thus, a transversely striated or beaded body is formed (Figs. 8-58 and 8-59).[99]

Clinical meaning: Out of 1,256 transthoracic fine needle aspirates, Leiman[95] found 56 specimens with asbestos bodies which were associated with malignancies in 30 cases, active or healed tuberculosis in 14 cases and abscess 5 cases regardless of positive evidence in the specimen. The presence of asbestos bodies is an important marker of underlying pathology. The choice of cytology modalities, exfoliative or aspiration, depends on localization of the lesion. Wheeler et al emphasize superiority of bronchial washings than sputum specimens and serial sections of different levels of cell blocks using iron stain in addition to Papanicolaou stain.[199]

■Proteinaceous material

This substance in sputum has been studied cytochemically,[28,106] since Rosen et al[151] described a peculiar and progressive disease which was termed alveolar proteinosis (Fig. 8-60). Macroscopically the sputum appears to be tenacious, amorphous and acellular like gelatin and grey-chalky in appearance. Microscopically it is rich in granular, amorphous and acellular substance which is acidophilic, sudanophilic, and positive for periodic acid-Schiff (PAS) reaction after diastase digestion. Crystal violet and Meyer's mucicarmine stains for amyloid and mucin are both negative.

■Charcot-Leyden crystals

The so-called Charcot-Leyden crystals are octahedral in shape, eosinophilic and are found, although rarely, with a number of eosinophilic leukocytes in the sputum of asthmatics (Fig. 8-61). The crystals are thought to be produced by crystallization of the dissolved substance of eosinophilic granules.

Eosinophilic pneumonia is found radiographically as

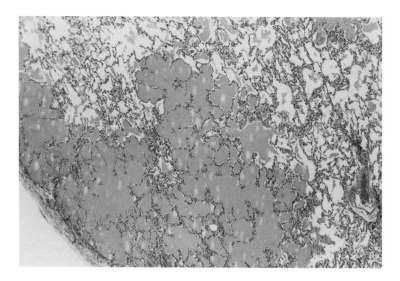

Fig. 8-60   Alveolar proteinosis showing fullness of alveolar ducts with PAS-positive, amorphous, proteinaceous material. This 16-year-old boy was pointed out abnormal flecked shadows bilaterally in chest X-ray films. Bronchial washing revealed slightly eosinophilic amorphous material without cellular elements.

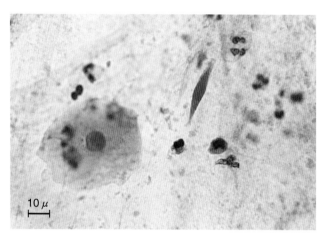

Fig. 8-61   Charcot-Leyden crystal in sputum from the patient with bronchial asthma. An octahedral crystal that is apparently acidophilic is often found in allergic and asthmatic patients. The crystals are associated.

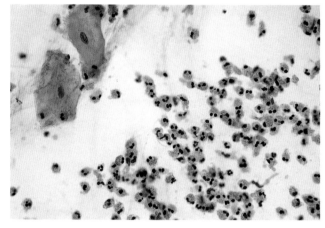

Fig. 8-62   Eosinophilic pneumonitis. Sputum specimen. Note major population of eosinophils which stain yellowish brown by Papanicolaou stain.

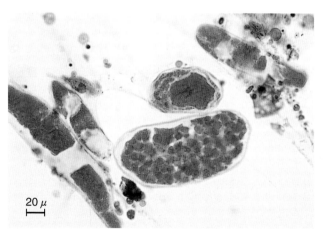

Fig. 8-63   Some plant expectorated in sputum. Note thick refractive cell wall. Bizarre shapes and variable sizes of these substances with odd interiors are indicative of food contaminants.

irregular opacities and recognized cytologically by predominance of eosinophils and occasionally Charcot-Leyden crystals (Fig. 8-62). This is not a disease entity but associated with asthma (allergic alveolitis), drug reaction, atopy and parasite infection.

■Elastic fiber
Elastic fibers which are highly refractile and delicate may be coughed up in cases of acute destruction of the pulmonary parenchyma such as in cavitation.

■Other non-cellular elements and infection
Since various kinds of bacterial flora are present in the mouth and the nasopharynx, the bacterial content of sputum is not unusual and not informative as to the etiologic organism of inflammation. A large number of pneumococci, hemolytic streptococci and *Klebsiella pneumoniae* may be pathognomonic. *Borrelia vincentii* and fusiform bacilli are occasionally found in sputum of the patients with pulmonary abscess, gangrene and long-standing bronchiectasis. Pollen is a common contaminant in sputum (Fig. 8-64). Some pollens resembling parasite ova can be distinguished by refractive cell walls and absence of miracidium in the interior. Plant cells

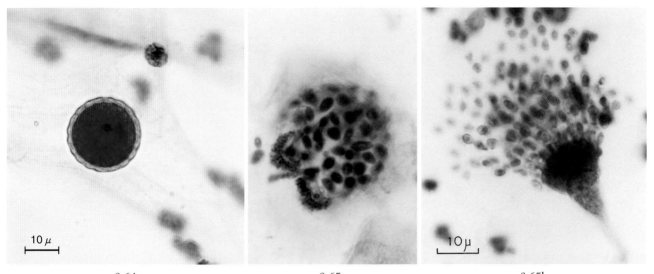

8-64 8-65a 8-65b

Fig. 8-64　Pollen in a sputum smear. Pollens are frequent contaminants in sputum and show various shapes. But the cell wall is thick and refractive. Although the interior is granular, no miracidium can be observed.

Fig. 8-65　*Aspergillus fumigatus* showing conidial production in a sputum smear. *Aspergillus fumigatus* is occasionally the cause of pulmonary infection. Note a characteristic conidiospore and sterigma showing chains of single-celled globose conidia. This structure of a bulbar fruiting body is similar to the conidiospore of penicillium fungi which are contaminants in sputum smears.

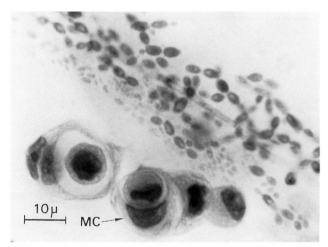

Fig. 8-66　Yeast-like fungi showing blastospores and pseudohyphae in a sputum smear. There are also malignant cells showing cannibalism and nuclear pyknosis (MC).

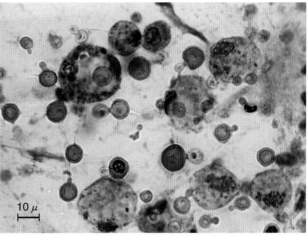

Fig. 8-67　Cryptococcosis in sputum from a patient who has received an immunosuppressive therapy. Note pinkish to bluish, spherical bodies. A transparent, mucinous capsule around the spherical body is noticed within the cytoplasm of a phagocyte.

derived from food are often mixed in sputum unless patients gargle before expectoration of sputum (Fig. 8-63).

Actinomycosis: Pulmonary actinomycosis results from aspiration of the organism in mouth. Detection of fungi in sputum may be valuable in diagnosis. The infection of *Actinomyces bovis* may produce grossly so-called "sulfur granules" in sputum. Microscopic examination of the fresh material shows a mass of delicate filaments in the center and "clubs" at the periphery.

Candidiasis: Detection of *Candida albicans* in sputum is not sufficient to establish its pathogenicity because the organism is often present in the mouth. Transbronchial specimens should be submitted to further mycological study when repeated tests for tuberculosis and cancer have been fruitless. The yeast cells and filaments of *Candida albicans* can be identified by Gram's stain as well

as PAS reaction (Fig. 8-66).

It is characterized by yellowish to brownish long hyphae and budding, round thick-walled conidia. Its growth in lung parenchyma occurs during the course of wide-spectrum antibiotics therapy. As the second opportunistic infection, aspergillosis should be considered. In case of aspergillosis the sputum may show fragments of septate hyphae, conidia or occasionally conidiophores (Fig. 8-65).

Cryptococcosis: Pulmonary cryptococcosis is secondary opportunistic infection in debilitated patients with Hodgkin's disease or allied reticuloendothelial abnormalities. The patient produces scanty, mucoid and rarely bloody-mucoid sputum in which ovoid or spherical, thick-walled organisms with occasionally single budding is found (Fig. 8-67). Faintly cyanophilic capsule that is positive for alcian blue or PAS stain is characteristic.[79]

191

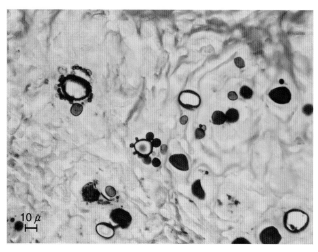

Fig. 8-68　*Blastomyces brasiliensis* showing single or tear-shaped buds from the surface of thick walled parent cells measuring 10 to 60 μm in diameter. The buds are as small as 1 to 2 μm in diameter. Note a characteristic marine pilot's wheel appearance (arrow). Gomori stain.

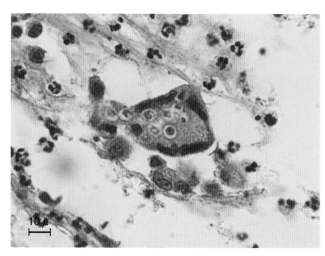

Fig. 8-69　Multinucleated histiocyte in sputum phagocytosing many organisms in infection of *Blastomyces brasiliensis.* The patient is productive of purulent sputum. Nodular lesions in the chest X-ray must be differentiated from pulmonary tuberculosis. HE stain. (By courtesy of Dr. E. M. Tani, Faculdade de Medicina de Botucatu, São Paulo, Brasil.)

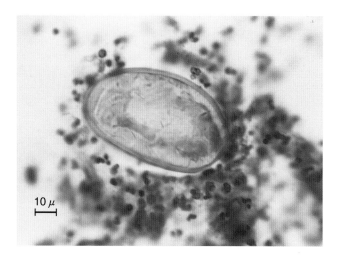

Fig. 8-70　Egg of *Paragonimus westermani* in sputum. The egg shows a thick shell and miracidium. Consolidated abscess formation associated with destruction of the bronchial wall may induce a productive cough containing the eggs.

North American blastomycosis: Blastomyces dermatitidis is the cause of North American blastomycosis and appears as a slightly orangeophilic yeast-like body measuring about 5 to 20 μm in diameter. The outer layer is composed of thick refractile cell wall. Single budding with broad attachment to the mother cell as pairs of cells is characteristic. *Blastomyces dermatitidis* gives rise to suppurative and granulomatous pneumonitis by inhalation infection. It may extend to visceral abdominal organs, brain and skeletal system like metastasis of malignant neoplasia.

South American blastomycosis: Paracoccidioides brasiliensis is the organism which causes South American blastomycosis. It is slightly larger than *Blastomyces dermatitidis* and measure from 5 to 30 μm in diameter. Multiple budding with a ship's wheel appearance is a characteristic feature (Fig. 8-68). The lung is the most commonly affected organ by inhalation. Confluent granulomatous lesions may develop to suppurative inflammation. Microscopical identification of paracoccidioidomycosis is easily made if the fungal nucleus is present. In the absence of fungal nucleus, the yeast form like a vacuole with refractile wall can be detected in the cytoplasm of multinucleated giant cells. The Papanicolaou's technique for sputum and bronchial brushings or washings is good enough for detection of pulmonary blastomycosis,[79] but cell-block preparation[76] is particularly useful for sputum specimens because of its accuracy and availability of specific methods such as Gomori, Grocott and PAS stains.

Paragonimiasis: In human infection with *Paragonimus westermani* that occurs most frequently in the Far East the adult worms settle down in peribronchial encapsulated cystic structures. The cyst wall is composed of granulomatous fibrous tissue with leukocytic infiltration. If communication of the cyst with respiratory tracts is formed, eggs will be expectorated with productive cough. The sputum is frequently blood tinged.[105] This hemorrhagic expectoration is called "endemic hemoptysis"; extrusion of the ova into the bronchioli following rupture of the peribronchial cyst causes cough, chest pain and hemoptysis.[203] A number of eosinophils and Charcot-Leyden crystals may be found. Paragonimiasis is also diagnosed by bronchial washing or bronchoscopic biopsy.[203] The egg is golden yellow and measures about

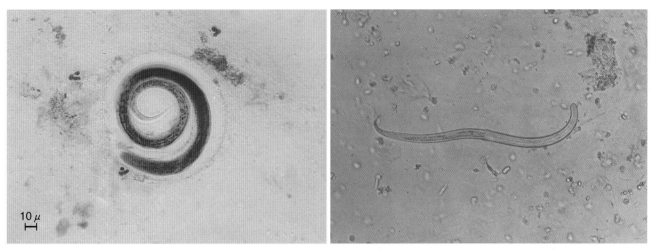

Fig. 8-71 *Strongyloides stercoralis* **coughed out in sputum.** Infective filariform larvae are able to penetrate the skin and are carried to the lungs via blood vessels. They burrow out of the capillaries into the alveolar ducts where hemorrhages and pneumonitis may follow.

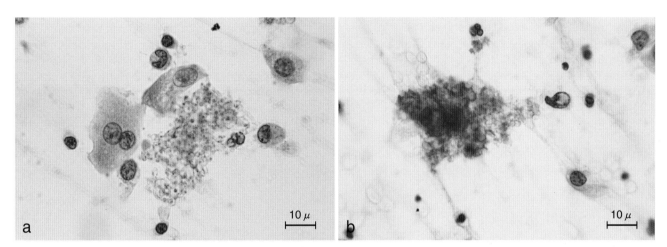

Fig. 8-72 *Pneumocystis carinii* **in a needle aspirate from the lung.**
a. The organism that affected the patient who had received intense immunosuppressive therapy displays a tiny, dot-like nucleus to be surrounded by a foamy mucus material.
b. PAS reaction stains the foamy mucus pinkish purple.

82-86 μm in length by 46-52 μm in width. It is operculated at the broader end (Fig. 8-70).

Schistosomiasis: The eggs of *Schistosoma* can be present in the lung with granuloma formation. The eggs of *Schistosoma japonicum*, for example, are 70-80 × 50-65 μm in size and have a thin shell without operculum or spine, but show a distinct miracidium in the interior. It must be emphasized that various kinds of pollen are expectorated or contaminated within sputum during flowering season. They have thick and refractive cell wall and granules in the interior, and may be often mistaken for ova of certain parasites.

Strongyloides stercoralis: The infective *filariform larvae* migrated into the cutaneous vessels are carried to the lung where fertilization of females occurs. The adolescent females swallowed reach the intestinal tract and mature followed by autoinfection without leaving the host. Fetal reinfection initiates in patients with lowered host resistance and *rhabditoid larvae* may recirculate into the venous route. Hemorrhagic sputum with respiratory infection is produced; *rhabditoid larvae* and occasionally *filariform larvae* are found (Fig. 8-71).

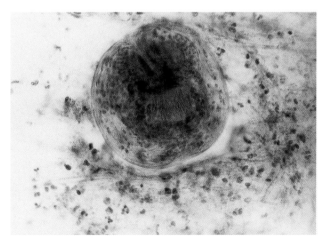

Fig. 8-73 **Invaginated protoscolex of *Echinococcus granulosus* in a sputum smear.** It measures approximately 150 μm in a greater diameter. (By courtesy of Dr. J. A. Tomb, American University, Beirut, Lebanon.)

193

Pulmonary hydatidosis: Echinococcus cysts may develop in the liver, lungs and brain in man or intermediate hosts during the cysticercal stage of the life cycle of *Echinococcus granulosus.* The cyst is composed of thick laminated coat (ectocyst) and thin nucleated germinal layer (endocyst) from which brood capsules containing scolices may develop. Echinococcus or hydatid cyst containing antigen occasionally ruptures into the bronchus. If the patient with the ruptured hydatid cyst did not succumb to anaphylactic shock, hydatid scolices coughed up can be diagnosed cytologically (Fig. 8-73). Tomb and Matossian[184] emphasize the value of sputum cytology, because the patients with pulmonary hydatidosis show relatively low positivity of hemagglutination[5] and complement fixation.[100]

Pneumocystis carinii: This is an organism which causes interstitial plasma cell pneumonia in premature infants or in adults who have received immunosuppressives or broad-spectrum antibiotics as an opportunistic infection. Acquired immune deficiency syndrome (AIDS) caused by HIV infection is the most grave state with *Pneumocystis carinii* pneumonia. Early in the course of AIDS lymphoproliferative disorders such as diffuse lymphoid hyperplasia, follicular bronchiolitis and lymphoid interstitial pneumonia (LIP) occur in not only adults but infants with maternal HIV transmission.[80,134] Although alveolar ducts are filled with honeycomb-like fibrillar material, spontaneous sputum does not contain the organism. Needle aspirates or imprint smears during open lung biopsy disclose tiny organisms measuring about 0.5 μm in diameter within the cysts (Fig. 8-72).

10 Cytological Typing of Lung Cancer

In cytodiagnosis an attempt to designate the tumor cell type has been successful. Four representative types, i.e., squamous cell or epidermoid, glandular and undifferentiated types, can be classified as in other organs. The last is further subdivided in lung cancer into undifferentiated large cell and undifferentiated small cell types (Tables 8-6 and 8-7). Although there may be several factors which are concerned with the prognosis of a neoplastic disease such as the presence of metastases to the regional and distant lymph nodes and/or organs, involvement of pleura, and resectability of the tumor, histological typing with reference to the grade of differentiation is the subject of discussion in exfoliative cytology.

It is generally accepted that epidermoid carcinoma has the most favorable prognosis while oat cell carcinoma is the worst from the clinical study of the five-year or longer survival rate.[44,121,195] The fact that the patient with localized adenocarcinoma had the best five-year survivals indicates the importance of stage and localization irrespective of histological types except for small cell carcinoma.[29,185] Since the histology of lung cancer is very variable and varies even in the same tumor from place to place, some investigators are conservative in their cell typing and affirmatory only for certain types which are limited to well differentiated squamous cell and undifferentiated small cell carcinoma.[207] Nevertheless, it is the opinion of many investigators that the majority of

lung cancers are able to be correctly typed.[87,141,189]

The discordance between histological and cytological typings is easily understood because histological classification depends on the most differentiated cell type, whereas cytological identification reflects the dominant cell type. A good example is shown in the "Histological Typing of Lung Tumours" by Kreyberg et al[92]: Some foci of the tumor display a high degree of differentiation, i.e., maturation to prickle cell type with pearl formation, whereas the greater part of the tumor is composed of small anaplastic cancer cells (see Fig. 8-91). According to the conventional histologic classification, the designation of squamous cell carcinoma may be possible. However, in exfoliative cytology, a chance to obtain well differentiated squamous cancer cells from such a case is very rare and it is a matter of course that cytological cell typing should rely on the predominant cell type of the tumor.

◾ Squamous cell carcinoma (Epidermoid carcinoma)

Although various degrees of differentiation are present in squamous cell carcinoma, it is histologically characterized by epidermis-like differentiation with keratinization and/or intercellular bridges. The inside of cancer cell nests that is the site of pearl formation tends to be necrotic and to from cavitation. If the cavitation is communicated to the air passages, it may produce a number of pyknotic cancer cells with cytoplasmic eosinophilia and odd deformation such as fiber-like, tadpole and snake cells.

Cytology: Squamous cancer cells spontaneously expectorated differ in shape and staining reaction according to the degree of differentiation and keratinization. The fiber cells and tadpole cells are the prototype of extreme differentiation (Fig. 8-76). Parabasal or intermediate cell-shaped cancer cells with bluish green stain are from the nonkeratinizing outer zone of cancer cell nests. These are easily identifiable (Figs. 8-75 and 8-77). However vigorously exfoliated cells by needle aspiration provides difficulty in differentiation from other types (Table 8-7).

Spindle cell squamous carcinoma is a variant of poorly differentiated squamous cell carcinoma defined by WHO classification: it locates peripherally with a propensity to involve the upper lobe.

Cytology of this tumor that is composed of spindly, sarcomatous cells leads to confusion with sarcoma (Fig. 8-78), carcinosarcoma, intermediate type of oat cell carcinoma (Fig. 8-79) and atypical carcinoid for interpretation. The case of spindle cell carcinoma admixed with areas of squamous cancer cells happens to be misinterpretated as carcinosarcoma. Characteristic features of squamous carcinoma in situ are summarized on page 213 (see Fig. 8-123).

◾ Adenocarcinoma

As a general rule, adenocarcinoma arises from the surface bronchial epithelium and is characterized by glandular structures occasionally with a papillary growth. It must be kept in mind that, even if only a small portion exhibits a gland-like differentiation and the remainder, i.e., the majority of the tumor is composed of solid

Table 8-6   Cell typing of malignant cells in sputum*

	Squamous cell carcinoma	Adenocarcinoma	Undifferentiated carcinoma	
			Small cell	Large cell
Cell clusters				
Specific feature	Keratinization	Gland-like pattern	Bare polygonal nuclei	None
Arrangement	Flat, cobblestone-like	Overlapping	Loose cluster or single	Loose cluster
Anisocytosis	Marked	Moderate	Slight to moderate	Slight to moderate
Cytoplasm				
Form	Variable and thick	Elliptic or oval	Scant or invisible	Indistinct
Staining reaction	Variable, green or orange	Bluish green	Pale blue	Pale blue
N/C ratio	Relatively small	Middle	Largest	Large
Nucleus				
Location	Central	Peripheral	Central	Commonly central
Form	Variable (pleomorphic)	Round or oval	Polygonal, relatively round	Round or oval, somewhat polyhedral
Chromatin pattern	Coarsely clumped and often pyknotic	Granular or moderate size	Coarsely granular	Granular of moderate size
Nucleolus				
Location	Indefinite	Central	Scattered	Indefinite
Size and shape	Variable, irregular	Large, round	Small, irregular	Variable, irregular
Number	Few, often invisible	Single or few	Multiple	Occasionally multiple

* These criteria are not decisive but are generally affirmatory. To attempt typing of malignant cells, one must carefully observe the entire field of the smear.

Table 8-7   Characteristics of cancer cell typing in needle aspirates

Squamous cell carcinoma	Adenocarcinoma	Small cell carcinoma	Large cell carcinoma
Frequently clustered with slight overlapping but tend to be arranged in parallel	Clustered with overlapping, but tubular pattern not so distinct as in sputum	Loosely clustered, occasionally in tandem	Loosely clustered without marked overlapping or any organoid pattern
Cellular border present but ill-defined	Cellular border well outlined	Cellular border poorly defined	Cellular border poorly defined
Moderate amount of cytoplasm with variable staining reaction according to degrees of keratinization	Moderate amount of cytoplasm, well stained by Light Green	Scant cytoplasm, pale and cyanophilic	Small to moderate amount of cytoplasm, pale and cyanophilic
Nuclear rims irregular but not so thickened as in sputum	Nuclear rims delicate and smooth	Nuclear rims distinct but not rigid as in sputum	Nuclear rims distinct but not thickened Nuclei large polymorphic
Nucleoli visible, one or two, irregular in shape	Nucleoli, centrally locating, enrounded and conspicuous	Nucleoli multiple and scattered throughout	Nucleoli multiple and prominent

masses of undifferentiated cancer cells, it is histologically designated as adenocarcinoma. A small proportion of adenocarcinoma may originate from the bronchial gland and its duct epithelium. This tumor locates relatively centrally and presents characteristic features such as the adenoid cystic type or the mucoepidermoid type which is commonly seen in carcinomas of the major salivary gland. Most adenocarcinomas are of peripheral type distal to segmental bronchi. Therefore, pathological changes secondary to obstruction of the large bronchi are seldom seen. Peripheral adenocarcinoma takes a direct spread along the visceral pleura producing mesothelioma-like desmoplasia. On the other hand, invasion to the chest wall is not so frequent as in epidermoid carcinoma such as Pancoast tumor. A close etiological relation of adenocarcinoma to scarring has been strongly stressed by many researchers; old infarction, tuberculosis and non-specific obsolescent fibrosis are concerned with scar cancer.

Histology: Characteristic features of adenocarcinoma are tubular and glandular (acinar) or papillary growth pattern irrespective of mucin production. Poorly differentiated adenocarcinoma is mainly composed of solid nests of polymorphic cancer cells with marked hyperchromatosis and high N/C ratio. The poorly differentiated type is hardly distinguished from large cell carcinoma with mucin content which was previously separated in WHO classification (1967). Some of the poorly differentiated adenocarcinomas may reveal bizarre, giant cells which are indistinguishable from giant cell carcinoma unless the remainder of the tumor reveals a glandular or papillary pattern (Fig. 8-81). The central type of adenocarcinoma often includes carcinoma of bronchial gland origin in which foci suggestive of mucin-containing mucoepidermoid or PAS-negative, clear serous cell differentiation are discernible.

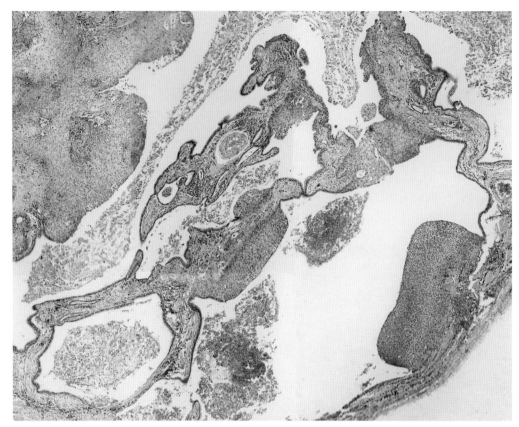

Fig. 8-74 Squamous cell carcinoma, early invasive and in situ carcinoma arising from bronchiectatic mucosa. Saccular bronchiectasis develops diffusely in the right lower lobe in association with multicentric cancerization. Chest X-ray shows only focal pneumonitis-like shadow. Selective bronchial washing was only successful to detect cancer.

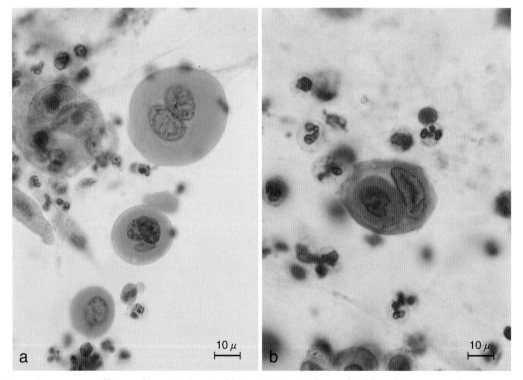

Fig. 8-75 Squamous cancer cells, nonkeratinizing and keratinizing, in bronchial brushing. From the same case as Fig. 8-74.
a. Nonkeratinizing epidermoid type is characterized by thickened, bluish green cytoplasm, centrally locating nuclei with occasional multiplication, and marked size and shape variation.
b. Keratinizing epidermoid type showing cannibalism. Intense orangeophilia suggests keratinization of cancer cells. Inclusion of one cancer cell by another that is called cannibalism is not infrequently encountered in squamous cell carcinoma.

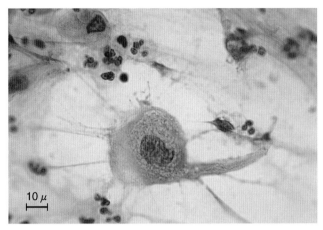

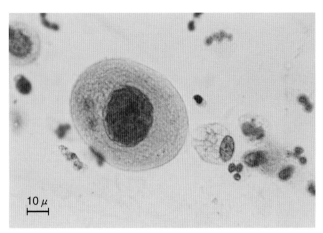

Fig. 8-76 Differentiated keratinizing squamous cancer cell in sputum. Note well preserved, thickened cytoplasm with intense orangeophilia that reflects keratinization.

Fig. 8-77 Differentiated nonkeratinizing squamous cancer cell in sputum. This well preserved cell of epidermoid type is characterized by a fibrillar structure of the cytoplasm, condensed chromatin, rigid nuclear contour and irregular nucleolus.

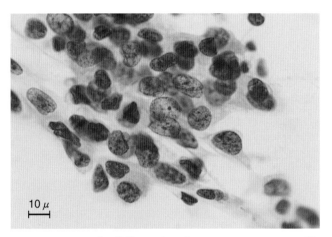

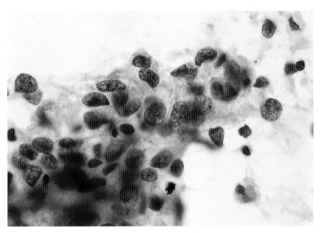

Fig. 8-78 Poorly differentiated squamous cancer cells seen in a cluster by needle aspiration. Cell typing for needle aspirates is based on the following diagnostic points: (1) ill-defined sheet of cells without organoid pattern, (2) poorly preserved indistinct cytoplasm, and (3) irregular contour of the nuclei are suggestive of poorly differentiated squamous cell carcinoma.

Fig. 8-79 Spindle cell variety of squamous cell carcinoma. Transbronchial needle aspirate. Note distinctive findings such as (1) marked variation of the nuclei in size and shape, (2) distinct nuclear rim, and (3) coarse chromatins to be distinguished from sarcoma or carcinoid.

Cytology: Cell typing of adenocarcinoma is established from formation of three dimensional cell clusters and its arrangement. Papillary adenocarcinoma will be based on papillary protrusion of cells with considerably different depths of focus (Fig. 8-82). Tubular adenocarcinoma reveals cell cluster with nuclear overlapping (Fig. 8-80). The depth of focus is not so marked as that of papillary carcinoma. Peripheral position of the nuclei is a valuable feature in cell typing, but becomes less prominent accordingly as histologic differentiation decreases. Cytoplasmic vacuolation, unless mucin content is verified, is not specific for adenocarcinoma but the result of cell degeneration.

■Bronchiolo-alveolar carcinoma
Bronchiolo-alveolar carcinoma or bronchiolar carcinoma is a peculiar type of adenocarcinoma that may grow without distortion of ordinary alveolar structures. The incidence of bronchiolo-alveolar carcinoma accounted

for 3% to 6% of pulmonary carcinomas tends to be increased in these 10 years[65,103] accordingly as the entity of the tumor has come to be clarified. This tumor is separated from adenocarcinoma because of distinctiveness of the pathologic picture and of the clinical behavior.

In its earliest stage the tumor grows as a solitary form and is resectable without pleural involvement and lymph node metastasis, thus more favorable prognosis can be expected than other types.[62] In the later stage this tumor reveals diffuse pneumonic extension or multinodular bilateral lesions by means of vascular, lymphatic and possibly endobronchial spread. The nomenclature of alveolar cell carcinoma derives from histologic findings that single layered tumor cells cover the alveolar wall without distortion of alveolar structures. The question may arise as to whether these tumor cells originate from endothelial-like membranous pneumocytes (type I) or cuboidal pneumocytes (type II) having an actively proliferating property in the alveolar ducts. Electron

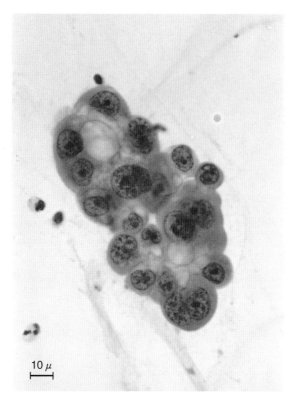

Fig. 8-80　Adenocarcinoma cells in sputum showing a characteristic tubular pattern. Note a tight cell cluster arranged around a lumen-like space is indication of well differentiated adenocarcinoma.

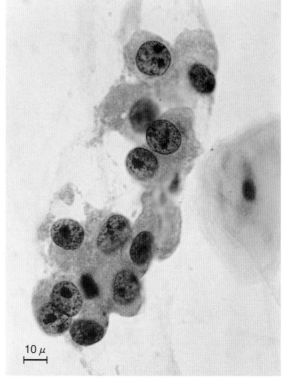

Fig. 8-81　Adenocarcinoma cells in sputum. The cell typing can be based on the following features: (1) Prominence of nucleoli, (2) round or oval nuclei with smooth contours, and (3) tendency to form a papillary cluster are suggestive of adenocarcinoma.

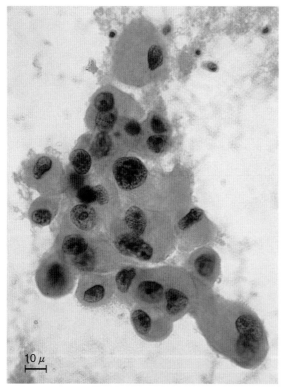

Fig. 8-82　Giant celled adenocarcinoma cells in needle aspirate. A variety of adenocarcinoma shows marked nuclear enlargement and anisonucleosis; this evidence supports the opinion that giant cell carcinoma is a form of adenocarcinoma originating in peripheral region. Immunocytochemically CEA is strongly positive.

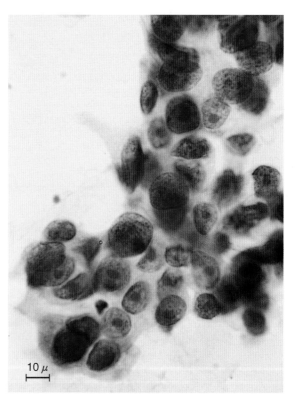

Fig. 8-83　Adenocarcinoma cells in needle aspirate. Moderately differentiated adenocarcinoma with ill-defined borders and pale stainability is hardly distinguished from squamous cell carcinoma especially in needle aspirates. Diagnostic points: (1) Marked overlapping of nuclei, (2) prominence of nucleoli, (3) fairly fine chromatin granules, and (4) smooth nuclear rims are favorable to adenocarcinoma.

198

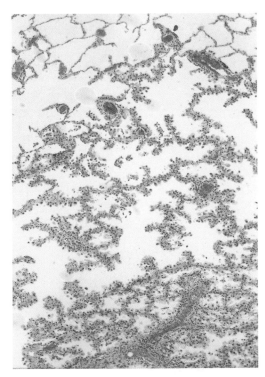

Fig. 8-84 Diffuse type of bronchiolo-alveolar carcinoma. Note alveolar walls left intact after neoplastic proliferation; bronchioles and alveolar ducts are not compressed or destroyed in spite of lining by cylindrical cancer cells.

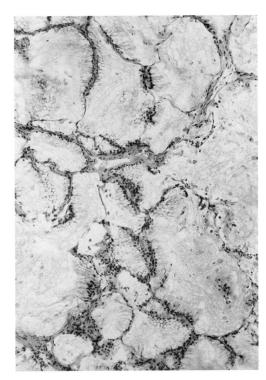

Fig. 8-85 Diffuse type of bronchiolo-alveolar carcinoma producing mucus. The alveolar ducts are filled with mucus secreted by tall cylindrical cancer cells. Desquamation of cancer cells into the alveolar lumen is seen.

microscopic studies supported the type II origin from the fact that lamellated bodies were found in cancer cells.[2,34,117]

The type II cell carcinoma can be proved by immunoperoxidase method using the antibody against surfactant-specific apoprotein.[163]

On the other hand, some investigators regard the mechanism of type II cell proliferation in the tumor as reactive proliferation as shown around the scar fibrosis of the tumor[62]; its histogenesis is suggested to be induced by mucous metaplasia of bronchiolar epithelium or by differentiation of bronchiolar stem cells to Clara cells.[77]

Clara cell bronchiolar carcinoma is a type of bronchiolo-alveolar carcinoma and forms about one third of peripheral adenocarcinoma.[24]

Histology: The tumor cells with basally placed nuclei proliferate as they creep up the inner wall of the alveoli. A well differentiated type that is also called pulmonary adenomatosis, from the meaning resembling "jaagsiekte" may display the alveolar wall to be lined with a single layer of uniform cylindrical tumor cells with abundant vacuolated cytoplasm which is frequently positive for mucicarmine or periodic acid-Schiff stain (Fig. 8-84).[176] Although there are no definite cytological criteria of bronchiolar carcinoma to be distinguished from conventional adenocarcinoma, the patient having a diffuse type of this tumor may expectorate a large amount of mucinous frothy sputum, in which tall columnar cells with vacuolated cytoplasm may be found arranged in a glandular or papillary pattern.[197]

Cytology: Production of large amounts of clear, watery, frothy or mucoid sputum is noted as a characteristic feature. Cancer cells are produced profusely and appear as acinar or papillary clumps (Fig. 8-86). It is open to discussion whether this tumor can be specifically indicated from the sputum cytology. Fairly small nuclei are basally situated in abundant clear cytoplasm with columnar shape. The nuclear size and shape variation is less marked than in adenocarcinoma (Figs. 8-82, 8-83, 8-88 and 8-90). The evidence of positive PAS stain in addition to the above features is favorable for this type of adenocarcinoma. They are similar to metastatic renal carcinoma cells; the columnar cytoplasm is as clear as what is commonly called "clear cell carcinoma."

Cytology of Clara cell bronchiolar carcinoma is characterized by dome- or club-shaped cytoplasm protruding at apical side, deep and homogeneous stain of the cytoplasm by Light Green and amphoteric prominent nucleolus (Fig. 8-89). For distinction of mucin-producing carcinoma PTAH stain at the free cell surface is strongly recommended.[43]

■Small cell carcinoma

This is the most aggressive tumor of bronchogenic carcinomas and arises in large bronchi proximal to segmental bronchi. The occurrence of this tumor is much more common in men than in women.[196] The majority of the patients with small cell carcinoma are heavy smokers.[6] Its incidence is increased correspondingly with increased amount of smoking habit in men of the sixth or seventh decade. In spite of the predominant localization of the tumor in hilar or perihilar region,[25] it seldom brings about obstruction of large bronchi, but extensive peribronchial infiltration to the parenchyma and distant or organ metastases (Fig. 8-96). The evidence that some of

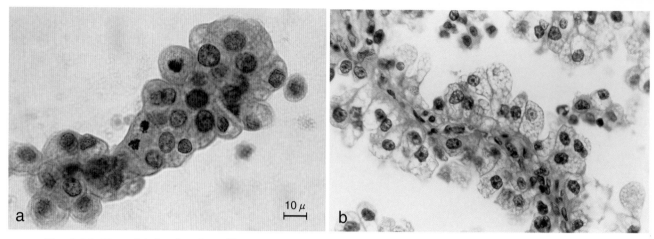

Fig. 8-86 Bronchiolo-alveolar cell carcinoma.
a. Sputum cytology characterized by a large number of clustered cells in frothy sputum. A mitotic figure is seen.
b. Higher magnification of histology. Note prominent nucleoli and clear, vesicular cytoplasm.

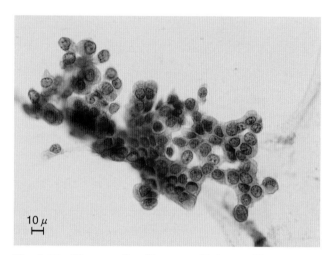

Fig. 8-87 Cluster of malignant cells in needle aspirates compatible with bronchiolar carcinoma. Note a characteristic feature displaying a papillo-tubular cell cluster. The nuclei are fairly small in size and uniform in shape.

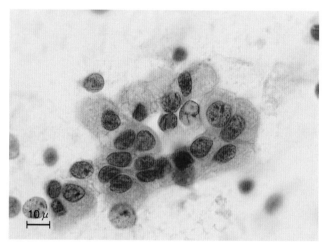

Fig. 8-88 Bronchiolar carcinoma cells with mucin content in a cluster. Higher magnification of the same smear as Fig. 8-87. Diagnostic points: (1) Basally situated small nuclei, (2) abundant, columnar clear cytoplasm, and (3) occasional mucin content which can be identified by PAS stain are suggestive of bronchiolar carcinoma.

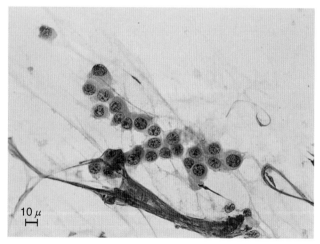

Fig. 8-89 Cluster of malignant cells in needle aspirates from bronchiolar carcinoma. Note a branching column of small cancer cells; this cell arrangement reflects the papillary growth of bronchiolar carcinoma. Amphoteric, prominent nucleoli, dome-shaped cytoplasm and deep greenish stain are suggestive of Clara cell adenocarcinoma.

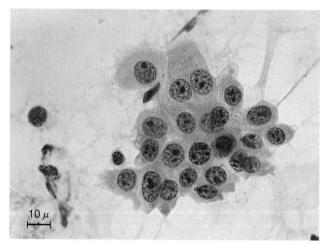

Fig. 8-90 Solid sheet of bronchiolar carcinoma cells in a needle aspirate. Higher magnification of the same smear as Fig. 8-89. Although the nuclei are uniform in shape and smaller in size than common adenocarcinoma, coarse clumping of chromatins and prominence of nucleoli are diagnostic of malignancy.

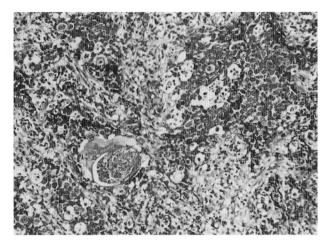

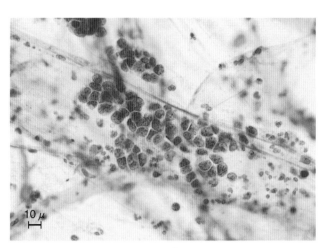

Fig. 8-91 Small cell carcinoma showing partial squamous cell differentiation. A proportion of abrupt differentiation represents 5% or less of the whole small cell population.

Fig. 8-92 Low power view of small carcinoma cells in sputum. Anaplastic small carcinoma cells are similar to lymphocytoid cells in sputum. If clustered cells with polyhedral nuclei are found, it is appropriate to consider small cell or oat cell carcinoma first of all.

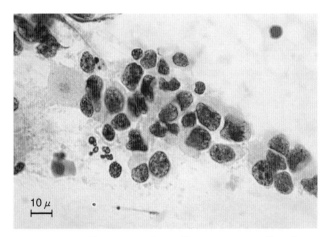

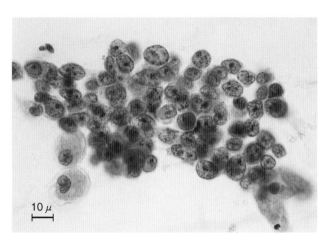

Fig. 8-93 Well preserved small carcinoma cells of oat cell type in bronchial brushing. Note almost stripped polyhedral nuclei with coarse but dispersed chromatins. Irregularity of nuclear shape is frequently encountered. Nucleoli are inconspicuous.

Fig. 8-94 Well preserved small carcinoma cells of intermediate type in bronchial brushing. In spite of well preservation of the nuclei, the cytoplasm is scarcely found. The intermediate type possesses larger nuclei than the oat cell type and reveals nucleoli identifiable. The tendency to form clusters is suggestive of partial glandular differentiation.

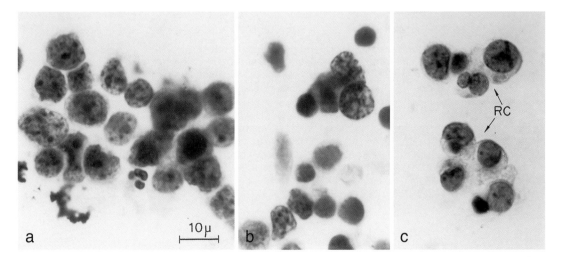

Fig. 8-95 Criteria for distinction between small, oat cell cancer cells and hyperplastic basal cells. The nuclear size in oat cell carcinoma (a, b) is about 8-10 μm in diameter and only slightly larger than that of hyperplastic reserve cells (RC). However, the nuclei of small cancer cells show moderate size variation and occasionally angular configuration. The most important criterion consists in their chromatin pattern: in small cancer cells coarsely clumped chromatin is intervened by clear spaces, while hyperplastic reserve cells showing hyperchromatism may reveal an opaque and homogeneous appearance.

201

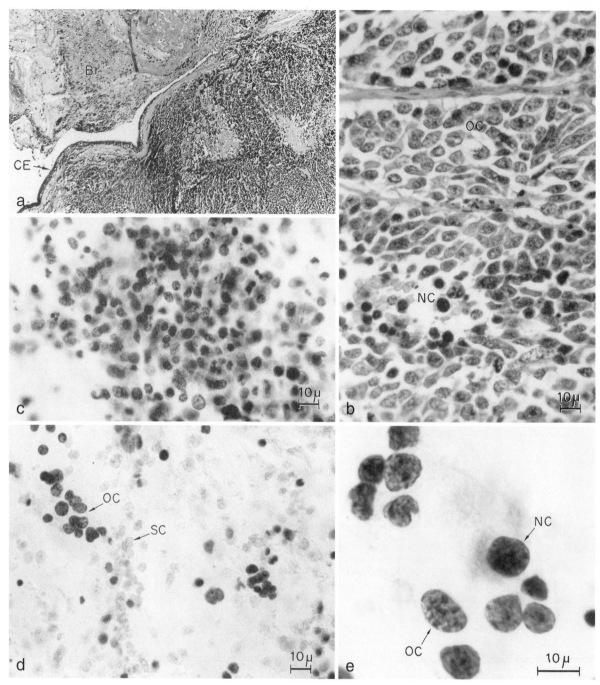

Fig. 8-96 Histology of oat cell carcinoma and its exfoliated cells in sputum.

a. Oat cell carcinoma or small cell carcinoma commonly arising from the main bronchus or stem bronchus infiltrates with massive sheets of closely packed cancer cells (Ca). The tumor occasionally exposes into the bronchial lumen (Br) and exfoliates numerous malignant cells. CE: ciliated columnar epithelium.

b. This highly anaplastic tumor tends to degenerate and show necrotic cells with nuclear pyknosis in the middle of cancer cell nests (NC). They are similar to large lymphocytes and may be missed in sputum. Actively growing oat cells, however, show angular or polygonal nuclei with coarse clumping of chromatin (OC).

c. Oat cells in sputum are produced massively and appear with loose cell clusters.

d, e. Once oat cell carcinoma is suspected, well preserved oat cells will be screened out with relative ease under low power field (OC). They retain the malignant chromatin pattern with coarse clumping. Diagnosis should not be based on necrotic cells similar to lymphoid cells (NC). Shadow cells are the result of advanced degeneration (SC). Note difference of oat cells with loose cluster formation from lymphoma cells singly isolated.

small cell carcinomas reveal a tubular pattern or abrupt keratinization induced contradictory opinions as to the parent cells. The concept that small-celled anaplastic carcinoma is composed of primitive reserve cells as first described as "Basalzellenkrebs" by Marchesani (1924) has sufficed the histological controversy, since the reserve cells are considered to be able to differentiate either columnar or squamous cells. Recent electron microscopic studies pointed out the similarity of oat cells having neurosecretory granules to Kulchitsky cells which are distributed in exocrine bronchial glands and terminal air ways. Bensch et al are the first to refer to the

Kulchitsky cells as the parent cells of carcinoid and small cell carcinoma. Hattori and his associates emphasized that oat cell carcinoma of the lung was special type of carcinoid because of argentaffine granules.[14,68] If it is assumed that oat cell carcinoma is all regarded as a malignant variety of carcinoid, the following facts must be fully explained: (1) Age distribution is higher for oat cell carcinoma than for carcinoid,[60] (2) male/female ratio for oat cell carcinoma is about four times as high as for carcinoid, (3) carcinoid producing distant metastases cannot be deemed either of oat cell carcinoma or of intermediate form between carcinoid and oat cell carcinoma, and (4) argentaffine granules are not all found in oat cell carcinoma.

Histologically, small cell carcinoma is characterized by massive infiltration of anaplastic cancer cells intersected into nests or cords by thin fibrovascular stromal tissue. No organoid pattern is visible, but a palisading pattern is often present at periphery of closely packed cell nests. Although degenerative changes occur microscopically, necrosis *en masse* giving rise to gross cavitation is unusual. Lymphatic invasion is pronounced and massive involvement of regional lymph nodes is rather more evident than at the primary site. The previous subtypes proposed by the WHO Committee (1981), fusiform cell carcinoma, polygonal cell carcinoma, lymphocyte-like carcinoma and others, are now classified into lymphocyte-like or oat cell type and intermediate type.

a) Lymphocyte-like type (oat cell type): Cancer cells of lymphocyte-like type possess small, hyperchromatic nuclei which vary considerably in size and shape. The cytoplasm is very scant. Several basophilic micronucleoli are present but often misinterpreted as chromocenters (Fig. 8-93).

b) Intermediate cell type: This type is composed of polygonal cells having somewhat larger, polyhedral, hyperchromatic nuclei in which one or two acidophilic, nucleoli are discernible. A fair amount of basophilic indistinctly bordered cytoplasm is present (Fig. 8-94).

c) Combined type: This third subtype is oat cell carcinoma including partial differentiation to epidermoid or adenocarcinoma. Abrupt epidermoid differentiation with keratinization is often encountered (Fig. 8-91). The combined type represents about 1% to 3% of small cell carcinomas.

In reference to prognosis of subtypes of small cell carcinoma, features of the nuclei, i.e., small round nuclei and densely stained coarse chromatins were described to be markers of poor prognosis.[116]

Hirsch and associates,[74] however, found no difference in survival between oat cell type and intermediate cell type of small cell carcinomas. Significance of cytology consists in detection of small cell carcinoma in a limited stage when a marked remission by combination chemotherapy can be expected.[57]

Cytology: Cancer cells are small, round, lymphocyte-like, and fusiform or polygonal in shape. As in histological subclassification of small cell carcinoma, at least two varieties can be classified cytomorphologically according to the nuclear size and prominence of nucleoli. The intermediate type will be distinguished by larger nuclei and prominent nucleoli, whereas lymphocyte-like type is characterized by small lymphocytoid nuclei with basophilic, karyosome-mimic nucleoli. They may appear in sputum as a mass of cells similar to lymphoid cells because of having pyknotic nuclei. One must notice variations in size and shape of the nuclei to differentiate them from lymphoid cells (Fig. 8-96).

■Large cell carcinoma

Large cell carcinoma consisting of large polyhedral cells with abundant cytoplasm and large vesicular nuclei locates peripherally as solitary, large masses. The diagnosis is made on transthoracic needle or transbronchial biopsy cytology.

This type of undifferentiated carcinoma is characterized by solid nests of cancer cells without evidence of epidermization or formation of glandular structures. The following subtypes are those classified by the WHO International Reference Center for the Histological Definition and Classification of Lung Tumors:

i) Solid tumors with mucin-like substances: Some concretions positive for periodic acid-Schiff (PAS) reaction, mucicarmine and alcian blue or alcian green stains are contained therein.

ii) Solid tumors without mucin-like substances.

iii) Giant cell carcinoma: This peculiar type was first designated as a type of undifferentiated carcinoma by Nash and Stout (1958) and is known to metastasize widely and to occur peripherally and in relatively younger ages.[13] *Histologically* the tumor is pleomorphic containing bizarre giant cells and resembles rhabdomyosarcoma. The tumor occasionally contains PAS-positive substances[73] and behaves clinically much like bronchogenic adenocarcinoma[53,92] (Figs. 8-97 to 8-99). Although this tumor is included in group 4, large cell carcinoma in the WHO classification, this is conventionally distinguished from large cell carcinoma.

Cytologically, giant tumor cells are shed singly or in loose cluster, and diagnosed accurately for their characteristic features.[23] The nuclei vary in size and shape, and are often multinucleated or multilobulated. The chromatin is increased in amount and distributed in a coarsely granular or reticular pattern. The cytoplasmic staining reaction is pale, cyanophilic and often vesicular. So-called phagocytic activity of the giant cell carcinoma that is lately interpreted as *tumor cell-tumor cell emperipolesis*[194] is not found in cytology smears. It should be kept in mind that giant cell formation occurs in some environmental conditions such as irradiation, virus infection and decreased oxygen tension.[36,194] Another peculiar giant cell tumor that secretes human chorionic gonadotropin (HCG) has been sporadically reported[37,69] since del Castillo (1941) described three patients with bronchogenic carcinoma with gynecomastia; histogenetically, disdifferentiation of multipotential cells in the distal bronchiole is suggested[194] (Figs. 8-100 and 8-101). Ectopic adrenocorticotropic hormone (ACTH) secretion also rarely occurs.

iv) Clear cell carcinoma composed of water-clear cells: Since this type of carcinoma shows no definite cytological features, it is indistinguishable from poorly differentiated adenocarcinoma (Fig. 8-105) or poorly differenti-

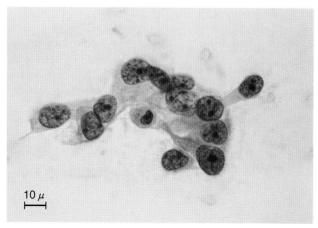

Fig. 8-97 Undifferentiated large carcinoma cells seen in a loose cluster. Diagnostic points: (1) Moderate amount of ill-defined, amphoteric cytoplasm, (2) high N/C ratio, (3) coarse clumping of chromatins, (4) prominent, acidophilic nucleoli, and (5) loose cell sheet are important features of large cell carcinoma.

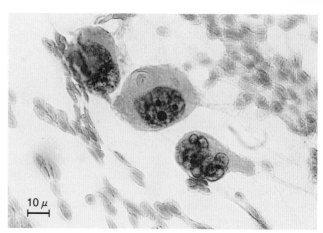

Fig. 8-98 Undifferentiated giant cell carcinoma in needle aspirates. Diagnostic points: (1) marked nuclear size variation with multinucleation, (2) multiple nucleoli, and (3) intracytoplasmic inclusion.

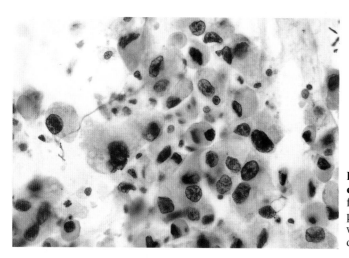

Fig. 8-99 Giant cell carcinoma in transbronchial needle aspirates. These are exhibited with a moderate magnification (40 × 3.3). Note strongly stained, highly polymorphic nuclei. Eccentrically locating nuclei and clear cytoplasm with brownish hue are indicative of the origin of adenocarcinoma.

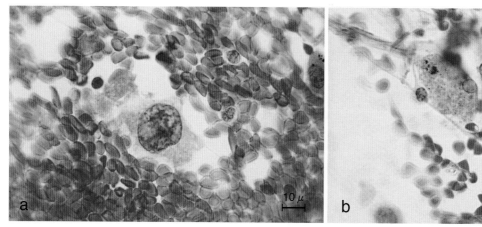

Fig. 8-100 Undifferentiated large carcinoma cells by needle aspiration from extragenital choriocarcinoma of the lung. The cytodiagnosis as undifferentiated large cell carcinoma was corrected by histologic examination after lobectomy as extragenital choriocarcinoma producing HCG.
a. Single trophoblastic malignant cell showing prominent nucleolus and poorly defined cytoplasm.
b. Syncytial trophoblastic malignant cell showing multinucleation. Scant cytoplasm is poorly outlined.

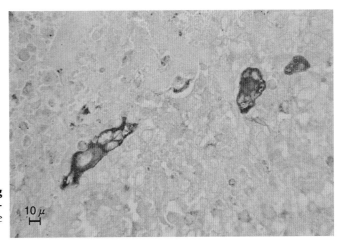

Fig. 8-101 Enzyme antibody technique demonstrating human chorionic gonadotropin, the same case as Fig. 8-100. It should be noted that not all cancer cells produce HCG.

10 μ

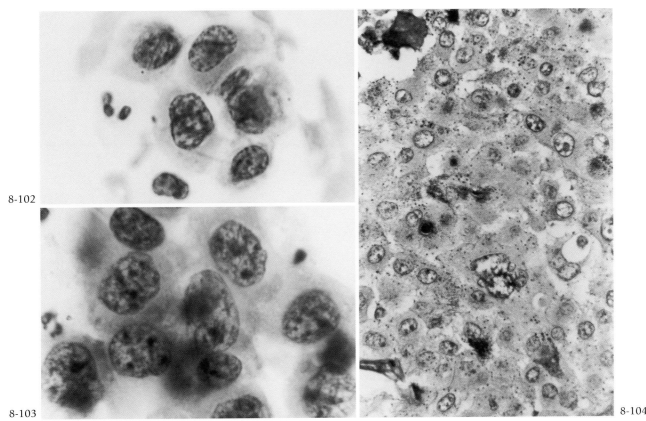

8-102

8-103

8-104

Fig. 8-102 Poorly differentiated large cell type of malignant cells in a scraping smear. These cells are from a case of poorly differentiated squamous carcinoma. The cytoplasm is pale, basophilic, scanty and ill-defined. The nuclei are irregularly outlined and densely stained. The chromatin is condensed in heavy strands.

Fig. 8-103 Poorly differentiated large cell type of malignant cells in a scraping smear. These cells are from a case of poorly differentiated adenocarcinoma. Nuclear overlapping and prominence of nucleoli are somewhat suggestive of glandular type, but these were classified into undifferentiated large cell type by such findings as pale basophilic cytoplasm, indistinct cellular borders and loose cell clusters.

Fig. 8-104 Large cell anaplastic carcinoma. This is a solid tumor cell nest occasionally with intracytoplasmic mucin-like substances positive for periodic acid-Schiff reaction.

Note: Cancer cells arising from such poorly differentiated squamous or adenocarcinomas will be frequently classified into the same category as the undifferentiated large cell type. This is an inevitable pitfall of cytology in cell typing.

ated squamous cell carcinoma. Some reliable criteria for this cell typing are loose texture of cell clusters, water-clear cytoplasm with indistinct cellular outlines, rich glycogen content and a high nuclear-cytoplasmic ratio (Fig. 8-106); the latter is important for distinction from metastatic renal cell carcinoma.

Reality of this tumor is rare incidence of pure clear cell carcinoma[110,208] and common occurrence of clear cell change in all types of carcinoma except for small cell carcinoma.[83] Of all clear celled carcinomas 55.2% were squamous cell carcinoma, 36.2% adenocarcinoma and 3.8% adenosquamous carcinoma, whilst only 4.8% were pure clear cell carcinoma.

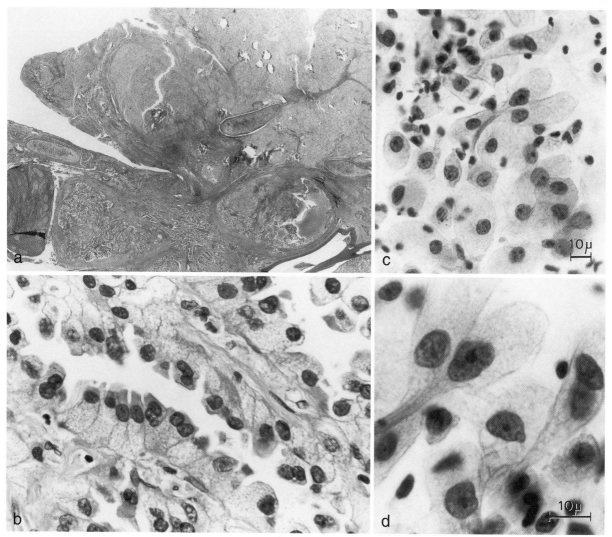

Fig. 8-105   Clear cell adenocarcinoma and its exfoliative cytology.
a. Exophytic tumor protruding into the stem bronchus. The tumor is thought to be originated from the bronchial gland or its duct epithelium.
b. Higher magnification of the tumor characterized by water-clear cancer cells arranged in a trabecula fashion.
c, d. Malignant cells with water-clear cytoplasm in sputum smears.

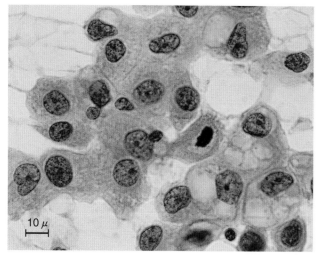

Fig. 8-106   Well preserved clear cell carcinoma in imprint. The cytoplasm is abundant and stains lightly by Light Green. Electron microscopically this clear cytoplasm was rich in glycogen. PAS stain with diastase digestion is important for cytochemical proof of glycogen.

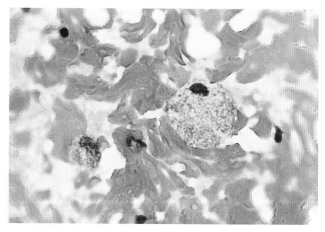

Fig. 8-107   Benign sugar cell in needle aspirate. From a coin lesion of a 21-year-old female. Note abundant foamy cytoplasm and a tiny eccentric nucleus.

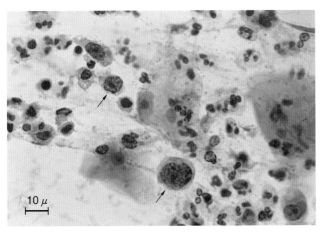

Fig. 8-108 Isolated malignant lymphoma cell in sputum. A large and smaller lymphoma cells are shed singly without any organoid pattern. Prominent nucleoli and indentation of nuclear rims are notable.

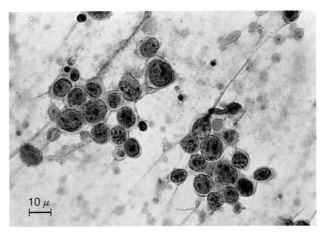

Fig. 8-109 Loosely aggregated lymphoma cells in sputum. Lymphoma cells aggregated may resemble anaplastic small cancer cells. However, multiple beads-like nucleoli and distinctly outlined but delicate nuclear membranes are characteristic of lymphoma cells.

■Clear cell tumor

Benign clear cell tumor first described as sugar tumor by Liebow and Castleman (1963) is a rare tumor of incidental finding as a coin lesion in chest X-ray. It occurs in all age groups without sex predominance.[4]

Cytologically this tumor is a target of fine needle aspiration. The tumor cells have abundant foamy cytoplasm and small ovoid nuclei (Fig. 8-107). Diastase digesting PAS stain accounts for much glycogen.[129]

■Malignant lymphoma

Extension of lymphoma to the lung frequently occurs from the mediastinal lymph nodes or as true metastasis. Primary malignant lymphoma has been rarely reported,[148,198,209] because the distinction from pulmonary involvement in systemic disease of lymphoma is difficult when lymph nodes of various parts of the body are generally involved. Of the three types of malignant lymphoma, lymphosarcoma is the most common in the lung[15,161] and is found to have a better clinical course than other types.[15,45] Since malignant lymphoma originates in the peribronchial lymphoid tissue, erosion of the bronchial mucosa may occur and makes cytological detection possible either by sputum or brushing cytology. According to Dawe et al[39] who studied 82 cases of histologically proven malignant lymphoma, only 9 cases (10.9%) were diagnosed. However, no interpretation on cytological characteristics of lymphoma cells was given. Of all types of lymphoma, though sputum cytology is beyond the reach to subclassification on non-Hodgkin's lymphoma, small celled lymphoma is the most common variety originated in mucosa-associated lymphoid tissue (MALT)[48,130]; lesions important for distinction from malignant lymphoma are pseudolymphoma (Saltzstein[161]) and lymphocytic interstitial pneumonia (Carrington and Liebow[30]) which are associated with dysproteinemia such as monoclonal macroglobulinemia or polyclonal gammopathy. As a causative factor smoking is important for bronchial MALT.[145] Bronchoscopic lavage and fine needle aspiration are indispensable procedures to confirm MALT lymphoma.[38,174]

Non-Hodgkin's lymphoma: Lymphoma cells shed in sputum are singly isolated and appear to be almost denuded; nuclear membranes are not thickened but distinct and partly folded. One or two prominent nucleoli with acidophilia locate paracentrally. Occasionally multiple nucleoli arranged in beads are characteristically encountered in lymphoreticular neoplasia (Fig. 8-108). The chromatin pattern consisting of dotted chromatin particles is not as condensed as that of small cell carcinoma; these are dispersed evenly (Figs. 8-109 and 8-110). Karyorrhexis and necrotic debris are not so pronounced as in small cell carcinoma (Fig. 8-96).

Differentiation between pseudolymphoma and lymphosarcoma: Since lymph follicles with true germinal centers are formed in pseudolymphoma, the presence of germinoblasts and reticulum cells showing phagocytosis (so-called starry sky pattern[78]) is a noteworthy finding. The mixture of variable numbers of inflammatory cells including plasma cells, eosinophils and histiocytes is suggestive of pseudolymphoma. However, these findings can be observed in transbronchial curettings or needle aspirates, because the bronchial or bronchiolar epithelium is preserved intact in spite of compression by subepithelial lymphocytic infiltrate. From a laboratory investigation, a diffuse form of pseudolymphoma or lymphocytic interstitial pneumonia (LIP) can be suggested by the association of dysproteinemia (Figs. 8-111 to 8-114).[9,96,118]

Infective follicular bronchiectasis: It should be kept in mind that follicular bronchiectasis is added to a variety of lymphoma-mimic cytopathology since massive subepithelial infiltrate by lymphoid cells may shed immature lymphoid cell to be distinguished from lymphoma cells. Chronic follicular bronchiectasis affects the youth before puberty with an episode of respiratory infection. The bronchial wall is thickened by massive lymphocytic infiltrate with hyperplasia of germinal centers (Fig. 8-111). Mature and immature lymphoid cells will be exfoliated admixed with neutrophils when erosion of the mucosa occurs.

Hodgkin's disease: The subclassification of Hodgkin's

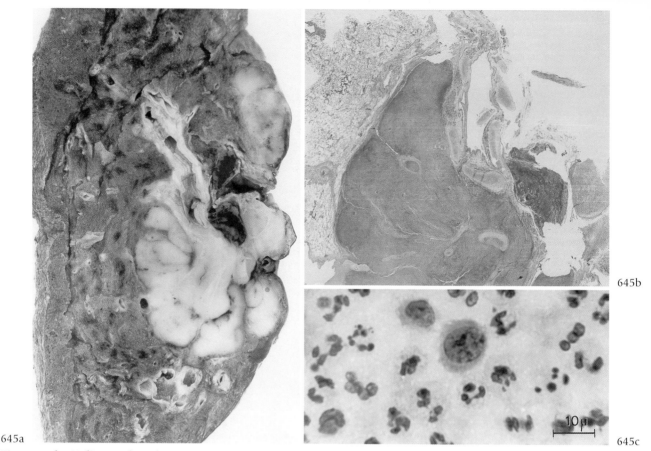

645a

645b

645c

10 μ

Fig. 110a, b Malignant lymphoma growing around the stem bronchus. Macroscopically, the tumor is homogeneous, pale pinkish grey and medullary in appearance. This tumor shows grossly an expansive growth.

Fig. 110c A single malignant cell suggestive of lymphosarcoma. This atypical cell in sputum has a relatively small, round and vesicular nucleus. The cytoplasm is sparse and indistinctly defined. Note a delicate distinct nuclear outline and coarse stippling of chromatin particles.

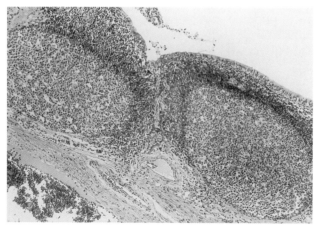

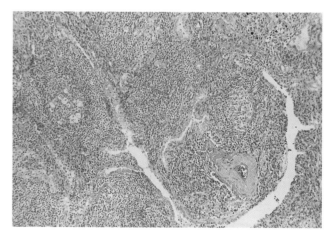

Fig. 8-111 Infective follicular bronchiectasis showing thickened bronchial wall by massive infiltration of lymphoid cells. This 37-year-old female has long been complained of productive cough of bloody sputum because of bronchiectasis. There is marked hypertrophy of germinal centers.

Fig. 8-112 Lymphoid interstitial pneumonitis by open lung biopsy. This 49-year-old male who has had skin eruption for 3 years was found to have diffuse nodular opacities in a chest roentgenogram. Hyperproteinemia of 10.8 g/dl with marked polyclonal gammopathy (IgG 5,540 mg/dl; IgA 1,380 mg/dl; IgM 960 mg/dl) was associated. Numerous nodules of a small finger tip size were found by open biopsy. Note massive infiltration of peribronchial and alveolar walls by lymphocytes and plasma cells in association with hyperplasia of germinal centers.

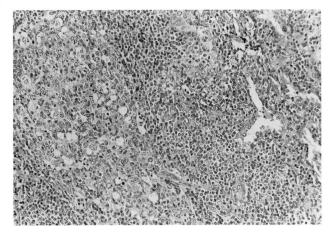

Fig. 8-113 Higher magnification of lymphoid interstitial pneumonitis. Massive cell infiltrate is composed of mature and immature lymphocytes and plasma cells. There is a hyperplastic germinal center showing many macrophages

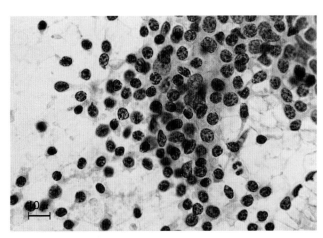

Fig. 8-114 Activated lymphocytes in bronchial brushing from pseudolymphoma. The cytology is characterized by the mixture of large lymphocytes with dotted chromatins and mature small pyknotic lymphocytes.

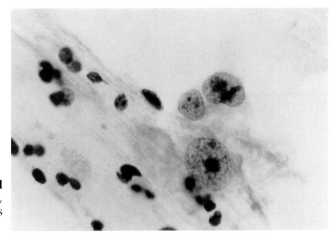

Fig. 8-115 Mononuclear Hodgkin's cell in bronchial scraping. Note a prominent acidophilic nucleolus and thin, delicate nuclear membrane. Nuclear details are the same as those of Reed-Sternberg cells.

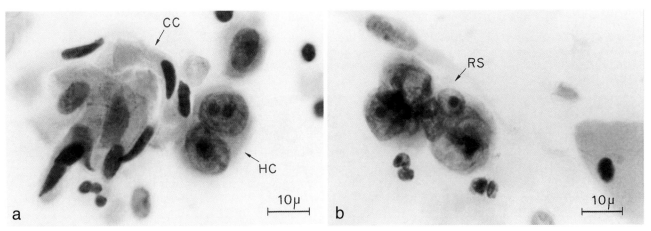

Fig. 8-116 Reed-Sternberg giant cells in bronchial scraping smears.
a. Binucleated Reed-Sternberg cell showing a typical "mirror image" arrangement. HC: Hodgkin's cell; CC: ciliated columnar cells.
b. A characteristic Reed-Sternberg giant cell (RS) showing multinucleation or multilobulation with prominent acidophilic nucleoli is diagnostic of Hodgkin's disease.

disease such as *lymphocyte predominance, mixed cellularity, nodular sclerosis* and *lymphocyte depletion* is that strictly based on histologic findings, Suprun and Koss[175] described a diagnostic rate of 29.7% and commented upon the distinctive features of Hodgkin's disease. The Reed-Sternberg cells that characterize Hodgkin's disease contain multilobulated or multinucleated nuclei with acidophilic nucleoli (owl-eyed appearance) (Figs. 8-115 and 8-116). The cytoplasm is small to moderate in amount, slightly basophilic and indistinctly bordered. The diagnosis of other lymphomas should be based on a general cellular patten, particularly on the predominant cells either of the lymphocytic type or histiocytic type.

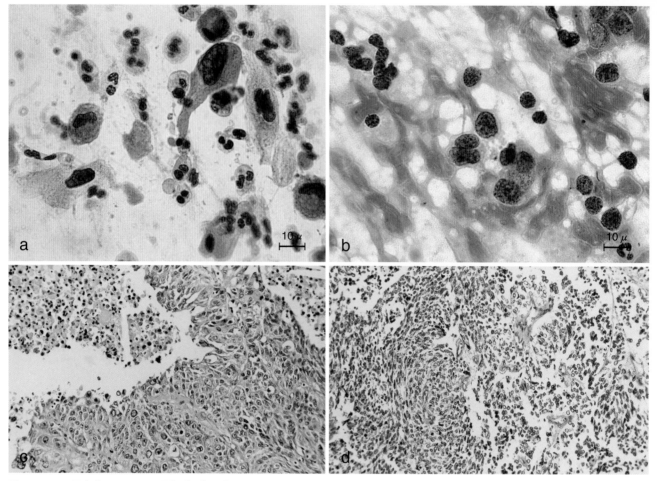

Fig. 8-117   Triple cancers with dual pulmonary cancers and synchronous gastric cancer.
a. Sputum cytology showing epidermoid carcinoma.
b. Sputum cytology showing oat cells that appeared metachronously.
c. Synchronous epidermoid carcinoma of the right lung. This 57-year-old male was done gastrectomy for intramucosal gastric adenocarcinoma two months before curable lobectomy of the right middle lobe, from which epidermoid cancer cells were expectorated (a).
d. Metachronous oat cell carcinoma of the right lung. One and a half years after previous operations (gastrectomy and right lobectomy) the patient has begun to produce bloody sputum, in which oat cells were found (b). Needle aspiration confirmed the third carcinoma in the right upper lobe. Autopsy revealed lymph node metastasis with different histologic types respectively.

■Combined carcinoma and dual carcinoma

The surface bronchial epithelium is endued with potential of metaplastic change. Squamous metaplasia of the surface epithelium and duct epithelium of the bronchial glands and adenomatous metaplasia of type II pneumocytes at the terminal bronchioles around scarring are quite common phenomena. It is not infrequent to see various degrees of differentiation or different kinds of differentiation in the same tumor; a combined form of cancer such as adenosquamous and small cell-epidermoid carcinoma is often encountered.

Strictly speaking dual carcinomas are defined based on the following criteria (Fig. 8-117): (1) definitely separated tumors with different histologic types, (2) metastasis formation of each histologic type, (3) absence of extrapulmonary primary carcinomas having the similar histologic types, (4) different anatomical location of either synchronous or metachronous carcinomas, and (5) different stem lines. From a cytological aspect, the mixture of cancer cells of different types in sputum is

not sufficient to distinguish the combined carcinoma from dual carcinoma, because dual carcinoma may occur simultaneously.[26,144] In order to confirm dual carcinomas, the specimen must be taken from both parts by means of selective brushing or needle aspiration to determine different malignant cell types. The difference of stem cell lines can be estimated by chromosomal analysis or DNA measurement using Feulgen microspectrophotometry.

■Pulmonary blastoma and carcinosarcoma

This rare tumor, pulmonary blastoma, originating from primitive multipotential cells, is characterized by epithelial cell element with a primitive tubular or glandular pattern and massive sarcomatous mesenchymal component of embryonal type. From this histological feature several authors regarded this tumor as a type of carcinosarcoma.[10]

Electron microscopic observation that confirmed the existence of primitive epithelial and mesenchymal cells

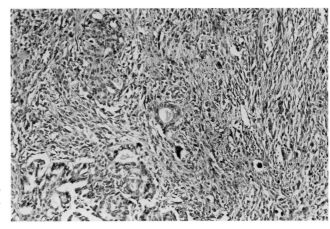

Fig. 8-118 Histology of pulmonary blastoma or embryoma. Fetal bronchiolar tissue is recognized in dense sarcomatous tissue.

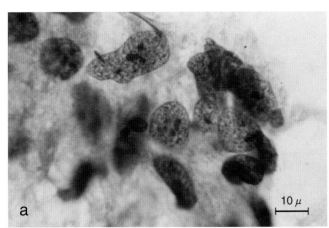
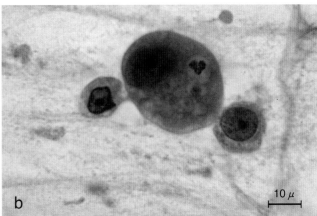

Fig. 8-119 Cytology from pulmonary blastoma by needle aspiration.
a. Mesenchymal component showing elongated nuclei with thin, delicate nuclear membranes and a smooth, lacy chromatin pattern. Since the mesenchymal component consists of a monotonous fibroblastic type, the differential diagnosis between pulmonary blastoma and carcinosarcoma cannot be made cytologically.
b. Glandular component showing round nuclei with prominent nucleoli.

with potential to differentiate into bronchial epithelial cells and cartilaginous cells may be favorable to the concept of carcinosarcoma.[107] On the other hand, adenomatous epithelial cells which are embedded as slits in dense sarcomatous stroma have vacuolated columnar cytoplasm. This feature is not likely carcinosarcoma but resembles fetal bronchiolar tissue as called embryoma.[85,210]

The biologic behavior of blastoma that occurs peripherally and undergoes a slow growth differs also from carcinosarcoma. Spencer[172] elicited the Waddell's concept (1949)[192] for explanation that the distal respiratory part of the lung originates from mesoblastic tissue that is inducible of the pulmonary counterpart of renal nephroblastoma. Although these forms are not criticized independently from clinicopathological aspects in WHO classification,[92] carcinosarcoma differs from embryonal blastoma in the location arising centrally, gross finding that grows as a polypoid lesion, and worse prognosis with distant metastases. Histologically the epithelial component in carcinosarcoma is commonly squamous cell in type, and the glandular type, if occurs, is not likely of fetal lung tissue. The concept of sarcomatous change of stroma in cancer for this tumor (Willis, 1960[204]) can be denied by the fact that both components of cancer and sarcoma were found in metastases.

Cytology: According to the reviews on 21 pulmonary blastomas by Karcioglu and Someren,[81] 15 cases showed symptoms of hemoptysis and/or productive cough, but no description was referred to its sputum cytology. Meinecke et al[112] described a case of positive cytology which was derived from adenocarcinoma component. Peripheral location of the tumor indicates applicability of fine needle aspiration (Fig. 8-119). A hypodermic needle is not suitable to obtain enough specimen because of the predominance of mesenchymal cell elements. Cytologically pulmonary blastoma is characterized by coexistence of fibroblast-like spindly nuclei with delicate unclear membrane and glandular epithelial cells, although cancer cells are rarely encountered.

■ Carcinosarcoma

Carcinosarcoma is a distinctive entity different from biphasic pulmonary blastoma that occurs peripherally beneath the visceral pleura. This tumor occurs mainly in men and arises from the lobar and segmental bronchus. There is an association with cigarette smoking. Histogenetically endodermal and mesodermal components are concerned.[173]

Carcinosarcoma is distinguished by two components of spindly fibrosarcomatous cells and squamous cancer cells regardless of keratinization. The definite diagnosis

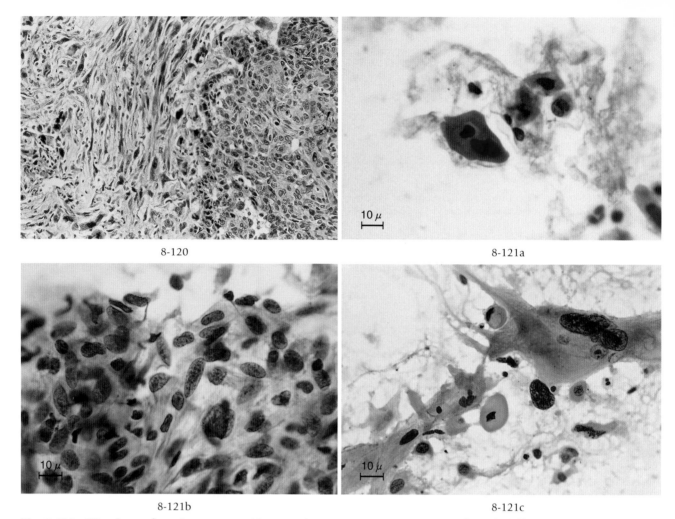

8-120

8-121a

8-121b

8-121c

Fig. 8-120   Histology of carcinosarcoma. Note two tissue components consisting of poorly differentiated epidermoid carcinoma in a solid cell nest and spindle celled sarcoma.

Fig. 8-121   Cytology from carcinosarcoma by transbronchial biopsy and imprint cytology. Carcinosarcoma from a polypoid tumor protruding into the segmental bronchus (right B$_6$). The tumor is composed of differentiated epidermoid cancer cells and pleomorphic sarcoma cells.

a. Epithelial component showing a keratinizing pyknotic cancer cell.

b. Mesenchymal component obtained as a tissue fragment. Note irregularly arranged spindly tumor cells. Thin nuclear rims and finely granular chromatins point out the nature of mesenchymal cells.

c. A keratinizing squamous cancer cell and sarcoma cells appear to be combined together.

(By courtesy of Dr. H. Kato, Tokyo Medical College, Tokyo, Japan.)

is only available when both squamous and mesenchymal malignant cells are found in the same smear (Fig. 8-121).

■Pulmonary hamartoma

Hamartoma is found as a well circumscribed lesion composed for the most part of cartilaginous tissue separated into lobules by fibroadipose tissue. Noteworthy findings are cleft-like spaces with lining of hyperplastic low columnar (respiratory) epithelium. Needle aspiration cytology may encounter hyperchromatic cuboidal cells mixed with cartilaginous cells either anucleated or mononucleated (Fig. 8-122).

11 Cytopathologic Changes in Heavy Smokers

It is known that the cigarette smoking habit induces goblet cell hyperplasia, basal cell hyperplasia, squamous

metaplasia and dysplasia of bronchial epithelium (Table 8-8). The grade of these changes with atypia is increased correspondently with increase in the amount of smoking, and carcinoma in situ of the bronchial epithelium would be induced by continuous heavy smoking.

The bronchiolar epithelium is also involved by heavy smoking. Ebert and Terracio[42] observed that Clara cells were decreased, whereas goblet cells producing tenacious acid mucopolysaccharide were remarkably increased in number. This evidence can be confirmed in sputum specimens as well as in bronchofiberscopic biopsy specimens. The Japan Lung Cancer Screening Program proceeds with cytologic screening of three day pool sputum of early morning cough for male heavy smokers over 50 of age. The grade of smoking habit is grouped by a cigarette index which is represented by the average number of cigarettes per day multiplied by years of smoking. The specimen can be satisfactorily obtained without aerosol induced technique. Increased goblet

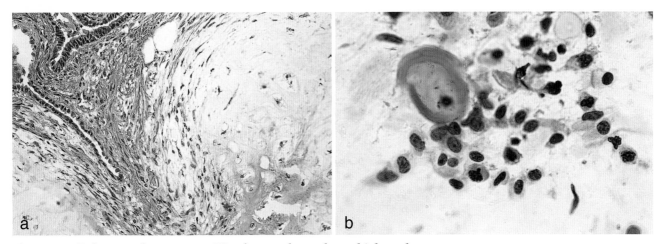

Fig. 8-122 Pulmonary hamartoma. Histology and transbronchial cytology.
a. Histology is characterized by lobulated cartilaginous tissue. Note cleft-like spaces lined by hypertrophic low columnar epithelium.
b. Transbronchial needle aspiration showed hyperchromatic cuboidal cells isolated and a cartilaginous cell with abundant, opaque cytoplasma and a peripherally locating small nuclear.

Table 8-8 Comparative histological changes of bronchial epithelium in heavy smokers* and non-smokers**

	Number of cases	Normal	Inflammation		Goblet metaplasia		Basal cell hyperplasia			Squamous metaplasia	
			Mild	Marked	Mild	Marked	Mild	Marked	Atypical	Benign	Atypical
Smoker	138	40	17	2	24	25	50	10	7	26	7
	(%)	(29.0)	(12.3)	(1.4)	(17.4)	(18.1)	(36.2)	(7.2)	(5.1)	(18.8)	(5.1)
Non-smoker	32	19	6	0	6	0	5	0	0	0	0
	(%)	(59.4)	(18.8)		(18.8)		(15.6)				

*Cigarette index: over 400.
**Patients with bronchitis are included.

cells, squamous metaplasia and dyskeratosis are found regardless of the cigarette index. The higher the index, the more marked cellular alteration such as squamous metaplasia with nuclear atypia and cytoplasmic dyskeratosis. Concerning a histologic type of lung cancer in relation to smoking habit, no clear-cut relationship is ascertained yet.[6]

12 Early Carcinoma and Carcinoma In Situ

■Goal of cytodiagnosis
Various kinds of techniques have been devised to detect occult carcinoma, i.e., X-ray negative and sputum-positive carcinomas without evidence of metastasis to the regional lymph nodes or distant organs (TX, N0, M0), for which a good 5-year-survival rate can be expected. In the report by Woolner and his associates[206] on curability of early pulmonary carcinoma 88% of their cases brought about 5-year-survival. Since carcinoma of the central type develops on the bronchial surface epithelium shedding of cancer cells into sputum may be continuous during the preclinical stage. In addition to the clinical staging and resectability of the tumor in relation to its location, cell typing is much concerned with prognosis. Search for cytological pictures peculiar to early carcinomas arising from the major bronchus was pursued on 22 cases collected by the Japanese project team for detecting early lung cancers, and came to obtain some

suggestive characteristics (Table 8-9).[178] Early carcinoma of central type is defined as a tumor arising from the stem or segmental bronchus regardless of histological patterns, and as a tumor strictly confined to the bronchial wall without metastasis to the regional lymph node.

Evaluations for lung cancer screening programs consisting of every four month examination with chest X-ray and sputum cytology for male smokers failed to expect the reduction of mortality by lung cancers even when these were detected in early stage (Fontana et al 1996,[50] Melamed and Flehinger 1987[113]). Since then mass screening using X-rays and sputum cytology has been abandoned. Sobue et al reported a case-control study choosing 273 cases and a total of 1,269 controls matched by sex, age and smoking status out of 136,860 males and 146,481 females in Japan. The Saccommano's 3-day-pool sputums were used in addition to 100 mm × 100 mm chest X-ray films for male heavy smokers. As a result of the case control study, the odds ratio of dying from lung cancer for those screened in 12 months versus those not screened was 0.72.[169] This result may suggest some benefits in reduction of lung cancer mortality.

■Cytologic pictures of early carcinoma
Although no definite pictures to determine early carcinoma are present, a background of the smear is clean without tumor diathesis. Frequent association with

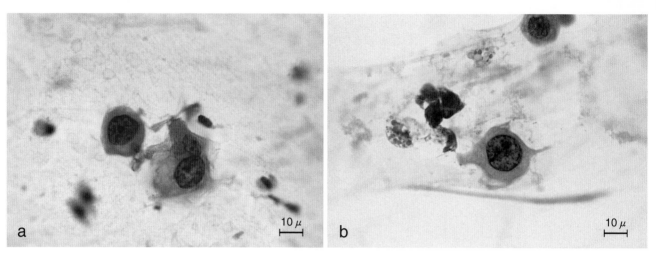

Fig. 8-123   Squamous cancer cells in sputum from a patient with cancer in situ and microinvasive carcinoma of the central type.
a. Parabasal type, keratinizing. **b.** Parabasal type, nonkeratinizing.
Clumping of chromatins, presence of nucleolus and high N/C ratio are the features distinguishable from metaplastic cells.

Table 8-9   Pulmonary cytology in 22 cases of early cancer arising from the major bronchus

Case	Age	Sex	Histology	Sputum	Cancer cells				Dysk. pykn.	Background				Cigar. index
					Ana.	Pab.	Int.	Sup.		Diat.	Spir.	Met.	CCP	
O.K. Natl. Tokyo Hosp.	71	M	Squamous †	Bloody	−	++	+	−	++	−	+	−	−	800
T.H. Tohoku Univ.	62	F	Poor squam.	Mucous	+++	+	−	−	+	−	−	−	−	0
S.S. Tohoku Univ.	56	M	Poor squam.	Bloody	+++	+	−	−	−	−	−	−	−	720
H.K. Natl. Kinki Hosp.	69	M	Squamous	Mucous	−	+	+	++	−	+	+	+	++	960
H.M. Natl. Kinki Hosp.	53	M	Squamous †	Mucous	−	++	+	−	−	−	−	−	−	600
Q.Z. Osaka Adult Ctr.	57	M	Squamous	Bloody	−	+	+	+	+	−	−	−	−	400
K.M. Osaka Adult Ctr.	58	M	Squamous	Bloody	−	++	+	−	+	−	+	+	−	900
I.T. Osaka Adult Ctr.	52	M	Squamous	-	−	+*	+*	−*	+*	**	**	+*	**	300
T.H. Tenri Hosp.	75	M	Squamous	Bloody	−	−*	+*	++**	−	**	**	−*	**	1,250
M.T. Tenri Hosp.	65	M	Squamous	Mucous	−	+	+	+	++	−	−	−	−	1,200
F. S. Tenri Hosp.	62	M	Squamous	Bloody	−	+	++	++	−	+	+	+	−	960
M.H. Tohoku Univ.	53	M	Squamous	Bloody	−	+	++	++	−	+	+	−	−	1,350
S.K. Mie University	65	M	Squamous	Mucous	−	+	+	++	+	−	−	−	−	800
Y.Y. Railway Hosp.	52	M	In situ Sq.	Bloody	−	+	+	−	+	−	−	+	−	600
K.S. Natl. Cancer Ctr.	47	M	Squamous	Bloody	−	−	−	−	+	−	−	+	−	420
B.M. Natl. Cancer Ctr.	55	M	Squamous	Bloody	−	+	−	+	++	−	−	+	+	2,310
N.K. Natl. Cancer Ctr.	64	M	Squamous	Bloody	−	+	+	+	++	−	+	+	−	2,000
T.T. Natl. Cancer Ctr.	60	M	Squamous	Bloody	−	+	+	+	++	−	+	−	+	925
K.O. Natl. Cancer Ctr.	44	M	Squamous	Bloody	+*	+*	++*	−*	−*	**	**	+*	**	550
N.N. Natl. Cancer Ctr.	46	M	Squamous	Bloody	−	−	−	+	+	−	+	−	+	700
K.T. Natl. Cancer Ctr.	56	M	Squamous	Bloody	−	−	+	++	+	−	+	+	−	1,480
S.Y. Tokyo Med. College	55	M	Oat cell	-	++*	−	−	−	−	+*	**	−*	**	0

Note: † Carcinoma in situ with microinvasive growth. *Findings in bronchial brushings. **Not evaluated because of brushing. Abbreviation: Ana. =anaplastic type; Pab. =parabasal type; Int. =intermediate type; Sup. =superficial type; Dysk. pykn. =dyskeratotic pyknotic cells; Diat. =diathesis; Spir. =curschmann's spirals; Met. =squamous metaplasia.

Curschmann's spirals may be the result of obliteration of small bronchi. As a direct finding, predominance of small parabasal type cancer cells shedding singly or in a loose small cluster with often refractive orangeophilia of the cytoplasm is suggestive of in situ carcinoma with or without early stromal invasion: Although the staining reaction of the cytoplasm is variable, from bluish green to intense orange, dyskeratotic cells with refractive cyto-plasm are frequent (Figs. 8-123 and 8-127). Since the cancer cells coughed out are from the fragile superficial layer, many are singly isolated and densely stained (Fig. 8-127). Selective brushing cytology is inevitably necessary before surgical procedure (1) to ascertain early epidermoid carcinoma characteristically comprised of small parabasal or basal type malignant cells and small dyskeratotic pyknotic cells, and (2) to make sure of localiza-

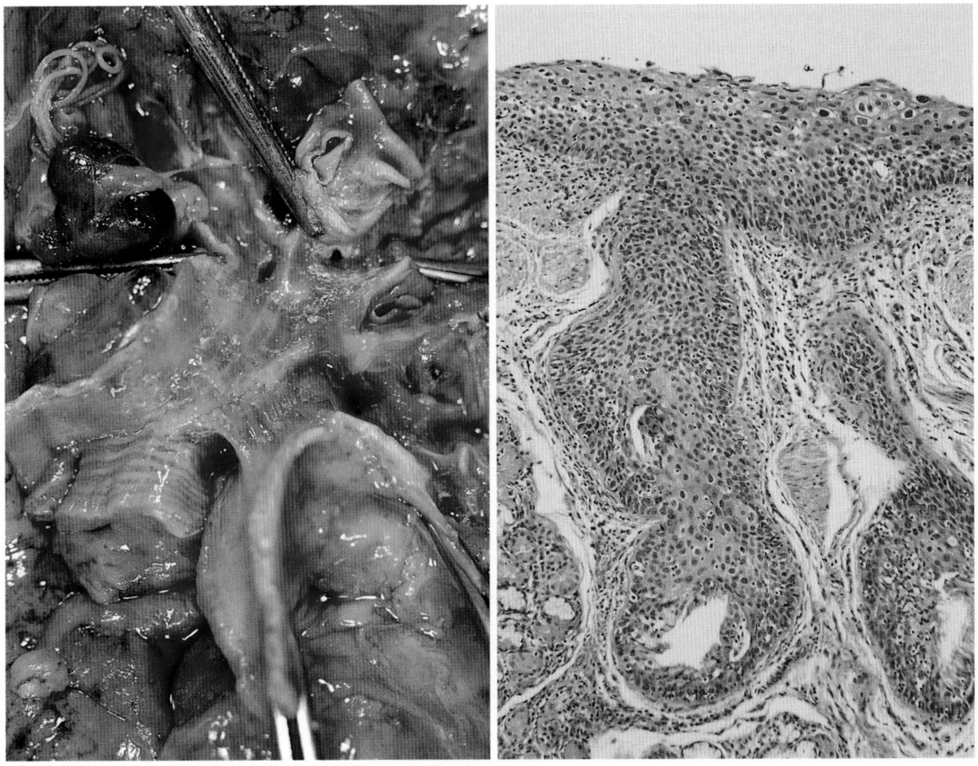

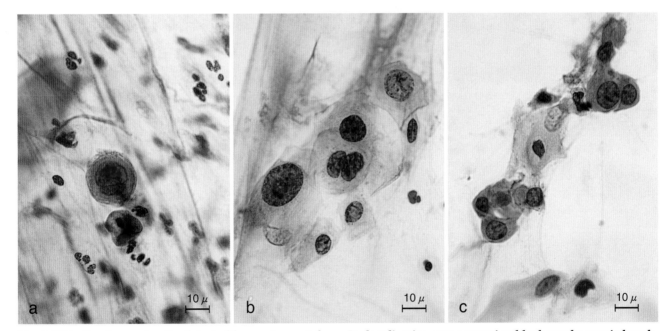

Fig. 8-124 X-ray negative sputum cytology positive early squamous cell carcinoma.

a. No evidence of macroscopic neoplasia except for dull thickened mucosa (B$_6$). Localization can be determined by selective bronchial brushing.

b. Early invasive squamous cell carcinoma. The mucosa is superficially replaced by cancer in situ involving the bronchial gland duct.

Fig. 8-125 Carcinoma in situ detected by sputum cytology. Its localization was ascertained by bronchoscopic brushing. A case of female without smoking habit. The same case as Fig. 8-124.

a. Keratinizing cancer cells of parabasal type scattered in a clear background. Endoscopic examination showed no evidence of tumor obstruction except for dull thickened mucosa of the stem bronchus of left upper lobe.

b, c. Superficial and parabasal type squamous cancer cells showing marked hyperchromatism, high N/C ratio and coarse chromatins.

tion of the cancerous lesion. Cytology is particularly important where the lesion is not visualized bronchoscopically and multiple punch biopsy is not available for localization (Fig. 8-126). Accordingly as invasive growth advances, cellular polymorphism becomes prominent, and tumor diathesis appears.

13 Bronchial Carcinoid and Bronchial Gland Carcinoma

General aspect

A group of slowly growing tumors which has been traditionally grouped as bronchial adenoma is comprised of carcinoid tumors, cylindromas, mucoepidermoid tumors, etc. The majority of bronchial adenomas, 85%

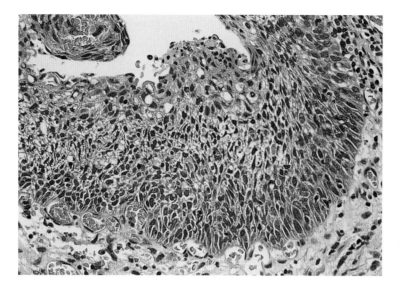

Fig. 8-126 Histology of squamous carcinoma in situ detected by mass screening program. Note replacement of mucosal membrane by squamous cancer cells showing nuclear hyperchromatism, anisonucleosis and irregular arrangement. Superficial differentiation is reflected in sputum cytology showing dyskeratosis (Fig. 8-127). Localization of carcinoma in situ spreading superficially around bifurcation between B_{1+2} and B_3 was proved by bronchofiberscopic cytology and biopsy. (By courtesy of Prof. H. Kato, Tokyo Medical College, Tokyo, Japan.)

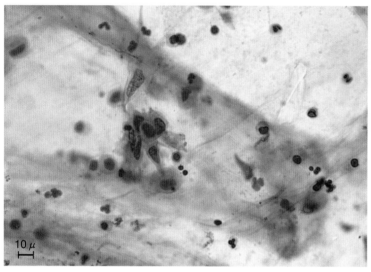

10 μ

Fig. 8-127 Keratinizing squamous carcinoma cells in sputum mass survey from a 57-year-old male with X-ray negative in situ carcinoma of the lung. Mass survey sputum placed in a mailable vinyl bag containing 1% formalin balanced solution preserves cells but tends to aggregate as mucin threads. Note fairly small cancer cells with dyskeratosis or parakeratosis in small groups. Tumor diathesis is absent. Many Curschmann's spirals were associated.

to 90%, are carcinoid tumors of the central type. Although carcinoid had been discussed under the same category with cylindroma, these are essentially different from cylindroma or adenoid cystic carcinoma that locates more proximally in the tracheobronchial tree and tends to metastasize more frequently than carcinoid.[183] Since these tumors growing as a circumscribed nodule with intact surface mucosal epithelium come to the same target in transbronchial cytology, they are described in this same chapter.

The occurrence of carcinoid syndrome and less frequent Cushing's syndrome is associated only with carcinoid. Ectopic endocrine syndrome produced by bronchogenic carcinomas such as carcinoid and oat cell carcinoma has led to the concept that argentaffine cells (Kulchitsky cells) in the bronchial glands and among the lining cells of the bronchioles become the parent cells of both carcinoid and oat cell carcinoma.[14] It is quite frequent that argentaffine reaction-negative carcinoid reveals positive argyrophil reaction. Therefore, argyrophil reaction is useful to determine the neurosecretory function of the tumor.

Although the tumor locates mainly in the main and segmental bronchi, occasional occurrence of peripheral carcinoid (15.7% of 217 tumors[133]) is consistent with

distribution of neurosecretory cells that belong to the APUD system in the bronchiolar epithelium.

■Cytopathology

1) *Cylindroma* (adenoid cystic carcinoma) is characterized by tight clusters of small hyperchromatic cells with scant cytoplasm. The nuclear feature resembles that of carcinoid tumor, however acinar structure around a lumen with often PAS-positive secrete in closely packed cluster is a diagnostic picture. The nuclei are uniformly round, nearly the same as carcinoid or oat cells in size and densely stained, but their chromatins are not coarse as for malignant cells (Figs. 8-128 and 8-129).

2) *Carcinoid* is grossly a well circumscribed, interbronchial tumor with covering of mucosal epithelium intact from ulceration. Hemoptysis is one of remarkable symptoms (42% of carcinoid tumors[133]) because of rich vascularity, but tumor cells exfoliate rarely into sputum, i.e., 5 of 133 cases by Okike et al.[133] These are recognizable by brushing or curetting. The tumor cells are small, uniform in size and shape, and appear in flat sheets. The nuclei are round or oval, moderately hyperchromatic and resemble those of undifferentiated small cancer cells or oat cells. The distinction of carcinoid to be differentiated from small cell carcinoma is in nuclear uniformity,

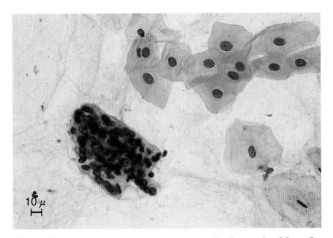

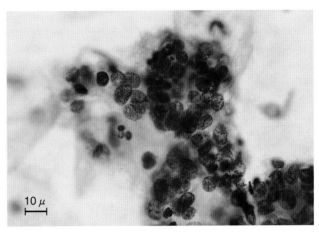

Fig. 8-128 Adenoid cystic carcinoma by bronchial brushing. The nuclei are nearly the same as those of carcinoid but they are tightly clustered and overlapped. Acinar structure can be recognized.

Fig. 8-129 Larger magnification of adenoid cystic carcinoma. By bronchial brushing from a polypoid exophytic tumor obliterating the bronchus. The nuclei are uniformly oval, generally hyperchromatic but fine in texture. Look at tight cluster with acinar pattern around lumens.

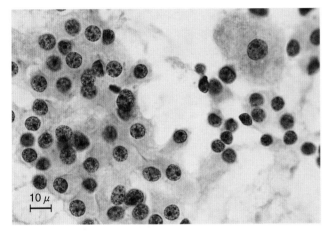

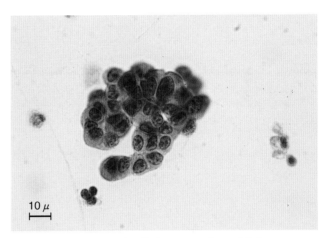

Fig. 8-130 Carcinoid by bronchial brushing. Well preserved carcinoid cells. Differential points from oat cells are based on (1) uniformity of nuclear size and shape, (2) smooth nuclear rims, and (3) the presence of moderate amount of pale finely granular cytoplasm.

Fig. 8-131 Carcinoid cells in a transbronchial curettage. These cells do not exfoliate into sputum because carcinoid is usually covered by mucosal epithelium. The tumor cells are clustered in a trabecular pattern. The nuclei are uniformly small, round and smoothly outlined.

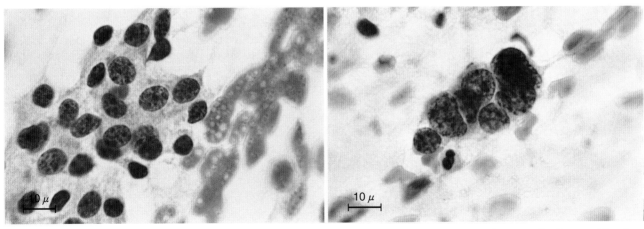

Fig. 8-132 Comparison between carcinoid and oat cell carcinoma. Illustration with the same high magnification. Carcinoid has uniformly oval nuclei and somewhat coarse but evenly distributed chromatins, whereas oat cell carcinoma varies in nuclear size and show much more coarsely granular chromatins.

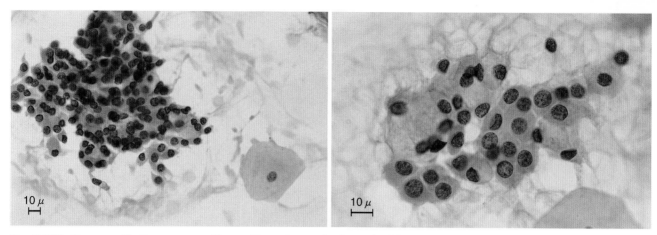

Fig. 8-133   Tumor cells from mucoepidermoid carcinoma by bronchial brushing. The nuclei are ovoid in shape and their rims are delicate. Prominence of nucleoli and finely granular chromatin pattern are rather in favor of the glandular type, while the cytoplasm is polyhedral, opaque and displays the flat sheet compatible with the epidermoid type.

smooth nuclear rim and even, finely granular chromatin pattern. The cytoplasm that stains feebly is ill-defined and somewhat granular when well preserved; the amount of cytoplasm is relatively small, but more abundant than that of oat cells (Figs. 8-130 to 8-132).

Pitfall of cytodiagnosis: Some anaplastic type of carcinoid (atypical carcinoid) which is histologically composed of ribbons or columns of hyperchromatic small cells displays cytology with oat cell-mimic feature such as polymorphic nuclei, high N/C ratio and irregular-coarse chromatins with perichromatin clearing. Pitfalls in the diagnosis of oat cell carcinoma are mentioned by Naib,[124] and others.[35,56,93] Because of morphological similarities between oat cells and carcinoid cells and of presence of argyrophil, neurosecretory granules in both tumors, oat cell carcinoma is often regarded as a malignant variant of carcinoid.[9]

Spindle cell carcinoid: Some of carcinoids arise from the periphery and are found in chest X-ray as a coin lesion. Peripheral carcinoids are distinctively composed of spindly cells with hyperchromatic elongated nuclei and conspicuous karyosomes. The tumor cells are arranged in a streaming fashion intersected by thin fibrovascular stroma. Thereby spindly tumor cells in needle aspirates may lead to misinterpretation as poorly differentiated squamous cell carcinoma[131] or small cell carcinoma.[46]

Cytological distinction from cancers can be based on the lack of necrosis, uniformity of the nuclei in size and shape and immunohistochemical approach for chromogranin A, neuron-specific enolase (NSE), etc.[46]

3) *Mucoepidermoid tumor* is recognized as a polypoid tumor with marked exophytic growth arising from bronchial glands within the major bronchi or its segments. Mucoepidermoid tumor when first described was thought to be a form of adenoid cystic carcinoma,[72] however, it has been separated from a group of adenoid cystic carcinoma, because its potential of malignancy is questionable; the case of mucoepidermoid carcinoma with invasive growth and distant metastases[167,183,193] is thought to be definitely a variety of epidermoid carcinoma. *Cytologically* the tumor cells are clustered but in a flat sheet; the cytoplasm is of moderate amount and polyhedral in shape like in metaplastic cells. Mucous cells with abundant goblet-like cytoplasm is distinctly admixed (Fig. 8-133).

14 Primary Sarcoma

Primary sarcoma of the lung is rare. Leiomyosarcoma is the most common in sarcomas and found as a localized, globular tumor protruding into the bronchial lumen. Since the surface is covered by an intact mucosal epithelium, sputum cytology is usually negative unless erosion occurs by vigorous scraping. Imprint cytology from a biopsy specimen may disclose elongated nuclei with smooth outlines and fine chromatins (Figs. 8-134 and 8-135).

15 Metastatic Carcinoma

The lung is an organ which is most commonly affected by hematogenous and lymphogenous metastases of malignant tumors from distant organs (Table 8-10). Initially, when the tumor cell lodgment is microscopical in blood vessels or lymphatic channels and its growth is confined to perivascular areas, exfoliation of tumor cells into the airways does not occur. Endobronchial metastatic foci producing intrabronchial protrusions are often shown in metastases from the rectum (Fig. 8-136) and the kidney, and occasionally from the breast; thus, they become the source of positive sputum cytology in a later stage. Since the needle aspiration cytology has come to be of common usage, we may encounter various types of carcinoma and sarcoma. It is an interesting subject to suggest the primary site of the metastatic carcinoma from a cytological aspect. Pulmonary metastasis is clinically suggested from X-ray findings such as single or multiple nodular shadows scattered throughout the bilateral lobes. As a general appearance of cytology, metastatic cells are found in a fairly clean background. In spite of the absence of tumor diathesis, hemoptysis may be present. Adenocarcinoma cells forming a tubular or acinar cell cluster and keratinizing epidermoid carcinoma cells are of the common type of primary carcinoma and not indicative of metastatic carcinoma. According to the report by Abrams

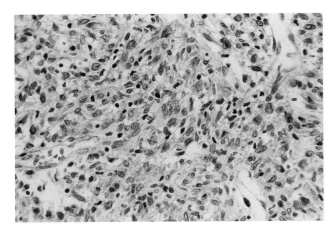

Fig. 8-134 Leiomyosarcoma growing as a submucosal globular tumor. Note irregular bundles of myogenic fibers, the nuclei of which vary in size and shape.

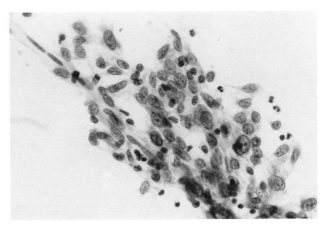

Fig. 8-135 Brushing cytology showing mesenchymal tumor cells from leiomyosarcoma. The nuclei are oval or elliptic and vary in size. Cytological differentiation from fibrosarcoma is difficult, but blunt nuclear poles are favorable to myogenic origin. (By courtesy of Dr. F. Yamamoto, Yamagata Adult Disease Center, Yamagata, Japan.)

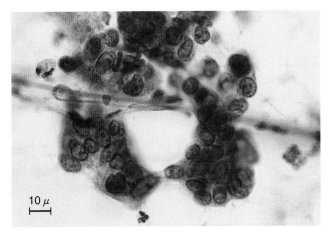

Fig. 136 Adenocarcinoma cells in sputum metastatic from rectum carcinoma. Although there is no specific feature to colon carcinoma, nuclear palisading in a cell cluster accounts for multilayered tubular adenocarcinoma.

Table 8-10 Cytodiagnosis of pulmonary metastatic carcinoma

Primary sites of tumors	Male		Female		Total	
	No. of cases	Posit.	No. of cases	Posit.	No. of cases	Posit.
Stomach	11	5	5	3	16	8
Colon	8	5	2	1	10	6
Breast	—	—	10	2	10	2
Esophagus	7	3	—	—	7	3
Liver	4	1	1	0	5	1
Pancreas	4	1	1	1	5	2
Thyroid	2	1	2	0	4	1
Kidney	3	2	1	0	4	2
Uterus	—	—	4	1	4	1
Prostate	3	0	—	—	3	0
Parotis	2	1	1	0	3	1
Bladder	1	0	1	1	2	1
Bile duct	0	0	2	0	2	0
Adrenal	2	0	0	0	2	0
Larynx	1	1	0	0	1	1
Thymoma	1	0	0	0	1	0
Lymphoma	7	2	1	0	8	2
Sarcoma	4	1	1	1	5	2
Myeloma	1	0	0	0	1	0

and his colleagues[1] on 1,000 autopsy cases with carcinoma, representative carcinomas that cause frequent pulmonary metastases are renal carcinoma in about 75%, malignant melanoma 65%, breast carcinoma 58%, sarcoma 40%, etc.

■Renal cell carcinoma

Renal cell carcinoma giving rise to endobronchial metastasis often sheds characteristic cell clusters in sputum. The cytoplasm is of fair amount and watery clear because of rich fatty substance. The nuclei are small rather for cytoplasm and possess single centrally locating distinctive nucleoli (Fig. 8-137). These cytological features resemble bronchiolo-alveolar carcinoma. However, the clear cytoplasm in bronchiolo-alveolar carcinoma is often positive for PAS reaction.

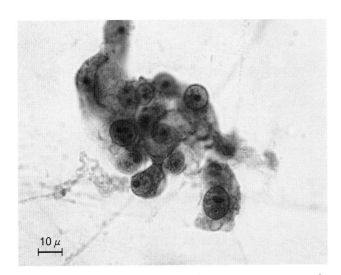

Fig. 8-137 Malignant tumor cells in sputum metastatic from renal cell carcinoma. Renal carcinoma cells have uniformly round and fairly small nuclei. Single, prominent nucleoli are locating centrally. The cytoplasm is of a moderate amount and clear in appearance.

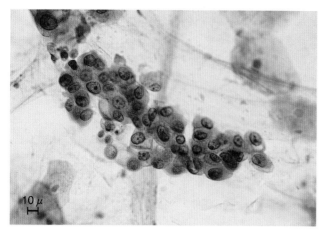

Fig. 8-138　Malignant tumor cells in sputum from hepatocellular carcinoma. Note irregular trabecular cluster of polyhedral cancer cells showing rounded, vesicular nuclei and centrally locating nucleoli. The cytoplasm is moderate in amount and clear.

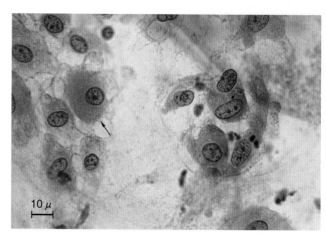

Fig. 8-139　Metastatic hepatoma cells in sputum. The characteristic feature that identifies liver cell carcinoma is the presence of production of bile pigments with yellowish green tint (arrow). They are from differentiated hepatoma with function to produce bile. (By courtesy of Dr. T. Satake, Dept. of Clinical Pathology, Nagoya Ekisaikai Hospital, Nagoya, Japan.)

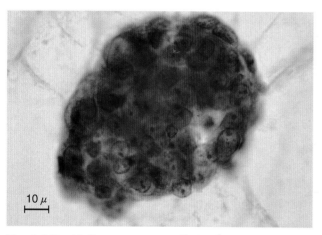

Fig. 8-140　Malignant tumor cells in sputum metastatic from mammary duct carcinoma. A large number of cancer cells are clumped tightly. Such a cell ball with the "morula pattern" reflects solid cancer cell nests of duct carcinoma.

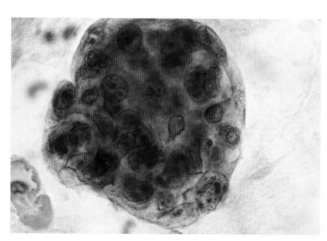

Fig. 8-141　Malignant tumor cells in sputum metastasized from papillary carcinoma of the thyroid. Three-dimensional cell ball formation in sputum specimens is shown in the case with carcinomas of the breast, thyroid and prostate. However, this is not a feature diagnostic of primary sites.

■Hepatoma (Hepatocellular carcinoma)

Hepatocellular carcinoma often metastasizing to the lung is notable in Asian countries. Enrounded cells in a flat sheet resemble renal carcinoma cells because of centrally locating, round nuclei and prominent single nucleoli (Figs. 8-138 and 8-139). Attention should be paid to bile pigmentation and α-fetoprotein (AFP) production. Although immunocytochemistry for AFP is diagnostic of hepatoma, the most poorly differentiated grade 4 carcinoma is not productive of AFP.

■Breast carcinoma

As one of the common types of infiltrating duct carcinoma, comedocarcinoma is histologically characterized by solid masses of cancer cells plugging the dilated ducts. It is known that these take a characteristic "morula pattern" in body cavity effusion. A similar type of cell balls is noted in sputum specimen (Fig. 8-140). Adenocarcinoma with apocrine metaplasia (apocrine carcinoma) is also a peculiar type to breast carcinoma (see breast carcinoma, Fig. 7-77).

■Thyroid carcinoma

Papillary carcinoma of the thyroid tends to invade blood vessels. The lung is the site of hematogenous metastasis. Three-dimensional cancer cells (cancer cell balls) shed in sputum is hardly distinguished from metastasized mammary duct carcinoma (Fig. 8-141).

■Other malignant tumors

Malignant melanoma is easily identified if melanin pigment is identified (Fig. 8-142). Sarcomas such as osteogenic sarcoma and fibrosarcoma metastasize to the lung and are found in sputum. *Osteosarcoma cells* are characterized by markedly polymorphic, large nuclei, the chromatin pattern of which is condensed by pyknosis

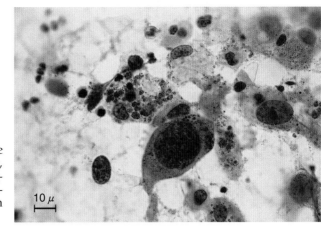

Fig. 8-142　Melanoma cells in sputum. Melanoma cells are easily distinguishabe by the cellular feature disclosing large, hyperchromatic nuclei and single macronucleoli. Melanin pigments are phagocytosed by macrophages. Cytochemical identification by bleaching is important to differentiate them from hemosiderin-laden macrophage.

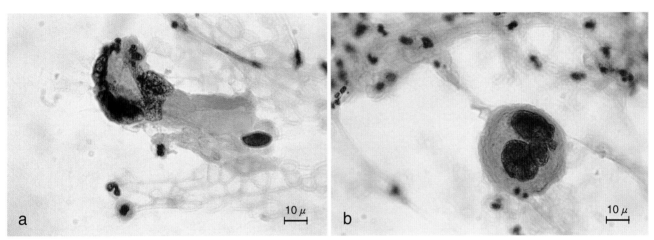

Fig. 8-143　Malignant mesenchymal tumor cells with poorly bordered polymorphic cytoplasm in sputum metastatic from osteogenic sarcoma of the femur. Metastatic sarcoma cells are found as isolated single cells. The nuclei are irregular in shape, multinucleated and delicately outlined.

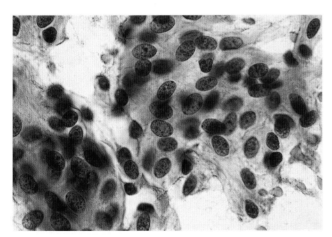

Fig. 8-144　Metastatic meningioma in transbronchial needle aspirates. Elongated cells with oval to elliptic nuclei with dense, homogeneous stain tend to aggregate in a whorl pattern. Refer to the feature of meningothelial meningioma (see page 443).

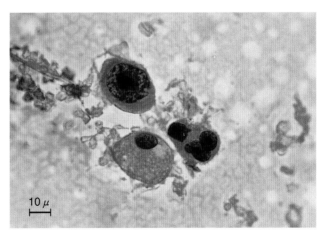

Fig. 8-145　Malignant germ cell by needle aspiration from metastatic seminoma. Mononuclear germ cell is characterized by enormously large, rounded nucleus and a centrally locating nucleolus.

and liquefaction. Prominent, acidophilic nucleoli are recognizable when the cells are not completely degenerated (Fig. 8-143). The nuclear rim of mesenchymal tumor cells is not thickened. *Fibrosarcoma cells* possess spindly, often pyknotic nuclei with variation of nuclear size. The nuclear polymorphism is not so marked as that in osteogenic sarcoma. Similar-looking *leiomyosarcoma*

cells may display blunt nuclear poles (Fig. 8-135). Other spindle shaped tumors that may metastasize to the lung have no specific cytomorphology. Immunocytochemistry is often adjunct to indicate the nature of mesenchymal neoplasias; factor VIII related antigen for angiogenic sarcoma, CD34 for hemangiopericytoma, S100 protein or vimentin for malignant schwannoma[159,186] are available.

221

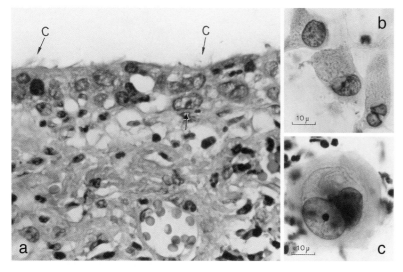

Fig. 8-146 Pitfall in cytodiagnosis by radiation therapy. Radiation effects on bronchial epithelium and its exfoliated cells.
a. The average tumor dosis by preoperative ^{60}Co teletherapy was 3,500 rad at the operation. The bronchial epithelium is almost denuded of cilia except for a small portion (arrow C), and shows marked variation in nuclear size. These altered epithelial cells showing marked nuclear enlargement (arrow) may be hardly identified if exfoliated.
b, c. The cytology specimen was taken at 2,260 rad of irradiation. Care should be taken for interpretation of these benign epithelial cell changes: (1) The chromatin is concentrated at the nuclear rim. (2) The nucleus presents a pale hazy appearance. (3) The N/C ratio is not much distorted, since the cytoplasm is also enlarged.

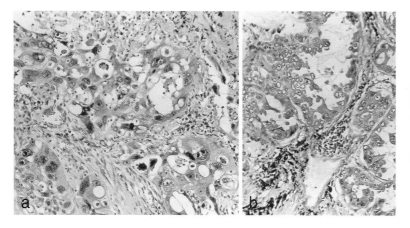

Fig. 8-147 Radiation effects on cancer cells by preoperative ^{60}Co teletherapy.
a. Epidermoid carcinoma that received irradiation of 3,500 rad. Note striking polymorphism and bizarreness of the nuclei.
b. Adenocarcinoma irradiated as much as 4,200 rad. Only moderate nuclear vacuolation is seen. Generally speaking, radiation response of adenocarcinoma is less marked histologically than epidermoid or oat cell carcinoma.

Metastasizing meningioma rarely displays benign-looking features consisting of elongated cells arranged in a whorl pattern[128] (Fig. 8-144). *Choriocarcinoma* also frequently metastasizes to the lung with multiple nodular masses, but is diagnosable only by needle aspiration. Germinoma as represented by seminoma exhibits distinctive features such as large, round to polygonal nuclei and moderate amount of cytoplasm. A prominent nucleolus centrally locating is very characteristic (Fig. 8-145).

16 Radiation Changes

The bronchial tree is an organ that is often affected by radiation therapy for malignant neoplasia of the lung, esophagus and mediastinum. Therefore, grave errors may occur if benign radiation changes of the bronchial mucosa are misinterpreted as recurrent or non-responsive carcinoma. It is also probable that recurrent carcinoma is overlooked as benign radiation change. Malignancy-mimic changes in irradiation are macrocytosis, bi- or multinucleation and pseudohyperchromasia because of condensation of chromatins. Differential points should be focused on (1) bizarre configuration of the cytoplasm, (2) enlargement of nucleus as well as cytoplasm without notable alteration of N/C ratio, (3) amphophilic and/or pseudo-eosinophilic staining reaction, and (4) loss of chromatin network, i.e., liquefaction or pyknosis. Irradiated cancer cells that retain increased N/C ratio and show variation in nuclear size in association with heavy condensation of chromatins can be identified without difficulty (Figs. 8-146 and 8-147).

Appendix. Histological Classification of Lung Tumours*

Epithelial tumours
A. Benign
   1. Papillomas
      a. Squamous cell papilloma
      b. "Transitional" papilloma
   2. Adenomas
      a. Pleomorphic adenoma (Mixed tumour)
      b. Monomorphic adenoma
      c . Others
B. Dysplasia
   Carcinoma in situ
C. Malignant
   1. Squamous cell carcinoma (epidermoid carcinoma)
      Variant:
      a. Spindle cell (squamous) carcinoma
   2. Small cell carcinoma
      a. Oat cell carcinoma
      b. Intermediate cell type

c . Combined oat cell carcinoma
3. Adenocarcinoma
   a. Acinar adenocarcinoma
   b. Papillary adenocarcinoma
   c . Bronchiolo-alveolar carcinoma
   d. Solid carcinoma with mucus formation
4. Large cell carcinoma
   Variants:
   a. Giant cell carcinoma
   b. Clear cell carcinoma
5. Adenosquamous carcinoma
6. Carcinoid tumour
7. Bronchial gland carcinomas
   a. Adenoid cystic carcinoma
   b. Mucoepidermoid carcinoma
   c . Others

Soft tissue tumours

Mesothelial tumours
A. Benign mesothelioma
B. Malignant mesothelioma
   1. Epithelial
   2. Fibrous (spindle-cell)
   3. Biphasic

Miscellaneous tumours
A. Benign
B. Malignant
   1. Carcinosarcoma
   2. Pulmonary blastoma
   3. Malignant melanoma
   4. Malignant lymphomas
   5. Others

Secondary tumours

Unclassified tumours

Tumour-like lesions
A. Hamartoma
B. Lymphoproliferative lesions
C. Tumourlet
D. Eosinophilic granuloma
E. "Sclerosing hemangioma"
F. Inflammatory pseudotumour
G. Others

*World Health Organization, 2nd ed., Geneva, 1981.

References

1. Abrams HL, Spiro R, Goldstein N: Metastases in carcinoma: Analysis of 1000 autopsied cases. Cancer 3: 74, 1950.
2. Adamson JS, Senior RM, Merrill T: Alveolar cell carcinoma: An electron microscopic study. Am Rev Resp Dis 100: 550, 1969.
3. Aisner SC, Gupta PK, Frost JK: Sputum cytology in pulmonary sarcoidosis. Acta Cytol 21: 394, 1977.
4. Anderson A, Mazzucco G, Gugliota P, et al: Benign clear cell (sugar) tumor of the lung: A light microscopic, histochemical and ultrastructural study with a review of the literature. Cancer 56: 2657, 1985
5. Apt W, Knierim F: An evaluation of diagnostic tests for hydatid disease. Am J Trop Med Hyg 19: 943, 1970.
6. Auerbach O, Garfinkel L, Parks VR: Histologic type of lung cancer in relation to smoking habits, year of diagnosis and sites of metastases. Chest 67: 382, 1975.
7. Baker RR, Marsh BR, Frost JK, et al: The detection and treatment of early lung cancer. Ann Surg 179: 813, 1974.
8. Barach AL, Bickernan HA, Beck GJ, et al: Induced sputum as a diagnostic technique for cancer of the lungs. AMA Arch Intern Med 106: 230, 1960.
9. Barden RP: The "oat-cell" tumors: A medical and biological challenge. Radiology 112: 743, 1974.
10. Bauermeister DE, Jennings ER, Beland AH, et al: Pulmonary blastoma, a form of carcinosarcoma. Am J Clin Pathol 46: 322, 1966.
11. Bayon MN, Drut R: Cytologic diagnosis of adenovirus bronchopneumonia. Acta Cytol 35: 181, 1991.
12. Beaver OL, Shapiro JL: A consideration of chronic pulmonary parenchymal inflammation and alveolar cell carcinoma with regard to a possible etiologic relationship. Am J Med 21: 879, 1956.
13. Bendel WL Jr, Ishak KG: Giant cell carcinoma of the lung: Report of two cases. Am J Clin Pathol 35: 435, 1961.
14. Bensch KG: What is the function of bronchial counterpart of the intestinal argentaffine (Kulchitsky) cell? Ann Thorac Surg 14: 568, 1972.
15. Berghuis J, Clagett OT, Harrison EG Jr: The surgical treatment of primary malignant lymphoma of the lung. Dis Chest 40: 29, 1961.
16. Berkheiser SW: The significance of bronchiolar atypia and lung cancer. Cancer 18: 516, 1965.
17. Berry G, Newhouse ML, Turok M: Combined effect of asbestos exposure and smoking on mortality from lung cancer in factory workers. Lancet 2: 476, 1972.
18. Bickerman HA, Sproul EE, Barach AL: An aerosol method of producing bronchial secretions in human subjects: A clinical technic for the detection of lung cancer. Dis Chest 33: 347, 1958.
19. Bonikos DS, Archibald R, Bensch KG: On the origin of the so-called tumorlets of the lung. Human Pathol 7: 461, 1976.
20. Borgeskov S, Francis D: A comparison between fine-needle biopsy and fiberoptic bronchoscopy in patients with lung lesions. Thorax 29: 352, 1974.
21. Botham SK, Holt PF: The development of glass-fibre bodies in the lung of guinea-pigs. J Pathol 103: 149, 1971.
22. Brenner SA, Lambert RL, Pablo GE: Superheated aerosol induced sputum in the cytodiagnosis of lung cancer. Acta Cytol 6: 405, 1962.
23. Broderick PA, Corvese NL, LaChance T, et al: Giant cell carcinoma of lung: A cytologic evaluation. Acta Cytol 19: 225, 1975.
24. Bunai Y: Studies on histogenetic classification of pulmonary peripheral adenocarcinoma: Especially on ultrastructural and cytological classification. 32: 829, 1984.
25. Byrd RB, Miller WE, Carr DT: Roentgenographic appearance of small cell carcinoma of the bronchus. Mayo Clin Proc 43: 333, 1968.
26. Caceres J, Felson B: Double primary carcinomas of the lung. Radiology 102: 45, 1972.
27. Campiche MA: Les inclusions lamellaires des cellules alvéolaires dans le poumon du raton. J Ultrastr Res 3: 302, 1960.
28. Carlson DJ, Mason EW: Pulmonary alveolar proteinosis: Diagnosis of probable case by examination of sputum.

Am J Clin Pathol 33: 48, 1960.

29. Carraga MT, Henson DE: The histologic grading of cancer. Cancer 75: 406, 1995.

30. Carrimgton CB, Liebow AA: Lymphocytic interstitial pneumonia. Am J Pathol 48: 36a, 1966.

31. Chan JKC, Tsang DNC, Wong DKK: Penicillium marneffei in bronchoalveolar lavage fluid. Acta Cytol 33: 523, 1989.

32. Chang SC, Russell WO: A simplified and rapid filtration technique for concentrating cancer cells in sputum. Acta Cytol 8: 348, 1964.

33. Churg A, Warnock ML: Pulmonary tumorlet: A form of peripheral carcinoid. Cancer 37: 1469, 1976.

34. Coalson JJ, Mohr JA, Pirtle SK, et al: Electron microscopy of neoplasms in the lung with special emphasis on the alveolar cell carcinoma. Am Rev Resp Dis 101: 181, 1970.

35. Collins BT, Cramer HM: Fine needle aspiration cytology of carcinoid tumors. Acta Cytol 40: 695, 1996.

36. Craig ID, Desrosiers P, Lefcoe MS: Giant-cell carcinoma of the lung: A cytologic study. Acta Cytol 27: 293, 1983.

37. Dailey JE, Marcuse PM: Gonadotropin secreting giant cell carcinoma of the lung. Cancer 24: 38, 1969.

38. Davis WB, Gadek JE: Detection of pulmonary lymphoma by bronchoalveolar lavage. Chest 91: 787, 1987.

39. Dawe CJ, Woolner LB, Parkhill EM, et al: Cytologic studies of sputum secretions and serous fluids in malignant lymphoma. Am J Clin Pathol 25: 480, 1955.

40. de Klerk NH, Musk AW, Eccles JL, et al: Exposure to crocidolite and the incidence of different histological types of lung cancer. Occup Environ Med 53: 157, 1996.

41. Dutra FR, Geraci CL: Needle biopsy of the lung. JAMA 155: 21, 1954.

42. Ebert RV, Terracio MJ: The bronchiolar epithelium in cigarette smokers. Observations with the scanning electron microscope. Am Rev Resp Dis 111: 4, 1975.

43. Ebihara Y, Sagawa H: Mucin-producing bronchioloalveolar cell carcinoma with special reference to a characteristic structure revealed by phosphotungstic acid-hematoxylin staining. Acta Cytol 30: 643, 1986.

44. Effler DB, Barr D: Five-year survival after surgery of bronchogenic carcinoma. Dis Chest 38: 417, 1960.

45. Ehrenstein F: Primary pulmonary lymphoma. Review of the literature and two case reports. J Thorac Cardiovasc Surg 52: 31, 1966.

46. Fekete PS, Cohen C, DeRose PB: Pulmonary spindle cell carcinoid: Needle aspiration biopsy, histologic and immunohistochemical findings. Acta Cytol 34: 50, 1990.

47. Fennessy JJ, Fry WA, Manalo-Estrella P, et al: The bronchial brushing technique for obtaining cytologic specimens from peripheral lung lesions. Acta Cytol 14: 25, 1970.

48. Fiche M, Capron F, Berger F, et al: Primary pulmonary non-Hodgkin's lymphomas. Histopathology 26: 529, 1995.

49. Flint A: Detection of pulmonary neoplasms by bronchial washings: Are cell blocks a diagnostic aid? Acta Cytol 37: 21, 1993.

50. Fontana RS, Sanderson DR, Woolner LB, et al: Lung-cancer screening: The Mayo Program. J Occup Med 28: 746, 1986.

51. Fraire AE, Underwood RD, McLarty JW, et al: Conventional respiratory cytology versus fine needle aspiration cytology in the diagnosis of lung cancer. Acta Cytol 35: 385, 1991.

52. Francis D, Borgeskov S: Progress in preoperative diagnosis of pulmonary lesions. Acta Cytol 19: 231, 1975.

53. Friedberg EC: Giant-cell carcinoma of the lung: A dedifferentiated adenocarcinoma. Cancer 18: 259, 1965.

54. Fry WA, Manalo-Estrella P: Bronchial brushing. Surg Gynecol Obstet 130: 67, 1970.

55. Fullmer CD, Short JG, Allen A, et al: Proposed classification for bronchial epithelial cell abnormalities in the category of dyskaryosis. Acta Cytol 13: 459, 1969.

56. Gal AA, Koss MN: Atypical carcinoid tumor vs typical carcinoid tumor. In: Differential Diagnosis: Pulmonary Disorders. p96. Baltimore: Williams and Wilkins, 1997.

57. Gatzemeier UK, Hossfeld DK, Love RR: Lung cancer. In: Love RR, et al (eds): Manual of Clinical Oncology. pp288-302. Geneva: UICC/Springer-Verlag, 1944.

58. Gilson JC: Asbestos cancer: Past and future hazards. Proc Roy Soc Med 66: 395, 1973.

59. Godwin JT: Aspiration biopsy: Technique and application. Ann NY Acad Sci 63: 1348, 1956.

60. Godwin JD, Brown CC: Comparative epidemiology of carcinoid and oat-cell tumors of the lung. Cancer 40: 1671, 1977.

61. Greaves TS, Strigle SM: The recognition of Pneumocystis carinii in routine Papanicolaou-stained smears. Acta Cytol 29: 714, 1985.

62. Greenberg SD, Smith MN, Spjut HJ: Bronchiolo-alveolar carcinoma: Cell of origin. Am J Clin Pathol 63: 153, 1975.

63. Grouls V: Über atypische Epithelhyperplasien (sog. "Tumorlets") bei chronischen Lungenerkrankungen. Prax Pneumol 26: 561, 1972.

64. Guyton AC: Textbook of Medical Physiology. Philadelphia: WB Saunders, 1958 (cited from Russell et al: Acta Cytol 7: 1, 1963).

65. Hackl H: Beobachtungen über eine Typenverschiebung des Lungenkrebses im letzen Jahrzehnt. Med Welt 24: 568, 1973.

66. Hattori S, Matsuda M, Sugiyama T, et al: Cytologic diagnosis of early lung cancer: Brushing method under X-ray television fluoroscopy. Dis Chest 45: 129, 1964.

67. Hattori S, Matsuda M, Sugiyama T, et al: Some limitations of cytologic diagnosis of small peripheral lung cancers. Acta Cytol 9: 431, 1965.

68. Hattori S, Matsuda M, Tateishi R, et al: Oat cell carcinoma of the lung: Clinical and morphological studies in relation to its histogenesis. Cancer 30: 1014, 1972.

69. Hayakawa K, Takahayashi M, Sasaki K, et al: Primary choriocarcinoma of the lung. Acta Pathol Jpn 27: 123, 1977.

70. Hayata T, Oho K, Ichiba M, et al: Percutaneous pulmonary puncture for cytologic diagnosis: Its diagnostic value for small peripheral pulmonary carcinoma. Acta Cytol 17: 469, 1973.

71. Hayata Y, Oho K, Ichiba M, et al: Percutaneous needle biopsy in the diagnosis of lung carcinoma. Lung Cancer (Japanese Lung Cancer Society) 9: 3, 1969.

72. Heilbrunn A, Crosby IK: Adenocystic carcinoma and mucoepidermoid carcinoma of the tracheobronchial tree. Chest 61: 145, 1972.

73. Hellstrom HR, Fischer ER: Giant cell carcinoma of lung. Cancer 16: 1080, 1963.

74. Hirsch FR, Osterlind K, Hansen HH: The prognostic significance of histopathologic subtyping of small cell carcinoma of the lung according to classification of the World Health Organisation: A study of 375 consecutive patients. Cancer 52: 2144, 1983.

75. Hoch-Ligeti C, Eller L: Significance of multinucleated epithelial cells in bronchial washings. Acta Cytol 7: 258, 1963.

76. Iwama DE, Mattos MC: Cell-block preparation for cyto-diagnosis of pulmonary paracoccidioidomycosis. Chest 75: 212, 1979.

77. Jacques J, Currie W: Bronchiolo-alveolar carcinoma: A Clara cell tumor. Cancer 40: 2171, 1977.

78. Jenkins BAG, Salm R: Primary lymphosarcoma of the lung. Br J Dis Chest 65: 225, 1971.

79. Johnston WW: Mycosis in cytopathology. In: Wied GL, Koss LG, Reagan JW (eds): Compendium on Diagnostic Cytology, 4th ed. pp299-314. Chicago: Tutorials of Cytology, 1979.

79a. Johnston WW: Fine needle aspiration biopsy versus sputum and bronchial material in the diagnosis of lung cancer. A comparative study of 168 patients. Acta Cytol 32: 641, 1988.

80. Joshi VV, Oleske JM, Minnefor AB, et al: Pathologic pulmonary findings in children with the acquired immunodeficiency syndrome: A study of ten cases. Hum Pathol 16: 241, 1985.

81. Karcioglu ZA, Someren AO: Pulmonary blastoma: A case report and review of the literature. Am J Clin Pathol 61: 287, 1974.

82. Karrer HE: An electron microscopic study of the fine structure of pulmonary capillaries and alveoli of the mouse. Bull Johns Hopkins Hosp 98: 65, 1956.

83. Katzenstein AA, Prioleau PG, Askin FB: The histologic spectrum and significance of clear-cell change in lung carcinoma. Cancer 45: 943, 1980.

84. Kern WH: Cytology of hyperplastic and neoplastic lesions of terminal bronchioles and alveoli. Acta Cytol 9: 372, 1965.

85. Kern WH, Stiles QR: Pulmonary blastoma. J Thorac Cardiovasc Surg 72: 801, 1976.

86. King EB, Russell WM: Needle aspiration biopsy of the lung: Technique and cytologic morphology. Acta Cytol 11: 319, 1967.

87. Kirsh MM, Orvald T, Naylor B, et al: Diagnostic accuracy of exfoliative pulmonary cytology. Ann Thorac Surg 9: 335, 1970.

88. Knudtson KP: Mucolytic action of hyaluronidase on sputum for the cytological diagnosis of lung cancer. Acta Cytol 7: 59, 1963.

89. Koss LG: Letters to the editors: Cancer of the lung. Acta Cytol 10: 308, 1966.

90. Koss LG: Diagnostic Cytology, 3rd ed. Philadelphia: JB Lippincott, 1979.

91. Koss MN, Hochholzer L, Langloss JM, et al: Lymphoid interstitial pneumonia: Clinicopathological and immuno-pathological findings in 18 cases. Pathology 19: 178, 1987.

92. Kreyberg L: Histological Typing of Lung Tumours. International Histological Classification of Tumours, No. 1. Geneva: WHO, 1967.

93. Kyriakos M, Rockoff SD: Brush biopsy of bronchial carcinoma: A source of cytologic error. Acta Cytol 16: 261, 1972.

94. Lanby VW, Efmory Burnett W, Rosemond GP, et al: Value and risk of biopsy of pulmonary lesions by needle aspiration. J Thorac Cardiovasc Surg 49: 159, 1965.

95. Leiman G: Asbestos bodies in fine needle aspirates of lung masses: Markers of underlying pathology. Acta Cytol 35: 171, 1991.

96. Liebow AA, Carrington CB: Diffuse pulmonary lymphoreticular infiltrations associated with dysproteinemia. Med Clin North Am 57: 809, 1973.

97. Lim GH, Ross P, Landman S: Bronchial brush biopsy and primary lung carcinoma. Med J Aust 2: 207, 1975.

98. Liu W: Concentration and fractionation of cytologic elements in sputum. Acta Cytol 10: 368, 1966.

99. Lopes Cardozo P: Atlas of Clinical Cytology, Targa, b.v.'s-Hertogenbosch, the Netherlands, 1975.

100. Lopes-Majano U: Biological diagnosis of pulmonary hydatidosis. Respiration 28: 471, 1971.

101. Low FN: The pulmonary alveolar epithelium of laboratory mammals and man. Anat Rec 117: 241, 1953.

102. Low FN, Daniels WC: Electron microscopy of the rat lung. Anat Rec 113: 437, 1952.

103. Ludington LG, Verska JJ, Howard T, et al: Bronchiolar carcinoma (alveolar cell), another great imitator: A review of 41 cases. Chest 61: 622, 1972.

104. Marchevsky A, Nieburgs HE, Olenko E, et al: Pulmonary tumorlets in cases of "tuberculoma" of the lung with malignant cells in brush biopsy. Acta Cytol 26: 491, 1982.

105. Markell EK, Voge M: Medical Parasitology. Philadelphia: WB Saunders, 1976.

106. Masin M, Masin F: Pulmonary alveolar proteinosis, cytologic and cytochemical observations on sputum specimens. Acta Cytol 6: 429, 1962.

107. McCann MP, Fu Y-S, Kay S: Pulmonary blastoma, a light and electron microscopic study. Cancer 38: 789, 1976.

108. McCormack LJ, Hazard JB, Effler DB, et al: Experiences with the cytologic examination of bronchial swabbings in the diagnosis of cancer of the lung. A study of 602 cases. J Thorac Surg 29: 277, 1955.

109. McDowell EM, McLaughlin JS, Merenyi DK, et al: The respiratory epithelium. V. Histogenesis of lung carcinomas in the human. J Natl Cancer Inst 61: 587, 1978.

110. McNamee CJ, Simpson RH, Pagliero KM, et al: Primary clear cell carcinoma of the lung. Respir Med 87: 471, 1993.

111. Meckstroth CV, Davidson HB, Kress GO: Muco-epidermoid tumor of the bronchus. Dis Chest 40: 652, 1961.

112. Meinecke R, Bauer F, Skouras J, et al: Blastomatous tumors of the respiratory tract. Cancer 38: 818, 1976.

113. Melamed MR, Flehinger BJ: Detection of lung cancer: Highlights of the Memorial Sloan-Kettering study in New York City. Schweiz Med Wschr 117: 1457, 1987.

114. Melamed M, Flehinger B, Miller D, et al: Preliminary report of the lung cancer: Detection program in New York. Cancer 39: 369, 1977.

115. Meyer EC, Liebow AA: Relationship of interstitial pneumonia honeycombing and atypical epithelial proliferation to cancer of the lung. Cancer 18: 322, 1965.

116. Miyamoto H, Inoue S, Abe S, et al: Relationship between cytomorphologic features and prognosis in small-cell carcinoma of the lung. Acta Cytol 26: 429, 1982.

117. Mollo F, Canese MG, Campobasso O: Human peripheral lung tumors: Light and electron microscopic correlation. Br J Cancer 22: 173, 1973.

118. Moran TJ, Totten RS: Lymphoid interstitial pneumonia with dysproteinemia. Am J Clin Pathol 54: 747, 1970.

119. Morrison R, Deeley TJ: Drill biopsy in intrathoracic malignant disease. Thorax 12: 87, 1957.

120. Moskowitz M, Freihofer A: Seldinger brush biopsy: A synthesis of techniques. Chest 57: 426, 1970.

121. Mountain CF, Carr DT, Anderson WAD: A system for

the clinical staging of lung cancer. Am J Roentgenol Nucl Med 120: 130, 1974.

122. Mylius EA, Gullvag B: Alveolar macrophage count as an indicator of lung reaction to industrial air pollution. Acta Cytol 30: 157, 1986.

123. Nagaishi C: Functional Anatomy and Histology of the Lung. Tokyo: Igaku-Shoin, 1972.

124. Naib ZM: Pitfalls in the cytologic diagnosis of oat cell carcinoma of the lung. Acta Cytol 8: 34, 1964.

125. Nasiell M: The general appearance of the bronchial epithelium in bronchial carcinoma: A histopathological study with some cytological viewpoints. Acta Cytol 7: 97, 1963.

126. Nasiell M: Metaplasia and atypical metaplasia in the bronchial epithelium: A histopathologic and cytopathologic study. Acta Cytol 10: 421, 1966.

127. Nasiell M: Diagnosis of lung cancer by aspiration biopsy and a comparison between this method and exfoliative cytology. Acta Cytol 11: 114, 1967.

128. Ng TH, Wong MP, Chan KW: Benign metastasizing meningioma. Clin Neurol Neurosurg 92: 152, 1990.

129. Nguyen G-K: Aspiration biopsy cytology of benign clear cell ("sugar") tumor of the lung. Acta Cytol 33: 511, 1989.

130. Nicholson AG, Wotherspoon AC, Diss TC, et al: Pulmonary B-cell non-Hodgkin's lymphomas. The value of immunohistochemistry and gene analysis in diagnosis. Histopathology 26: 395.1995.

131. Nishikawa A, Nakamura A, Nishikawa H, et al: Exfoliative cytopathology of atypical carcinoid: A case report. Jpn J Clin Oncol 16: 413, 1986.

132. O'Donnell WE, Day E: Early cancer: Its detection, diagnosis and management. Med Clin North Am 40: 591, 1956.

133. Okike N, Bernatz PE, Woolner LB: Carcinoid tumors of the lung. Ann Thorac Surg 22: 270, 1976.

134. Oldham SAA, Castillo M, Jacobson FL, et al: HIV-associated lymphocytic interstitial pneumonia: Radiological manifestation and pathologic correlation. Thorac Radiol 170: 83, 1989.

135. Oswald NC, Hinson KFW, Canti G, et al: The diagnosis of primary lung cancer with special reference to sputum cytology. Thorax 26: 623, 1971.

136. Papanicolaou GN: Degenerative changes in cilliated cells exfoliating from the bronchial epithelium as a cytologic criterion in the diagnosis of diseases of the lung. NY J Med 56: 2647, 1956.

137. Papanicolaou GN, Bridges EL, Railey C: Degeneration of the ciliated cells of the bronchial epithelium (ciliocytophthoria) in its relation to pulmonary disease. Am Rev Resp Dis 83: 641, 1961.

138. Papanicolaou GN, Liebow AA: Exfoliative cytology in diagnosis of lung cancer. In: Mayer E, Maier HC (eds): Pulmonary Carcinoma. New York, New York University Press, 1956.

139. Pfitzer P, Wehle K, Blanke M, et al: Fluorescence microscopy of Papanicolaou-stained bronchoalveolar lavage specimens in the diagnosis of Pneumocystis carinii. Acta Cytol 33: 557, 1989.

140. Pharr SL, Farber SM: Cellular concentration of sputum and bronchial aspirations by tryptic digestion. Acta Cytol 6: 447, 1962.

141. Philips FR: The identification of carcinoma cells in the sputum. Br J Cancer 8: 67, 1954.

142. Pierce CH, Knox AW: Ciliocytophthoria in sputum from patients with adenovirus infections. Proc Soc Exp Biol Med 104: 492, 1960.

143. Rannels DE, Rannels SR: Influence of the extracellular matrix on type 2 cell differentiation.Chest 96: 165, 1989.

144. Razzuk MA, Pockey M, Urschel HC, et al: Dual primary bronchogenic carcinoma. Ann Thorac Surg 17: 425, 1974.

145. Richmond I, Pritchard GE, Ashcroft T, et al: Bronchus-associated lymphoid tissue (BALT) in human lung: Its distribution in smokers and non-smokers. Thorax 48: 1130, 1993.

146. Roberts TW, Pollak A, Howard R, et al: Tracheobronchial cytology utilizing an improved tussilator (cough machine). Acta Cytol 7: 174, 1963.

147. Rome DS, Olson KB: A direct comparison of natural and aerosol produced sputum collected from 776 asymptomatic men. Acta Cytol 5: 173, 1961.

148. Rose AH: Primary lymphosarcoma of the lung. J Thorac Surg 33: 254, 1957.

149. Rosen P, Hadju SI: Visceral herpesvirus infections in patients with cancer. Am J Clin Pathol 56: 459, 1971.

150. Rosen P, Melamed M, Savino A: The "ferruginous body" content of lung tissue: A quantitative study of eighty-six patients. Acta Cytol 16: 207, 1972.

151. Rosen SH, Castleman B, Liebow AA: Pulmonary alveolar proteinosis. N Engl J Med 258: 1123, 1958.

152. Rosenberg LM: Multiple tracheobronchial melanomas with ten-year survival. JAMA 192: 717, 1965.

153. Rosenblatt MB, Trinidad S, Lisa JR, et al: Specific epithelial degeneration (ciliocytophthoria) in inflammatory and malignant respiratory disease. Dis Chest 43: 605, 1963.

154. Ryan RF, McDonald JR, Glagett OT: Histopathologic observations on bronchial epithelium with special reference to carcinoma of the lung. J Thorac Surg 33: 264, 1957.

155. Sabour MS, Osman LM, LeGolvan PC, et al: Needle biopsy of the lung. Lancet 2: 182, 1960.

156. Saccomanno G, Archer VE, Auerbach O, et al: Development of carcinoma of the lung as reflected in exfoliated cells. Cancer 33: 256, 1974.

157. Saccomanno G, Saunders RP, Archer VE: Cancer of the lung: The cytology of sputum prior to the development of carcinoma. Acta Cytol 9: 413, 1965.

158. Saccomanno G, Saunders RP, Ellis H, et al: Concentration of carcinoma or atypical cells in sputum. Acta Cytol 7: 305, 1963.

159. Saleh HA, Haapaniemi J: Aspiration biopsy cytology of malignant hemangiopericytoma metastatic to the lungs: Cytomorphologic and immunocytochemical study of a case. Acta Cytol 41: 1265, 1997.

160. Salm R: A primary malignant melanoma of the bronchus. J Pathol Bacteriol 85: 121, 1963.

161. Saltzstein SL: Pulmonary malignant lymphomas and pseudolymphomas: Classification, therapy, and prognosis. Cancer 16: 928, 1963.

162. Sato M, Saito Y, Nagamoto N, et al: Diagnostic value of differential brushing of all branches of the bronchi in patients with sputum positive or suspected positive for lung cancer. Acta Cytol 37: 879, 1993.

163. Singh G, Katyal SL, Torikata C: Carcinoma of type II pneumocytes: Immunodiagnosis of a subtype of "bronchioloalveolar carcinoma." Am J Pathol 102: 195, 1981.

164. Skitarelic K, von Haam E: Bronchial brushings and washings: A diagnostically rewarding procedure? Acta Cytol 18: 321, 1974.

165. Smith P, Heath D, Moosavi H: The Clara cell. Thorax 29: 147, 1974.
166. Smith MJ, Naylor B: A method for extracting ferruginous bodies from sputum and pulmonary tissue. Am J Clin Pathol 58: 250, 1972.
167. Sniffen RC, Soulter L, Robbins LL: Muco-epidermoid tumors of the bronchus arising from surface epithelium. Am J Pathol 34: 671, 1958.
168. Solomon DA, Solliday NH, Gracey DR: Cytology in fiberoptic bronchoscopy. Comparison of bronchial brushing, washing and post-bronchoscopy sputum. Chest 65: 616, 1974.
169. Sobue T, Suzuki T, Matsuda M, et al: Sensitivity and specificity of lung cancer screening in Osaka, Japan. Jpn J Cancer Res 82: 1069, 1991.
170. Sobue T, Suzuki T, Naruke T, Japanese Lung-Cancer-Screening Research Group: A case-control study for evaluating lung-cancer screening in Japan. Int J Cancer 50: 230, 1992.
171. Spain DM: The association of terminal bronchiolar carcinoma with chronic interstitial inflammation and fibrosis of the lungs. Am Rev Tuberc 76: 559, 1957.
172. Spencer H: Pathology of the Lung, 3rd ed. Oxford: Pergamon Press, 1977.
173. Spencer H: Pathology of the Lung, 3rd ed. Vol 2, pp834-835, 990. Oxford: Pergamon Press, 1977.
174. Sprague RI, Deblois GG: Small lymphocytic pulmonary lymphoma. Diagnosis by transthoracic fine needle aspiration. Chest 96: 929,1989.
175. Suprun H, Koss LG: The cytological study of sputum and bronchial washing in Hodgkin's disease with pulmonary involvement. Cancer 17: 674, 1964.
176. Swan LL: Pulmonary adenomatosis of man. Arch Pathol 47: 517, 1949.
177. Takahashi M: Pulmonary cytology and its clinicopathological bases. J Jpn Soc Clin Cytol (Tokyo) 7: 58, 1968.
178. Takahashi M: Cytology of pulmonary cancer. In: Ikeda S (ed): Atlas of Early Cancer of Major Bronchi. pp21-26. Tokyo: Igaku-Shoin, 1976.
179. Takahashi M, Hashimoto K, Osada H: Parenteral administration of chymotrypsin for the early detection of cancer cells in sputum. Acta Cytol 11: 61, 1967.
180. Takinen E, Tukiainen P, Renkonen R: Bronchoalveolar lavage: Influence of cytologic methods on the cellular picture. Acta Cytol 36: 680, 1992.
181. Tang C-S, Kung ITM: Homogenization of sputum with dithiothreitol for early diagnosis of pulmonary malignancies. Acta Cytol 37: 689, 1993.
182. Theodos PA, Allbritten FF Jr, Breckenridge RL: Lung biopsy in diffuse pulmonary disease. Dis Chest 27: 637, 1955.
183. Tolis GA, Fry WA, Head L, et al: Bronchial adenomas. Surg Gynecol Obstet 134: 605, 1972.
184. Tomb JA, Matossian RM: Diagnosis of pulmonary hydatidosis by sputum cytology. Johns Hopkins Med J 139: 38, 1976.
185. Travis WD, Travis LB, Devesa SS: Lung cancer. Cancer 75: 191, 1995.
186. Traweek ST, Kandalaft PL, Metha P, et al: The human hematopoietic cell antigen (CD34) in vascular neoplasia. Am J Clin Pathol 96: 25, 1991.
187. Tsuboi E: Atlas of Transbronchial Biopsy: Early Diagnosis of Peripheral Pulmonary Carcinomas. Tokyo: Igaku-Shoin, 1970.
188. Tsuboi E, Ikeda S, Tajima M, et al: Transbronchial biopsy smear for diagnosis of peripheral pulmonary carcinoma. Cancer 20: 687, 1967.
189. Umiker W: Typing of bronchogenic carcinoma. Arch Pathol 71: 295, 1961.
190. Vainio H, Boffetta P: Mechanisms of the combined effect of asbestos and smoking in the etiology of lung cancer. Scand J Work Environ Health 20: 235, 1994.
191. Valentine EH: Squamous metaplasia of bronchus: Study of metaplastic changes occurring in epithelium of major bronchi in cancerous and non cancerous cases. Cancer 10: 272, 1957.
192. Waddell WR: Organoid differentiation of the fetal lung: A histologic study of the differentiation of mammalian fetal lung in utero and in transplants. Arch Pathol 47: 227, 1949.
193. Walls WJ, Thornbury JR, Naylor B: Pulmonary needle aspiration biopsy in the diagnosis of Pancoast tumors. Radiology 111: 99, 1974.
194. Wang N-S, Seemayer TA, Ahmed MN, et al: Giant cell carcinoma of the lung. A light and electron microscopic study. Hum Pathol 7: 3, 1976.
195. Watson WL: Ten-year survival in lung cancer. Cancer 18: 133, 1963.
196. Watson WL, Berg JW: Oat cell lung cancer. Cancer 15: 759, 1962.
197. Watson WL, Smith RR: Terminal bronchiolar or "alveolar cell" cancer of the lung: Report of 33 cases. JAMA 147: 7, 1951.
198. Weissman I, Christie JM: Primary lymphosarcoma of the lung. Arch Surg 62: 129, 1951.
199. Wheeler TM, Johnson EH, Coughlin D, et al: The sensitivity of detection of asbestos bodies in sputa and bronchial washings. Acta Cytol 32: 647, 1988.
200. Whitwell F: Tumourlets of the lung. J Pathol Bacteriol 70: 529, 1955.
201. Whitwell F, Newhouse ML, Bennett DR: A study of the histological cell types of lung cancer in workers suffering from asbestosis in the United Kingdom. Br J Indust Med 31: 298, 1974.
202. WHO: Advisory committee on asbestos cancers. Ann Occup Hyg 16: 9, 1973.
203. Willie SM, Snyder RN: The identification of Paragonimus westermani in bronchial washings: Case report. Acta Cytol 21: 101, 1977.
204. Willis RA: Pathology of Tumours, 3rd ed. London: Butterworth, 1960.
205. Witschi H: Proliferation of type II alveolar cells: A review of common responses in toxic lung injury. Toxicology 5: 267, 1976.
206. Woolner LB, David E, Fontana RS, et al: In situ and early invasive bronchogenic carcinoma: Report of 28 cases with postoperative survival date. J Thorac Cardiovasc Surg 60: 275, 1970.
207. Woolner LB, McDonald JR: Carcinoma cells in sputum and bronchial secretions: Study of 150 consecutive cases in which results were positive. Surg Gynecol Obstet 88: 273, 1949.
208. Yamamoto T, Yazawa T, Ogata T, et al: Clear cell carcinoma of the lung. A case report and review of the literature. Lung Cancer 10: 101, 1993.
209. Yardumian K, Myers L: Primary Hodgkin's disease of the lung. Arch Intern Med 86: 233, 1950.
210. Yokoyama S, Hayashida Y, Nagahama J, et al: Pulmonary blastoma: A case report. Acta Cytol 36: 293, 1992.
211. Zaman MB, Hadju SI, Melamed MR, et al: Transthoracic aspiration cytology of pulmonary lesions. Semin Diagn Pathol 3: 176, 1986.

212. Zaman MK, Wooten OJ, Suprahmanya B, et al: Rapid non-invasive diagnosis of Pneumocystis carinii from induced liquified sputum. Ann Intern Med 109: 7, 1988.

9 Thyroid

1 Introduction

Nodular struma is the target of aspiration biopsy cytology (ABC), because it is palpable and can be aimed with ease by fine needle aspiration. However, laboratory examinations of the thyroid function give a clue to clinical diagnosis of thyroid disorders and should be performed prior to ABC. The majority of thyroxine (T4) and 3, 5, 3'-triiodothyronine (T3) are present in plasma bound with protein. A small amount of the hormone, approximately 0.3% of T3 and 0.03% of T4 are present as free T3 and free T4 which are biologically active fractions of the thyroid hormone under a regulatory feedback mechanism between the thyroid and the hypophysis; TSH plays a regulatory role dependent on either increase or decrease of T3 and T4 in amount. While the hypophysis is functioning in a physiological condition, FT3, FT4 and TSH are good indicators to determine the state of thyroid function (Table 9-1). Instead of sophisticated measurement of FT3 and FT4, T3 and T4 are commonly measured in laboratories to conjecture the thyroid function. Thyroglobulin, a major constituent of the follicle colloid that is the storage form containing thyroxine has no value of disease specificity, however, its elevation over 30 ng/ml will be induced by TSH stimulation or derived from release owing to destruction of thyroid tissues by malignant neoplasia or thyroiditis. In increased FT3 and FT4, an inhibition of TSH secretion occurs in Basedow disease and early stage of subacute thyroiditis. Toxic struma may accelerate ^{125}I intake and reveal a "hot nodule."

In normal ranges of FT3 and FT4 in association with palpable nodules, various disorders including chronic thyroiditis, follicle adenoma, carcinoma and lymphoma are targets of further investigations.

2 Anatomy and Histology

The thyroid locates in the midportion of the neck in front of the larynx and trachea at the level of the cricoid cartilage. It consists of two lateral lobes and connecting narrow isthmus. Although there is some variation of both lobes in size, they are usually symmetric and weigh about 20 g. The internal carotid veins are present laterally in both sides. The recurrent laryngeal nerves run in the left side between the trachea and esophagus. Thin fibrous tissue covers the organ as capsule and enters into the parenchyma as septa.

Histologically the parenchyma consists of cyst-like follicles lined by a single layer of cuboidal or flattened epithelium. The height of epithelium varies according to the state of thyroid function. The lumen contains eosinophilic, gelatinous colloid, stainability of which varies also by the change of function; maximum storage of thyroglobulin strengthens homogeneous, eosinophilic stain. These follicles are surrounded by well vascularized fibrous stroma.

Cytologically follicular cells are scarcely obtained by fine needle aspiration from the thyroid gland within normal limits. The cells are arranged in sheet and display orderly arrangement. Their nuclei are small, round

Table 9-1 Values of T3, T4 and free forms and disorders

Over reference interval T4: over 13.0 μg/dl FT4: over 1.9 ng/dl T3: over 180 ng/dl FT3: over 5.8 pg/ml	*Hyperthyroidism* 　　　　　　Plumer disease Low TSH　Basedow disease 　　　　　　Subacute thyroiditis 　　　　　　(early stage)
Within reference interval T4: 4.5-13.0 μg/dl FT4: 0.8-1.9 ng/dl T3: 84-180 ng/dl FT3: 3.0-5.8 pg/ml	*Euthyroidism* 　　　　　　Follicle adenoma 　　　　　　Chronic thyroiditis 　　　　　　Colloid goiter 　　　　　　Carcinoma 　　　　　　Lymphoma
Under reference interval T4: under 4.5 μg/dl FT4: under 0.8 ng/dl T3: under 84 ng/dl FT3: under 3.0 pg/ml	*Hypothyroidism* 　　　　　　Hashimoto disease High TSH　After thyroidectomy 　　　　　　Subacute thyroiditis 　　　　　　(late stage)

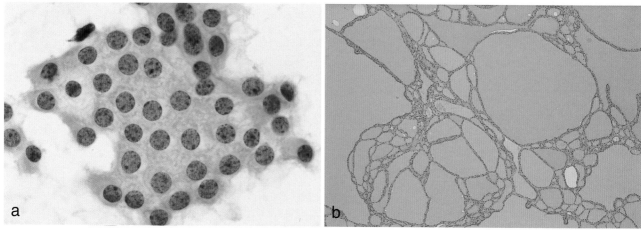

Fig. 9-1a,b Adenomatous goiter. ABC cytology and histology. Nuclei of epithelial cells are slightly enlarged, but uniform in size and shape. Their arrangement is regular in flat sheet and the distance from nucleus to nucleus appears equal. Chromatins are smoothly distributed.

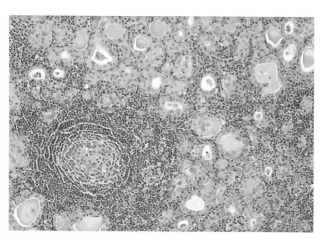

Fig. 9-2 Histology of chronic thyroiditis. Note heavy infiltration of lymphocytes in stroma forming a lymphoid follicle with a prominent germinal center. Epithelial cells possess broad eosinophilic cytoplasm (Askanazy cells).

or ovoid and uniform in size. The cytoplasm is even but vary in amount within the normal range.

3 Procedure of Thyroid Needle Aspiration

The patient is laid supine putting a pillow under the patient's shoulder. A position stretching the neck is appropriate for ABC. A 22 to 23 gauge needle is recommended because of vascularity of the organ. During the procedure of needle aspiration, swallow, cough and speech are prohibited.

1) Fix a nodule of the thyroid between the thumb and index finger of a physician.
2) Stick the needle to upper lateral direction from the midst and aspirate. Minute up and down movements rotating the needle after returning gently the negative pressure within the tube are desirable in order to obtain sufficient aspirates. If pure blood is aspirated, a press with fingers for 5 to 10 min is necessary to avoid hematoma formation.
3) Place a small amount of isotonic saline into the syringe and make smears onto glass slides.

There is no special contraindication except for hemor-rhagic diathesis. Vigorous palpation may cause hemorrhage and/or infarction.[11]

4 Adenomatous Goiter (Nodular Goiter)

Deficiency of dietary iodine intake or ingestion of the substance that blocks thyroid hormone synthesis is the cause of adenomatous goiter. Diffuse colloid goiter derived from a long term iodine deficiency is out of the focus of ABC, however, nodular goiter is occasionally examined by fine needle aspiration. Since hyperactive nodules show excessive iodine uptake, "hot nodules" will be distinguished clinically from "cold nodules" of malignant neoplasia. Since blood chemistry of T3 and T4 is often kept in the low normal range, TSH is stimulated.

Cytology: Compensatory hyperplasia of columnar follicle cells by TSH stimulation may often lead to overdiagnosis. In addition, follicular adenoma cannot be definitely distinguished (Fig. 9-1).

It must be kept in mind that there is age-related changes in follicular nuclei of the thyroid with nodular goiter; nuclear volumes from nodular goiters are significantly larger in an older age group than in a younger age group.[19]

5 Chronic Thyroiditis (Hashimoto's Disease)

This is much more common in females than in men and its first peak of occurrence is seen at menopause and the second larger peak is seen at the age over 40. The disease belongs to autoimmune diseases because of the presence of antithyroid antibodies which are responsible for damages of the parenchyma. Grossly it enlarges symmetrically unless other neoplasia are associated. Coexistence with other autoimmune diseases such as lupus erythematosus, Sjögren's syndrome and rheumatoid arthritis supports the entity of immunologic disorder. Histologically it is characterized by heavy lymphocytic infiltration occasionally admixed with plasma cells. Formation of follicles with prominent germinal centers is not infrequent (Fig. 9-2).

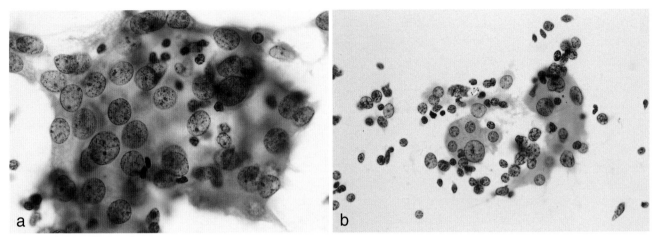

Fig. 9-3a,b Chronic thyroiditis characterized by clustered oxyphilic epithelial cells with broad amphoteric cytoplasm (Askanazy cells). Some nuclear overlapping and anisonucleosis are present, but they are uniform in shape and vesicular. Note some neutrophils within the cell cluster.

Cytology: Since follicles are reduced in number and size, epithelial cells are not much obtained by fine needle aspiration. As a characteristic finding enlarged oxyphilic epithelial cells similar to Hürthle cells (Askanazy cells) appear. Nuclear enlargement and overlapping should not be referred to as malignant. Lymphocytes and some plasma cells are indispensable constituents in ABC (Fig.9-3).

6 Subacute Thyroiditis (Granulomatous Thyroiditis)

Subacute thyroiditis first described by de Quervain in 1936 occurs as a febrile thyroid swelling accompanying insidious weight loss and general lassitude. Middle age females are affected. When localized swelling occurs, ABC becomes indicative to rule out neoplasia. Initially the pathology is localized within follicles such as follicular destruction, disappearance of colloid and occurrence of multinucleated giant cells (Fig. 9-4). Lately the lesion is replaced by granulation tissue with infiltration of histiocytes and lymphocytes.

7 Graves' Disease (Basedow's Disease)

This is a form of primary hyperthyroidism that causes symmetric diffuse enlargement of the thyroid. Exophthalmos is often associated but not invariably seen as disease entity. Graves' disease is not a target of ABC, because high vascularity in hyperthyroidism has a risk of bleeding during needle aspiration. Nodular localized swelling of the thyroid most likely caused by fibrosis and/or inflammatory infiltration may come to be an indication of ABC. Administration of antithyroid drugs, i.e., thiouracil and thiamazole may inhibit thyroid hormone synthesis and stimulates TSH. As a result, follicular cells appear hyperplastic.

Cytology: Follicular epithelial cells become diffusely enlarged[25] with slight variation in nuclear size(Fig.9-5). Chromatins are bland. Association with lymphoid cells not as much as chronic thyroiditis occurs occasionally in needle aspirates.

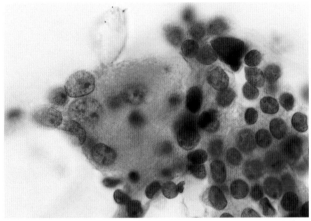

Fig. 9-4 Cytology of needle aspirates in subacute thyroiditis showing a multinucleated giant cell with abundant cytoplasm. The surrounding follicular cells are uniformly round in shape.

8 Follicular Adenoma

Follicular adenoma develops as a solitary nodule encapsulated by thin fibrous tissue. Over one third of thyroidectomy cases are benign adenomas which are histologically classified into trabecular (embryonal), microfollicular (fetal), colloid (Fig. 9-6) and oxyphilic cell (Hürthle cell, Fig. 9-9) types. Atypical adenoma is another variety. Although histological classification is based on the structure of follicles, cytology is not able to define the structure but to distinguish between ordinary follicle cells (Fig. 9-7) and oxyphilic cells (Fig. 9-8). The oxyphilic cell type was formerly called "Hürthle cell" type. Distinction of follicular adenoma from well differentiated follicular carcinoma requires experience and should be focussed on:

1) less cellularity in aspiration,
2) uniformity of nuclei in shape and size,
3) bland chromatin distribution,
4) regular sheet-like arrangement.

Oxyphilic adenoma is of a special form showing sheets of cells with broad, finely granular cytoplasm.

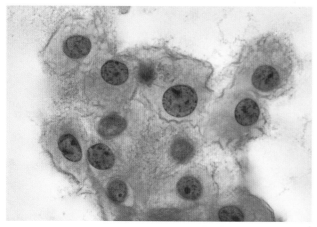

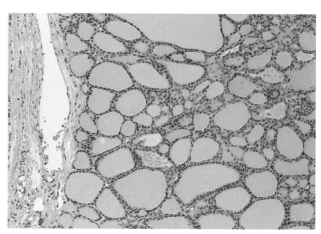

Fig. 9-5 Basedow's disease treated with antithyroid drug thiamazole for five years. Note enlarged cells with enlarged round nuclei showing slight variation in size. Characteristically, the cytoplasm is abundant and shows marginal vesiculation.[27]

Fig. 9-6 Follicular adenoma. The simple type consists of macro- and micro-follicles containing homogeneously stained colloid. The lining epithelium is cuboidal in shape.

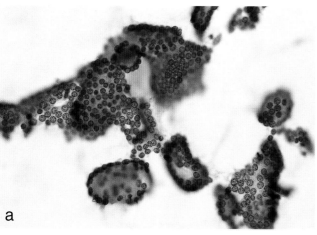

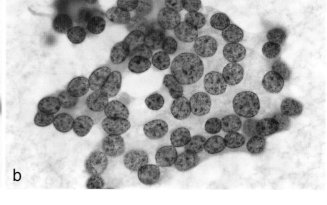

Figs 9-7 Follicular adenoma by ABC.
a. Small round cells taken from micro-follicles exhibit the pattern of follicles varying in size. Colloid contents are occasionally visible. Follicular cells in flat sheet are most likely derived from macro-follicles. Chromatins are finely granular and evenly distributed.
b. Higher magnification shows uniformity of nuclei in shape. Chromatins are finely granular and evenly distributed. Nuclear rims are delicate. Note inspissated colloid that stains darkly brown.

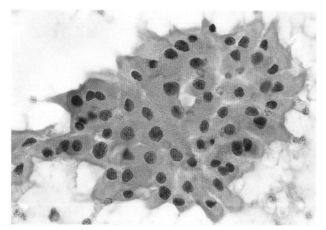

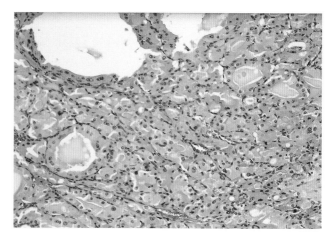

Fig. 9-8 Oxyphilic adenoma by ABC. Low magnification. Note a sheet of oxyphilic cells or oncocytic cells which are characterized by large polyhedral cells with abundant, finely granular, eosinophilic cytoplasm. Their nuclei tend to be pyknotic.

Fig. 9-9 Histology of oxyphilic adenoma. Note densely packed trabeculae lined with tall columnar cells with abundant, waxy cytoplasm and small, pyknotic nuclei.

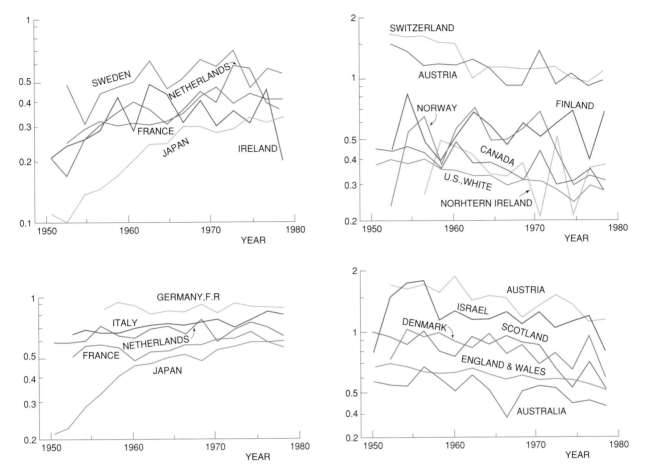

Fig. 9-10   Trends in age-adjusted death rates for malignant neoplasm of thyroid gland from 1950-51 to 1978-79 in 21 countries. (From Kurihara M, Aoki K, Tominaga S(eds): Cancer Mortality Statistics in the World. p81. Nagoya: The University of Nagoya Press, 1984.)

Eosinophilic stain is also a characteristic feature (Figs. 9-8 and 9-9).

9 Thyroid Carcinoma

Thyroid carcinoma accounts for about 1% of all cancers with a male to female ratio 1 to 2. Epidemiologically there are two peaks of incidence; the first minor peak is at young ages under 20 years and the second major peak is shown over 40 years old.[4] The death rates from thyroid cancer is low about 0.3 to 0.8 per 100,000 population in males and 1.0 to 1.5 in females (Fig. 9-10). However, there is some geographical difference of the mortality in Austria and Switzerland that is about two time higher than that of other countries. Japan is taking an increasing trend. We must note an etiological factor such as irradiation to the neck and thymus in infants and exposure to radioiodine. Clinically attention should be paid for the state with persistent high level of TSH and antecedence of hyperplastic lesions. The age of patients with thyroid cancer has a significant prognostic factor; patients older than 60 have a worse survival rate than those who are younger than 40 regardless of histologic types, i.e., either follicular carcinoma or papillary carcinoma.[29] According to Hay, association of papillary carcinoma with benign disorders is not infrequent such as 41% with nodular goiter and 20% with chronic thyroiditis.[14] Whereas the association had nothing to do with the outcome of cancer, noteworthy reports[2,9] mentioned that thyroid carcinoma associated with Basedow's disease tended to develop to lymph node involvement, distant metastasis and invasion beyond the thyroid.

As a prognosis-related factor quantitative measurement of DNA content of tumors has an independent value for survival; correlation between DNA aneuploidy and high mortality is strongly suggested.

10 Follicular Carcinoma

The average age of occurrence of follicular carcinoma is about 10 years older than that of papillary carcinoma.[20] Capsular invasion and intravascular invasion are encountered histologically and become a reliable criterion for distinction from follicular adenoma (Fig. 9-12). Clinically distant metastasis to the lung and bones is occasionally initial manifestation.

Cytology of follicular carcinoma varies according to the grade of differentiation; a differentiated type that retains a follicular structure resembles the benign counterpart and a less differentiated type shows trabecular clusters with nuclear overlappings. It is difficult to distinguish well differentiated follicular carcinoma from follicular adenoma. In carcinoma, nuclei are overlapping and disorderly arranged. Presence of a distinctive nucleolus is also a target of differentiation(Fig. 9-13). The case showing mixture of follicular and papillary features is

233

Table 9-2   Diagnostic accuracy of ABC for thyroid lesions

Authors	Total cases	Malig cases	Benign cytology malig/total	Susp cytology malig/total	Malignant cytology malig/total	Above susp %
Löwhagen, et al	412	96	9/226	24/123	63/63	90.6%
Bugis, et al	197	30	8/126	9/55	13/16	73.3%
Hawkins, et al	415	72	10/336	15/28	47/51	86.1%
Klemi, et al	186	22	3/107	7/67	12/12	86.4%
Altavilla, et al	233	46	8/160	18/53	20/20	82.6%
Mandreker, et al	213	30	9/163	10/38	11/12	70.0%

Malig : malignant, Susp : suspicious

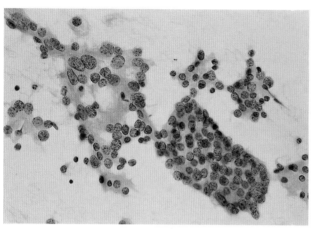

Fig. 9-11   Follicular carcinoma by ABC. A closely aggregated tubule and ill-defined loose clusters of ovoid cancer cells are illustrated. Their nuclei are uniformly shaped but vary in size. Chromatins are somewhat coarsely granular. These findings can be distinguished from those of adenoma (see Fig. 9-7)

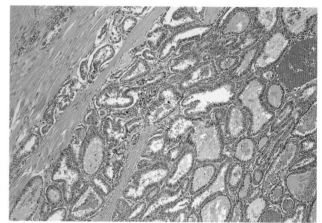

Fig. 9-12   Follicular carcinoma consisting of irregular follicles varying in size and shape. Some follicles are devoid of colloid content. Note capsular invasion of cancer cells. The lining cells of follicles are cuboidal and hypertrophic with densely staining nuclei.

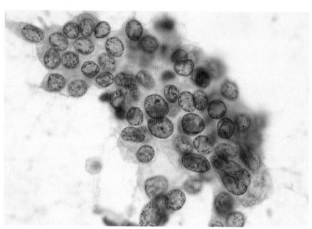

Fig. 9-13   Follicular carcinoma by ABC. An irregular follicular cluster of cells with hyperchromatic nuclei which are disorderly overlapping. Note distinct nucleoli.

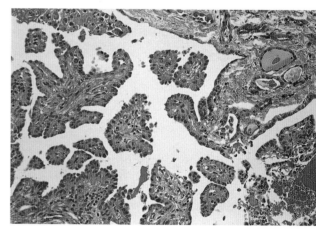

Fig. 9-14   Papillary carcinoma. Large papillary processes of tumor cells growing with fibrovascular stalks. The tumor cells are cuboidal and possess ovoid hyperchromatic nuclei.

classified as papillary carcinoma because of the behavior of growth. It has been described as the follicular variant of papillary carcinoma.[7,17,21]

11 Papillary Carcinoma

Cytology of papillary carcinoma is characterized by papillary growth pattern lining on long fibrovascular stalks. Irregular branching or anastomosing patterns can be observed under a low power field of histology. (Fig. 9-14). Nuclear features are represented as indentations,

pseudoinclusions, ground glass or powdery chromatin appearance and formation of nuclear grooves (Fig. 9-15). Papillary carcinoma is often associated with squamous metaplasia [6,24](Fig. 9-16).

However, local occurrence of squamous metaplasia is not the diagnostic target in cytology but significant when the association is detected. Association with calcification forming psammoma bodies (calcospherites) is not specific for papillary carcinoma, but distinct from other types. The frequency of psammoma bodies is reported about 40% to 60%[6,10,24] in the usual type papillary car-

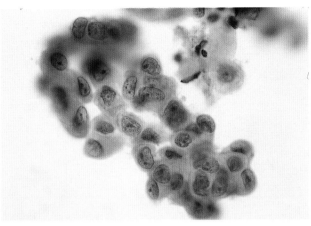

Fig. 9-15   Papillary carcinoma in needle aspirates. Polyhedral nuclei reveal occasionally indentations. Some show a ground glass appearance. The nuclei are disorderly arranged. Note prominence of nucleolus and nuclear crescent.

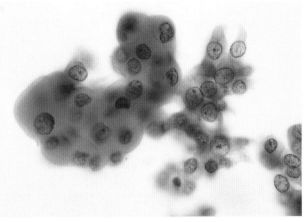

Fig. 9-16   Papillary carcinoma showing branching or finger-like projections in needle aspirates. Note three-dimensional cell overlappings with different levels of focus. Broad, thickened cytoplasm that is intensely stained with Light Green is the result of squamous metaplasia.

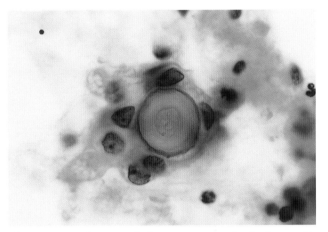

Fig. 9-17   A characteristic psammoma body in needle aspirate from papillary carcinoma. The surrounding cancer cells have hyperchromatic nuclei forming the shape of crescent.

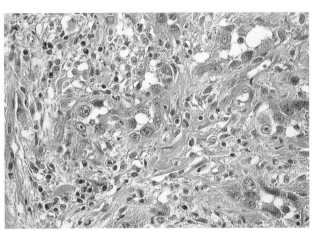

Fig. 9-18   Anaplastic carcinoma showing indistinct follicular spaces. Note markedly large follicular cells with prominent nucleoli which are surrounded by spindle-shaped cells. No colloid content.

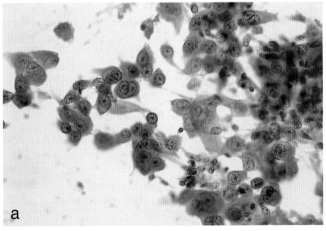

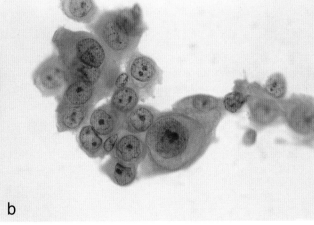

a   b

Fig. 9-19a,b   Giant cell carcinoma by imprint cytology during operation. These are giant follicular cells having large, round nuclei admixed with some spindly cells. Prominent acidophilic nucleoli are noticed.

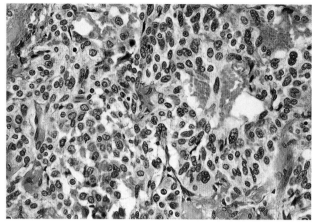

Fig. 9-20 Medullary carcinoma consisting of solid nests of polyhedral cancer cells. The cancer cells are surrounded by amyloid-containing stroma that is positive for Congo Red stain.

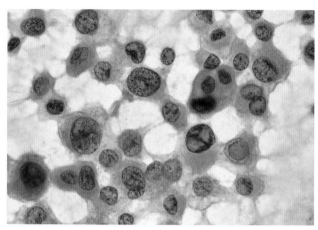

Fig. 9-21 Cytology of medullary carcinoma by ABC. Isolated cancer cells form a loose aggregation. Markedly enlarged cells are occasionally present. The cytoplasm is of moderate amount and homogeneous in stain. Note intranuclear cytoplasmic inclusion.

cinoma. Since these are exceptionally rare in benign diseases, cytological detection of psammoma bodies in ABC is indicative of malignancy (Fig. 9-17).

Besides the usual type of papillary carcinoma, histological variants comprise the follicular variant, tall cell variant and encapsulated variant. The follicular variant could be interpreted as the follicular carcinoma; in Kini's 379 follicular lesions 28 (7.4%) are of follicular variant. Out of general cytologic findings, intranuclear pseudoinclusions[17] and nuclear grooves appearing along the long axis of nuclei are very distinctive features. The tall cell variant is clinically important because of aggressive course.[5,16] The tall cell is characterized by abundant, elongated, eosinophilic cytoplasm with a height about twice the width.

12 Anaplastic Carcinoma

Small cell carcinoma which was defined as a form of anaplastic or undifferentiated carcinoma is currently regarded as malignant lymphoma.[28] Immunohistochemistry is needed for differentiation of small cell carcinoma from lymphoma. As a distinctive feature of anaplastic carcinoma a remarkable component of the tumor is polymorphic, giant cells having enlarged nuclei and prominent nucleoli (Figs. 9-18 and 9-19). This type is the most malignant thyroid carcinoma that affects elderly individuals over 60 years and more commonly women than men as in other types of thyroid carcinoma. It invades rapidly the adjacent tissues and causes compression signs.[15,26]

13 Medullary Carcinoma

Medullary carcinoma first described by Horn (1951) and subsequently named by Hazard (1959) is a peculiar type of thyroid carcinoma that is associated with carcinoid-like syndrome. Histogenetically this tumor is referred to as parafollicular C cell origin,[31] whereas the other types of thyroid cancers arise from follicular epithelium. It accounts for 5% to 10% of all thyroid cancers and

occurs sporadically in about two third of the tumor. The sporadic form is seen in middle-aged adults (mean age; 36-51 years) with a slight female predominance.[8]

The remainder is of familiar occurrence in association with adrenal medullary and parathyroid proliferative disorders (multiple endocrine neoplasia syndrome type II) that is inherited in an autosomal dominant pattern.

Cytology: The cancer cells are singly isolated or loosely aggregated in sheets. Characteristically scattered are markedly enlarged cells with large, peripherally locating, hyperchromatic nuclei (Figs. 9-20 and 9-21).

References

1. Altavilla G, Pascale M, Nenci I: Fine needle aspiration cytology of thyroid gland diseases. Acta Cytol 34: 251, 1990.
2. Belfiore A, Garofalo MR, Giuffrida D: Increased aggressiveness of thyroid cancer in patients with Graves' disease. J Clin Endocrinol Metab 70: 830, 1990.
3. Bugis SP, Young JEM, Archibald SD, et al: Diagnostic accuracy of fine needle aspiration biopsy versus frozen sections in solitary thyroid nodules. Am J Surg 152: 411, 1986.
4. Caldwell CB, Sherman CD Jr: Thyroid carcinoma. In: Love RR, et al (eds): Manual of Clinical Oncology. pp 279-287. Berlin: Springer-Verlag, 1994.
5. Cameselle-Teijeiro J, Febles-Perez C, Cameselle-Teijeiro JF, et al: Cytologic clues for distinguishing the tall cell variant of thyroid papillary carcinoma: A case report. Acta Cytol 41: 1310, 1997.
6. Carcangiu ML, Zampi G, Pupi A: Papillary carcinoma of the thyroid: A clinicopathologic study of 241 cases treated at the University of Florence Italy. Cancer 55: 805, 1985.
7. Chen KTK, Rosai J: Follicular variant of thyroid papillary carcinoma: A clinicopathologic study of six cases. Am J Surg Pathol 1: 123, 1977.
8. Collins BT, Cramer HM, Tabatowski K, et al: Fine needle aspiration of medullary carcinoma of the thyroid: Cytomorphology, immunocytochemistry and electron microscopy. Acta Cytol 39: 920, 1995.
9. Farbota LM, Calandra DB, Lawrence AM, et al: Thyroid carcinoma in Graves' disease. Surgery 98: 1148, 1985.

10. Franssila KO: Is the differentiation between papillary and follicular thyroid carcinoma valid? Cancer 32: 853, 1973.

11. Gordon DL, Gattuso P, Castelli M, et al: Effect of fine needle aspiration biopsy on the histology of thyroid neoplasms. Acta Cytol 37: 651, 1993.

12. Harach HR, Zusman SB: Cytologic findings in the follicular variant of papillary carcinoma of the thyroid. Acta Cytol 36: 142, 1992.

13. Hawkins F, Bellido D, Bernal C, et al: Fine needle aspiration biopsy in the diagnosis of thyroid disease. Cancer 59: 1206, 1987.

14. Hay ID: Papillary thyroid carcinoma. Endocrinol Metab Clin North Am 19: 545, 1990.

15. Jensen MH, Davis RK, Derrick L: Thyroid cancer: A computer assisted review of 5287 cases. Otolaryngol Head Neck Surg 102: 51, 1990.

16. Johnson TL, Lloyd RV, Thompson NW, et al: Prognostic implications of the tall cell variant of papillary thyroid carcinoma. Am J Surg Pathol 12: 22, 1988.

17. Kini SR, Miller JM, Hamburger JI, et al: Cytopathology of follicular lesions of the thyroid gland. Diagn Cytopathol 1: 123, 1985.

18. Klemi PJ, Heikki J, Nylamo E: Fine needle aspiration biopsy in the diagnosis of thyroid nodules. Acta Cytol 35: 434, 1991.

19. Klencki M, Slowinska-Klencka D, Sporny S, et al: Age-related changes in the size of thyrocyte nuclei in aspirates from nontoxic nodular goiter. Acta Cytol 38: 524, 1994.

20. Lang W, Choritz H, Hundeshagen H: Risk factors in follicular thyroid carcinomas: A retrospective follow-up study covering a 14-year period with emphasis on morphological findings. Am J Pathol 10: 246, 1986.

21. Leung C-S, Hartwick RWJ, Bedard YC: Correlation of cytologic and histologic features in variants of papillary carcinoma of the thyroid. Acta Cytol 37: 645, 1993.

22. Löwhagen T, Willems JS, Lundell G, et al: Aspiration biopsy cytology (ABC) in nodules of the thyroid suspected to be malignant. Surg Clin North Am 59: 3, 1979.

23. Mandreker SRA, Nadkarni NS, Pinto RGW, et al: Role of fine needle aspiration cytology as the initial modality in the investigation of thyroid lesions. Acta Cytol 39: 898, 1995.

24. Meissner WA, Warren S: Tumors of the Thyroid Gland. Atlas of Tumor Pathology. Washington, DC: Armed Forces Institute of Pathology, 1969.

25. Myren J, Sivertssen E: Thin-needle biopsy of the thyroid gland in the diagnosis of thyrotoxicosis. Acta Endocrinol 39: 431, 1972.

26. Nel CJ, van Heerden JA, Goellner JR: Anaplastic carcinoma of the thyroid: A clinicopathologic study of 82 cases. Mayo Clin Proc 60: 51, 1985.

27. Nilsson G: Marginal vacuoles in fine needle aspiration biopsy smears of toxic goiters. Acta Pathol Microbiol Scand 80: 289, 1972.

28. Schmid KW, Kroll M, Hofstadter F, et al: Small cell carcinoma of the thyroid: A reclassification of cases originally diagnosed as small cell carcinoma of the thyroid. Pathol Res Pract 181: 540, 1986.

29. Simpson WJ, McKinney SE, Carruthers JS: Papillary and follicular thyroid cancer: Prognostic factors in 1,578 patients. Am J Med 83: 479, 1987.

30. Venkatash YS, Ordonez NG, Schultz PN, et al: Anaplastic carcinoma of the thyroid: A clinicopathologic study of 121 cases. Cancer 66: 321, 1990.

31. Williams ED: Histogenesis of medullary carcinoma of the thyroid. J Clin Pathol 19: 114, 1966.

10 Thymus and Mediastinum

Masayoshi Takahashi and Toshiro Kawai

Imaging studies by chest X-ray, ultrasonography and computed tomography assess accurate localization of tumors and allied diseases in the mediastinum. Main tumors and cysts are arranged in Table10-1 in order of incidence; thymoma, germ cell tumors and neurogenic tumors are representative in relatively rare tumors of the mediastinum. Thymic tumors are the target easily accessible by fine needle aspiration.

During embryogenesis, the thymus plays an important role as a primary lymphoid organ, into which stem cells derived from the bone marrow migrate. The thymus concerns with maturation of the precursor cells which become T lymphocytes in cell-mediated immune response. The T lymphocytes migrated into blood stream may populate in the peripheral lymphoid tissues, predominantly in the paracortical area. Thymoma is a fairly rare tumor situated in the anterior mediastinum, and grows as a well encapsulated tumor mass. If it is diagnosed early, the tumor is resectable and curable. Needle aspiration cytology can be possible, but its localization must be carefully examined beforehand roentgenologically, because it is in a close connection with the pericardium and great vessels. Practically needle aspiration is performed during an operative procedure as adjunct to biopsy.

Although needle aspiration is easily accessible to tumors in the anterior mediastinum, cytologists will be lost in a maze when they are asked for making a report on the biological malignancy grade of thymoma, because an attempt to predict the outcome of thymoma based on only morphological grounds is futile. Preservation of the capsule and lack of blood vessel invasions are obviously informative rather than cytological atypicality.[7] Shimosato and Mukai, however, mentioned that predominantly polygonal epithelial cell type was more worse than the spindle cell type in their clinicopathological studies.[11] we must take into account that needle aspiration represents only a part of the tumor that may admix both cell types.

1 Thymocyte Maturation

Regarding the maturation of thymocytes, surface antigenic phenotypes and the rearrangement and expression of the genes encoding their surface receptors for antigen demonstrate the process of maturation such as prothymocytes to be matured to cortical lymphocytes acquiring many cell surface protein/glycoprotein molecules; CD1 is expressed only on the surface of cortical thymocytes whereas CD2 and CD5 appear on all cortical and medullary thymocytes. CD8 that appears early in T lymphocyte development may become double positive for CD4 and CD8 cortical thymocytes. A minority of these double-positive cells give rise to the functionally mature, single positive CD4+ or CD8+ lymphocytes of the medulla. The enzyme terminal deoxyribonucleotidyl transferase (TdT) is present within prothymocytes and cortical thymocytes. It is during the development of thymocytes in the cortex that expression of T cell receptor genes occurs.

2 Tumors of the Thymus and Mediastinum

Thymoma locating ordinarily in the anterior mediastinum can be approached by needle aspiration, however, one must note that it may not always be located in the normal place of the thymus. The tumor may develop in ectopic thymic tissue such as in the neck or in a lower part of the mediastinum close to the diaphragm.

Generally speaking, differentiation between thymoma and lymphoma is based on cell population in smears; thymoma contains epithelial cells to some extent, whereas non-Hodgkin's lymphoma shows a monomorphic cell population.

Thymic carcinoma is mostly nonkeratinizing squa-

Table 10-1 Tumors and allied lesions in the mediastinum

Anterior mediastinum	Middle mediastinum	Posterior mediastinum
Thymoma	Mediastinal goiter	Neurogenic tumor
Cancer (primary and metastatic)	Bronchial cyst	neurinoma
Germinoma		neurofibroma
Mediastinal goiter		ganglioneuroma
Thymic cyst		Lymphoma
Lymphoma		Mesenchymal tumor
Carcinoid		

238

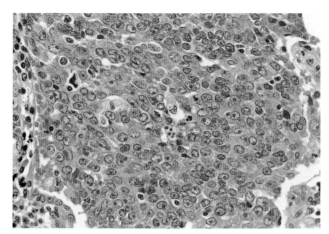

Fig. 10-1 Thymic carcinoma showing a solid mass of large polygonal cells. The nuclei are round to polygonal with prominent nucleoli. Note the presence of "starry sky-like" macrophages.

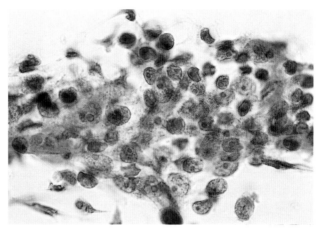

Fig. 10-2 Thymic carcinoma in fine needle aspirates. Note enlarged, round nuclei, high N/C ratio and prominent, amphoteric nucleoli. The cytoplasm is scant and ill-defined.

mous cell variety, and is distinguished by the presence of cells with marked atypia such as high N/C ratio, nuclear hyperchromatism and nucleolar prominence. In short, a resemblance to nasopharyngeal carcinoma is a noteworthy finding (Figs. 10-1 and 10-2).

Squamous cell carcinoma in the mediastinum is often metastatic spread from pulmonary carcinoma. Exclusion of secondary carcinoma is needed in case of thymic carcinoma. The primary thymic carcinomas studied by Shimosato and his associates[11] have shown fairly good survivals longer than five years after surgery.

Germ cell tumors include seminoma-like germinoma (Fig. 10-3), yolk sac tumor, embryonal carcinoma and choriocarcinoma. Germinoma or germ cell tumor consisting of large, enrounded tumor cells and lymphocytes is apt to be misinterpreted as thymoma in needle aspirates because two cell pattern of histology is occasionally obscure. Recognition of large, round tumor cells having a centrally locating macronucleolus is a key point of diagnosis. Taking account of not infrequent occurrence of germinoma in the anterior mediastinum may lead to accurate diagnosis.

■Thymoma

Shimosato and Mukai[12] have recently described the classification of thymic tumors from various aspects as follows:

by histology; lymphocyte predominant, mixed lymphocytic and epithelial, epithelial cell predominant,

by cell type; spindle cell, mixed spindle cell and polygonal cell, polygonal-oval cell,

by cell atypia; absent, slight, moderate, and marked.

From the aspect of cytology, Tao and associates introduced subclassification; (1) small epithelial cell type, (2) intermediate epithelial cell type, (3) large epithelial cell type, (4) pleomorphic epithelial cell type, and (5) spindle epithelial cell type.[13] This Color Atlas adopts the basic classification of three types according to the cell feature of thymoma. The epithelial thymoma is subclassified to polygonal or spindle cell type.

1) Cytology of thymoma

Principal cell constituents of thymoma are composed of two different elements, epithelial cells and lymphoid

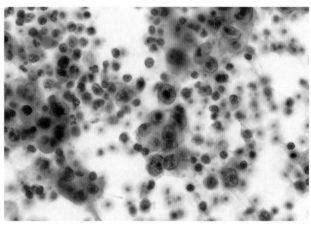

Fig. 10-3 Germinoma arising in the anterior mediastinum. Markedly enlarged, round nuclei, scant cytoplasm and centrally locating prominent nucleoli are characteristic features of germinoma. Noteworthy finding is the presence of lymphoid cells around loose clusters of tumor cells.

cells. The tumor was simply classified into (1) lymphocytic thymoma, (2) epithelial thymoma, and (3) mixed lymphoepithelial thymoma based on cell components.[3,15]

Lymphocytic thymoma is characterized by small thymic lymphocytes or thymocytes with a few scattered plump epithelial cells. Lymphocytic thymoma that dominates the smear resembles lymphoma (Fig. 10-4). The lack of nuclear convolution and the lack of prominent nucleolus are important points of differentiation of thymoma from lymphoma. General appearance of low power magnification is useful to grasp a general pattern of cell constituents. In lymphocytic thymoma there are two varieties, i.e., a type consisting of mature small lymphocytes and another type comprised of larger active lymphocytes which are called "stimulated" cells[7](Figs. 10-5 and 10-6).

Epithelial thymoma is composed of atypical epithelial cells arranged in islands or trabeculae, which are surrounded by small thymocytes (Fig. 10-7 to 10-14). The epithelial cells in thymoma possess oval, vesicular nuclei with smooth nuclear membrane. The cytoplasm of the epithelial tumor cells is more scant than that of mature

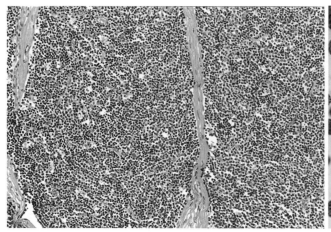

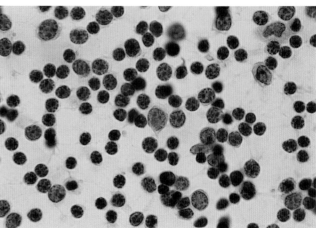

Fig. 10-4 Histology of lymphocytic thymoma. Low magnification. The majority of the tumor consists of small thymic lymphocytes. Very few polygonal epithelial cells are admixed.

Fig. 10-5 Lymphocytic thymoma. From a 65-year-old male without myasthenia gravis. The most of tumor cells consist of mature thymic lymphocytes and a few scattered larger cells having faint vesicular nuclei. Imprint smear.
(By Courtesy of Dr. S. Hanzawa, Department of Thoracic Surgery, Hamamatsu Medical Center, Hamamatsu, Japan.)

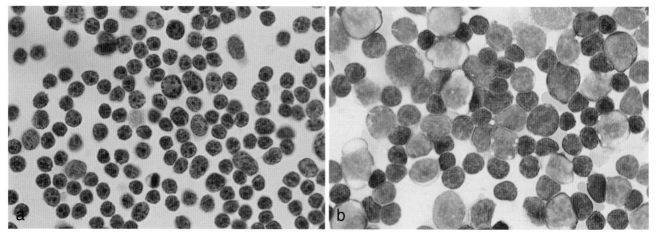

Fig. 10-6 Lymphocytic thymoma. From a 57-year-old male without myasthenia gravis. A characteristic lymphocyte predominance type to be composed of "stimulated" lymphocytes of intermediate size. Imprint smear.
a. Papanicolaou stain showing active thymic lymphocytes of intermediate size with dotted chromatins. A few larger cells are scattered.
b. Giemsa stain showing mature thymic lymphocytes and larger cells with vesicular nuclei in which prominent nucleoli are invisible.
(By Courtesy of Dr. S. Hanzawa, Department of Thoracic Surgery, Hamamatsu Medical Cnter, Hamamatsu, Japan.)

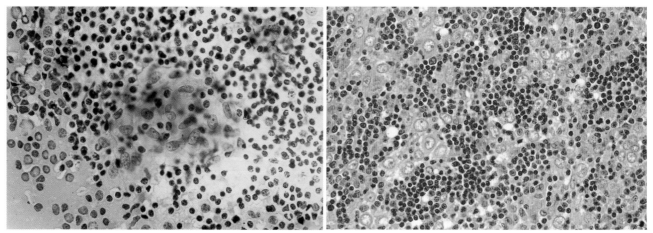

Fig. 10-7a Histology of thymoma of mixed lympho-epithelial variety. Note an admixture with a large number of thymic lymphocytes and small group of epithelial cells with pale adequate cytoplasm.

Fig. 10-7b Histology of mixed lymphoepithelial thymoma showing broad bands or nests of polyhedral epithelial cells. Small thymic lymphocytes appear to be compressed among epithelial cells.

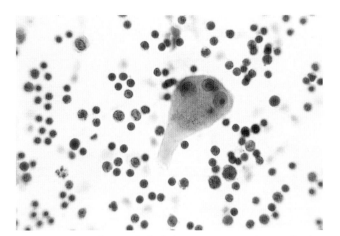

Fig. 10-8 **Lymphoepithelial thymoma in aspirates during thoracotomy.** A multinucleated cell with thickened homogeneous cytoplasm is consistent with Hassall's corpuscle.

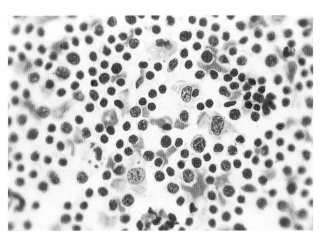

Fig. 10-9 **Thymoma of mixed lymphoepithelial type in imprint smear.** Several epithelial cells having large vesicular nuclei are mixed with a number of mature thymic lymphocytes.

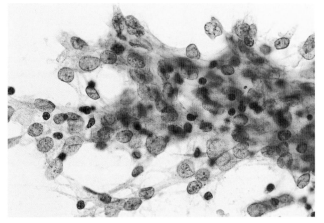

Fig. 10-10 **Cytology of epithelial thymoma.** Fine needle aspiration showing dominant polyhedral epithelial cells aggregated in irregular sheets. Note normochromic nuclei and abundant, ill-defined cytoplasm. Small lymphocytes are intermingled within the tumor cell nest.

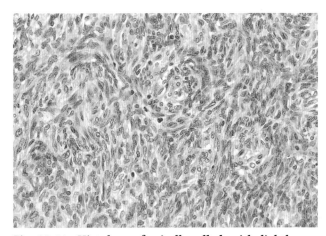

Fig. 10-11 **Histology of spindle-celled epithelial thymoma.** The tumor is composed of interlacing bundles of spindle-shaped epithelial cells. Note a few lymphocytes within the tumor mass.

epithelial cells, but is recognizable with basophilic tint. The grade of atypicality and the proportion of epithelial cell components were referred to as invasion capability by Tao et al.[13]

From a view of differential diagnosis, spindle-celled thymoma looks like neurogenic tumor that arises from the posterior mediastinum. Location of the tumor and admixture with small thymocytes in and around the tumor cell aggregates are in favor of thymoma (Fig. 10-12). Germinoma that is combined with lymphocyte background resembles lymphoepithelial thymoma. Attention for differentiation should be paid to large round nuclei with centrally locating macronucleolus of germinoma (Fig. 10-3).[10]

2) Immunocytochemistry

Application of immunocytochemistry/electron microscopy is worthy for differential diagnosis of thymoma from thymoma-mimic neoplasms.[1,4,6,14] Demonstration of cytokeratin for epithelial cells and T6/Tdt for cortical thymic lymphocytes is valuable,[5,9] since most of small cell lymphomas, lymphocytic thymoma resembling lym-

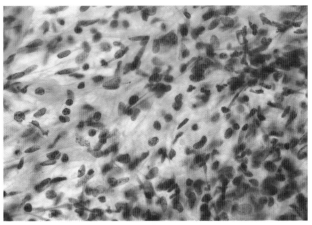

Fig. 10-12 **Cytology of spindle-celled epithelial thymoma.** Fine needle aspiration. Spindle-shaped nuclei vary in size. There is no organoid pattern. Note small thymic lymphocytes admixed with spindle epithelial cells.

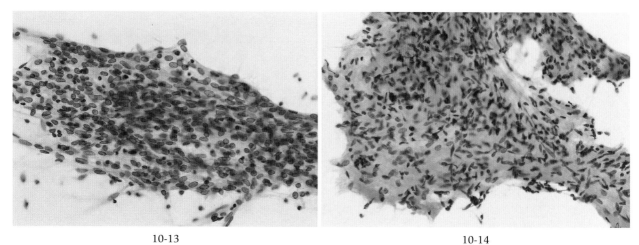

10-13

10-14

Figs. 10-13 and 10-14 Comparative demonstration between epithelial thymoma and neurinoma. The epithelial cells of thymoma(Fig. 10-13) show elongated, elliptic nuclei with blunt nuclear poles, while neurinoma (schwannoma) (Fig. 10-14) reveals spindly and wavy nuclei. Nuclear size variation is more marked in neurinoma than in thymoma. Nuclear palisading or fasciculation is occasionally noticeable in cytology.

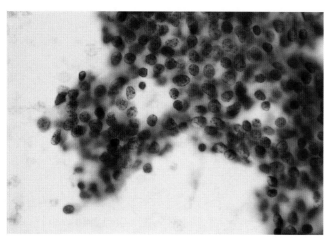

Fig. 10-15 Carcinoid in the anterior mediastinum. From a 76-year-old male by fine needle aspiration.
Overlapping cell clusters arranged in a trabecular fashion. The ovoid nuclei are fairly uniform in size and shape and show finely granular chromatins. The cytoplasm is scant and ill-defined.

phoma can be confirmed by the presence of T6-positive and Tdt-positive cells.[10]

3 Carcinoid Tumor

Carcinoid tumor of the mediastinum (Fig. 10-15) as ever thought to be parathyroid mediastinal tumor or metastatic endocrine tumor must be strictly defined. This tumor is somewhat lobulated and encapsulated like thymoma.

Microscopically a solid tumor is comprised of anastomosing trabeculae or ribbons of small round or ovoid cells. The cytoplasm is scant and ill-defined. The nuclei are vesicular but finely granular. Grimelius and Fontana-Masson stains are useful but not always positive for carcinoid tumor.[2]

References

1. Finley JL, Silverman JF, Strausbauch PH, et al: Malignant thymic neoplasms: Diagnosis by fine-needle aspiration biopsy with histologic, immunocytochemical, and ultrastructural confirmation. Diagn Cytopathol 2: 118, 1986.
2. Kawai T, Tsunoda N, Kubono S, et al: Three cases of thymic carcinoid. J Jpn Soc Clin Cytol 27: 564, 1988.
3. Lewis JE, Wick MR, Scheithauer BW, et al: Thymoma: A clinicopathologic review. Cancer 60: 2727, 1987.
4. Millar J, Allen R, St John Wakefield J, et al: Diagnosis of thymoma by fine-needle aspiration cytology: Light and electron microscopic study of a case. Diagn Cytopathol 3: 166, 1987.
5. Mokhtar N, Hsu S, Lad RP, et al: Thymoma: Lymphoid and epithelial components mirror the phenotype of normal thymus. Hum Pathol 15: 378, 1984.
6. Ritter MA, Lampert IA: The thymus. In: McGee J O'D, et al (eds): Oxford Textbook of Pathology. pp1807-1821. Oxford: Oxford Med Publ, 1992.
7. Rosai J, Levine GD: Tumors of the Thymus. Atlas of Tumor Pathology (ed by Firminger HL). Bethesda: Armed Forces Institute of Pathology, 1976.
8. Salyer WR, Eggleston JC: Thymoma: A clinical and pathological study of 65 cases. Cancer 37: 229, 1976.
9. Sato Y, Watanade S, Mukai K, et al: An immunohistochemical study of thymic epithelial tumors: II. Lymphoid component. Am J Surg Pathol 10: 862, 1986.
10. Sherman ME, Black-Schaffer S : Diagnosis of thymoma by needle biopsy. Acta Cytol 34: 63, 1990.
11. Shimosato Y, Kameya T, Nagai K, et al: Squamous cell carcinoma of the thymus. An analysis of eight cases. Am J Surg Pathol 1: 109, 1977.
12. Shimosato Y, Mukai K: Atlas of Tumor Pathology: Tumors of the Mediastinum. Washington, DC: Armed Forces Institute of Pathology, 1997.
13. Tao L-C, Pearson FG, Cooper JD, et al: Cytopathology of thymoma. Acta Cytol 28: 165, 1984.
14. Vengrove MA, Schimmel M, Atkinson BF, et al: Invasive cervical thymoma masquerading as a solitary thyroid nodule. Acta Cytol 35: 431, 1991.
15. Wick MR, Scheithauer BW, Weiland LH, et al: Primary thymic carcinomas. Am J Surg Pathol 6: 613, 1982.

11 Salivary Glands

Masayoshi Takahashi and Toshiro Kawai

Tumors of the major salivary glands are relatively uncommon and mostly arise from the parotid gland. The parotid gland is the major target of aspiration biopsy cytology (ABC) of all salivary gland lesions in 125 of 151 cases (82.8%) by Cajulis et al.[2] Approximately three quarters of tumors are benign and histologically pleomorphic adenoma (mixed tumor), Warthin's tumor, myoepithelioma, etc. The site predilection of tumors for submandibular gland tumors in Uganda,[6] sublingual tumors in West Indies,[12] and malignant salivary gland tumors in Canadian Eskimos.[19] The location of salivary gland tumors are correctly identified by imaging studies, thereby the tumor is accessible with ease by ABC and surgical procedures can be indicated preoperatively.[1,3,4]

1 Procedure of Aspiration and Preparation

Fine needle aspiration is performed using 21-22 gauge needles mounted on previously heparinized 20 ml syringes. The method is based on the technique described by Zajicek (1974).[20] Preparation of smears is the same as shown in ABC of other organs. When a rash report is required, air-dried smears and Diff-Quik stain are recommended.[2]

2 Pleomorphic Adenoma

Pleomorphic adenoma or mixed tumor comprises around 50% of salivary gland tumors diagnosed by ABC; Cajulis et al[2] diagnosed 33 of 61 tumors (54.1%) and Cristallini et al[5] 38 of 87 tumors (43.7%). Pleomorphic adenoma affects a middle age group and more frequently women than men.[9] The tumor is well-defined and encapsulated by thin fibrous membrane. Histologically the tumor is composed of two components of epithelial and mesenchymal tissues irrespective of predominance of either component (Fig. 11-2).

Cytology: In ABC the epithelial component may predominate and appears to be closely packed with arrangements suggestive of acinar or trabecular aggregates, Their nuclei are spindle or ovoid in shape and uniform in size. The stromal component that is gelatinous, myxoid or chondroid in matrix can be easily recognized. Observation of these cell mixture may lead to correct interpretation (Figs. 11-3 and 11-4); complexity in the texture of pleomorphic adenoma will be contrastive with monotony of normal gland aspirates (Fig. 11-1).

3 Warthin's Tumor (Adenolymphoma) (Papillary Cystadenoma Lymphomatosum)

Warthin's tumor constitutes 11.5% of salivary tumors by Cajulis et al and 10.3% by Cristallini et al, respec-

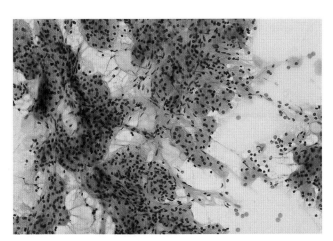

Fig. 11-1 Normal parotid gland aspirates. Acinar parenchyma arranged in trabeculae or balls of epithelial cells. Note uniform small nuclei and abundant cytoplasm.

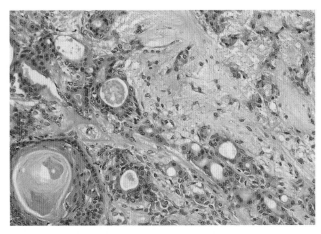

Fig. 11-2 Pleomorphic adenoma of the parotis of a 48-year-old male. The same case of Fig. 11-3. Variously sized tubules showing squamous metaplasia are surrounded by chondro-myxoid stroma.

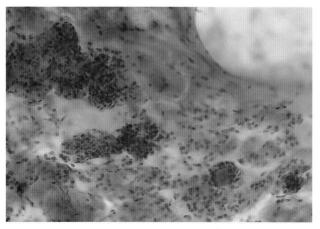

Fig. 11-3　Pleomorphic adenoma in ABC showing epithelial cell clusters or balls surrounded by gelatinous material. Brownish mucoid stroma overshadows compact epithelial cell aggregates.

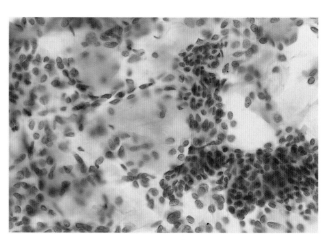

Fig. 11-4　Pleomorphic adenoma in ABC. From a parotid gland tumor of a male aged 35. Irregular epithelial cell clusters and amphoteric stromal matrix suggestive of chondromyxoid substance.

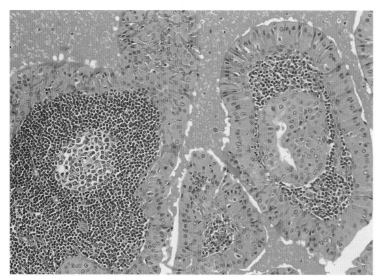

Fig. 11-5　Warthin's tumor characterized by eosinophilic, columnar cells and dense lymphocytic cuffs. The same case of Fig. 11-6 from a male aged 57.

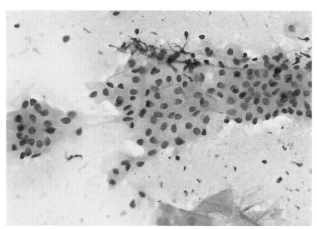

Fig. 11-6　Oncocytes aggregated in sheet. Needle aspirates of the parotid gland. Epithelial cells predominate. Uniform small nuclei and abundant, opaque cytoplasm are characteristic.

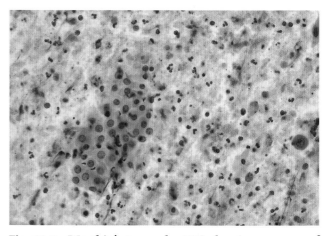

Fig. 11-7　Warthin's tumor by ABC showing a group of columnar epithelial cells, lymphocytes and desquamated amorphous substances. Components of cell debris and some leukocytes may mislead interpretation to sialadenitis.

tively. The tumor has a predilection of site for the parotid gland that is rich in lymphoid tissues. An important clinical behavior is a simultaneous or asynchronous bilateral occurrence. Histologically, the tumor is characterized by papillary cystadenomatous structures of tall columnar cells which are finely granular and eosinophilic. Noteworthy finding is the presence of small, round nuclei near the free surface. A mass of lymphoid tissue is encompassed by papillary epithelial growths (Fig. 11-5).

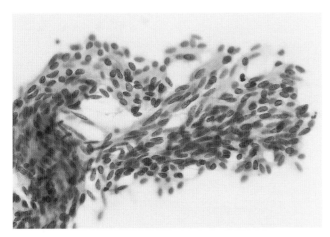

Fig. 11-8 Spindle celled myoepithelioma by ABC. Dense bundles of spindly cells. Their nuclei appear ovoid or elliptic in shape and uniform in size. Chromatins are bland.

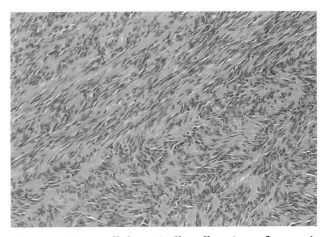

Fig. 11-9 Hypercellular spindle cell variety of myoepithelioma. From the parotid gland of a 40-year-old female.

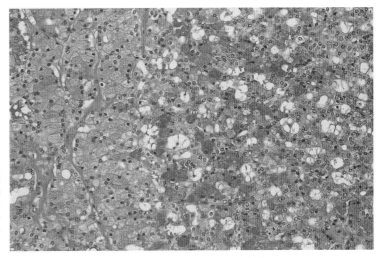

Fig. 11-10 Acinic cell carcinoma of the parotid gland. From a 19-year-old female. The same case as Figs. 11-11 and 11-12. The cancer cells resembling normal acinar cells are massively, closely packed. Note a mosaic pattern with mixture of darkly staining cells.

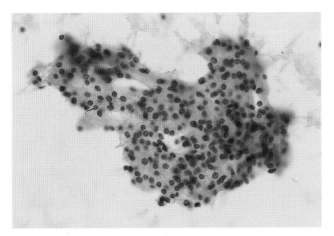

Fig. 11-11 Acinic cell carcinoma in needle aspirates. General appearance with low power magnification showing irregular sheet of epithelial cells and distortion of an alveolar structure.

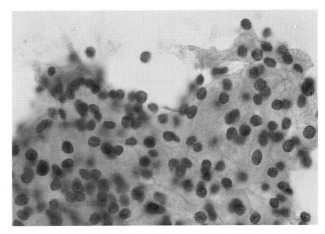

Fig. 11-12 Acinic cell carcinoma with a higher magnification. The cytoplasm is broad, clear and very finely granular like benign acinar cells. Their nuclei are hyperchromatic, variable in size and disorderly arranged.

Cytology: Aspirates do not always include both cellular elements under the same microscopic field. Although epithelial cells having abundant, finely granular cytoplasm will be the key points of diagnosis (Fig. 11-6), mucoid material containing amorphous substance or cellular debris is frequently encountered (Fig. 11-7).[16]

Their nuclei are small, round or ovoid in shape and have bland chromatins. Since cytologic features of epithelial cells are similar to oncocytes, a case with preponderance of oncocytic epithelial cells cannot be distinguished from monomorphous oncocytoma[16] and uncommon oncocytic salivary duct carcinoma.[10, 11]

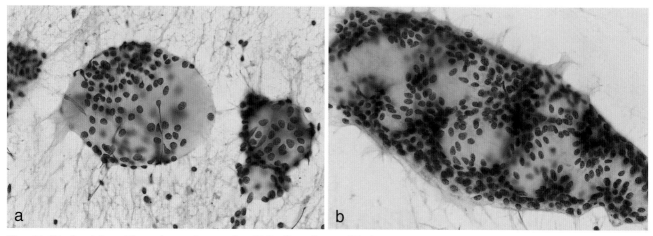

Fig. 11-13a,b Adenoid cystic carcinoma by ABC of a submandibular tumor. From a 67-year-old woman. Note characteristic cytology showing three-dimensional clusters arranged in globules (a) and cylinders (b) encompassing amorphous, opaque or mucoid material. The nuclei are basophilic, round or oval, and suggestive of basal cell origin.

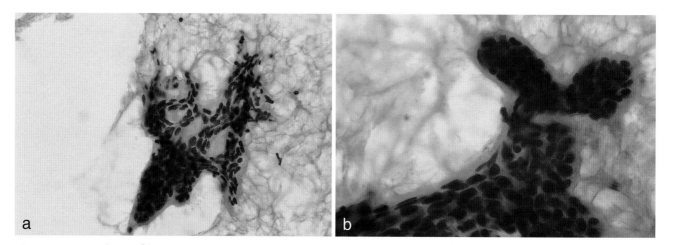

Fig. 11-14 Cytology of basal cell adenoma resembling adenoid cystic carcinoma.
a. A cluster of basal cells under low magnification is hardly distinguished from adenoid cystic carcinoma.
b. Club-shaped clusters with some papillary projections are similar to low-grade papillary adenocarcinoma.[17] Note the periphery lined by low columnar cells in row.

4 Myoepithelioma

Myoepithelioma is adopted in classification of the salivary tumor of the third edition of *Atlas of Tumor Pathology* by AFIP[7] and defined as a monomorphous myoepithelial component without duct formation. The frequency of the tumor constitutes 1.5% of salivary gland tumors in the AFIP files.[7] Histologically, the tumor consists of bundles or nests of spindle cells, polygonal epithelioid cells, or plasmacytoid cells. Immunohistochemical reactivity for cytokeratin and S 100 protein is helpful for diagnosis in histology but not contributory in ABC. Cytologically, a characteristic feature is a monomorphous constituent of spindly or epithelioid cells (Figs. 11-8 and 11-9). However, determination either monomorphous or polymorphous based on only ABC is difficult unless aspirated cells represent the whole histopathology of the tumor.

5 Acinic Cell Carcinoma

Acinic cell carcinoma is a relatively uncommon tumor and accounts for about 17% of all salivary gland malignant tumors according to the AFIP series[7] and 13% by Eneroth's report.[8] The tumor is a slow growing, solid tumor of low-grade malignancy and affects mostly the parotid gland. Histologically, the tumor is composed of closely packed, acinus-like or trabecular cells that resemble normal salivary glands. Occasionally darkly staining cells are admixed most likely due to zymogen granules. In comparison with normal acinar or lobular cell aggregates, regular cell arrangements are distorted and alveolar structures are lost in observation with low power magnification (Fig. 11-10).

Cytology: The cytoplasm is abundant, foamy in appearance and very finely granular. Occasionally dark cytoplasm is demonstrated but not distinctive as in tissue section of H.E. Stain. Their nuclei are uniform in shape but vary slightly in size and disorderly arranged. The importance of interpretation consists in grasp of a whole texture or cell clusters (Figs. 11-11 and 11-12).

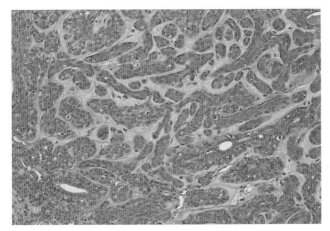

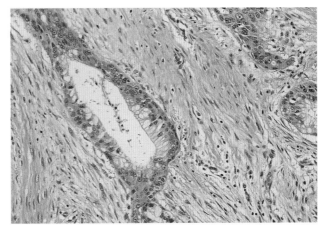

Fig. 11-15 Histology of basal cell adenoma. From a parotid tumor of a woman aged 34. The tumor is composed of solid tubules or trabeculae varying in size, which are separated by small amount of stroma.

Fig. 11-16 Mucoepidermoid carcinoma showing infiltrative growth with desmoplasia. Variously sized tubules are lined with disorderly arranged squamous cells and mucus-containing cells.

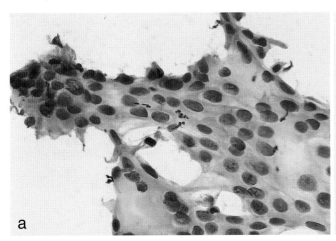

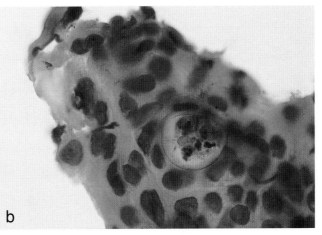

Fig. 11-17 Mucoepidermoid carcinoma in needle aspirates of submandibular tumor. From a 42-year-old woman.
a. A large irregular sheet of squamous cells having oval to elliptic nuclei that vary slightly in size. Chromatins are bland.
b. A higher magnification displays cohesive clusters of squamous cells with obvious nuclear atypicality. Note a mucus concretion with brownish hue.

6 Adenoid Cystic Carcinoma

Adenoid cystic carcinoma has a site predilection for the minor salivary glands especially the submandibular gland.[14] The frequency is about the same as mucoepidermoid carcinoma; i.e., 23% of 528 malignancies by Eneroth.[8] This slowly growing carcinoma has a proclivity to invade locally around the nerve sheaths and to insidiously recur.[18] Histologically, the tumor forms compact nests or columns of basophilic cells with scant cytoplasm. A distinctive cart wheel or cribriform pattern can be recognized.

Cytology: Tumor cells form closely packed globules or balls of small basophilic cells in which acellular mucoid materials are recognizable.[14] Spherical pinkish purple globules can be identified by Giemsa stain[14, 16] (Fig. 11-13a, b).

7 Basal Cell Adenoma

Benign monomorphous basal cell adenoma mimics adenoid cystic carcinoma because of its cell compo-

nents.[13, 15] These are densely gathered in solid, trabecular or tubular pattern with scant cytoplasm (Figs 11-14 and 11-15). Cytology in ABC is seldom depicted.[7]

8 Mucoepidermoid Carcinoma

Mucoepidermoid carcinoma arises commonly in the parotid gland without age and sex predilections. This tumor is the most common in the malignant tumors of the salivary glands and constitutes 23% in the report by Eneroth.[8] Histologically, this is a solid tumor comprised of sheets or cords of squamous cells of low atypicality containing mucus-producing cells (Fig. 11-16).

Aspiration cytology demonstrates loose cell clusters of squamous cells with varying degrees of atypia. The majority of the tumor lack obvious atypia; thereby the cytoplasm is broad, opaque in appearance, and greenish blue by Papanicolaou stain. The nuclei are oval or round and slightly vary in size. The N/C ratio is not much distorted. Mucus-producing cells or mucoid material are admixed (Fig. 11-17a,b).

References

1. Bhatia A: Fine needle aspiration cytology in the diagnosis of mass lesions of the salivary gland. Ind J Cancer 30: 26, 1993.
2. Cajulis RS, Gokaslan ST, Yu GH, et al: Fine needle aspiration biopsy of the salivary glands. A five-year experience with emphasis on diagnostic pitfalls. Acta Cytol 41: 1412, 1997.
3. Chan MKM, McGuire LJ, King W, et al: Cytodiagnosis of 112 salivary gland lesions. Correlation with histologic and frozen section diagnosis. Acta Cytol 36: 353, 1992.
4. Cramer H, Layfield L, Lampe H: Fine-needle aspiration of salivary glands: Its utility and tissue effects. Ann Otol Rhinol Laryngol 102: 483, 1993.
5. Cristallini EG, Aseani S, Farabi R, et al: Fine needle aspiration biopsy of salivary gland, 1985-1995. Acta Cytol 41: 1421, 1997.
6. Davies JNP, Dodge OG, Burkitt DP: Salivary-gland tumors in Uganda. Cancer 17: 1310, 1964.
7. Ellis GL, Auclair PL: Tumors of the Salivary Glands, 3rd ed. Atlas of Tumor Pathology. Washington, DC: Armed Forces Institute of Pathology, 1996.
8. Eneroth CM: Salivary gland tumors in the parotid gland, submandibular gland, and the palate region. Cancer 27: 1415, 1971.
9. Everson JW, Cawson RA: Salivary gland tumours. A review of 2410 cases with particular reference to histological types, site, age and sex distribution. J Pathol 146: 51, 1985.
10. Fyrat P, Cramer H, Feczko JD, et al: Fine-needle aspiration biopsy of salivary duct carcinoma. Report of five cases. Diagn Cytopathol 16: 526, 1997.
11. Gilerease MZ, Guzman-Paz M, Froberg K, et al: Salivary duct carcinoma. Is a specific diagnosis possible by fine needle aspiration cytology? Acta Cytol 42: 1389, 1998.
12. Gore DO, Annamunthodo H, Harland A : Tumors of salivary gland origin. Surg Gynecol Obstet 119: 1290, 1964.
13. Hood IC, Qizilbash AH, Salama SS, et al: Basal cell adenoma of parotid. Difficulty of differentiation from adenoid cystic carcinoma on aspiration biopsy. Acta Cytol 27: 515, 1983.
14. Kline TS: Handbook of Fine Needle Aspiration Biopsy Cytology. St Louis : CV Mosby, 1981.
15. Layfield LJ: Fine needle aspiration cytology of a trabecular adenoma of the parotid gland. Acta Cytol 29: 999, 1985.
16. Löwhagen T, Tani EM, Skoog L: Salivary glands and rare head and neck lesions. In: Bibbo M (ed) : Comprehensive Cytopathology. pp621-648. Philadelphia: WB Saunders, 1991.
17. Pisharodi LR: Low grade papillary adenocarcinoma of minor salivary gland origin. Diagnosis by fine needle aspiration cytology. Acta Cytol 41: 1407, 1997.
18. Ranger D, Thackray AC, Lucas RB: Mucous gland tumours. Br J Cancer 10: 1, 1956.
19. Wallace AC, MacDougall JT, Hildes JA, et al: Salivary gland tumors in Canadian Eskimos. Cancer 16: 1338, 1963.
20. Zajicek J: Aspiration Biopsy Cytology. Part 1. Cytology of Supradiaphragmatic Organs. Basel: S Karger, 1974.

12 Esophagus and Intestines

ESOPHAGUS

The major risk factors of esophageal cancer are heavy alcohol ingestion, heavy smoking and hot, spicy food intake. Geographically esophageal cancers are common in the Middle East and Far East. The prognosis of esophageal cancer is poor and its overall 5-year survival rates are low, about 5% to 10% in spite of easy endoscopic accessibility for esophageal lesions, because the esophagus has not been the target organ of cancer screening programs. Intraluminal ultrasound will be valuable in defining the depth of cancer. Brushing cytology for esophageal lesions remains not only for detection of early cancers but for differential diagnosis of cancer from esophagitis, leukoplakia, benign erosion, Barrett's esophagus, etc.

1 Collection of Specimens and Preparation of Smears

■Esophageal washings
The patient should not be given a breakfast on the day of cytologic examination. The stomach is emptied prior to washing. The lesion is washed with 20 ml balanced saline under direct vision by esophagoscopy. Washing the lesion blindly through a Levin tube (No. 12, 14 or 16) can also be used to obtain a specimen. In this procedure, a Levin tube with length markings is gently introduced above the lesion, the level of which must be previously ascertained roentgenologically. Aspirated fluid should be immediately centrifuged[27] or placed in an equal amount of the pretreatment solution, which has been prepared with 2 parts of formalin and 98 parts of 40% alcohol diluted with balanced saline solution. Koss recommends placing the material in 50% alcohol without delay.

■Esophageal brushings
It is accepted that forceps biopsy is most common and reliable as a single technique for diagnosis of malignancy of the esophagus.[20,23]

Since the esophageal wall is thin, biopsy may risk perforation. Scraping the lesion with a small brush is by far more effective and safer than endoscopic biopsy. A small brush is passed through a Levin tube (Fig. 12-1). Where the stenotic lesion is located must be determined beforehand roentgenologically. Otherwise, a rough, blind technique may risk perforation. Cytological examination should not be performed within 48 hours after X-ray examination, because barium contamination interferes with microscopic study.

Endoscopic brushing cytology prevails over brushing through a Levin tube, since flexible endoscopy have become a very common technique. As an adjunct to biopsy the postbiopsy brushing cytology is strongly recommended.[20]

Procedure:
1) A Levin tube is inserted to above the lesion, then the brush is gently extended.
2) After scraping, the brush is retracted into the Levin tube and withdrawn with the tube.
3) The cells are transferred from the brush onto a clean glass slide by tapping lightly, then fixed immediately in balanced 95% ethanol. Then the brush is rinsed with 10 ml of balanced saline. The rinsings are centrifuged at 1,500 rpm for five minutes. The sediment is smeared on glass slides and fixed.

■Capsulated sponge method
A unique sponge to brush the lesion of the esophagus was designed by Onozawa and Nabeya.[25] The framework of the sponge is made of polyurethane foam measuring 25 mm in diameter to which a string is attached. The outer layer of the elastic sponge is made of a rough net of polypropylene fibers that is suitable for brushing the mucosal surface. The sponge is compressed and packed within a soluble capsule that is used commonly for medication (Fig. 12-2). The patient is able to swallow the capsulated sponge like a medicine with a little water. Since the attached string has length marks, the depth of insertion can be ascertained. The advantage of the sponge technique consists in not only appropriateness to detect esophageal cancer but in accessibility to cardiac cancer by a swallow down to the cardia to which biopsy forceps hardly approach.

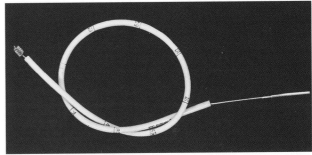

Fig. 12-1 Levin tube with length markings and a brush for scraping.

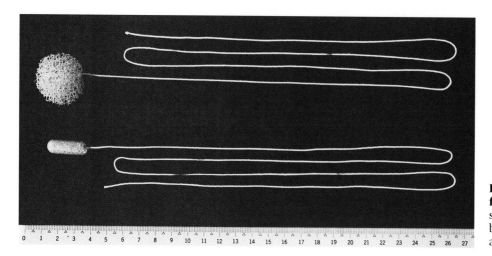

Fig. 12-2 Capsulated sponge for esophageal brushing. A sponge placed in the capsule to be digested swells up in the stomach.

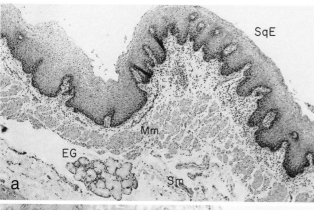

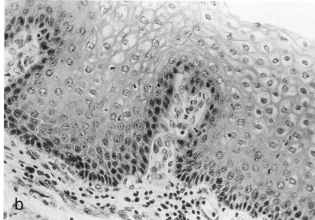

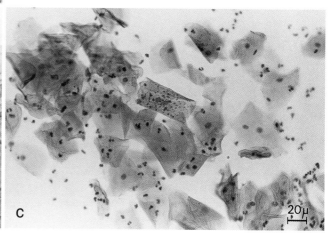

Fig. 12-3 Histology of esophageal mucosa and its normal cellular constituents.

a. Histology of the esophagus covered with stratified squamous epithelium. The esophageal glands proper (EG) locate in the submucosa (Sm). SqE: stratified squamous epithelium, Mm: muscularis mucosae.

b. Higher magnification of stratified squamous epithelium. Note absence of cornification.

c. Benign cellular components in esophageal washings. Superficial and intermediate squamous cells are seen.

Procedure: The gelatine made capsule is solved within the stomach in several minutes after swallowing. Then the elastic sponge blown like a ball should be gently taken out, especially when it passes through the esophageal-cardiac junction with some resistance. Smears are made by pressing the sponge on clean glass slides. Rinsing the sponge with saline and centrifugation may ensure cell collection.

2 Histology of the Esophagus

The mucous membrane is covered with stratified squamous epithelium which abruptly changes to simple columnar epithelium as it meets the gastric cardia. The superficial layer does not undergo true cornification. There are two kinds of glands in the esophagus: (1) The esophageal glands proper, i.e., tubuloalveolar mucous glands located in the submucosa, and (2) the cardiac glands, which are located at levels below the larynx and above the esophageal-gastric junction, are lined with low columnar cells and present in the *lamina muscularis mucosa*. Solitary lymph follicles are present in the *tunica propria mucosa*.

3 Benign Cellular Components

The normal cells shed by esophageal washing or scraping are squamous cells of various layers, although superficial cells and intermediate cells are more common than cells from the deeper layers. Since the squamous epithelium lacks true cornification, these cellular components are quite similar to those in vaginal smears (Fig. 12-3c).

Fig. 12-4 Benign hyperplasia in chronic esophagitis.
a. Basal cell and columnar cell hyperplasia in association with chronic esophagitis. Note multilayered basal cells with nuclear enlargement and nucleolar prominence.
b. Hypertrophic basal cells seen in a sheet from brushing cytology. Oval or elliptic nuclei are enlarged and possess prominent nucleoli.
c. Sheet of hypertrophic columnar cells from brushing cytology. The nuclei are overall oval in shape but vary in size. Broad, vesicular cytoplasm arranged in a honeycomb appearance will be suggestive of columnar epithelium origin. Columnar epithelial-lined esophagus originated from the stomach (Barrett's esophagus) may shed columnar cells in sheet.

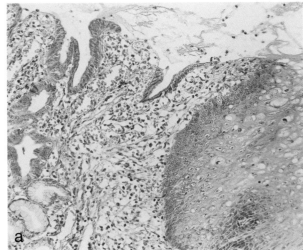

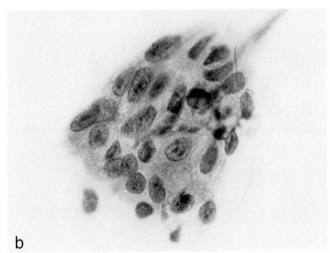

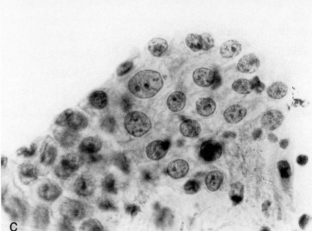

4 | Benign Lesions

■ Esophagitis (Reflux esophagitis)

Smears prepared from lesions of ulcerative esophagitis show no diagnostic cellular features except for general appearance of the smear such as abundance of degenerative leukocytes or cellular debris, admixture with histiocytes and presence of microorganism. There may appear some hypertrophic prickle cells and/or cardiac columnar cells in a cluster.

■ Hyperplasia (Prickle cell hyperplasia)

Multilayering of deep prickle cells with hypertrophic nuclei is known as prickle cell hyperplasia. It may occur as its own disease entity or in association with esophagitis and pernicious anemia. Hyperplasia of columnar epithelium at the esophageal-cardiac junction is associated.

Cytologically, sheets of parabasal cells showing hypertrophy and hyperchromatism of nuclei may mislead unexperienced observer to the diagnosis of cancer. Differentiation of benign hyperplasia from carcinoma is based on regularity of cell arrangement and uniformity of nuclear size and shape. In spite of nuclear hyperchromatism and nucleolar prominence, uneven distribution of chromatins is not present and nucleoli are usually single and centrally located (Fig. 12-4).

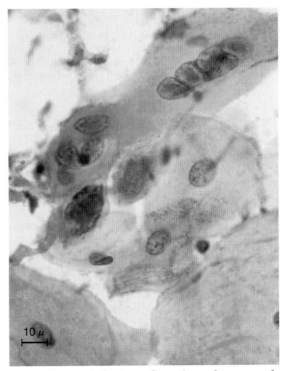

Fig. 12-5 Herpes infection of esophageal mucosa showing acidophilic intranuclear inclusions and multinucleation. This 61-year-old patient is debilitated because of irradiation (3,000 rad) and bleomycin (15 mg) therapy. Nuclear molding is characteristic of herpes virus infestation.

251

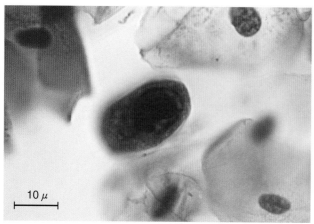

Fig. 12-6 Cytomegalovirus infection affecting the cardiac mucosa in esophageal brushings. An enlarged cell having a giant intranuclear inclusion with "owl-eyed appearance" is derived from the affected surface or foveolar epithelium of the stomach. This 48-year-old women who has been intensively treated with prednisone and antibiotics showed multiple stress ulcers.

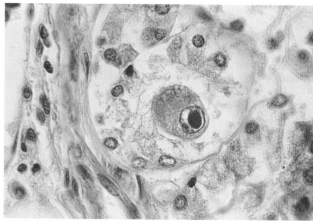

Fig. 12-7 Histological proof of cytomegalovirus infection affecting the glandular epithelium of the cardia. Columnar epithelial cells of the glandular or parenchymatous organ are apt to be affected. Note an "owl-eyed appearance" around the large amphoteric inclusion body in association with pronounced cell enlargement. (By courtesy of Miss K. Tsuzaki, C.T., IAC, and Dr. K. Kotoh, Osaka Prefectural Hospital, Osaka.)

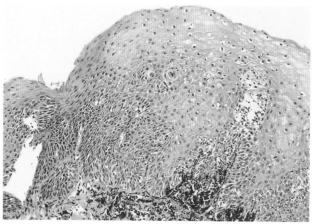

Fig. 12-8 Dysplasia of esophageal mucosa. Note thickening of the mucosal epithelium by hypertrophy of the deep prickle cells (parabasal cells). The superficial layer with slight dyskaryosis will be submitted to scraping cytology.

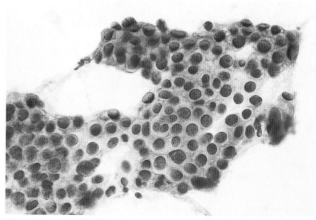

Fig. 12-9 Mild dysplastic cells in large sheet obtained by esophageal brushing. The nuclei are enlarged and increased in chromatin amount. However, these are regularly arranged and display a honeycomb appearance.

■Herpetic esophagitis

Herpetic infection is often diagnosed at postmortem examination in patients who have been treated with immunosuppressive therapy. Antemortem diagnosis can be made by brushing cytology[19] or biopsy[36] during esophagoscopic examination. Recently attention for mucosal vesicular lesions in association with skin manifestation of herpes zoster infection was strongly emphasized by Wisloff et al.[37] Endoscopic examination for the patients with non-radicular thoracic pain should be performed to diagnose mucosal involvement of the digestive tract by herpetic infection.[2] Brush cytology is worthy to be applied from two reasons: (1) Sampling affected epithelial cells can be done more widely with brush than with biopsy forceps. (2) Gentle maneuver of brushing does not give rise to a grave side effect like perforation of thin esophageal wall.

Cytology: Enlargement of nuclei showing multinucleation with molding, intranuclear inclusion and smoky or ground glass appearance of the nuclei are exactly the

same irrespective of affected sites of herpes infection (Fig. 12-5).

■Cytomegalovirus infection

Cytomegalovirus infection occasionally involves epithelial cells of the gastrointestinal tracts[6,26] especially in deteriorated patients with immunosuppressive state (Figs. 12-6 and 12-7). These are also found in the endothelial cells or mesenchymal cells at the site of gastric ulcer[6,26]; the presence of macrophages containing the inclusions after blood transfusion may explain their localization in ulcers. It should be kept in mind that whereas herpes virus inclusions are usually found in the epithelial cells of stratified squamous epithelium, cytomegalovirus inclusions can be found in vascular endothelial and mesenchymal cells.

■Candidiasis

Pseudomycelia of candida species are commonly seen in the immunosuppressive patients who complains of

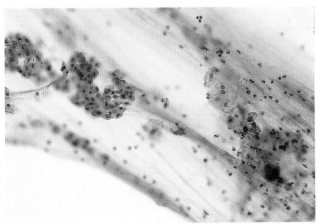

Fig. 12-10 Well differentiated squamous cell carcinoma shed as single cells and in cell sheets. Capsulated sponge method. Note a keratinizing single cell with orangeophilic cytoplasm. Nonkeratinizing cells are arranged side by side.

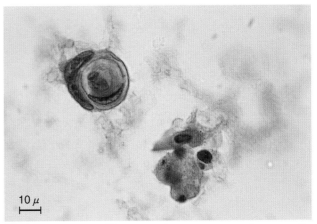

Fig. 12-11 Well differentiated squamous carcinoma cells with pearl formation. The cancer cells in the left upper corner are arranged in a whorl pattern. The chromatin is increased in amount and coarsely clumped. A keratinizing cancer cell with pyknosis is in the right lower corner.

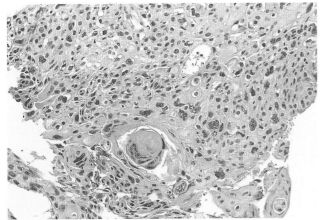

Fig. 12-12 Keratinizing squamous cell carcinoma consisting of large solid nest of polymorphic cancer cells. Note keratinizing pearl formation.

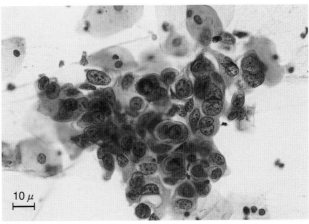

Fig. 12-13 Well differentiated nonkeratinizing squamous cell carcinoma cells shed in a cluster by esophagoscopic scraping. Note anisonucleosis, distinct nuclear membrane and marked hyperchromatism with coarse chromatin clumping. The cytoplasm is intensely stained by Light Green.

dysphagea.[9] Pinkish stained filamentous pseudohyphae that resemble bamboo canes and budding spores can be found alongside squamous cells.

5 Precancerous Lesions of the Esophagus

Since the majority of esophageal carcinomas are of the epidermoid type, Bowenoid change and carcinoma in situ (intraepithelial carcinoma) may be precursors of invasive squamous carcinoma, as is the case in other sites lined with stratified squamous epithelium. It has been reported that carcinoma in situ is present adjacent to invasive carcinoma,[24,30-32] but few studies have been carried out on precancerous dysplasia or carcinoma in situ without association of invasive carcinoma (Figs. 12-8 and 12-9).[3,17,35]

6 Malignant Cells

Cancer of the esophagus develops insidiously at the onset without symptoms. When the patient complains of dysphagia, the tumor is already in invasive stage. Rapid invasion beyond the esophageal wall is enhanced by the lack of serosa. Berry et al[4] found 15 cases with cancer for 500 asymptomatic persons using inflatable balloon catheter.

Epidemiologically cancer of the esophagus accounts for about 2% of malignant neoplasia and fairly common in northern China, northeast of Caspian countries and Southeast of Africa.[29]

Carcinoma of the esophagus occurs with about equal frequency in the areas of physiological narrowing; at the beginning of the esophagus, the bronchial bifurcation and the esophageal cardiac junction. Grossly, the tumor has a polypoid, penetrating or infiltrative form. About 95% or more of esophageal carcinomas are of epidermoid type microscopically. Adenocarcinoma accounts for about 3% to 10%[7] and most are located at the cardia, where the gastric glands are present normally or as heterotopic tissue. The developmental variety known as Barrett's esophagus is the site of higher frequency of adenocarcinoma. Some adenocarcinoma reveals metaplastic differentiation to a squamous variety, and is referred to

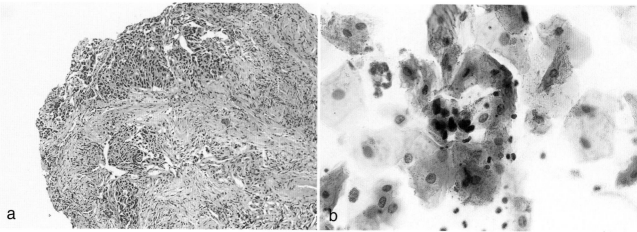

Fig. 12-14   Poorly differentiated squamous cell carcinoma.
a. Hyperchromatic, polymorphic cancer cells with increased N/C ratio infiltrate deeply as solid nests or cords. Dense desmoplasia is associated.
b. The nuclei are markedly hyperchromatic like Indian ink droplets. The cytoplasm is scanty. Capsulated sponge method prior to surgery.

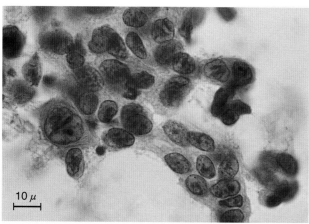

Fig. 12-15   Poorly differentiated squamous cancer cells in sheet by esophageal brushing. The cytoplasm is scant and indistinctly bordered because of poor differentiation. Diagnostic point: (1) These cells are arranged in a flat sheet. (2) The nuclear rims are thickened and rigid.

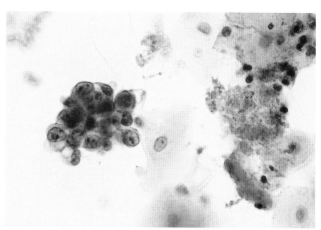

Fig. 12-16   Adenocarcinoma cells aggregated in tight cluster. Capsulated sponge method. Note three dimensional cell cluster showing prominent nucleoli and cytoplasmic vacuolation.

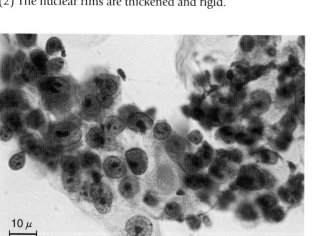

Fig. 12-17   Poorly differentiated adenocarcinoma cells seen in clusters by esophagoscopic brushing. The nuclei are enlarged, hyperchromatic and possess prominent nucleoli. The cytoplasm is scant and ill-defined. Diagnostic points: (1) Cell overlapping, (2) thin nuclear rim, and (3) prominence of nucleoli with acidophilic hue are suggestive of the glandular type.

as adenoacanthoma.

■ Squamous cell type (Epidermoid type)

Differentiated epidermoid type is characterized by tadpole cells, fiber cells, snake cells and third-type cells termed by Graham[12] (Figs. 12-10 to 12-13). These characteristic cells are exactly the same as those described in the chapters on carcinoma of the uterine cervix and carcinoma of the lung. The poorly differentiated type reveals distinctive features such as scantiness and pale cyanophilia of cytoplasm, marked increase in N/C ratio, and marked hyperchromatism of nuclei with a coarsely granular chromatin pattern (Figs. 12-14 and 12-15).

■ Glandular type

The glandular type of malignant cells (adenocarcinoma cells) tends to form cell clumps with nuclear overlapping, eccentric position of nuclei and cytoplasmic vacuolation. The nuclei are round to oval with delicate, smooth outlines and possess prominent nucleoli (Figs. 12-16 and 12-17). Chromatins in adenocarcinoma are not so coarse as in epidermoid carcinoma. The majority

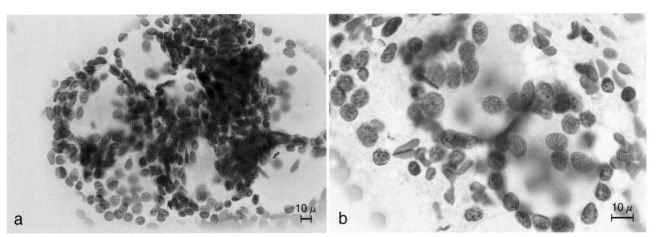

Fig. 12-18 Cylindroma arising from the upper part of esophagus.
a. There is a characteristic cell arrangement with cribriform texture; equally sized and shaped nuclei are arranged in such circular strands that luminal spaces are formed.
b. Higher magnification of cylindroma. These denuded nuclei which are uniformly oval and homogeneously stained resemble basalioma of the skin. In spite of scarce cell atypism this type tumor undergoes recurrent and invasive growth.

Fig. 12-19 Adenoacanthoma in the upper portion of the esophagus.
a. Cytology of esophagoscopic scraping smear. Whereas enrounded nuclei having droplet-like eosinophilic nucleoli are likely to be of adenocarcinoma, thickened cytoplasm with amphoteric staining reaction and slit-like intercellular space are suggestive of epidermoid type.
b, c. Histology of adenoacanthoma. A glandular pattern of cancer cells is shown in the left upper side (Ad), while distinctive squamous differentiation with intercellular bridges is shown in solid cancer cell nest (Sq).

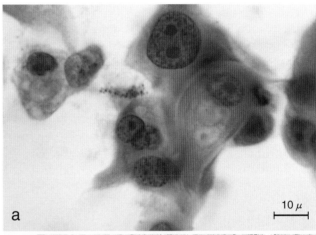

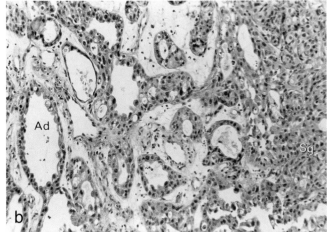

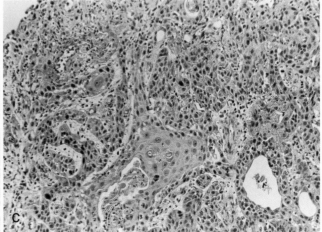

of adenocarcinomas are derived from the distal thoracic segment and the esophageal cardiac junction; they are considered to be of gastric origin and to be subepithelially extended.

Occasionally columnar cells of Barrett's esophagus is the origin of adenocarcinoma.[14] Papillary type adenocarcinoma originating from Barrett's esophagus is known to have cilia at the cellular surface.[28]

Adenoid cystic carcinoma or cylindroma is of a rare occurrence in the esophagus and shows histologic features identical with those seen in the salivary gland (Fig. 12-18a,b).

■Adenoacanthoma

This type is metaplastic in nature. Foci of epidermoid differentiation are found in adenocarcinoma cell nests. Cytology of adenoacanthoma is characterized by differentiation towards both epidermoid and adenocarcinomas. The cytoplasm becomes broad and thickened owing to squamous metaplastic change, while its nucle-

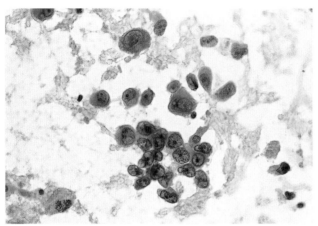

Fig. 12-20   Melanoma cells obtained by esophagoscopic scraping. The cytoplasm intensely stained by Light Green reveals brownish tint. Note characteristic prominent nucleoli.

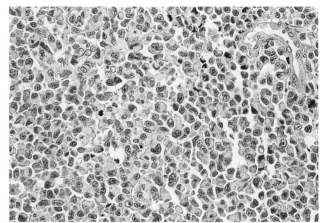

Fig. 12-21   Melanoma consisting of polymorphic melanoma cells in sheet. Light brownish melanin pigments can be recognized in moderate amount of cytoplasm. Prominent nucleoli are characteristic.

us retains the feature as shown in adenocarcinoma; the nuclei are round to oval, smoothly outlined and possess single, centrally locating prominent nucleoli (Fig. 12-19).

■Small cell carcinoma

Argyrophilic, Grimelius-positive small cell carcinoma is a rare carcinoma that is similar to oat cell carcinoma of the lung.[16] The patient with small cell carcinoma takes a fetal downhill course. The median survival was reported only 4 months in the report of 87 cases by Tennvall et al.[33]

7 Malignant Melanoma

The primary melanoma has been doubted because of the lack of melanoblasts in the esophagus. The rare melanoma shows a polypoid tumor in the lower part of the esophagus in elderly persons. Cytology shows the same feature as in other sites of the body (Figs. 12-20 and 12-21). Absence of metastatic melanoma must be carefully proved.

SMALL AND LARGE INTESTINES

Out of three segments of the small intestine, the duodenum is the target organ of exfoliative cytology because excretes from the papilla Vateri may contain diagnostic cells from the biliary ducts and the pancreatic ducts. Percutaneous needle aspiration prevails conventional methods to diagnose the tumor of the pancreatic head (see Chapter 13). However, pancreatic excretions are still worthy of use.

Colorectal cancers are common and mostly preceded by villous or tubular adenomas. Histological types with marked grades of atypicality and the size over 2.0 cm have more malignant potential. Screening using Guaiac-based fecal tests are widely used followed by further investigations such as double-contrasted barium enema

and colonoscopy. Biopsy histology is the most common procedure to confirm the malignancy.

1 Collection of Specimens and Preparation of Smears

■Choledocho-pancreatic duct cytology

1) Pancreozymin secretin method

Duodenal excrete induced by intravenous injection of Pancreozymin is useful as a routine technique of choledocho-pancreatic duct cytology. The procedure is performed as follows: (1) In early morning before breakfast, let the patient lie down on the right side and intubate a double lumen gastroduodenal tube, which is designed to obtain gastric and duodenal juice separately. (2) Inject slowly Pancreozymin (1 unit/kg body weight) intravenously and collect duodenal juice for 10 minutes (No. 1 tube). (3) Inject secretin (1 unit/kg) intravenously and aspirate the juice for 10 minutes (No. 2 tube). (4) Collect duodenal juice continuously into a tube for another 10 minutes (No. 3 tube) and into two more tubes each for 20 minutes (No. 4 and No. 5 tubes). These tubes must be placed in a basin with ice cubes during collection. (5) Immediately after collection, centrifuge the specimen at 1,500 rpm for 5 minutes and make smears for cytology.

2) Endoscopic aspiration and washing

Duodenal juice excreted after intravenous injection of secretin (1 unit/kg) is aspirated directly through an intubated vinyl tube under endoscopic vision.

Washing the biliary duct and/or pancreatic duct through the intubated vinyl tube with about 10 ml physiologic saline solution is a useful technique to obtain fresh specimen from the lesion.

3) Method combined with cholangiography

Cytology combined with endoscopic retrograde cholangio-pancreatography (ERCP)[8,15] and with percutaneous transhepatic cholangiography bring forth satisfactory results[1] since these are able to obtain mucosal cells directly from the bile duct without contamination from other parts of the digestive tract.

■Colon cytology

1) Endoscopic scraping

Mucosal secretions obtained at proctosigmoidoscopy are excellent specimens.[38] Although there is a limit to the reach of the sigmoidoscope, nearly 70% of colon carcinomas are located in the portion of the rectosigmoid within the reach of sigmoidoscopic examination. The lesion is scraped with a cotton swab applicator. The cellular elements are transferred onto clean glass slides by carefully rolling the cotton swab across the slides or lightly tapping the cotton swab against the slides.

2 Histology of the Intestine and Its Benign Cellular Components

■Histology of the intestine

1) Small intestine

The small intestine is composed of the duodenum, jejunum and ileum. The histological structure is somewhat different in each part but essentially the same. The mucosal surface discloses delicate processes called intestinal villi, to the roots of which the intestinal glands open (crypts of Lieberkühn). The Brunner's glands in the duodenum are located in the submucosa and their ducts pass through the *muscularis mucosae* into the pits. In jejunum and ileum, the intestinal glands, 320-450 μm long, are present in the mucosa and lined by low columnar cells producing chiefly mucus and some enzymes. The *tunica propria* is composed of a reticular framework of connective tissue with lymphatic channels and shows some infiltration of polymorphonuclear leukocytes, plasma cells, lymphocytes, etc. The lymphoid apparatus is well developed and solitary follicles and Peyer's patches are formed. In the bottom of the crypts of Lieberkühn are the Paneth cells with eosinophilic granules. The surface epithelium is composed of goblet cells and slender columnar cells with striated borders which may function in the absorption of water and digested products. Alkaline phosphatase activity is strongly positive at striated borders.

2) Large intestine

The mucosa of the large intestine lacking villi is quite smooth, but retains crypts. The surface epithelium is lined with a single layer of columnar cells with striated borders intermingled with goblet cells. The crypts are straight, perpendicular to the surface and longer than those of the small intestine. There are so many goblet cells that they are called intestinal mucous glands (crypts of Lieberkühn). The Paneth cells are not seen.

■Benign cellular components

1) Duodenal epithelial cells

The duodenal mucosal cells are shed in regular sheets with a honeycomb appearance if seen from above. In lateral aspect, the cells are columnar with parallel or polar-positioned nuclei. Isolated columnar cells have slender cytoplasm that stains green with Papanicolaou stain. Striated borders are distinctive at the broad cellular boundary. The nuclei are oval and quite uniform in shape and size. Although the chromatin of well preserved cells is finely granular and evenly distributed, degenerate epithelial cells which are frequent in duodenal aspirates reveal liquefied or fragmented pyknotic nuclei.

2) Colonic epithelial cells

Although the colonic mucosal cells obtained by the washing technique are poorly preserved, denuded or liquefied, the cells obtained by scraping under endoscopy are shed in sheets or clusters (Fig. 12-22a). The cytoplasm has a long slender shape. Basally situated nuclei are oval with fairly uniform size and possess finely granular chromatins.

3) Goblet cells

The cytoplasm is distended with mucus, and appears transparent. The nuclei are extremely peripheral in position. Usually they are not found singly but recognized within clusters of columnar epithelial cells.

4) Biliary tract epithelial cells

The mucosal epithelial cells from the gallbladder and biliary ducts disclose a characteristic long slender cytoplasm which is lacking in the striated border. The cells tend to degenerate by bile, but well preserved cells taken by endoscopic aspiration retain transparent columnar cytoplasm and basal ovoid nuclei. The cells derived from the pancreatic duct have low columnar cytoplasm. Otherwise they are hardly differentiated from the biliary duct cells (Figs. 12-23 and 12-24).

■Non-cellular components

Contaminants from food particles are found in aspirates or washings of the digestive tract. Protozoa and eggs of intestinal parasites are also encountered in cytology.

1) *Giardia lamblia*[21]

Giardia lamblia are of common protozoal infection of the intestinal tract. It is usually nonpathogenic but occasionally gives rise to mild symptoms such as diarrhea, anorexia and cramp-like abdominal pain. In children malabsorption syndrome with steatorrhea and hypoproteinemia may occur. The trophozoite that appears in biliary drainage is pear-shaped, measuring about 10-20 μm in length by 5-10 μm in width, and possesses two nuclei, two median bodies and four pairs of flagella. The nuclei locate in the area of the sucking disk. The cyst measures about 8-14 μm in length by 7-10 μm in width and possesses four prominent nuclei eccentrically and four median bodies (Fig. 12-25).

2) Eggs of liver fluke[21]

Eggs in infection of various types of liver fluke are found in biliary drainage and feces; representative liver flukes are *Clonorchis sinensis* (*Opisthorchis sinensis*) occurring in Asia (Fig. 12-26), *Opisthorchis viverrini* occurring in Thailand, *Dicrocoelium dendriticum* occurring in many parts of the world, etc. Human infection occurs by way of fresh-water fish containing the encysted metacercariae. The egg of *Clonorchis sinensis* measuring about 29 by 16 μm has a moderately thick light yellowish brown shell and a distinct convex operculum. Its abopercular end is provided with a small comma-shaped process. The egg of *Dicrocoelium dendriticum* measuring about 40 by 25 μm shows a thick golden brown shell and a large operculum.

Cytopathologically it should be kept in mind that marked hyperplasia of the small mucinous glands of the duct occurs in infection in association with dilatation

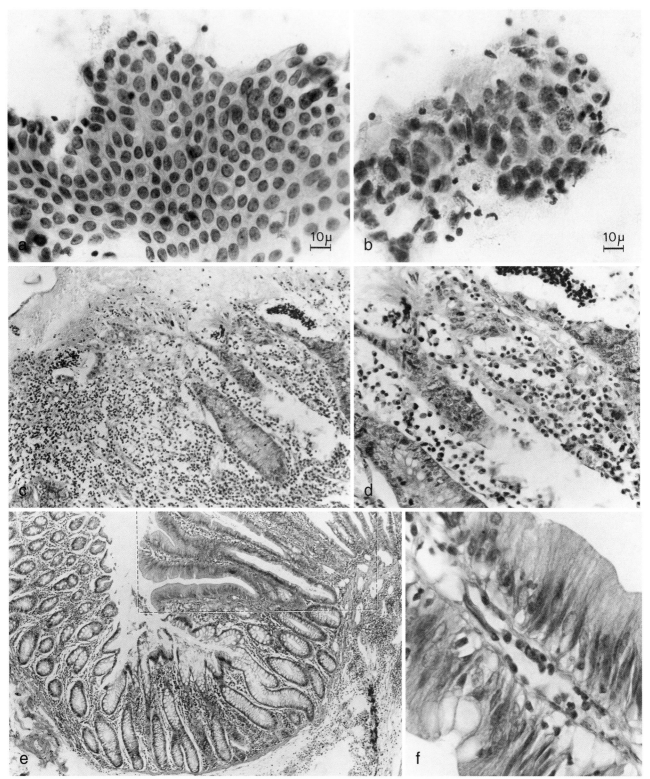

Fig. 12-22　Normal and benign epithelial changes of the colon.
a. Sheet of normal mucosal epithelial cells with a regular honeycomb appearance in a scraping smear of the sigmoid colon. The nuclei are round or oval, uniform in size, and show evenly smooth distribution of chromatins.
b. Sheet of hyperplastic mucosal epithelial cells from a case with nonspecific ulcerative colitis. The nuclei are pleomorphic, irregular in arrangement, and densely stained.
c, d. Histology of nonspecific ulcerative colitis. Necrotic foci with edema and infiltration by neutrophils, mononuclear cells and plasma cells are seen. The glands adjacent to the ulcer show frequent mitotic figures and multilayered nuclei.
e, f. Benign adenomatous polyp (villous adenoma) of the colon. The mucosal epithelial cells of the lesion are taller, slender and more crowded than normal epithelial cells. The "needle cells" may exfoliate from such lesion. Goblet cells may be absent or few in polyp.

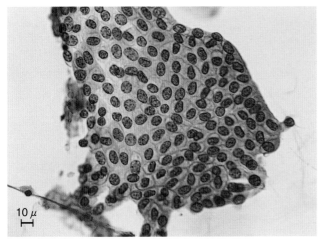

Fig. 12-23   Well preserved mucosal cells of the common biliary duct in duodenal aspirates under direct vision. Normal columnar cells seen from the above are arranged in a honeycomb appearance. The nuclei are uniform in size and arranged regularly.

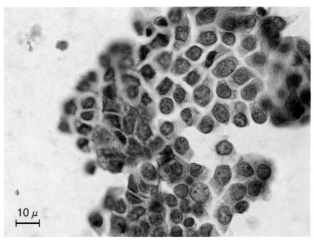

Fig. 12-24   Hyperplastic mucosal cells with atypia of the biliary duct. Aspiration smear through the Vater's papilla. The nuclei are enlarged and vary in size, but arranged orderly. From a dysplastic lesion.

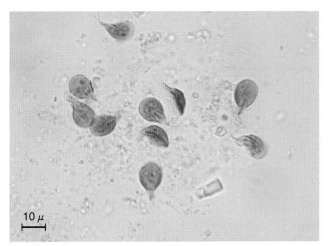

Fig. 12-25   Trophozoites of *Giardia lamblia* in bile by Lugol stain. Pear-shaped trophozoites have two nuclei lying in the area of the sucking disk. Four pairs of flagella are recognizable in well preserved trophozoites.

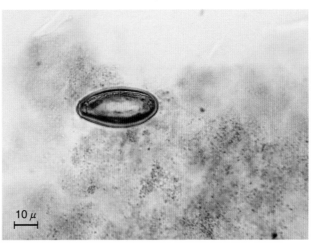

Fig. 12-26   Egg of *Clonorchis sinensis* in duodenal aspirates. Note a thickened yellowish brown shell and distinct operculum.

and thickening of the biliary duct.

3) *Entamoeba histolytica*[21]

Infection of *Entamoeba histolytica* is widespread in the world, especially in unsanitary areas. The large intestine is inhabitable; the trophozoites multiply in the crypts and give rise to ulceration which is referred to as amebic dysentery. The trophozoites are variable in size and measure about 12-60 μm in diameter. A histiocyte-like ill-defined trophozoite contains small ovoid nucleus with thick chromatin attached to the nuclear membrane and a prominent karyosome. Erythrocytes and bacteria are often ingested within the cytoplasm. Cysts are usually spherical in shape, measuring about 10-20 μm in diameter, and provided with hyaline, refractile walls. They contain one to four round nuclei and often chromoidal bars which stain bluish by hematoxylin like chromatin. The cytoplasm stains light yellowish brown with iodine staining (Fig. 12-27).

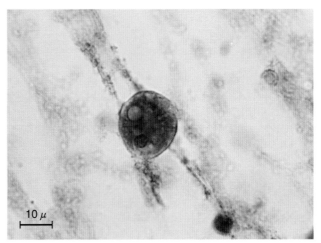

Fig. 12-27   Trophozoites of *Entamoeba histolytica* in a resistant pre-cyst stage. A scraping smear from ulcerative colon mucosa. The pre-cyst is characterized by enrounded form, round single nucleus, absence of hyaline cyst wall, etc. There is degenerative vacuolation in the cytoplasm. This 57-year-old male has been feverish and has had liver abscess. (By courtesy of Dr. K. Kotoh, and Miss K. Tsuzaki, CT, IAC, Osaka Prefectural Hospital, Osaka, Japan.)

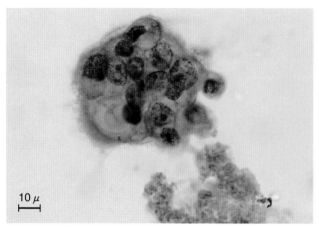

Fig. 12-28　Adenocarcinoma cells lying in cluster in duodenal aspirates by intubation. The nuclei are enlarged, enrounded and eccentric in columnar cytoplasm, which is clear and often vacuolated. Bile pigments are seen. Note prominent acidophilic nucleolus.

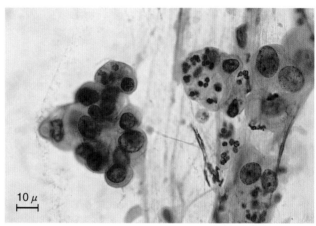

Fig. 12-29　Adenocarcinoma cells seen in a tight cluster arising from the pancreatic duct. Aspiration smear through the endoscopic intubation. These well preserved cancer cells are markedly overlapping, tightly clustered and vacuolated.

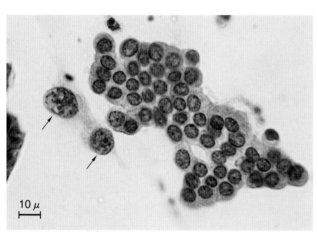

Fig. 12-30　Poorly differentiated carcinoma cells isolated singly and a sheet of normal surface mucosal cells. Normal mucosal cells are arranged in a regular honeycomb pattern and retain well preserved cytoplasm. Note undifferentiated cancer cells shedding singly and showing scant cytoplasm (arrows).

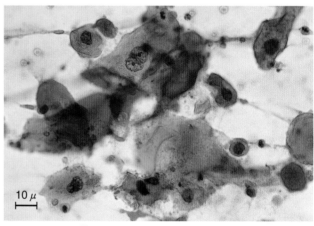

Fig. 12-31　Differentiated epidermoid cancer cells arising from the gallbladder. In about 5% of carcinoma of the biliary tract the tumor growth takes a squamous metaplastic pattern. Note keratinizing epidermoid cancer cells with India ink droplet-like pyknotic nuclei.

3　Benign Atypical Cells

Benign atypical cells often originate from the lesions of diverticulitis, chronic ulcerative colitis and adenomatous polyp (tubular adenoma). The columnar cells are hyperplastic with slight nuclear enlargement and hyperchromatism. The N/C ratio is not markedly changed and the nucleoli are usually not prominent.

■Chronic ulcerative colitis

In advanced cases, longitudinal ulcers are adjacent to hyperplastic mucosa, occasionally adenomatous hyperplasia. Galambos and Klayman[10,11] described two kinds of cells characteristic of the disease; large active cells and large bland cells. The large active cells originating in actively proliferating cells at the base of crypts of Lieberkühn show an increase in nuclear size and chromatin content and coarse distribution of chromatin; however, the cytoplasm is also swollen and the N/C ratio is not greatly altered. The bland cells, which are

referable to hydropic degeneration,[34] have vesicular, pale and homogeneous nuclei with vacuolated cytoplasm.

■Adenomatous polyp

There are two of cells arising from benign polyps[34]: (1) needle type with centrally located, hyperplastic nuclei and slender cytoplasm (tubular adenoma), and (2) fan type, less common than the former, with a broad vesicular cytoplasm and basal nuclei. The fan type epithelial cells are considered to be derived from mucus-secreting cells.

4　Malignant Cells

■Carcinoma of biliary tract and pancreatic duct

Carcinomas of the bile duct are relatively rare compared with carcinomas of the gallbladder and pancreas that often involves the common duct. Most of carcinomas arising from the choledocho-pancreatic duct are tubular

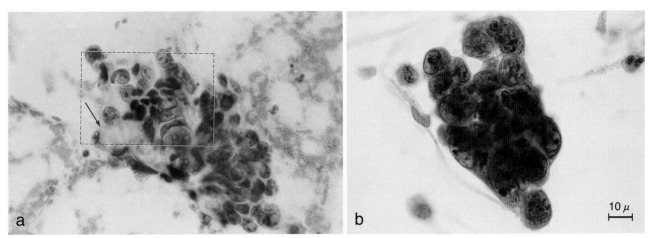

Fig. 12-32   Large clusters of tubular adenocarcinoma from the rectum.
a. There is marked variation in nuclear size and shape. Note prominence of nucleoli and hyperchromatism with coarse clumping. A few malignant signet ring cells are intermingled within the cell cluster (arrow). An arrangement of cancer cells side by side is characteristic of the tubular type adenocarcinoma (frame).
b. Well preserved cancer cells tightly clustered and disorderly arranged. Note high N/C ratio, anisonucleosis and hyperchromatism.

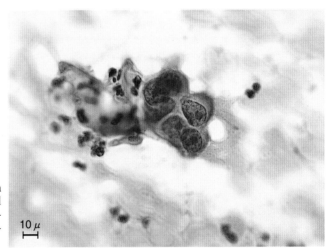

Fig. 12-33   Tubular adenocarcinoma in scrapings from the rectum under sigmoidoscopy. Formation of tight cell cluster, high N/C ratio, hyperchromatism with coarse chromatin pattern and prominent acidophilic nucleoli are diagnostic of common tubular adenocarcinoma.

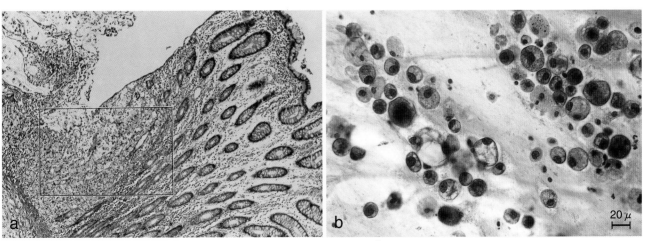

Fig. 12-34   Histology and cytology of mucinous adenocarcinoma of rectum.
a. Histology. See abrupt change of normal mucosa to mucocellular adenocarcinoma massively infiltrating (frame). Although this rare type of carcinoma in the colon is functionally differentiated, distant metastases are frequently encountered.
b. Cytology. Well preserved signet ring cancer cells are seen singly in sediments of rinsings of a brush. The nuclei varying in size are displaced to the periphery by full of mucous content.

adenocarcinomas (Figs. 12-28 to 12-30), whereas adenoacanthoma and epidermoid carcinomas occasionally occur presumably as a result of squamous metaplasia (Fig. 12-31). According to the report by Gupta et al[13] on 328 cases with carcinoma of the gallbladder, epidermoid carcinoma was seen only in 6.4% and adenosquamous carcinoma in 2.4%. It should be remembered that carcinoma of the gallbladder is about 3-4 times as frequent in women as in men,[13,22] but carcinomas of the bile duct and the pancreas are slightly more frequent in men than women.

■Carcinoma of intestines

Carcinoma of the small intestine is rare and accounts for about 2.5% of all intestinal carcinomas.[18] It is located almost in the duodenum or upper portion of the jejunum. On the other hand, carcinoma of the large intestine is quite common and comprises about 10% of all malignant tumors. According to the study by Buser et al[5] concerning the relative frequency of carcinoma in various segments of the colon, 71.1% of 478 tumors were located in the rectosigmoid, 22 cases (4.6%) in the ascending colon, 33 cases (6.9%) in the cecum, 11 cases (2.3%) in the hepatic flexura, 18 cases (3.8%) in the transverse colon, 8 cases (1.7%) in the lienal flexura, 45 cases (9.4%) in the descending colon, 65 cases (13.6%) in the sigmoid, and 276 cases (57.7%) in the rectosigmoid and anus.

Histologically, carcinoma of the colon is fairly well differentiated with a tubular or glandular pattern lined by tall columnar cancer cells (adenocarcinoma tubulare). Another common type is mucinous or colloid carcinoma characterized by marked production of mucus intracellularly (adenocarcinoma mucocellulare) as well as extracellularly (adenocarcinoma muconodulare).

Cytologically, malignant cells form tubular cell clusters with irregular nuclear overlappings (Figs. 12-32 and 12-33). A typical picture of this type shows a large trabecular clumping of cells, in which multiplied but parallel arrangement of the nuclei can be characteristically recognized. The nuclei are elongated, enlarged (high nuclear-cytoplasmic ratio), hyperchromatic, and vary in size. The shape variation is not so marked as size variation. They are usually elliptic in shape. Two or three prominent nucleoli are also distinctive in adenocarcinoma cells. It is also a noteworthy finding that these cells are interwoven with some malignant signet ring cells. Shed from mucinous carcinoma are many scattered, signet ring cells in isolated form or tightly clustered form. The latter form is commonly observed in scraping smears (Fig. 12-34).

References

1. Akashi M, Uchimura M, Hayashida Y: Studies on the technique and the results of the cytologic examination of bile. J Jpn Soc Clin Cytol 10: 43, 1971.
2. Artigas JMG, Saumell CB, Faure RA, et al: Herpes Zoster of upper gastrointestinal tract. Lancet 2: 43, 1980.
3. Auerbach O, Stout AP, Hammond EC, et al: Histologic changes in esophagus in relation to smoking habits. Arch Environ Health 11: 4, 1965.
4 Berry AV, Baskind AF, Hamilton DG: Cytologic screen-

ing for esophageal cancer. Acta Cytol 25: 135, 1981.
5. Buser JW, Kironer JB, Palmer WL: Carcinoma of the large bowel: Analysis of clinical features in 478 cases, including eighty-eight five-year survivors. Cancer 3: 214, 1950.
6. Campbell DA, Piercey JRA, Shnitka TK, et al: Cytomegalovirus-associated gastric ulcer. Gastroenterology 72: 533, 1977.
7. Dee AL, Hanssen B: An experience in the cytopathologic diagnosis of carcinoma of the esophagus. Acta Cytol 7: 236, 1963.
8. Endo Y, Morii T, Tamura H, et al: Cytodiagnosis of pancreatic malignant tumors by aspiration, under direct vision, using a duodenal fiber scope. Gastroenterology 67: 944, 1974.
9. Eras P, Goldstein MJ, Sherlock P: Candida infection of the gastrointestinal tract. Medicine 51: 367, 1972.
10. Galambos JT, Klayman MI: The clinical value of colonic exfoliative cytology in the diagnosis of cancer beyond the reach of the proctoscope. Surg Gynecol Obstet 101: 673, 1955.
11. Galambos JT, Massey BW, Klayman MI, et al: Exfoliative cytology in chronic ulcerative colitis. Cancer 9: 152, 1956.
12. Graham RM: The Cytologic Diagnosis of Cancer, 2nd ed. Philadelphia: WB Saunders, 1963 (3rd ed, 1972).
13. Gupta S, Udupa KN, Gupta S: Primary carcinoma of the gallbladder: A review of 328 cases. J Surg Oncol 14: 35, 1980.
14. Haggit RC, Tryzelaar J, Ellis FH, et al: Adenocarcinoma complicating columnar epithelium lined (Barrett's) esophagus. Am J Clin Pathol 70: 1, 1978.
15. Hatfield ARW, Smithies A, Wilkins R, et al: Endoscopic retrograde cholangiopancreatography (ERCP) and pure pancreatic juice cytology: A combined diagnostic approach in pancreatic disease. J Br Soc Gastroenterol 16: 405, 1975.
16. Hoda SA, Hajdu SI: Small cell carcinoma of the esophagus: Cytology and immunohistology in four cases. Acta Cytol 36: 113, 1992.
17. Imbriglia JE, Lopusniak MS: Cytologic examination of sediment from the esophagus in a case of intra-epidermal carcinoma of the esophagus. Gastroenterology 13: 457, 1949.
18. Karsner HT: Human Pathology. p590. Philadelphia: JB Lippincott, 1949.
19. Lightdale CJ, Wolf DJ, Marucci RA, et al: Herpetic esophagitis in patients with cancer: Ante mortem diagnosis by brush cytology. Cancer 39: 223, 1977.
20. Malhotra V, Puri R, Chinna RS, et al: Endoscopic techniques in the diagnosis of upper gastrointestinal tract malignancies: A comparison. Acta Cytol 40: 929, 1996.
21. Markell EK, Voge M: Medical Parasitology. Philadelphia: WB Saunders, 1976.
22. Nevin JE, Moran TJ, Kay S, et al: Carcinoma of the gallbladder. Cancer 37: 141, 1976.
23. O'Donoghue JM, Horgan PG, O'Donohoe MK, et al: Adjunctive endoscopic brush cytology in the detection of upper gastrointestinal malignancy. Acta Cytol 39: 28, 1995.
24. O'Gara RW, Horn RC Jr: Intramucosal carcinoma of the esophagus. Arch Pathol 60: 95, 1955.
25. Onozawa K, Nabeya K: Abrasive cytology with capsel for esophageal cancer. Jpn J Gastroenterol Surg 10: 1, 1977.
26. Rosen P, Armstrong D, Rice N: Gastrointestinal

cytomegalovirus infection. Arch Intern Med 132: 274, 1973.

27. Rubin CE: Newer advances in the exfoliative cytology of the gastrointestinal tract. Ann NY Acad Sci 63: 1377, 1956.

28. Rubio CA, Aberg B, Stemmermann G: Ciliated cells in papillary adenocarcinomas of Barrett's esophagus. Acta Cytol 36: 65, 1992.

29. Sherman CD Jr, Caldwell CB, Kim J-P: Cancer of esophagus and stomach. In: Love RR, et al (eds): Manual of Clinical Oncology, UICC. pp310-329, Berlin: Springer-Verlag, 1994.

30. Stout AP, Lattes R: Tumors of the Esophagus. Atlas of Tumor Pathology. Washington, DC: Armed Forces Institute of Pathology, 1957.

31. Suckow EE, Staley CJ, Brock DR: Extensive intra-epithelial carcinoma of the esophagus with multiple invasive sites. Quart Bull Northwestern Univ Med Sch 35: 45, 1961.

32. Suckow EE, Yokoo H, Brock DR: Intraepithelial carci-noma concomitant with esophageal carcinoma. Cancer 15: 733, 1962.

33. Tennvall J, Johansson L, Albertsson M: Small cell carci-noma of the esophagus: A clinical and immuno-histopathologic review. Eur J Surg Oncol 16: 109, 1990.

34. Thabet RJ, Macfarlane EWE: Cytological field patterns and nuclear morphology in the diagnosis of colon pathology. Acta Cytol 6: 325, 1962.

35. Ushigome S, Spjut HJ, Noon GP: Extensive dysplasia and carcinoma in situ of esophageal epithelium. Cancer 20: 1023, 1967.

36. Weiden PL, Schuffler MD: Herpes esophagitis compli-cating Hodgkin's disease. Cancer 33: 1100, 1974.

37. Wisloff F, Bull-Berg J, Myren J: Herpes zoster of the stomach. Lancet 2: 953, 1979.

38. Wisseman CL, Lemon HM, Lawrence KB: Diagnosis of malignant lesions of the rectum and sigmoid colon by cytological studies of exfoliative cells. Bul Am Coll Surg 32: 243, 1947.

13 Liver, Bile Ducts and Pancreas

Kunio Mizuguchi

LIVER

The liver is the largest organ in human body; the parenchyma is frequently affected by diffuse pathologic changes such as viral hepatitis, cirrhosis, alcoholic and drug-induced disorders, immunologic and metabolic diseases, etc. In these diffuse disorders biopsy histology is preferred to cytology. However, accurate localization of small nodular lesions by skill image analysis may bring about successful cytodiagnosis. Among the hepatic malignancies, metastatic cancers are the most frequent tumors and accounts for about 75%[39] of overall malignant tumors in the liver. In order to determine whether the tumor is primary, metastatic, or to clarify the exact nature of the tumor, fine needle aspiration cytology under ultrasound guide is now widely performed aside from other imaging techniques such as CT scan, angiography and MRI. Aspiration cytology technique is mainly focused onto the space-occupying lesions in the liver, such as tumorous or cystic lesions. These lesions include liver cysts, cystic tumors, tumor-like lesions, benign or malignant solid tumors.

1 Collection of the Specimen and Preparation of the Smears[39]

■Procedure of liver aspiration
There are two methods of fine needle aspiration. The first procedure is to use an aspiration needle and syringe under ultrasonic guide. The second is an aspiration method under fluoroscopic guide. However, the latter method is not frequently used because of its inaccuracy and danger. Both methods use a special type of syringe with 21 gauge needle (Fig. 13-1). Aspiration under laparoscopy or during operation may occasionally be performed. Among these methods, aspiration under ultrasonic guide is most widely applied, because of its technique with ease, safe and accurate access to the lesion of the liver. In order to avoid a complication of intraperitoneal bleeding, hemorrhagic diathesis with measurement of a platelet count and a prothrombin should be examined preoperatively. A care to the presence of the tumors such as hemangioma and angiosarcoma which are easy to bleed should be paid. Another caution for liver aspiration by this method is the possibility of tumor cell dissemination through the needle tract by this method. However, the frequency of dissemination is very low.

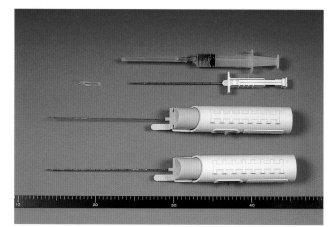

Fig. 13-1 Apparatus for aspiration cytology of the liver. Different types of syringe are shown.

■Preparation of the aspirated material[39]
Tiny liver tissue fragments are picked up by the forceps and dipped in formalin fixative. These tissues are used for further histologic examination. A preparation technique of needle aspirates is processed in the same way as for fine needle aspiration of other organs. The specimen present in the needle lumen is pushed out onto a clean glass slide after detaching the syringe to introduce air. Another technique is placing the specimen that remains in the needle lumen into balanced saline. The saline is centrifuged and the sediment is used for preparing cytologic smear on glass slides.

2 Normal Cell Components[7,9,11,34,38,44]

The normal liver tissue is composed of liver cell cords, central veins and the Glisson's sheath, including veins, arteries and bile ducts. Aspiration cytology from the normal liver shows mainly liver cells and bile duct epithelial cells with a few mesenchymal cells such as endothelial cells and lymphocytes. The normal liver cells are polygonal with a relatively small nucleus and finely granular broad cytoplasm. These cells are cohesive with a plane arrangement. Liver cells are fairly uniform in size and shape: centrally locating small nuclei possess finely distributed chromatins (Fig. 13-2). The bile duct epithelium is composed of uniform columnar cells arranged in acinar pattern with no cellular pleomorphism (Fig. 13-3).

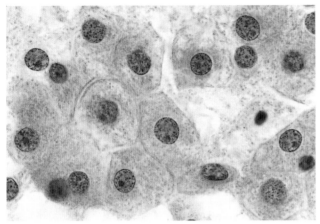

Fig. 13-2 Normal liver cells with uniform nuclei and finely granular cytoplasm. High magnification.

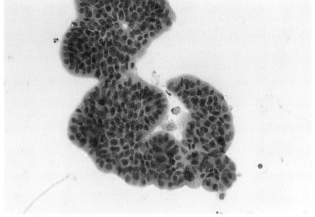

Fig. 13-3 Bile duct epithelium obtained by PTCD. The nuclei are slightly hyperplastic and hyperchromatic. Note well preserved columnar cytoplasm. Low magnification.

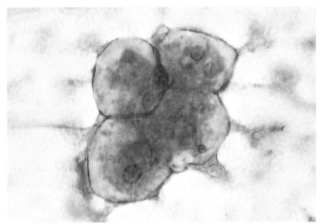

Fig. 13-4 Ameba from the liver abscess. Erythrophagocytosis is seen. Cytoplasm is vacuolated. Moderate magnification.

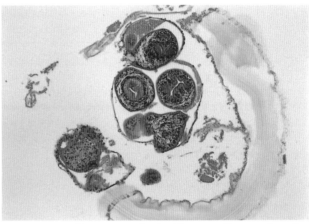

Fig. 13-5 *Echinococcus* hydatid cyst of monolocular type. Daughter cyst and hydatid sands are identified. HE stain. Low magnification.

3 Non-neoplastic Lesions[31]

■Abscess including amebic abscess

Liver abscess due to bacterial infection is characterized by numerous polymorphonuclear leukocytes (pus). Occasionally, bacterial colonies may be identified within the abscess. In an amebic abscess, cysts or trophozoite forms are diagnostic of the disease. Trophozoite form of the ameba shows active erythrophagocytosis which may mimic the feature of histiocyte. The cytoplasm of ameba is usually foamy in appearance (Fig. 13-4). Pyogenic abscess that must be differentiated from amebic abscess occurs occasionally as a result of ascending cholangitis or in association with immune deficiency state.

■Cyst

Cysts are frequently found in the liver. Usually the cystic content is clear serous fluid. The inner surface of the cyst is lined by low cuboidal cells resembling bile duct epithelium. If cystic content is mucinous, mucinous cystadenoma or carcinoma should be considered.

■Hydatid cyst

Another rare cystic lesion of the liver is a hydatid cyst which is caused by *Echinococcus granulosus* or *Echinococcus*

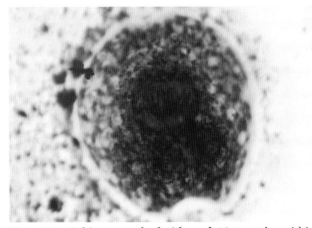

Fig. 13-6 *Echinococcus* hydatid sand. Note scolex within granule. Giemsa stain. High magnification.

multilocularis infestation. Man is infected by ingestion of the ova that lodges in the liver to develop to the larvae and form two types of hydatid cyst which are as monolocular and multilocular. Cystic content is composed of daughter cysts, hydatid sand and hook of scolex as well as necrotic debris (Figs. 13-5 and 13-6).

4 Benign Tumors and Tumorous Lesions

■Regenerative nodule

This lesion may also be called as adenomatous or adenomatoid hyperplasia. Regenerative nodule is seen in the cirrhosis of the liver in which the nodule is larger than other nodules and grossly resembles hepatocellular carcinoma. There are Glisson's sheaths in the nodule. Histology of this lesion, however, shows hyperplastic change. Occasionally, the cells in the nodule show slight to moderate atypia. The nuclei vary in size and the N/C ratio is not so high as in hepatocellular carcinoma. Cytological diagnosis is often difficult because of the presence of nuclear atypia[8] (Fig. 13-7).

■Focal nodular hyperplasia[10]

Focal nodular hyperplasia (FNH) is a hyperplastic nodular lesion which arises in a non-cirrhotic liver and is usually solitary. Macroscopically, there are central stellate scars which characterize this type of the lesion. There are various numbers of bile ducts adjacent to the fibrous scar. There is no capsule surrounding the FNH lesion. The liver cells in FNH do not show atypia but may have clear cytoplasm. The sinusoidal structure are well preserved. Cytological features of FNH are almost similar to those of normal liver cells.

■Liver cell adenoma

Liver cell adenomas arise in the non-cirrhotic liver. It may relate to the administration of oral contraceptives. Most of liver cell adenoma originate as a solitary lesion. The tumor arises in subcapsular region resulting in hemispherical protrusion. The liver cells in this tumor are quite uniform in size and shape without atypia. Therefore, cytological differentiation between adenoma cells and normal liver cells is very difficult. Also, the adenoma cells resemble tumor cells of well differentiated type of hepatocellular carcinoma. There are some reports trying to clear the cytologic characteristics of liver cell adenoma though complete criteria have not been assessed because the examined cases were very limited in number. There is no bile duct in the tumor.

■Hemangioma

Usual hemangiomas of the liver are small in size and composed of a cavernous structure. It can be found by ultrasonography or computerized tomography. This tumor may bleed easily, therefore, fine needle aspiration should not be performed. In rare case, however, cavernous hemangioma may develop a large mass with marked sclerosis. Such a case is quite difficult to differentiate from other tumors. Aspiration can collect a few cells including red blood cells, endothelial cells, fibroblasts and some inflammatory cells.

■Inflammatory pseudotumor

Inflammatory pseudotumor is a rare tumor of the liver characterized by well discrete mass. This mass is composed of abundant fibrous collagen and infiltration of inflammatory cells such as mature plasma cells, lym-

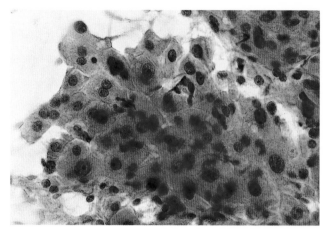

Fig. 13-7 Regenerative nodule from liver cirrhosis. Mild nuclear atypia is present. The nuclei are centrally located within broad cytoplasm. Moderate magnification.

phocytes and histiocytes. The exact cause of this lesion is still unclear. This tumor is also called as plasma cell granuloma, when plasma cells are predominant or xanthogranuloma just as described in other organs such as the lung and intestinal tracts. Cytologically, many mature plasma cells are seen as well as some lymphocytes, histiocytes and fibroblasts.

■Angiomyolipoma

Angiomyolipoma is a rare tumor of the liver which was first described in the kidney.

This lesion is sometimes associated with multiple sclerosis. The tumor is composed of three main structures; fatty tissue, blood vessels and smooth muscle fibers. The ratio of these components differ case by case. In case that the tumor consists of merely smooth muscle fibers, it may give rise to the feature of leiomyoma. Occasionally, smooth muscle cells show moderate atypia. This tumor may be found incidentally using ultrasonographic examination and also at the time of autopsy. Diagnosis is not difficult, however, some cases show only compactly arranged epithelioid cells without fat or blood vessels. Cytologically, these different types of cells may be found with or without some nuclear atypia.

5 Malignant tumors (Table 13-1)[18]

■Hepatocellular carcinoma[16,23,32]

Hepatocellular carcinoma (liver cell carcinoma, malignant hepatoma) is composed of tumor cells resembling liver cells in both cytological feature and structural architecture. Formerly, Edmondson's classification based on grades of differentiation has been widely used. However, recently WHO classification has come to illustrate structural and cellular changes; these include (1) trabecular form, (2) pseudoglandular form, compact form, (4) scirrhous form, and (5) others (pleomorphic type, clear cell type, small cell type, spindle cell type). Aside from these classifications, there is another classification that depends on the degree of differentiation. Cytological diagnosis is effective for determination of cell atypia and structural differentiation. Usually the degree of differentiation is as follows:

Table 13-1 Histological classification of tumours of the liver (WHO, 1994)

1	**Epithelial Tumours**		3.2	*Teratoma*
1.1	*Benign*		3.3	*Yolk sac tumour (endodermal sinus tumour)*
1.1.1	Hepatocellular adenoma (liver cell adenoma)		3.4	*Carcinosarcoma*
1.1.2	Intrahepatic bile duct adenoma		3.5	*Kaposi sarcoma*
1.1.3	Intrahepatic bile duct cystadenoma		3.6	*Rhabdoid tumour*
1.1.4	Biliary papillomatosis		3.7	*Others*
1.2	*Malignant*		**4**	**Unclassified Tumours**
1.2.1	Hepatocellular carcinoma (liver cell carcinoma)		**5**	**Haemopoietic and Lymphoid Tumours**
1.2.2	Intrahepatic cholangiocarcinoma (peripheral bile duct carcinoma)		**6**	**Metastatic Tumours**
			7	**Epithellal Abnormalities**
1.2.3	Bile duct cystadenocarcinoma		7.1	*Liver cell dysplasia*
1.2.4	Combined hepatocellular and cholangiocarcinoma		7.2	*Bile duct abnormalities*
1.2.5	Hepatoblastoma		**8**	**Tumour-like Lesions**
1.2.6	Undifferentiated carcinoma		8.1	*Hamartomas*
2	**Nonepithelial Tumours**		8.1.1	Mesenchymal hamartoma
2.1	*Benign*		8.1.2	Biliary hamartoma (microhamartoma, von Meyenburg complex)
2.1.1	Angiomyolipoma			
2.1.2	Lymphangioma and lymphangiomatosis		8.2	*Congenital billary cysts*
2.1.3	Haemangioma		8.3	*Focal nodular hyperplasia*
2.1.4	Infantile haemangioendothelioma		8.4	*Compensatory lobar hyperplasia*
2.2	*Malignant*		8.5	*Peliosis hepatis*
2.2.1	Epithelioid haemangioendothelioma		8.6	*Heterotopia*
2.2.2	Angiosarcoma		8.7	*Nodular transformation (nodular regenerative hyperplasia)*
2.2.3	Undifferentiated sarcoma (embryonal sarcoma)			
2.2.4	Rhabdomyosarcoma		8.8	*Adenomatous hyperplasia*
2.2.5	Others		8.9	*Focal fatty change*
3	**Miscellaneous Tumours**		8.10	*Inflammatory pseudotumour*
3.1	*Localized fibrous tumour (localized fibrous mesothelioma, fibroma)*		8.11	*Pancreatic pseudocysts*
			8.12	*Others*

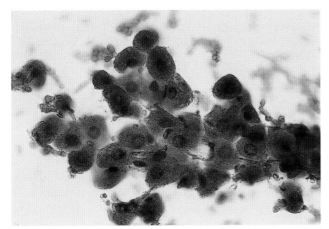

Fig. 13-8 Hepatocellular carcinoma, well differentiated. Atypia is slight. Pseudogland is present. HE stain. Moderate magnification.

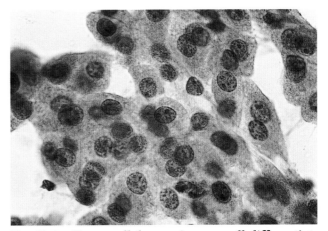

Fig. 13-9 Hepatocellular carcinoma, well differentiated. Cellularity is slightly increased. Compare with normal hepatocytes (Fig. 13-2). High magnification.

Well differentiated type: Histologically this is very close to a normal liver. Histopathological and cytological diagnosis is very difficult. Cellular atypia is not so helpful (Figs. 13-8 and 13-9). Structural changes of liver cell cords are more diagnostic, and are easily distinguished by silver-impregnation method. The N/C ratio is slightly increased.

Moderately differentiated type: Both histology and cytology of this type are intermediate between well and poorly differentiated type (Figs. 13-10 and 13-11).

Poorly differentiated type: In this type, both structural and cellular atypia are marked. The N/C ratio is increased. The cytoplasm is scanty. The nuclei are predominantly prominent, and nuclear chromatins are increased in amount. Cohesiveness of the tumor cells are lost and individual cells are separately isolated (Figs. 13-12 and 13-13).

◼Cholangiocellular carcinoma (CCC)

This carcinoma arises from the bile duct epithelium, therefore, histologic and cytologic features are almost similar to those of carcinoma of the common bile duct

267

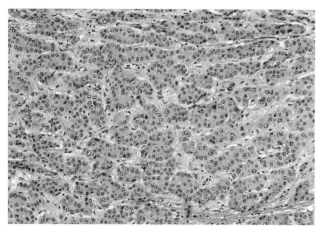

Fig. 13-10 Hepatocellular carcinoma, moderately differentiated. Trabecular pattern is prominent. HE stain. Low magnification.

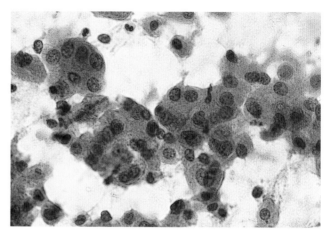

Fig. 13-11 Hepatocellular carcinoma, moderately differentiated. Showing increased cellularity and nuclear disarrangement. Moderate magnification.

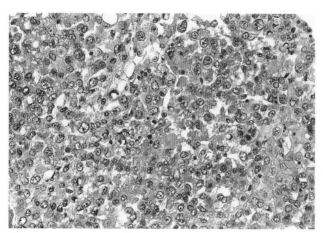

Fig. 13-12 Hepatocellular carcinoma, poorly differentiated. Cellular atypia is marked. HE stain. Low magnification.

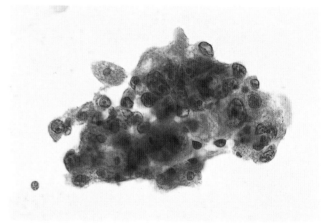

Fig. 13-13 Hepatocellular carcinoma, poorly differentiated. Cellular and structural atypia is seen. Moderate magnification.

or gallbladder. The only difference lies in that CCC may form larger mass in the liver associated with dense fibrous stroma. The CCC is classified as follows:

1) Adenocarcinoma

Well differentiated type: Histologically the tumor is composed of well preserved glandular (tubular) or papillary structure associated with various amount of fibrosis. Cellular and nuclear atypia are slight to moderate. Nucleoli are usually prominent. The N/C ratio is not so high. The nuclear arrangement is slightly disoriented (Fig. 13-14).

Moderately differentiated: The structural and cellular atypia are moderate and show intermediate patterns between well and poorly differentiated type (Fig. 13-15).

Poorly differentiated type: Poorly differentiated CCC shows marked anisocytosis, and less cohesiveness than well or moderately differentiated type; a papillary or tubular pattern is indiscernible and nucleoli are prominent and irregular in size. The cytoplasm is scanty (Fig. 13-16).

2) Squamous cell carcinoma (SCC)[37]

Incidence of pure type SCC is very low. Usually a mixed type of SCC and cholangiogenic adenocarcinoma is more common than pure SCC. If the lesser component comprises more than 20%, it is defined as adenosqua-

mous carcinoma. Cytologically, both adenocarcinoma and squamous cell carcinoma can be distinguished (Fig. 13-17).

3) Mucus-secreting carcinoma

This group consists of mucinous carcinoma and mucinous cystadenocarcinoma. The former shows relatively uniform cell clusters that appear to be floating in a mucus lake. The mucinous background is a clue to the cytological diagnosis. The latter is made of mucus-rich cell clusters. The mucus content can be observed within the cytoplasm. Therefore, the cytological feature is absolutely different. It should be noted that nuclear atypia of both tumors may be not so prominent (Fig. 13-18).

4) Others

Undifferentiated carcinoma does not show differentiation to any specific structure and is hardly distinguished from sarcoma. Other rare tumors are carcinosarcomas. Most of these cases are poorly differentiated adenocarcinomas with sarcomatoid change (Fig. 13-19).

■Hepatoblastoma

This rare tumor occurs mainly in infants around one year of age. Most cases may reveal markedly elevated α-fetoprotein in the serum. The following classification

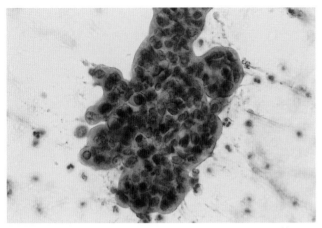

Fig. 13-14 Cholangiocellular carcinoma, well differentiated. Cellular arrangement is preserved. Moderate magnification.

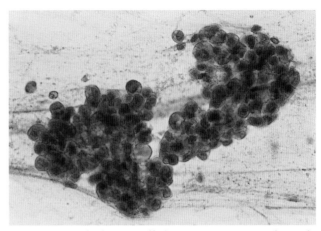

Fig. 13-15 Cholangiocellular carcinoma, moderately differentiated. Moderate atypia is present. Moderate magnification.

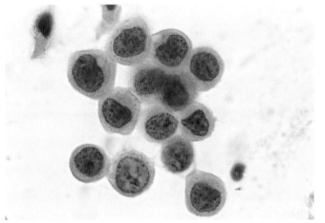

Fig. 13-16 Cholangiocellular carcinoma, poorly differentiated. Cytoplasm is scant. Cohesiveness is decreased. High magnification.

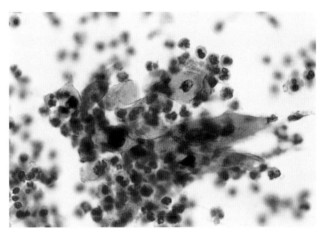

Fig. 13-17 Squamous cell carcinoma of the bile duct. Atypical keratinizing cells are seen. The squamous cell component is demonstrated. Moderate magnification.

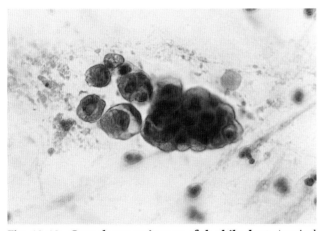

Fig. 13-18 Cystadenocarcinoma of the bile duct. Atypical cells in bile. Note mucin content with eosinophilic tint in the cytoplasm. Moderate magnification.

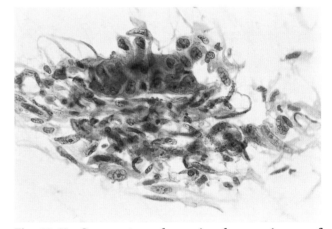

Fig. 13-19 Sarcomatous change in adenocarcinoma of the gallbladder. Spindle-shaped cells and adenocarcinoma cells are present. Low magnification.

depends on the cellular components and their immaturity. Most of these tumors show mixture of epithelial malignant cells with a resemblance to embryonal or fetal hepatocytes with primitive mesenchymal cells.

Well differentiated type: Tumor cells are polygonal to cuboidal in shape and smaller than normal liver cells. They resemble fetal liver tissue. The nuclei are small and

homogeneous in appearance. Cytological diagnosis is very difficult and clinical information is inevitable .

Poorly differentiated type: This tumor is composed of spindle cells with scanty cytoplasm. The N/C ratio is very high. The nuclei are rich in chromatin and mitoses are frequently seen. Cohesiveness of epithelial tumor cells is poorly preserved but rosette, ribbon, or tubular arrange-

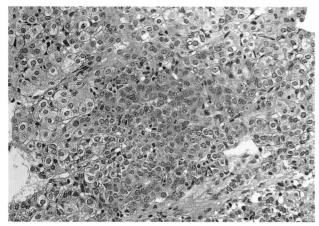

Fig. 13-20 **Hepatoblastoma.** Mixture of well differentiated and poorly differentiated patterns. HE stain. Low magnification.

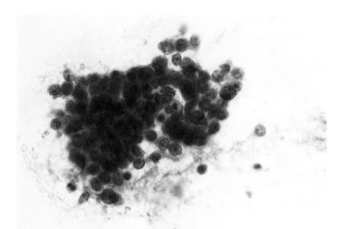

Fig. 13-21 **Hepatoblastoma. Immature cells are seen in cluster.** Nuclear chromatin is rich. Moderate magnification.

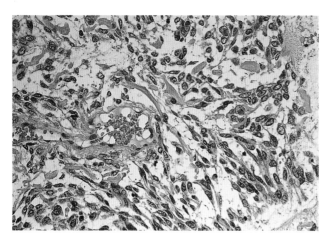

Fig. 13-22 **Angiosarcoma.** Proliferation of spindle-shaped cells with vessel formation is present. HE stain. Low magnification.

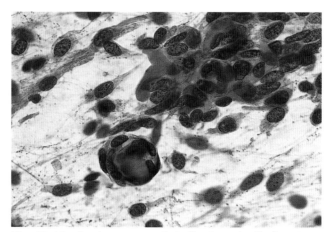

Fig. 13-23 **Angiosarcoma.** Spindle-shaped cells forming a vessel-like structure. Moderate magnification.

ments are occasionally discernible (Figs. 13-20 and 13-21).

Immature type: The main tumor cells are round to spindle-shaped with scarce cytoplasm and hyperchromatic nuclei and resemble those of malignant lymphoma or neuroblastoma. Careful observation on the nuclei, cellular arrangement and background is important for differential diagnosis.

■ Other rare tumors

There are other rare tumors in the liver which include; malignant lymphoma, angiosarcoma, undifferentiated embryonal sarcoma, germ cell tumor, lymphangioma, epithelioid hemangioendothelioma and hamartoma.

Malignant lymphoma: This is a rare primary tumor of the liver. However, the liver is often involved as a part of systemic disease. Histologic and cytologic patterns are the same as malignant lymphoma in other organs. Giemsa stain is necessary for the diagnosis of hematologic neoplasia including malignant lymphoma.

Hemangiosarcoma[22,29]: This is a malignant counterpart of hemangioma. Increased vascularity is the main finding. Cellular atypia of endothelium which is located along the slit-like space of blood vessels, may vary from slight atypia suggestive of increased sinusoids to marked

atypia constituting large, bizarre giant cells. In the case of slight atypia, it may lead to a misdiagnosis as benign neoplasia. Attention is very important not to overlook low-grade hemangiosarcoma: typical case of hemangiosarcoma is composed of atypical non-epithelial cells, which are spindle to plump in shape with hyperchromatic irregular nuclei (Figs. 13-22 and 13-23). A bloody background accompanying hemosiderin deposition is helpful for diagnosis. The tumor is macroscopically spongy and filled with abundant blood. Special caution is required in aspiration biopsy cytology since severe bleeding after aspiration may occur.

■ Metastatic tumors

The liver is the organ which is most frequently affected by metastasized tumors from many other primary sites. At times it is hard to distinguish the metastatic adenocarcinoma from primary cholangiocellular carcinoma and even hepatocellular carcinoma. It may show the similar features in aspiration cytology. One of the reason for this event is that metastatic tumors often change their structure to those resembling liver tissue (so-called "organ mimic"). Therefore, clinical data are very important for differential diagnosis, such as the presence of primary tumor in other organs, multicentricity, elevated

Table 13-2 Histological classification of tumours of the gallbladder and extrahepatic bile ducts (WHO, 1991)

1	**Epithelial Tumours**		3.1.5	Haemangioma
1.1	*Benign*		3.1.6	Lymphangioma
1.1.1	Adenoma		3.1.7	Neurofibroma
1.1.1.1	Tubular		3.2	*Malignant*
1.1.1.2	Papillary		3.2.1	Rhabdomyosarcoma
1.1.1.3	Tubulopapillary		3.2.2	Kaposi sarcoma
1.1.2	Cystadenoma		3.2.3	Leiomyosarcoma
1.1.3	Papillomatosis (adenomatosis)		3.2.4	Malignant fibrous histiocytoma
1.2	*Dysplasia*		3.2.5	Angiosarcoma
1.3	*Malignant*		**4**	**Miscellaneous Tumours**
1.3.1	Carcinoma in situ		4.1	*Carcinosarcoma*
1.3.2	Adenocarcinoma		4.2	*Malignant melanoma*
1.3.3	Papillary adenocarcinoma		4.3	*Malignant lymphomas*
1.3.4	Adenocarcinoma, intestinal type		**5**	**Unclassified Tumours**
1.3.5	Mucinous adenocarcinoma		**6**	**Secondary Tumours**
1.3.6	Clear cell adenocarcinoma		**7**	**Tumour-like Lesions**
1.3.7	Signet-ring cell carcinoma		7.1	*Regenerative epithelial atypia*
1.3.8	Adenosquamous carcinoma		7.2	*Papillary hyperplasia*
1.3.9	Squamous cell carcinoma		7.3	*Adenomyomatous hyperplasia*
1.3.10	Small cell carcinoma (oat cell carcinoma)		7.4	*Intestinal metaplasia*
1.3.11	Undifferentiated carcinoma		7.5	*Pyloric gland metaplasia*
2	**Endocrine Tumours**		7.6	*Squamous metaplasia*
2.1	*Carcinoid tumour*		7.7	*Heterotopias*
2.2	*Mixed carcinoid-adenocarcinoma*		7.8	*Xanthogranulomatous cholecystitis*
2.3	*Paraganglioma*		7.9	*Cholecystitis with lymphoid hyperplasia*
3	**Non-epithelial Tumours**		7.10	*Inflammatory polyp*
3.1	*Benign*		7.11	*Cholesterol polyp*
3.1.1	Granular cell tumour		7.12	*Malacoplakia*
3.1.2	Ganglioneurofibromatosis		7.13	*Congenital cyst*
3.1.3	Leiomyoma		7.14	*Amputation neuroma*
3.1.4	Lipoma		7.15	*Primary sclerosing cholangitis*

serum tumor markers like α-fetoprotein and CEA, and imaging or graphic characteristics. The cytological diagnosis of metastatic cancer will be made by the same methods as in other primary sites.

GALLBLADDER AND EXTRAHEPATIC BILE DUCTS

■ Cells and material

The gallbladder and the common bile duct are composed of low columnar or cuboidal cell lining in mucosa. Although non-epithelial lesions are present in these bile tract system, the neoplastic lesions are mainly derived from these epithelium. Specimen for cytological examinations is usually bile juice.[30] Bile is obtained by either endoscopic retrograde cholangiopancreatography (ERCP) or percutaneous transhepatic cholangiodrainage (PTCD). Occasionally bile is directly and percutaneously aspirated from the gallbladder.[1,21] Recent studies reported are of bile duct brushing cytology and also fine-needle aspiration for gallbladder carcinoma.[13,14,20,25] Sometimes the imprint cytology during operation is helpful for the case indistinguishable between benign and malignant lesions of the gallbladder.[41] Histological and cytological features are almost similar to those of intrahepatic cholangiocellular carcinoma. Differences between aspiration and bile cytology are cellular degeneration and cluster formation in bile

juice. Desquamated epithelial cells tend to form atypical cell cluster or balls (Fig. 13-24). It is important not to misdiagnose these clusters as carcinoma. Other types of carcinoma of the common bile duct and gallbladder are similar to CCC (see the description on page 268). The cytological pictures of different grade of cellular differentiation of adenocarcinomas from the bile duct system are shown in Figs. 13-25 to 13-28. Table 13-2 is a classification of carcinoma of the gallbladder and extrahepatic bile ducts.[2]

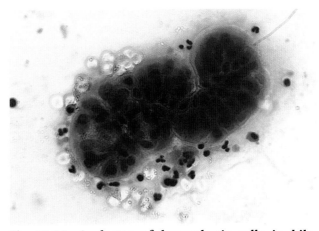

Fig. 13-24 A cluster of hyperplastic cells in bile. Cellularity is increased but nuclear arrangement is still preserved. Moderate magnification.

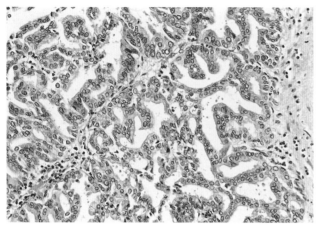

Fig. 13-25   Well differentiated adenocarcinoma of the gallbladder. Papillary growth is seen. HE stain. Low magnification.

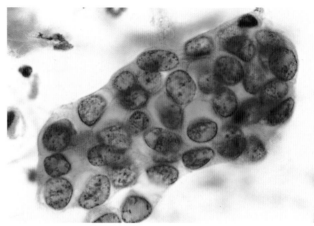

Fig. 13-26   Well differentiated adenocarcinoma of the gallbladder. Cohesiveness is well preserved. High magnification.

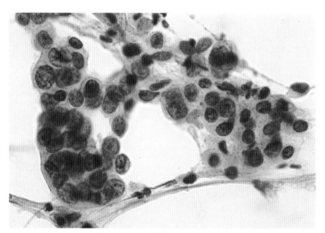

Fig. 13-27   Moderately differentiated adenocarcinoma of the bile duct. Cellular atypia is moderate. Moderate magnification.

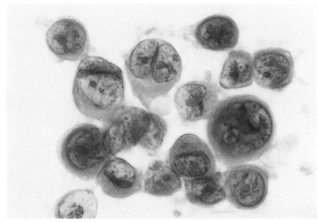

Fig. 13-28   Poorly differentiated adenocarcinoma of the bile duct. Decreased cohesiveness and nuclear atypia are marked. High magnification.

PANCREAS

1 Diagnostic Procedures

Many kinds of diagnostic procedures (e.g., ultrasound, computed tomography scan, angiography, endoscopic cholangiopancreatography and laparoscopy), are available for detection of pancreatic tumors. In these methods, on the economic and safety standpoint, helical CT scan may provide the best overall assessment of the patient with pancreatic head lesions. However, for the decision of the therapy, histological or cytological proof is necessary. The cytological specimens from the pancreatic tumors are obtained usually by ERCP or percutaneous aspiration.[3,4,12,33] The aspiration method is changing from fluoroscopic guide to CT or ultrasonic guide. Accuracy of cytologic diagnosis has markedly improved up to 96%.[39] Recent studies indicate the endoscopic aspiration cytology is very effective for detection of early or small cancers of the pancreas.[27,28,40] Intraoperative fine needle aspiration cytology of the pancreas is also helpful for the diagnosis.[35]

2 Pancreatic Tumors

It is well known that an early diagnosis of the pancreatic tumor is clinically very hard, because the organ is located in the retroperitoneum and clinical symptoms occur in a late stage. However, recent development of image analysis makes it possible to find the pancreatic tumor though early diagnosis remains difficult. When the tumor is found in the pancreas, both aspiration cytology and pancreatic juice cytology are very important to determine the exact nature of the tumor.[4,12] The pancreas has three main cellular components which include columnar duct cells, acinar cells (Fig. 13-30) and islet cells. Most pancreatic tumors are derived from pancreatic duct cells. Tumors arising from acinar cells and islet cells are very few in number. In Table 13-3, classification of pancreatic carcinoma is presented.

■ Duct cell carcinoma
This is the common group of carcinoma simply originating from the pancreatic duct cells.
1) Papillary adenocarcinoma
Papillary growth is characteristic. Often the pancreatic duct is dilated by abundant mucin production.

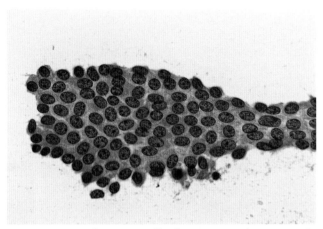

Fig. 13-29 Columnar cells from normal pancreatic duct. The cells are uniform in size showing regular arrangement. Moderate magnification.

Table 13-3 Histological classification of pancreatic carcinoma (modified Japan Pancreas Society)

Ⅰ. Duct cell carcinoma
1. Papillary adenocarcinoma
2. Tubular adenocarcinoma
 · Well differentiated type
 · Moderately differentiated type
 · Poorly differentiated type
3. Cystadenocarcinoma
4. Squamous cell carcinoma
5. Adenosquamous carcinoma
6. Mucinous adenocarcinoma
7. Others
 · Cribriform adenocarcinoma
 · Mucoepidermoid carcinoma
 · Signet-ring cell carcinoma
Ⅱ. Acinar cell carcinoma
Ⅲ. Islet cell carcinoma
Ⅳ. Combined carcinoma
Ⅴ. Undifferentiated carcinoma
 · Round cell type
 · Spindle cell type
 · Pleomorphic type
   (include osteoclastic type of giant cell tumor)
Ⅵ. Unclassified
Ⅶ. Miscellaneous
 · Pancreatoblastoma
 · Carcinosarcoma
 · Solid and cystic tumor
 · Carcinoid tumor

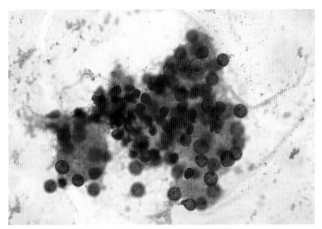

Fig. 13-30 Acinar cells from non-neoplastic pancreas. Well preserved acinar patterns are present. Moderate magnification.

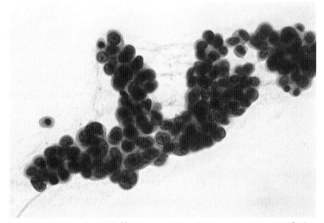

Fig. 13-31 Well differentiated adenocarcinoma of the pancreatic duct. Irregular arrangement of duct cells is seen. Moderate magnification.

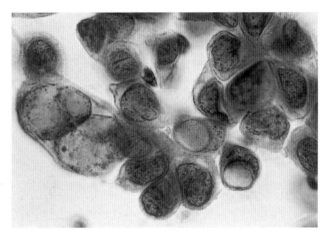

Fig. 13-32 Poorly differentiated adenocarcinoma of the pancreatic duct. Highly atypical cells with intracytoplasmic lumina are present. High magnification.

Cytologically, papillary clusters of the tumor cells are prominent. If the mucinous background is marked, differentiation from mucinous carcinoma is difficult. Occasionally, psammoma bodies may be seen.

2) Tubular adenocarcinoma
This is the most common type of pancreatic carcinomas.

Tumors are sub-classified as well differentiated, moderately differentiated and poorly differentiated in type. The criteria for these types are almost similar to those of cholangiocellular carcinoma (CCC). In the case of moderately differentiated or poorly differentiated type, intracytoplasmic lumina (ICL) may be seen. The typical pic-

273

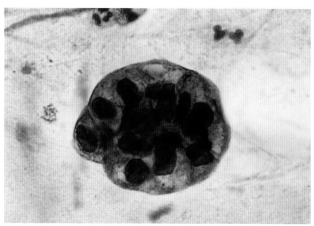

Fig. 13-33 Cystadenocarcinoma of the pancreas. Showing less atypical cell cluster with clear cytoplasm. Moderate magnification.

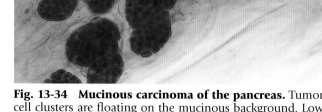

Fig. 13-34 Mucinous carcinoma of the pancreas. Tumor cell clusters are floating on the mucinous background. Low magnification.

tures of moderately differentiated tubular type of carcinoma and moderately differentiated type with ICL can be figured out (Figs. 13-31 and 13-32).

3) Mucinous cystadenocarcinoma

This is a tumor forming monolocular or multilocular cysts. This tumor is mostly characterized by abundant mucus production. This is the reason why this particular tumor is called "mucinous" cystadenocarcinoma. Cytological features such as intracytoplasmic mucin, well preserved cohesiveness and less nuclear atypia are distinctive of this tumor. The cellular differentiation is usually developed (Fig. 13-33), although it is often difficult to differentiate cystadenocarcinoma from cystadenoma. Usually enlarged hyperchromatic nuclei with coarse chromatin and irregular nuclear membrane are seen in cystadenocarcinoma.[15,42]

4) Squamous cell carcinoma and adenosquamous carcinoma

The tissue from these tumors contain various degrees of squamous cancer cells. If all the tumor cells show squamous cell differentiation, it can be called squamous cell carcinoma. Adenosquamous carcinoma is made of both adenocarcinoma and squamous cell carcinoma components.[45] Cytological characteristics of squamous cancer cells are the same as those of SCC of other organs. For example, individual keratinization, thick cytoplasm with distinct borders, irregular and hyperchromatic nuclei, and presence of tadpole or snake cells are representative.

5) Mucinous adenocarcinoma

In these tumors, mucus production is prominent resulting in mucus-lake formation. The tumor cell clusters are floating on the mucus (Fig. 13-34). The cell clusters may be more irregular than papillary adenocarcinoma. Mucus background is very important for the diagnosis of these tumors.

6) Others

These include special histologic types of tumor such as cribriform adenocarcinoma, mucoepidermoid carcinoma and signet ring cell carcinoma.

■Acinar cell carcinoma[36]

The tumor cells and structures are similar to the pancre-

atic acinar cells. For exact diagnosis, zymogen granules should be proved by histochemical and ultrastructural examination. Compared with duct cell carcinoma, the cytologic features of acinar cell carcinoma are not well described. The tumor cells have smooth-contoured nuclei containing one or two prominent nucleoli. Cytoplasm may contain granules which are positive for non-digestive type PAS reaction. Supplemental studies including immunocytochemistry, cytochemistry and electron microscopy are important in facilitating their identification.

■Islet cell carcinoma

The tumor cells resemble those of islet cells in structure. In cytological specimen, tumor cells show flat sheets or cord-like arrangements with less cellular atypia. Rosette may be seen. Like other endocrine tumors, it is not difficult to suspect islet cell tumor. However, general cytological criteria for malignant tumor are not a good indicator for the islet cell tumors.[5,6,17,38] The cytological points for malignant islet cell tumor include the feature of endocrine cells, in addition to increase in nuclear size, well defined cytoplasm, eccentrically located nuclei, and the presence of multinucleated tumor cells and necrotic debris (Figs. 13-35 and 13-36).

■Undifferentiated carcinoma

This tumor group does not show any differentiation to duct, acinar or islet cells. There are three histologic types in this group which are round cell type, spindle cell type and pleomorphic type. The osteoclastic type of giant cell tumor is included in the pleomorphic type. In this tumor, multinucleated giant cells of osteoclastic type as well as interstitial spindle cells are characteristic (Figs. 13-37 and 13-38). There are many studies which indicate this tumor is epithelial in origin.[24,26,43]

■Miscellaneous

This group includes very rare tumors such as pancreatoblastoma, carcinosarcoma, solid and cystic tumors. In these rare tumors, solid and cystic tumor (also called as papillary cystic or solid and papillary epithelial neoplasm) is seen in young women. This tumor is mostly

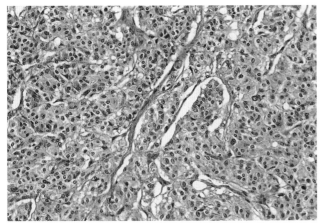

Fig. 13-35 Islet cell carcinoma of the pancreas. Endocrine cell nests with fine capillary network are present. This tumor metastasized to regional lymph node. HE stain. Low magnification.

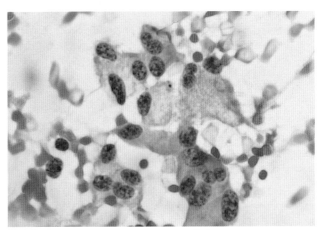

Fig. 13-36 Islet cell carcinoma of the pancreas. Tumor cells have clear cytoplasm. Nuclear atypia is slight. Moderate magnification.

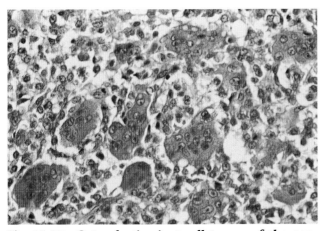

Fig. 13-37 Osteoclastic giant cell tumor of the pancreas. Multinucleated osteoclastic cells are characteristic. HE stain. Low magnification.

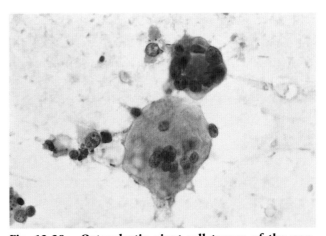

Fig. 13-38 Osteoclastic giant cell tumor of the pancreas. Multinucleated giant cells are seen in aspirated material. Low magnification.

benign but can be malignant. Medium-sized ovoid cells form pseudorosettes and pseudopapillary structures[19] (Fig. 13-39). Cytological features show those of between endocrine cells and ductal cells. Alpha 1 anti-trypsin is frequently positive in the tumor cell cytoplasm.

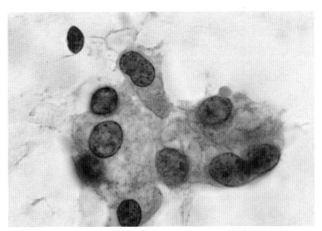

Fig. 13-39 Solid and cystic tumor of the pancreas. Slightly atypical cells forming rosette-like structure. High magnification.

References
1. Akosa AB, Barker F, Desa L, et al: Cytologic diagnosis in the management of gallbladder carcinoma. Acta Cytol 39: 494, 1995.
2. Albores-Saavedra J, Henson DE, Sobin LH (eds): Histological Typing of Tumours of the Gallbladder and Extrahepatic Bile Ducts, 2nd ed. International Histological Classification of Tumours (WHO). Berlin: Springer-Verlag, 1991.
3. Al-Kaisi N, Siegler EE: Fine needle aspiration cytology of the pancreas. Acta Cytol 33: 145, 1989.
4. Alpern GA, Dekker A: Fine-needle aspiration cytology of the pancreas. Acta Cytol 29: 873, 1985.
5. Banner BF, Myrent KL, Memoli VA, et al: Neuroendocrine carcinoma of the pancreas diagnosed by aspiration cytology. Acta Cytol 29: 442, 1985.
6. Bell DA: Cytologic features of islet-cell tumors. Acta Cytol 31: 485, 1987.
7. Bell DA, Carr CP, Szyfelbein WM: Fine needle aspiration cytology of focal liver lesions: Results obtained

with examination of both cytologic and histologic preparations. Acta Cytol 30: 397, 1986.

8. Berman JJ, McNeill RE: Cirrhosis with atypia: A potential pitfall in the interpretation of liver aspirates. Acta Cytol 32: 11, 1988.

9. Bognel C, Rougier P, Leclere J, et al: Fine needle aspiration of the liver and pancreas with ultrasound guidance. Acta Cytol 32: 22, 1988.

10. Casarella WJ, Knowles DM, Wolff M, et al: Focal nodular hyperplasia and liver cell adenoma: Radiologic and pathologic differentiation. AJR Am J Roentgenol 13: 393, 1978.

11. Civardi G, Fornari F, Cavanna L: Ultrasonically guided fine needle aspiration biopsy: A useful technique for the diagnosis of abdominal malignancies. Eur J Cancer Clin Oncol 22: 225, 1986.

12. Das DK, Bhambhani S, Kumar N, et al: Ultrasound guided percutaneous fine needle aspiration cytology of pancreas: A review of 61 cases. Trop Gastroenterol 16: 101, 1995.

13. Dodd LG, Moffatt EJ, Hudson ER, et al: Fine-needle aspiration of primary gallbladder carcinoma. Diagn Cytopathol 15:151, 1996.

14. Ferrari JAP, Lichenstein DR, Slivka A, et al: Brush cytology during ERCP for the diagnosis of biliary and pancreatic malignancies. Gastrointest Endosc 40:140, 1994.

15. Gupta RK, alAnsari AG: Needle aspiration cytology in the diagnosis of mucinous cystadenocarcinoma of pancreas: A case of five cases with an emphasis on utility and differential diagnosis. Int J Pancreatol 15:149, 1994.

16. Gupta SK, Das DK, Rajwanslin A, et al: Cytology of hepatocellular carcinoma. Diagn Cytopathol 2: 291, 1986.

17. Hsiu JG, D'Amato NA, Sperling MH: Malignant islet-cell tumor of the pancreas diagnosed by fine needle aspiration biopsy: A case report. Acta Cytol 29: 576, 1985.

18. Ishak KG, Anthony PP, Sobin LH (eds): Histological Typing of Tumours of the Liver, 2nd ed. International Histological Classification of Tumours (WHO). Berlin: Springer-Verlag, 1994.

19. Jagannath P, Bhansali MS, Murthy SK, et al: Solid and cystic papillary neoplasm of pancreas: A report of seven cases. Indian J Gastroenterol 13: 112, 1994.

20. Kocjan G, Smith AN: Bile duct brushing cytology: Potential pitfalls in diagnosis. Diagn Cytopathol 16: 358, 1997.

21. Layfield LJ, Wax TD, Lee JG, et al: Accuracy and morphologic aspects of pancreatic and biliary duct brushings. Acta Cytol 39: 11, 1995.

22. Ludwig J, Hoffman HN II: Hemangiosarcoma of the liver: Spectrum of morphologic changes and clinical findings. Mayo Clin Proc 50: 255, 1975.

23. Lundqvist A: Fine-needle aspiration biopsy of the liver. Acta Med Scand [Suppl] 520: 1, 1971.

24. Manci EA, Gardner LL, Pollock WJ, et al: Osteoclastic giant cell tumor of the pancreas: Aspiration cytology, light microscopy, and ultrastructure with review of the literature. Diagn Cytopathol 1: 105, 1985.

25. McGuire DE, Venu RP, Brown RD, et al: Brush cytology for pancreatic carcinoma: An analysis of factors influencing results. Gastrointest Endosc 44: 300, 1996.

26. Mullick SS, Mody DR: "Osteoclastic" giant cell carcinoma of the pancreas: Report of a case with aspiration cytology. Acta Cytol 40: 975, 1996.

27. Nakaizumi A, Tatsuta M, Uehara H, et al: Effectiveness of the cytologic examination of pure pancreatic juice in the diagnosis of early neoplasia of the pancreas. Cancer 76: 750, 1995.

28. Nakaizumi A, Tatsuta M, Uehara H, et al: Usefulness of simple endoscopic aspiration cytology of pancreatic study. Dig Dis Sci 42: 1796, 1997.

29. Nguyen GK, McHattie JD, Jeannot A: Cytomorphologic aspects of hepatic angiosarcoma: Fine-needle aspiration biopsy of a case. Acta Cytol 26: 527, 1982.

30. Nilsson B, Wee A, Yap I: Bile cytology: Diagnostic role in the management of biliary obstruction. Acta Cytol 39: 746, 1995.

31. Perry MD, Johnston WW: Needle biopsy of the liver for the diagnosis of nonneoplastic liver diseases. Acta Cytol 29: 385, 1985.

32. Piloti S, Rilke J, Claren R, et al: Conclusive diagnosis of hepatic and pancreatic malignancies by fine needle aspiration. Acta Cytol 32: 27, 1988.

33. Pinto MM, Avila NA, Criscuolo EM: Fine needle aspiration of the pancreas. Acta Cytol 32: 39, 1988.

34. Ramsey WH, Wu GY: Hepatocellular carcinoma: Update on diagnosis and treatment. Dig Dis 13: 81, 1995.

35. Saez A, Catala I, Brossa R, et al: Intraoperative fine needle aspiration cytology of pancreatic lesions: A study of 90 cases. Acta Cytol 39: 485, 1995.

36. Samuel LH, Frierson HF Jr: Fine needle aspiration cytology of acinar cell carcinoma of the pancreas: A report of two cases. Acta Cytol 40: 585, 1996.

37. Shinagawa T, Tadokoro M, Takagi M, et al: Primary squamous cell carcinoma of the liver: A case report. Acta Cytol 40: 339, 1996.

38. Soda K, Yamada S, Yamanaka T, et al: Minute malignant islet cell tumor of the pancreas: Report of a case. Surg Today 25 :444, 1995.

39. Tao LC: Liver and pancreas. In: Bibbo M (ed): Comprehensive Cytopathology. pp822-859. Philadelphia: WB Saunders, 1991.

40. Uehara H, Nakaizumi A, Tatsuta M, et al: Diagnosis of carcinoma in situ of the pancreas by peroral pancreatoscopy and pancreatoscopic cytology. Cancer 79: 454, 1997.

41. Valliengua C, Rodriguez OJC, Proske SA, et al: Imprint cytology of the gallbladder mucosa: Its use in diagnosing macroscopically inapparent carcinoma. Acta Cytol 39: 19, 1995.

42. Vellet D, Leiman G, Mair S, et al: Fine needle aspiration cytology of mucinous cystadenocarcinoma of the pancreas. Acta Cytol 32: 43, 1988.

43. Walts AE: Osteoclast-type giant cell tumor of the pancreas. Acta Cytol 27: 500, 1983.

44. Whitlach S, Nunez C, Pitlik DA: Fine needle aspiration biopsy of the liver: A study of 102 consecutive cases. Acta Cytol 28: 719, 1984.

45. Wilcznski SP, Valente PT, Atkinson BF: Cytodiagnosis of adenosquamous carcinoma of the pancreas: Use of intraoperative fine needle aspiration. Acta Cytol 28: 733, 1984.

14 Effusions in Body Cavities

1 Preparation of Smears

■Centrifugation method

This is the most common and simplest technique for cytology of serous effusions. Dry smear may lose a cellular detail for Papanicolaou stain. Wet preparation and fixation are indispensable to represent good nuclear features (Fig. 14-1). To prevent fibrin formation, anticoagulants must be added during or immediately after thoracentesis and paracentesis; (1) aspirate fluid with a heparinized syringe, (2) add balanced sodium citrate solution (3.8%) or oxalate mixture solution (1.2 g of crystalline ammonium oxalate and 0.8 g of crystalline potassium oxalate in a total amount of 100 ml distilled water) to the fluid to a final concentration of one-fifteenth. It is desirable to send the fluid specimen to the cytology laboratory immediately, but if some delay is unavoidable, the specimen should be kept in the ice box; cellular components are well preserved even if the specimen is left overnight. Papanicolaou has recommended the addition of an equal amount of 50% ethyl alcohol to fluid specimens. A lower concentration of ethyl alcohol (20-30%) with the addition of little formalin is preferable to 50% alcohol. The pretreatment solution is made at the ratio of 2 ml formalin, 68 ml balanced electrolyte solution and 30 ml of 95% ethyl alcohol. The centrifugation method with pretreatment has the advantages that cellular elements are well preserved and the effects of centrifugation are increased. Disadvantages are that protein-rich exudates are coagulated by alcohol and often cellular elements are lost during the wet preparation procedures.

■Cell block method

The cell block method is carried out in the same manner as the histological method; fixing, dehydrating and then embedding the sediment into paraffin later to be sectioned[14,34,120](Fig. 14-2).

Serial sections or step serial sections cut about 5 μm thick are preferable to examine a whole specimen. To obtain step serial sections, we pick up a ribbon with two or three sections every five or more cuttings. Although the preparation of slides is time-consuming, the details of cells in large clumps with heavy overlappings will be clearly demonstrated as in standard histological methods, to which pathologists are accustomed (Fig. 14-3).

■Membrane filter method

Membrane filtration with use of Millipore filter, Nuclepore filter, etc. is theoretically satisfactory for cell concentration from fluids, however, with serous effusions containing a large number of cells, we have found the conventional centrifugation method is more satisfactory than membrane filtration. The use of membrane is limited to an occasion when the specimen is too small in amount to centrifuge.

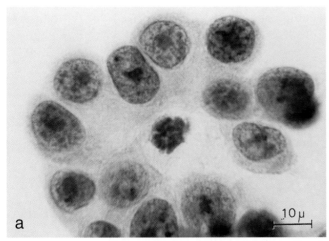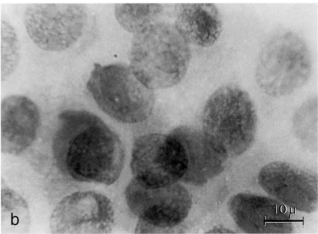

Figs. 14-1 Comparison of smears prepared by wet fixation method and air dry method.
a. Wet preparation. Immediate preparation after paracentesis with the wet preparation technique preserves malignant features such as coarse chromatin clumping and prominent nucleoli or karyosomes.
b. Air dry preparation. A dry smear shows loss of nuclear details. The nuclei are enlarged and their chromatin network is obscure.

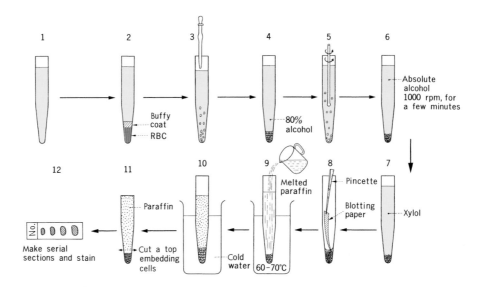

Figs. 14-2　Schema of rapid cell block method.
1, 2: Centrifuge at 1,500 rpm for 5 minutes. 3: Pipette buffy coat, drop it into the fixative and leave for 5 minutes. 4: Decant the fixative and pour 80% alcohol. 5: Stir with a glass bar and change alcohol (80%-95%-100%). 6: Finish dehydration. 7: Clear with xylol. 8: Decant and blot the remainder of xylol. 9: Pour melted paraffin. 10: Place the tube into cold water. 11: Cut a top of paraffin bar to be sectioned. (By courtesy of Dr. N. Fukushima, Doai Memorial Hospital, Tokyo.)

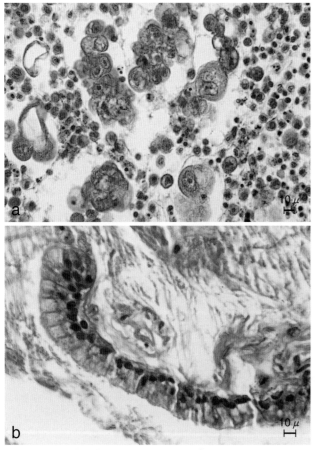

Fig. 14-3　Cell block preparation of ascitic sediment.
a. From a case with embryonal carcinoma. Note a tissue section-like feature showing mixture of inflammatory cells with malignant cells forming a tubular cluster.
b. From a case with pseudomyxoma peritonei. A glandular structure that is lined by tall columnar cells with mucin content accords with the same feature as histology of pseudomyxoma peritonei.

The ThinPrep Processor (Cytyc Corp., Boxborough, Massachusetts, U.S.A.) is a unique membrane filtration device that is useful for effusions and liquid-based samples scraped from lesions.[32] According to the study by Papillo and Lapen,[83] concentration of epithelial cells in comparison with the Cytospin method (Shandon Southern Instruments, Pittsburgh, Pennsylvania, U.S.A.) has resulted in about three times more epithelial cell collection than the cytocentrifugation method. Thus the ThinPrep preparation is practically worthy of use when clinically diagnostic cells are very small in number in fluids and when diagnostic cells have to be absolutely recovered from rinsing samples.

■Thin layer preparation method

Thin layer preparation is processed semi-automatically using the AutoCyte preparation processor (TriPath, Burlington, North Carolina, U.S.A.) through cell sedimentation and Papanicolaou stain. Before the automated preparation, preparatory steps for fluid samples are as follows; (1) Mix body fluids well and place about 50 ml into a centrifuge tube and centrifuge for 10 minutes at 800 ×g. (2) Mix the sediments with 10 ml alcohol-based prediluted CytoRich preservative (after 2×dilution of the preservative by physiologic saline) and vortex the sample for 10 seconds until the cell button appears to be mixed homogeneously. (3) Leave the specimen at room temperature for about 30 minutes. (4) Recentrifuge for 10 minutes at 800 ×g. (5) Resuspend with saline and vortex for 5 to 10 seconds before loading onto the AutoCyte PREP System.

Cellular components stained are settled thinly onto 13 mm circle. The smear has a clean background and displays well preserved nuclear detail.[6] For hemorrhagic sample red cell-lytic CytoRich Preservative Red is recommended.[118]

2 Macroscopic Examination

It is important to examine the macroscopic and physical characteristics, e.g., appearance of fluids, specific gravity and protein content prior to microscopic examination. Coloration of the fluid, for example, may give some indications concerning the pathological process. If the fluid is a transudate with few cellular elements it is usually watery clear and its coloration will depend on its bilirubin content. The fluid may be tinged dark red to reddish brown by erythrocytes. A high white cell count often gives a yellowish-white tint. Chylous fluid which is milky-white or milky-green is suggestive of a traumatic rupture and obstruction of the thoracic duct or main lymphatic vessels such as by filariasis or tumor infiltration. Similar to chylous fluid is chyliform fluid which is rich in fat droplets produced by cell degeneration. Turbidity of fluid largely depends on the amount of suspended formed particles, but mostly on cellularity. Coagulability of the fluid depends on the amount of fibrin content and permits the differentiation of exudates from transudates.

■Transudate

Accumulation of transudate is mostly due to decreased colloid osmotic pressure, sodium retention and circulatory disturbance such as venous outflow block; common causes of transudation are hepatic cirrhosis, cardiac insufficiency, renal insufficiency, hypoproteinemia, adhesive pericarditis, Meigs' syndrome, Chiari's syndrome, etc. It should be kept in mind that metastasis of malignant tumor to the liver or periportal lymph nodes gives rise to a transudate type of ascites without malignant cells. Its specific gravity is less than 1.018 and the protein content is less than 2.3 g/dl. The Rivalta test for presence of proteinaceous material by dropping two or three drops of glacial acetic acid is usually negative. Cellular elements are few and the gross appearance is watery clear.

■Exudate

Exudation occurs with inflammation and malignant neoplasm. The specific gravity is generally over 1.018; there is a high protein content (over 3 g/dl) and a high cellularity (more than 100/mm³ not including erythrocytes). The glucose content is usually lower in exudates than in transudates; while the transudates retain almost the same qualities as those of their parent plasma, glycolysis in inflammatory process is considered to give rise to decrease in glycogen content. The appearance of fibrin clot is often encountered. Light and his associates[64] emphasize that LDH value is useful to discriminate exudate from transudate; LDH activity higher than 200 units (Wroblewski) is seen in about 70% of exudates.

3 Histology of Mesothelium and Benign Cellular Components from Serous Cavities

■Histology of mesothelium

The mesothelium covering the peritoneum, pleura and pericardium is composed of two parts; the parietal and

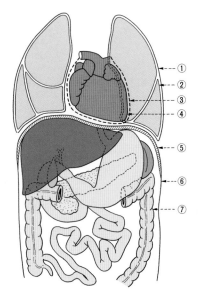

Fig. 14-4 Schema of serous body cavities. The pleural, the pericardial and the peritoneal cavities which are formed between the parietal and the visceral layers.
1. parietal pleura, 2. visceral pleura, 3. pericardium, 4. epicardium, 5. diaphragm, 6. parietal peritoneum, 7. visceral peritoneum.

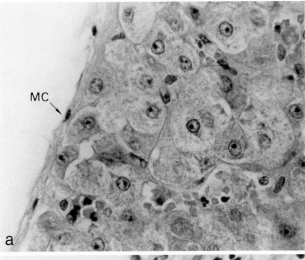

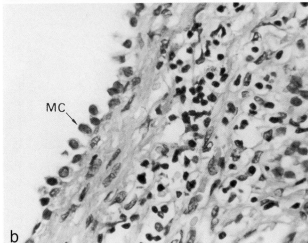

Fig. 14-5 Histology of mesothelium in normal and activated states.
a. Normal liver covered with serosa with lining of endothelium-like flat cells.
b. Cirrhotic liver covered with proliferating mesothelium. Note cuboidal cells with enlarged, round or oval nuclei.

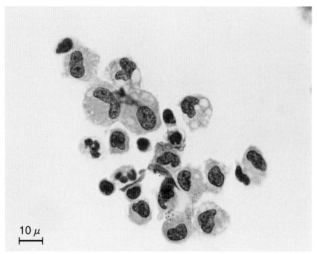

Fig. 14-6　Typical macrophages in ascites. Oval or bean-shaped nuclei are peripheral in position. Chromatins are finely and evenly distributed. The cytoplasm is abundant and vacuolated.

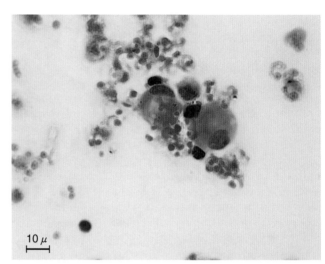

Fig. 14-7　Macrophages phagocytosing destroyed erythrocytes. Note phagocytosis of red cell debris ingested into the cytoplasm.

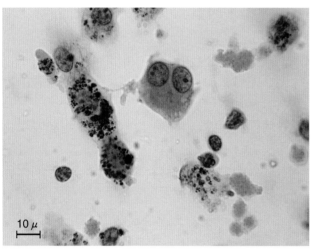

Fig. 14-8　Melanophages and a binucleated mesothelial cell in ascites from a case with disseminated melanoma. The cytoplasm is filled with dark brown pigments; melanin can be identified by a bleaching technique. Compare with melanoma cells in Fig. 14-119; oxidized, phagocytosed pigments are more darker than those produced within the melanoma cells.

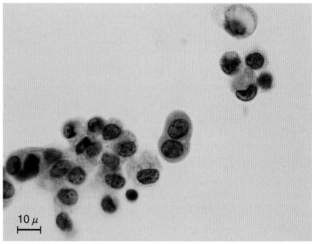

Fig. 14-9　Macrophages phagocytosing bile pigments and mesothelial cells in ascites associated with biliary cirrhosis. Mesothelial cells showing round, central nuclei locate in the middle. Light yellow bile pigments are phagocytosed by some macrophages. Diagnostic points: Prussian blue to identify hemosiderin-laden macrophages and Fontana-Masson or bleaching method to identify melanophages are useful for differential diagnosis. These techniques are rather simple than unstable Gmelin reaction.

visceral membranes. Histologically, it is simple squamous epithelium which is equivalent to the endothelium of blood and lymph vessels. This flat epithelium becomes cuboidal and richly cellular when stimulated by inflammation or other causes (Fig. 14-5). Physiologically a small amount of serous fluid is present in the serous cavities, but it is not discernible.

■Benign cellular components

Benign cellular components contained in the serous effusions are as follows:

1) Macrophages (Histiocytes)

Since the origin of this cell either from tissue or blood has not been clarified, the appropriateness of this term has long been debated. Free macrophages in the peritoneal cavity are considered to originate in the milky

spots of the omentum; however it may also be possible that macrophages are transformed monocytes which have migrated from the blood vessels, since these wandering mononuclear cells increase in volume with a striking development of the Golgi areas and also they appear to acquire phagocytic properties during inflammatory stimulation.[28,39,70,73,113,116]

Differentiation or characterization of blood monocytes to macrophages was demonstrated by introduction of various techniques such as supravital staining (Seemann[100]), tissue culture (Ehrich[27]) and experimental injection of Latex particles (Joos et al[51]). The theory of blood monocyte origin of mononuclear macrophages was pointed out from the fact that most of intraperitoneal phagocytes showed positive oxidase and peroxidase reactions. As to the origin of some phagocytic cells

Fig. 14-10 Electron micrograph of free macrophages in pericardial fluid. The nuclei are irregularly outlined and peripherally located. Lysosomal dense bodies (L) are numerous. Flask-shaped caveolae are recognized on the cell surface. Numerous vacuoles or vesicles varying in size may result from degeneration. There are many finger-like cytoplasmic processes (c). (original magnification ×4,000)

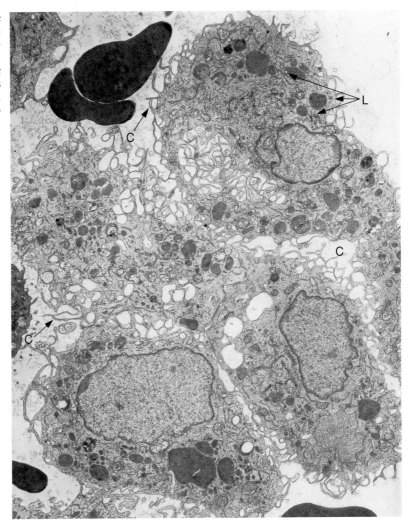

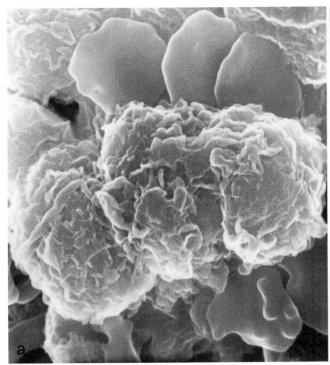

Fig. 14-11 Scanning electron microscopic view of macrophages showing irregular cytoplasmic processes. Plate-like flat cells are red cells. (original magnification ×7,500)

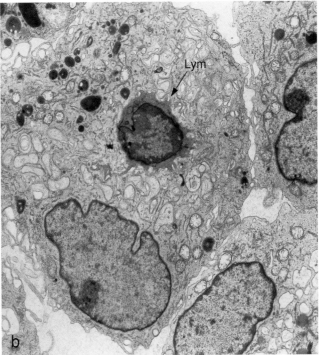

Fig. 14-12 Electron micrograph of macrophages in pericardial fluid. The nuclei are characteristically indented and peripherally located. Electron-dense granules are most likely secondary lysosomes. A pyknotic lymphocyte (Lym) is included showing bridge-like processes. It is not certain whether or not macrophage-lymphocyte interaction is illustrated. (original magnification ×4,000)

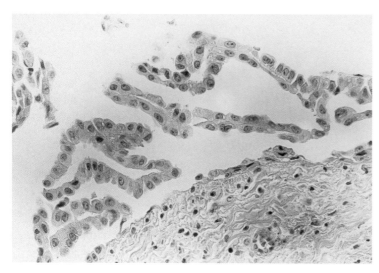

Fig. 14-13 Hyperplastic mesothelial cells in a single layer. The nuclei are round or oval with prominent one or two nucleoli. The cytoplasm is cuboidal to columnar in shape. Double nucleation is occasionally seen (Figs. 14-8 and 14-17d). See a tubular pattern of cytology (Fig. 14-17c, e) that is correspondent with this histologic feature.

Table 14-1 Differentiation between mesothelial cells and macrophages in Papanicolaou stain

	Mesothelial cells	Macrophages (histiocytes)
Shape of nuclei	Round or oval	Kidney-shaped and rarely lobulated
Location of nuclei	Nearly central	Peripheral
Chromatin pattern	Finely granular and occasionally with karyosomes	Finely granular and smoothly distributed
Nucleolus	Distinct	Indistinct
Cytoplasm	Intensely stained with acidophilic perinuclear halo (basophilic in Giemsa stain)	Weakly cyanophilic and lacy with vesicular degeneration
Cellular border	Clearly defined	Ill-defined
Grouping of cells	Distinct within syncytial cytoplasm	Singly isolated

showing negative peroxidase reaction and prominent neutral red granules, these were regarded as degenerated or aged blood monocytes. Joos et al[51] demonstrated that abrupt increase in number of peritoneal macrophages in 16 to 24 hours after injection of Latex particles was not due to local proliferation of histiocytic cells but to migration of mononuclear cells from blood into the peritoneal cavity cells.

Accordingly as immunological studies progress, the precursor of blood monocytes which may differentiate or transform to macrophages is considered to be small lymphocytes. [47] Small lymphocytes are multipotent cells which can be transformed to plasmacytoid mononuclear cells and to phagocytes in immune reaction; immune macrophages stimulated by specific antigens are enrounded with smooth outer surface and rich in lysosomes.

Macrophages, as a term in exfoliative cytology, are used to represent the cell function rather than its origin. Macrophages in malignant effusion apposed to neoplastic cells present many finger-like cytoplasmic processes (Fig. 14-10); Carr[10] interpreted this evidence as the ingestion and digestion of the tumor cells with cytotoxic factors emitted from macrophages. Distinction of macrophages from mesothelial cells is practically important.

Nucleus: The nuclei are oval, kidney-shaped and occasionally lobulated. They vary markedly in size. The eccentric position of the nuclei is specific for this type. The chromatin is finely granular and smoothly distributed. The nucleoli are invisible.

Cytoplasm: The cytoplasm is abundant, poorly bordered and faintly cyanophilic. Distinctive features are vesicular vacuolation and intracytoplasmic phagocytosis of erythrocytes and nuclear debris in various stages of digestion (Figs. 14-6 to 14-9). Occasionally shown are signet ring cells with extremely eccentric nuclei which should be differentiated from mucus-secreting cancer cells (see Figs. 14-31 and 14-32). Differentiation can be based on the following findings: Malignant signet ring cells vary in size from cell to cell and are often found in cell clusters in which the evidence of malignancy is discernible. In supravital staining, the neutral red granules stain quite rapidly and appear concentrated in the Golgi areas or scattered around the nucleus.

With Giemsa stain, the nucleus shows a lacy pattern of pale-staining chromatin. Degenerate macrophages have distorted or irregular nuclei with a loose reticular chromatin pattern. Although the enzyme content of macrophages varies with their physiological state, such readily demonstrable enzymes as acid phosphatase, β-glucuronidase, esterase of various types and peroxidase are characteristically increased in activity in macrophages of serous effusions.

2) Mesothelial cells

The mesothelium is a variety of simple squamous epithelium which is composed of flat plate-like cells joined together with desmosomes. Mesothelial cells are shed from the lining epithelium of the serous cavities, and those which are found in serous effusions are considered to have been in various stages of degeneration. Accordingly, it is important to identify the origin of mesothelium based on well preserved cells (Fig. 14-13).

Nucleus: The nuclei are round or oval and paracentral-

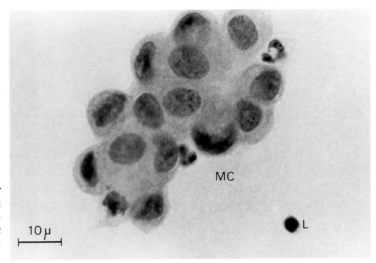

Fig. 14-14 Mesothelial cells seen in a large cluster showing an acinar pattern. The nuclei are oval and generally uniform in size except for peripherally situated pyknotic cells. The chromatin particles are smoothly and evenly distributed. MC: mesothelial cells, L: lymphocyte.

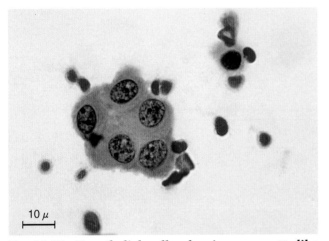

Fig. 14-15 Mesothelial cells showing a rosette-like arrangement. The nuclei are equally round or oval. The cytoplasm is syncytial and each is indistinguishable. Papanicolaou stain.

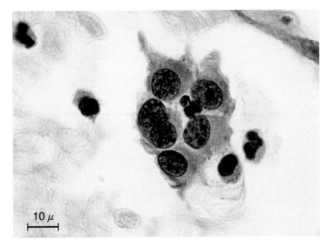

Fig. 14-16 Mesothelial cells appeared in a loose cell cluster. A flat sheet arrangement is in favor of mesothelial origin. Note equally round nuclei with fine, evenly distributed chromatins.

ly located. Occasionally, however, cuboidal mesothelial cells show an eccentric nuclear position. The size variation is more marked than shape variation. The nuclear rim is delicate, smooth but distinct. Slight indentation is occasionally noticeable; it is likely cerebral convolution. The chromatin is finely granular and smoothly distributed in a resting stage, but the chromatin pattern becomes coarsely stippled with nuclear enlargement in case of mesothelial reaction. Prominent nucleoli, occasionally indistinguishable from karyosomes, are seen (Fig. 14-17a, b). Multinucleation composed of two or three nuclei is frequently found (Fig. 14-17f, g). According to the study by Luse and Reagan,[68] multinucleation was encountered in 26% of 396 cases. Mitoses of the nuclei have nothing to do with a diagnosis of malignancy.[97]

Cytoplasm: The cytoplasm is round and of a fair amount. The cytoplasmic border, though relatively well defined, is not sharply outlined. Well preserved cells display an opaque appearance and a basophilic staining reaction with an acidophilic tinge around the nuclei. Mesothelial cells tend to form groups of several cells, usually not more than ten cells in each group (Figs. 14-15 and 14-16). They are arranged flatly with a cobble-

stone appearance (*Placards endothéliaux*) and sometimes aggregate in a rosette formation. Acinus-like arrangement as well as rosette formation may give rise to confusion with adenocarcinoma cells. The mesothelial cells, however, have nuclei of uniform size and shape (Figs. 14-14 and 14-16). There is no marked nuclear overlapping (morula-formation)[42] which is one of the distinctive criteria of adenocarcinoma.

In Giemsa stain, the cytoplasm stains deeply blue and the nucleus exhibits a compact stippling of fine chromatins. There is occasionally a perinuclear halo with a pinkish hue. Functionally mesothelial cells are able to ingest such small particles as thorotrast, but this micropinocytosis differs from true phagocytosis of macrophages. The difference of the function can be proved by India ink phagocytosis with ease.

Special stain and immunocytochemistry for cell identification: The periodic acid-Schiff (PAS) reaction is valuable for identification of mesothelial cells.[13,36,105] PAS-positive fine granules are concentrated at the periphery along the cytoplasmic border, whereas phagocytes are usually negative or weakly positive with only a few fine granules (Fig. 14-22). The behavior of malignant cells for PAS reaction is variable. Although there have been some

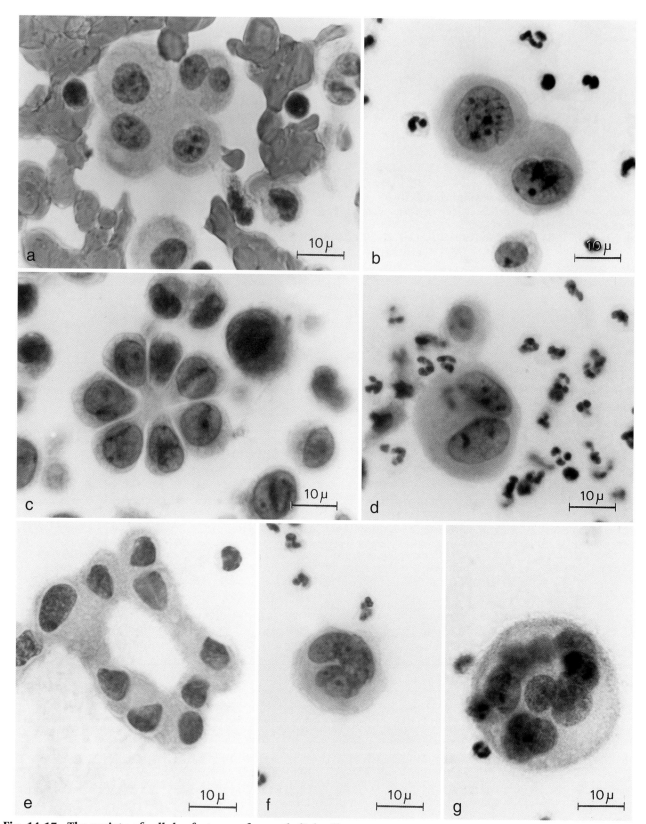

Fig. 14-17 The variety of cellular features of mesothelial cells.

a. Typical mesothelial cells with central, round nuclei and opaque cytoplasm. The cytoplasm appears to merge together. This is the most common type of mesothelial cells.

b. Large active mesothelial cells showing nuclear enlargement, hyperchromatism and some granular chromatins. Note a ground glass appearance and central position of the nuclei. Nucleoli are visible but hardly distinguished from karyosomes.

c. Group of mesothelial cells showing a rosette-like arrangement. Cerebral convolution-like indentation of nuclear outlines is also a characteristic feature of mesothelial cells.

d. Double-nucleated mesothelial cell with bland, lightly stained nuclei.

e. Group of mesothelial cells with acinar arrangement.

f. Mesothelial cell with multilobulated or multinucleated nuclei, the chromatin particles of which are evenly and smoothly distributed.

g. Multinucleated mesothelial cell with a bland chromatin pattern.

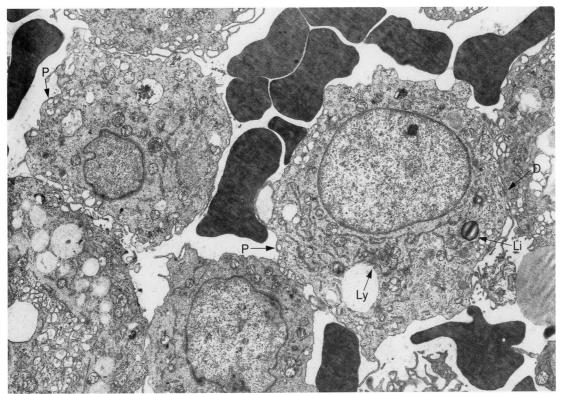

Fig. 14-18   Electron micrograph of well preserved mesothelial cells which are freely exfoliated in pericardial fluid. Pinocytotic vesicles (P) are seen at periphery. Cytoplasmic processes are not distinctive. The nuclei are oval and smoothly outlined. (original magnification ×4,000) D: desmosome, Ly: lysosome, Li: lipid.

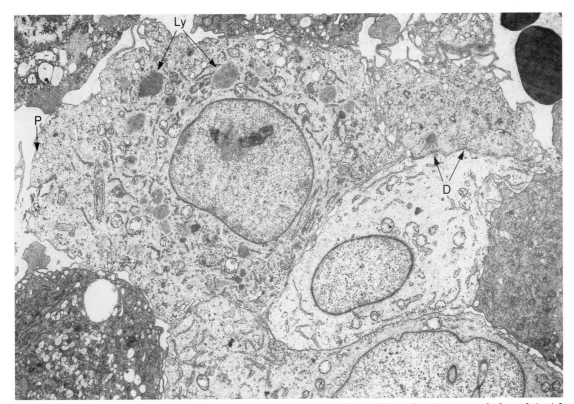

Fig. 14-19   Electron micrograph of mesothelial cells arranged in a dense sheet showing rounded nuclei with smooth nuclear envelopes. Pinocytotic vesicles and rough endoplasmic reticulum are well developed. Distinctive desmosomal junction is recognized. (original magnification ×4,000) P: pinocytotic vesicles, D: desmosome, Ly: lysosomes.

285

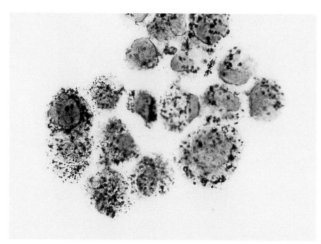

Fig. 14-20 Cytochemical LDH stain of gastric adenocarcinoma cells in ascites. Intensity of LDH is quantified by the amount of nitro-blue tetrazolium formation.

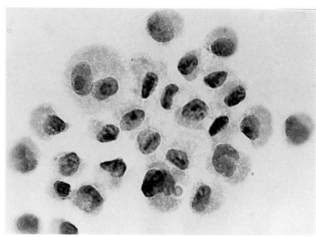

Fig. 14-21 Cytochemical quantitation of LDH M fraction by urea treatment. The same specimen as Fig. 14-20. Loss of LDH stainability by the addition of 2.6 mol urea in comparison with total LDH stain is indicative of the M fraction.

Table 14-2 Special stains for identification of cells

	Mesothelial cells	Macrophages (histiocytes)	Malignant cells (adenocarcinoma)
Periodic acid-Schiff (PAS)	Fine granules concentrated at periphery (diastase digestible)	None or few fine lacy granules at periphery	Variable (negative or intensely positive; with droplet-like or diffuse positive reaction)
Mucicarmine stain	Negative	Negative	Occasionally positive
Peroxidase reaction	Negative	Occasionally positive	Negative
Phagocytosis*	Negative	Positive	Negative
Neutral red supravital stain	A few scattered fine granules	Rapidly stain and concentrate in a rosette-like pattern	Scattered or rosette-like pattern
Sudan Black B stain[4]	Negative	Scattered granules or negative	Negative
LDH isoenzyme stain**	H subunit predominant	H subunit predominant	M subunit predominant

  * India Ink Phagocytotic Test (Takahashi and Urabe)
  1. To examine the function of phagocytosis of cells in effusions, take a part of the sediment with a capillary and mix with 1 ml of the supernatant in a small test tube. The fluid specimen should have been treated with anticoagulant. For peripheral blood, take 1 ml of blood with a heparinized syringe.
  2. Filtrate the Pelikan Drawing Ink, Günther Wagner, Germany, with a paper filter (No. 4 or 6, Toyo Roshi Co., Tokyo) twice or three times without changing the filter. Add a drop of this filtrate into the above specimen and mix gently.
  3. Stand still in an incubator at 37°C for two hours.
  4. Mix the specimen and make a thin smear.
  5. Fix and stain with Giemsa method. Papanicolaou stain can be also available.
  ** Isoenzyme fractions of lactate dehydrogenase (LDH) in effusions[107] reflect in part those of blood serum, but the fraction of such a released enzyme as LDH depends on the cell constituents in effusions; selecting the specimen in which single type of the cell either mesothelial cell or macrophage or cancer cell is preponderant, we found the M fraction to be distinctively elevated in cancer cases and characteristically inhibited by 2.6 mol urea treatment. Mesothelial cells and lymphoid cells revealed the highest third fraction that is resistant to 2.6 mol urea biochemically as well as cytochemically.

reports[60,61] that cancer cells do not present PAS-positive granules or droplets, many adenocarcinoma cells are evidently positive with large droplet-like or diffuse staining reactions (Table 14-2). In other words, negative or weakly positive PAS reaction does not indicate any specific type of cells, but strongly positive reaction is indicative of adenocarcinoma cells (Figs. 14-23 and 14-24). The PAS reaction is commonly applied to hematology to identify lymphocytic leukemic and lymphosarcoma cells. While polymorphonuclear leukocytes and megakaryocytes are diffusely stained, lymphoblasts and lymphosarcoma cells reveal perinuclear, coarsely granular positivity.[75]

Other special techniques such as India ink phagocyto-sis for identification of macrophages (Fig. 14-25) and cytochemical verification of lactic dehydrogenase for demonstration of M fraction-rich cancer cells are applied to differential diagnosis (Table 14-2; Figs. 14-20 and 14-21).

Immunocytochemistry for cell identification: Immunocytochemistry is useful for distinction of reactive mesothelial cells from cancer cells. Reactive mesothelial cells and mesothelioma cells are positive for vimentin whereas normal mesothelium does not react.[1,15] Carcinomas originally negative for vimentin may be coexpressed with cytokeratin in certain cases that take a worse clinical course.[5] Epithelial membrane antigen (EMA) is the most useful marker for distinguishing can-

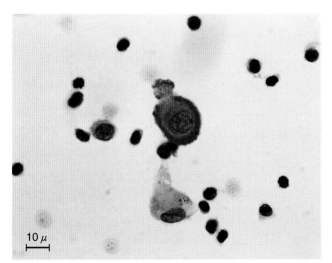

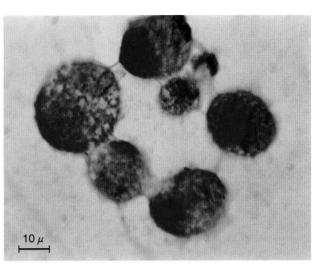

Fig. 14-22 A PAS-positive mesothelial cell and a PAS-negative macrophage. Fine PAS-positive granules at periphery of the cytoplasm of mesothelial cells are digestible by diastase.

Fig. 14-23 Large PAS-positive droplets in single cancer cells derived from gastric adenocarcinoma. The PAS stain as droplet in cancer cells is regarded as mucus production.

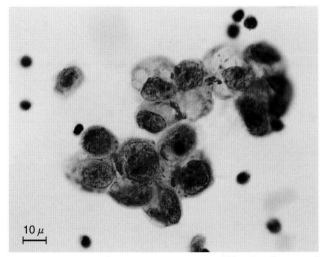

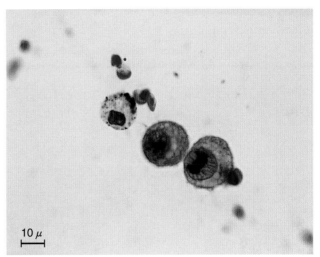

Fig. 14-24 Irregular PAS-positive speckles in clustered cancer cells from gastric adenocarcinoma. The PAS stain resistant to diastase digestion demonstrates mucus products.

Fig. 14-25 India ink phagocytosis of a macrophage. A macrophage (upper left) reveals phagocytosed ink droplets at the periphery of the cytoplasm. Two cancer cells are free from phagocytosis.

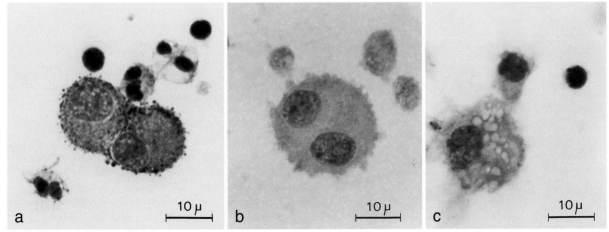

Fig. 14-26 Periodic acid-Schiff reaction for identification of various types of cells in body effusions.
a. Mesothelial cells showing fine PAS-positive granules aggregated at the periphery of the cytoplasm.
b. Mesothelial cells showing loss of PAS reaction after diastase digestion.
c. A vacuolated macrophage negative for PAS reaction. A common type of macrophage.

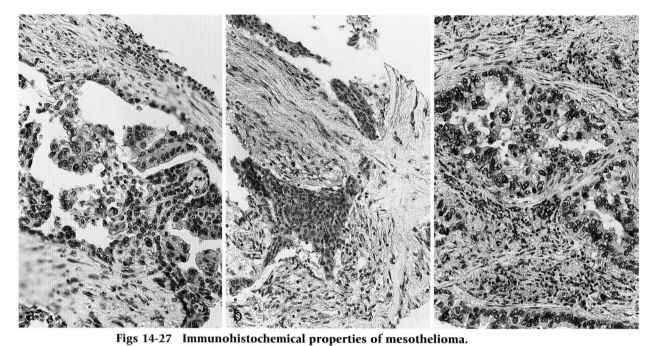

Figs 14-27   Immunohistochemical properties of mesothelioma.
a. Sporadic stain of epithelial (papillary) mesothelioma for EMA.
b. Diffuse stain of epithelial mesothelioma for EMA.
c. Positive stain of pulmonary adenocarcinoma for MOC 31.
(By courtesy of Dr. Taiying Chen, BioGenex, San Ramon, California, U.S.A.)

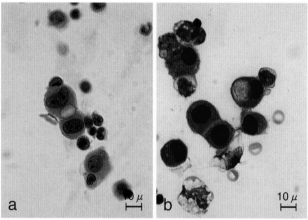

Fig. 14-28   Degenerate mesothelial cells showing vacuolation in ascites caused by liver cirrhosis. Uniform enrounded nuclei are those of mesothelial cells. A degenerative vacuole is transparent and sharply outlined. These are often hardly distinguishable from macrophages. Compare with Fig. 14-32. Note the same property of basophilic cytoplasm. a: Papanicolaou stain, b: Giemsa stain.

Table 14-3   Immunocytochemical reactivity in mesothelial cells and cancer cells

	Mesothelial cells reactive	Mesothelioma cells	Cancer cells
CEA	− , (+)	− , (+)	+, ++ **
EMA	−	− , +	++, +++
MOC-31	−	−	+, ++
Vimentin	++	++	− , +
Fibronectin	++		−
Keratin	++	+++	+, ++ *
Ber-Ep4	−	−	+

* Note 1: Adenocarcinoma positive for low molecular weight. Squamous carcinoma positive for low and high molecular weight (Bedrossian, 1994). Mesotheliomas express more high molecular weight keratin than adenocarcinomas

** Note 2: Some hyaluronate-rich mesotheliomas and reactive mesothelial cells may show false positive staining for CEA. Hyaluronidase treatment prior to staining if necessary.[91]

cer cells from reactive mesothelial cells. However, it must be kept in mind that EMA is reactive in malignant mesothelioma (Fig. 14-22).[110] Vimentin and EMA are good opposite markers between reactive mesothelial cells and cancer cells, but these are not absolute clues for distinction (Table 14-3). Vimentin-positive adenocarcinomas reveal its expression towards the basal pole of cancer cells, while mesotheliomas are positive homogeneously in the cytoplasm.[4] Bedrossian describes that EMA positivity ranges from 44% to 98% in mesothelioma and 62% to 100% in cancer.

Immunocytochemistry for Ber-Ep4 that demonstrates positivity along the surface of cancer cells and negative reactivity of mesothelium and mesothelioma are wor-

thy for the differential diagnosis.[22,61]

Monoclonal antibody MOC-31 (ESA: epithelial specific antigen) recognizes a membrane glycoprotein of 40-kd molecular weight and shows positive stain on adenocarcinoma cells but not on either mesothelial ceflls or mesothelioma cells (Fig. 14-27c).[94]

CEA is selectively positive for adenocarcinoma arising from the gastrointestinal tract but low positive in cancers of other organs. Bedrossian has stressed the texture of CEA stain: Adenocarcinomas showing CEA secretory differentiation illustrate finely stippled or flocculent pattern, whilst poorly differentiated carcinomas reveal a compact, diffuse staining pattern of the cytoplasm.

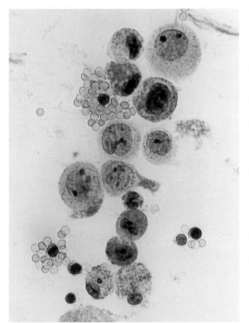

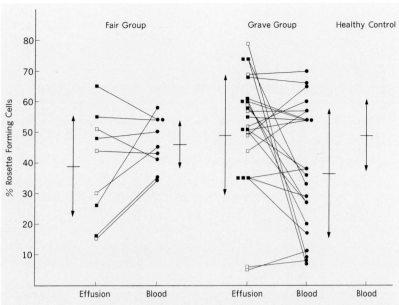

Fig. 14-29 E-rosette forming lymphocytes in ascites with gastric adenocarcinoma. E-rosette forming lymphocytes are seen in two of six lymphocytes.

Fig. 14-30 Percentage of rosette forming lymphocytes in effusions and peripheral blood of cancer patients. The ranges shown with arrowhead bars represent the mean ± 1 SD in each group.(From Tani EM, Takahashi M: J Jap Soc Clin Cytol 19: 534, 1980.) ■Effusion without cancer cells, □Effusion with cancer cells, ●Peripheral blood.

3) Degenerate mesothelial cells

Mesothelial cells in degenerating process show vesicular or vacuolar degeneration of the cytoplasm and displacement of the nuclei to an eccentric position. The chromatin is not increased in quantity but shows a roughly stippled pattern because of degenerative condensation. They are often hardly discriminated from macrophages merely from the morphological standpoint (Fig. 14-28).

4) Inflammatory cells

Neutrophils, eosinophils, lymphocytes, plasma cells and erythrocytes are recognizable in effusions. The proportion and types of blood cells depend on the nature and stage of the inflammation. The cellularity and cytogram are important for interpretation of a pathophysiological state in body cavities. The preponderance of erythrocytes indicates hemorrhage. A mixture with phagocytes and reactive mesothelial cells may be suggestive of chronic extravasation. Marked lymphocytosis in tuberculosis, plasmacytosis in monoclonal gammopathy and leukocytosis in acute inflammation are of a distinctive pattern (see Fig. 14-37).

5) Lymphocytes with reference to phenotyping

It is frequent that we encounter a large number of lymphocytes in excess of fluid in body cavities secondary to malignancy including chronic lymphatic leukemia and lymphoma. There may occur simple questions: (1) Which lymphocytes predominate, T cells or B cells? (2) What role do these lymphocytes play in malignant effusion? i.e., are they concerned with production of immunoglobulins or active defense reaction against cancer cells? Domagala and his associates[25] have found major population of lymphocytes in leukemia or lymphoma to be composed of B cells whereas in inflammatory process to be comprised of T cells. In another word,

the case with B cell predominance is indicative of lymphoma.

It is postulated that cell-mediated immune mechanism plays an important role to prevent the spread of cancer cells; this evidence is proved from the fact that there occurs hyperplasia of paracortical areas in lymph nodes draining the locus of malignant neoplasm.[112] There is a possibility that long-lived T lymphocytes play an afferent immune reaction producing lymphotoxin to inhibit the mitotic activity and specific macrophage-arming factor to induce phagocytic activity of macrophages.[29,30] In regression of neoplasms, lymphoreticular reaction against tumor cells is manifested by close action of activated macrophages on the tumor cells: Observations on neoplastic cells encircled by many flap-like and finger-like cytoplasmic processes of macrophages as a cytotoxic phenomenon have been reported.[11,52] (see Fig. 14-11). The release of cytotoxic factors by macrophages and phagocytosis of tumor cells followed by lysosomal digestion are possibly induced by the effector lymphocytes.

Practically, spontaneous rosette formation of lymphocytes with sheep erythrocytes and blastoid transformation against PHA are simple and reliable techniques to determinate the classification of T cells.[127] Several reports have referred to the fact that the patients with various types of cancers are accompanied by abnormalities either in absolute levels or in the relative proportions of circulating T and B cells.[40,44,62] The lymphocyte count and its T cell population of peripheral blood are considered to be prognostically significant because a cell mediated immune response plays an important role in the defense against malignancy: Mitogen reactivity of T cells[44] and the number of rosette forming cells[40] are depressed in patients with advanced carcinoma, whereas

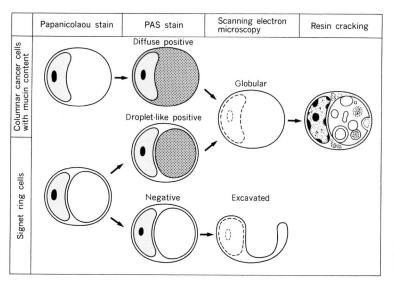

	Papanicolaou stain	PAS stain	Scanning electron microscopy	Resin cracking

Fig. 14-31 Entity of signet ring cells. A PAS-negative vacuole that is linearly outlined is due to cave-like excavation of the cytoplasm.

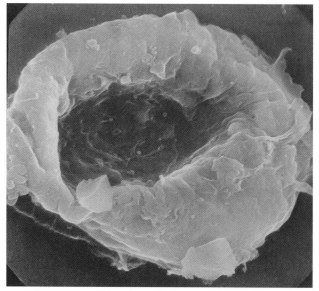

Fig. 14-32 Scanning electron micrograph of a benign signet ring form of histiocyte. A doughnut-like shape of the cell does not represent a hole but excavation. (original magnification ×6,000)

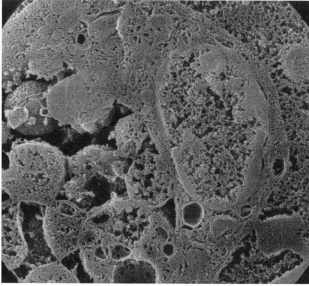

Fig. 14-33 Scanning electron micrograph of a columnar cancer cell after resin crack. This columnar cancer cell is from adenocarcinoma of the stomach that contained PAS-positive droplet in the cytoplasm light microscopically. Note an eccentric nucleus and irregular cisterns. (original magnification ×10,000)

they are not impaired in the patient with localized growth. Specific immunoglobulins in effusions are not excessive and almost equal in amount as those of blood serum.[65] However, there have appeared a few reports concerning the T cell population in effusions from a prognostic aspect. Tani and Takahashi[108] investigated on the T cell population in body fluid comparing with that of peripheral blood (Figs. 14-29 and 14-30). Although the patients who have pleural or peritoneal effusion caused by malignancy are already in the advanced stage of malignancy, some informative difference was suggested between two subjects having been divided into a grave or fair group from their physical states. Whereas not much difference of T cell population between effusion and peripheral blood was found in the fair group, marked decrease in number of circulating T cells comparing with those of effusions was noted in the grave group. In other words, the population of E-rosette forming cells and PHA responsiveness of circulating lymphocytes are more informative prognostically rather than local population of T cells in body cavity fluids.

6) What does signet ring cell mean ?

Signet ring cells do not mean a specific cell type but include benign and malignant cells which are characterized by eccentric nuclei replaced by intracellular vacuoles or vacuolar configurations. A malignant ring cell filled with mucus content is possessed of characteristic malignant nucleus with hyperchromatism and coarse chromatins. The vacuolar mucus products show some opaque density and ill-defined boundaries; PAS stain often indicates a droplet-like mucus content[128] (Fig. 14-23). Benign signet ring cells are derived mostly from macrophages and occasionally from mesothelial cells as the result of degeneration. The degenerative vacuoles are

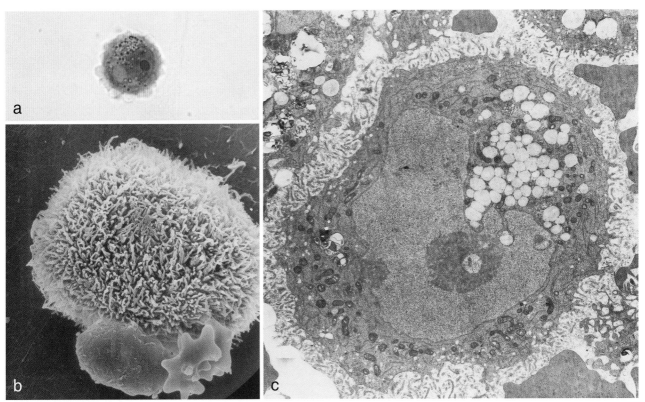

Fig. 14-34   Single cancer cell in ascites secondary to gastric adenocarcinoma.
a. Sternheimer-Malbin supravital stain showing a prominent nucleolus and many refractive granules alongside of the nucleus.
b. Scanning electron micrograph showing numerous microvilli. A lymphocyte and a red cell are attached to it. (original magnification ×4,000)
c. Electron micrograph with well developed microvilli. Note a prominent nucleolus with compact texture. (original magnification ×5,400)

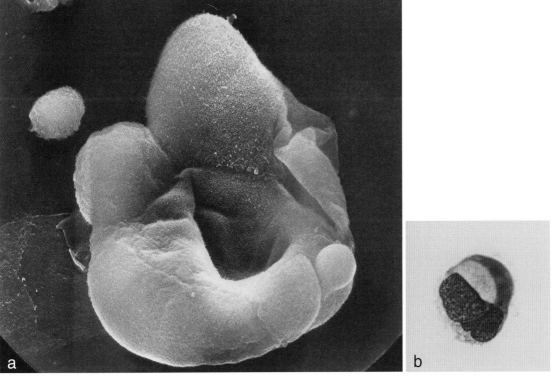

Fig. 14-35   Clustered adenocarcinoma cells varying in size. Note marked excavation of cytoplasm that is correspondent to signet ring-like vacuolation under light microscopy.
a. Scanning electron micrograph. (original magnification ×2,000)
b. Diff-Quik stained cancer cell with marked vacuolation. Diff-Quik is a commercial rapid stain with similar stainability to the Giemsa method.

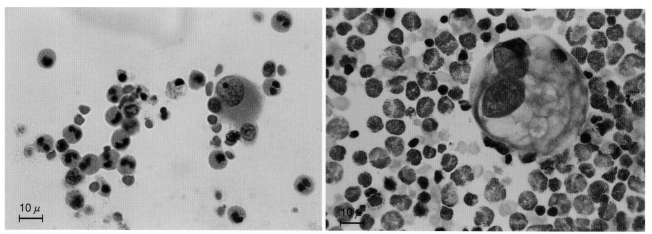

Fig. 14-36   Pleural eosinophilia associated with malignancy. Tremendous eosinophilia in pleural effusion occurred temporarily in association with shedding of adenocarcinoma cells from pulmonary carcinoma. Papanicolaou-stained eosinophilic leukocytes display fine yellowish particles.

light microscopically distinguished by sharp contours on focusing in and out and transparency. Scanning electron microscopy demonstrates a deep cave-like hollow which is apparently responsible for signet ring formation in light microscopy. The evidence that the vacuolar appearance is not due to intracytoplasmic vacuole but a hole-like concavity is proved by the presence of cell surface structures at the bottom of a hollow (Figs. 14-32 to 14-35).

4 Effusions in Benign Pathological Conditions

Eosinophilia in effusions

Preponderance of eosinophils in effusions is informative of its causative conditions such as traumatic or spontaneous pneumothorax,[102] pneumonia, pulmonary infarction, malignant neoplasm,[26,82,114,129] parasitic infestation (amebiasis,[18,21] ascariasis,[89] ruptured hydatid cyst,[93] etc) and allergic diseases.[115] The allergic conditions are those of rheumatic pericarditis[84,90] and allergic vasculitis.[45,46,79,120] According to a study by Spriggs, the causative diseases of 37 cases with pleural eosinophilia (exceeding 20% eosinophils) were pneumothorax and hemothorax (32.4%), pneumonia (13.5%), malignant neoplasm (13.5%)(Fig. 14-36), and pulmonary infarction (10.8%). Concerning eosinophilia in pleural effusions that occurs in association with pneumothorax, histological examination has proved that proliferative pleural reaction with marked eosinophilic and histiocytic infiltration is consistent with eosinophilic granuloma. Mesothelial reaction with multinucleation is often associated with this pathologic change in the localized lesion.[3]

Acute suppurative inflammatory effusions

Acute suppurative peritonitis and pleuritis are caused by perforating lesions of various organs into the serous cavities and consequent bacterial infection. Acute inflammation also occurs without perforation probably by extension along the lymphatics. Migratory peritonitis[121] or transdiaphragmatic peritonitis (*Durchwanderungsperitonitis*) is the typical case, which may result from

inflammation close to the peritoneum. Migratory pleuritis may occur *vice versa*. Etiological microorganisms for suppurative inflammation are as follows: Streptococci, staphylococci, *Escherichia coli*, pneumococci (in nephrotic children), gonococci and anaerobic bacilli are the common agents for suppurative peritonitis, while streptococci, pneumococci and staphylococci are most commonly implicated for suppurative pleuritis. Neutrophilic polymorphonuclear leukocytes are predominant in acute suppurative inflammation. Mesothelial cells, more or less degenerated, are present with neutrophils for a short period of initial stage of inflammation.[103] They soon disappear and are replaced by numerous neutrophils. Other cellular components, macrophages and lymphocytes are few and less than 10% of total cell populations. The presence of cellular debris is characteristic.

Acute non-suppurative inflammatory effusions

An excess of pleural effusion may occur secondary to pneumonia, influenza, pulmonary infarction, subphrenic abscess and hepatic abscess. Retention of fluid is generally small in amount. Cellular components are variable and include polymorphonuclear neutrophils, lymphocytes in large numbers, macrophages and active mesothelial cells showing multinucleation and mitoses. The cytology of effusions *per se* is not diagnostic. Awareness of clinical findings is quite important when considering the cellular pattern. As a general pattern, parapneumonic effusions are abundant in polymorphonuclear leukocytes.

Effusions secondary to pulmonary infarction will be often associated with a sudden pleuritic pain, hemoptysis and a slight fever. The cellular pattern is similar to that found in postpneumonic pleurisy, but there is a striking predominance of active mesothelial cells and erythrocytes. The number of mesothelial cells occasionally may account for up to 20% to 30%.[101]

Tuberculous effusions

Effusions due to tuberculous inflammation are serofibrinous and may even be fibrinopurulent in appearance with a bloody tinge and are persistently lymphocytic

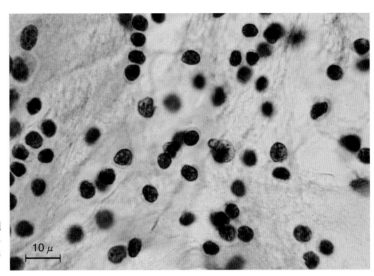

Fig. 14-37 Marked lymphocytosis of pleural fluid secondary to pulmonary tuberculosis. Tuberculous effusion reveals such marked lymphocytosis in association with fibrin exudation.

except for the early stages of acute inflammation when variable cellular patterns are noticeable; there may be 10% to 30% polymorphonuclear neutrophils and some mesothelial cells in addition to the lymphocytes. Predominance of lymphocytes admixed with activated lymphocytes may lead to a misdiagnosis of lymphoma (Fig. 14-38). Tuberculous lymphocytosis can be distinguished from lymphoma by the uniformity and/or preponderance of mature lymphocytes. A predominance of T cells is indicative of benignancy.[25] Fibrin-rich background is often favorable to tuberculosis (Fig. 14-37). In tuberculous peritonitis, loops of small intestine are matted together by adhesion, so that only a part of the effusions can be obtained by paracentesis. In the chronic healing stage of tuberculous pleurisy, fibrous adhesions may result in formation of sacculated spaces accumulating a clear transudate.

Tuberculous empyema is a rare condition at present. If it develops, the cytology of the effusions abruptly changes from lymphocytosis to neutrophilic preponderance.[78] When it lasts for a long time, the fluid occasionally turns milky white in appearance (chyliform effusion) and may even contain cholesterin crystals (cholesterol effusion).[20] Since the fat content is generally low, it differs from chylous ascites. The highly refractile cholesterin crystals may be stained by Schultz's method or identified by differential interference microscopy.[99]

■Effusions secondary to liver cirrhosis

A decrease in colloid osmotic pressure and portal hypertension are responsible for ascites in cases of liver cirrhosis. The cause of ascites may be easily established by clinical features and laboratory findings. A dynamic mechanism of ascites formation in liver cirrhosis should be considered on intrahepatic (postsinusoidal) and prehepatic (presinusoidal) flow hypertension as well as decreased osmotic pressure (hypoproteinemia); in early cirrhosis, postsinusoidal obstruction by regenerative nodular hyperplasia promotes the formation of excessive lymph and leakage of fluid primarily from the liver, while in late cirrhosis, presinusoidal obstruction by scarring produces inflow block.[125] The cytology of ascites is particularly important from two aspects (see Fig. 14-9):

1) Active mesothelial cells in jaundiced ascites are apt to be misinterpreted as malignant cells particularly hepatoma cell.

2) It may happen rarely that the cytology is indicative of malignant transformation from hepatitis-associated cirrhosis to hepatoma. In such a case immunological quantitation of alpha-fetoprotein is confirmative in determination of hepatoma.

Active mesothelial cells which shed in acinus-like aggregates and possess enlarged nuclei, showing somewhat granular karyosomes and one or two prominent nucleoli can be distinguished from adenocarcinoma cells by the following cellular features; the nuclei of reactive mesothelial cells are fairly uniform in size and shape, and their nuclear overlappings are less prominent than in adenocarcinoma cells. In addition to the above mentioned mesothelial cells, varying numbers of active macrophages, neutrophils and erythrocytes are associated particularly in posthepatitic cirrhosis.

■Effusions secondary to congestive heart failure

Transudates are common in peritoneal and pleural cavities secondary to congestive heart failure. The cellular pattern is similar to that of the effusion caused by hepatic cirrhosis of Laennec's type, but the sediments are so few that cellular elements are scarcely recognizable by ordinary centrifugation.

5 Malignant Effusions

■General consideration for malignant effusions

It should be noted that malignant tumors of the intrathoracic and intra-abdominal organs give rise to effusions in two ways:

a) Firstly, when the serous membranes are directly involved by tumor invasion, malignant cells will be detectable. However, there are two kinds of cases with malignant effusions: (1) Malignant tumors develop to dissemination on serous membranes by implantation metastases, and yield exfoliation of numerous malignant cells into the body cavities as shown in animal trans-

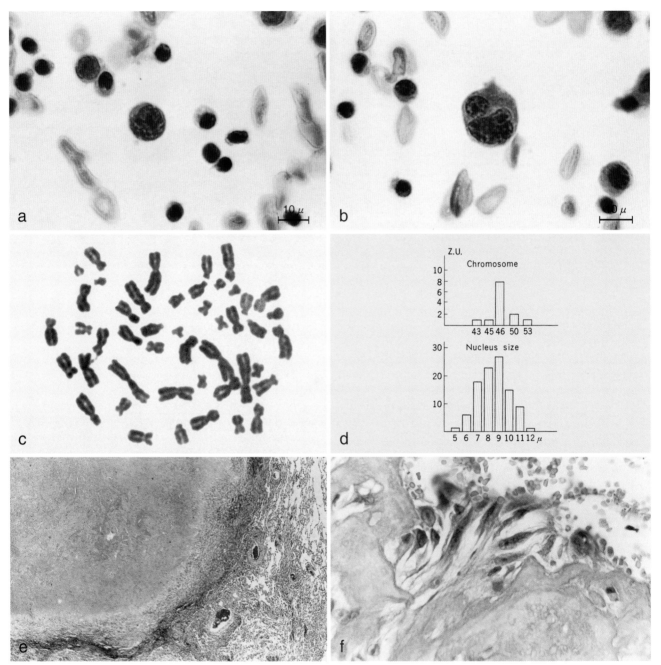

Fig 14-38 Tuberculous pleural effusion of a 49-year-old man showing lymphocytosis and scattered atypical cells misinterpreted as malignant.

a, b. Fused atypical cells with hyperchromatic multinucleation. A single atypical cell is most probably activated lymphoid cell (a). These are likely to be activated mesothelial cells showing syncytial aggregation (b).

c, d. Direct chromosomal analysis after thoracentesis. Note predominance of diploid cells with normal karyotypes. Several cells in near diploidy showed no evidence of abnormal chromosomes.

e. Low power view of the resected lung. Lobectomy of the right upper lobe was performed because of suspicious cytology in small amount of pleural fluid and a localized tumor-like density in the apex. Histologically, the lesion was the so-called tuberculoma in subpleural parenchyma showing massive caseous necrosis. It is demarcated by fibrous wall with epithelioid cell infiltration.

f. High power magnification of the visceral pleura showing hyperplastic mesothelial cells with round or spindly hyperchromatic nuclei. They may be responsible for atypical cytology.

Summary: There is no case of malignant effusion with lymphoma or leukemia showing only a few abnormal immature lymphoid cells mixed with large population of mature lymphocytes. In another word, lymphoma or leukemia shows monomorphous, monoclonal population of abnormal lymphoid cells.

plantable ascitic tumor. Implantation peritoneal carcinomatosis by ovarian serous papillary cystadenocarcinoma is a typical case. (2) Malignant tumors growing in intra-thoracic and intra-abdominal organs progressively involve their serous membranes, from which relatively small numbers of malignant cells may shed into the body cavities admixed with activated mesothelial cells. Hepatoma (primary hepatocellular carcinoma) is a good example of this type of tumor infiltration.

b) Secondly, serous effusions may occur by secondary inflammation or circulatory disturbance. For instance, pleural effusion will be caused by bronchopneumonia secondary to bronchogenic carcinoma. In Budd-Chiari syndrome occasionally caused by occlusion of hepatic veins by hepatoma and in metastasis of malignant tumor to the liver that causes portal obstruction, a transudate form of ascites without shedding of malignant cells may occur.

As an important sign of malignancy-associated effusion, appearance of serosanguinous effusion is significantly frequent in cancer cases regardless of positivity of cytology. Broghamer et al found serosanguinous pleural fluid in 104 of 145 patients who were histologically confirmed to have intrathoracic malignancies. It is most likely due to tumorigenesis promoting capillary growths and permeability on the serosal membrane involved by cancer cell infiltration.[8]

Since varying types of cancer cells and sarcoma cells are observed in malignant serous effusions, cell typing of the malignant cells is important and interesting for the possible interpretation of their primary sites. Cell typing should proceed as follows; (1) whether the malignant cells are of epithelial or mesenchymal origin, i.e., carcinomatous or sarcomatous, (2) whether the cancer cells are of glandular, epidermoid or anaplastic type, and (3) whether the malignant cells have specific cell features indicative of primary neoplasms or not. For example, pseudociliation and psammoma bodies in ovarian serous papillary cystadenocarcinoma, melanin pigment in malignant melanoma and azurophilic granules in myelogenous leukemia are of diagnostic value (Figs. 14-47 to 14-49, 14-86 and 14-119).

■Features of malignant cells

Malignant cells in serous effusions mostly derive from adenocarcinoma and exhibit the typical features of glandular type cells as follows:

1) Adenocarcinoma cells

Globular type: This type shows a characteristic cell cluster as called cell ball or morula-like cell aggregate. A large number of cells are closely packed inside the ball, and usually arranged in parallel at the boundary (Figs. 14-45, 14-69 and 14-79). Vacuole formation is usually absent. This type is most commonly found in the case with duct carcinoma of the breast; it is most likely to be originated from medullary tubular carcinoma that is composed of solid masses of cancer cells intervened by stromal connective tissue. Included in this type are cell balls with petal-like cell arrangement at periphery; petal-like cancer cell projections outwards are indicative of papillary tubular pattern of adenocarcinoma (Figs. 14-45 and 14-46).

Tubular type: This is the common type of adenocarcinoma in serous effusions and is characterized by cell clusters varying in size and appearance; a large trabecular or tubular column consisting of columnar cancer cells is a prototype (Fig. 14-43). Great variation in size and arrangement of cancer cells is characteristic of common tubular adenocarcinoma. Cell vacuolation is often associated (Fig 14-39).

Mucus-secreting type: This type of malignant cell shows intracytoplasmic vacuolation filled with mucus. The vacuoles which are sharply outlined are due to degenerative change. A signet ring cell which is the extreme form of

Table 14-4 Primary tumors responsible for malignant body cavity effusions

Primary tumor	Johnson Patients	%	Murphy and Ng Patients	%
Carcinoma of lung	80	23.2	16	16.5
Carcinoma of breast	51	14.8	37	38.1
Carcinoma of ovary	45	13.0	23	23.7
Lymphoma and leukemia	39	11.3	0	·
Carcinoma of colon	21	6.1	4	4.1
Carcinoma of stomach	19	5.5	5	5.2
Carcinoma of endometrium	–		6	6.2
Carcinoma of pancreas	–		2	2.1
Carcinoma of gallbladder	–		1	1.0
Renal cell carcinoma	–		1	1.0
Carcinoma of thyroid	–		1	1.0
Carcinoma of cervix	–		1	1.0
Others (various sites)	57	16.5		
Unknown sites	33	9.6		
Total	345		97	

this type should be defined by PAS or alcian blue stains (Figs. 14-23, 14-31 and 14-41).

Single columnar type: Single malignant cells isolated from poorly differentiated adenocarcinoma show a columnar or cuboidal shape with peripheral nuclei (Fig. 14-40).

Since the above subtypings are common in adenocarcinomas of various organs such as the ovary, the uterine corpus, the digestive tract, the breast, the lung and the prostate, the primary sites of carcinomas cannot be determined only from the cellular features.

2) Epidermoid carcinoma cells

The esophagus, the lung and the uterine cervix are the sites involved by epidermoid carcinoma, however exfoliation of the cancer cells into body cavities is rarely encountered; 6.8% in 176 cases of malignant pleural effusions and 0.9% in 213 cases of malignant ascites may exhibit epidermoid cancer cells (Table 14-5). The epidermoid cancer cells exfoliate usually as single cells and occasionally as clustered cells. Involvement of the serosal membrane is not likely disseminated serosal implantation, therefore epidermoid cancer cells exfoliated are small in number (Fig. 14-67). These single cancer cells shed into fluid specimens are not so pleomorphic as those directly scraped from epidermoid carcinoma and tend to be enrounded (Figs. 14-72 and 14-73). The nuclear rims are rigid and the cytoplasm is thick and distinctly bordered. Fine lamellar fibrils are diagnostic of the epidermoid type. Clustered cells are arranged in a flat sheet with a cobblestone appearance. The cancer cells arising from the cancer pearls are arranged in a lily bulb-like pattern.

3) Background of malignant effusion and its significance

1. Malignant effusions are hemorrhagic either macroscopically or microscopically.

2. Lymphocytosis of effusions is associated with various diseases including malignancy and is especially copious in tuberculous effusions. In case of malignancy the preponderance of T lymphocytes means host's local

Table 14-5   Primary sites and cell types in malignant effusions

A: *Pleural effusion*

Primary tumor	Male					Female				
	Gland. type	Squam. type	Undiff. type	Others	No (%)	Gland. type	Squam. type	Undiff. type	Others	No (%)
Ca. of lung	36	4	8	0	48(49.0)	19	2	0	0	21(26.9)
Ca. of stomach	14	0	0	0	14(14.3)	7	0	0	0	7(9.0)
Ca. of breast	0	0	0	0	0(0)	19	0	0	0	19(24.4)
Ca. of esophagus	1	4	0	0	5(5.1)	0	1	0	0	1(1.3)
Ca. of pancreas	3	0	0	0	3(3.1)	2	0	0	0	2(2.6)
Ca. of colon	3	0	0	0	3(3.1)	2	0	0	0	2(2.6)
Ca. of ovary	−	−	−	−	−	3	0	0	0	3(3.8)
Ca.of bile duct	1	0	0	0	1(1.0)	1	0	0	0	1(1.3)
Ca.a of uterus	−	−	−	−	−	0	1	0	0	1(1.3)
Unknown site	7	0	0	3	10(10.2)	9	0	0	2	11(14.1)
Lymphoma, Leukemia	−	−	−	10	10(10.2)	−	−	−	5	5(6.4)
Multiple myeloma	−	−	−	1	1(1.0)	−	−	−	1	1(1.3)
Seminoma, dysgerminoma	−	−	−	1	1(1.0)	−	−	−	1	1(1.3)
Malignant teratoma	−	−	−	1	1(1.0)	−	−	−	1	1(1.3)
Neuroblastoma	−	−	−	0	0(0)	−	−	−	1	1(1.3)
Malignant thymoma	−	−	−	1	1(1.0)	−	−	−	0	0(0)
Sarcoma	−	−	−	0	0(0)	−	−	−	1	1(1.3)
Total	65	8	8	17	98	62	4	0	12	78

B: *Ascites*

Primary tumor	Male					Female				
	Gland. type	Squam. type	Undiff. type	Others	No (%)	Gland. type	Squam. type	Undiff. type	Others	No (%)
Ca. of stomach	63	0	0	0	63(55.8)	44	0	0	0	44(44.0)
Ca. of colon	9	0	0	0	9(8.0)	5	0	0	0	5(5.0)
Ca. of ovary	−	−	−	−	−	12	0	0	0	12(12.0)
Ca. of bile duct	7	0	0	0	7(6.2)	4	0	0	0	4(4.0)
Ca. of pancreas	5	0	0	0	5(4.4)	6	0	0	0	6(6.0)
Ca. of liver	5	0	1	0	6(5.3)	2	0	0	0	2(2.0)
Ca. of uterus	−	−	−	−	−	3	1	0	0	4(4.0)
Ca. of lung	0	0	0	0	0(0)	3	0	0	0	3(3.0)
Ca. of bladder	0	0	0	2	2(1.8)	0	0	0	0	0(0)
Ca. of esophagus	0	0	0	0	0(0)	0	1	0	0	1(1.0)
Unknown site	10	0	0	3	13(11.5)	9	0	0	1	10(10.0)
Lymphoma, Leukemia	−	−	−	6	6(5.3)	−	−	−	1	1(1.0)
Multiple myeloma	−	−	−	1	1(0.9)	−	−	−	1	1(1.0)
Seminoma, dysgerminoma	−	−	−	0	0(0)	−	−	−	1	1(1.0)
Malignant teratoma	−	−	−	0	0(0)	−	−	−	2	2(2.0)
Granulosa cell tumor	−	−	−	−	−	−	−	−	1	1(1.0)
Neuroblastoma	−	−	−	0	0(0)	0	−	−	1	1(1.0)
Sarcoma	−	−	−	1	1(0.9)	−	−	−	1	1(1.0)
Melanoma	−	−	−	0	0(0)	−	−	−	1	1(1.0)
Total	99	0	1	13	113	88	2	0	10	100

immune response against cancer cells (Fig. 14-30), but the number does not parallel with a total T cell population of peripheral blood, that is important from a prognostic aspect.

3. Histiocytes or macrophages frequently increase in number and their phagocytic activity and neutral red stainability are intensified.

4. Many lipophagocytes may be demonstrated by Sudan III or other fat stains. It is known that actively proliferating malignant neoplasms tend to degenerate quite rapidly. Therefore, Sudan III-positive granules are shown in degenerative cancer cells as well as in macrophages surrounding the necrotic tumor tissues. The triphenyltetrazolium (TPT) reaction,* owing to its liposolubility, is available for demonstration of macrophages filled with coarse TPT granules and cancer

---

*TPT reaction is simply performed as follows: (1) Centrifuge serous effusion at 1,000 rpm for 5 minutes, (2) place a drop of sediment on a clean glass slide to which add 1% TPT saline solution to be mixed, (3) put a cover slip sealing with paraffin and keep it in an incubator at 37 ℃ for 30 minutes.

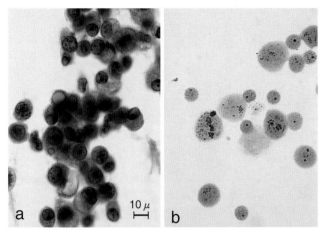

Fig. 14-39 Tubular cluster of gastric adenocarcinoma cells.
a. Papanicolaou stain. Although the nuclei are hyperchromatic and possess prominent nucleoli, they are rather small for cancer cells. Single columnar cancer cells are also seen.
b. Ag-As stain for nucleolar organizers. Whereas mature lymphocytes disclose a single spot nucleolar organizer region, cancer cells can be distinguished by increase in number and irregular dispersion of the nucleolar organizer regions. Application of Ag-As stain is simple and valuable to account for cytogenetic abnormality.

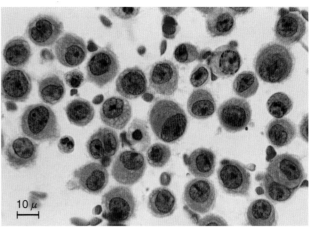

Fig. 14-40 Single adenocarcinoma cells in ascites derived from gastric carcinoma. Large hyperchromatic nuclei are eccentric in position and prominent nucleoli are present. Exfoliation of single type of cancer cells as many as in this smear is frequent in malignant effusion secondary to gastric carcinoma.

cells with fine granules. This reaction depends on the reduction of colorless TPT by succinic dehydrogenase to the reddish triphenylformazan which is soluble in lipid and thus easily discernible of sudanophilic substance.[2]

6 General Cellular Pattern in Carcinomas of Various Organs

■Carcinoma of the stomach

The majority of gastric carcinomas are derived from foveolar epithelium and reveal various degrees of differentiation even in the same tumor. Commonly shown in effusions are tubular and/or isolated columnar types which may be referable to as tubular and poorly differentiated adenocarcinomas respectively (Figs. 14-39 and 14-40). Signet ring cells are also frequently encountered singly or in small clusters (see Figs. 14-23 and 14-41), although the term signet ring cells can be used in a broad sense not only for malignant cells but for benign macrophages when large vacuoles replace the nuclei towards periphery. Signet ring cells with mucus products which are positive for PAS stain usually originate from mucinous carcinoma (signet ring cell carcinoma). Thus signet ring cell denotes a form of mucinous adenocarcinoma in cytology.

Cell typing for malignant cells in ascites secondary to gastric carcinoma disclosed a tubular type to be most common (38.8%) and a signet ring type to be the next (32.7%).

■Carcinoma of the colon

Histologically most of carcinomas of the colon are composed of atypical tubular structures multilayered by tall columnar cells (*adenocarcinoma tubulare cylindrocellulare*) (Fig. 14-43). Thus the tubular type of malignant cells

with nuclear palisading is often suggestive of colon carcinoma. There is also an apparently mucus-producing carcinoma; signet ring cells occur singly or often forming aggregates into the peritoneal cavity.

■Carcinoma of the ovary

It is serous cystadenocarcinoma that frequently produces malignant effusions caused by ovarian cancers in the peritoneal cavity. In this tumor numerous cancer cells of the papillo-tubular type will be found. Vacuolation in these papillo-tubular cell clusters is in favor of ovarian carcinoma (Fig. 14-44). Psammoma bodies in cell clumps are rarely encountered but, if present, are indicative of ovarian papillary cystadenocarcinoma (Fig. 14-47).[57] They may also occur in association with benign papillary proliferation of the ovary[55,85]. Spriggs[101] described a specific cellular feature as called pseudociliation which corresponds to *cystadenoma cilioepitheliale*: ciliated cellular features are more clearly demonstrated by Giemsa stain than by Papanicolaou stain (Fig. 14-49). Scanning electron microscopy demonstrates the characteristic cilia-like structure as abnormal specialization (Fig. 14-50).

An unusual occurrence in the case of mucinous cystadenoma is the spillage of tumor cell secreting mucinous material into the peritoneal cavity (2.9% of 102 cased by Woodruff et al[126]). This may develop into so grave condition that the bowels are embedded in thick mucinous material (*pseudomyxoma peritonei*). The tumor cells originated from mucinous cystadenoma will be scarcely found as clusters in mucinous fluid; thick needle as used for lumbar puncture (18- to 20-gauge) is needed for paracentesis (Fig. 14-52). They are arranged in sheets with a cobblestone appearance or in cell balls with elongated clear cytoplasm. The nuclei are ovoid and fairly uniform in size. The PAS-positive amorphous material surrounds these cell clusters (Fig. 14-54). There is an occasion that elongated, spindly tumor cells look free-floating within mucus (Fig. 14-55). If the cell block preparation is applied, these cytologic features are comparable with characteristic, histologic pattern in muci-

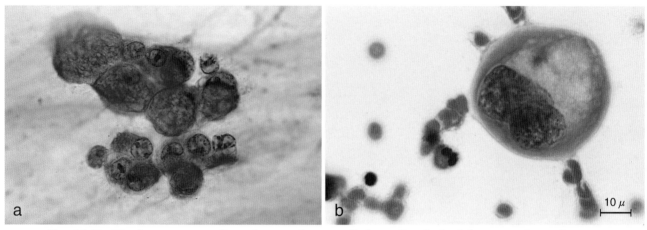

Fig. 14-41 Single mucus-secreting adenocarcinoma cells in ascites derived from gastric carcinoma.
a. Note signet ring cells filled with alcian blue positive mucus. Eccentric nuclei possess prominent nucleoli.
b. A perinuclear halo with pinkish hue is due to mucus content.

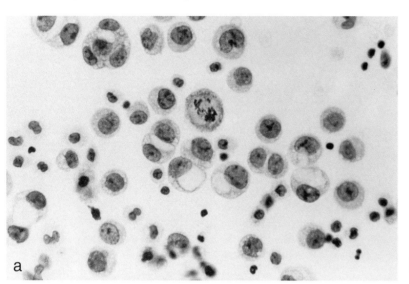
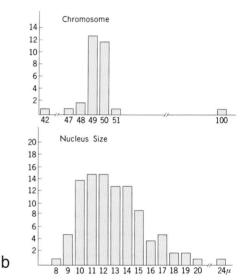

Fig. 14-42 Malignant cells of single columnar type in ascites caused by gastric adenocarcinoma and their chromosome analysis.
a. The malignant cells are numerous like a pure cell suspension in ascitic fluid. Coarse chromatin clumping, irregular karyosomes and abnormal mitosis meet the malignant criteria. However, the variation in nuclear size and in chromatin content is not so pronounced as in other polyploid and aneuploid cancer cells.
b. Chromosomal studies showed two hyperdiploid modal cells with 49 and 50 chromosomes. The variation in nuclear size of these cells revealed a relatively narrow range from 10 to 15 μm in diameter. Nuclear size and nuclear-cytoplasmic ratio of normal somatic cells will be kept quite constant according to their maternal tissues from which they are derived. There have been reports that pointed out the parallel relationship among the nuclear size, DNA content and chromosomal number.

nous cystadenoma (Figs. 14-3b and 14-52b, c). It must be taken into account that rupture of mucocele of the appendix also becomes the cause of gelatinous implants and pseudomyxoma peritonei.

■Specific tumors of the ovary

Specific ovarian tumors associated with excess of ascitic fluid are embryonal carcinoma, dysgerminoma, mesonephroma, etc.

Granulosa cell tumor rarely accompanies peritoneal effusion; the tumor cells in cluster take a loose trabecular, column-like or acinar arrangement. The nuclei of granulosa cell tumor are fairly small, oval or round, uniform in size and shape, and disclose a bland chromatin pattern (Figs. 14-56 and 14-57). Well preserved cytoplasm is scanty and cuboidal in shape (Fig. 14-57a). There may be denuded cells in ascites (Fig. 14-57b). The

diffuse type that forms densely packed masses of tumor cells intervened by scanty stroma may shed isolated or loose tumor cells with the similar features of the common follicular type (Fig. 14-58).

Table 14-6 Cell types of cancer cells in ascites secondary to gastric carcinoma

Cell type	No. of cases	Percent
Single columnar	9	18.4
Signet ring	16	32.7
Tubular	19	38.8
Vacuole formation	(3)	
Non-vacuolated	(16)	
Mixed	5	10.2
Total	49	

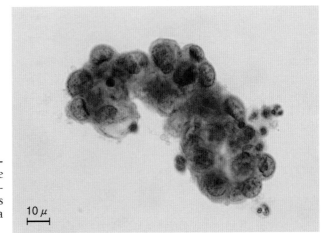

Fig. 14-43 Adenocarcinoma cells in ascites with a characteristic tubular arrangement. A case with carcinoma of the rectum. Hyperchromatic cancer cells are displayed in an elongated tubular fashion with marked nuclear overlappings; this pattern corresponds to the histology of tubular adenocarcinoma of the gastrointestinal tract.

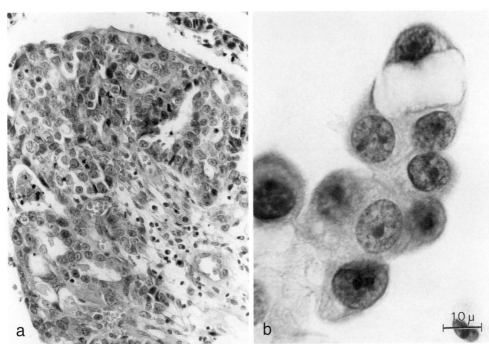

a   b

Fig 14-44 Common tubular type of cancer cells in ascites shed from serous cystadenocarcinoma of the ovary.
a. Histology of papillary serous cystadenocarcinoma. The cancer cells tend to pile up and to form a glandular arrangement. This structural pattern illustrates the exfoliation of the tubular type of malignant cells (b).
b. Cytology showing eccentric position of the nuclei, prominence of the nucleoli, cytoplasmic vacuolation and tubular arrangement of the cells with nuclear overlapping. Shedding of these cancer cells is large in number because of peritoneal implantation.

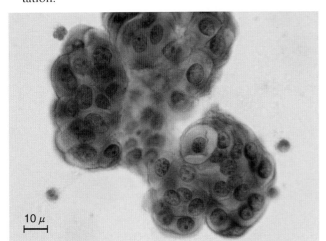

Fig. 14-45 Morula-like cell balls derived from duct carcinoma of the breast. Morula-like cell balls with smooth outlines are suggestive of common infiltrating duct carcinoma of the breast. Linear sharp outline differs from petal-like arrangement of the papillary cell ball.

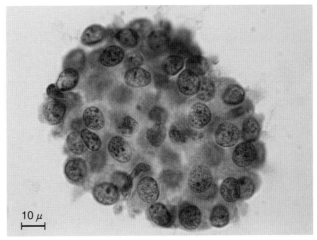

Fig. 14-46 Tightly clustered cell ball derived from serous papillary adenocarcinoma of the ovary. Note marked nuclear overlapping. This type cell ball formation with petal-like arrangement at periphery is indicative of papillary tubular adenocarcinoma.

299

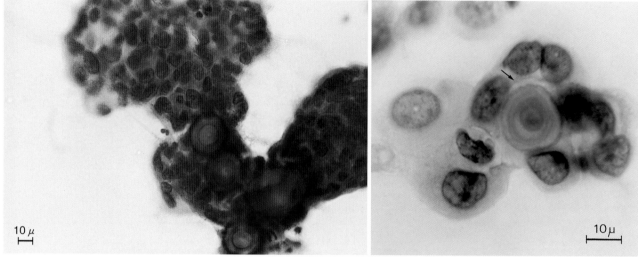

Fig. 14-47　Papillary cell cluster with psammoma body formation derived from ovarian serous papillary adenocarcinoma.
Diagnostic points: (1) typical papillary arrangement with nuclear overlapping, and (2) multiple laminated psammoma bodies.

Fig. 14-48　Psammoma body in serous cystadenoma of the ovary. Note a lamellar basophilic concretion within a tumor cell clump.

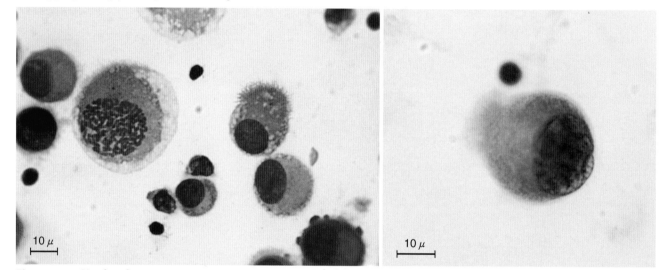

Fig. 14-49　Single adenocarcinoma cells with ciliation in ascites derived from serous papillary adenocarcinoma of the ovary. Giemsa stain demonstrates ciliated columnar cancer cells more clearly than Papanicolaou stain.

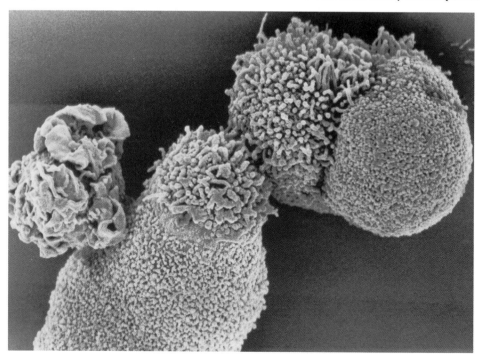

Fig. 14-50　Scanning electron micrograph of ciliated serous cystadenocarcinoma of the ovary. The cancer cells covered with dense microvilli are provided with cilia-like configurations. (By courtesy of Dr. K. Tsuji, Dept. of Central Diagnostic Laboratory, Oita Prefectural Hospital, Oita, Japan.)

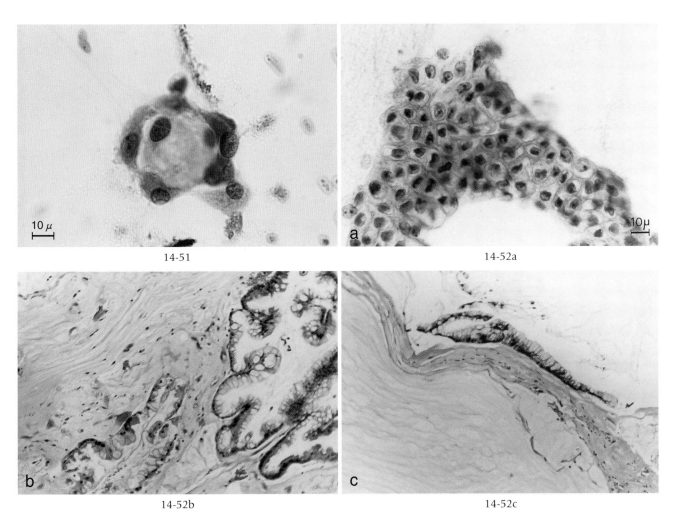

14-51

14-52a

14-52b

14-52c

Fig. 14-51 Pseudomyxoma peritonei arising from mucinous cystadenoma of the ovary. Small clustered cells with mucinous material in the center show small, round or oval nuclei and tapered cytoplasm. Such tailed or fiber-like cells are frequently encountered.

Fig. 14-52 Pseudomyxoma peritonei arising from mucinous cystadenoma of ovary.

a. Tissue fragment-like large cell clusters in gelatinous ascites. Abundant and clearly vacuolated cytoplasm around the nuclei gives a honeycomb appearance when they are seen from above. These epithelial cells are scarcely seen in huge amounts of gelatinous material and no inflammatory cells are found.

b, c. Histology of peritoneal implants of ovarian mucinous cystadenoma showing secretory acini embedded in thick gelatinous material. The tumor cells are tall columnar with darkly staining nuclei at the base of vacuolated cytoplasm. This case is that of 59-year-old female who had an episode of oophorectomy several years ago because of mucinous cystadenoma. Since then, accumulation of thick mucinous material has been undergoing by transplanted tumor cells.

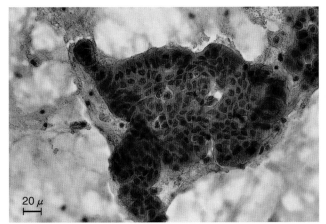

Fig. 14-53 Pseudomyxoma peritonei arising from mucocele of the appendix: large columnar cell cluster in abundant mucus lake. The nuclei are small and their abnormality is out of the grade of malignancy. However, mucinous background with light brownish or bluish hue points out pseudomyxoma peritonei.

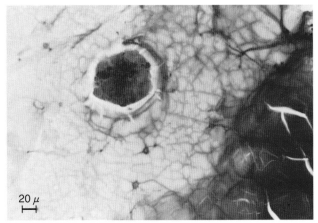

Fig. 14-54 Low power view of pseudomyxoma peritonei. Note PAS-positive mucinous background and a cluster of columnar cells. Alcian blue stain is also available.

301

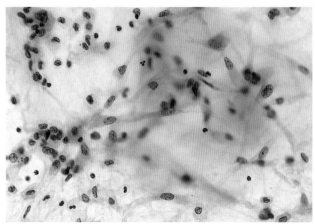

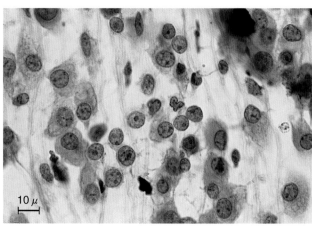

Fig. 14-55 Spindly cell components in pseudomyxoma peritonei. Somewhat stout, supranuclear translucent cytoplasm with brownish tint is in favor of epithelial origin, although description of fibroblastic origin are present.[19,63)]

Fig. 14-56 Granulosa cell tumor of the ovary in ascites. From a diffuse type. Uniform oval nuclei are characteristically smoothly outlined and contain single nucleolus. Note a bland chromatin pattern.

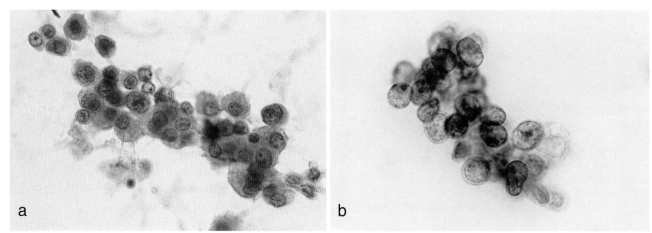

a

b

Fig. 14-57 Granulosa tumor cells with trabecular and follicular patterns in ascites. Hemorrhagic fluid contains nearly naked and uniformly oval cells arranged in a folliculoid or tubular fashion. Single nucleoli, bland chromatin pattern, uniformity of nuclear size, and smooth distinctive nuclear rim are in favor of granulosa cell tumor. **a:** trabecular pattern, **b:** follicular pattern

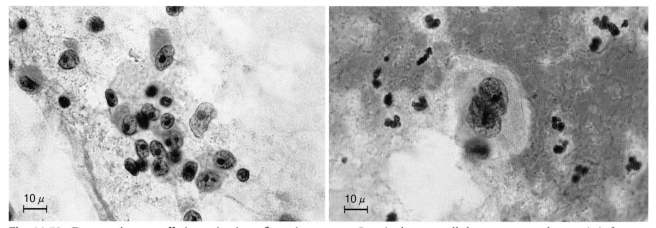

Fig. 14-58 Dysgerminoma cells in aspiration of ovarian tumor. Germinal tumor cells have common characteristic features such as (1) large, round nucleus with vesicular chromatin pattern, (2) usually single, prominent nucleolus in central position, and (3) moderate amount of clear, polygonal cytoplasm. Association with lymphoid cells in and around the tumor cells is diagnostic of germinoma.

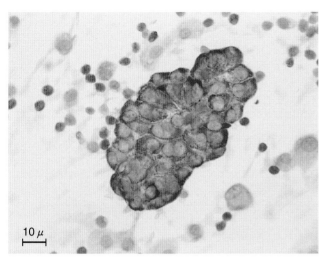

Fig. 14-59 PAS-positive dysgerminoma cells seen in a cluster. Germinal tumors are positive for PAS reaction. Note scattered lymphoid cells.

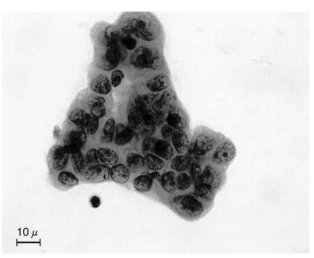

Fig. 14-60 Seminoma from hydrocele of a 29-year-old male. Cytology of seminoma is comparable to dysgerminoma. Note intermingled lymphoid cells in and around the cluster of epithelial cells.

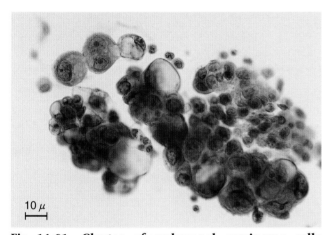

Fig. 14-61 Cluster of embryonal carcinoma cells arranged in a papillotubular fashion in ascites arising from testicular tumor. High N/C ratio, marked hyperchromatism, cytoplasmic vacuolation and a papillotubular pattern are of the features characteristic of embryonal carcinoma.

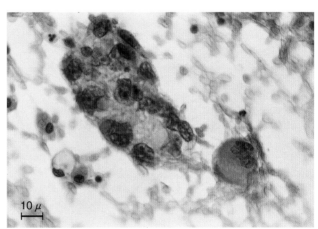

Fig. 14-62 Malignant cells seen in a tubular cluster arising from ovarian embryonal carcinoma. Cytologically only tubular adenocarcinoma is assessed. The cytochemical finding of alpha-fetoprotein by immunofluorescence and clinical evidence of ovarian mass are suggestive of embryonal carcinoma.

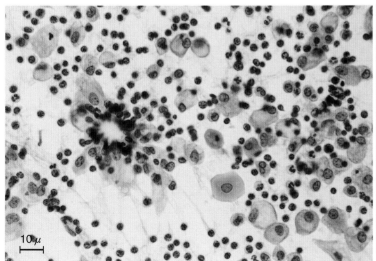

Fig. 14-63 Pleural gliomatosis in a 14-year-old girl in excess of pleural fluid associated with ovarian teratoma. A characteristic "Meigs' syndrome." Diagnostic points: (1) Unusual neuroglial cells in a cluster, and (2) a number of squamous cells are indicative of teratoma.

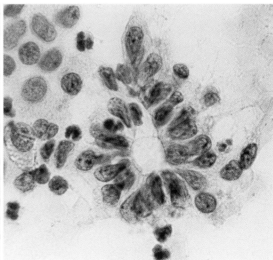

Fig. 14-64 Pleural gliomatosis showing characteristic "rod and cone cells" which are indicative of neural origin. These are of high power magnification from the same smear as Fig. 14-63.

303

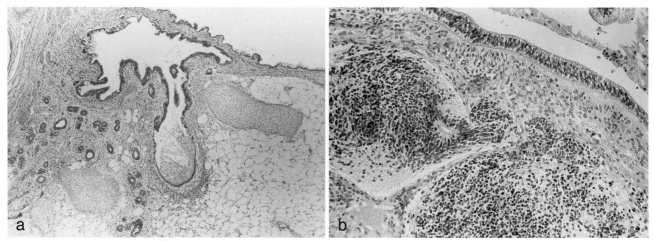

Fig. 14-65   Histology of differentiated cystic teratoma of the ovary. This is the same case as Figs. 14-63 and 14-64 that showed loss of pleural effusion after oophorectomy. The respiratory epithelium, adipose tissue and cartilage are illustrated. Only immature tissue element is neuroglial tissue as shown in (b).

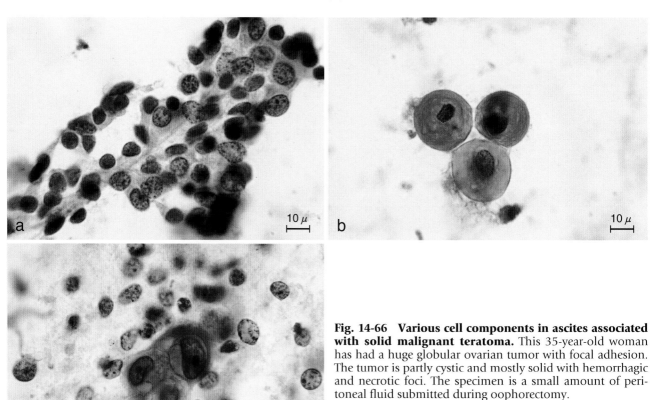

Fig. 14-66   Various cell components in ascites associated with solid malignant teratoma. This 35-year-old woman has had a huge globular ovarian tumor with focal adhesion. The tumor is partly cystic and mostly solid with hemorrhagic and necrotic foci. The specimen is a small amount of peritoneal fluid submitted during oophorectomy.
a. Immature neural tissue element consisting of hyperchromatic ovoid cells with a neurofibril-like structure.
b. Dyskaryotic parabasal squamous cells.
c. Malignant cells with a glandular pattern showing eccentric position of nuclei and prominence of nucliodi.

Dysgerminoma shows direct peritoneal dissemination and sheds characteristic large, polyhedral cells having centrally located, round, vesicular nuclei. These cells are arranged in a flat sheet with scarce overlapping (Fig. 14-60). The mixture of lymphocytes in and around the cell clusters is occasionally present and informative (Figs. 14-58 and 14-59).

Embryonal carcinoma is frequently associated with effusion and sheds tubular cell clusters with prominent vacuolation. The nuclei are markedly pleomorphic, hyperchromatic and possess macronucleoli (Fig. 14-62). An occasional production of alpha-fetoprotein by the tumor

can be detected by immunocytochemical technique. This type of germinal tumor is that defined for a variety with a predominance of glandular structures and closely related to the other germinal tumors which undergo trophoblastic differentiation (choriocarcinoma) or somatic differentiation (teratocarcinoma or malignant teratoma).

Solid malignant teratoma reveals histologically various types of tissue components in varying stages of differentiation; there may be embryonic stellate cells including various sarcomatous areas, primitive neural tissues and immature glandular structures in addition to islands of cartilaginous and bony tissues.

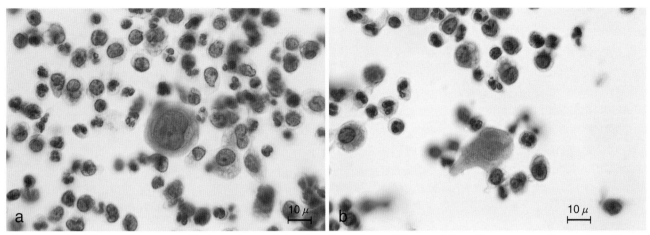

Fig. 14-67 Epidermoid cancer cells in ascites originating from dermoid cyst with malignant transformation. One to two per cent of benign cystic teratomas developing to malignant tumor is of epidermoid variety.
a. Parabasal type cell with a large hyperchromatic nucleus and perinuclear fibrillar structure.
b. Keratinizing epidermoid cell with nuclear liquefaction.

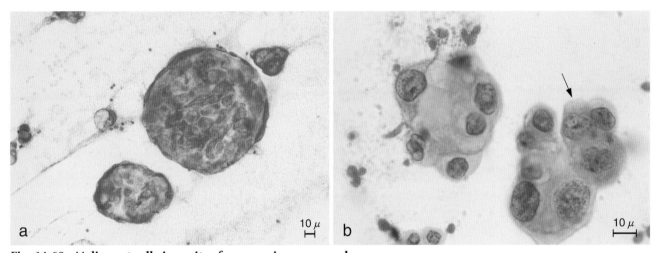

Fig. 14-68 Malignant cells in ascites from ovarian mesonephroma.
a. Cell balls with clear cytoplasm that is strongly positive for PAS reaction.
b. High power magnification. The nuclei are hyperchromatic and vary in size. Multiple small nucleoli are present. Diagnostic points: (1) abundant, transparent cytoplasm, (2) intense positive PAS stain, and (3) occasional cell protrusion in a cell cluster (arrow) that is correspondent to hobnail cells in histology.

Cystic teratoma of the ovary is of benign variety, the lumen of which is lined by epidermis combined with various skin appendages (Fig. 14-65). As this tumor is called dermoid cyst, various kinds of tissue such as nerve tissue, teeth, thyroid, bronchial epithelium and cartilage are associated. It should be noted that enormous ovarian teratoma in a young child often includes immature neuroglial elements and accompanies an excess of body cavity effusion; the presence of neuroglial cells and benign squamous cells in effusions despite wall encapsulation of the tumor could not be explained (Figs. 14-63 and 14-64). Fox and Langley[38] have proffered three possibilities, i.e., (1) deposition of immature neural tissue which undergoes maturation from the teratoma on to the serosal membrane, (2) lymphatic metastasis, and (3) extrusion of neuroglial tissue through a defect of the teratoma capsule. These unusual teratoma with *peritoneal gliomatosis* accompanies small nodular implants on the serosal membrane[37,92,124](Fig. 14-63). A malignant behavior is infrequent and fairly good prognosis can be expected when the primary tumor and serosal implants are matured. A malignant transformation of benign cystic teratoma occurs rarely (1.7% by Kelly and Scully[54]) and major parts of the tumor are of squamous cell carcinoma (Fig. 14-67).

Clear cell tumor of the ovary which is also called *mesonephric tumor* (Teilum[109]) or adenocarcinoma with clear cells (Saphir and Lackner[98]) grows in a papillary or solid pattern of cuboidal cells, the cytoplasm of which is clear like hypernephroma because of rich glycogen. The tumor cells in peritoneal effusion aggregate as cell balls or clusters with strongly PAS-positive clear cytoplasm (Fig. 14-68).

■Carcinoma of the lung

Although pulmonary carcinoma is very variable histologically, tubular and isolated columnar types are very common in serous effusion, because peripheral carcinomas which tend to extend along the visceral pleura are frequently of adenocarcinoma type and give rise to mesothelioma-like pleural involvement. It is rare to see metastasis or seeding of differentiated squamous cell car-

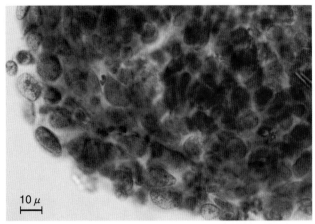

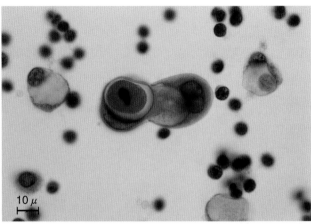

Fig. 14-69   Huge cell ball in pleural fluid originating from implants of thymoma. Occurrence of thymoma cells in pleural effusion is rarely encountered and small in number. Note cell constituents of pale-staining epithelial cells with thin nuclear membranes. These are benign-looking, however, serosal seeding is one of the behaviors of aggressive growth.

Fig. 14-70   Malignant pearl in pericardial fluid derived from pulmonary adenosquamous carcinoma. The nuclei, constituting the pearl are elongated and arranged in a scroll-work. Note keratinization in the cancer of the pearl. (By courtesy of Dr. R. Yatani, Dept. of Pathology, Mie University School of Medicine, Tsu, Japan.)

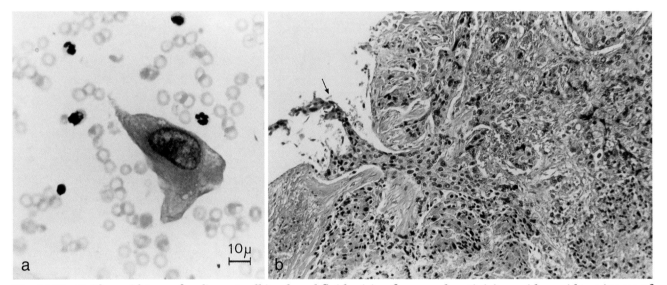

Fig. 14-71   Epidermoid type of malignant cell in pleural fluid arising from nonkeratinizing epidermoid carcinoma of the lung.
a. Giemsa stained cell showing abundant thickened cytoplasm with an odd shape and distinct cellular border is from squamous cell carcinoma. Exfoliation of epidermoid cancer cells is seldom encountered in effusion.
b. Histology. Infiltration of solid nests of squamous cancer cells breaks into the pleural cavity (arrow). There is no pearl formation within the cancer cell nests. It is rare that malignant cells of epidermoid type occur in serous effusions.

cinoma into the serous cavities. When the exception occurs, pleomorphic cells with voluminous and thick cytoplasm will be seen as isolated single cells (Fig. 14-71). Oat cells are apt to be overlooked owing to their similarity in size to lymphocytes. They may be differentiated from lymphocytes, however, in the points that they form aggregations of cells with considerable variations from cell to cell in nuclear size and shape; an "Indian-file" or "tandem" arrangement with molding of the nuclei against one another is often recognized (Figs. 14-76 and 14-77). Occasionally nucleolar prominence in these cells is discernible.

Active mesothelial cells are the problematical cells which must be distinguished from cancer cells. Identification of active mesothelial cells may be made based on the following findings: (1) Mesothelial cells do not display either absolute increase in the chromatin content or abnormally coarse chromatin aggregation, (2) thickening and irregularity of nuclear rims are not so pronounced as in cancer cells, (3) the active nuclei, though moderately enlarged, are associated with a fairly abundant cytoplasm and are centrally located, and (4) the nucleolo-nuclear ratio (Quensel) remains within benign limits[76,88]; in non-neoplastic cells the n/N ratio lies between 0.15 and 0.20, while in malignant cells it ranges from 0.25 to 0.40.[35] Immunocytochemistry plays a great role to characterize mesothelial cells (Table 14-3). Table 14-7 discloses the frequency of malignant pleural effusion according to the cell types. Comparing cytology of pleural effusion with histology of the lung cancers submitted to the same Laboratory of Clinical Pathology during the same period, from 1961 to 1975,

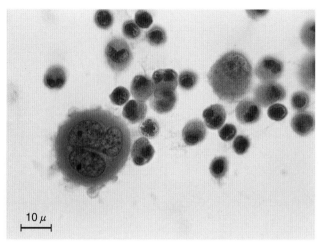

Fig. 14-72 Nonkeratinizing epidermoid cancer cells in pleural fluid. Note thickened cytoplasm with distinct cellular border and bizarre nuclei with coarsely reticular chromatins.

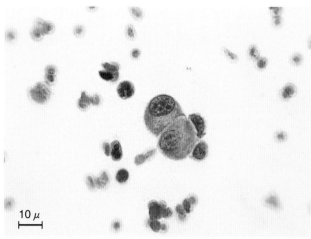

Fig. 14-73 Epidermoid cancer cells derived from esophageal carcinoma. Note a typical cobblestone arrangement. A straight slit-like space between two cells is of distinctive feature of epidermoid cancer cells.

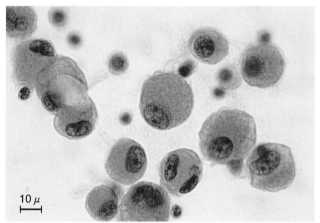

Fig. 14-74 Singly isolated cancer cells arising from pulmonary adenocarcinoma. Note eccentric nuclei and finely vesicular cytoplasm that is characteristic of adenocarcinoma. Such a single cell population is not usual in effusion secondary to lung cancer.

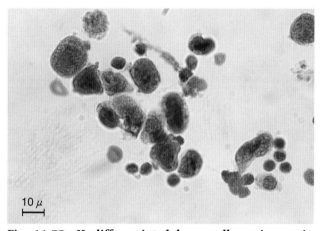

Fig. 14-75 Undifferentiated large cell carcinoma in pleural fluid arising from pulmonary large cell carcinoma. Diagnostic points: (1) Very scant, ill-defined cytoplasm, (2) high N/C ratio, (3) polymorphic nuclei varying in shape and size, and (4) prominent nucleoli are indicative of undifferentiated large cell carcinoma.

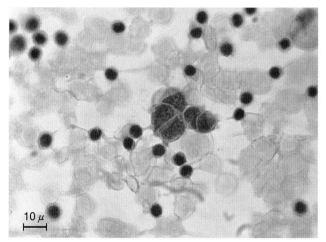

Fig. 14-76 Undifferentiated small cell carcinoma in pleural fluid originating from small cell carcinoma of the lung. Small cell carcinoma tend to form a loose cluster although they shed singly in sputum. The cytoplasm is very scant and poorly preserved.

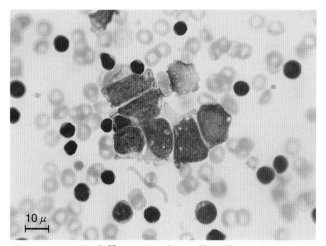

Fig. 14-77 Undifferentiated small cell carcinoma by Giemsa stain. An arrangement in tandem is a distinctive feature of epithelial nature. Small cell carcinoma shows more scant cytoplasm than lymphocytic leukemic cells.

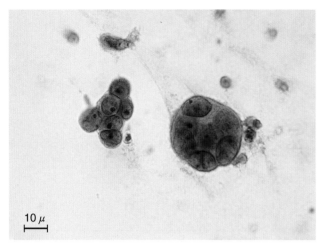

Fig. 14-78 Small adenocarcinoma cells in pleural fluid associated with breast carcinoma. These small cancer cells resemble active mesothelial cells. Note following features as differential points: (1) single but prominent nucleoli, (2) high N/C ratio though nuclei are fairly small, and (3) nuclear overlapping in a cell cluster that is different from flat cell sheet of mesothelium.

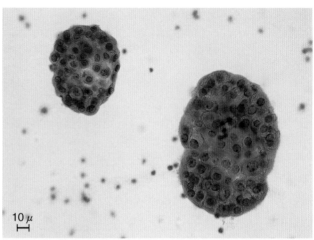

Fig. 14-79 Adenocarcinoma cells in pleural fluid originating from duct carcinoma of the breast showing characteristic cell ball formation. Cell constituents of the ball are fairly uniform in size and shape. Single nucleoli are locating centrally. Note a characteristic morula-like clump of cells.

Table 14-7 Incidence of lung cancer and malignant pleural effusion submitted to the Department of Clinical Pathology, JNR Hospital during the same period from 1961 to 1975*

Lesion \ Type	Adenoacanthoma Case	Adenoacanthoma Total	Epidermoid ca. Case	Epidermoid ca. Total	Adenoca. Case	Adenoca. Total	Large cell ca. Case	Large cell ca. Total	Small cell ca. Case	Small cell ca. Total	Total cases
Central	0		19		18		1		11		
Peripheral	0	2	17	36	62	90	5	6	10	22	156
Undetermined	2		0		10		0		1		
Excess of malignant effusion	0		6 (16.7%)		52 (57.8%)		1 (16.7%)		7 (31.8%)		66 (42.3%)

*Note: Since the inpatients in JNR Hospital are limited to the employee of JNR and their families, long term follow-up observation are possible. Thus data on the frequency of association of lung cancer with excess of malignant effusion, either synchronous or metachronous, is reliable.

one may recognize the highest frequency to accompany pleural effusion in adenocarcinoma; the ratio is 1.0 to 0.58 in adenocarcinoma, 1.0 to 0.17 in epidermoid carcinoma and large cell carcinoma, and 1.0 to 0.32 in small cell carcinoma (Table 14-7).

■Carcinoma of the breast

The breast is one of the most important organs that produce malignant effusion. The nuclei are fairly small, fairly uniform in shape and only two-fold or less the size of active mesothelial nuclei. The chromatins are fine and bland.[69] These features are subjected to false negatives when the cancer cells shed as single cells or small cell clumps. The presence of macronucleolus is valuable to be distinguished from active mesothelial cells (Fig. 14-78). Characteristic of breast carcinoma is shedding of cancer cells as closely packed cell balls varying in size; the ductal medullary-tubular carcinoma, the most common type of mammary carcinoma, will be the source of such a distinctive cell cluster (Fig. 14-79). In infiltrating ductal carcinoma solid masses of cancer cells plugging the ductal lumina invade the stroma with some productive fibrosis. In some instances or in some areas of ductal carcinoma, one sees elongated strands or columns of cancer cells to be disposed irregularly in dense fibrous tissue (scirrhous carcinoma); this type of histologic feature is responsible for a typical "Indian-file" arrangement in serous effusions. Cells from advanced lobular carcinoma tend to show small "Indian file" chains resembling oat cell carcinoma. Bland chromatins and visible cytoplasm are distinguishable from oat cell carcinoma.

Summarizing the characteristic profile of breast cancer cells in effusions, we often encounter (1) morula-like cell balls (cannonball) that may be derived from medullary tubular carcinoma, and (2) small groups of cells lying in row, either in single file or in tandem, that may originate from scirrhous tubular carcinoma or lobular carcinoma.

■Carcinoma of the liver

There have been scarce cytological descriptions on detection of hepatoma cells in the literatures, probably because of very low mortality by hepatoma in Europe and in the United States and of rare incidence of exfoliation of the cancer cells from hepatoma which locates within the liver parenchyma. The study of geographic pathology has shown its higher mortality in South Africa and Orient, and has suggested the relation of

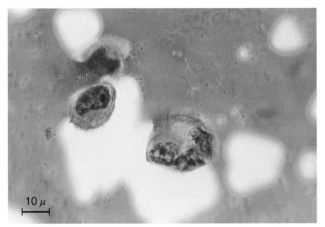

Fig. 14-80 Hepatoma cells in bloody ascites. Although ascites with hepatoma is often purely bloody, exfoliation of hepatoma cells is not so pronounced as in peritoneal carcinomatosis. The cytoplasm is fairly abundant and clearly bordered. Binucleation or trinucleation is of common occurrence.

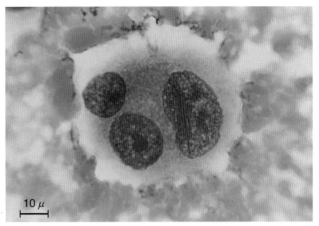

Fig. 14-81 Giemsa stained hepatoma cells showing syncytial trinucleation. The cytoplasm is fairly abundant and nucleoli are prominent. The same case as Fig. 14-80.

dietary habits to liver cirrhosis and hepatocellular carcinoma; food-borne toxins such as aflatoxin and sterigmatoxin are the subject to carcinogenesis.[48] A recent virological study ascertained the frequent association of hepatoma with macronodular cirrhosis (post hepatitic and postnecrotic cirrhosis) to be related to precedence of hepatitis B infection; Prince et al[87] in Senegal reported that the frequency of hepatitis B surface antigen was 61.2% of 165 patients with hepatoma as compared to 11.7% of 154 patients with other site cancers. Nishioka and his associates[81] also described the relationship in Orientals introducing their sensitive immune adherence hemagglutinin method (Table 14-8).

Many cases hepatoma have been found to have antibodies to non-A, non-B virus (hepatitis C virus), while a few of these have the additional HBV infection. Long-lasting HCV infection in association with chronic hepatitis in thought to precede hepatoma.

Since hepatoma grows within the liver parenchyma as a fused mass (massive form) or nodules (nodular form), hepatoma cells are seldom exfoliated in ascites. When cancer cells are shed, the ascites is usually hemorrhagic, often purely bloody in appearance, and the cell constituents are scarce probably because of hemorrhage. Although no specific features diagnostic of hepatoma are present except for bile production in the cytoplasm (Fig. 14-83), hepatoma cells are (1) arranged in flat sheets like mesothelial cell clusters and possessed of fair amount of cytoplasm and centrally locating, enrounded, large nuclei. They have one or two prominent nucleoli (Figs. 14-81 and 14-82). In liver cirrhosis that is a frequent precursor of hepatoma, reactive mesothelial cells shed in large numbers into peritoneal fluids should not be overdiagnosed as hepatoma cells.[31] Positive cytology of hepatoma in effusion specimens is the late evidence. Fine needle aspiration is useful in early malignant transformation (see Chapter13). The shedding hepatoma cells are few in number. If hepatoma is suspected clinically, immunological quantitation of alpha-fetoprotein from blood serum as well as effusions is important rather than cytology only. Takahashi and his co-work-

Table 14-8 Hepatitis BS antigen positive rates in patients with hepatocellular carcinoma (Courtesy of Dr. K. Nishioka, The Tokyo Metropolitan Institute of Medical Sciences, Tokyo)

Country or area	Hepatocellular carcinoma	Healthy control
Africa		
Mozambique	62.4 (%)	10.7(%)
Uganda	44.6	4.0
Zambia	52.6	5.0
Senegal	70.7	10.3
Mali	47.6	5.0
Ethiopia	50.0	3.3
Ghana	80.0	-
Kenya	54.8	4.7
South Africa (Bantu)	59.5	9.0
Nigeria	40.0	6.3
Europe		
Greece	38.9	4.8
United Kingdom	16.0	0.2
France	4.7	0.01
North America		
United States	26.0	0.3
Pacific		
Papua New Guinea	71.9	18.0
Asia		
Burma	33.3	9.5
China	47.8	4.5
Hong Kong	45.0	12.2
India	41.2	3.2
Indonesia	52.4	3.0
Japan	36.7	2.7
Philippines	62.5	16.1
Thailand	71.4	9.8
Vietnam	70.0	19.0
Taiwan	54.8	12.2
Singapore	35.3	7.5

Note: The data are based on IAHA, RPHA and RIA methods.

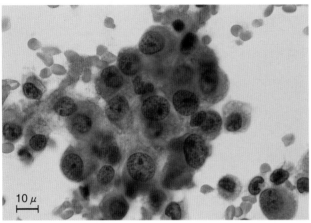

Fig. 14-82 Hepatoma cells in ascites arranged in a flat cell sheet. Enlarged, round nuclei are locating centrally and possess one or two prominent nucleoli. The cytoplasm is moderate in abundant.

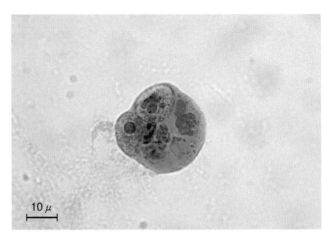

Fig. 14-83 Single hepatoma cell producing bile pigments in the cytoplasm. Note a large round nucleolus and an enormously large nucleolus. Bile production, though specific, is seen in a fairly differentiated type.

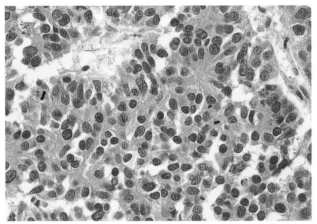

Fig. 14-84 Moderately differentiated hepatoma composed of massive cancer cells arranged in a trabecular pattern. Note marked hyperchromatism, high N/C ratio and variation in nuclear size.

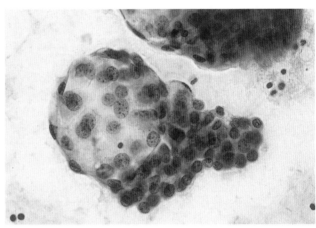

Fig. 14-85 Adenocarcinoma cells in ascites arising from biliary duct carcinoma. Tubular adenocarcinoma that is the common type of the biliary duct or gallbladder origin tends to form such a cell ball in effusions.

ers[106] have introduced anti-fetoprotein immunofluorescent technique for the identification of hepatoma cells. It must be kept in mind, however, that hepatoma cells are not all producing alpha-fetoprotein (AFP); its function is related to the grade of differentiation of hepatoma. The most poorly differentiated hepatoma cells are not productive of AFP.

■Leukemia and malignant lymphoma

Although serous cavities may be involved in malignant lymphomas and leukemias, few references have been made to their exfoliative cytology in textbooks. As to the proportion of lymphoma in malignant effusion Johnson reported positive cytology with lymphomas and leukemias in 39 of 345 cases (11.3%),[50] Ceelen in 6 of 72 cases (8.3%),[13] and our laboratory in 22 of 389 cases (5.7%). As a staining technique, air-dried Giemsa stain has been particularly recommended for the identification of lymphoma cells or leukemic cells by Spriggs,[101] Grunze[43] and Lopes Cardozo.[66]

1) Leukemia

Myeloid leukemic cells in various stages of maturation can be identified with ease by their specific granulation

within the cytoplasm which are poorly demonstrated by Papanicolaou stain (Fig. 14-86). Lopes Cardozo[67] emphasizes worthy cytochemical approaches such as Sudan black B stain for demonstration of Auer's bodies and alkaline phosphatase reaction that becomes negative in myelogenous leukemia. Lymphocytic leukemic cells are hardly distinguishable from lymphosarcoma cells unless systemic pathological and hematological studies are carried out. The number of leukemic cells is variable from case to case and not so as many as lymphosarcoma cells with a monomorphic pattern.

2) Lymphoma

Lymphoma affects pleural and peritoneal cavities as a part of manifestation of generalized lymphoma. The most universally used classification of lymphomas was that of Rappaport (1996) based on morphological features. It has been greatly modified after phenotypic and cytogenetic analyses. Although the updated Kiel classification of non-Hodgkin's lymphomas is nowadays commonly accepted for histopathological diagnosis, small lymphocytic lymphoma of low grade malignancy and large cell lymphoblastic lymphoma of high grade malignancy as proposed by the NC I International Working

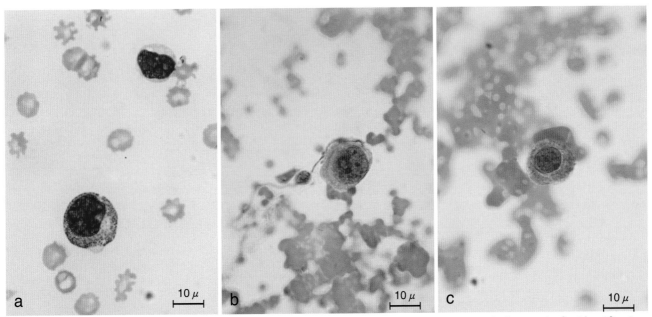

Fig. 14-86 Myelogenous leukemic cells in pleural fluid. Giemsa stain is superior to Papanicolaou stain for identification of leukemic cells.
a. Promyelocyte showing characteristic granules in the cytoplasm. Giemsa stain.
b, c. Papanicolaou-stained immature myeloid cell showing multiple nucleoli and cyanophilic, somewhat granular, hue at periphery of the cytoplasm.

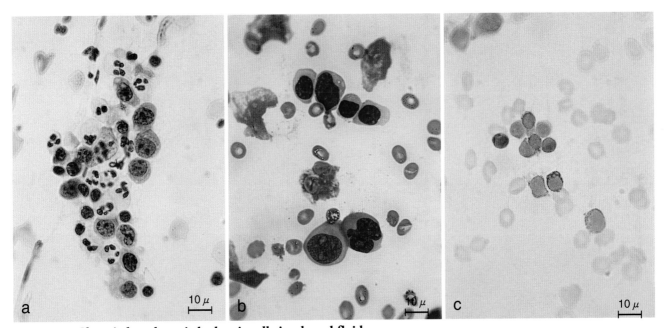

Fig. 14-87 Chronic lymphocytic leukemic cells in pleural fluid.
a. Papanicolaou-stained leukemic cells. Leukemic cells are few in number and admixed with small, mature lymphocytes. These are consistent with small lymphocytic lymphoma cells. Lymphoma cells that involved the pleural cavity are few but exhibit distinctive features such as coarse chromatins, prominence of nucleoli and indentation of nuclear rims.
b. Giemsa-stained lymphocytic leukemic cells. Note an admixture of immature lymphocytic leukemic cells with prominent nucleoli and three mature lymphocytes.
c. PAS reaction for chronic lymphocytic leukemic cells. Lymphocytic leukemic cells show characteristically coarse PAS-positive granules whereas myelocytic and monocytic leukemic cells reveal slight and diffuse stainability.

Formulation (see Table 17-5) may suffice routine fluid cytologic classification.

Small lymphocytic lymphoma cells or chronic lymphocytic leukemic cells are often hardly distinguished from activated lymphocytes. However, lymphocytosis induced as inflammatory reaction is of T cell origin. They can be differentiated from chronic lymphocytic leukemic cells derived from B lymphocytes.[5,80]

Large lymphoblastic lymphoma affecting body cavities can be recognized with ease by a monomorphic population of large or medium sized lymphoid cells having coarse chromatins and prominent nucleoli. Cytomorphological features are well studied by Giemsa or Wright's stain (Figs. 14-89 and 14-90c)

311

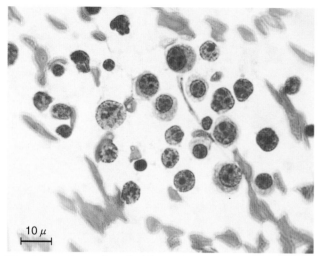

Fig. 14-88   Large lymphoblastic lymphoma cells in pleural fluid from a 6-year-old boy. The patient has been hospitalized with acute onset complaining of dysphagia and dyspnea in association with lymphadenopathy and hepatosplenomegaly. Note scant cytoplasm, prominent nucleoli and coarse chromatins.

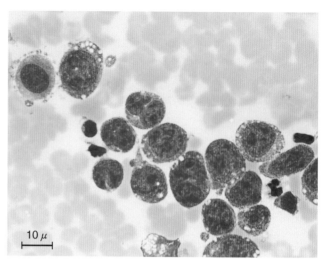

Fig. 14-89   Large lymphoblastic lymphoma cells in ascites by Giemsa stain. Note enrounded, hyperchromatic nuclei with delicate nuclear membranes and prominent nucleoli. The basophilic cytoplasm is scant and vacuolated. A mesothelial cell showing an oval nucleus and homogeneous cytoplasm (left) and a few lymphocytes are seen.

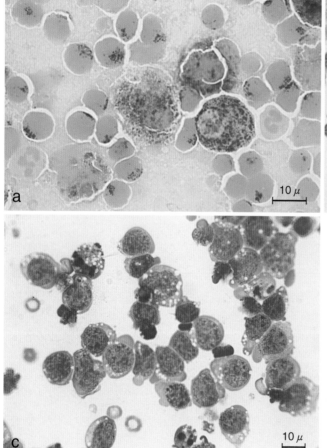

Fig. 14-90   Large cell lymphoblastic lymphoma cells in ascites from a 12-year-old girl.
a. Acid phosphatase stain. Focal or single-spot positive stain for acid phosphatase is indicative of T cell variety of acute lymphoma. Some macrophages diffusely stained are distinguishable.
b. Monomorphic population of lymphoblastic lymphoma cells: These are singly isolated, nearly denuded or possessed of scant cytoplasm that is indistinctly bordered, and show a chromatin pattern stippled with coarse particles.
c. Giemsa stain shows abnormal chromatin pattern more in details than in Papanicolaou stain. One or two prominent nucleoli and coarsely stippled chromatins are evident. Scant basophilic cytoplasm shows fine vacuolations (immunoblastic lymphoma suggested).

3) Histiocytoid lymphoma

This type, which has been designated as reticulum cell sarcoma, is actually few in incidence and classified in the miscellaneous group (see Table 17-5). Histiocytoid cells with broader cytoplasm vary more markedly in size than lymphosarcoma cells. Cytoplasmic vacuolation is remarkable and demonstrable evidently by Giemsa stain. The chromatin structure is finely reticular or finely granular. The nucleoli are prominent, occasionally multiple, and show bead-like irregular shapes (Fig. 14-91, Table 14-9).

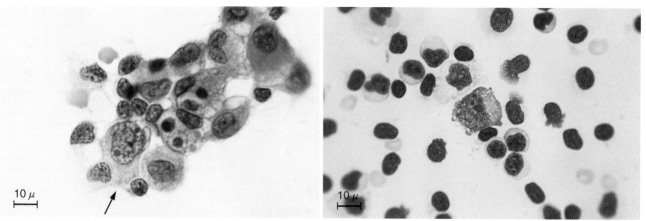

Fig. 14-91 Histiocytioid lymphoma or reticulosarcoma cells in ascites. A typical lymphoma cell of histiocytoid type (arrow) is found to be surrounded by mature macrophages and lymphocytes. The nucleus of histiocytic type varies in shape rather than in size; note polyhedral and indented nucleus and moderate amount of cytoplasm.

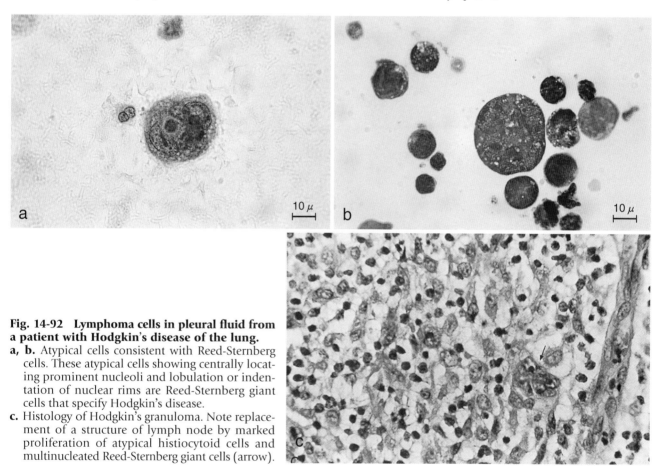

Fig. 14-92 Lymphoma cells in pleural fluid from a patient with Hodgkin's disease of the lung.
a, b. Atypical cells consistent with Reed-Sternberg cells. These atypical cells showing centrally locating prominent nucleoli and lobulation or indentation of nuclear rims are Reed-Sternberg giant cells that specify Hodgkin's disease.
c. Histology of Hodgkin's granuloma. Note replacement of a structure of lymph node by marked proliferation of atypical histiocytoid cells and multinucleated Reed-Sternberg giant cells (arrow).

Table 14-9 Cytologic characteristics of lymphoma cells

Features	Small lymphocytic lymphoma cells	Lymphoblastic lymphoma cells	Histiocytoid lymphoma cells
Cell size	10-15 μm; small size variation	15-25 μm; moderate size variation	10-30 μm; large size variation
Cytoplasm size/ stain	Narrow Light basophilia	Narrow Intense basophilia	Broad Light basophilia
Cytoplasmic vacuolation*	Rare	Frequent small vacuolization	Frequent vacuolization
Nuclear shape**	Round or oval	Round or oval	Polyhedral
Chromatins	Stippled	Stippled/coarse	Reticular
Nucleolus	Indistinct	Distinct	Distinct and prominent

*Vacuolization can be depicted by Giemsa stain.
**Shapes of nuclear membrane such as indentation, cleavage and convolution can be recognized by Giemsa stain.

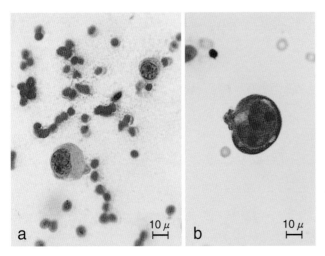

Fig. 14-93   Myeloma cells in ascites derived from Bence-Jones myeloma. Myeloma cells are isolated singly and characterized by (1) eccentric enrounded nuclei, (2) perinuclear halo, (3) basophilic stain at the periphery, and (4) prominent single nucleolus. **a.** Papanicolaou-stained myeloma cells. **b.** Giemsa-stained myeloma cells.

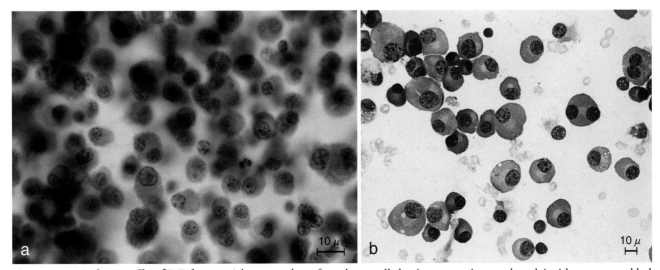

Fig. 14-94   Myeloma cells of IgD kappa. A large number of myeloma cells having eccentric round nuclei with coarse speckled chromatins and prominent nucleoli. Note perinuclear halo and intense basophilic stain at periphery.
a. Papanicolaou stain, **b**. Giemsa stain.

4) Hodgkin's disease

Although multinucleated Reed-Sternberg giant cells are diagnostic of the disease, they are rarely encountered in exfoliative cytology of serous effusions. It is the mononucleated form (Hodgkin's cell) having the same characteristic cellular features as the Reed-Sternberg giant cell that exfoliates frequently in effusions. The nuclei of these cells, regardless of their multinucleation or multilobulation, have distinct nuclear rims and acidophilic or amphoteric prominent nucleoli, around which clear zones are characteristically seen (Fig. 14-92).

■Multiple myeloma

Extramedullary lesions in multiple myeloma are frequently encountered (73% by Churg[16] and 51% by Innes[49]), but an appearance of myeloma cells in serous effusions rarely occurs; some reported cases are on pleural effusions and/or pericardial fluids with involvement of serous membranes by myelomatosis. Identification of myeloma cells that produce abnormal M-proteins and characterize plasma cell dyscrasias is occasionally difficult only from the morphological base, because well differentiated neoplastic cells are normal-looking plasma cells. A preponderance of plasma cells and admixture of immature plasma cells necessitate further investigations biochemically as well as clinically to establish the diagnosis of multiple myeloma. The myeloma cells are singly isolated and possess eccentrically located round or oval nuclei, the chromatins of which are condensed towards the nuclear membrane. One or two nucleoli are recognizable in immature cells. The cytoplasm is basophilic with perinuclear halo; it is intensely stained at periphery and distinctly bordered (Fig. 14-94). Protein aggregates are often demonstrated as acidophilic *Russel bodies*. The myeloma cells are classified into several variants according to five types of major classes and two types of light chains. The myeloma cells usually produce a single specific type of immunoglobulins as referred to as monoclonal gammopathy; IgG, IgA, IgM, IgD, IgE and Bence Jones types are ordinarily classified.

■Malignant mesothelial cells

The majority of mesothelioma are pleural in origin and known to be related to asbestos exposure.[74] Thus its existence was even doubted in the past.[122] Macroscopically, pleural involvement of peripheral adenocarcinoma of the lung is strikingly similar to mesothelioma described as diffuse type. As primary site of malignant

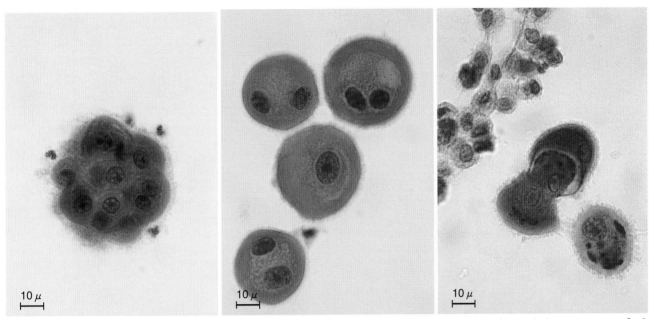

Fig. 14-95 Malignant peritoneal mesothelioma of epithelial type. As a general appearance of mesothelioma one may find various kinds of cells transitional from activated mesothelial cells to mesothelioma cells in association with enlargement and multiplication of nuclei. Acidophilic stainability around the nucleus is also one of the characteristic properties of mesothelial cells. Irregular granules or flecks positive for PAS reaction account for deviated cell differentiation of mesothelioma. Note fine, chest nut bur-like processes at the cellular border. (By courtesy of Dr. T. Motoyama, Dept. of Pathology, Niigata University School of Medicine, Niigata, Japan.)

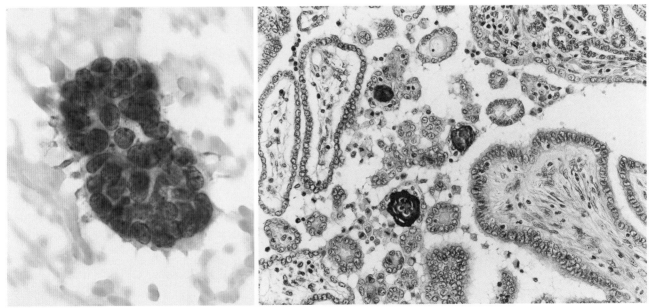

Fig. 14-96 Peritoneal mesothelioma cells aggregated in a papillary cluster. PAS-positive stain.

Fig. 14-97 Malignant peritoneal mesothelioma showing a characteristic papillotubular growth. Psammoma bodies are associated. These growth patterns may lead to mistake of adenocarcinoma. Rare occurrence of psammoma bodies in papillotubular mesothelioma is described in Atlas of Tumor Pathology (AFIP).[4]

pleural effusions, the frequency of pleura is variable; 26 out of 143 malignant tumors confirmed at autopsy (18.2%) by DiBonito et al[23], and 4 out of 126 malignant neoplasms (3.2%) by Monte et al.[77] Triol and associates[111] have described a noteworthy findings on 75 histologically confirmed malignant mesothelioma cases which were first reported to be mesothelioma or probably mesothelioma in 42 (56%). Five cases were originally suspected to have adenocarcinoma (6.7%). Mesotheliomas have been classified by Stout[104] as follows:

Benign mesothelioma
    solitary; fibrous or tubular
    diffuse; fibrous or tubular
Malignant mesothelioma
    solitary; fibrous or tubular
    diffuse; fibrous or tubular
Koss[58] classifies mesothelioma into those originating from (1) mesothelial lining cells (epithelial mesothelioma), (2) mesothelium supporting connective tissue (fibrous mesothelioma) and (3) mixed cellularity show-

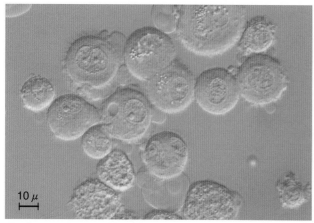

Fig. 14-98 Peritoneal epithelial mesothelioma under differential interference microscopy. Note enlarged, occasionally binucleated nuclei in a paracentral position. Fine cytoplasmic processes can be seen in supravital observation.

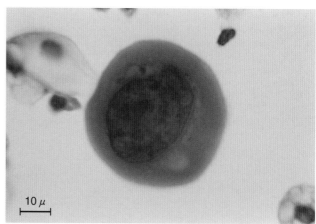

Fig. 14-99 Single malignant mesothelioma cell in ascites. This type of cell should be differentiated from hepatoma cells and epidermoid cancer cells. (1) Differentiation from epidermoid cancer; syncytial binucleation differs from cannibalism in epidermoid carcinoma. Fibrillar structure is not seen in the thickened cytoplasm of mesothelioma. (2) Hepatoma cells are occasionally clustered in sheet but not syncytial. Note distinctive perinuclear vacuolation.

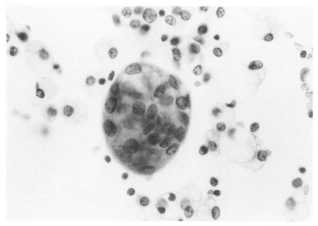

Fig. 14-100 Morula-like clustered cells. These nuclei are round or oval and fairly uniform in size. Note a bland chromatin pattern. These benign-looking mesothelioma cells coexist with malignant cells as shown in the preceding color figures.

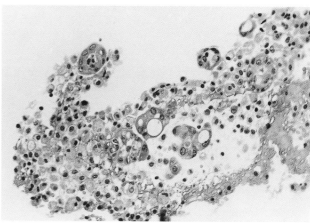

Fig. 14-101 Malignant mesothelioma of diffuse tubular type. This case is that of 61-year-old woman who has had abdominal distension over one and a half years. Necropsy showed nodular tumors scattered on the surface of the peritoneum, mesenterium and intestinal serosa. Histologically, the tumor shows nodular or papillary projections with loose fibrous stalks. The tubular spaces are lined with atypical mesothelial cells with pleomorphic nuclei and prominent nucleoli. (By courtesy of Dr. N. Fukushima, Doai Memorial Hospital, Tokyo.)

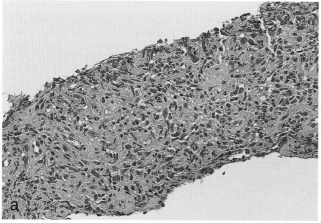

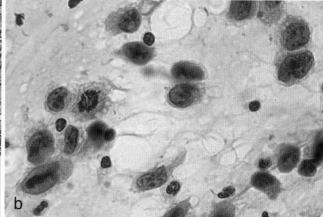

Fig. 14-102 Fibrous mesothelioma from the pleural cavity.
a. Transthoracic biopsy showing diffuse, massive infiltration of spindly cells. No organoid pattern is discernible.
b. Isolated sarcomatous cells with elongated but plump nuclei.

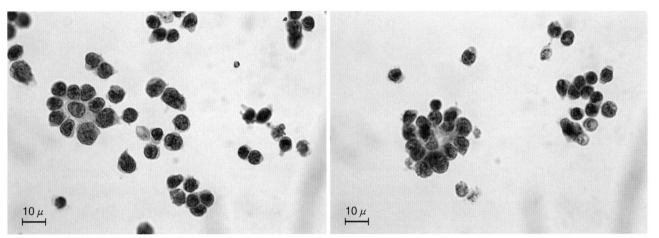

Fig. 14-103 Sympathicoblasts in ascitic fluid from a 2-year-old girl. These small lymphoid cells arising from neuroblastoma should be differentiated from lymphoma and oat cell carcinoma. Diagnostic points: (1) Formation of cell clumping, occasionally in a rosette pattern, differs from isolated lymphoma cells. (2) Fine fibrillar structures are occasionally noticeable as a characteristic feature of neural origin, although not visible in this case.

ing synovioma-like both cellular components. Battifora and McCaughey describe four main histologic types: epithelial (tubulopapillary and epithelioid), sarcomatous (desmoplastic), biphasic (mixed) and poorly differentiated.[4] Desmoplastic mesothelioma, a variety of diffuse sarcomatous mesothelioma, is poorly cellular with dense fibrous stroma.[123]

Solitary fibrous mesothelioma, though more common in the pleural cavity than other types, is not easily diagnosed in exfoliative cytology (Fig. 14-102). Tubular mesothelioma with papillary growth frequently shows multiple nodulations or large conglomerations. Histologically papillary projections and slit-like spaces are lined by cuboidal or low columnar cells with round nuclei and prominent nucleoli. This type of mesothelioma may cause an accumulation of somewhat viscid fluid in which a large number of atypical mesothelial cells are contained. Cytology of mesothelioma cells in body cavities is of epithelial origin.

Malignant mesothelial cells show distinctive features of malignancy such as hyperchromatism and coarse chromatin clumping of nuclei, prominence of nucleoli, multinucleation and marked variation in cellular size. The identification of mesothelial tumor cells will be successful based on a cytochemical property that contains hyaluronic acid[117] and a pattern of general cellular components; because malignant mesothelial cells retain some similar features to benign mesothelial cells. Positive alcian blue stain becomes negative following treatment with hyaluronidase. There may be single cells showing fairly broad, opaque cytoplasm and enlarged round nuclei with single prominent nucleoli. Cell clumps showing several round nuclei in fused cytoplasm that resemble syncytial mesothelial cells are also encountered (Fig. 14-100). Cytoplasmic granules or flecks in the PAS reaction digestible by diastase are occasionally useful for the identification of mesothelioma cells[33,53]; irregular distribution of PAS-positive granules is of importance in cytochemical findings (Fig. 14-96). The staining reaction often displays a characteristic tinctorial feature that, the periphery being densely green, is gradually converted lighter and eosinophilic towards the center. Occasionally observed are brush-like borders or cytoplasmic blebs,[51,101] though they may be formed during preparation.[101]

In some cases, particularly of the undifferentiated type, the cellular pattern is represented by a predominance of large morula-like cell balls with closely packed nuclei. Aggregation of these cells in a cluster resembles acinar or papillary cancer cells. Implantation metastasis of ovarian serous cystadenocarcinoma is the representative case that may be confused with malignant mesothelioma. One must seek for single cells or small cell clumps that retain the cellular features of mesothelial origin. Perinuclear vacuolization is described as a characteristic feature of mesothelial cells.[7] Klempman states that copious exfoliation of small and medium sized atypical mesothelial cells occurs singly and in aggregations.[56]

■ Rare tumors

Other rare tumor cells found in ascitic and pleural fluids are often concerned with retroperitoneum and mediastinum in origin.

a) Tumors of epithelial origin are from the kidney, urinary bladder, uterus, etc. (Figs. 14-106 to 14-112). The appearance of malignant cells from renal cell carcinoma, urothelial carcinoma and endometrial or cervical carcinoma is infrequent, but attention should be paid for the possibility of exfoliation and for their distinctive features. Renal carcinoma cells are characterized by abundant cytoplasm and round nuclei with single prominent nucleoli (Figs. 14-109 to 14-112). Squamous cancer cells in ascites are considered to be originated from squamous cell carcinoma of the uterine cervix, epidermoid component of ovarian teratoma (Fig. 14-66) and much more rarely from squamous carcinoma of the lung and the pancreas. Transitional carcinoma cells seen in a cluster can be suggestive from the heaping pattern in a unipolar position and abundant cytoplasm resembling that of squamous cancer cells (Fig. 14-106).

Apocrine carcinoma arising from the breast and apoc-

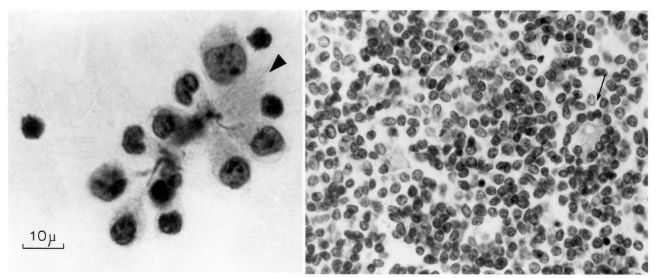

Fig. 14-104   Sympathicoblastoma arising from the posterior mediastinum. This highly cellular tumor resembles a collection of lymphoid cells or lymphosarcoma cells. However, careful examination points out the distinctive feature of the tumor that is composed of sympathicoblasts with vesicular nuclei and scarcely demonstrable cytoplasm. The tumor cells tend to aggregate around a fine fibrillar network (arrowhead). Pseudorosette formation is shown with arrow.

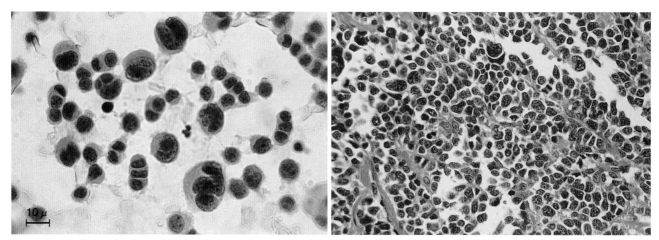

Fig. 14-105   Oat-celled carcinoid cells in ascites and its histology originated from the stomach. This clinically malignant and histologically anaplastic carcinoid resembles oat cell carcinoma of the lung. The nuclei are hyperchromatic with densely packed chromatin particles and generally enrounded, although some large polymorphic nuclei are admixed. Note a characteristic epithelial arrangement in tandem.

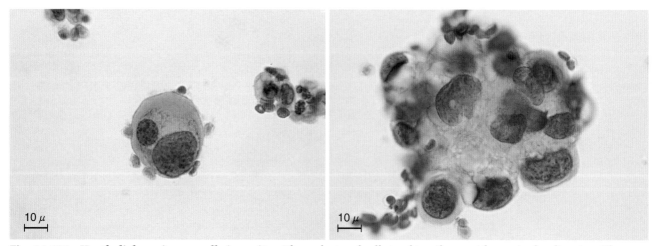

Fig. 14-106   Urothelial carcinoma cells in ascites. These clustered cells tend to pile up with a unipolar direction. The cytoplasm is not as thick as that of epidermoid cancer cells; perinuclear clearing is also a distinctive feature.

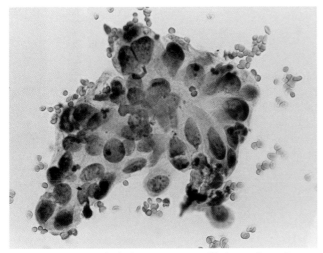

Fig. 14-107 Urothelial carcinoma cells in ascites. Cancer cells seen in sheet tend to form a parallel arrangement. An irregular cell cluster differs from glandular clumping.

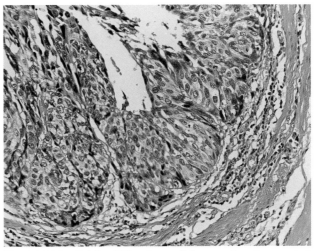

Fig. 14-108 Original histology of the previously resected carcinoma. A grade III transitional carcinoma that is composed of multilayered bands of urothelial cancer cells. Polarity is distorted but remains.

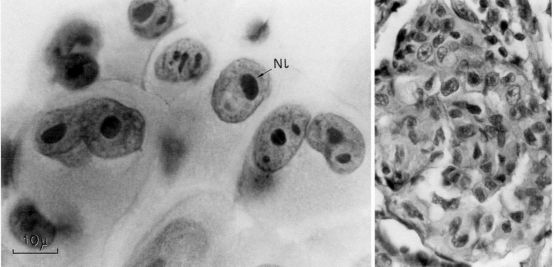

14-109

14-110

Fig. 14-109 Clear celled malignant cells in ascites peculiar to renal cell carcinoma. A large amount of clear cytoplasm with indistinct cellular border and India ink droplet-like nucleoli (Nl) are characteristic of this type. Although hematuria is a frequent and early symptom of this tumor, these malignant cells are rarely detected in urine.
Fig. 14-110 Histology of renal cell carcinoma. The tumor cells have abundant, clear, ill-defined or vacuolated cytoplasm. The nuclei are polyhedral, varying in size, vesicular, and show prominent nucleoli. The property of vascular invasion is peculiar to this tumor and develops early metastasis to lungs, liver and bones.

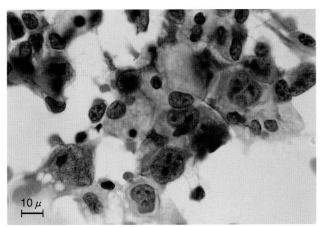

Fig. 14-111 A loose cluster of derived clear cancer cells is from poorly differentiated adenocarcinoma of the kidney. Note clear cytoplasm and polymorphic nuclei having one or two prominent nucleoli.

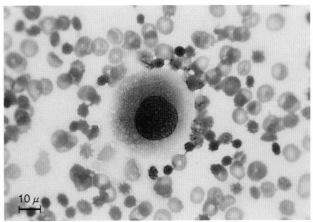

Fig. 14-112 Giemsa stained renal adenocarcinoma cell in ascites. A round nucleus contains a prominent central nucleolus. Its hyperchromasia is due to densely packed chromatin particles. The abundant cytoplasm is vesicular and ill-defined.

319

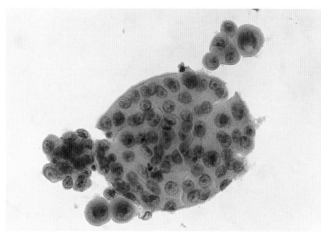

Fig. 14-113a Apocrine adenocarcinoma in pleural fluid. Cancer cells shed as cell balls or tight cluster in effusions. The tumor shows fair amount of finely granular or homogeneous cytoplasm. Prominent nucleoli are also characteristic of this tumor.

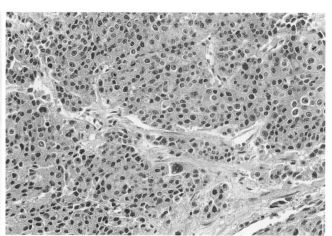

Fig. 14-113b Apocrine adenocarcinoma of the skin involving the lung and pleura. A 46-year-old male developed visceral dissemination after excision of skin tumor at axilla. (By courtesy of Drs. M. Otani and M. Miura, Toho University, School of Medicine, Tokyo.)

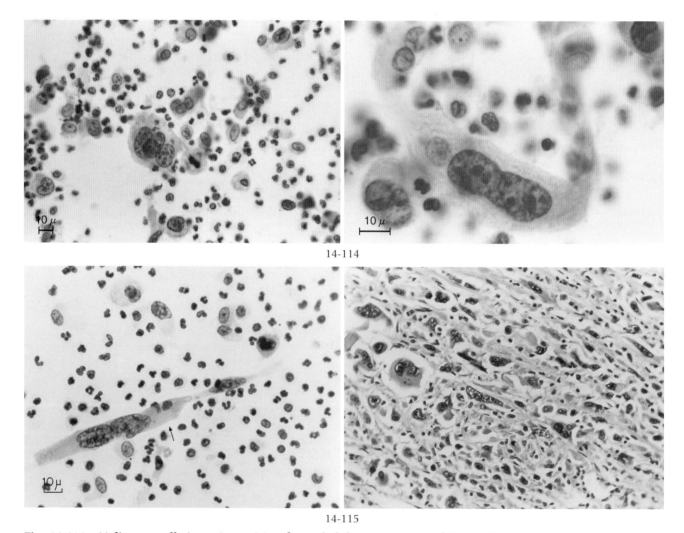

14-114

14-115

Fig. 14-114 Malignant cells in ascites arising from rhabdomyosarcoma of the gallbladder. Unless cross striation is found, no definite diagnosis can be made. However, several characteristic features are summarized as follows: (1) Bizarre cellular feature with indistinct borders, (2) odd shapes and multiplication of the nuclei, and (3) prominence of multiple, eosinophilic nucleoli.

Fig. 14-115 Cytology and histology of the original tumor, rhabdomyosarcoma arising from the gallbladder. This is a loosely textured malignant neoplasm showing diffuse infiltration of enormously pleomorphic sarcoma cells; there are spindly cells, round cells, multinucleated giant cells with eosinophilic finely granular cytoplasm and strap-shaped cells with two or three nuclei arranged in tandem (arrow).

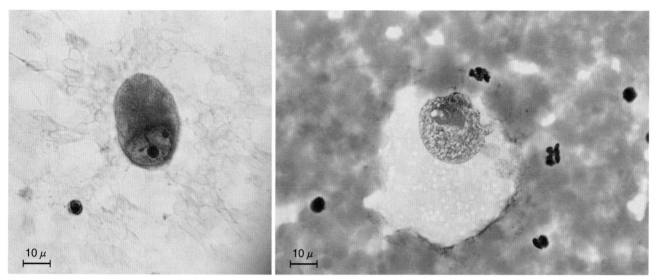

Fig. 14-116 Malignant cells arising from undifferentiated liposarcoma of the retroperitoneum. There is no diagnostic feature of liposarcoma except for a few findings that characterize mesenchymal origin: (1) prominent acidophilic nucleoli, (2) thin nuclear rim, (3) abundant, opaque cytoplasm with fine vesicles, and (4) sudanophilia though not illustrated in these figures. (By courtesy of Cytology Laboratory, Teikyo University Hospital, Tokyo.)

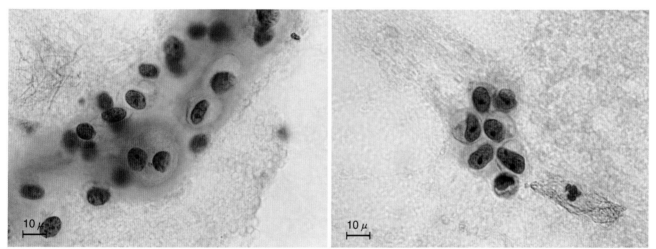

Fig. 14-117 Malignant cells in pleural fluid appearing by recurrence of chondrosarcoma of the lung. In excess of fluid after resection of pulmonary chondrosarcoma, single or loosely clustered sarcoma cells were found. Diagnostic points: (1) binucleation with single prominent nucleoli, (2) opaque, waxy cytoplasm with distinctive perinuclear halo, (3) ill-defined cellular border, and (4) positive PAS reaction though not illustrated.

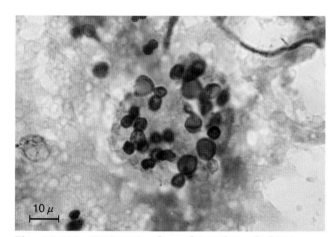

Fig. 14-118 PAS-stained multinucleated histiocyte including *Cryptococcus neoformans*. A 72-year-old deteriorated man was examined with abdominal paracentesis because of double cancers of hepatoma and colon carcinoma. Attention must be paid for opportunistic fungus infection in abnormally multinucleated histiocytes.

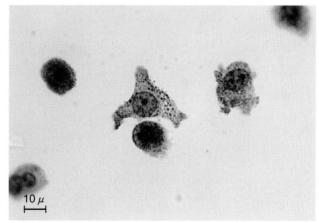

Fig. 14-119 Melanoma cells in ascites form a patient with disseminated malignant melanoma. The nuclei of melanoma cells are enrounded and possess large, droplet-like nucleoli. Melanin pigments can be identified by special stains. Compare with melanophages (Fig. 14-8).

321

rine glands of the skin develops occasionally visceral spread and results in excess of effusion of body cavities (Fig. 14-113). This tumor displays characteristic cytologic features such as fair amount of homogeneous or finely granular cytoplasm and prominent, amphoteric nucleoli. Diastase-resistant, PAS-positive granules may reveal granular cytoplasm.[17]

b) Tumors of mesodermal origin are from the adipose tissue (Fig. 14-116), smooth and striated muscle, blood or lymph vessels, bone, cartilage (Fig. 14-117), etc. Sarcoma cells, generally speaking, are polymorphic and varying in size. Rhabdomyosarcoma is a good example of extremely polymorphic mesenchymal tumor. The nuclear rims are distinct but delicate. They have prominent, often multiple nucleoli (Figs. 14-114 and 14-115).

c) Tumors of neurogenous origin include the neoplasm arising from nerve sheaths, sympathetic nervous tissue and paraganglionic tissue (Figs. 14-103 and 14-104). The most malignant neuroblastoma is rarely encountered in cytology of effusions. These are small, lymphocyte-like, pyknotic cells with scant cytoplasm. Although they resemble oat cell carcinoma, malignant carcinoid and lymphosarcoma cells, the nuclei are polygonal or polyhedral, smaller than those of oat cells, and aggregate into cell groups showing a rosette-like pattern (Fig. 14-103). It should be kept in mind that neuroblastoma occurs in childhood under three years of age.

d) Body cavities are involved by metastatic melanoma. In serous effusion detection of melanophages is occasionally the first in diagnosis. Melanoma cells are characterized by a centrally locating prominent nucleolus and round nucleus regardlcss of the pigment (Fig. 14-119). The Fontana-Mason is stressed to be an important cytochemical stain (Bedrossian).[4]

References

1. Armstrong GR, Raafat F, Ingram L, et al: Malignant peritoneal mesothelioma in childhood. Arch Pathol Lab Med 112: 1159, 1988.
2. Apffel CA, Baker JR: Lipid droplets in the cytoplasm of malignant cells. Cancer 17: 176, 1964.
3. Askin FB, McCann BG, Kuhn C: Reactive eosinophilic pleuritis. Arch Pathol Lab Med 101: 187, 1977.
4. Battifora H, McCaughey WTE: Tumors of the Serosal Membranes, Atlas of Tumor Pathology. p39. Washington, DC: Armed Forces Institute of Pathology, 1995.
5. Bedrossian CWM: Malignant Effusions: A Multimodal Approach to Cytologic Diagnosis. New York: Igaku-Shoin, 1994.
6. Bishop JW, MacFarlane K, Cheuvront D, et al: Cell recovery and appearance in thin-layer preparations in nongynecologic cytology. Anal Quant Cytol Histol 20: 229, 1998.
7. Boon ME, Kwee HS, Alons CL, et al: Discrimination between primary pleural and primary peritoneal mesotheliomas by morphometry and analysis of the vacuolization pattern of the exfoliated mesothelial cells. Acta Cytol 26: 103, 1982.
8. Broghamer WL, Richardson ME, Faurest SE: Malignancy-associated serosanguinous pleural effusions. Acta Cytol 128: 46, 1984.
9. Carr I: The Macrophage: A Review of Ultrastructure and Function. London: Academic Press, 1973.
10. Carr I, Underwood JCE: The ultrastructure of the local cellular reaction to neoplasia. Int Rev Cytol 37: 329, 1974.
11. Carr I, Underwood JCE, McGinty F, Wood P: The ultrastructure of the local lymphoreticular response to an experimental neoplasm. J Pathol 113: 175, 1973.
12. Cawson RA: The cytological diagnosis of oral cancer. Br Dent J 108: 294, 1960.
13. Ceelen GH: The cytologic diagnosis of ascitic fluid. Acta Cytol 8: 175, 1964.
14. Chapman CB, Whalen EJ: The examination of serous fluids by the cell block technic. N Engl J Med 237: 215, 1947.
15. Churg A: Immunohistochemical staining for vimentin and keratin in malignant mesothelioma. Am J Surg Pathol 9: 360, 1985.
16. Churg J, Gordon AJ: Multiple myeloma: Lesions of the extraosseous hematopoietic system. Am J Clin Pathol 20: 934, 1950.
17. Conangla M, Gimferrer E, Combalia N, et al: Fine needle aspiration biopsy in the diagnosis of apocrine sweat gland adenocarcinoma. Acta Cytol 34: 282, 1990.
18. Cordier V, Morenas L: Eosinophilie pleurale symptomatique d'un abcès hépatique amibien évacué par vomique. Constatation d'Entamoeda dysenteriae dans l'expectoration. Comp Rend Soc Biol 104: 198, 1930.
19. Costa M, Oetel YC: Cytology of pseudomyxoma peritonei: Report of two cases arising from appendiceal cystadenomas. Diagn Cytopathol 6: 201, 1990.
20. Curran TM: Cholesterol pleural effusion. Edin Med J 55: 252, 1948.
21. de Lavergne V, Abel E, Debenedetti R: Eosinophilie pleurale au cours d'un abces amibien du poumon. Bull et Mem Soc Méd Eop d Paris 54: 593, 1930.
22. Diaz-Arias AA, Loy TS, Bickel JT, et al: Utility of BER-EP4 in the diagnosis of adenocarcinoma in effusions: An immunocytochemical study of 232 cases. Diag Cytopathol 9: 516, 1993.
23. DiBonito L, Falconieri G, Colautti I, et al: The positive pleural effusion: A retrospective study of cytopathologic diagnoses with autopsy confirmation. Acta Cytol 36: 329, 1992.
24. Djeu JY, McCoy JL, Cannon GB, et al: Lymphocytes forming rosettes with sheep erythrocytes in metastatic pleural effusions. J Natl Cancer Inst 56: 1051, 1976.
25. Domagala W, Emerson EE, Koss LG: T and B lymphocyte enumeration in the diagnosis of lymphocyte-rich pleural fluid. Acta Cytol 25: 108, 1981.
26. Dyment L-J: Les pleurésies cancéreuses à éosinophiles. Thesis, Paris, 1931.
27. Ehrich WE: Die Leukozyten und ihre Entstehung. Erg Allg Path 29: 1, 1934.
28. Ehrich WE: Die Entzündung. Im: Handbuch der Allgemeinen Pathologie. Berlin-Heidelberg-New York: Springer-Verlag, 1956.
29. Evans R, Alexander P: Cooperation of immune lymphoid cells with macrophages in tumour immunity. Nature 228: 620, 1970.
30. Evans R, Alexander P: Mechanism of immunologically specific killing tumour cells by macrophages. Nature 236: 168, 1972.
31. Falconieri G, Zanconati F, Colautti I, et al: Effusion cytology of hepatocellular carcinoma. Acta Cytol 39: 893, 1995.
32. Fischler DF, Toddy SM: Nongynecologic cytology utilizing the ThinPrep processor. Acta Cytol 40: 669, 1996.

33. Fisher ER, Hellstrom HR: The periodic acid-Schiff reaction as an aid in the identification of mesothelioma. Cancer 13: 837, 1960.

34. Foot NC: The identification of tumor cells in sediments of serous effusions. Am J Pathol 13: 1, 1937.

35. Foot NC: The identification of neoplastic cells in serous effusions: Critical analysis of smears from 2,029 persons. Am J Pathol 32: 961, 1956.

36. Foot NC: The identification of mesothelial cells in sediments of serous effusions. Cancer 12: 429, 1959.

37. Fortt RW, Mathie IK: Gliomatosis peritonei caused by ovarian teratoma. J Clin Pathol 22: 348, 1969.

38. Fox H, Langley FA: Tumours of the Ovary. London: William Heinemann, 1976.

39. Frerichs JB: Influence of monocytosis of peripheral blood stream upon cellular character of acute inflammation. Bull Johns Hopkins Hosp 74: 49, 1944.

40. Fudenberg HH, Wybran J, Robbins D: T-rosette-forming cells, cellular immunity and cancer. N Engl J Med 292: 475, 1975.

41. Garrioch DB, Good RA, Gatti RA: Lymphocyte response to P.H.A. in patients with nonlymphoid tumours. Lancet 1: 618, 1970.

42. Giese W: Zytodiagnostik der Pleuratumoren. Lehrbuch der Speziellen Pathologischen Anatomie II, 3 Teil. Berlin: Walter de Gruyter, 1960.

43. Grunze H: The comparative diagnostic accuracy, efficiency and specificity of cytologic technics used in the diagnosis of malignant neoplasm in serous effusions of the pleural and pericardial cavities. Acta Cytol 8: 150, 1964.

44. Han T, Takita H: Immunologic impairment in bronchogenic carcinoma: A study of lymphocyte response to phytohemagglutinin. Cancer 30: 616, 1972.

45. Harkavy J: Vascular allergy: Pathogenesis of bronchial asthma with recurrent pulmonary infiltrations and eosinophilic polyserositis. Arch Intern Med 67: 709, 1941.

46. Harkavy J: Vascular allergy. J Allergy 14: 507, 1943.

47. Holub M: Potentialities of the small lymphocyte as revealed by homotransplantation and autotransplantation experiments in diffusion chambers. Ann NY Acad Sci 99: 477, 1962.

48. Hutt MSR, Anthony PP: Tumours of liver, biliary system and pancreas. Recent Result in Cancer Res 41: 57, 1973.

49. Innes J, Gordon AJ: Myelomatosis. Lancet 1: 239, 1961.

50. Johnson WD: The cytological diagnosis of cancer in serous effusions. Acta Cytol 10: 161, 1966.

51. Joos F, Roos B, Bürki H, et al: Umasatz, Proliferation und Phagozytosetätigkeit der freien Zellen im Peritonäalraum der Maus nach Injektion von Polystyren-Partikeln. Z Zellforsch 95: 60, 1969.

52. Journey LJ, Amos DB: An electron microscope study of histiocyte response to ascites tumor homografts. Cancer Res 22: 998, 1962.

53. Kasdon EJ: Malignant mesothelioma of the tunica vaginalis propria testis. Cancer 23: 1144, 1970.

54. Kelley RR, Scully RE: Cancer developing in dermoid cysts of ovary: A report of 8 cases, including a carcinoid and a leiomyosarcoma. Cancer 14: 989, 1961.

55. Kern WH: Benign papillary structures with psammoma bodies in culdocentesis fluid. Acta Cytol 13: 178, 1969.

56. Klempman S: The exfoliative cytology of diffuse pleural mesothelioma. Cancer 15: 691, 1962.

57. Koss LG: Diagnostic Cytology and Its Histopathologic Bases. Philadelphia: JB Lippincott, 1968.

58. Koss LG: Diagnostic Cytology and Its Histopathologic Bases, 3rd ed. p911. Philadelphia: JB Lippincott, 1979.

59. Kotani H, Sakashita Y, Hayashi W, et al: A case of peritoneal mesothelioma. J Jpn Soc Clin Cytol 19: 226, 1980.

60. Lapes M, Rosenzweig M, Barbieri B, et al: Cellular and humoral immunity in non-Hodgkin's lymphoma. Am J Clin Pathol 67: 347, 1977.

61. Latza U, Niedobitek G, Schwarting R, et al: Ber-Ep4: New monoclonal antibody which distinguishes epithelia from mesothelial cells. J Clin Pathol 43: 213, 1990.

62. Lee Y-T N, Marshall GJ, Weiner J, et al: Pleural B- and T-lymphocyte counts in patients with sarcoma and breast carcinoma. Cancer 40: 667, 1977.

63. Leiman G, Goldberg R: Pseudomyxoma peritonei associated with ovarian mucinous tumors: Cytologic appearance in five cases. Acta Cytol 36: 299, 1992.

64. Light RW, Macgregor MI, Luchsinger PC, et al: Pleural effusions: The diagnostic separation of transudates and exudates. Ann Intern Med 77: 507, 1972.

65. Lopes Cardozo E, Harting MC: On the function of lymphocytes in malignant effusions. Acta Cytol 16: 307, 1972.

66. Lopes Cardozo P: Advantages and disadvantages of the May-Grünwald Giemsa procedure in exfoliative cytology. Acta Cytol 2: 284, 1958.

67. Lopes Cardozo P: Atlas of Clinical Cytology. Targa b. v.'s-Hertogenbosch, the Netherlands, 1975.

68. Luse SA, Reagan JW: A histocytological study of effusions. I. Effusions not associated with malignant tumors. Cancer 7: 1155, 1954.

69. Mallonee MM, Lin F, Hassanein R: A morphologic analysis of the cells of ductal carcinoma of the breast and of adenocarcinoma of the ovary in pleural and abdominal effusions. Acta Cytol 31: 441, 1987.

70. Masugi M: Über die Beziehungen zwischen Monocyten und Histiocyten. Beitr Pathol Anat 76: 396, 1927.

71. Mavrommatis FS: The identification on mesothelial cells by the technic of Mowry. Acta Cytol 6: 443, 1962.

72. Mavrommatis FS: Some morphologic features of cells containing PAS positive intracytoplasmic granules in smears of serous effusions. Acta Cytol 8: 426, 1964.

73. Maximow A: Cultures of blood leucocytes: From lymphocyte and monocyte to connective tissue. Arch Exper Zellforsch 5: 169, 1928.

74. McDonald AD, McDonald JC, Pooley FD: Mineral fibre content of lung in mesothelial tumours in North America. Ann Occup Hyg 26: 417, 1982.

75. McDonald GA, Dodds TC, Cruickshank B: Atlas of Haematology. London: Churchill Livingstone, 1978.

76. McDonald JR, Broders AC: Malignant cells in serous effusions. Arch Pathol 27: 53, 1939.

77. Monte SA, Ehya H, Lang WR: Positive effusion cytology as the initial presentation of malignancy. Acta Cytol 31: 448, 1987.

78. Morrison JB: Pleurisy with effusion during pneumothorax therapy. Am Rev Tuberc 57: 598, 1948.

79. Nagel O: Ein Beitrag zur Pathologie der flüchtigen Lungenfiltrate mit Bluteosinophilie. Beitr Klin Tuberk 96: 185, 1941.

80. Naylor B: Pleural, peritoneal and pericardial fluids. In: Bibbo M (ed): Comprehensive Cytopathology. pp541-614. Philadelphia: WB Saunders, 1991.

81. Nishioka K, Hirayama T, Sekine T, et al: Australia antigen and hepatocellular carcinoma. Gann Monogr

Cancer Res 14: 167, 1973.

82. Page DS: Pleural effusion, its cytology and the results of paracentesis. Lancet 198: 584, 1920.

83. Papillo JL, Lapen D: Cell yield ThinPrep vs. Cytocentrifuge. Acta Cytol 38: 33, 1994.

84. Perkins JJ, Dudgeon LS: A case of eosinophilic pleurisy. Trans Pathol Soc London 58: 119, 1907.

85. Picoff RC, Meeker CI: Psammoma bodies in the cervicovaginal smear in association with benign papillary structures of the ovary. Acta Cytol 14: 45, 1970.

86. Piéron R, Fernet M, Cosset JM, et al: Maladie de kahler avec plasmacytome sous-costal, pleurèsie et péricardite a plasmocytes. Sem Hop Paris 48: 2953, 1972.

87. Prince AM, Szmuness W, Michon J, et al: A case/control study of the association between primary liver cancer and hepatitis B infection in Senegal. Int J Cancer 16: 376, 1975.

88. Quensel U: Zur Frage der Zytodiagnostik der Ergüsse seröser Höhlen. Methodologische und pathologisch-anatomische Bemerkungen. Acta Med Scand 68: 427, 1928.

89. Reinikanien M: Exudative pleurisy with eosinophilia. Ann Med Int Fennioe: 14, 1947.

90. Rist E, Kindberg L: Deux cas d'eosinophilie pleurale. Bull et Mem Soc Med Hop d Paris 28: 693, 1909.

91. Robb JA: Mesothelioma versus adenocarcinoma: False positive CEA and Leu-Ml staining due to hyaluronic acid [Letter]. Hum Pathol 20: 400, 1989.

92. Robboy SJ, Scully RE: Ovarian teratoma with glial implants on the peritoneum. Hum Pathol 1: 643, 1970.

93. Rolleston HD: A case of eosinophile ascites: With remarks. Br Med J 1: 238, 1914.

94. Ruitenbeck T, Couw ASH, Poppema S: Immunocytology of body cavity fluids: MOC-31, a monoclonal antibody discriminating between mesothelial and epithelial cells. Arch Pathol Lad Med 118: 265, 1994.

95. Safa AM, Ordstand HS: Pleural effusion due to multiple myeloma. Chest 64: 246, 1973.

96. Sandkühler ST, Doemheld L: Myelome mit plasmazytärem Pleuraergu β. Acta haematol 18: 403, 1957.

97. Saphir O: Cytologic diagnosis of cancer from pleural and peritoneal fluids. Am J Clin Pathol 19: 309, 1949.

98. Saphir O, Lackner JE: Adenocarcinoma with clear cells (hypernephroid) of ovary. Surg Gynecol Obstet 79: 539, 1944.

99. Schultz A, Löhr G: Zur Frage der Spezifität der mikrochemischen Cholesterinreaktion mit Eisessig-Schwefelsäure. Zbl Allg Path Path Anat 36: 529,1925.

100. Seemann S: Über die Beziehungen zwischen Lymphocyten und Histiocyten insbesondere bei Entzündung. Ziegler Beitr Path Anat 85: 303, 1930.

101. Spriggs AI: The Cytology of Effusions in the Pleural, Pericardial and Peritoneal Cavities. London: William Heinemann, 1957.

102. Spriggs AI: Pleural eosinophilia due to pneumothorax. Acta Cytol 23: 425. 1979.

103. Steinberg G: Peritoneal exudate. JAMA 116: 572, 1941.

104. Stout AP: Mesothelioma of the pleura and peritoneum: cited from Tumors of the Retroperitoneum, Mesentery and Peritoneum. Washington, DC: Armed Forces Institute of Pathology, 1954.

105. Tajima M: Cytological diagnosis of pleuroperitoneal effusion. Proceedings of 17th General Assemb Jap Med Congr 2: 995, 1967.

106. Takahashi M, Kawaguchi A, Kawamata K, et al: α-Fetoprotein by the immunofluorescent antibody technique in primary liver carcinoma and its histopathological basis. Gann Monogr Cancer Res 14: 219, 1973.

107. Takahashi M, Miwa K, Kunizane H: Isoenzyme fractions of lactic dehydrogenase in effusions (in Japanese). J Med Technol 20: 292, 1976.

108. Tani EM, Takahashi M: Prognostic evaluation of T lymphocyte population in body cavity effusions secondary to malignancy. J Jpn Soc Clin Cytol 19: 534, 1980.

109. Teilum G: Histogenesis and classification of mesonephric tumors of the female and male genital system and relationship to benign so-called adenomatoid tumors (mesotheliomas). Acta Pathol Microbiol Scand 34: 431, 1954.

110. Tickman RJ, Cohen C, Varma VA, et al: Distinction between carcinoma cells and mesothelial cells in effusions: Usefulness of immunohistochemistry. Acta Cytol 34: 491, 1990.

111. Triol JH, Conston AS, Chandler SV: Malignant mesothelioma: Cytopathology of 75 cases seen in a New Jersey Community Hospital. Acta Cytol 28: 37, 1984.

112. Turk JL, Oort J: The production of sensitized cells in cell mediated immunity. In: Studer A, Cottier H (eds): Handbuch der Allgemeinen Pathologie, Bd 7, Teil 3, Immunreaktionen/Immune Reactions. pp392-435. Berlin-Heidelberg-New York: Springer-Verlag, 1970.

113. van Furth R: Origin and kinetics of monocytes and macrophages. Semin Hematol 7: 125, 1970.

114. Vargas-Suárez J: Über Ursprung und Bedeutung der in Pleuraergüssen vorkommenden Zellen. Beitr Klin Tuberk 2: 201, 1904.

115. Veress JF, Koss LG, Schreiber K: Eosinophilic pleural effusions. Acta Cytol 23: 40, 1979.

116. Volkman A: The origin and turnover of mononuclear cells in peritoneal exudates in rats. J Exp Med 124: 241, 1966.

117. Wagner JC, Munday DE, Harington JS: Histological demonstration of hyaluronic acid in pleural mesothelioma. J Pathol Bacteriol 84: 73, 1962.

118. Weidmann J, Chaubal A, Bibbo M: Cellular fixation: A study of CytoRich Red and Cytospin collection fluid. Acta Cytol 41: 182, 1997.

119. Whitaker D, Shilkin KB: The cytology of malignant mesothelioma in Western Australia. Acta Cytol 22: 67, 1978.

120. Wihman G: A contribution to the knowledge of the cellular content in exudates and transudates. Acta Med Scand (Stockholm) 130 (Suppl): 205, 1948.

121. Wile SA, Saphir O: Migratory peritonitis: A clinicopathologic study of so-called hematogenous peritonitis in children. Am J Dis Child 43: 610, 1932.

122. Willis RA: Pathology of Tumours. London: Butterworth, 1948.

123. Wilson GE, Hasleton PS, Chatterjee AK: Desmoplastic malignant mesothelioma: A review of 17 cases. J Clin Pathol 45: 295, 1992.

124. Wisniewski M, Deppisch LM: Solid teratomas of the ovary. Cancer 32: 440, 1973.

125. Witte CL, Witte MH, Cole WR, et al: Dual origin of ascites in hepatic cirrhosis. Surg Gynecol Obstet 129: 1027, 1969.

126. Woodruff JD, Bie LS, Sherman RJ: Mucinous tumors of the ovary. Obstet Gynecol 16: 699, 1960.

127. Wybran J, Fudenberg HH: Thymus-derived rosette-forming cells in various human disease states: Cancer, lymphoma, bacterial and viral infections, and other diseases. J Clin Invest 52: 1026, 1973.

128. Zach J: Praktische Zytologie für Internisten. Stuttgart: Georg Thieme, 1972.
129. Zadek I: Die Zytologie der Exsudate und Transudate. In: Hirschfeld H, Hittmair A (eds): Handbuch der allgemeinen Hämatologie. Berlin: Urban, 1933.

15 Urinary Tract

1 General Aspect of Urinary Cytology

Cytologists have devoted themselves to the detection of malignant neoplasms of the urinary tract because of the ease of sampling specimens since Papanicolaou and Marshall (1945)[86] referred cytologists to detection of cancer, but they have not paid much attention to qualitative and quantitative examinations of the urinary sediment as evaluated in urinalysis that examines casts, blood cells, renal epithelial cells which are derived from the nephron. It is because they have not examined unstained fresh sediments. Identification of various kinds of the urinary casts and measurement of their numbers and dimensions are of diagnostic value for intrinsic parenchymal diseases of the kidney.

On the other hand, laboratory technologists who are engaged in urinalysis are not familiar with cytological technique using Papanicolaou stain. During a microscopic examination of the fresh sediment without stain, they may overlook incipient carcinoma of the urinary tract, if they are not familiar with small atypical cells which shed singly in urine without tumor diathesis. These pitfalls in both sides will be covered by the use of supravital stains such as Sternheimer-Malbin stain and toluidine blue stain, and by the introduction of various techniques of special microscopy. Phase contrast microscopy and differential interference microscopy are worthy of visualization of cells and casts. Fluorescence microscopy is available for acridine orange supravital or permanent stain, immunofluorescence antibody method,

and for the detection of a substance that yields a primary fluorescence.

Routine urinalysis and cytology

In urinalysis we must deal with a large number of urine specimens soon after micturition, thus the sediments are routinely examined without stain; a delay for a couple of hours after collection of the specimen permits the growth of contaminated bacteria and promotes alkalization of urine which may dissolve cellular elements. In cases where the detailed structure should be observed, differential interference microscopy enables unstained material visible with a distinctive color contrast; chromatin threads and cytoplasmic granules can be observed. The use of polarized light illumination is valuable to demonstrate doubly refractive substances; cholesterol and leucine crystals disclose a characteristic "Maltese cross" appearance (Fig. 15-1). However, examination with stain is important to demonstrate microorganisms and to distinguish renal epithelial cells from deep transitional epithelial cells. As a usual laboratory examination, the procedure for urinalysis will be summarized as follows:

First step: Examination of fresh sediments with condensed, proper illumination.
　Use of green filter that is occasionally valuable to enhance a contrast.
Second step: Phase contrast microscopy for fresh sediments (Fig. 15-2).
　Differential interference microscopy for fresh sediments; more useful than phase contrast microscopy.
　Supravital stain such as Sternheimer,[82] Sternheimer-Malbin (Fig. 15-3) and acridine orange (see Fig. 15-77).
　Rapid stain as a permanent smear. Diff-Quik stain (Baxter Co., Miami, Florida, U.S.A.) is a good example.
Third step: Papanicolaou or Shorr stain as a permanent smear.
　Other special stains added if necessary.

Urinalysis consists of analysis of cellular components and organic or inorganic crystals, determination of the number and nature of casts in the urinary sediments and grading of albuminuria, glycosuria, bilirubinuria, hematuria (hemoglobinuria), bacteriuria, etc. Cytologic studies were mainly focused on qualitative and occasionally quantitative alterations of epithelial cells exfoliated from

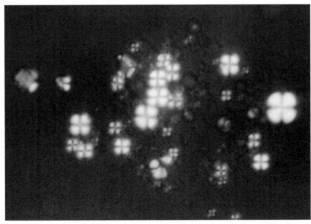

Fig. 15-1　Maltese cross configuration of lipid bodies. Under polarized light, a typical "Maltese cross" configuration can be observed on lipid bodies.

326

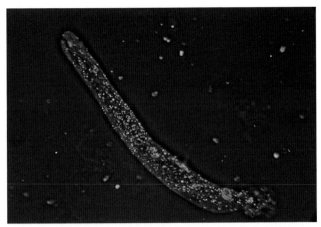

Fig. 15-2 Phase Contrast microscopy with bright contrast for fine granular cast. The outline of the cast is clearly demonstrated. Note beautifully illuminated fine granules which are scattered in glassy matrix.

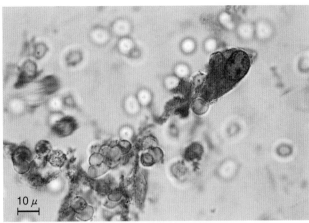

10 μ

Fig. 15-3 Transitional cancer cell in urinary sediment by supravital stain. The Sternheimer astrablue stain is appropriate to visualize cells and casts with ease and rapidity. Initial stain ability of cells relates to their viability; devitalized cells stain immediately in supravital stain.

the urinary tract below the pelvis. However, pathologic changes of the nephron will be reflected in the formation of various kinds of casts as well as on degenerative cell changes of tubular cells.

Gross appearance of urine

Appearance of urine becomes a clue to pathological conditions of the urinary tract. It is normally clear and straw-colored because of urochrome. The grade of coloration depends upon urine concentration. Attention should be paid to turbidity and abnormal coloration.

1) Turbidity

Pathologically turbidity is caused by much content of bacteria, fat droplets, and pus cells. Chylous color (chyluria) is due to either obstruction of lymphatic channels or to fistula formation between lymphatic channels and the urinary tract. Cloudy smoky urine is caused by much content of red blood cells. Physiologically, urate in acidic urine may give rise to turbidity, whereas phosphate is responsible for turbidity in alkaline urine. If urine is placed in a refrigerator, clear urine becomes milky because of urate crystals.

2) Abnormal coloration

Bilirubinuria or biliuria is characterized by yellowish brown or greenish brown color; shaking the urine in a tube, one sees bubbles to be yellow. Abnormally reddish hue is obtained in various conditions such as hematuria, hemoglobinuria, porphyrinuria, and medication of senna, santonin, sulfa drugs etc. Hematuria is distinguishable from hemoglobinuria by smoky turbidity and sedimentation of red cells. Hemoglobinuria, unaccompanied by red cells, is determined by occult blood test. Porphyrinuria may shed purple wine fluorescence by ultraviolet illumination.

3) Passage of gas

Pneumaturia is noted during micturition as a result of fistula formation between the bowel and urinary bladder. Rectosigmoid carcinoma, diverticulitis and gas-forming anaerobic bacterial infection are responsible for pneumaturia.

Sensitivity dependent on cell typing and efficacy of serial urinary cytology

The fact that the sensitivity of urinary cytology depends on the grade of malignancy of urothelial tumors and the repeat of examinations has strongly emphasized by de Voogt, et al.[26] There is an increase of diagnostic rates from only single specimen to three repeated specimens in each grade of urothelial tumors in such degrees as from 34% to 79% in the grade II tumor, from 68% to 85% in the grade III tumor, and from 92% to 98% in the grade IV tumor respectively. It has been seen also in the grade I urothelial tumor (Table 15-1).

There appeared several reports[55,81] regarding the sensitivity increased with malignancy grading. It is noteworthy that very high sensitivity could be shown for patients with carcinoma in situ (Table 15-2).

2 Collection of Specimen

Voided urine and catheterized urine are ordinarily used. A midstream morning specimen is suitable for routine urinary cytology, because the initial portion of the urine may contain extraneous bacteria and contaminants. Random specimens also offer satisfactory results in symptomatic patients.

Three-glass urine: Separate specimens obtained by fractional voiding that is known as the three-glass urine are available for urinary cytology. The first glass collected from the initial urination of about 10 to 30 ml contains washings from the urethra, whereas the second glass from the midstream is the sample from the urinary bladder. The third glass urine usually 10 to 20 ml is that collected at the terminatory urination and may contain secretions from the prostate. Cytology by the three-glass technique is informative for localization of the lesion.

Ureteric catheterization is of value to examine a selective site of the ureter or renal pelvis; the site of the lesion should be clinically determined beforehand. Cystoscopy is detectable of the bleeding site, and excretory urography is able to discover the dysfunction of urinary excre-

Table 15-1 Relationship between positive cytological report and histological grading and differentiation (Leiden, 1970-1975)

	Total cases	1st spec. pos.	2nd spec. pos.	3rd spec. pos.	>3rd spec. pos.
Grade 0-I tumor	47	0	0	0	0
Grade II tumor					
without infiltration	3	3	–	–	–
with infiltration	35	10	14	1	2
Grade III tumor	53	36	11	2	–
Grade IV tumor	41	35	3	2	–
Carcinoma in situ	16	11	4	1	–
Squamous carcinoma	8	8	–	–	–
Adenocarcinoma	4	3	–	–	–
Total	207	106	32	6	2

The data of the first two columns were subjected to a χ^2 test for the contingency table: $\chi^2=34.8$ DF=6. The adenocarcinomas and squamous carcinomas were grouped together. The result is highly significant (p<0.0001). The rate of detection of tumors grade II, grade III and grade IV differ. The difference between grade IV carcinomas and the combined cases of carcinoma in situ, squamous cell carcinoma and adenocarcinoma (all with a known high exfoliation rate) was not significant. (From de Voogt HJ, et al: Urinary Cytology. Berlin: Springer-Verlag, 1977.)

tion on the affected side. Retrograde pyelography is also necessary to find the filling defects; compression, obliteration, displacement and dilatation of the urinary tract can be proved with ease. Although both sites are usually examined separately, the specimen should be taken first from an intact site to avoid cell contamination from the diseased site. Washing the urinary bladder with isotonic saline or electrolyte solutions after voiding is suitable for obtaining an adequate number of cells. The irrigation method with isotonic saline solution containing a small amount of proteolytic enzyme that may enhance exfoliation of cells is also recommended. The osmotic pressure of urine is said to vary physiologically from 50 to 1,200 mOsm/kg (500 to 800 mOsm/kg on the average) to keep the blood osmotic pressure constant. Hypertonicity of urine may influence the feature of cells exfoliated; red blood cells will be characteristically crenated. Therefore, the irrigation method with balanced saline is valuable, not only to promote exfoliation of cells, but to

preserve cells well. External massage by hand pressure on the lower abdomen is expected to bring about a good exfoliation of urothelial cells.

3 Preparation of Smears

Routine centrifugation method

1) Centrifuge a urine specimen at 1,500 rpm for five minutes. If enough sediment cannot be obtained, decant the supernatant, pour urine into the same centrifugation tube and recentrifuge.

2) Make smears on thinly albuminized glass slides with the pull-apart method. Instead of using the albuminized slides, put a drop of egg albumin onto the urine sediment, mix lightly with a glass bar and make smears.

3) Place the slides soon into 95% ethanol when the edges of the smears begin to dry. The use of egg albumin has advantages, not only for adhering cells to slides, but also for preventing cells from complete drying.

***Addendum:* Special caution for urinary cytology**
Cytological follow-up after transurethral resection is indispensable to examine the recurrence of cancer. Laboratories may deal with specimens which are scanty in cellularity. Fluid specimens which have a small number of cells and low content of protein may lose diagnostic cells or expel the cells towards the periphery of smears in wet preparation. Cautions for minimizing cell loss or concentrating cells in a defined area on the glass slide are important.

1) Addition of coating adhesives after centrifugation. A drop of bovine albumin, fetal calf serum or 2% carbowax will be effective to recover cells on glass slides.

2) Cytocentrifugation by Cytospin Mark 1 (Shandon Southern Products, Runcorn, Cheshire, U.K.) is suitable for collecting cells in a defined area. However, this method is not suitable for highly cellular samples.

3) Cytosedimentation by AutoCyte PREP (TriPath) is suitable for collecting cells from urine samples with low cellularity by using the preservative and the semiautomated processing system (see page 278). Thin layer films smeared in a small circle (13 mm diameter) can appreciate screening cancer cells either primary or recurrent with ease (Fig. 15-4).

Table 15-2 Relationship between urinary cytology and histological grading and typing

Tumor type	No. of cases	Cytologic classification					
		Negative		Suspicious		Positive	
		No.	%	No.	%	No.	%
Transitional cell carcinoma, in situ*	62	1	2	21	33	40	65
Transitional cell carcinoma, grade I	152	56	37	52	34	44	29
Transitional cell carcinoma, grade II	248	52	21	91	37	105	42
Transitional cell carcinoma, grade III or IV	237	14	6	54	23	169	71
Squamous cell carcinoma, bladder	15	1	7	3	20	11	73
Adenocarcinoma, bladder	15	2	13	4	27	9	60
Carcinoma, urethra	5	0	–	1	20	4	80
Adenocarcinoma, kidney	36	29	81	7	19	0	–
Adenocarcinoma, prostate	62	35	56	14	23	13	21
Metastatic carcinoma	28	6	21	8	29	14	50
Total	860	196	23	255	30	409	47

* 1975-1986 From Kern WH: Acta Cytol 32: 651-654, 1988.

■Centrifugation after pretreatment

Smears should be made while the specimen is fresh. If it is impossible to send the specimen to the cytology laboratory soon after voiding or to prepare the smears soon after they are received, pretreatment of the specimen will preserve cells adequately.

1) Addition of an equal amount of 95% ethanol to the urine specimen (Papanicolaou).

2) Addition of formalin to the specimen in the proportion 1:9 (Vincent Memorial Hospital Staff).

3) Addition of an equal amount of the "pretreatment solution" to the specimen. The pretreatment solution is prepared as stock solution in the proportion of 2 ml of formalin and 98 ml of 40% ethanol diluted with balanced electrolyte or balanced saline solution (in the author's laboratory).

■Irrigation method

1) Insert a catheter, evacuate the bladder, and inject 50 ml of the saline or enzyme-containing irrigation solution. The irrigation solution should be prepared prior to the procedure by resolution of 5 mg of alpha-chymotrypsin (Kimopsin one vial, Eisai Co., Ltd., Tokyo, Japan) per 100 ml of a balanced electrolyte solution.

2) Pump four or five times and collect the solution.

3) Repeat the irrigation with another 50 ml of the irrigation solution and collect it.

4) Centrifuge the collected solution at 1,500 rpm for five minutes.

■Modified method with proteolytic enzyme saline solution

1) Insert a catheter and evacuate the urinary bladder.

2) Inject 30 ml of saline solution with Elase containing 15,000 units DNase and 25 units fibrinolysin (Parke-Davis Co., Detroit, Michigan, U.S.A).

3) Remove the catheter and let the solution stand for 10 minutes.

4) Collect the voided specimen in a urine cup, then add an equal amount of 2% formalin · 40% ethanol · balanced electrolyte solution (pretreatment solution).

5) Centrifuge the collected solution at 1,500 rpm for five minutes.

4 Staining Technique

Microscopic examinations with staining are important to clarify microorganisms and to distinguish the types of urothelial cells. Various kinds of supravital stains are applied to urinalysis in addition to routine Papanicolaou stain.

■Sternheimer-Malbin stain

This staining solution, which was first reported to be applied for diagnosis of pyelonephritis, has come to be very useful to distinguish various types of urothelial cells. A commercial stain "Uri-Cel" (Cambridge Chemical Products, U.S.A.) is thought to have the constituents similar to the Sternheimer-Malbin solution and convenient for a practical use; a drop of urinary sediments and a drop of the solution put beside are mixed well with a cover glass edge. It is ready for observation a

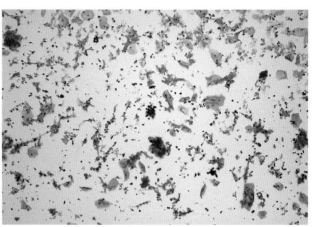

Fig. 15-4 Monolayer smear of urine prepared by AutoCyte PREP processor. Low power magnification from a field of the 13 mm smear. Cells are concentrated and evenly smeared.

few minutes later. Afterwards, stainability increases gradually; nuclei and cytoplasm come to be diffusely stained by devitalization.

Staining method:
First stock solution
Crystal violet ······························3.0 g
95% ethyl alcohol·····················20.0 ml
Ammonium oxalate·····················0.8 g
Distilled water ·····························80.0 ml
Second stock solution
Safranin O ·······························0.25 g
Ethyl alcohol (95%) ·····················10.0 ml
Distilled water·····························100.0 ml
Staining procedure:
1. Mix the first and second solutions in the proportion of 3:97. This mixture should be renewed every three months.
2. Filter prior to use.
3. Place one drop of the staining solution on fresh urine sediment on slide.

■Sternheimer stain

A mixture of aqueous solutions of National fast blue (Allied Chemical Corp., No. 1946P) that is a copper-phthalocyanine dye and pyronin B (Matheson Coleman & Bell, No. Pb 17) facilitates recognition of wide range of formed elements such as benign or malignant cells and various types of casts. Instead of National fast blue, alcian blue or Astrablau of histological use as a strong basic dye is available. Pyronin B is a basic dye of lower molecular weight interacts with polynucleotides. An affinity of copper-phthalocyanine to acid mucopolysaccharides is proper to stain the matrix of casts which are rich in Tamm-Horsfall protein.

1) Make 2% National fast blue solution.
2) Make 1.5% pyronin B aqueous solution. It is advised by Sternheimer that the commercial pyronin should be purified by alcohol extraction.
3) Mix equal parts of both solutions and store.
4) Mix two drops of urinary sediments and a drop of the stain. Examine immediately under light microscope. Devitalized cells stain promptly.

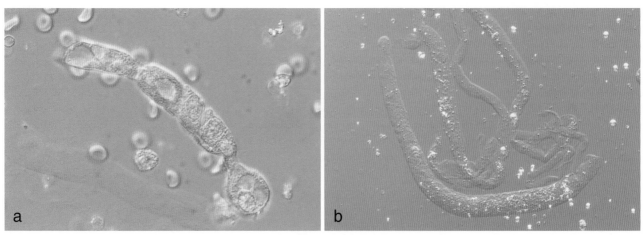

Fig. 15-5　Differential interference microscopy on epithelial-granular cast and many hyaline casts. Differential interference microscopy has advantages in (1) rapid observation without staining, and (2) three dimensional visualization of casts and cells.
a. Fine granular cast containing tubular epithelial cells and red cells scattered.
b. Hyaline casts composed of homogeneous matrix.

■Iodine stain

Iodine reaction is applicable to identify the cells rich in glycogen and vegetable contaminants; glycogen content in urothelial cells reflects cytohormonal alterations in reproductive females. Smears fixed with alcohol are stained with Lugol solution of three time dilution in 5 minutes (Fig. 15-36); the Lugol solution is made with composition of iodine 1 g, potassium 2 g, and distilled water 300 ml.

■Acridine orange stain

A 0.01% acridine orange phosphate buffer solution (pH 6.0) is a good supravital stain, although fluorescence microscope is needed. Bacteria, hyaline casts and various kinds of cells will be easily demonstrated (Figs. 15-34 and 15-77b).

■Instant Giemsa stain

In urinalysis unstained smears or supravital stains are used for microscopical observations. However, rash preparation and interpretation of cellular components or their derivatives are often required in cytology laboratory. The instant alcohol fixed Diff-Quik stain can be utilized with ease, because the stain exhibits features similar to Giemsa stain.[40]

5 Differential Interference Microscopy for Urinalysis

Differential interference microscopy (Nomarski) is able to visualize formed structures such as casts and cells without staining. Phase changes produced by any given component of a specimen can be observed with a high accuracy. Proteinaceous material of hyaline casts with little difference of refractive index from the background and degenerated cellular constituents of cell casts can be clearly identified (Fig. 15-5).

Prompt examination of live unstained material and cells in urinary sediments gives important clue to further clinical examination.

6 Polarized Light Illumination for Urinalysis

Insertion of a polarizer just below the substage condenser and the analyzer adjacent to the eyepiece; the polarizer and the analyzer are placed so that the axes are perpendicular. Whereas amorphous subjects are not visible, crystalline such as lipid droplets and uric acid crystals can be identified[106](Fig. 15-1).

7 Standardization of Urinalysis

Urinalysis including microscopic counts of red blood cells, white blood cells, casts and various types of epithelial cells gives important clues to intrinsic glomerular and tubular diseases, and infectious and neoplastic diseases of the lower urinary tract. For example, white blood cell casts are indicative of pyelonephritis, whereas the presence of a large number of neutrophils, particularly pale, viable neutrophils without cast may reflect infections of the lower urinary tract. Quantitative counts of these blood cells, epithelial cells and casts per either 10×10 or 10×40 microscopic fields are meaningful to consider severity of the disease (Table 15-3). The Addis count technique was employed for quantitation of absolute numbers of cells and casts. However, we must consider various factors to be concerned with standardization of urinalysis. Variations in centrifugation speed, amount of supernatant for resuspension of the packed sediment, and technique to place a cover slip over the drop become factors to report different values from laboratory. For the purpose of standardization of urinalysis, Kova system distributed by ICL Scientific (Fountain Valley, California, U.S.A.) is a device to make a uniform, reproducible procedure using a disposable centrifuge tube with marks of urine amount and a disposable transfer pipette with built-in plastic disc that facilitates discarding of centrifuged urine. The disc is designed to hold back 1 ml of urine with packed sediment. An optically clear plastic slide possesses four

Table 15-3 Normal range of urinary sediments

Costituents	Magnification	Description	Normal value
RBC and WBC	×400	number/every field* many/every field (in case over 30) number/several fields number/total fields	RBC: 1/several fields WBC: 1/every field
Casts	×100 or (×400)	number/total fields (identification is possible under ×400)	1 or 2/total fields (hyaline casts are found in a healthy person)
Epithelial cell	×100 or (×400)	number/total fields (identification is possible under ×400)	None or 1/10 fields (×100)**
Crystal and bacteria	×100 or (×400)	+ or – (nature is described if identifiable)	Abnormal crystals are absent such as cystine, tyrosine, leucine, cholesterol, bilirubin, sulfonamides, etc.

* Observe at least 30 fields and describe with average.

** Squamous cells will be contaminated in females.

Table 15-4 Reference values for urinary sediment

Red blood cells	0-2/hpf
White blood cells	0-5/hpf
Hyaline casts	0-2/lpf
Renal epithelial cells	few/hpf
Transitional cells	few/hpf
Squamous cells	few/lpf
Bacteria	neg/hpf
Abnormal crystals	none/lpf

Note : hpf: high power(40×objective) field
lpf: low power (10×objective) field
few means some are present

Table 15-5 Alterations of tubular epithelial cells

1) Hydropic degeneration in proximal tubules in case of osmotic nephrosis.
2) Protein droplets; hyaline droplets (PAS-positive) in proximal tubules in case of renal disease which is related to excessive proteinuria.
3) Fat droplets; appear in nephrotic syndrome originally described as lipid nephrosis.
4) Glycogen deposits; appear in proximal tubules in case of glycogen storage disease (von Gierke).
5) Calcium deposits; appear either as dystrophic or as meta-static calcification.
6) Multinucleation; occurs as a degenerative phenomenon.
7) Cytomegalic inclusion; intranuclear inclusion in association with nuclear enlargement.

chambers which are able to contain equal amounts of the specimen by capillary action. Thus the counts of cells and casts under microscope are accurate and valuable for interpretation of pathological conditions of the urinary tract.

■NCCLS recommendation[82](Table 15-4)

A first-morning, midstream urine is preferable, although any freely voided urine collection is acceptable.

(1) Urine volume: 12 ml is recommended as the standardized volume. Specimens should be centrifuged immediately after collection. They can be refrigerated and examined within 2 hours.

(2) Centrifugation: A relative centrifugal force (RCF) of 400 g or 400-500 g for 5 minutes are recommended.

(3) Concentration factor of the sediment: This is based on the volume of well-mixed urine centrifuged and the final volume of sediment remaining after decant of supernatant.

8 Nephron and Its Derivatives

■Histology and function of the nephron

The nephron is a functioning unit of the kidney and composed of a glomerulus, proximal convoluted tubule, loop of Henle, distal convoluted tubule and a collecting tubule. The glomerulus is a capillary network which is ensheathed by characteristic epithelial cells (Figs. 15-6 to 15-8); these are called podocytes since their cytoplas- mic processes cover the outer layer of glomerular capillaries. The cytoplasmic processes do not possess organella but contain many microfilaments. Although the physiological role of podocytes cannot be fully understood, the function is referable to as filtration; normally the primary urine filtered by hydrostatic pressure which is opposed by plasma colloid osmotic pressure and intracapsular pressure is acellular and contains small molecular substances such as glucose, amino acids, creatinine, electrolytes, albumin, urea, uric acid, etc. The mesangial cells among capillaries act as a supporting element of the capillaries. The proximal convoluted tubule is lined by a low columnar or cuboidal epithelium having granular, eosinophilic cytoplasm. The cytoplasmic border faced to the lumen possesses the so-called brush border which is composed of numerous microvilli (Fig. 15-9); these are directly concerned with active absorption of glucose, albumin, amino acids, ascorbic acid and with transfer of creatinine and foreign exogenous substances. On the other hand, the distal convoluted tubule is lined by a single layer of cuboidal epithelium having the pale cytoplasm without brush border. This portion is responsible for exchange of electrolytes; sodium is reabsorbed whereas potassium ions, hydrogen ions and urate ions are excreted.

■Tubular epithelial cells

Tubular cells are usually isolated and occasionally grouped in a small, loose cell cluster. There are round or cuboidal, small cells with eccentrically or paracentrally

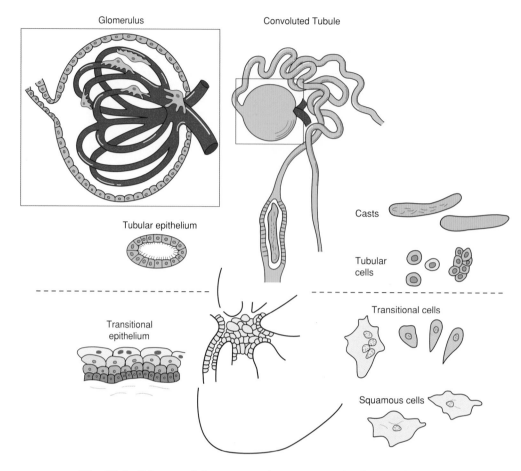

Fig. 15-6 Schema of the upper urinary tract. Histology and cells.

situated oval nuclei, as ever called "small round cells". The cytoplasm stains feebly and shows often vesicular vacuolation. The oval fat body is a degenerative tubular cell with refractable fat droplets; thus it appears as a round globule which is rich in refractive, lightly brown vacuoles in an unstained smear. Polarized light microscopy demonstrates a characteristic "Maltese cross" pattern (Fig. 15-1). Sudan III stain reveals strongly positive droplets (Figs. 15-15 to 15-17). The appearance of oval fat bodies is the result of tubular degeneration (nephrotic syndrome). Other various alterations of tubular epithelial cells are summarized in Table 15-5.

■Cast formation and pathognomonic evaluation

Although urinalysis has been used since antiquity as one of the most common medical laboratory tests, not much attention has been paid to the evaluation of noncellular subjects such as casts and crystals in cytodiagnosis. Casts are formed in urinary sediments as the mold of lumens of renal tubules and collecting ducts. In healthy persons only hyaline casts are exceptionally found in the case when the urine is highly concentrated and/or acidic. The presence of other kinds of casts becomes an important clue to intrinsic pathologic changes of the nephron (Tables 15-6 and 15-7).

Although identification of various types of urinary casts in routine urinalysis is performed in smears without special stain or with supravital stain, their charac-

teristics by Papanicolaou stain should be understood (Table 15-8).[103]

1) Hyaline or proteinaceous cast

The condition for formation of hyaline casts includes abnormal filtration of albumin through the glomerular basement membrane and impairment of reabsorption at the tubular epithelium. In a physiological state, a small amount of albumin not over 100 to 150 mg per day is thought to be within normal limits. Acute glomerulonephritis gives rise to increased permeability of albumin, and pyelonephritis or tubular necrosis produces renal tubular damage. A sulfated mucopolysaccharide liberated from impaired tubular epithelial cells is a constituent of a hyaline cast. A pure hyaline cast is a straight cylinder which is composed of translucent, nonrefractive, colorless mucoid substance. The use of a green filter or differential interference microscopy for observation of fresh sediments is suitable to visualize a hyaline cast (Fig. 15-5b). Hyaline granular and hyaline fatty casts are often formed by the mixture of cellular debris and fat droplets as a result of tubular epithelial degeneration. The PAS stain is positive because of content of mucopolysaccharide that is called Tamm-Horsfall mucoprotein.[96]

2) Myeloma cast

The myeloma cast is formed with a large content of 7S globulin light chain subunits which are known as Bence Jones proteins. Since this monoclonal, light chain has a small molecular weight in the range of 22,000, it will

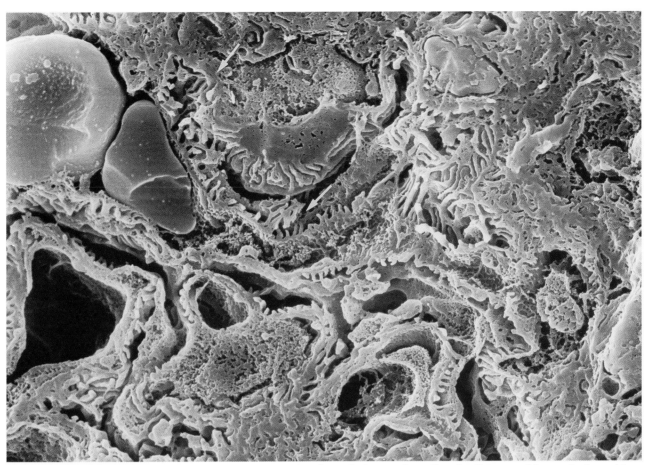

Fig. 15-7 Scanning electron microscopic observation of renal glomerulus of rat by freeze etching technique. Note fenestrated capillary loops, to which branched foot processes of podocytes are attached. Interrelation between blood capillary and podocyte (arrows) should be referred to a transmission electron micrograph (Fig. 15-8). (By courtesy of JEOL Ltd., Akishima, Tokyo)

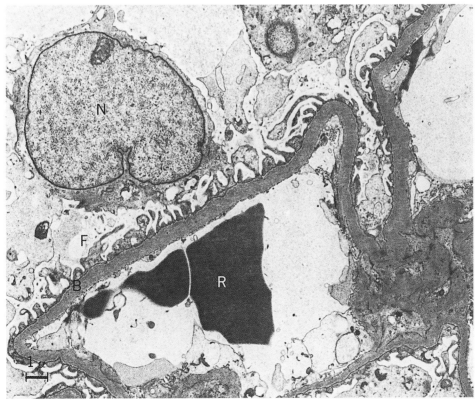

Fig. 15-8 Glomerular capillary loop grasped by podocytes. Club-shaped foot processes (F) extend around the basement membrane (B) of blood capillaries. (original magnification ×6,000) N: nucleus of a podocyte, R: RBC.

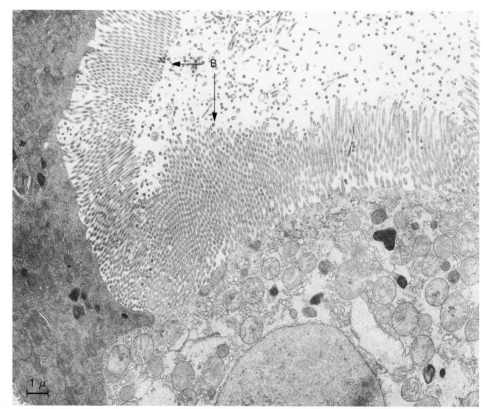

Fig. 15-9 Proximal tubular epithelium of the kidney showing well developed microvilli which are called brush borders. A large number of mitochondria are present. (original magnification ×6,000) B: brush borders.

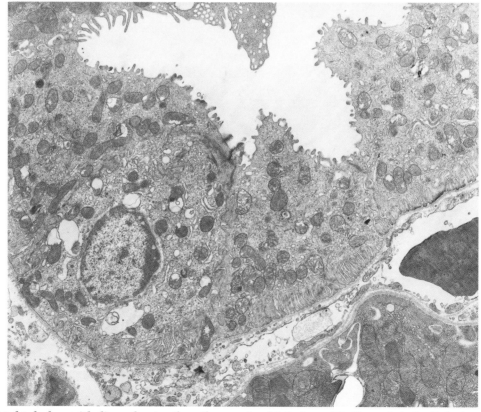

Fig. 15-10 Distal tubular epithelium shorter than the proximal tubular epithelium. Apical cellular border reveals short microvilli. (original magnification ×6,000)

Table 15-6 Classification of casts by Lippmann[63]

(1) Hyaline cast	(2) Cellular cast
Hyaline cast(simple)	Epithelial cast
Hyaline cast containing organized structures	White blood cell cast
Hyaline red blood cell cast	Coarse granular cast
	Fine granular cast
Hyaline white blood cell cast	Waxy cast
	Fatty cast
Hyaline epithelial cast	(3) Blood cast
Hyaline granular cast	Red blood cell cast
Hyaline fatty cast	Blood cast (true)
(Hyaline bacterial cast)	Fibrinoid or fibrinous cast

Table 15-7 Classification of casts

1. Hyaline cast	a) Hyaline simple cast
	b) Hyaline mixed cast
	Hyaline red cell cast
	Hyaline white cell cast
	Hyaline epithelial cast
	Hyaline fatty cast
2. Cellular cast	a) Epithelial cell cast
	b) Degenerative cell cast
	Coarse granular cast
	Fine granular cast
	Waxy cast
	Fatty cast
3. Blood cast	a) Red blood cell cast
	b) Hemoglobin cast*
	c) White blood cell cast (leukocyte cast and lymphocyte cast)
4. Pigment cast	a) Bilirubin cast
	b) Melanin cast
5. Miscellaneous form	a) Myeloma cast
	b) Amyloid cast
	c) Bacterial cast

* Hemoglobin cast is often included in the pigment cast.
(From Takahashi M, Ito, K: Color Atlas of Urinary Sediments. Tokyo: Uchudo-Yagishoten, 1979.)

be excreted through the glomerular basement membrane. Immunoelectrophoresis of urine following concentration and dialysis is a proper technique to identify the specific protein in the urine. Immunofluorescence antibody method with use of fluorescent, specific antibody visualizes the proteinaceous cast containing a monoclonal kappa or lambda light chain (Fig. 15-12). Since multiple myeloma is frequently associated with tubular epithelial damage (Fig. 15-13), the myeloma casts are of a mixed type such as a proteinaceous epithelial cast and take a form of broad cast (Fig. 15-11).

3) Cellular cast and blood cast

a) Tubular epithelial casts are those which are composed of desquamated epithelial cells with a little protein matrix. The appearance of epithelial cell casts is suggestive of intrinsic pathology of renal tubules (see Table 15-5). In acute tubular necrosis as in nephrotoxic acute nephropathy involving the proximal convoluted tubules, epithelial cell cast formation is distinctive. Since the cells are more or less degenerative, details of cellular structure are obscure unless either supravital or permanent stainings are performed (Figs. 15-13 and 15-15). The presence of fat droplets in the cell cast, i.e., formation of fatty cast, is indicative of lipid nephrosis or diabetic nephropathy (Fig. 15-16).

As one variety of acute tubular necrosis, shock kidney or lower nephron nephrosis that produces segmental necrosis in the distal convoluted tubules discloses hyaline cellular casts and/or pigment casts.

b) Leukocyte cast: When leukocytes are embedded in a protein matrix as a cluster, this cast is called leukocyte cast (Fig. 15-15). As degeneration progresses, the distinction of leukocyte cast from the epithelial cell cast becomes difficult. It should be also distinguished from the artificial clump of leukocytes after centrifugation; in the true leukocyte cast its contour is sharp and leukocytes are present as entrapped in it. The pseudocast is a tight cluster of leukocytes with distinctive nuclear features. The use of staining technique is valuable for the differentiation. The pseudocast will be formed from inflammatory leukocytes in the lower urinary tract, whereas the leukocyte cast is indicative of infection of the nephron such as acute pyelonephritis and occasionally acute glomerulonephritis.

c) Erythrocyte cast: This cast is made of red blood cells. Characteristic features of erythrocytes are discernible, but a granular cell debris with a hemoglobin

Table 15-8 Types of urinary casts and their characteristics for accurate identification (by Papanicolaou stain)

1. Hyaline	Pale blue or occasionally orange transparent cylinder without inclusions.
2. Erythrocytic	Dark red to red-orange cylinder, consisting of a blue cylinder filled with intact erythrocytic stroma.
3. Blood	Same as erythrocytic cast but intact stroma not seen. Granular appearance.
4. Leukocytic	Pale blue, gray or orange cylinder filled with segmented leukocytes.
5. Epithelial cell	Blue, gray or orange transparent or granular cylinder filled with renal tubular epithelial cells.
6. Granular	Blue to gray matrix containing fine or coarse refractile granules.
7. Waxy	Homogeneous orange or blue cylinder containing a highly refractive material. The borders have cracks and indentations.
8. Fatty	Blue, orange, gray or brown cylinder filled with large vacuoles.
9. Mixed cell	Any combination of the above.
10. Broad	Present in all types of casts, but the waxy type is most common. Two-six times larger than normal.

(From Schumann GB, et al: Am J Clin Pathol 69: 18, 1978.)

hue will be mixed according to the stage of cell disintegration (Figs. 15-19 and 15-20).

Hemoglobin cast is the end product of disintegrated erythrocytes in massive hematuria; acute hemorrhagic glomerulonephritis, polyarteritis nodosa, crush syndrome and trauma including needle aspiration are responsible for the red cell cast formation. Blood casts occur in acute allograft rejection in association with appearance of epithelial casts.[103]

335

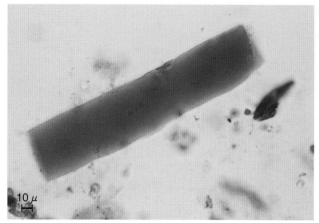

Fig. 15-11 Broad cast by toluidine blue supravital stain from a case with multiple myeloma. A proteinaceous cast with high concentration of globulin fragments is characterized by densely stained, metachromatic matrix and sharply defined straight borders.

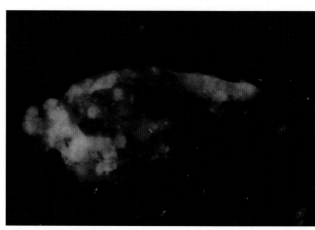

Fig. 15-12 Myeloma cast containing a monoclonal light chain by immunofluorescence antibody technique. A coarse granular cast made of degenerative cells in myeloma nephropathy can be verified to have the Bence Jones protein by such a FITC-fluorescent antiserum.

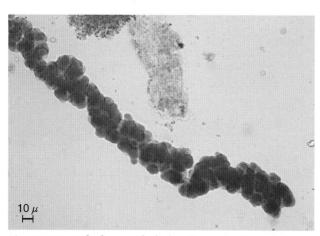

Fig. 15-13 Tubular epithelial cell cast by Sternheimer-Malbin stain from a patient with multiple myeloma. This cast is composed of sloughed tubular cells as a result of severe degenerative change of tubular epithelium.

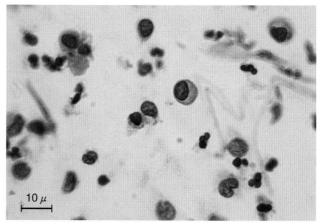

Fig. 15-14 Plasmacytoid cell in urine that is positive for Bence Jones protein. An ovoid hyperchromatic nucleus is surrounded by halo. Furthermore this patient was found to excrete myeloma casts and epithelial cell in urinary sediments. Papanicolaou stain.

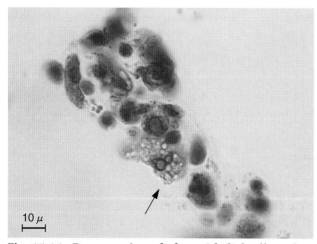

Fig. 15-15 Degenerative tubular epithelial cells and an oval fat body from a patient with nephrotic nephropathy. Note an oval corpuscle filled with tiny refractile droplets or vesicles (arrow).

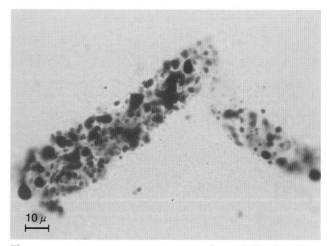

Fig. 15-16 Fatty cast containing sudanophilic droplets. Fatty degeneration of tubular epithelial cells gives rise to refractile droplets; these are found in coarse granular casts. Fatty casts are often associated with oval fat bodies.

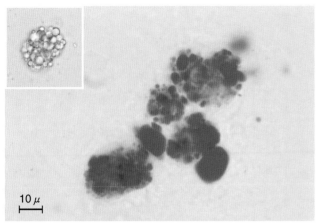

Fig. 15-17 **Oval rat bodies from a patient with nephrotic nephropathy.** Fatty degenerative tubular cells which are known as oval fat bodies in urinalysis. Non-stained oval fat body from the urinary sediments (inset).

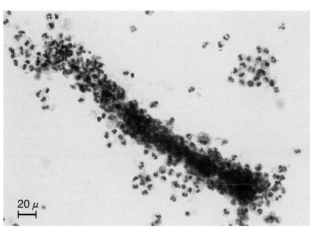

Fig. 15-18 **Leukocyte cast by Papanicolaou stain forming a long cylinder consisting of WBC.** This cast differs from a leukocytic pseudocast that is a simple aggregate of neutrophil leukocytes. The presence of leukocyte cast is suggestive of inflammation of the lower nephron such as acute pyelonephritis.

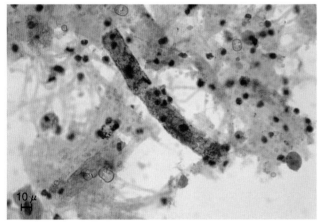

Fig. 15-19 **Erythrocyte cast by Papanicolaou stain from a case with acute glomerulonephritis.** Reddish-brown blood casts are composed of degenerate red cells and their fragments. Note an association with leukocytes and degenerate tubular epithelial cells.

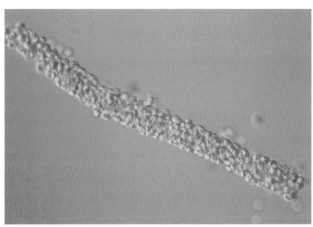

Fig. 15-20 **Erythrocyte cast or red cell cast under differential interference microscopy.** This cast is formed by closely packed red blood cells. The lack of other casts or cell derivatives indicates of bleeding to the nephron. This was seen after renal needle aspiration.

4) Pigment cast

The hemoglobin cast as above described is one of the pigment casts. The hepatorenal syndrome with acute renal insufficiency following the failure of liver parenchyma or biliary tract surgery is attended with bile pigmentation of renal tubular cells and formation of bilirubin or bile casts. Melanin and its colorless precursor, melanogen, pass through the glomeruli when exceed the renal threshold and bring about pigmentation of degenerate epithelial cells (melanin cast) (Fig. 15-21). Melanophages may also appear as free cells in the late stage of malignant melanoma, and should not be regarded as metastatic melanoma cells, which possess large, enrounded nuclei with single macronucleoli and thickened nuclear membrane.[120]

5) Granular cast

Granular casts are considered to be elaborated from degenerated cells or cell debris and various kinds of matrix substances such as protein and lipid; thus the entity of pure granular cast does not exist. They are divided into coarse granular and fine granular casts according to the stage of disintegration of cellular elements (Figs. 15-2, 15-22 and 15-23). Waxy casts are those which are composed of amorphous, refractive mass as the end products of degenerating granular casts (Fig. 15-24). The presence of waxy casts is often observed in chronic renal failure and advanced allograft renal rejection as a result of chronic stagnation of flow in impaired tubules.[103]

9 Urinary Tract and Its Benign Cellular Components

■ Histology of the urinary tract

The urinary tract consists of the renal pelvis, ureters, urinary bladder and urethra. From the renal pelvis to the bladder, the mucosal membrane is lined with transitional epithelium, although its thickness differs from place to place. The pelvis and the ureters are lined with 3 or 4 layers, while the bladder is lined with 6 to 8 layers. In

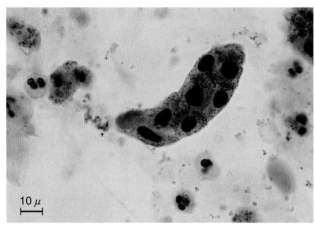

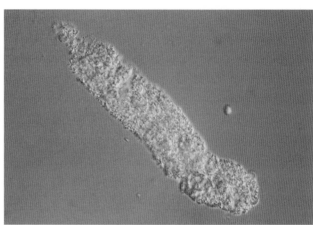

Fig. 15-21 Tubular cell-melanin mixed cast from a patient with advanced generalized melanoma. Melanin pigments deposit on the matrix of tubular epithelial cell cast.

Fig. 15-22 Coarse granular cast under differential interference microscopy. From a case of rapidly progressive glomerulonephritis. Such granular casts are composed of desquamated tubular cells in various grades of degeneration, blood cells and mucoprotein. This form shows sharply defined contours and ill-defined or broken end.

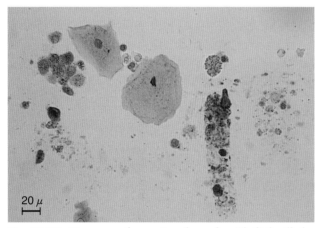

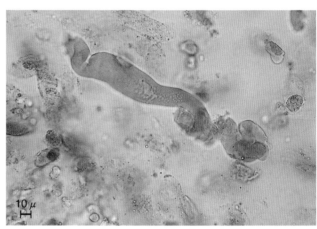

Fig. 15-23 Fine granular cast and renal epithelial cells by Sternheimer-Malbin stain. From a case with lupus nephritis. The granular cast displays sharply defined, straight contours. The fine granules in the cast are considered to be made of disintegrated cells. Note tubular epithelial cells, squamous cells and an oval fat body.

Fig. 15-24 Convoluted waxy cast by Sternheimer-Malbin stain. Note characteristic waxy appearance looking like a thin paraffin section and distinct outlines with deep incisions. Waxy casts are well preserved either in acidic or alkaline urine.

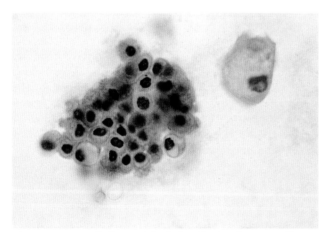

Fig. 15-25 Tubular epithelial cells in sheet. Well preserved low columnar cells are arranged in a flat sheet. From a case of female with eclampsia. Uniform small round or oval nuclei locate basally in low columnar cytoplasm. Compare with a transitional cell for cell size.

the vicinity of the urethral orifice, the urinary bladder shows small invaginations of the mucous membrane (crypts), in which intraepithelial aggregations of mucus-secreting cells can be found. The male urethra to which the reproductive system opens consists of different types of epithelium as follows: the *pars prostatica* lined with transitional epithelium, the *pars membranacea* and the *pars cavernosa* lined with stratified columnar epithelium, and the *fossa navicularis* lined with stratified squamous epithelium. The mucous membrane has many recesses, to which the branching tubular glands of Littré open. The female urethra reveals individual differences in its lining epithelium; mostly it is lined with stratified squamous epithelium and occasionally with pseudostratified columnar epithelium.

■Benign cellular components in urine
1) Tubular epithelial cells
The epithelial cells from renal tubules are those which are referred to as small round cells in urinalysis (Figs.

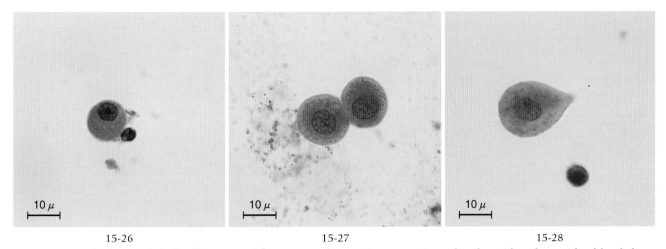

15-26   15-27   15-28

Fig. 15-26 Tubular epithelial cell aspirated from a renal retention cyst. Note a basal ovoid nucleus with a bland chromatin. Papanicolaou stain.
Fig. 15-27 Supravital stain of a tubular cell. The nucleus is round and smoothly outlined. The cytoplasm is also enrounded and vesicular. Note the different configuration from the transitional cell (Fig. 15-28). Sternheimer-Malbin supravital stain.
Fig. 15-28 Pyriform transitional cell with tail-like process. The cytoplasm is sharply outlined. A small ovoid nucleus shows a bland chromatin. Sternheimer-Malbin supravital stain.

15-26 and 15-27). The nuclei take a paracentric position in the round cytoplasm. The cytoplasm stains weakly by Light Green and shows often fine vesicular vacuolation. These are distinguished from deep transitional cells with a caudal process by the shape (Fig. 15-28). Parabasal squamous cells are larger in size and have more thickened cytoplasm than renal tubular cells.

Cell clusters arranged in sheet display a stone pavement or honeycomb pattern (Fig. 15-25). Identification of low columnar or cuboidal cell feature is possible at cell boundaries. Schumann et al mention the importance of observing cytoplasmic pigmentation and lipid droplet deposition.[104]

2) Transitional epithelial cells

Superficial transitional cells: The superficial transitional cells vary in size and shape. The nuclei differ from triangular or polyhedral form to oval form. Their shape variation depends largely on the influence of the hypertonicity of the urine. Well preserved cells have oval or round nuclei with finely granular and evenly distributed chromatin particles, while dehydrated cells have speckled condensations of chromatin.

The cytoplasm is fairly abundant, polyhedral or pear-shaped with tail-like processes (Fig. 15-28). Cytoplasmic vacuolation is occasionally seen (Fig. 15-29c, d). The staining reaction is from bluish green to brownish green with a frosted-glass like appearance; its cellular border has an intense hue, whereas perinuclear zone appears clear.

Multinucleated transitional cells: The occurrence of multinucleation with two or three oval, uniform nuclei is not uncommon in the superficial transitional epithelial cells (umbrella cells)(Fig. 15-30).[19,31,46,53] It is also not rare to see multinucleated syncytial giant cells with innumerable nuclei in the broad syncytial cytoplasm (Fig. 15-31). Concerning the nature of these cells, no consensus is obtained yet. Whereas Harman and Hogan[34] emphasize tubular epithelial origin from proximal convoluted tubules, reactive modification of the

transitional epithelium in response to the presence of large aggregates of mucin is suggested as a causative agent (Dorfman and Monis[27]); it is frequent to see association of multinucleation with cytoplasmic inclusions composed of acid mucopolysaccharide. Although multinucleation is still a matter of conjecture, some investigators attribute it to irritation or inflammation,[77] administration of some laxative[67] and catheterization of the upper urinary tract. There has been no evidence whether or not neoplastic growth is related.

Deep transitional cells: The nuclei are oval or elliptic with finely granular chromatin. The nuclear rims are occasionally accentuated because of condensation of chromatin. The deep transitional cells have scant, spindly or columnar cytoplasm with tail (Fig. 15-29f). The cytoplasm appears to be translucent or vacuolated, and stains pale bluish green around the nucleus (Fig. 15-32).

3) Squamous epithelial cells

Squamous epithelial cells arising from the urethra are of normal cellular components. The urine of the female frequently contains a considerable number of squamous cells by contamination from a vaginal flora. Shedding of squamous cells occurs in males who have had estrogen therapy for carcinoma of the prostate. The urocytogram is used as a tool to evaluate ovarian hormonal states of adult females, because there is a parallelism between urinary squamous cells and vaginal cells.[6,24,59,61] The urocytograms of pregnant women from the 20th to the 37th week are known to indicate increased progestational activity[2,61]: Low values in the eosinophilic and karyopyknotic indices, formation of folding and clustering of cells, and an increase in the number of navicular cells are the representative changes.

4) Red blood cells and neutrophil leukocytes

In general, red blood cells are not found under normal conditions (see Table 15-3). Microscopic hematuria can be found as a tiny blood button in the centrifuged tube. Hematuria is an important clue to manifestation of

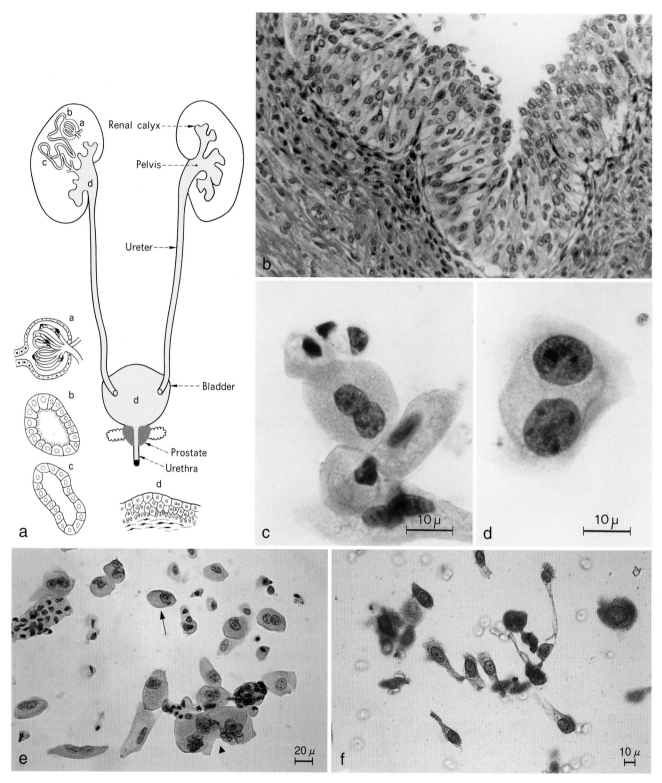

Fig. 15-29 Histology and cytology of the lower urinary tract.
a. Anatomical diagram of the lower urinary tract.
b. Transitional epithelium of the urinary bladder. The lining epithelium is composed of several layers of rectangular cells with clear cytoplasm.
c. d. Superficial transitional cells in urine. Binucleation and variation in size of nuclei are fairly common findings. The cytoplasm is adequate and polyhedral.
e. Superficial, often multinucleated (arrowhead), and deep transitional cells in catheter urine. Note pyriform cytoplasm with perinuclear clearing (arrow).
f. Deep transitional cells with tails originating from renal pelvis. These cells should not be misinterpreted as columnar cells. Sternheimer-Malbin stain.

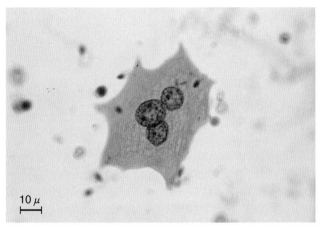

Fig. 15-30 Multinucleated transitional cells. Superficial transitional cells are often multinucleated. The nuclei are uniform in size and shape. The cytoplasm is abundant and polyhedral. Some superficial cells called "umbrella cells" have a number of round nuclei.

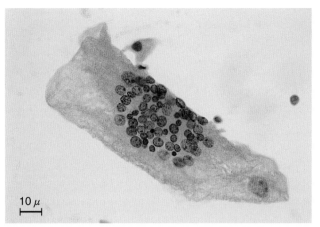

Fig. 15-31 Multinucleated giant cells in vesical washings. This syncytial broad cytoplasm is well defined and different from that of a histiocyte with foamy appearance although the texture is edematous. The round or oval nuclei are equal in size and shape. There is no evidence of malignancy.

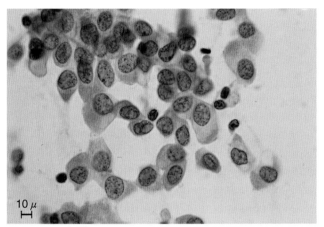

Fig. 15-32 Deep transitional cells by direct scraping of the urinary bladder. The nuclei are uniformly round or oval. Note caudate or pyriform shape and perinuclear transparency of the cytoplasm.

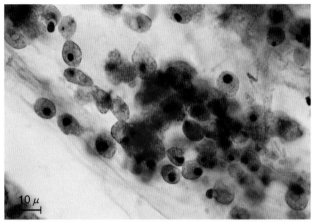

Fig. 15-33 Degenerate transitional cells in voided urine from a patient with colo-vesical plasty. These unfamiliar cells with nuclear pyknosis and cytoplasmic eosinophilia are formed as a result of degenerative change of urothelium.

Table 15-9 Causes of hematuria

Diagnosis	Note
1) Systemic disease	
Purpura	–
Hemophilia	–
Leukemia	–
2) Upper urinary tract	
Neoplasm	unilateral, often gross hematuria
Pyelonephritis	pus cells and WBC cast found
Tuberculosis	about 30% of cases, unilateral
Cystic kidney	about 20% of cases, either unilateral or bilateral
Trauma	unilateral
Acute glomerulonephritis	bilateral
Infarction	unilateral, transient
Malignant nephrosclerosis	bilateral
Nephrolithiasis	unilateral with colic pain
3) Lower urinary tract	
Hemorrhagic cystitis	pus cells found
Neoplasm	often gross hematuria
Bladder stones	with colic pain
Diverticulum	–

intrinsic renal pathology, systemic hemorrhagic disorders and extrarenal diseases (Table 15-9). Gross hematuria occurs often in malignant neoplasm, hemorrhagic glomerulonephritis, hemorrhagic cystitis, trauma, hemorrhagic diathesis, etc. Bleeding into the tubules is responsible for erythrocyte casts (see Fig. 15-19).

White blood cells, more than one to five pus cells per high power field, are indicative of an inflammatory process such as pyelonephritis, ureteritis, cystitis and urethritis. The significance of leukocytes in urine does not always consist in the quantity but in the form of appearance. For example, leukocyte casts or combined hyaline white cell casts reflect intrinsic renal inflammatory disease (see Fig. 15-18).

5) Other non-human organism in urine

Trichomonas urethritis and cystitis occur more commonly in females than in males. Men are also affected a few days after intercourse. The protozoa will be found in urinary sediments (Fig. 15-34). The ova of *Schistosoma haematobium* can be found in urine is often hemorrhagic. Schistosomiasis or bilharziasis is particularly important because of the relationship to carcinogenesis[29](see Fig.

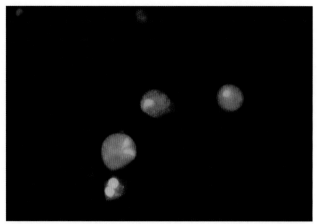

Fig. 15-34 Protozoan *Trichomonas vaginalis* in urinary sediment of a male by acridine orange supravital stain. Note a characteristic pear shape of the protozoa and longitudinal binary fission. A serological test by means of hemagglutination is important to differentiate from fecal contaminant, *Trichomonas hominis*.

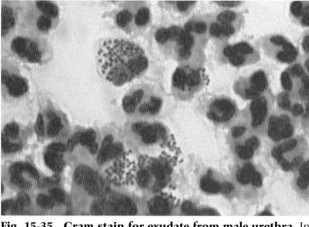

Fig. 15-35 Gram stain for exudate from male urethra. In acute stage of gonococcal infection, Gram-negative intracellular diplococci are demonstrated.

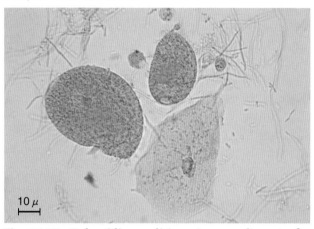

Fig. 15-36 *Balantidium coli* in urinary sediment of a female with urethritis. This largest protozoa is covered with cilia and shows an inverted conical depression, that is called cytosome. Lugol stain.

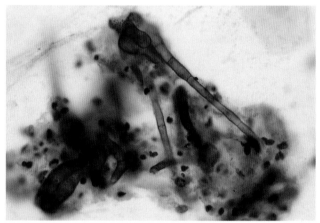

Fig. 15-37 *Alternaria* in urine. A common non-pathogenic contaminant fungus. A brownish conidia is muriform having transverse and longitudinal cross walls.

15-87). Acute stage of gonorrhea becomes a target of microscopical examination using a conventional Gram stain for purulent exudate from the urethra (Fig. 15-35). There are non-pathogenic contaminants such as *Alternaria* (Fig. 15-37), pollen, etc.

10 Benign Atypical Cells

Exfoliation of papillary and ball-shaped cell clusters is strongly pointed out as catheter artifact by Kannan and Bose. Catheterized urine specimens often lead to false positive interpretation as papillary tumors. Definite cell boundaries (cytoplasmic collar), flattened outer cells and uniform oval nuclei shown in catheterized urine are unfavorable to papillary carcinoma.[47] Desquamation of many superficial transitional cells after catheterization and cystoscopy should be kept in mind.

■Urolithiasis

Urothelial lithiasis is associated with complex epithelial changes, i.e., repeat of erosion and repair, and superimposition of chronic inflammation. Thus, malignancy-

mimic cell changes occur in the urothelium of the pelvis and ureter[26,39,90]; deep or intermediate transitional cells with high N/C ratio and hyperchromasia exfoliate singly as well as in clusters (Figs. 15-38 to 15-40). The cellular border is ill-defined. The nuclei are disorderly arranged. Dyskaryosis and presence of nucleolus in cell clusters are hardly distinguished from atypicality in papilloma or grade 1 papillary carcinoma. An admixture with red cells particularly following colic pain is a noticeable finding.

As a chronic tissue response squamous metaplasia superimposed on reparatory hyperplasia becomes a pitfall in urinary cytology (Figs. 15-41 and 15-42).

■Benign tumor cells

Angiomyolipoma is uncommon benign tumor and comprises about 0.7 to 2.0% of renal tumors. Patients with angiomyolipoma are commonly associated with tuberous sclerosis (80%), however, less than half of them are asymptomatic and found by various imaging studies.[80]

Out of benign tumors angiomyolipoma is of clinical importance, because massive growth within the kidney resembles renal cell carcinoma. Fine needle aspiration is

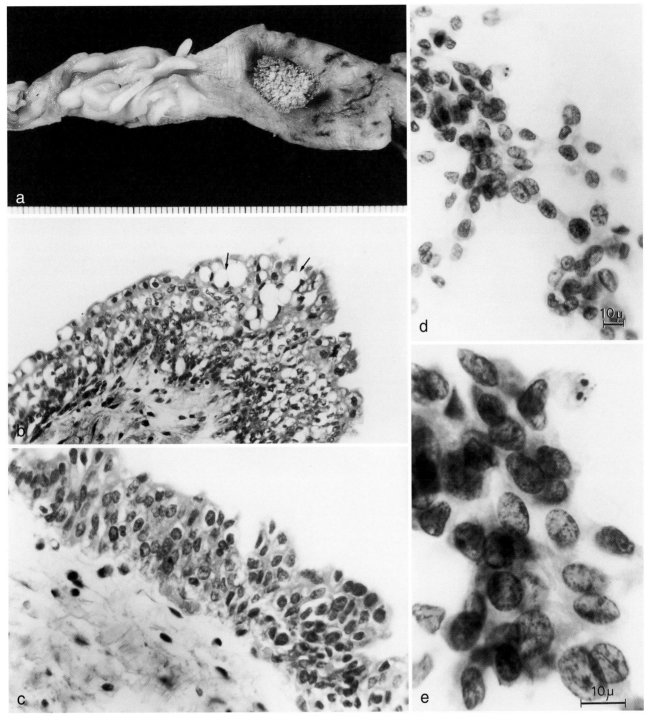

Fig. 15-38 Benign atypical hyperplasia from a case with villous polyp and ureterolithiasis.
a. Incarcerated stone with spiny processes are surrounded by thickened, rough mucosa.
b, c. Regenerative hyperplasia of the transitional epithelium. Note enlargement of nuclei, variation in nuclear size and irregular cell arrangement. Some mucinous cells are appearing as a result of chronic inflammation (arrows).
d, e. Clusters of dyskaryotic transitional cells in catheter urine. Loose cell clusters show distortion of arrangement that is a different finding from benign papilloma. Although nuclear enlargement and hyperchromatosis are present, the nuclei are uniform in shape and bland in chromatin pattern.

a useful method to distinguish from renal cell carcinoma.

This tumor is characterized by smooth muscle fibers and mature adipose tissue growing around vessels. These tissue components are distinguishable cytologically (Fig. 15-43).

■Tuberculosis

A small percentage of patients with pulmonary tubercu-losis, less than 10%, are affected with renal tuberculosis; caseous tubercles in the medulla of the kidney are pathologic findings of common occurrence. There may be discharge of the content of the tubercle if liquefaction of coalesced lesions extends down to the surface of the papilla (Figs. 15-44 and 15-45); epithelioid cells desqua-mate in urine forming simple groups or tubercles. Microscopic hematuria is found in about 30% of the

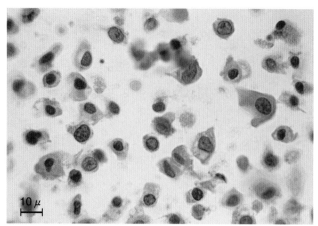

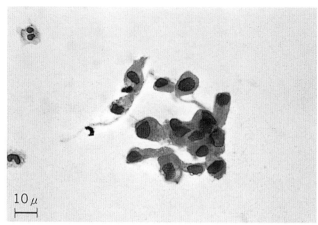

Fig. 15-39 Hyperplastic transitional cells from a case with bladder diverticulum. Benign regenerative hyperplasia is virtually the same as that encountered in nephrolithiasis. Note high cellularity and admixture of regenerative cells having large hyperchromatic nuclei and degenerative pyknotic cells are unfavorable to carcinoma in situ.

Fig. 15-40 Hyperplastic transitional cells associated with calculi of the lower urinary tract. Note singly shed, hypertrophic, hyperchromatic degenerative cells. Differentiation from similar looking in situ cancer cells should be focused on (1) lack of nuclear atypism, (2) inadequacy of increase in N/C ratio, and (3) preservation or tail-like cytoplasm.

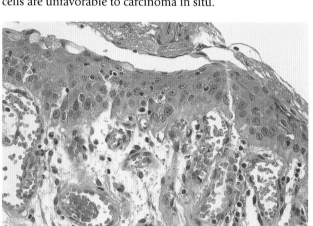

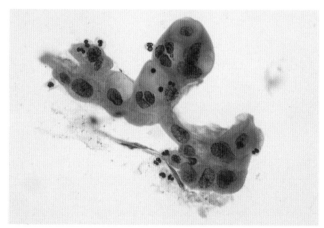

Fig. 15-41 Pronounced squamous metaplasia in urolithiasis. Urothelium of the pelvis with stones exhibits extensive squamous metaplasia.

Fig. 15-42 Cell group showing marked squamous metaplasia. From the same case of Fig. 15-41. Condensed hypertrophic nuclei and opaque, thickened cytoplasm may lead to overinterpretation.

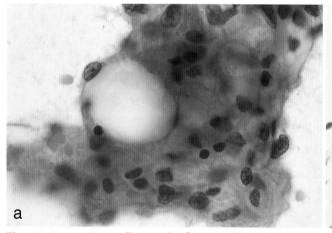

Fig. 15-43 Angiomyolipoma by fine needle aspiration.
a. A large vacuole derived from adipose tissue component is surrounded by bundles of spindly smooth muscle cells.
b. Dense fascicular bundles consisting of spindly muscle cells.

patients with tuberculosis.

The urothelium of the lower urinary tract is also affected by tuberculosis in association with urothelial hyperplasia and squamous metaplasia. Hyperplasia is a reversible reactive change.[87]

Cytologically detection of epithelioid cells and/or Langhans' giant cells is diagnostic of tuberculosis (Figs. 15-44 and 15-45). Regarding the frequency of these cells, Piscioli et al have found Langhans' cells and epithelioid cells in 11 and 5 respectively of 13 cases with

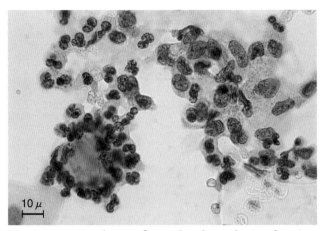

Fig. 15-44 **Cytology of renal tuberculosis showing epithelioid cells and necrotic debris surrounded by leukocytes like a cell ball.** This 47-year-old female has had dull back pain and persistent hematuria. Epithelioid cells of a mesenchymal origin are characterized by elliptic nuclei with delicate nuclear rim, lacy chromatins and indistinct cytoplasm. Orangeophilic necrotic debris may be derived from caseous necrosis.

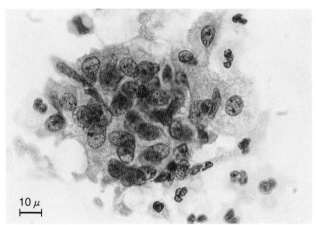

Fig. 15-45 **Cytology of renal tuberculosis showing epithelioid cells appeared in a tubercle.** The same case as Fig. 15-44. Enrounded epithelioid cells may become a pitfall of false positive cytology because of compact cluster formation. Differential points: Thin nuclear rim, lacy chromatin pattern and poorly preserved pale cytoplasm may argue for mesenchymal origin.

histologically proven tuberculosis.

11 Malignant Cells

Urinalysis is important not only for diagnostic procedure of urothelial tumors but for screening of precancerous lesions or carcinoma in situ from high risk groups. Carcinoma of the urothelium shows a high incidence in the industrial workers in the textile, printing, plastic and rubber factories who have exposed to the dyestuffs; 2-naphthylamine, 4-aminobiphenyl, benzidine and 4-nitrobiphenyl have been identified as bladder carcinogens (Table 15-10). Also remarkable is a high incidence of bladder carcinoma in the inhabitants with exposure to endemic infection of *Schistosoma haematobium*[68](see Fig. 15-87); a proportion of epidermoid carcinoma is greater than transitional carcinoma in bladder carcinoma induced by schistosomiasis.[16,29]

■Staging and grading[100]
After surgical resection and pathologic examination, staging and grading are accomplished according to Marshall's modification of the Jewett-Strong system.

Stage 0:   Mucosal involvement only, including both papillary (Ta) and carcinoma in situ (Tis)
Stage A: Penetration into the lamina propria (T1)
Stage B1: Focal invasion into the muscle (T2)
Stage B2: Deep muscle invasion, but not beyond (T3a)
Stage C: Involvement of prevesical fat (T3b)
Stage D1: Direct invasion into other pelvic structures, or nodal spread below the bifurcation of the aorta (T4) (Aorta)
Stage D2: Spread to distant organs to nodes above the bifurcation of the aorta

A three-grade system of scoring the degree of anaplasia has generally been adopted[55,105]:

Table 15-10 Urinary cytology for dyestuff workers

Cytology / Year	1973	1974	1975	1976	Total specimen	Case	Percent
Normal	1	24	21	30	76	11	20%
Mild atypia	9	19	24	24	76	38	69%
Dysplasia (suspicious)	1	4	4	2	11	5	9%
In situ carcinoma	1	1	0	0	2	0	2%
Invasive carcinoma	0	0	0	0	0	0	0%

(By courtesy of Dr. K. Sasaki, Kitasato University School of Medicine, Sagamihara, Japan.)

Grade I: Tumors have the least degree of anaplasia compatible with diagnosis of malignancy
Grade II: Tumors have a degree of anaplasia between grades I and III
Grade III: Tumors have the most severe degree of anaplasia

Dealing with urine specimens in malignancy of the urothelium, one must notice that either macroscopic or microscopic hematuria is an important symptom that persists and increases in severity as the tumor progresses. In the presence of a tumor, initial hematuria in three glass urine collection is suggestive of the growth in proximity to the internal meatus, whereas terminatory hematuria indicates it to be induced from the bladder tumor by contraction. Red blood cells originating from the urinary tract below the pelvis are mixed uniformly in the urine; red blood cell casts are not formed.

As many as 90% of neoplasms of the urinary bladder and ureters are of the transitional cell type. Histologically there are many benign-looking neoplasms which may be classified as benign transitional cell papillomas.[25,28,57,70,83] However, most papillomas repeat recurrence after cauterization and develop into invasive carcinomas.[70,92] Therefore, well differentiated papillary neo-

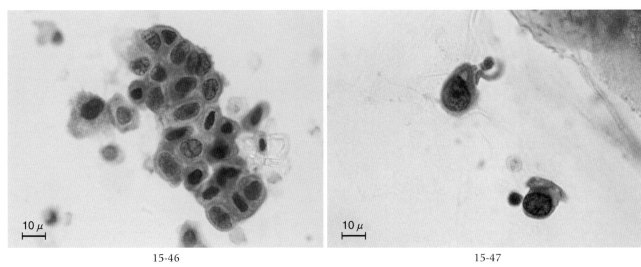

15-46   15-47

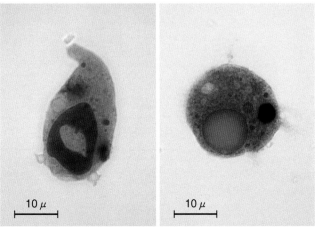

15-48

Fig. 15-46 Dysplastic or atypical hyperplastic cells in a cluster from an asymptomatic dyestuff factory worker. Although cell atypism is present, a cobblestone pavement arrangement side by side is not likely to be malignant. No tumor diathesis is associated.

Fig. 15-47 Malignant cells exfoliated singly from in situ carcinoma of the bladder of a dyestuff factory worker. Fairly small single atypical cells with scant cytoplasm and hyperchromatic enrounded nuclei are similar to in situ carcinoma cells of the uterine cervix. The tumor diathesis is not accompanied.

Fig. 15-48 Odd-shaped, degenerate transitional cells in urine from a man who has been exposed to alpha-naphthylamine. Neither tumor cells nor inflammatory cells are associated. It is conjectured that degenerative process persists in dyestuff workers.

plasm has been regarded biologically as grade I transitional cell carcinoma.[10] The majority of tumors are papillary and composed of villous projections of multilayered tumor cells with stalks of vascular connective tissue. Based on the anaplasia of tumor cells, i. e., cytologic grading by Broders,[15] the neoplasms are graded into four groups as described below.

Adenocarcinoma of the urinary tract is relatively rare but of academic interest in cytopathology. Remnants of the cloaca or competence of the vesical transitional epithelium to preserve the property of primitive cloaca are prevailing for the morphogenesis. The frequent occurrence of glandular or mucinous cells in pyelitis and ureteritis (Fig. 15-38) may also postulate the antecedence of metaplastic adenocarcinoma.

Broders' classification

This classification dependent on the grade of cell differentiation or maturation is widely used in histopathology of carcinoma. The report of the registry of genito-urinary pathology to the American Urological Association used a similar classification combined with structural pattern for the total of 1,004 cases and proved a reliable prognosis in relation to the grading of cell differentiation. In histological studies, grading of cell differentiation is represented by size variation, chromatin content and mitosis of nuclei in addition to distortion of cell polarity. From the point of exfoliative cytology, the classification

is more rudely available as follows: (a) papilloma grade 0 or papillary carcinoma grade I (papillary transitional cell carcinoma grade I), (b) papillary carcinoma grade II (papillary transitional cell carcinoma grade II), (c) transitional cell carcinoma grade III. Cytologically it is possible to suspect flat carcinoma in situ regardless of the extent of superficial spread (Fig. 15-49).

■Transitional cell carcinoma
1) Cytology of papilloma grade 0 or papillary carcinoma grade I

Voided urinary cytology is not effective in the detection of low grade papillary tumor, because tumor cells are unlikely to shed in voided urine and their identification seems to be difficult. The appearance of small epithelial cells or elongated cells is mentioned to be a signal of bladder papilloma.[119]

From a cytomorphological view, no definite criterion to identify benign papilloma is present. Exfoliation with large clumps of cells (papillary fronds) is significant, because it suggests detachment of epithelial fragments from the tumor (Fig. 15-57). There may be a unipolar arrangement of the cells on the lateral aspect of cell clusters.

Although there is no cytodiagnostic basis for distinction of papilloma from papillary carcinoma grade I except for some difference of nuclear atypia, a suggestive

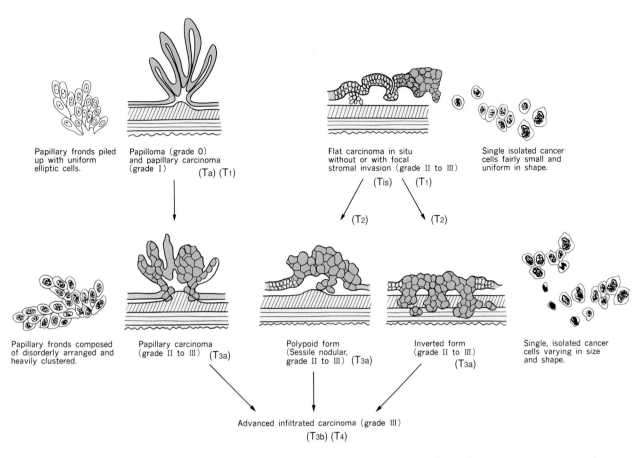

Papillary fronds piled up with uniform elliptic cells.

Papilloma (grade 0) and papillary carcinoma (grade I) (Ta) (T1)

Flat carcinoma in situ without or with focal stromal invasion (grade II to III) (Tis) (T1)

Single isolated cancer cells fairly small and uniform in shape.

(T2)   (T2)

Papillary fronds composed of disorderly arranged and heavily clustered.

Papillary carcinoma (grade II to III) (T3a)

Polypoid form (Sessile nodular, grade II to III) (T3a)

Inverted form (grade II to III) (T3a)

Single, isolated cancer cells varying in size and shape.

Advanced infiltrated carcinoma (grade III) (T3b) (T4)

Fig. 15-49   Schematic classification of urothelial tumors and their cytology. (see Figs. 15-50 to 15-56.)

Table 15-11   Cytopathology of urothelial tumors

	Papilloma (grade 0) or papillary carcinoma (grade I)	Papillary or nonpapillary transitional cell carcinoma (grade II)	Transitional cell carcinoma (grade III)
Gross findings	Fern-like projections with delicate fibrous stalks.	Papillary or sessile nodular tumor with blunt, confluent, short villous growth.	Sessile, exophytic or flat inverted tumor often with ulceration.
Microscopic findings	Piled-up transitional cells of normal thickness with uniform nuclei. Polarity remains.	Thick cell bands with hypertrophy, hyperchromatism and moderate size variation of nuclei. Polarity is distorted.	Thick cell bands with disorderly arranged anaplastic cells. Marked nuclear hyperchromatism, polymorphism, multinucleation and frequent mitoses are seen.
Exfoliative cytology	1) Shedding of normal-looking cells in a tight cluster, in which nuclei regularly and centrally placed. Low cellularity. 2) Microscopic hematuria. 3) Cell clusters consist of deep transitional cells with slight nuclear hypertrophy; polarity is present. 4) A fair amount of cytoplasm preserved.	1) Shedding of clumps of atypical transitional cells with enlarged, hyperchromatic, polymorphic nuclei. 2) Unipolar arrangement of nuclei in cell cluster remains, though apparently distorted; thus irregular nuclear budding is seen in the cluster. 3) A fair amount of cytoplasm.	1) Shedding of isolated anaplastic cells with high N/C ratio 2) Marked nuclear variation in size and shape. 3) Giant multinucleated cells occasionally intermingled. 4) Scant, pale, indistinctly outlined cytoplasm.

information for papillary outgrowth can be obtained from the shape and arrangement of tumor cells (Table 15-11).

a) Spontaneous desquamation of papillary fronds with distinctive linear outline is a favorable finding to neoplasm.[41] We must remember that cellular desquamation by catheterization or in the presence of stones is more pronounced than shedding into voided urine in low grade papillary tumors.

b) Papillary fronds of papillary urothelial tumors are

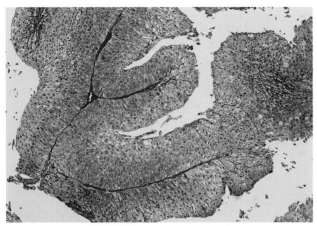

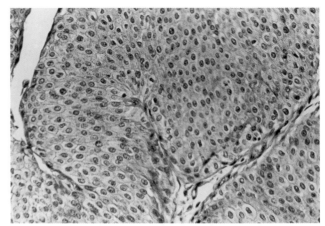

Fig. 15-50   Papilloma of the urinary bladder showing typical villous proliferation accompanied by thin fibrous stalks (grade 0). Note the lining epithelium of papillary fronds of almost equal thickness.

Fig. 15-51   Differentiated papillary carcinoma (grade I). Papillary fronds growing with thin fibro-vascular stalk are lined with multilayered transitional cells. Note uniform nuclear size and shape despite the marked thickness.

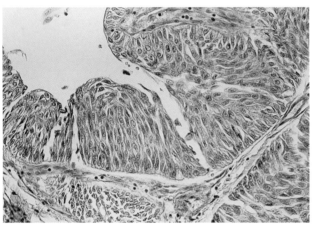

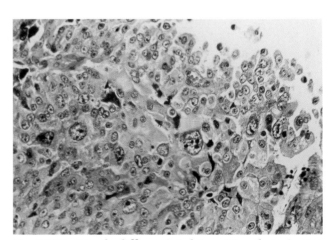

Fig. 15-52   Moderately differentiated papillary carcinoma (grade II). The nuclei are more enlarged and more hyperchromatic than in grade I carcinoma of the previous Fig. 15-51.

Fig. 15-53   Poorly differentiated transitional carcinoma (grade III). Solid nest is composed of polymorphic cancer cells with multiple prominent nucleoli.

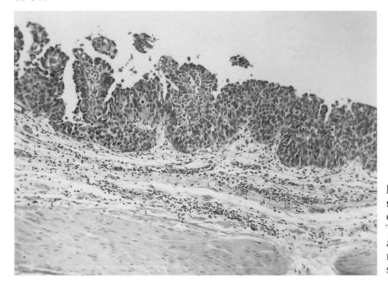

Fig. 15-54   Superficial spreading carcinoma in situ involving ureter and pelvis. The mucosal epithelium is entirely replaced by small cancer cells. The cancer cells are generally monotonous in shape and possess round to polyhedral, hyperchromatic nuclei with high N/C ratio. The cellular atypia corresponds to grade II. Polarity and stratification are lost.

lined by several layers of elongated transitional cells on thin fibro-vascular stalk (Figs. 15-50 and 15-51). A palisade pattern of nuclei in the cell cluster that resembles schooling of fish is also significant for interpretation. The outer surface of spontaneously exfoliated papillary fronds in urine is enrounded and lined with flattened cells (Fig. 15-57). On the contrary, nuclear budding with irregular outline in the papillary cluster depicts the configuration of more poorly differentiated papillary carcinoma (Figs. 15-58 and 15-59).

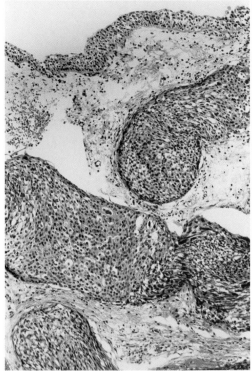

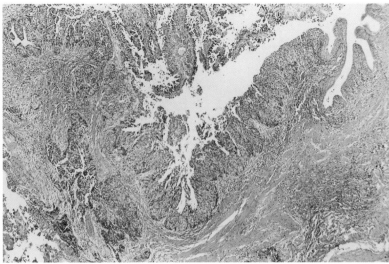

15-55

Fig. 15-55 Inverted form of infiltrating transitional cell carcinoma of the urinary bladder (grade III). The tumor is characterized by massive downgrowth of anastomosing solid nests or trabeculae composed of poorly differentiated transitional cancer cells.

Fig. 15-56 Sessile nodular form of infiltrating transitional cell carcinoma of the urinary bladder (grade III). This advanced tumor shows a nodular outgrowth with broad base and deep downgrowth.

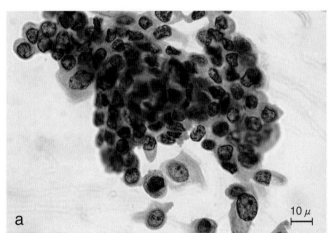

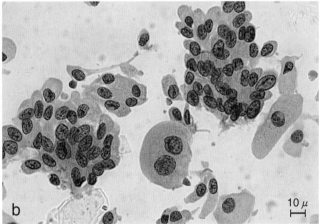

Fig. 15-57 Papillary fronds illustrative of papilloma.
a. Catheterized urine containing papillary clusters, the nuclei of which are shrunken. Diagnostic points of papilloma: (1) shedding of the cells as papillary fronds being composed of cells uniform in shape and arrangement, and (2) absence of tumor diathesis despite of microscopic hematuria.
b. Vesical washings showing well preserved papillary clusters by Shorr stain.

2) Cytology in advanced transitional cell carcinoma

The more undifferentiated the more shedding of single malignant cells because of decreased mutual cohesiveness. Grade II malignant cells possess fairly abundant cytoplasm with distinctive borders; a greenish or cyanophilic staining reaction will be obtained (Fig. 15-60). Undifferentiated malignant cells of grade III malignancy exfoliate singly or in loose cell clusters and show marked variation in nuclear size, cellular shape and chromatin content (Figs. 15-61 and 15-62). The chromatin is unevenly distributed and condensed at the nuclear rim because of cell degeneration. There may be heavy clumps or strands of chromatin material. The nuclear outlines are irregularly indented or notched. The cytoplasm is scant and ill-defined. Denuded nuclei are frequently present. The nucleoli, often multiple, tend to be basophilic; these are irregular in shape and often indistinguishable from large karyosomes. Irregularity of nuclear and nucleolar features is exaggerated by degenerative process within the non-isotonic voided urine.

3) Carcinoma in situ and its cytology

Flat carcinoma in situ develops in multicentric foci without remarkable macroscopic changes; slightly elevated, granular, reddish foci are often visible cystoscopically.[115] Histologically, the thickened mucosa layered by more than four or five cells is entirely replaced by round hyperchromatic cancer cells which have lost polarity (Fig. 15-54). Vousta and Melamed[117] studied 20 cases of carcinoma in situ and emphasized its specific pattern in exfoliative cytology. They mentioned that the malignant

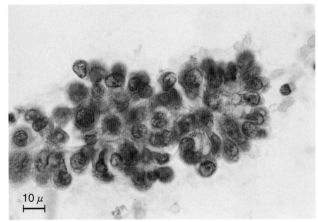

Fig. 15-58　Grade II papillary carcinoma showing a typical papillary pattern with long cytoplasmic tails. There is moderate anisonucleosis, although karyopyknosis exaggerates nuclear size variation.

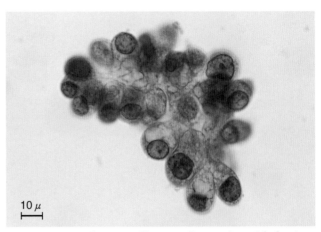

Fig. 15-59　Grade II papillary carcinoma in voided urine. Note a papillary arrangement with tails and cytoplasmic vacuolation that represents the clear cell variant.

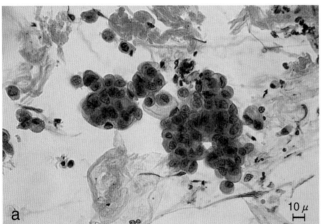

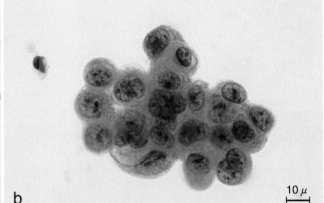

Fig. 15-60　Grade II papillary carcinoma cells in clusters in voided urine. The tumor cells shed in urine retain a papillary pattern in the grade II carcinoma, although the nuclei are varying in size and disorderly arranged. The cytoplasm is of a moderate amount and well stained by Light Green.

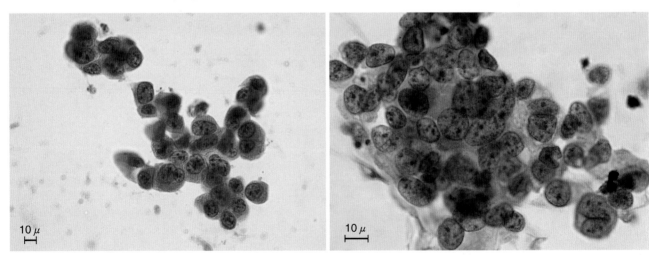

Fig. 15-61　Grade III transitional carcinoma cells forming an irregular cluster. Marked nuclear enlargement, high N/C ratio and coarse chromatin clumping are seen. Irregular papillary projections reflect a papillary growth of the tumor.

cells in carcinoma in situ are relatively small, about the size of normal bladder epithelial cells or slightly larger. As a rule, the nuclei are rather moderately enlarged for this stage and rounded, although they tend to be irregular or angular in shape because of degeneration. They are always hyperchromatic with a coarse chromatin

structure. The cytoplasm is usually thin, small to moderate in amount, and not greatly different from that of a normal deep transitional cell (see Fig. 15-47). Its staining reaction, a hazy cyanophilic opacity, differs from transparency in benign cells. Some enormously enlarged nuclei are admixed within a fairly uniform cellularity

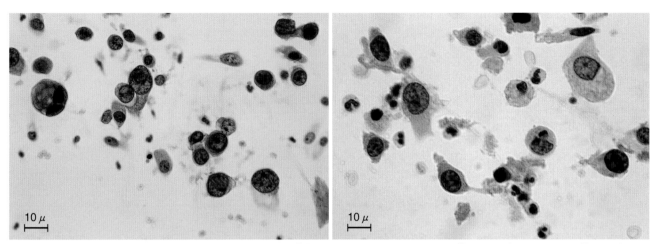

Fig. 15-62 Grade III transitional carcinoma cells exfoliated singly. The most undifferentiated carcinoma cells are characterized by scant cytoplasm, ill-defined cellular borders and lack of cluster formation.

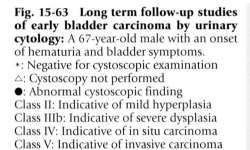

Fig. 15-63 Long term follow-up studies of early bladder carcinoma by urinary cytology: A 67-year-old male with an onset of hematuria and bladder symptoms.
∗: Negative for cystoscopic examination
△: Cystoscopy not performed
●: Abnormal cystoscopic finding
Class II: Indicative of mild hyperplasia
Class IIIb: Indicative of severe dysplasia
Class IV: Indicative of in situ carcinoma
Class V: Indicative of invasive carcinoma
(Note: The transitional Papanicolaou's classification had been used)

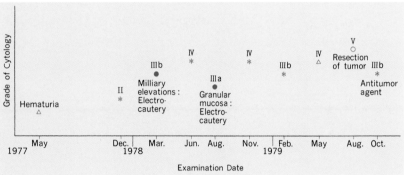

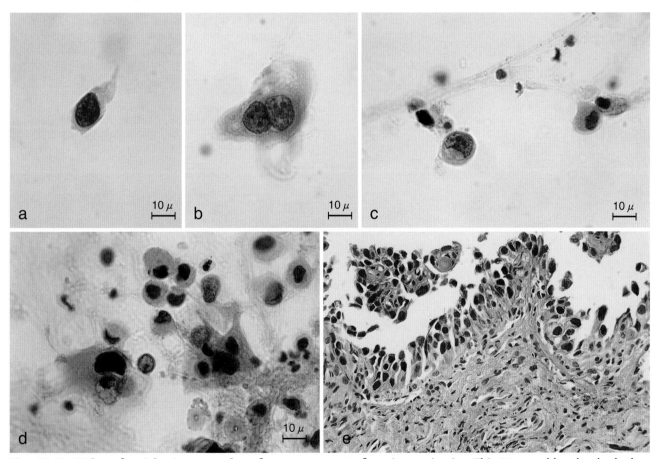

Fig. 15-64 Value of serial urinary cytology for management of carcinoma in situ. This 67-year-old male who had an episode of cauterization for multiple miliary elevations of the mucosa has been examined every month for two years. Although no macroscopical tumors were found cystoscopically, a small number of single small malignant cells were continuously found (a-c). One and half year later he showed apparent positive cytology (d) by development of early invasive carcinoma with extensive intraepithelial carcinoma.

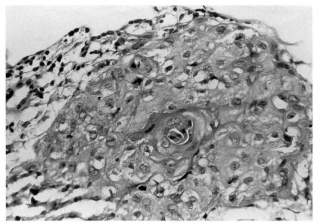

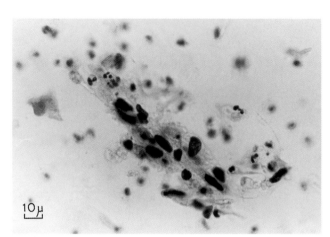

Fig. 15-65 Squamous cell carcinoma of the urinary bladder. A solid nest of cancer cells shows characteristic epidermoid differentiation; the cancer cells possess broad cytoplasm and tend to form a pearl in the middle of the cancer cell nest.

Fig. 15-66 Clusters of spindly cancer cells in voided urine secondary to squamous cell carcinoma of the uterine cervix. Malignant cells with spindle nuclei and elongated cytoplasm are arranged like bundles of fibroblasts.

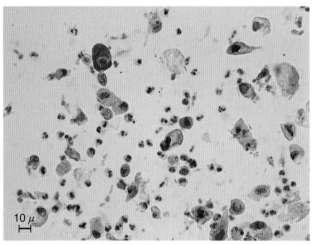

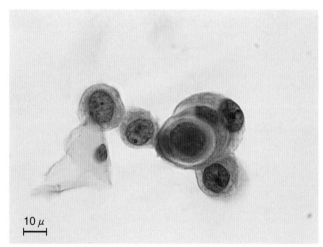

Fig. 15-67 Squamous cancer cells exfoliated singly in voided urine. Cancer cells are variable in shape and in staining reaction.

Fig. 15-68 Squamous cancer cells in voided urine showing cannibalism, i.e., cancer cell inclusion in another cancer cell, that is often encountered in the epidermoid type. Note parabasal type cancer cells.

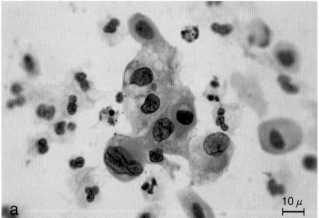

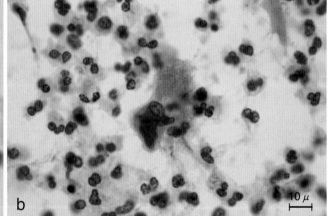

Fig. 15-69 Squamous carcinoma cells secondary to cervical carcinoma.
a. Nonkeratinizing cancer cells in a loose cluster. The nuclei are markedly polymorphic and densely stained. These cells are arranged in a flat sheet.
b. A keratinizing cancer cell is characterized by a bizarre hyperchromatic nucleus and intensely orangeophilic cytoplasm.

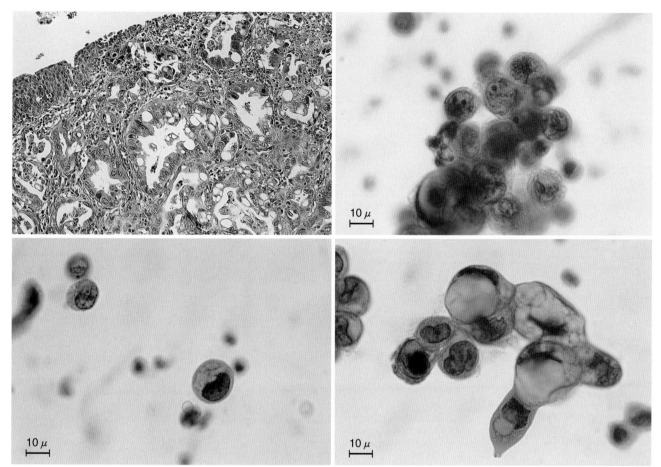

Fig. 15-70 Histology and cytology of adenocarcinoma growing in the dome of a 72-year-old male. Primary adenocarcinoma showing marked downgrowth is histologically characterized by atypical glandular structures varying in size. Note cytoplasmic vacuolation. Cytology shows cancer cells either singly or in cluster. Single cells show eccentric nuclei and often cytoplasmic vacuolation. The pattern showing marked nuclear overlapping in a tight cluster is also favorable to a glandular type of malignancy.

(Fig. 15-48). As a general appearance, these malignant cells are small in number and tend to shed singly or in a small cluster. Also characteristic in carcinoma in situ is a clean background; neither necrotic cell debris nor pus cells are found unless infection is superimposed (Figs. 15-63 and 15-64).

■Squamous cell carcinoma

Squamous cell carcinoma (epidermoid carcinoma) occurring as a pure form is about 5% of tumors of the urinary bladder. It is a grossly non-papillary and slightly elevated tumor that is found in the anterior wall or the dome. It tends to infiltrate into the bladder wall through the stage of carcinoma in situ like in other organs. Microscopically, the tumor is composed of anastomosing sheets or bands of polygonal cancer cells with abundant cytoplasm and intercellular bridges. The center of cancer cell nests shows keratinization (pearl formation) in well differentiated type of squamous cell carcinoma (Fig. 15-65). Since the urothelium is susceptible to metaplastic change, carcinoma of metaplastic form may occur in focal areas of transitional cell carcinoma.

The exfoliative cytology is the same as that of squamous cell carcinoma in other organs (Figs. 15-67 and 15-68): Tadpole cells, fiber cells, snake cells and third type cells are found in cases of well differentiated squamous cell carcinoma in the advance stage. It must be mentioned that squamous carcinoma cells are derived not only from primary carcinoma of the urinary tract, but also from metastatic lesions or secondary extensions of epidermoid cancer from the adjacent organs. Squamous cell carcinoma of the uterine cervix frequently invades the urinary bladder (Figs. 15-66 and 15-69).

■Adenocarcinoma

Adenocarcinoma of the transitional epithelium is rarely seen. According to the review on adenocarcinoma of the urinary bladder by Jaske et al,[45] it accounts for only 2.5 %, 18 or 715 vesical tumors. The morphogenesis of adenocarcinoma is attributed to neoplastic manifestation from glandular metaplastic cells[65] or from intraepithelial cysts of mucous cells which are similar to the Littré glands. Some adenocarcinoma of the bladder may be of the urachal origin. The cephalic segment of the urogenital sinus that is the derivative with lower intestine from the cloaca forms the fetal bladder and its tapered urachus. It is possible that the remnant of mucus-producing glandular epithelium of enteric type accompanies along the urachal cord. Urachal adenocarcinoma that accounts for about 30% of vesical adenocarcinoma locates at the bladder dome or anterior wall in the midline and grows intramurally. Histologically the tumor shows all variants

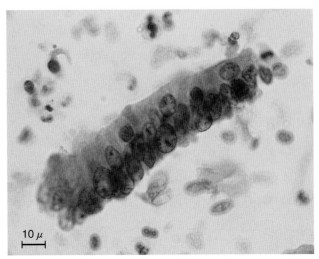

Fig. 15-71 Adenocarcinoma cells in a cluster in voided urine developed from rectum carcinoma. Such a tubular type of adenocarcinoma is suggestive of colon carcinoma. Diagnostic points: (1) Columnar cancer cells with basally locating nuclei, and (2) marked overlapping of the nuclei are typical features of tubular adenocarcinoma.

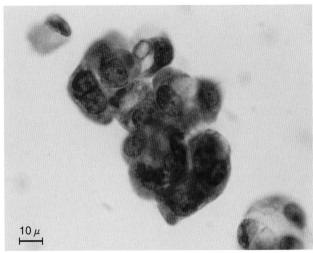

Fig. 15-72 Adenocarcinoma cells shed in an overlapping cell cluster secondary to endometrial carcinoma. It is impossible to determine cytologically whether these are from primary or secondary adenocarcinoma. Only cell typing is made as follows. (1) Nuclear overlapping is more marked than in transitional cell carcinoma, and (2) PAS-positive droplets within the cytoplasm are distinctive of adenocarcinoma.

Table 15-12 Malignant tumors of kidney

Pathology	Origin	Incidence	Cytology
Renal cell carcinoma (Grawitz's tumor)	Tubular epithelium	75% of renal tumors frequent in 5th, 6th decade	Rare in urine Broad, clear cytoplasm Fairly small nuclei Prominent nucleoli
Nephroblastoma (Wilms' tumor)	Wolffian body or embryonic nephrogen tissue	20% of all infant malignant tumors	Embryonal epithelial cells: small, oval or elongated cells with scant cytoplasm Mesenchymal cells; sarcoma cells
Transitional tumor	Transitional cells of pelvis	5 to 10% of renal malignant tumors	Malignant cells of transitional cell type

of enteric adenocarcinoma, especially of mucus-producing tubular and signet ring cell types (Fig. 15-70). A rareness of positive urinary cytology of urachal adenocarcinoma consists in the behavior of the tumor growth; it grows intramurally beneath the intact urothelium. However, the fact that hematuria and passage of necrotic debris or mucous material are virtually frequent should lead to the attention to this tumor by cytologists.[64]

Histology of the tumor that is composed of columnar mucus-secreting cells arranged in tubules or alveoli cannot be distinguished from invasive colonic adenocarcinoma without appropriate clinical information.

Cytology: Exfoliated malignant cells of this type reveal the ordinary glandular pattern as characterized by clumping of cells with nuclear overlappings, columnar or cuboidal form with eccentric position of nuclei and cytoplasmic vacuolation due to mucus production. It is not rare to see the admixture of the glandular type with the ordinary transitional type malignant cells occurring as the result of adenocarcinomatous metaplasia of transitional cell carcinoma.

Attention should be paid in adenocarcinoma to exclude secondary carcinoma which are spread from the adjacent organs such as rectum and uterus (Figs. 15-71 and 15-72).

■ **Renal cell carcinoma** (Table 15-12)

Renal cell carcinoma (adenocarcinoma, clear cell carcinoma, hypernephroma, Grawitz's tumor) is the most common in relative frequency of malignant neoplasms of the kidney. This tumor forms a spheroidal mass frequently associated with necrosis and hemorrhage. Hematuria is an important clinical symptom of this tumor as a result of tumor invasion into the pelvis.

Tweeddale[113] mentions that the one-half of renal cell carcinomas show direct involvement of the renal pelvis. However, detection rates of malignant cells from the bloody urine are not as frequent as expected,[71] and its cytology is said to be of questionable value.[53] On the other hand, there have been several descriptions reported fairly high diagnostic rates: Harrison[36] found malignant cells in 12 of 15 cancers, Deden[23] in 11 of 15, and Meisels[74] in all of 9 cases. Detectability of cancer cells from urine depends on the stage whether the tumor is invading the pelvis or not. Hajdu and his associates[33] mentioned the appearance of Oil-red O positive gran-

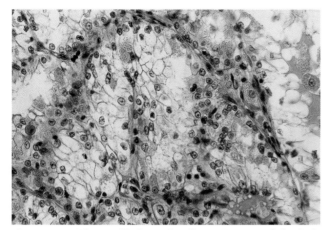

Fig. 15-73   Clear celled renal cell carcinoma of the kidney. This highly cellular tumor is composed of tubules which are lined with tall columnar cells having small round nuclei and clear columnar cytoplasm. Note single prominent nucleoli which are centrally locating.

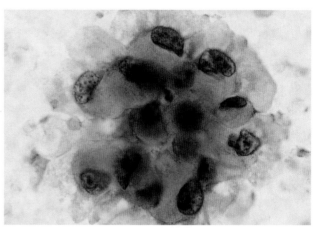

Fig. 15-74   Clear celled renal cell carcinoma with three-dimensional cell cluster. From voided urine by Millipore filter method. Note characteristic clear cytoplasm and prominent nucleoli.

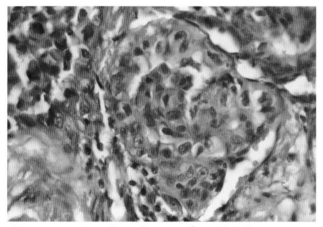

Fig. 15-75   Granular cell type of renal cell carcinoma. Solid alveolar cell nests are composed of polyhedral cells staining strongly eosinophilic. Note prominent nucleolus.

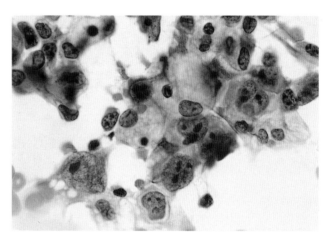

Fig. 15-76   Well preserved renal cell carcinoma showing a trabecular arrangement. From fine needle aspirate. The polyhedral cytoplasm shows Light Green stain by Papanicolaou stain. Note anisonucleosis and prominent nucleoli.

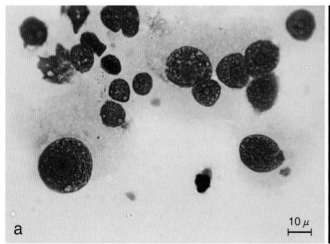

10 μ

a

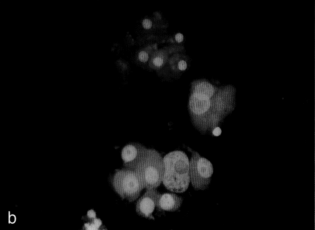

b

Fig. 15-77   Renal cell carcinoma in catheter urine.
a. In Giemsa stain cancer cells appear to be more larger than in Papanicolaou stain. Note enormously large, central nucleoli.
b. In acridine orange stain the cytoplasm and macronucleoli emit reddish orange fluorescence, while the nuclei rich in DNA yield intense green to yellow fluorescence.

ules in the cytoplasm to be valuable as a diagnostic criterion of renal cell carcinoma. However, lipid stain is not specific for carcinoma but suggestive of degenerative process of both malignant cells and tubular epithelial cells; the latter is known as oval fat bodies in nephrotic syndrome. Concerning this problem, Milsten et al[79] placed a diagnostic value only for large, single cells laden with fine lipid granules. The tumor is histologically composed of two kinds of cells, clear cells and granular cells arranged in sheets, cords or tubules: (1) clear cells with abundant foamy cytoplasm separated by delicate connective tissue and relatively small pyknotic nuclei and (2) smaller cells with a darkly stained, granular cytoplasm. Either type may be predominant or combined together in this tumor. In addition to two main types, there is an anaplastic type which is characterized by polymorphic cancer cells with indistinctly bordered cytoplasm and pleomorphic nuclei.

Subtypes of renal cell carcinoma according to the Armed Forces Institute of Pathology (Washington, D.C.) are classified to clear cell (hypernephroid), papillary, granular cell, chromophobe cell, sarcomatoid and collecting duct types.[80] Clear cell type (Fig. 15-73) is the most common variant and accounts for about 70% of renal cell carcinomas. Papillary type is the second frequent variant of multifocal occurrence accounting for about 10%.[56] This type has a distinctive papillary growth pattern of small cuboidal cells. Granular cell type (Fig. 15-75) shows a similar growth pattern to that of clear cell type; abundant cytoplasm is characterized by densely stained, eosinophilic, somewhat granular appearance. Chromophobe type is a peculiar variety having broad granular or transparent cytoplasm and small round nuclei. This histologic pattern is called to be "plant like." Sarcomatoid type pursuing aggressive course was ever called anaplastic type that is comprised of spindle-shaped cells.[12,93] There may be mixture with characteristic renal cancer cells which can be identified with ease. Collecting duct type is a rare variant which presents histologic similarity to collecting tubules of the kidney.

Accurate fine needle aspiration assessment is reported by Renshaw et al; 74% of 34 renal carcinomas are correctly diagnosed with satisfactory subtyping results.[89]

Cytologically, the clear cells exhibit characteristic features, such as an abundant and delicately vacuolated cytoplasm, polygonal nuclei which are relatively small for the cytoplasm (slight alteration of N/C ratio) and prominent nucleoli[42,53] (Fig. 15-74). Differential cytodiagnosis is important between reactive histiocytes and clear carcinoma cells with small nuclei, and between transitional carcinoma cells and anaplastic renal carcinoma cells. Prominent nucleoli and central round nuclei with distinctive nuclear membrane are found in renal cell carcinoma and not peculiar to histiocytes (Figs. 15-76 and 15-77). Anaplastic renal carcinoma cells are hardly differentiated from grade III transitional carcinoma cells except for the prominence of nucleoli that is characteristic of hypernephroma; poorly differentiated transitional carcinoma cells may possess basophilic, more than single nucleoli which resemble karyosomes.

■Nephroblastoma (Wilms' tumor)

Nephroblastoma is one of the most common malignant tumors in childhood.[1,21]

It forms a large spherical tumor, which is demarcated from the renal parenchyma in the early stage and extends diffusely toward the surrounding tissues in the later stage. The tumor often shows invasion of the renal vein and downward extension to the ureter. Hematuria is seen in about 10 to 20% of the cases because of the spread to the pelvis. Marked variation in the histology of the tumor depends on the preponderance and/or grade of differentiation either of the embryonal epithelial element or of the mesoblastic stromal element. The epithelial cells are small embryonal nephroblasts arranged as curved tubules or rosettes. Tubules and glomerulus-like structures are similar to the developing nephrons in the fetus. The mesenchymal elements undergo frequent differentiation to smooth or striated muscle, cartilage, bone, fatty tissue, etc.

Cytological description is very rare. Characteristic tubular features shown in a cell block preparation were obtained by Kmetz and Newton[51] from pleural fluid. Koss described a single case showing "clumps of very small anaplastic cells" in urinary sediment.[53] In our experience from an infant, a few clumps of fairly small cells without cytoplasm or with scant indefinite cytoplasm were encountered in the urinary sediment. These anaplastic small cells were considered to be of embryonal epithelial origin, because of the tendency to compose cell clumping suggestive of primitive tubules (Fig. 15-78d, f). Another type of cell population comprised of spindle-shaped cells shed in sheets is most likely to be of the mesenchymal constituent (Fig. 15-78c).[32] The importance of observation of two cell populations is emphasized by Hajdu.[32] However identification of the tumor cells whether of epithelial or nonepithelial origin is not as easy as in adult carcinosarcoma, since these primitive cells resembling neuroblastoma and Ewing's tumor have a bipotentiality of differentiation.

■Malignant lymphoma

Primary malignant lymphoma affecting the urinary tract is rare. Cytological detection of renal lymphoma was ever described by Sano and Kaprowska,[99] but most of lymphomas involving the renal parenchyma are of the secondary form. Primary lymphoma of the bladder is mentioned to be associated with follicular cystitis. Since the transitional epithelium adjacent to the tumor is often hyperplastic, dyskaryotic epithelial cells may take the leading part of cytological interpretation; lymphoma cells liable to degrade may be overlooked. Cytological distinction whether of primary neoplasm or of secondary manifestation of generalized lymphoma is impossible unless systemic examinations are available for the problem case.

Cytologically lymphoma cells are shed singly and characterized by almost denuded, mononuclear, hyperchromatic cells. Smooth nuclear membranes with occasional indentation and one or two distinctive nucleoli are favorable findings to lymphoma (Figs. 15-79 and 15-80).[54]

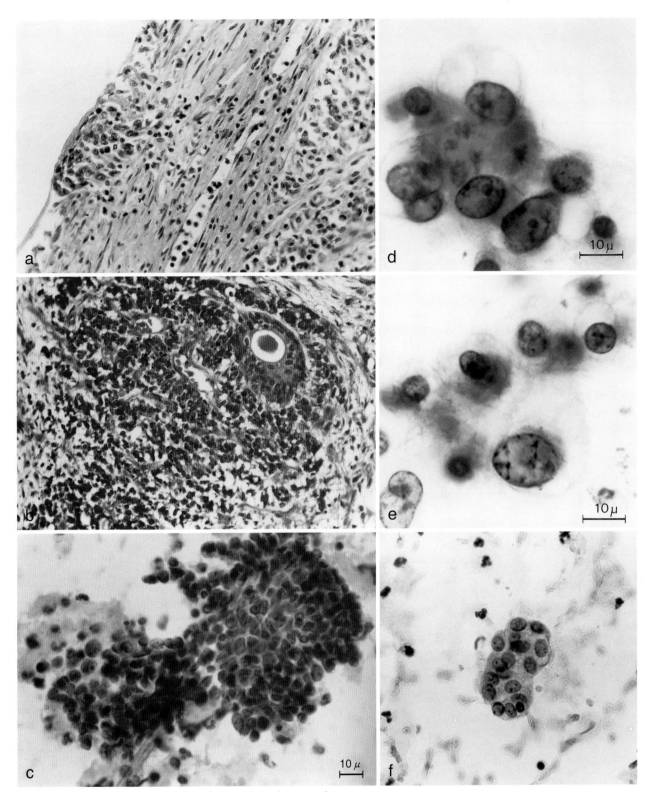

Fig. 15-78 Nephroblastoma (Wilms' tumor) and its cytology.

a. Low power magnification of the pelvis showing the mucosal epithelium to be eroded by tumor cell infiltration (left side).

b. Color figure showing differentiation to renal tubules and mesenchymal elements. There is only a small proportion and scant, ill-defined cytoplasm.

c. Clustered malignant cells from nephroblastoma of an infant. Massive cluster of undifferentiated nephroblasts which are hyperchromatic and stripped of the cytoplasm may be derived from mesenchymal component of the tumor.

d, e. Clumped or single malignant cells in urine. The nuclei are almost denuded and show irregular condensation of chromatin particularly at the nuclear border. The background behind condensed chromatin appears hazy and homogeneous as a result of liquefaction degeneration. The cytoplasm is scant, basophilic and indistinctly outlined.

f. Tumor cells showing tubular differentiation. Cluster formation suggestive of a tubular pattern is reflective in histology (b). Note preservation of cytoplasm.

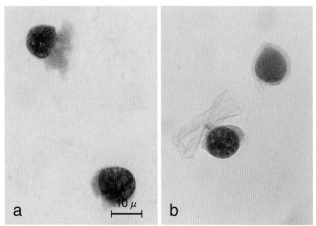

Fig. 15-79 Lymphosarcoma cells in voided urine from a patient with generalized lymphadenopathy. Single mononucleated cells reflect the histology of malignant lymphoma. Diagnostic points: Although chromatin pattern is vague by degeneration, (1) shedding as single mononuclear cells, (2) high N/C ratio owing to scant cytoplasm, (3) one or two prominent nucleoli meet the criteria of malignant lymphoma irrespective of either primary or secondary variety. (By courtesy of Mr. Fujinaga, CT, IAC, Hyogo Prefectural Amagasaki Hospital, Amagasaki, Japan.)

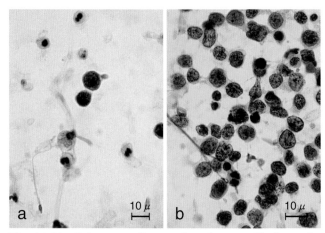

Fig. 15-80 Primary malignant lymphoma cells from the urinary bladder. a: in voided urine, **b:** from imprints after resection of a tumor. Note delicate nuclear membrane with occasional indentations.

■Sarcoma

Nonepithelial tumors are of rare occurrence comprising less than 1% of bladder tumors. Fibrous, myogenic, angiogenic and neurogenic tumors are more common than heterotopic varieties. Rhabdomyosarcoma, the most malignant heterotopic sarcoma, infiltrates diffusely within the bladder wall and extends extravesically.

Cytologically fibrosarcoma is characterized by spindly cells varying in size. Rhabdomyosarcoma cells are markedly polymorphic, bizarre in shape especially in pleomorphic type, and distinguished from other types of sarcoma by the detection of cross striation (Figs. 15-82 and 15-83).

12 Cytomegalic Inclusion Disease

Salivary gland virus (cytomegalovirus) transmitted by a mother to her fetus is known to affect not only epithelial cells of the salivary gland but also various organs; liver cells, bile ducts, renal tubules, bronchial epithelium, etc. This inclusion body contains DNA.[98] The affected cells are enlarged and contain large basophilic, occasionally acidophilic, inclusions in their nuclei (Figs. 15-84 and 15-85). The inclusion is situated in the center of nuclei and surrounded by clear halo. The nuclear membrane is thickened by chromatin condensation. Bolande[11] states that the inclusion is acidophilic in the early stage and becomes basophilic as cell infection progresses. In fresh urinary sediment one may encounter the affected cells in as many as 28% of normal children,[94] since the renal tubular epithelial cells are frequently affected subclinically by cytomegalovirus at an early age.[4,7,9,30,69,94] The high percentage of newborn infants with complement-fixing antibodies to cytomegalovirus (71%) may suggest transplacental transfer of maternal antibody.[95] Warner et al[118] found them in sputum specimens, Blanc in gastric washings,[8] and McElfresh et al in cere-

brospinal fluid.[73]

Adult patients who have received an immunosuppressive therapy are susceptible to the cytomegalovirus infection.

13 Polyomavirus Infection

Polyomavirus infection may cause similar intranuclear inclusion which is large, centrally located and densely basophilic. Periinclusion halo is not so distinctive as that in the cytomegalovirus infection. Furthermore, we must note a number of epithelial cells are affected in polyomavirus infection[17](Fig. 15-86).

We have experienced pronounced degenerative and swollen urothelial cells in urinary sediments of infants with upper respiratory infection often associated with conjunctivitis. The representative figures of infected nuclei are illustrated as densely granular or homogeneously condensed basophilic mass. A basophilic intranuclear inclusion resembles a Cowdry type A inclusion (Fig. 15-86). The former is compatible with "smudge cell" that was described in adenovirus infection.[109] The cause and the mechanism of such nuclear changes in urine have not been described as far as we know except for the report by Coleman[17,18] in polyomavirus infection in renal allograft recipients and by Horiuchi et al[43] in viral infection associated with leukemia. Our observation on peculiar basophilic intranuclear inclusions filled with virions consistent with B.K. polyomavirus particles is that suggested the pathogenicity of polyomavirus in human childhood infection.

14 *Schistosoma Haematobium* and Carcinogenesis

Bilharziasis manifests with passages of blood in urine and occasionally associated with temporary dysentery. In a later stage the patient begins to have a burning sensation at the time of micturition with so-called bladder colic. The eggs migrated into the vesical wall give rise to hyperplasia of mucosal epithelium and multiple minute abscesses. It is occasionally associated with malignant

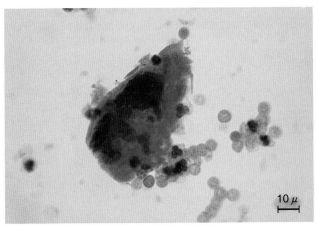

Fig. 15-81 Syncytial trophoblastic cell in catheter urine. A syncytial multinucleated giant cell in hemorrhagic urine is derived from choriocarcinoma involving the bladder wall.

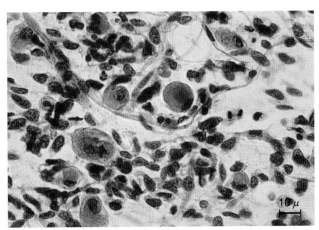

Fig. 15-82 Pleomorphic sarcoma cells in imprint smears of juvenile rhabdomyosarcoma of the bladder. A highly cellular mesenchymal tumor consisting of spindly cells and enrounded, racket-shaped cells. Some tapered cells are intermingled. Cross striations could be demonstrated by PTAH stain.

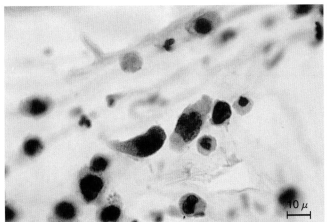

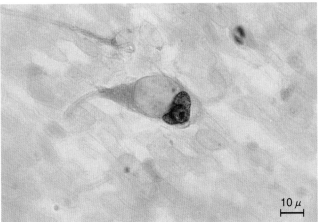

Fig. 15-83 Fibrosarcoma cells in urine. The bladder is involved by mesenterial fibrosarcoma. The tumor cells are spindly and their nuclei are polymorphic. There is no thickening of nuclear rim.

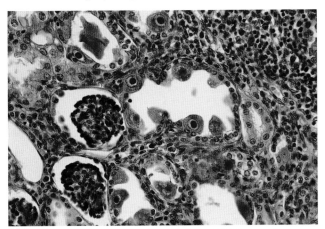

Fig. 15-84 Cytomegalovirus-infected kidney of a newborn infant. Note many intranuclear inclusions surrounded by clear halos in the renal tubular cells. (By courtesy of Prof. M. Montenegro and Dr. E. Tani, Faculdade de Medicina, Universidade de Botucatu, São Paulo, Brazil.)

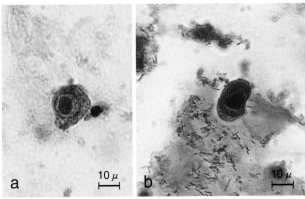

Fig. 15-85 Cytomegalovirus inclusion in an enlarged tubular cell. Diagnostic points: An infected epithelial cell is characterized by (1) a marked nuclear enlargement, (2) cyanophilic to basophilic intranuclear inclusion with halo, and (3) thickening of a nuclear membrane. (By courtesy of Dr. C.R. Vieira e Silva, São Paulo, Brazil.)

neoplasm. There has been the question that the infection with *Schistosoma haematobium* giving rise to inflammatory growth of urothelium solely induce invasive cancer.[58] It is postulated by Hicks et al that endemic schistosomiasis supplies the proliferative stimulus to acceler-ate cancer growth in patients affected with low doses of urothelial carcinogen.[37] The evidence that the cancer initiator nitrosamine was detected in the urine of Egyptians and Europeans with bacteriuria is also interesting in carcinogenesis of the urothelium.[38] The eggs can be found

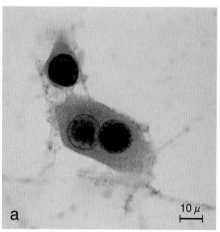
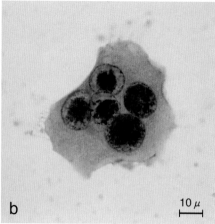

Fig. 15-86 Malignancy-mimic urothelial cells containing large, basophilic intranuclear inclusions. From voided urine of a 3-year-old girl who has been hospitalized because of acute respiratory infection with fever. In spite of intense nuclear stain by hematoxylin, loss or reduction of Feulgen reaction was referred to as infection by numerous virions consistent with B.K. polyomavirus by electron microscopy.
a. So-called smudge cells filled with granular or diffusely dispersed basophilic mass.
b. Intranuclear inclusion consistent with Cowdry type A inclusion.

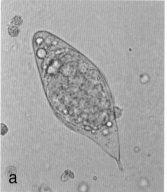
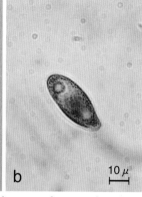

Fig. 15-87 Mature egg of *Schistosoma haematobium*.
a. Fresh sediment showing a viable ova that is characterized by a distinct terminal spine and fully embryonated miracidium.
b. Papanicolaou stained *Schistosoma haematobium* in urine from a 25-year-old African male who has had persistent hematuria. Probably not viable because of invisibleness of the miracidium.

Table 15-13 Frequency of eosinophilic inclusions in urinary sediments by random screening at Pediatric Clinic[48]

Non-viral disease		Viral disease	
Examined cases 1,650		99	
Inclusion-positive 133 (8.1%)		71 (71.7%)	
Clinical diagnosis	cases	Clinical diagnosis	cases
Acute bronchitis	28	Mumps	63
Acute pharyngitis	27	Roseola infantum	3
Nephritis	22	Measles	2
Acute tonsillitis	16	Rubella	1
Acute enteritis	11	Herpes stomatitis	1
Streptococcal infection	7	Polyomavirus infection**	1
Hemorrhagic cystitis	5		
Atypical pneumonia*	3		
Hydronephritis	2		
Vulvitis	1		
Pertussis	1		
Periodic vomiting	1		
Malignant lymphoma	1		
Allergic disorder	1		

* One of these cases was suspicious for adenovirus infection.
** Basophilic intranuclear inclusions were characteristic (see Fig. 15-86).

in urinary sediments; they are pale yellow and elongated-oval in shape with a distinct terminal spine (Fig. 15-87). It measures 110 to 170 by 40 to 70 μm and possesses fairly refractive wall.

15 Cytoplasmic Inclusion Bodies

Various kinds of intracytoplasmic inclusions should not be mistaken for those of cytomegalovirus. Mucin-containing inclusions found in transitional epithelial cells and multinucleated giant cells are globular in shape with delimited halos and appear grey to brown with Papanicolaou stain.[27] They are confirmed cytochemically positive for alcian blue stain or periodic acid-Schiff reaction after diastase digestion.

There may be also non-specific acidophilic inclusion in degenerating epithelial cells,[76] often in association with various viral diseases such as measles and chicken pox (Table 15-13); they are single or multiple, round, intensely eosinophilic inclusions that resemble ciliocytophthoria of the respiratory epithelial cells. There are negative for PAS reaction and reddish in Azan-Mallory stain. Since the presence of these eosinophilic inclusions is associated with karyorrhexis, it is considered to be the result of the degenerative process (Fig. 15-88).

16 Malakoplakia

Intracytoplasmic Michaelis-Guttmann bodies[64,89] in large macrophages (von Hansemann cells) of the urinary sediment are diagnostic of malakoplakia (Figs. 15-89 and 15-90). They are spherical, 2-10 μm in diameter, concentrically laminated and hematoxylin-staining concretions with centrally pale figures.[75] Electron microscopically Michaelis-Guttmann bodies are found to be located and formed in the phagosomes; bacteria incorporated within the phagosomes persist as dense amorphous aggregates because of impairment of enzymatic digestion. These become later mineralized. Thorning and Vracko[112] referred to it as the block of intralysosomal acidification by sulfonamide therapy. Recently there have appeared other reports[52,62] suggesting that immune suppressed states are related to the genesis of malakoplakia. Viana de Camergo et al[18] reported a case of

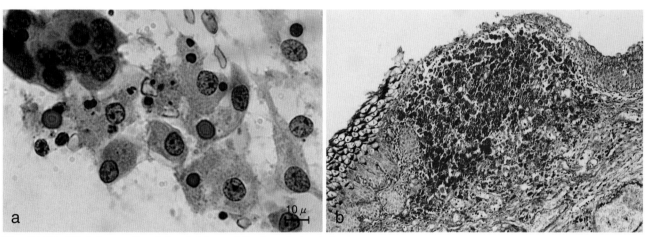

Fig. 15-88 Intracytoplasmic acidophilic inclusions resembling respiratory ciliocytophthoria from a patient with chicken pox. The patient suffering from acute viral infections such as measles, variola and chicken pox produces frequently such non-specific inclusion cells. a: Azan-Mallory stain, b: Papanicolaou stain.

Fig. 15-89 Histology of malakoplakia and its imprint cytology.
a. See characteristic Michaelis-Gutmann bodies in the cytoplasm of macrophages. These are from a small yellowish nodule of the urinary bladder by imprinting. Laminated bodies were positive for Von Kossa's calcium and Perl's iron stains. (By courtesy of Dr. J.L. Viana de Camargo, Faculdade de Medicina, Universidade de Botucatu, São Paulo, Brazil.)
b. Malakoplakia cystitis showing sessile plaque in the bladder mucosa. A slightly elevated plaque is composed of aggregates of macrophages which are intensely stained by PAS reaction. Michaelis-Gutmann bodies are included in these macrophages.

young girl for whom malnutrition was responsible. Michaelis-Guttmann bodies are positively stained with PAS reaction, Sudan black B and Oil Red O solutions. In addition, they are positive for von Kossa's calcium stain and for Perl's iron stain.[3] Malakoplakia is a rare lesion characterized by yellowish grey plaque 0.5 to 1.0 cm in diameter in the mucosa of the urinary bladder, ureters and renal pelvis. Histologically, there is massive infiltration by numerous macrophages with or without Michaelis-Guttmann bodies.

17 Comet Cells

Comet cells (Papanicolaou) or decoy cells (Ricci) are cancer-mimic degenerative cells which are often found in voided urine of dyestuff workers. They are characterized by round or oval nuclei with coarse chromatins or homogeneously dark chromatins (Fig. 15-91). The cytoplasm is absent or small in amount with a little tag-like processes.[20]

18 Glitter Cells in Urine

There are polymorphonuclear leukocytes in urinary sediment showing Brownian movement of intracytoplas-

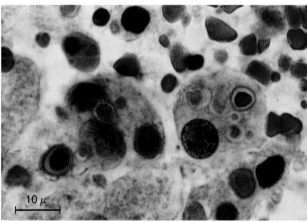

Fig. 15-90 Malakoplakia and its associated inclusions. Imprint smears from a yellowish plaque show multiple basophilic round inclusions in the cytoplasm of macrophages. Some inclusions are accompanied by halo formation. These are positive for PAS reaction. (By courtesy of Dr. J. Kawachi, Miyazaki Prefectural Hospital, Miyazaki, Japan.)

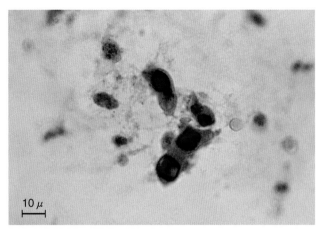

Fig. 15-91　Decoy cells exfoliated singly in urine. Note irregular nuclei with condensed chromatins and elongated cytoplasm with tails.

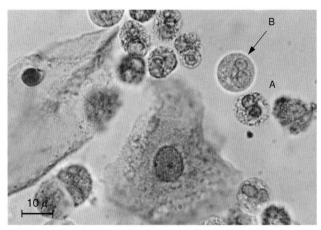

Fig. 15-92　Glitter cells in urine from a case with acute pyelonephritis. Small superannuated leukocytes containing immobile coarse granules (A) and glitter cell (B) with fine granules showing Brownian movements in the colorless cytoplasm.

mic granules under the dark field microscopy. These leukocytes named "glitter cells" by Schilling (1908)[101] are easily distinguished with supravital staining by gentian violet-safranin O solution (Sternheimer and Malbin[108]). Two kinds of leukocytes are identified by this stain: (1) Intensely stained small leukocytes with reddish violet nuclei and faintly purple cytoplasm, in which immobile granules are scattered. They are regarded as superannuated leukocytes; (2) Glitter cells which are larger than the former cells. They have pale bluish nuclei and colorless or faintly bluish cytoplasm in which fine granules undergo active Brownian movements (Fig. 15-92). Some of these larger pale leukocytes show no evidence of movement of intracytoplasmic granules, and they are seen in hypertonic urine with a specific gravity greater than 1,020.[121]

Brownian movement of intracytoplasmic granules is influenced by the concentration of electrolytes or osmotic pressure; it is conspicuous and lasts for a long time in hypotonic urine around 1,015 specific gravity.

■Clinical significance of glitter cells

The appearance of glitter cells was attributed to pyelonephritis[108] rather than to lower urinary tract inflammation. Poirier and Jackson[88] carried out a comparative study between the needle biopsy of kidneys and urinalysis, and confirmed that the more inflammatory leukocytes infiltrate the renal stroma, the more glitter cells appear in urinary sediment. It was also recognized that glitter cells were not found in the healing stage of pyelonephritis, particularly after treatment with antibiotics. At present a large number of leukocytes with nuclei stained blue are considered to be indicative of active inflammation in the urinary tract, irrespective of site of the lesion.[35] Although there has been some agreement concerning the frequent incidence of glitter cells in pyelonephritis, they may also occur in various other conditions such as glomerulonephritis, Kimmelstiel-Wilson disease[5] and cystitis.[91] It should not be considered as specific for pyelonephritis.

In our laboratory, we have come to the conclusion

that these are frequently encountered in hypotonic urine that is less than 1,020 in specific gravity and less than 700 mOsm/liter. The case showing glitter cells more than 10% and viable leukocytes with pale blue hue in Sternheimer-Malbin stain more than 50% is highly suggestive of manifestation of infectious disease of the urinary tract.

19 Allograft Rejection Cytology

In acute allograft rejection beginning with clinical signs of localized abdominal pain, fever and increased serum creatinine, urinary sediments disclose exudation of a large number of lymphocytes and neutrophils, exfoliation of tubular cells and appearance of granular cast in a week after operation[36,80]; more than 15 tubular cells per 10 high-power fields are indicative of rejection. Granular cast and erythrocyte cast formation with a dirty background is also an important finding in rejection. Lymphocyturia and exfoliation of tubular cells tend to be decreased in effective response to the therapy. Polyoma virus infection following renal transplantation has been emphasized by Coleman.[17] It is often to see, in urine of renal allograft recipients, degenerative epithelial cells having characteristic features such as basophilic, ground glass nuclei without haloing, increased N/C ratio and occasional bird's eyed inclusion. These were regarded as polyoma virus affected cells.[17]

■Rejection of renal transplantation

Both cell-mediated and humoral immunities are concerned with rejection mechanisms. Sensitized T-cells, granulocytes with lysosomal enzymes and macrophages are all involved in a cell-mediated mechanism.

Hyperacute rejection that occurs within hours after transplantation is developed with the presence of pre-existing antibodies; failure of the graft in vascularization, fibrin thrombi in the recipient following adherence of neutrophils and platelets, ischemic degeneration of tubules and infiltration by lymphocytes, neutrophils and macrophages are noticed.

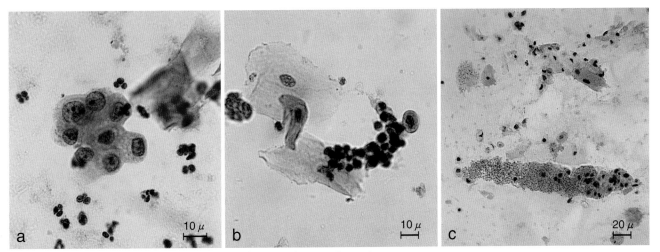

Fig. 15-93　Acute rejection crisis 10 days after renal transplantation.
a. Desquamation of swollen tubular cells. The nuclei and nucleoli are enlarged, but normochromic nuclei with finely granular chromatin, and adequate, edematous cytoplasm are rather suggestive of early degenerative change.
b. Lymphocyte cast composed of two or three columns of lymphocytes.
c. Fine granular cast containing degenerate tubular cells.

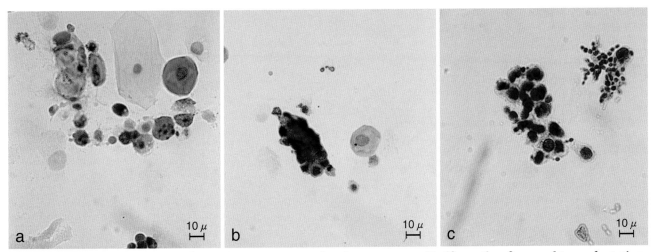

Fig. 15-94　Acute rejection crisis during immunosuppressive therapy one and half months after renal transplantation.
a: Degenerate tubular and transitional epithelial cells. **b:** Degenerate epithelial cell cast. **c:** Candidiasis caused by immunosuppressive therapy. Degenerate cells are also seen.

Acute rejection that occurs within several days is mediated by both cellular and humoral mechanisms; direct killing action of T-cells and reaction of macrophages activated by lymphokines are considered to be predominant. Acute rejection can occur much later in individuals who have been receiving immunosuppressive therapy. The clinical evidence of acute rejection is suggestive by gradual decrease of renal function after initial good urinary output.

Chronic rejection is characterized by gradual loss of renal function in association with proteinuria and hypertension over months to years after transplantation.

■Cytologic profile of urine after renal transplantation
1) Healthy recipient
Urinary sediments of the recipient in a week after transplantation is characterized by desquamation of transitional and tubular cells accompanied by a small number of leukocytes and granular casts. These cellular elements

gradually decrease in number and the background becomes clean in a month. Thus the urinary profile is not much different from normal individuals except for a mixture with some tubular cells. The tubular epithelial cells appear somewhat active in the recipient free from rejection; these tend to form an acinar cluster and display enlargement of cells as a whole. The enlarged nuclei are round and smoothly outlined.
2) Patient with rejection crisis
Output of urine declines and fever, weight gain and hypertension may begin to follow. Such a rejection crisis is mostly reversible by increasing corticosteroid dosage in combination with azathioprine (antimetabolite) or cyclophosphamide (alkylating agent). The criteria of urinary cytology suggestive of rejection are summarized as follows:
1) Increase in red blood cells,[13]
2) Increase in lymphocytes, occasionally with formation of lymphocyte casts[13](Fig. 15-93b).
3) Appearance of many degenerative tubular cells

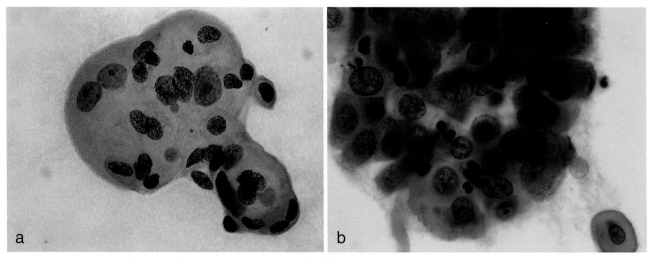

Fig. 15-95   Radiation effect on the urothelium by ^{60}Co irradiation for carcinoma of the uterine cervix.
a. Bizarre enlargement of cells with multinucleation after 5,000 rad irradiation.
b. Marked nuclear enlargement of deep transitional cells with engulfment of leukocytes.

with pyknotic or smudgy chromatins and liquefied or acidophilic cytoplasm,[85]
4) Presence of dirty granular background containing cellular debris,[13]
5) Admixture with various types of casts[13](Fig. 15-93c).

3) Associated changes in immunosuppressive therapy

The cytologic profile of rejection crisis may improve by intensive therapy. On the other hand opportunistic infection such as fungus and viral infection may be superimposed in urinary sediments. Lately it is emphasized that patient survival in such infection is more important than graft survival[97](Fig. 15-94c).

20 Radiation Cystitis

Radiation cystitis following radiation therapy not only for bladder tumors but for tumors of the adjacent tissue such as uterine cervix gives rise to some degree of cytological effects on the urothelium. There may occur bizarre cellular enlargement, chromatin condensation, multinucleation, vacuolization, abnormal stain of the cytoplasm, etc. (Fig. 15-95). In acute cystitis there is mucosal ulceration associated with hemorrhage. Therefore, clinical informations are indispensable in urinary cytology.

References

1. Arei JB: Malignant neoplasms in early life. J Pediat 35: 776, 1949.
2. Arrighi AA, Terzano G: Urinary cytology during pregnancy. Acta Cytol 3: 298, 1959.
3. Ashton PR, Lambird PA: Cytodiagnosis of malakoplakia. Report of a case. Acta Cytol 14: 92, 1970.
4. Bancroft J, Seybolt, Windhager HA.: Cytologic diagnosis of cytomegalic inclusion disease. Acta Cytol 5: 182, 1961.
5. Berman LB, Schreiner GE, Feys JO: Observations on the Glitter-cell phenomenon. N Engl J Med 255: 989, 1956.
6. Biot R, Bertran Nuñez R: Modificaciones periódicas del sedimento urinario en relación con el ciclo menstrual. Su posible aplicación como test de ovulación. Semana Med (Buenos Aires) 2: 532, 1944.
7. Birdsong M, Smith DE, Mitchell FN, et al: Generalized cytomegalic inclusion disease in newborn infants. JAMA 162: 1305, 1956.
8. Blang WA: Cytologic diagnosis of cytomegalic inclusion disease in gastric washings. Am J Clin Pathol 28: 46, 1957.
9. Blang WA, Gaetz R: Simplified millipore filter technique for cytomegalic inclusion disease in examination of the urine. Pediatrics 29: 61, 1962.
10. Bloom NA, Vidone RA, Lytton B: Primary carcinoma of the ureter: A report of 102 new cases. J Urol 103: 590, 1970.
11. Bolande RP: Cellular Aspects of Developmental Pathology. Philadelphia: Lea & Febiger, 1967.
12. Bonsib SM, Fischer J, Plattner S, et al: Sarcomatoid renal tumors. Cancer 59: 527, 1987.
13. Bossen EH, Johnston WW, Amatulli J, et al: Exfoliative cytopathologic studies in organ transplantation. III. The cytologic profile of urine during acute renal allograft rejection. Acta Cytol 14: 176, 1970.
14. Bredin HC, Daly JJ, Prout GR, Jr: Lactic dehydrogenase isoenzymes in human bladder cancer. J Urol 113: 487, 1975.
15. Broders AC: Epithelioma of the genitourinary organs. Ann Surg 75: 574, 1922.
16. Cole P: Lower urinary tract. In: Schottenfeld D (ed): Cancer Epidemiology and Prevention. pp233-262. Springfield, Ill: Charles C Thomas, 1975.
17. Coleman DV: The cytodiagnosis of human polyomavirus infection. Acta Cytol 19: 93, 1975.
18. Coleman DV, Gardner SD, Field AM: Human polyomavirus infection in renal allograft recipients. Br Med J 3: 371, 1973.
19. Crabbe JGS: Exfoliative cytological control in occupational cancer of the bladder. Br Med J 2: 1072, 1962.
20. Crabbe JGS: 'Comet' or 'Decoy' cells found in urinary sediment smears. Acta Cytol 15: 303, 1971.
21. Dargeon HW: Cancer in children from birth to fourteen years of age. JAMA 136: 459, 1948.
22. DeCamargo JLV, Gregório EA, Rodrigues MAH, et al:

Malcoplasia em "face precoce" na infancia: Relato de caso com observações ultra-estruturais. Revta Paul Med 94: 34, 1979.

23. Deden C: Cancer cells in urinary sediment. Acta Radiol Suppl Stockh 115: 1, 1954.

24. Del Castillo EB, Argonz J, Galli Mainini C: Cytological cycle of urinary sediment and its parallelism with vaginal cycle. J Clin Endocrinol 8: 76, 1948.

25. Deming CL: The biological behavior of transitional cell papilloma of the bladder. J Urol 63: 815, 1950.

26. de Voogt HJ, Rathert P, Beyer-Boon ME: Urinary Cytology: Phase Contrast Microscopy and Analysis of Stained Smears. Berlin: Springer-Verlag, 1977.

27. Dorfman HD, Monis B: Mucin-containing inclusions in multinucleated giant cells and transitional epithelial cells of urine: Cytochemical observations on exfoliated cells. Acta Cytol 8: 293, 1964.

28. Dukes CE: Tumours of the Bladder. Edinburgh: E & S Livingstone, 1959.

29. El-Bolkainy MN, Chu EW, Ghoneim MA, et al: Cytologic detection of bladder cancer in a rural Egyptian population infested with schistosomiasis. Acta Cytol 26: 303, 1982.

30. Fetterman GH: A new laboratory aid in the clinical diagnosis of inclusion disease of infancy. Am J Clin Pathol 22: 424, 1952.

31. Graham RM with members of The Vincent Memorial Laboratory Staff: The Cytologic Diagnosis of Cancer. Philadelphia: WB Saunders, 1954.

32. Hajdu SI: Exfoliative cytology of primary and metastatic Wilms' tumors. Acta Cytol 15: 339, 1971.

33. Hajdu SI, Savino A, Hajdu EO, et al.: Cytological diagnosis of renal cell carcinoma with the aid of fat stain. Acta Cytol 15: 31, 1971.

34. Harman JW, Hogan JM: Multinucleated epithelial cells in the tubules of the human kidney. Arch Pathol 47: 29, 1949.

35. Harris DM: Staining of urinary leucocytes as an aid to the diagnosis of inflammation in the urinary tract. J Clin Pathol 22: 492, 1969.

36. Harrison JH, Botsford TW, Tucker MR: The use of the smear of the urinary sediment in the diagnosis and management of neoplasm of the kidney and bladder. Surg Gynecol Obstet 92: 129, 1951.

37. Hicks RM, James C, Webbe G: Effect of Schistosoma haematobium and N-butyl -N-(4-hydroxy-butyl) nitrosamine on the development of urothelial neoplasia in the baboon. Br J Cancer 42: 730, 1980.

38. Hicks RM, Walters CL, Elsebai I, et al: Demonstration of N-nitrosamines in human urine. Proc R Soc Med 70: 413, 1977.

39. Highman W, Wilson E: Urine cytology in patients with calculi. J Clin Pathol 35: 350, 1982.

40. Hirschowitz SL, Mandell D, Nieberg RK, et al: The alcohol fixed Diff-Quik stain: A novel rapid stain for the immediate interpretation of fine needle aspiration specimens. Acta Cytol 38: 499, 1994.

41. Holmquist ND: Diagnostic Cytology of the Urinary Tract in Monographs in Clinical Cytology. Basel: S Karger, 1977.

42. Hopman BC: Clinical Cytology and Cytologic Research. Miami: Miami Post Publishing Co, 1960.

43. Horiuchi H, Takeda B, Ihara S, et al: Studies of cellular morphology suggesting viral infection in the urine of patients with malignant lymphoma and leukemia. J Jpn Soc Clin Cytol 17: 196, 1978.

44. Hrushesky W, Sampson D, Murphy GP: Lymphocyturia in human renal allograft rejection. Arch Surg 105: 424, 1972.

45. Jaske G, Schneider HM, Jacobi GH: Urachal signet-ring cell carcinoma, a rare variant of vesical adenocarcinoma: Incidence and pathological criteria. J Urol 120: 764, 1978.

46. Johnson WD: Cytopathological correlations in tumors of the urinary baldder. Cancer 17: 867, 1964.

47. Kannan V, Bose S: Low grade transitional cell carcinoma and instrument artifact: A challenge in urinary cytology. Acta Cytol 37: 899, 1993.

48. Kasai A, Kubo I, Takahashi M: Detection of inclusion bearing cells in the urine. J Med Technol 24: 820, 1980.

49. Kern WH: The diagnostic accuracy of sputum and urine cytology. Acta Cytol 32: 651, 1988.

50. Kern WH, Webster WW: Combined lactic acid dehydrogenase and cytologic examinations of urine in the diagnosis of cancer of the urinary tract. Acta Cytol 8: 302, 1964.

51. Kmetz DR, Newton WA Jr: The role of clinical cytology in a pediatric institution. Acta Cytol 7: 207, 1963.

52. Konnak JW, Hart WR: Malakoplakia of the prostate in an immunosuppressed patient. J Urol 116: 830, 1976.

53. Koss LG: Diagnostic Cytology and Its Histopathologic Bases. Philadelphia: JB Lippincott , 1968.

54. Koss LG: Tumors of the Urinary Bladder. Atlas of Tumor Pathology. Bethesda: Armed Forces Institute of Pathology, 1975.

55. Koss LG, Deitch D, Ramanathan R, et al: Diagnostic value of cytology of voided urine. Acta Cytol 29: 810, 1985.

56. Kovacs G, Kovacs A: Parenchymal abnormalities associated with papillary renal cell tumors: A morphologic study. J Urol Pathol 1: 301, 1993.

57. Kretschmer HL, Stika EA: Papilloma of the bladder, life after fulguration. JAMA 141: 1039, 1949.

58. Kuntz RE, Cheever AW, Bryan GT, et al: Natural history of papillary lesions of the urinary bladder in schistosomiasis. Cancer Res 38: 3836, 1978.

59. Lencioni LJ: Comparative and statistical study of vaginal and urinary sediment smears. J Clin Endocrinol 13: 263, 1953.

60. Lencioni LJ: El Urocitograma. Buenos Aires: Editorial Médica Panamericana, 1963.

61. Lencioni LJ, Martinez Amezaga LA, Lo Biango VS: Urocytogram and pregnancy. 1. Methods and normal values. Acta Cytol 13: 279, 1969.

62. Lewin KJ, Fair WR, Steigbigel RT, et al: Clinical and laboratory studies into the pathogenesis of malakoplakia. J Clin Pathol 29: 354, 1976.

63. Lippmann RW: Urine and the Urinary Sediment: A Practical Manual and Atlas, 2nd ed. Springfield, Ill: Charles C Thomas, 1957.

64. Loening SA, Jacobo E, Hawtrey CE, et al: Adenocarcinoma of the urachus. J Urol 119: 68, 1978.

65. Lucke B, Schlumbuger HG: Tumors of the Kidney, Renal Pelvis and Ureter. Atlas of Tumor Pathology. Washington, DC: Armed Forces Institute of Pathology, 1957.

66. Macfarlane EWE: Some pathologic conditions affecting urine cytology. Acta Cytol 7: 196, 1963.

67. Macfarlane EWE: The appearance of multinucleated cells in the urine after purgation. Acta Cytol 10: 104, 1966.

68. Makhyoun NA, El-Kashlan KM, Al-Ghorab MM, et al: Aetiological factors in bilharzial bladder cancer. J Trop

Med Hyg 74: 73, 1971.

69. Margileth AM: The cytodiagnosis and treatment of generalized cytomegalic inclusion disease of the newborn. Pediatrics 15: 270, 1955.

70. Marshall VF: Current clinical problems regarding bladder tumors. Cancer 9: 543, 1956.

71. McDonald JR: Exfoliative cytology in genitourinary and pulmonary diseases. Am J Clin Pathol 24: 684, 1954.

72. McDonald JR, Priestiey JT: Carcinoma of renal pelvis. Histopathologic study of seventy-five cases with special reference to prognosis. J Urol 51: 245, 1944.

73. Mcelfresh AE, Arey JB: Generalized cytomegalic inclusion disease. J Pediat 51: 146, 1957.

74. Meisels A: Cytology of carcinoma of the kidney. Acta Cytol 7: 239, 1963.

75. Melamed MR: The urinary sediment cytology in a case of malakoplakia. Acta Cytol 6: 471, 1962.

76. Melamed MR, Wolinska WH: On the significance of intracytoplasmic inclusions in urinary sediment. Am J Pathol 38: 711, 1961.

77. Melamed MR, Koss LG, Ricci A, et al: Cytohistological observations on developing carcinoma of the urinary bladder in man. Cancer 13: 67, 1960.

78. Michaelis L, Guttmann C: Über Einschlüsse in Blasentumoren: Über Malakoplakie der Harnblase. Virchows Arch Pathol Anat 173: 302, 1903.

79. Milsten R, Frable WJ, Texter JH, et al: Evaluation of lipid stain in renal neoplasms as adjunct to routine exfoliative cytology. J Urol 10: 169, 1973.

80. Murphy WM, Beckwith JB, Farrow GM: Tumors of the Kidney, Bladder, and Related Urinary Structures. Washington, DC: Armed Forces Institute of Pathology, 1994.

81. Murphy WM, Soloway MS, Jukkola AF, et al: Urinary cytology and bladder cancer: The cellular features of transitional cell neoplasms. Cancer 53: 1555, 1984.

82. National Committee for Clinical Laboratory Standards: Routine urinalysis and collection, transportation, and preservation of urine specimens, tentative guidelines. Villanova, PA: 12 (26) GP16-T, P 9, 1992.

83. Nichols JA, et al: Treatment of biologically benign papilloma of the urinary bladder by local excision and fulguration. Cancer 9: 566, 1956.

84. Nieburgs HE: Diagnostic Cell Pathology in Tissue and Smears. New York: Grune & Stratton, 1967.

85. O'Morchoe PJ, Erozan YS, Cooke CR, et al: Exfoliative cytology in the diagnosis of immunologic rejection in the transplanted kidney. Acta Cytol 20: 454, 1976.

86. Papanicolaou GN, Marshall VF: Urine sediment smears as diagnostic procedure in cancers of urinary tract. Science 101: 519, 1945.

87. Piscioli F, Pusiol T, Polla E, et al: Urinary cytology of tuberculosis of the bladder. Acta Cytol 29: 125, 1985.

88. Poirier KP, Jackson GG: Characteristics of leukocytes in the urine sediment in pyelonephritis. Am J Med 23: 579, 1957.

89. Renshaw AA, Lee KR, Madge R, et al: Accuracy of fine needle aspiration in distinguishing subtypes of renal cell carcinoma. Acta Cytol 41: 987, 1997.

90. Renshaw AA, Nappi D, Weinberg DS: Cytology of grade 1 papillary transitional cell carcinoma: A comparison of cytologic, architectural and morphometric criteria in cystoscopically obtained urine. Acta Cytol 40: 676, 1996.

91. Reubi F, Goodgold A, Schmid A: La présence de cellules de Sternheimer Malbin dans le sédiment urinaire est-elle liée a l'existence d'une pyélonephrite? Helvet Med

Acta 20: 392, 1953.

92. Riches EW, Griffiths IH, Thackray AC: New growths of the kidney and ureter. Br J Urol 23: 297, 1951.

93. Ro JY, Ayala AG, Sella A, et al: Sarcomatoid renal cell carcinoma: Clinicopathologic: A study of 42 cases. Cancer 59: 516, 1987.

94. Rowe WP, Hartley JW, Cramblett HG, et al: Detection of human salivary gland virus in the mouth and urine of children. Am J Hyg 67: 57, 1958.

95. Rowe WP, Hartley JW, Waterman S, et al: Cytopathogenic agents resembling human salivary gland virus recovered from tissue cultures of human adenoids. Proc Soc Exp Biol Med 92: 418, 1956.

96. Rutecki GJ, Goldsmith C, Schriener GF: Characterization of proteins in urinary casts. N Engl J Med 284: 1049, 1971.

97. Salvatierra O, Feduska NJ: Renal transplantation. In: Smith DR (ed): General Urology, 9th ed. Los Altos, CA: Lange, 1978.

98. Sandritter W, Müller D, Mantz O: Zur Histochemie der Cytomegalie. Frankfurt Z Pathol 70: 589, 1960.

99. Sano ME, Koprowska I: Primary cytologic diagnosis of a malignant renal lymphoma. Acta Cytol 9: 194, 1965.

100. Sarosdy MF: Genitourinary cancer. In: Weiss GR (ed): Clinical Oncology. pp180-193, Norwalk: Appleton & Lange, 1993.

101. Schilling V: Lebende weisse Blutkörperchen im Dunkelfeld: Beiträge zur normalen und degenerativen Struktur, besonders der Neutrophilen. Folia Haematol 6: 429,1908.

102. Schumann GB, Burleson RL, Henry JB, et al: Urinary cytodiagnosis of acute renal allograft rejection using the cytocentrifuge. Am J Clin Pathol 67: 134, 1977.

103. Schumann GB, Harris S, Henry JB: An improved technic for examining urinary casts and a review of their significance. Am J Clin Pathol 69: 18, 1978.

104. Schumann GB, Johnston JL, Weiss MA: Renal epithelial fragments in urine sediment. Acta Cytol 25: 147, 1981.

105. Schumann GB, Schumann JL, Marcussen N: Cytodiagnostic Urinalysis of Renal and Lower Urinary Tract Disorders. New York: Igaku-Shoin, 1994.

106. Setterington R: Microscopy. In: Coleman DV, Chapman PA (eds): Clinical Cytotechnology. pp125-136. London: Butterworths, 1989.

107. Sternheimer R: A supravital cytodiagnostic stain for urinary sediments. JAMA 231: 826, 1975.

108. Sternheimer R, Malbin B: Clinical recognition of pyelonephritis, with a new stain for urinary sediments. Am J Med 11: 312, 1951.

109. Strano AJ: Light microscopy of selected viral diseases (morphology of viral inclusion bodies). Pathology Annual 11: 53, 1976.

110. Sunderland H, Lederer H: Prostatic aspiration biopsy. Br J Urol 43: 603, 1971.

111. Takahashi M, Ito K: Color Atlas of Urinary Sediments. Tokyo: Uchudo-Yagishoten, 1979.

112. Thorning D, Vracko R: Malakoplakia: Defect in digestion of phagocytized material due to impaired vacuolar acidification. Arch Pathol 99: 456, 1975.

113. Tweeddale DE: Urinary Cytology. Boston: Little Brown, 1977.

114. Umiker W, Lapides J, Sourenne R: Exfoliative cytology of papillomas and intraepithelial carcinomas of the urinary bladder. Acta Cytol 6: 255, 1962.

115. Utz DC, Hanash KA, Farrow GM: The plight of the patient with carcinoma in situ of the bladder. J Urol

103: 160, 1970.
116. Von Hansemann: Über Malakoplakie der Harnblase. Virchows Arch Pathol Anat Physiol 173: 302, 1903.
117. Voutsa NG, Melamed MR: Cytology of in situ carcinoma of the human urinary bladder. Cancer 16: 1307, 1963.
118. Warner NE, McGrew EA, Nanos S: Cytologic study of the sputum in cytomegalic inclusion disease. Acta Cytol

8: 311, 1964.
119. Wolinska WH, Melamed MR, Klein FA: Cytology of bladder papilloma. Acta Cytol 817, 1985.
120. Woodard BH, Ideker RE, Johnston WW: Cytologic detection of malignant melanoma in urine: A case report. Acta Cytol 22: 350, 1978.
121. Yanaka M, Eto M, Kurosawa A, et al: Studies on glitter cells in urine. J Jpn Soc Clin Cytol 6: 181, 1967.

16 **Prostate**

Cancer of the prostate is one of the most frequent cancer in western countries. Approximately 7% of clinically overt tumors are of the prostate. The mortality accounts for 4% of all cancer deaths.[5] The incidence is low in the Orient. It occurs with an increasing frequency in older age groups. The majority of carcinoma of the prostate locate at the periphery of the organ, whereas glandular or glandular fibromuscular hypertrophy involves the central zone. Diagnostic procedures consist of digital rectal palpation, transrectal ultrasound, transrectal fine needle aspiration and measurement of prostatic specific antigen of blood serum. Zone-dependent carcinogenesis should match the diagnostic procedure.

1 Structure of Prostatic Glands

Prostatic glands are divided into three groups. The innermost "mucosal glands" locate in fibromuscular tissues around the urethra. The outer glands are composed of main glands that occupy the major part of the prostate and submucosal glands that surround ejaculatory ducts. The main glands contribute most to the volume of prostatic secretion. Tubulo-alveolar glands with lining of single columnar cells assemble at excretory ducts and open to the urethra. We must note that prostatic carcinoma arises from the main gland at periphery (Fig. 16-1).

2 Collection of Specimens

■Prostatic massage
Prostatic fluid obtained by gentle massage throughout the entire lobes of the prostate should be smeared and immersed immediately in 95% ethanol solution. Massage is performed by pressing the prostate with an index finger pad from its lateral edge towards the middle in order to express secretion out of the urethra (Fig. 16-2). Specimens ejaculated into a condom are also useful for cytology (Fig. 16-3). The secretions usually contain many cells of the lining epithelium of the tubulo-alveolar glands and their ducts. The danger of dissemination of cancer by massage may be suggestive. But the procedure has nothing much to do with tumor extension comparing with a surgical procedure like transurethral resection.[10]

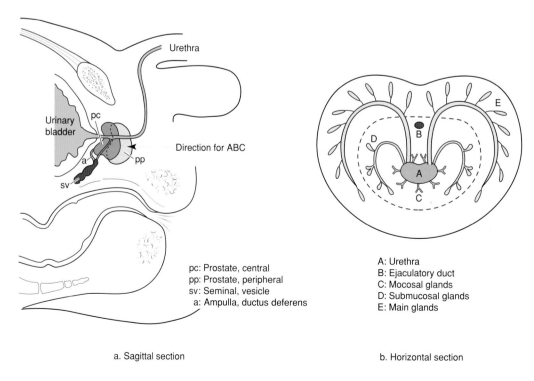

pc: Prostate, central
pp: Prostate, peripheral
sv: Seminal, vesicle
a: Ampulla, ductus deferens

A: Urethra
B: Ejaculatory duct
C: Mocosal glands
D: Submucosal glands
E: Main glands

a. Sagittal section

b. Horizontal section

Fig. 16-1 Anatomical diagrams of the prostate.

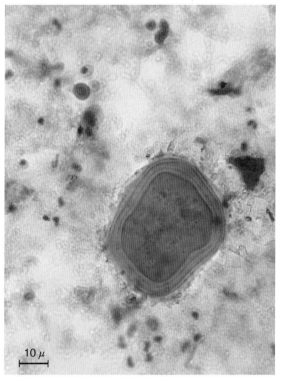

Fig. 16-2 **Corpora amylacea in a prostatic fluid.** Prostatic concretions are polygonal, sharply outlined, and occasionally laminated. A small number of sperma are present.

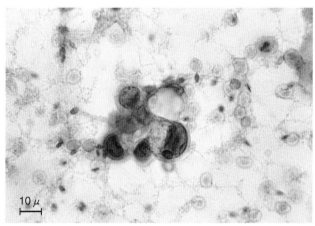

Fig. 16-3 **Adenocarcinoma cells from prostatic gland in bloody semen.** Note variation of the nuclei in size and cytoplasmic vacuolation. These malignant cells of a glandular type were found in admixture with spermatozoa in bloody semen.

■Prostatic needle aspiration

Needle aspiration is an excellent tool to detect occult prostatic carcinoma. Transrectal aspiration using a flexible fine-bore needle is a standardized technique since Franzen's first description in 1960. Williams et al[13] and Sunderland and Lederer[11] confirmed the high accuracy of cytodiagnosis and low false positives. The Franzen's instrument consists of an outer metal guide tube and inner needle to get specimen. Grip the guide tube with a gloved hand and fix the direction to the prostate using an index finger. Thereafter introduce the fine needle 20cm (22 gauge) long and insert through the rectal mucosa to aspirate. As an adjunct diagnostic procedure, measurement of serum prostate-specific antigen is valuable because it may elevate according as development of cancer.

3 Prostatic Cytology

The procedure of prostatic massage is performed with ease, although it is not so effective as transrectal needle aspiration using the Franzen needle. A successful massage yields varying types of formed elements; columnar and squamous epithelial cells, macrophages, blood cells, corpora amylacea, spermatozoa and others.

Transrectal needle aspiration using the Franzen's aspiration instrument is more reliable in cytodiagnosis than examination of urinary sediments and prostatic fluid; it facilitates collecting specimens from tumors as well as non-neoplastic lesions of the prostate (Fig. 16-1a).
  1) tuberculosis that is characterized by epithelioid cells,

  2) granulomatous eosinophilic prostatitis showing infiltration of eosinophils, plasma cells and giant cells of Touton type,
  3) glandular hyperplasia that is composed of sheets of large but uniform columnar cells having abundant cytoplasm.

■Value of prostatic cytology

If appropriate needle aspirates were submitted to a cytology laboratory positive diagnosis would assure the neoplasia. However, negative results may include inadequate sampling, minute carcinoma and unexperienced interpretation, etc.[8] Regarding cytological interpretation in prostatic cytology, the diagnosis may rely on structural atypia such as overlapping cell clusters, disorderly cell arrangement, scattered single atypical cells and absence of myoepithelial cells rather than cellular atypia. Even in histology of well differentiated adenocarcinoma that is composed of fairly uniform cells, presence of mitoses and bizarre or giant cells have little value in the assessment of malignancy. Contamination with cells from adjacent tissues such as seminal vesicles and rectum during aspiration may lead to misinterpretation if this event passed unnoticed; recognition of cytoplasmic lipochrome granules of seminal cells and distinctive nuclear palisading at the base of rectal columnar cells may prevent the error.[7]

While there are reports of very high diagnostic rates for carcinoma of the prostate as high as 97.6% (in 205 of 210 cases) by Esposti[4] and 93.2% (in 41 of 44 cases) by Koss and his associates,[8] Kaufman et al[6] have mentioned 60% accuracy. It is notable that diagnostic rates varied according as dedifferentiation of carcinoma, although poorly differentiated carcinoma was diagnosed in 92.6%, well differentiated carcinoma could not be clearly identified (Table 16-1).

■Prostatitis

Although acute and subacute prostatitis are contraindication of needle aspiration biopsy,non-specific chronic

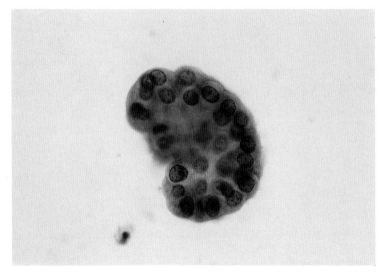

Fig. 16-4 **Prostatic glandular epithelium within normal limits.** Uniform low columnar epithelial cells are arranged in glandular pattern with polarity.

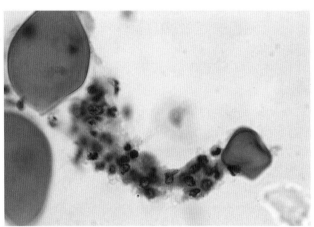

Fig. 16-5 **Nonspecific chronic prostatitis from prostatic fluid.** Epithelial cells are deformed and pyknotic as a result of reactive and degenerative changes. Note some neutrophils intermingled within epithelial cell cluster. Corpora amylacea are seen.

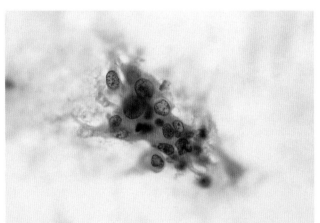

Fig. 16-6 **Nonspecific chronic prostatitis from a needle aspirates.** Chronic inflammatory response brings about nuclear hypertrophy and deterioration (pyknosis). Chromatins are finely granular. Metaplastic change is associated.

Table 16-1 Comparison of fine needle aspirations with histopathological diagnoses (January 1982-April 1984, Gifu University Hospital)*

Fine needle aspiration	No of cases	Pathologic findings			
		Benign	Well diff ca	Mod diff ca	Poor diff ca
Negative	92	68	12 (44.4%)	10 (40.0%)	2 (7.4%)
Atypical	5	1	3 (55.6%)	1 (60.0%)	0 (92.6%)
Positive	53	2	12	14	25
Total	150	71	27	25	27

*Takahashi M, Sugie S: Needle aspiration cytology of the prostate (in Japanese). Pathol Clin Med 3 (Suppl) : 131-151, 1985.

prostatitis develops a firm enlargement of the prostate. Clinically neoplasia-suspicious lesions are of indication of needle aspiration cytology.[7-9] Cytologically the nuclei are slightly enlarged but retain regularity in arrangement (Figs. 16-5 and 16-6).

■Nodular hyperplasia
Benign hypertrophy affects fibrous tissue, smooth muscle and glands at the central zone. However, hyperplasia is not uniform and predominates either in fibromuscular

(stromal nodule) or glandular components (glandular nodule). Thereby cellularity in needle aspirates depends on the site of aspiration. When needle aspirates are obtained from glandular nodules, a moderate to large number of epithelial cells can be examined.

Cytologically epithelial cells are demonstrated in sheets (Figs. 16-7 to 16-9). The nuclei are round or oval, and uniform in size. Uniformly arranged cells display a "honeycomb" appearance. The cytoplasm of moderate amount is lacy and poorly stains.

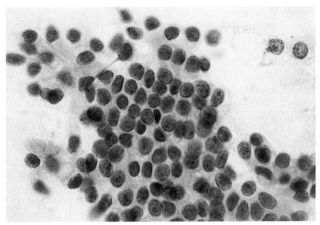

Fig. 16-7 **Benign hyperplastic epithelial cells in sheet from needle aspirates.** Glandular epithelial cells showing a "honeycomb" appearance are regularly arranged. The nuclei are normochromic and bland in chromatin distribution.

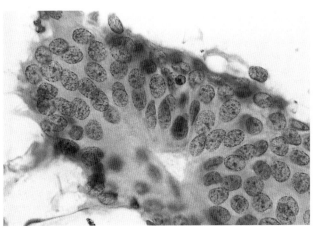

Fig. 16-8 **Benign granular hyperplasia from needle aspirates.** Epithelial cells are observed from above. The nuclei are enlarged and multilayered with regular polarity. Note evenly distributed chromatins.

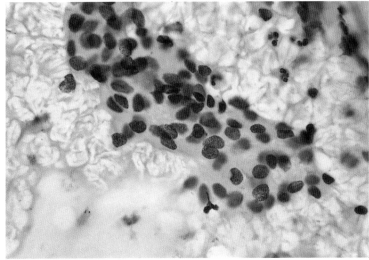

Fig. 16-9 **Atypical epithelial hyperplasia from needle aspirates.** Interpretation was first considered to be suspicious because of anisonucleosis and disorderly arrangement. Bland nuclear chromatins, absence of nucleoli and accompaniment of small pyknotic cells (myoepithelial) are in favor of benign hyperplasia.

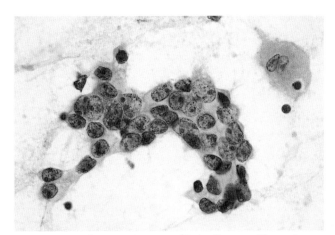

Fig. 16-10 **Adenocarcinoma cells in fine needle aspirates of the prostate.** A tubular aggregate of cancer cells with prominent nucleoli. Anisokaryosis and coarse chromatins are distinctive.

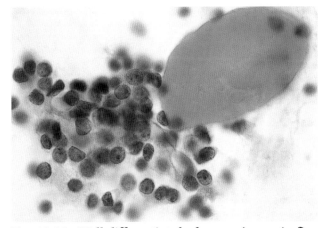

Fig. 16-11 **Well differentiated adenocarcinoma in fine needle aspirates.** The nuclei are fairly uniform in size but exhibit indistinctive cell cluster. Note distortion of nuclear arrangement and prominence of nucleoli.

■Carcinoma

Most prostatic carcinomas arise from the main glands at periphery, while these acinar glands undergo involution with increasing ages. Transurethral resection is not likely to be reliable for detection of cancer in an early stage. In the stage of carcinoma of the prostate involving the urinary bladder, adenocarcinoma can be detected in urinary cytology (Fig. 16-13). The majority of carcinoma are adenocarcinomas showing wide spectrum of histological differentiation (Figs. 16-10 to 16-12).

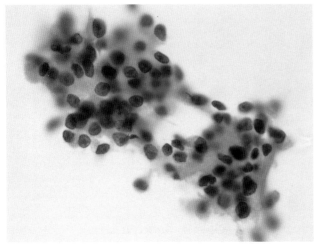

Fig. 16-12 Undifferentiated prostatic carcinoma consisting of small cells. Small hyperchromatic cancer cells disorderly arranged tend to form acinar pattern. The cytoplasm is ill-defined and scanty. Note presence of nucleoli.

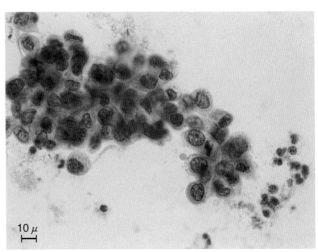

Fig. 16-13 Adenocarcinoma cells in urine from carcinoma of the prostate. Prostatic cancer cells often shed in the late micturition. Note disorderly arrangement of polymorphic cells in a papillo-tubular fashion.

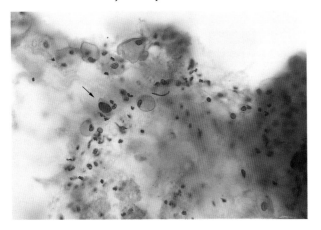

Fig. 6-14 A "monstrous" cell in prostatic fluid by massage. Note a single, enormously large, hyperchromatic cell (arrow). Mucus and some deep transitional cells are seen.

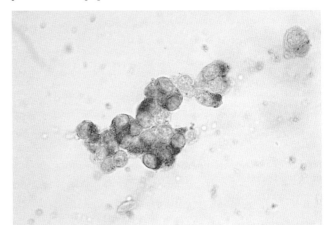

Fig. 16-15 Adenocarcinoma cell cluster stained with acid phosphatase stain. Strong stain of the cytoplasm is indicative of prostate origin. This was L-tartrate sensitive in reaction.

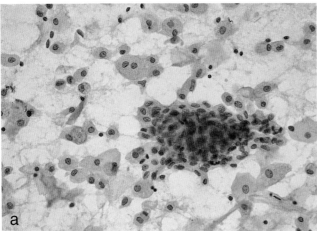 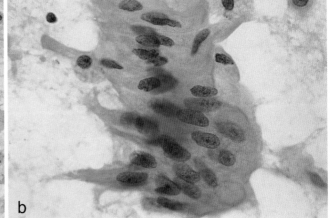

Fig. 16-16 Iatrogenic cell changes by stilbestrol administration.
a. Cancer cell cluster showing degeneration and eosinophilic tint. Squamous metaplasia is associated.
b. Enormous swelling of cells with vesicular degeneration of the cytoplasm.

Cytologically malignancy grading is classified into well differentiated, moderately differentiated and poorly differentiated carcinomas.

Well differentiated and moderately differentiated car-

cinomas have features as follows:
1) cell aggregates varying in size and shape and suggestive of three dimensional acinar structures,
2) anisonucleosis, and

3) macronucleoli.

Poorly differentiated carcinoma is characterized by

1) loose aggregates of cells which are disorderly arranged,

2) high N/C ratio regardless of nuclear size,

3) scant cytoplasm which is poorly defined, and

4) macronucleoli

Pitfall of interpretation is the appearance of abnormally large ("monstrous") cells which derive from the lining epithelium of the seminal vesicles[1](Fig. 16-14). The nuclear polymorphism such as hyperchromasia, enlargement and irregular shape becomes predominant with aging.[1] Singly appearing large nucleus is a cause of false positive. In advanced stage beyond the capsule, adenocarcinoma cells are hardly distinguished in urinary sediments whether they are from the urinary bladder or from the prostate. Intense reaction of acid phosphatase is indicative of the prostate origin (Kline)[7](Fig. 16-15).

■Therapeutic effect

The best treatment for old patients with localized cancer of the prostate is still in debate whether prostatectomy or external beam radiotherapy is preferable. In more advanced stage beyond stage C surgery is not considered. Radiotherapy or hormone therapy using estrogens or lutenizing hormone releasing hormone analogues, and antitumor chemotherapy are adopted. Böcking and Auffermann described cytologic assessment of therapeutic effect by scoring morphological alterations such as vacuolization and loss of cytoplasm, shrinkage and pyknosis of nuclei, and reduction and loss of nucleoli.[2]

■Stilbestrol therapy

Stilbestrol administration with consecutive measurements of serum alkaline phosphatase is a way of management for prostatic carcinoma with bone metastases. Cancer cells responsive to hormone therapy swell enormously. Squamous metaplasia is often associated during estrogen therapy (Fig. 16-16).

References

1. Ansell ID: The male generative system: Prostate. In: McGee J O'D, et al (eds): Oxford Textbook of Pathology, Vol 2a. Oxford: Oxford Univ Press, 1992.
2. Böcking A, Auffermann W: Cytological grading of therapy-induced tumor regression in prostatic carcinoma: Proposal of a new system. Diag Cytopathol 3: 108, 1987.
3. Coleman DV, Gardner SD, Field AM: Human polyomavirus infection in renal allograft recipients. Br Med J 3: 371, 1973.
4. Esposti PL: Aspiration Biopsy Cytology in the Diagnosis and Management of Prostatic Carcinoma. Stockholm: Stahl and Accidenstryct, 1994.
5. Jones WG, Smith PH: Cancer of the prostate. In: UICC Manual of Clinical Oncology. pp409-418. Berlin: Springer-Verlag, 1994.
6. Kaufman JJ, Rosenthal M, Goodwin WE: Methods of diagnosis of carcinoma of the prostate: A comparison of clinical impression, prostatic smear, needle biopsy, open perineal biopsy and transurethral biopsy. J Urol 72: 450, 1954.
7. Kline TS: Handbook of Fine Needle Aspiration Biopsy Cytology. St Louis: CV Mosby, 1981.
8. Koss LG, Woyke S, Olszewski W: Aspiration Biopsy: Cytologic Interpretation and Histologic Bases. New York: Igaku-Shoin, 1984.
9. Linsk JA, Franzen S: Clinical Aspiration Cytology. London: JB Lippincott, 1983.
10. Nieburgs HE: Diagnostic Cell Pathology in Tissue and Smears. New York: Grune & Stratton, 1967.
11. Sunderland H, Lederer H: Prostatic aspiration biopsy. Br J Urol 43: 603, 1971.
12. Takahashi M, Sugie S: Needle aspiration cytology of the prostate (in Jpn). Pathol Clin Med 3 (Suppl): 131, 1985.
13. Williams JP, Still BM, Pugh RCB: The diagnosis of prostatic cancer: Cytological and biochemical studies using the Franzen biopsy needle. Br J Urol 39: 549, 1967.

17 Lymph Node

1 Collection of Specimens and Preparation of Smears

Lymph node is the most appropriate organ for cytology by needle aspirations and imprint smears. Since a diagnosis by lymph node puncture was first described by Guthrie in 1921, there have been many descriptions concerned with adjunct procedures to hematological and pathological diagnoses for hematopoietic disorders such as various types of lymphoma and leukemia and for superficial lymphadenopathies.[23,53,73,105,114,127,128,150,153,170] Simplicity of the technique of aspiration cytology without leaving a scar is the main advantage. In addition, determination whether or not cancer spread is present into the regional lymph nodes by imprint smear during surgery is important to evaluate the prognosis and to determine the operating procedure.

■Technique

The techniques of collecting specimens are exactly the same as those described in the chapter on the breast: (1) Lymph node puncture with a hypodermic, silicone-coated needle of suitable length and a 20 ml or 10 ml syringe (2.5 to 6 cm long and 0.5 to 0.9 mm in outer diameter), and (2) preparation of imprint smears of a lymph node are performed as follows.

Lymph node aspiration can be performed even at the outpatient clinic with local anesthesia. The syringe can be operated with one hand, while the other hand is used to hold the swollen lymph node (Franzén)(Fig. 17-1).[42] Thin hypodermic needles are sufficient merely to obtain specimens from the aimed lymph node that is diffusely affected, however, needle aspirates may fail to represent the entire histological changes and may overlook small metastatic foci of a malignant tumor. Negative cancer cytology by needle aspiration does not assure a negative result. On the other hand, imprint smears are possible to diagnose cancer metastasis as well as lymphoma and its allied diseases even when small amounts of specimen are submitted. The fresh unfixed lymph node is immediately transected and its cut surface is pressed against a clean glass slide. Two or three imprints on a slide, each imprint by each press, are preferable to single imprint because one of these imprints provides a satisfactory film for observation of structural alteration. Pressing the slide over the cut surface of a lymph node which was fixed by forceps is also a good technique to obtain a good smear. Rubbing the cut surface on a slide brings about deformity of cellular structures. This simple pressing maneuver is the best technique to reflect histological changes of a lymph node as it is. As a staining technique, May-Grünwald-Giemsa stain is important for identification of lymphoma cells and leukemic cells in addition to the Papanicolaou stain. Rash stain of air-dried smear by Diff-Quik is adopted in American laboratories.[62]

2 Histology of Lymph Nodes

Several afferent lymphatic vessels enter the marginal sinuses through the convex surface of the organ. The lymph flows out through the efferent vessels at the hilus. The sinus is composed of networks of reticuloendothelial or retothelial cells. The parenchyma is divided into two compartments, the cortical region and the medullary region. The cortical region contains lymphatic nodules which are the sites of closely packed small lymphocytes; while the primary follicles are simply composed

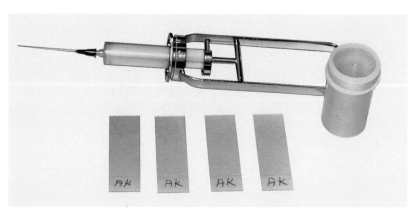

Fig. 17-1 Franzen-Stormby's disposable apparatus for needle aspiration.

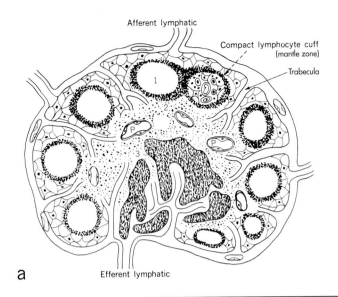

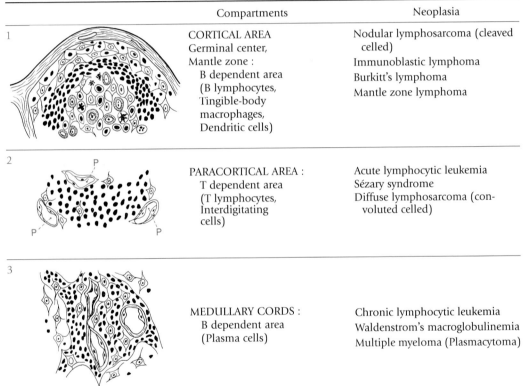

	Compartments	Neoplasia

1	CORTICAL AREA
	Germinal center,
	Mantle zone :
	   B dependent area
	   (B lymphocytes,
	   Tingible-body
	   macrophages,
	   Dendritic cells)
		Nodular lymphosarcoma (cleaved
		   celled)
		Immunoblastic lymphoma
		Burkitt's lymphoma
		Mantle zone lymphoma

2	PARACORTICAL AREA :
	   T dependent area
	   (T lymphocytes,
	   Interdigitating
	   cells)
		Acute lymphocytic leukemia
		Sézary syndrome
		Diffuse lymphosarcoma (con-
		   voluted celled)

3	MEDULLARY CORDS :
	   B dependent area
	   (Plasma cells)
		Chronic lymphocytic leukemia
		Waldenstrom's macroglobulinemia
		Multiple myeloma (Plasmacytoma)

Fig. 17-2　Schema of a normal lymph node and its diagrammatic compartments.
Anatomical knowledge of a lymph node from immunological view points is necessary to understand a functional role of lymphocytes and classification of non-Hodgkin's lymphomas; the lymph node is divided into the B cell and T cell areas. Within the cortex there are aggregates of lymphocytes mixed with some dendritic reticulum cells which are called primary follicles. The antigen-stimulated lymph node reveals enlarged germinal centers (secondary follicles) which are composed of immature lymphoid cells to be differentiated to B cells. The germinal centers are surrounded by a corona of dense lymphocytes that is called mantle zone. The paracortical areas contain a large number of thymus-dependent cells which are supposed to have emigrated through the postcapillary venules (P) in the paracortical areas.

of small lymphocytes, the secondary follicles possess lightly staining germinal centers (Fig. 17-2). The globular germinal centers are composed of variable cell components as described below and considered to be the center of lymphocytopoiesis as well as immune reaction. The paracortical zone is not distinctly figured but functionally different from the B-dependent germinal center. The medullary region consists of medullary sinuses and medullary cords containing many plasma cells.

3　Normal Cellular Components

Over 90% of the cells of normal lymph nodes are lymphocytes. A small number of prolymphocytes, histiocytes or reticulum cells, plasma cells and neutrophils are present. In inflammation some type of cells may be in excess in the differential cell count; monocytosis in infectious mononucleosis (Pfeiffer's glandular fever);

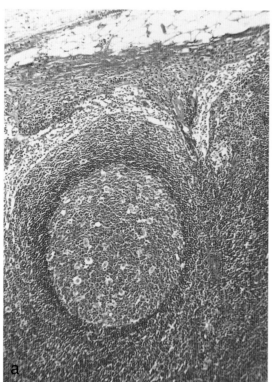

**Fig. 17-3 Histology of a lymph node with reactive lymphoreticu-
lar hyperplasia.**
a. Lower magnification. The peripheral sinus is widened and the germinal center of a secondary follicle is remarkably hypertrophic. Clear
spots in the germinal center represent macrophages (tingible body
macrophages).
b, c. Higher magnifications. Hyperplastic germinal center shows various
cell constituents such as reticulum cells, lymphoblasts or lymphogonia, prolymphocytes and some mature lymphocytes. Macrophages
with phagocytosed inclusions can be seen. RC: reticulum cells, LG:
lymphogonia, PL: prolymphocytes, LC: lymphocytes.

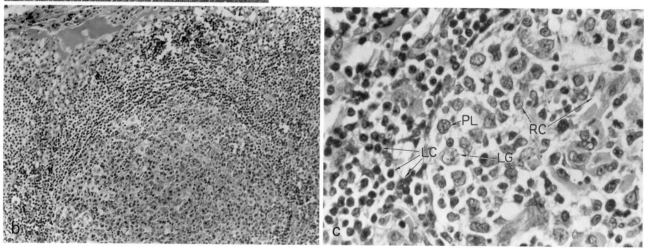

Table 17-1 Differential counts from normal lymph node
smears

Cell	Percent
Reticulum cells	0 – 0.1
Mast cells	0 – 0.5
Lymphoblasts	0.1 – 0.9
Prolymphocytes	5.3 – 16.4
Lymphocytes	67.8 – 90.0
Monoblasts	0 – 0.5
Promonocytes	0 – 0.5
Monocytes	0.2 – 7.4
Plasmoblasts	0 – 0.1
Proplasmocytes	0 – 0.5
Plasma cells	0 – 4.7
Neutrophils	0 – 2.2
Eosinophils	0 – 0.3
Basophils	0 – 0.2

(Modified from Lucas PF: Blood 10: 1030, 1955. Used by
permission from the publishers.)

plasmacytosis in chronic polyarthritis[87] and cirrhosis of
the liver,[17] etc. However, it does not indicate specific
inflammation. The presence of specific giant cells of
Reed-Sternberg type is sufficient for the diagnosis of
Hodgkin's disease, although some exceptional cases with
giant cells resembling Reed-Sternberg cells were pointed
out.[149] A diagnosis of malignant lymphoma can be
made if 80% or more of the cells are abnormal. It is
important to know the differential cell counts or
cytogram from normal lymph node smears (Table 17-1).
The following descriptions of morphology of the cells
are based on Giemsa stain, which is indispensable for
studies of lymph node smears. Then the nuclear size
measures larger in Giemsa stain than in Papanicolaou
stain.

■ Lymphocytes

There is a moderate size variation of cells (6-16 μm)
according to their maturation. The nucleus is round or
oval with some indentation of the nuclear rim. The chro-

Table 17-2 Differential counts from benign lymphadenopathy

	Reactive hyperplasia 15 cases	Toxoplasmosis 4 cases	Necrotizing lymphadenitis 4 cases	Infectious mononucleosis 2 cases
Small lymphocytes	82.5 (97.0-63.4)	87.0 (93.4-81.4)	79.4 (91.6-68.8)	73.4 (90.2-56.6)
Medium-sized lymphocytes*	10.5 (22.6-2.6)	5.3 (6.2-3.8)	9.2 (17.4-2.7)	12.5 (16.5-8.4)
Lymphoblasts	0	0	0.7 (1.6-0)	7.5 (14.9-0)
Large lymphoid cells**	5.2 (15.4-0)	4.6 (9.6-0.4)	5.0 (7.2-2.4)	5.0 (8.9-1.0)
Immunoblasts	0	0.4 (1.0-0)	0.6 (2.2-0)	0.7 (1.4-0)
Plasma cells	0.1 (0.8-0)	0.3 (0.4-0.2)	0.4 (0.8-0)	0
Plasmacytoid cells	0.2 (2.2-0)	0.8 (2.4-0)	0.5 (1.5-0)	0.1 (0.2-0)
Neutrophils	0.1 (1.0-0)	0.1 (0.4-0)	0.2 (0.4-0)	0.4 (0.6-0.2)
Eosinophils	0.2 (1.4-0)	0.1 (0.2-0)	0	0
Monocytes/ histiocytes	0.4 (1.0-0)	0.9 (1.6-0)	3.2 (6.4-0.8)	0.3 (0.3-0.2)
Others	0.8 (3.6-0)	0.5 (1.6-0)	0.9 (2.2-0)	0.3 (0.6-0)

*synonym: prolymphocytes
**synonym: large cleaved, large noncleaved cells
Differential counts: percent per 500 cells by May-Giemsa stain
(From S. Kurita in Malignant Lymphoma Cytology Atlas, Nagoya: Nagoya Univ Press, 1994.)

matin shows thick strands. Nucleoli are absent. The cytoplasm is scant and stains sky blue. Occasional azurophilic granules are seen.

Normal lymphocytes are classified into two distinct population, thymus-derived (T) and bursa-equivalent element or bone marrow-derived (B) cells by different cell surface markers (see Table 17-3 and Fig. 17-4). The B cells are identified by the presence of immunoglobulins on cell surface and receptor for the Fc portion of aggregated immunoglobulin. The T cells are characterized by the ability to form rosettes with sheep erythrocytes.

■Prolymphocytes (Medium-sized lymphocytes, Small cleaved cells)

The term of prolymphocyte is used for lymphocytic cells in the transitional stage of maturation from lymphoblasts to mature lymphocytes. Activated or irritated lymphocytes in a retrograde maturation belong to this type morphologically. Small cleaved cells (Lukes) and centrocytes (Lennert) of germinal center orgin are also included. Prolymphocytes are larger than lymphocytes (10-18 μm). The nucleus is round or oval and slightly indented. The chromatin pattern is slightly coarse. No definite nucleolus is usually seen, but there may be a single nucleolus as called "shadow nucleolus."[104] The cytoplasm is slightly more abundant than in mature lymphocytes, basophilic and may contain azurophilic granules.

Feature in Papanicolaou stain: In Papanicolaou stain, the nuclei of small lymphocytes are pyknotic with dense chromatin bands. These nuclei appear to be denuded or framed with scanty cytoplasm. Large lymphocytes and prolymphocytes reveal larger nuclei with somewhat stippled chromatins. Nucleoli are invisible by Papanicolaou stain. A small amount of cytoplasm is slightly stained with Light Green.

■Lymphoblasts

Synonyms; lymphogonia (Amano), large lymphoid reticulum cell (Rohr, Moeschlin), immunoblasts (Lukes and Tindal) and germinoblasts (Lennert). The size variation ranges from 20 to 30 μm. The large centrally locating nucleus has fine stipplings or thin strands (reticular network) of chromatin. The nuclear rim is thin but definite. Usually no indentation is seen. One or two pale nucleoli are present. The cytoplasm is sparse and basophilic (deep blue) with occasionally a lighter perinuclear halo. Immunoblasts of B cell origin is more intensely basophilic than those of T cell origin.

Papanicolaou stain shows a large, round nucleus, in which one or two prominent, centrally or paracentrally locating, cyanophilic nucleoli are present. A perinuclear halo is often recognizable. The term "immunoblast" has been employed recently to the immature-looking cells that resemble blastoid transformed cells with a meaning concerned with hypersensitive immune reaction. In WHO International Histological Classification of Tumours, the immunoblast is designated as a large lymphoid cell with pyroninophilic, very basophilic cytoplasm.

■Large lymphoid cells

Large lymphoid cells measure 15 to 25 μm and have polyhedral, cleaved nuclei with one or two nucleoli which are closed to nuclear rims. These cells locate in germinal centers and are called large cleaved cells (Lukes) or centroblasts (Lennert). The cytoplasm is small to moderate in amount and slightly basophilic.

■Reticulum cells

Since reticulum cells are originally fixed histiocytic cells which are present as a meshwork of parenchyma, they occupy only a small proportion of the normal cellular components in lymph node smears. Gall[46] simplified the definition and classification of reticulum cells and included several synonyms and subtypes as a whole; some subtypes such as histiocytes, basophilic stem cells

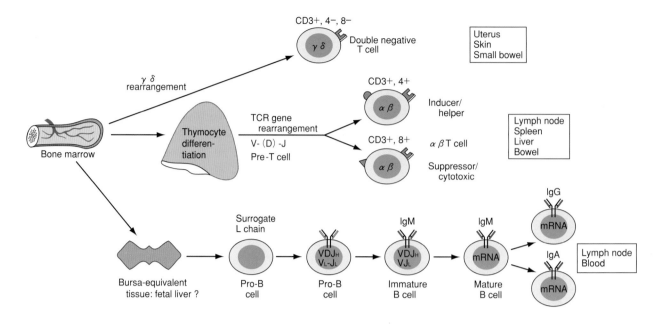

Fig.17-4 Hypothetical diagram of lymphocyte differentiation.

and germinoblasts have been traditionally classified. Distinct identification of reticulum cells may be made only for those cells having an oval or slightly lobulated nucleus with finely reticular chromatin and abundant cytoplasm with frequent vacuolation. Phagocytosis is occasionally discernible. One or two small colorless or basophilic nucleoli are present.

1) Follicular dendritic cells
Follicular dendritic cells form a meshwork and play a role in transportation of messages concerned with B cell activity.[67] These cells are large polyhedral cells with cytoplasmic processes. The nuclei are large, round and show a lacy chromatin pattern.

2) Interdigitating cells
Interdigitating cells are distributed in the paracortical area and demonstrate histiocyte-like features. Nuclear chromatins are lacy in appearance. The cytoplasm is of fair amount and stains lightly. Immunocytochemically these are positive for S100 protein and negative for lysozyme.[166]

3) Endothelial cells or histiocytes
The eccentric nucleus varies in shape; oval, elliptic or bean-shaped. The cytoplasm is of a fair amount, grey or pale blue and very often vacuolated. This type of reticuloendothelial cell occurs in predominance as represented as sinus histiocytosis. Phagocytosis is frequently seen; the function can be examined by the India ink method as described in the chapter of effusion (see Table 14-2 on page 286).

■Plasma cells
The nucleus is round or oval and eccentrically located. The chromatin is coarsely clumped. Nucleoli are invisible in normal plasma cells. The cytoplasm stains brilliant blue with light perinuclear zone.

4 T Cell and B Cell Population of Lymphocytes

The bursa fabricius is a mass of lymphoid tissue in birds and immunologically equivalent to bone marrows and gastrointestinal tracts of mammals. The B cells are the origin of plasma cells which produce specific immunoglobulins against the antigens. The B cells comprise 20 to 30% of circulating lymphocytes and have a short life span. A large number of microvilli are present in the B cells, while the surface of the T cells is smooth and deficient in microvilli.[130,132] The thymus is an organ in which cell mediated lymphocytes are induced; the undifferentiated stem cells (prothymic stem cells) derived from bone marrows differentiate to the T cells in the thymus; Prethymic thymocytes may transform through phenotypic differentiation (C_3, C_4 and C_8) and T cell receptor rearrangement to posthymic cells that migrate into the periarterial sheaths of the white pulp of the spleen, paracortical zones of lymph nodes, etc. (Fig. 17-4). The T cells have a long life span and comprise about 60% of circulating lymphocytes. Functionally the T cells act as a helper as well as suppressor to the B cells in addition to taking part in a cellular immunity. In addition to subpopulation of two lymphocytic classes, current studies[19] on surfacemarkers of lymphoma cells have found that lymphoma cells respond simultaneously to multiple markers. Distribution of these cells with double and triple markers in lymphoma is mentioned to be distinguishable from benign lymphadenopathy.

■Identification of human B cells
(Table 17-3 and Fig. 17-5)
B cells can be identified by the presence of immunoglobulin determinants particularly IgM and receptors for the Fc fraction of aggregated immunoglobulin G[9,38] and for a modified complement component (C_3 receptor).[11,85]

Table 17-3 Different characteristics between T cells and B cells

	T cells	B cells
Origin	Thymus	Bone marrow
Life	Long span	Short span
Function		
Antibody production	Helper and suppressor to B cells	Antibody (immunoglobulin) producer
Cellular immunity	Delayed type	−
Lymphokine production	+	−
Surface immunoglobulin*	−	+
Antigen θ [99]	+	−
Ly (A, B, C)	+	−
Receptor C$_3$ [11]	−	+
Fc	−	+
SRBC [7, 16, 84]	+	−
Blast formation		
LPS	−	+
PHA	+	−
ConA	+	−
PWM	+	+
Electron microscopy	Scant microvilli	Well developed microvilli
Cytochemistry		
Acid α-naphthyl acetate esterase [82]	High activity	Low activity
β-glucuronidase [120, 151]	High activity	Low activity
Acid phosphatase [25, 151]	High activity	Low activity
ATPase [78]	Low activity	High activity
Terminal deoxynucleotidyl transferase [12]	High activity	Low activity

* Surface immunoglobulin that characterizes B lymphocytes is mainly 7S IgM. [102]

The surface immunoglobulins, IgM and/or IgD, are detectable with immunofluorescence antibody technique using FITC-conjugated goat antiserum. IgM is the first immunoglobulin that appears on the surface of B lymphocytes in immune response (Figs. 17-7 and 17-8).

EAC rosette formation

A certain percentage of lymphocytes contain surface receptors for complement. This is a B cell marker easily demonstrated by an erythrocyte-amboceptor-complement (EAC) technique.

1. Isolation of lymphocytes
1) Take blood 3.0 ml with heparinized sterile syringe, add 6.0 ml physiologic saline and then mix gently.
2) Pour 3.0 ml Conray 400-Ficoll solution (Pharmacia, Sweden) beneath the above described blood.
3) Centrifuge at 400 ×g (1,550 rpm) for 30 minutes.
4) Replace an upper lymphocyte layer into another centrifuge tube to add an equal volume of physiologic saline, and then mix.
5) Centrifuge at 40 ×g (1,550 rpm) for 5 minutes.
6) Add barbital-buffer and centrifuge at 400 ×g for 5 minutes twice, then pure lymphocytes 95 ± 2.65%, 7-8 × 10$^{6-7}$/ ml will be obtained. The use of sodium metrizolate Ficoll is rapid and reliable technique to separate lymphocytes. [14]

2. Rosette formation
1) Incubate equal volumes of a 5% suspension of washed sheep erythrocytes and a 1:1,000 dilution of a 2-mercaptoethanol sensitive rabbit antiserum against sheep erythrocytes at 37 ℃ for 30 minutes (19 S fraction from a rabbit antiserum is also available by column chromatography).

2) Wash the above sensitized cells (EA) and prepare a 5% EA suspension in Veronal-buffered saline (pH 7.4).
3) Incubate equal volumes of a 5% EA suspension and a 1:10 dilution of mouse complement at 37 ℃ for 30 minutes (EAC). The complement should be supplied from a mouse without C5. Wash the EAC suspension 3 times and resuspend to a concentration of 0.5% in PBS.
4) Incubate equal volumes of isolated lymphocytes (0.1 ml) and 0.5% EAC suspension, and then leave for one hour at a room temperature. Count lymphocytes showing a rosette formation with more than four red cells as positive.

■Identification of human T cells
(Table 17-3 and Fig. 17-6)
The best means to distinguish long-lived T cells from B cells in peripheral blood is to use surface markers or receptors. For human T cell identification the common and easy techniques are detection of a surface receptor to sheep erythrocytes [56] and response to a phytohemagglutinin or concanavalin A (Fig. 17-10).

E rosette formation (Fig. 17-9)
1. Isolation of lymphocytes is the same as described above.
2. Rosette formation
1) Wash sheep erythrocytes which have been preserved in Alsever solution with PBS (pH 7.4) three times.
2) Make 0.5% sheep erythrocyte fetal calf serum. Practically, place 0.1 ml sheep erythrocytes into 2.0 ml Eagle medium containing fetal calf serum.
3) Mix lymphocytes 0.1 ml and 0.5% sheep erythrocytes 0.1 ml, and then incubate at 37 ℃ for 15 minutes.

Differen-tiation	Pro-B cell	Pre-pre B cell	Pre-B cell	Immature B cell	Mature B cell	Active B cell	Plasma cell
TdT	TdT						
Phenotype	anti HLA-DR						
		CD19					
			CD10			CD10	
			CD20				
				CD22			
				CD21			
Immuno-globulin			cμ			cIgM, G, A	
				sIgM, sIgM/D, sIgM/G/A			
Gene re-arrangement		IgH chain gene rearrangement					
			IgL chain gene rearrangement				
Leukemia Lymphoma	AUL		cALL	B-ALL	B-CLL		
						B-lymphoma	Myeloma

17-5

Differen-tiation	Pre-thy-mocyte	Early T-cyte*	Common T-cyte	Mature T-cyte	T cell	Active T cell
TdT	TdT					
Phenotype	CD7					
		CD5				
		CD2				
			CD1			
			CD4/8	CD4 / CD8		
				CD3		
Gene re-arrangement		TCR beta chain gene rearrangement				
			TCR alpha chain gene rearrangement				
Leukemia Lymphoma	T-ALL		Lymphoblastic T-lymphoma		T-CLL T cell	ATL lymphoma

17-6

Fig. 17-5 B cell differentiation and B cell neoplasia.
TdT: terminal deoxynucleotidyl transferase, AUL: acute undifferentiated leukemia, cALL: common acute lymphoblastic leukemia, CD: cluster determinant.

Fig. 17-6 T cell differentiation and T cell neoplasia.
*T-cyte: Thymocyte

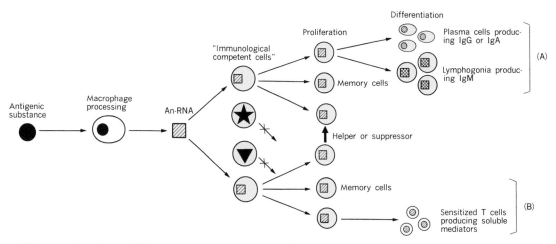

Fig. 17-7 Diagram showing cellular components concerned with immune response.
A: Gut-dependent immunocytes [121,145] responsible for humoral immunity (production of immunoglobulin and antibodies).
B: Thymus-dependent immunocytes [31,164] responsible for cellular immunity (delayed hypersensitivity, graft rejection, resistance to neoplasm and viral infection, etc.).

4) Centrifuge at 1,000 rpm for 10 minutes and place the sediment in iced bath at 0 ℃ for one hour.
5) Take lymphocytes with a capillary and count lymphocytes showing a rosette formation with more than four red cells as positive.

■**Cytochemical property as a T cell marker**
The morphological difference of the cell surface between T and B lymphocytes is still under debate. Although it has been proved that thymocytes have smoother surface than peripheral circulating lymphocytes irrespective of preparative techniques,[140] different features of the surface do not represent their own properties either of T or of B cells but are related to the functional state and environmental response. For example, low temperature state as produced in E rosette formation gives rise to retraction of microvilli, whereas cell to cell contact causes redistribution of microvilli.[142] As a T cell marker, cyto-

chemical alpha-naphthyl acetate esterase activity is mentioned to be useful for separation from B cells[64,136]; a few dots are observed as reaction products in the cytoplasm of T cells. Histiocytes or macrophages which are intensely and diffusely positive for this enzymatic activity can be distinguished from T cells from a morphological standpoint. Chloroacetate esterase by the method of Yam et al is mentioned to be positive as fine granular cytoplasmic stain in T cell lymphomas.[37,168] Although the specificity of enzymatic reactivity is still open to discussion, strongly positive single-spot acid phosphatase and beta-glucuronidase are characteristically activated in T cells and T lymphoma cells (Figs. 17-11 and 17-12). In B cells, on the other hand, a strong membrane-bound ATPase activity can be recognized.

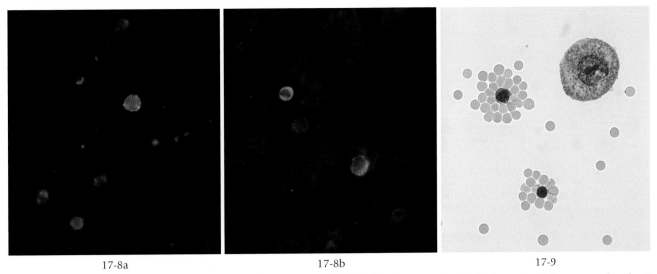

17-8a   17-8b   17-9

Fig. 17-8 B lymphocytes demonstrating surface immunoglobulin by fluorescein-labeled goat anti-human polyvalent immunoglobulin antisera. See distribution of fluorescence as ring or patch along the cellular border. The patch fluorescence means aggregation of the complex to be coalescent as a cap. Peripheral lymphocytes are those separated by Ficoll-Hypaque Gradient method.[19]

Fig. 17-9 T lymphocytes showing spontaneous E rosette formation. Lymphocytes are from ascites containing cancer cells derived from gastric carcinoma.

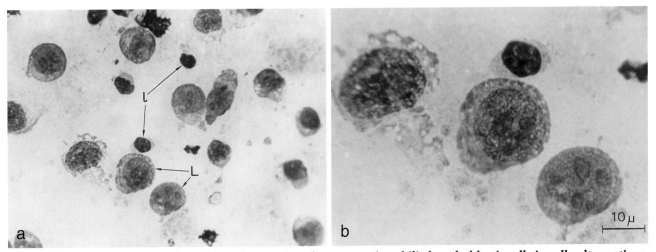

Fig. 17-10 Transformation of small lymphocytes into large pyroninophilic lymphoblastic cells in cell culture stimulated by phytohemagglutinin.
a. Low power magnification of a smear. l: small lymphocytes, L: large pyroninophilic cells.
b. Higher magnification of lymphoblastic cells. Note enlarged hyperchromatic nuclei, multiple nucleoli and deeply basophilic cytoplasm.

■Clinical application

At present, leukemia and malignant lymphoma arising from lymphoid cells are classified into B cell and T cell varieties. Since a dissociation of surface marker properties occurs occasionally, laboratory examinations on multiple surface markers will be necessary. For example, B cell chronic lymphatic leukemia with EAC rosette formation may lack surface immunoglobulins. Even in the case showing no surface markers it should not be simply regarded as a null cell variety,[10] because the neoplastic cell may lose their own properties during dedifferentiation and may not derive from lymphocytes. The significance of immunological classification of the lymphoma cell type consists in the difference of clinico-pathological manifestation and refractoriness to aggressive chemotherapy.[37,158]

5 Blastic Transformation of Lymphocytes by Mitogens

In recent years, the concept that lymphocytes have potentialities to differentiate to other cell types and to relate to immunological reactions has been sustained by blood cultures, in which lymphocytes underwent lymphoblastic transformation during first three days by addition of phytohemagglutinin (PHA), an extract of the bean (*Phaseolus vulgaris*).[6,22,60,113] It was considered to be the result of antigenic stimulation of the lymphocytes by phytohemagglutinin (Fig. 17-10).[6,39] From the same immune reaction, tuberculin also induces blastic transformation of lymphocytes which have ever been sensitized by BCG inoculation or tuberculous infection.[33,126]

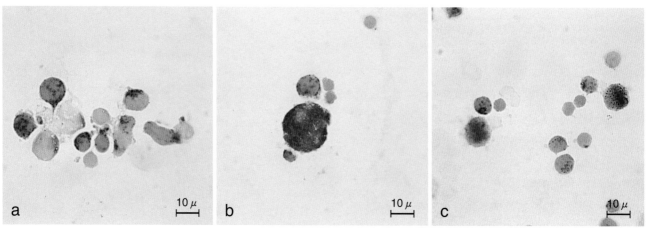

Fig. 17-11 Cytochemistry of lymphocytes in PHA stimulation.
a, b. Acid phosphatase activity shown in enlarged blastoid cells. Small PHA-nonresponsive lymphocytes are negative. A large macrophage or histiocyte is diffusely positive.
c. Beta-glucuronidase activity shown in blastoid cells. Small nonresponsive lymphocytes are negative.

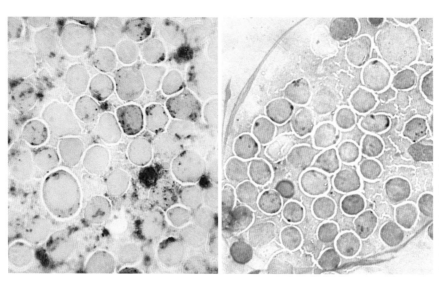

Fig. 17-12 Acid phosphatase- and beta glucuronidase-positive T cells in imprints of a lymph node. From a case with immunoblastic lymphadenopathy of rare T cell origin; E rosette-forming cells account for 66% while surface immunoglobulin-positive B cells are only 5% in the study for surface markers of abnormal cells from the lymph node.

MacKinney[100] observed the appearance of PHA cells with basophilic cytoplasm and prominent nucleoli in healthy persons as many as 85% of lymphocytes.

Cytologically, the PHA cells possess fairly broad, basophilic cytoplasm with irregular cellular borders and small vacuoles. The nuclei are round or oval with delicate strands of chromatin and prominent nucleoli. The content of ribosomal particles in such lymphoblasts increases electron microscopically,[28] and numerous electron-dense granules appear, which are presumably lysosomal in nature.[123]

■Clinical application

Some specific systemic diseases involving lymphoid tissues such as Hodgkin's disease, chronic lymphatic leukemia and sarcoidosis with impairment of delayed hypersensitivity do not respond well to phytohemagglutinin[5,10,29,58,61,83,115,121,135]: Whilst the function of immunoglobulin producing system is usually intact, some defect of thymus-dependent lymphocytic function occurs.[36] It may be possible that peripheral blood with mixed lines of lymphocytes is manifested by overpro-

duction of the PHA nonresponsive lymphocytes in lympho-proliferative disorders.

■Procedure

Cultures of peripheral lymphocytes are performed according to the method of Moorhead usually used for the chromosome analysis of leukocytes. After 72-hour-incubation without colchicinization, smears of cell deposit obtained by light centrifugation at 500 rpm for 5 minutes are stained by Giemsa method. The proportion of lymphocytes having the property of "blast" transformation determined with percentage by 500 cell differential counts. In cytology morphological account of blastoid response seems to be preferable to isotopic quantification of DNA synthesis.

Measurement of DNA synthesis is accomplished by pulse-labeling the cultures with tritiated thymidine (^{3}H-Tdr), a nucleoside precursor which will be incorporated into synthesized DNA. Autoradiography and grain counts or by scintillation counting in a liquid scintillation spectrophotometer will be applied in laboratories.

Table 17-4 Traditional classification of malignant lymphoma

Jackson and Parker (1947)[72]	Rappaport (1966)[137]	Nomenclature Committee Lukes et al (1966)[96]
	Nodular (Follicular) → Diffuse	
Reticulum cell sarcoma	Undifferentiated (Stem cell lymphoma)	
Lymphocytoma	Histiocytic (Reticulum cell sarcoma)	
Lymphoblastoma	Mixed cell (Histiocytic-lymphocytic)	
Lymphosarcoma	Lymphocytic, poorly differentiated (Lymphoblastic lymphosarcoma)	
Giant follicle lymphoma	Lymphocytic, well differentiated (Lymphocytic lymphosarcoma)	
Hodgkin's disease:	Hodgkin's disease:	Hodgkin's disease:
Paragranuloma	Hodgkin's paragranuloma	Lymphocyte predominance
Granuloma	Hodgkin's granuloma	Nodular psclerosis Mixed cellularity
Sarcoma	Hodgkin's sarcoma	Lymphocyte depletion

6 Malignant Lymphoma

General concept on classification

Malignant lymphoma has been traditionally divided into three main groups; lymphosarcoma, reticulum cell sarcoma and Hodgkin's disease (Table 17-4). The so-called follicular lymphoma is not a specific subtype but signifies an architectural pattern of neoplastic cell infiltration or the stage of the neoplasm.[138] The subtypes of these neoplasms are described later.

Recent immunological studies on non-Hodgkin's lymphoma have made the classification of T- and B-types. The major parts of lymphosarcoma are of B cell origin arising from the secondary follicles which are regarded as the sites of immunoblastic activity. T cell lymphoma with a diffuse pattern takes an aggressive course frequently involving the mediastinum, the central nervous system, the skin, etc. T lymphoma cells possess medium-sized nuclei with considerable size variation and convoluted contours. The characteristic nuclear feature is called "*baseball catcher's mit appearance.*"[51] B cell lymphoma takes a follicular or nodular pattern in an early stage. Although the recent concept[90,95,139] includes plasmacytoid, small cleaved (germinocytic) or large cleaved (germinoblastic), non-cleaved (germinoblastic) and immunoblastic types cytologically, light microscopic observation does not give definite classifications as yet, unless determination of surface (phenotypic) markers is introduced. Immunocytochemically T cell and B cell lymphomas can be distinguished by means of specific antisera conjugated with a fluorochrome (Figs. 17-14 and 17-15). Generally speaking, the nuclei of B cell lymphomas are small or medium-sized, round to oval, and cleaved with finger-like clefts. Nucleoli are small and indistinct in the small cleaved type, while multiple in the large cleaved or non-cleaved type. A group of hyperplasia of immunoblasts with a plasmacytoid pyroninophilic cytoplasm in association with proliferation of arborizing small vessels is called "*immmunoblastic lymphadenopathy*" mainly of B cell variety. The presence of PAS-positive amorphous material[77] as first described by Lukes and Tindle is considered to be an inconstant finding.[97]

Classification of lymphomas based on immunological cell typing by Lukes and Collins[95]

The Lukes-Collins classification is still now worthy for cytomorphological interpretation in which the concept of B cell and T cell categories was introduced (Figs. 17-16 and 17-17). More recent classifications such as International Working Formulation[111] and Kiel[88] indicating clinical behavior (malignancy grading) can be useful in further examinations based on histological as well as cytological criteria (Tables 17-5 to 17-7).

Key points in cytologic interpretation

Cytologic assessment for lymphoma and allied diseases by fine needle aspiration and imprint preparation must be based on population of cells (differential cell counts) and general cell features whether the cells are small, intermediate or large in size, whether the cells are cleaved, convoluted or smooth in shape, and whether the nuclei possess prominent nucleoli or not. Predominant cell population unusual in normal cell population or mixture of abnormal cells (Table 17-8) such as eosinophils, epithelioid cells and necrotic cell debris may point out a diagnostic key.

T cell lymphoma: It is composed of small lymphocytes having round nuclei and larger lymphocytes with irregular convoluted nuclei; deep subdivisions of the nuclear rim or even lobation are distinctive in electron micrograph (Fig. 17-18). T cell lymphomas are much less than B cell lymphomas, comprising about 10% to 15% of non-Hodgkin's lymphomas; a group lymphomas developing cutaneous manifestations such as Sézary syndrome and mycosis fungoides are of T cell origin. Adult T cell leukemia induced by retrovirus HTLV-1 infection

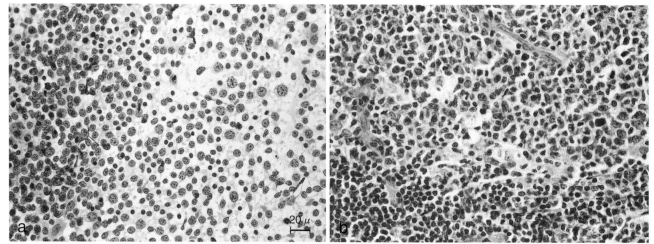

Fig. 17-13 Histology and cytology of follicular lymphoma.
a. If good imprint smears are taken, the architecture of lymph node can be observed (see Fig. 17-23).
b. Histological pattern of follicular or nodular pattern of follicle centers derived from lymphosarcoma is reflected in the smear showing a mosaic that is made of loosely textured lymphosarcoma cells and densely packed lymphocytes.

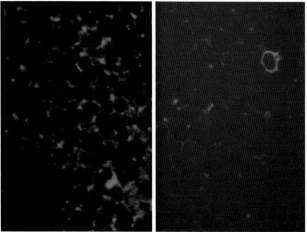

Fig. 17-14 Imprint smear of T cell lymphoma diffusely fluorescent by anti-T cell FITC-labeled antiserum. The smear is negative for anti-B cell antiserum. (Figs. 17-14 to 17-20 are presented by Prof. K. Kikuchi, Department of Pathology, Sapporo Medical College, Sapporo, Japan.)

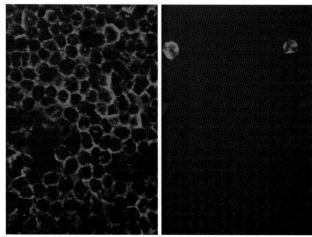

Fig. 17-15 Imprint smear of B cell lymphoma intensely positive for anti-B cell FITC-labeled antiserum. Note negative fluorescence for anti-T cell antiserum except for two cells.

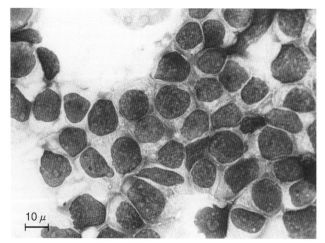

Fig. 17-16 T cell lymphoma in imprints of a lymph node. The smear is entirely replaced by immature lymphoid cells with angulated nuclei. Two or three pale nucleoli are present. The T cell variety was confirmed by anti-T cell antiserum.

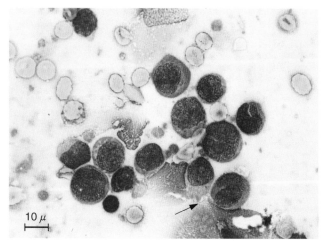

Fig. 17-17 B cell lymphoma in imprints of a lymph node. Immature lymphoid cells possess rounded and occasionally cleaved (arrow) nuclei. The cytoplasm is basophilic. The B cell variety was confirmed by anti-B cell antiserum. Note mixed proliferation of cleaved and non-cleaved cells.

Table 17-5 Classification of non-Hodgkin's lymphoma for clinical usage by NCI International Working Formulation*

Low-grade

A. Malignant lymphoma
　　Small lymphocytic
　　　Consistent with CLL
　　　Plasmacytoid
B. Malignant lymphoma, follicular
　　Predominantly small cleaved cell
　　　Diffuse areas
　　　Sclerosis
C. Malignant lymphoma, follicular
　　Mixed, small cleaved and large cell
　　　Diffuse areas
　　　Sclerosis

Intermediate-grade

D. Malignant lymphoma, follicular
　　Predominantly large cell
　　　Diffuse areas
　　　Sclerosis
E. Malignant lymphoma, diffuse
　　Small cleaved cell
　　　Sclerosis
F. Malignant lymphoma, diffuse
　　Mixed, small and large cell
　　　Sclerosis
　　　Epithelioid cell component

G. Malignant lymphoma, diffuse
　　Large cell
　　　Cleaved cell
　　　Noncleaved cell
　　　Sclerosis

High-grade

H. Malignant lymphoma
　　Large cell immunoblastic
　　　Plasmacytoid
　　　Clear cell
　　　Polymorphous
　　　Epithelioid cell component
I. Malignant lymphoma
　　Lymphoblastic
　　　Convoluted cell
　　　Nonconvoluted cell
J. Malignant lymphoma
　　Small noncleaved cell
　　　Burkitt's
　　　Follicular areas
K. Miscellaneous
　　Composite
　　Mycosis fungoides
　　Histiocytic
　　Extramedullary plasmacytoma
　　Unclassifiable
　　Other

*The Non-Hodgkin's Lymphoma Pathologic Classification Project: National Cancer Institute sponsored study of classifications of non-Hodgkin's lymphoma. Summary and description of a Working Formulation for Clinical Usage. Cancer 49: 2112-2135, 1982.

Table 17-6 Revised European American Lymphoma (REAL) Classification

B cell neoplasms	T cell and natural killer cell neoplasms
Precursor B cell nesplasm	*Precursor T cell neoplasm*
Precursor B lymphoblastic leukemia/lymphoma	Precursor T lymphoblastic leukemia/lymphoma
Peripheral B cell neoplasms	*Peripheral T cell and NK cell neoplasms*
B cell chronic lymphocytic leukemia/prolymphocytic leukemia/small lymphocytic lymphoma	T cell chronic lymphocytic leukemia/ prolymphocytic leukemia
Immunocytoma/lymphoplasmacytic lymphoma	Large granular lymphocyte leukemia (LGL)
Mantle cell Lymphoma	Mycosis fungoides/Sézary syndrome
Follicular center lymphoma, follicular	Peripheral T cell lymphomas, unspecified
Marginal zone B cell lymphoma	Angioimmunoblastic T cell lymphoma (AILD)
Hairy cell leukemia	Angiocentric lymphoma
Plasmacytoma/plasma cell myeloma	Intestinal T cell lymphoma
Diffuse large B cell lymphoma	Adult T cell lymphoma/leukemia (ATL/L)
Burkitt's lymphoma	Anaplastic large cell lymphoma (ALCL)

(From Harris NL, Jaffe ES, Stein H, et al: A revised European-American classification of lymphoid neoplasms: A proposal from the International Lymphoma Study Group. Blood 84: 1361, 1994.)

Table 17-7 Updated Kiel classification of non-Hodgkin's lymphomas (1988) with a few additions

B cell	T cell
Low-grade malignant lymphomas	
Lymphocytic	Lymphocytic
Chronic lymphocytic leukaemia	Chronic lymphocytic leukaemia
Prolymphocytic leukaemia	Prolymphocytic leukaemia
Hairy-cell leukaemia	
	Small cell, cerebriform
	Mycosis fungoides, Sézary syndrome
Lymphoplasmacytic/-cytoid (immunocytoma)	Lymphoepithelioid (Lennert's lymphoma)
Plasmacytic	Angioimmunoblastic (AILD, LgX)
Centroblastic-centrocytic	T-zone lymphoma
follicular ± diffuse	
diffuse	
Centrocytic (mantle cell)	Pleomorphic, small cell (HTLV-1±)
Monocytoid, including marginal zone cell	
High-grade malignant lymphomas	
Centroblastic	Pleomorphic, medium-sized and large cell
	(HTLV-1±)
Immunoblastic	Immunoblastic (HTLV-1±)
Burkitt's lymphoma	
Large cell anaplastic (Ki-1+)	Large cell anaplastic (Ki-1+)
Lymphoblastic	Lymphoblastic
Rare types	*Rare types*

Table 17-8 Differential counts from lymphoma

B cell					*T cell*	
	Diffuse small lymphocytic 2 cases %	Follicular small cleaved 5 cases %	Follicular large cell 3 cases %	Diffuse large cell 18 cases %	Polymorphic medium-large 8 cases %	Anaplastic large cell 3 cases %
Small lymphocytes	36.3	10.5	14.1	6.6	7.8	76.1
Medium-sized lymphocytes	58.4	86.0	30.2	14.6	39.8	3.3
Lymphoblasts	0	0.1	0	0	0	0
Large lymphoid cells	1.4	2.6	51.9	75.8	38.7	15.1
Immunoblasts	0.5	0.1	1.2	0.2	0.3	0
Plasma cells	0	0	0.4	0.1	0.1	0
Plasmacytoid cells	0.2	0	0.3	0.3	9.1	0.5
Neutrophils	0.1	0.1	0.5	0.2	0.6	1.4
Eosinophils	0	0.2	0	0	0.1	0.5
Monocytes/histiocytes	0.2	0.1	1.3	1.1	1.0	2.5
Others	2.9	0.3	1.1	1.1	2.5	0.7

(Adapted from S. Kurita in Malignant Lymphoma Cytology Atlas, Nagoya: Nagoya Univ Press, 1994.)

occurs not infrequently in southwestern Japan. Angio-immunoblastic lymphadenopathy with dysproteinemia is also T cell derived.

B cell lymphoma: Variable cell types are included as follows:

1) Small lymphocytic lymphoma
2) Plasmacytoid lymphocytic lymphoma
3) Follicular center cell derived lymphoma
   a. Small cleaved
   b. Large cleaved
   c. Small non-cleaved
   d. Large non-cleaved
4) Immunoblastic sarcoma

Small lymphocyte type or chronic lymphocytic leukemia is composed of mainly small cells with round nuclei showing compact chromatin and indistinct nucleoli. There are some larger lymphocytes having single central nucleoli and finely reticular chromatin networks.

This type is of low grade malignancy and occurs in old age groups above 60 years. For differentiation from reactive hyperplasia, a monotonous cellular constituent and lack of macrophages are the key point.

Plasmacytoid lymphocyte type (immunocytoma) is similar to the small lymphocyte type except for intranuclear PAS-positive inclusions and plasmacytoid cytoplasm.

About 20% to 30% of the case are associated with monoclonal immuno-gammaglobulinemia. The case with monoclonal IgM globulinemia is referred to as primary macroglobulinemia.[54]

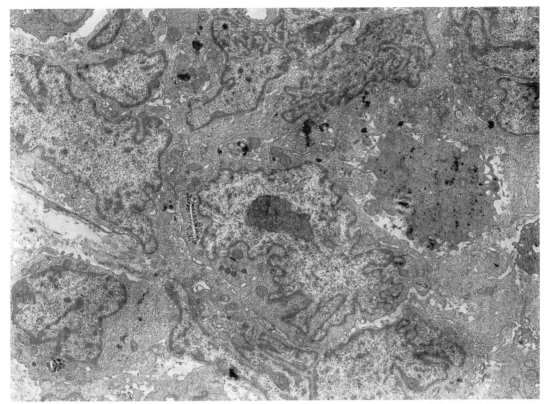

Fig. 17-18 Electron micrograph of T cell lymphoma showing complex convolution of the nuclei. Note poor development of rough endoplasmic reticulum. (original magnification ×6,000)

Table 17-9 WHO classification for lymphoid tumor*

A. Lymphosarcomas
  1. Nodular lymphosarcoma
  2. Diffuse lymphosarcoma
    (a) Lymphocytic
    (b) Lymphoplasmacytic
    (c) Prolymphocytic
    (d) Lymphoblastic
    (e) Immunoblastic
    (f) Burkitt's tumour
B. Mycosis fungoides
C. Plasmacytoma
D. Reticulosarcoma
E. Unclassified malignant lymphomas
F. Hodgkin's disease
  1. With lymphocyte predominance
  2. With nodular sclerosis
  3. With mixed cellularity
  4. With lymphocyte depletion
G. Others
  1. Eosinophilic granuloma
  2. Mastocytoma

*From International Histological Classification of Tumours, No. 14. Geneva, 1976.

Note: WHO Classification for Lymphoid Tumors[162]
After the 1976 WHO classification for lymphoid tumors, some modification was added to the conventional nomenclature in which the terms such as lymphosarcoma and reticulosarcoma or histiocytic lymphoma have been employed. Prolymphocytic and lymphoblastic diffuse lymphosarcomas are mentioned to be used according to the resemblance to hematological prolymphocytes and lymphoblasts. Lymphoplasmacytic lymphosarcoma is a variety of lymphosarcoma admixed with plasmacytoid cells in association with monoclonal gammopathy. Immunoblastic lymphosarcoma is a newly added variety containing large basophilic (pyroninophilic) cells with prominent nucleoli. Histiocytic lymphoma (Rappaport) is of common usage because of the resemblance to histiocytes; the large nuclei, either cleaved or uncleaved, reveal evenly dispersed chromatins and prominent nucleoli. The lack of distinctive cytochemical or immunochemical cell markers and the evidence producing reticulin fibers lead us to use the term reticulosarcoma or reticulum cell sarcoma.[101,162]

■Follicle center cell-derived follicular lymphoma (Fig. 17-13)

This type is characterized by nodular aggregates of lymphoma cells having phenotypic properties positive for CD19, CD20, CD22 and CD10, and negative for CD5.
Small cleaved cell type is characterized by the presence of cleavage in the nuclei. Nucleoli are small and indistinct. The chromatin is finely distributed (Fig. 17-19).

Large cleaved cell type is variable in size and configuration of the nuclei; mixed proliferation with small cleaved and non-cleaved cells is characteristic of this type.
Non-cleaved cell type is subdivided into small non-cleaved and large non-cleaved types respectively. The nuclei are rounded and smoothly bordered. The chromatin is finely distributed and two or three nucleoli are

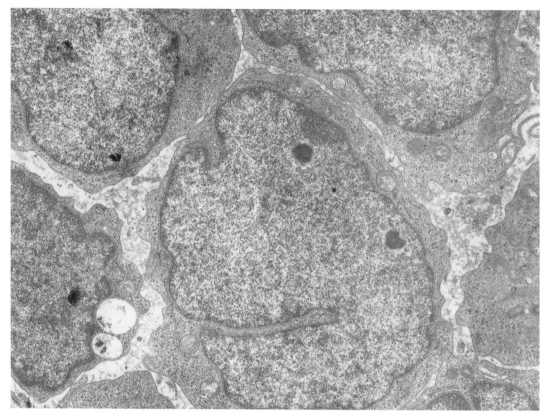

Fig. 17-19   Electron micrograph of B cell lymphoma showing a characteristic cleaved cell. Note a linear cleavage and increased polysomes in the cytoplasm. (original magnification ×16,600)

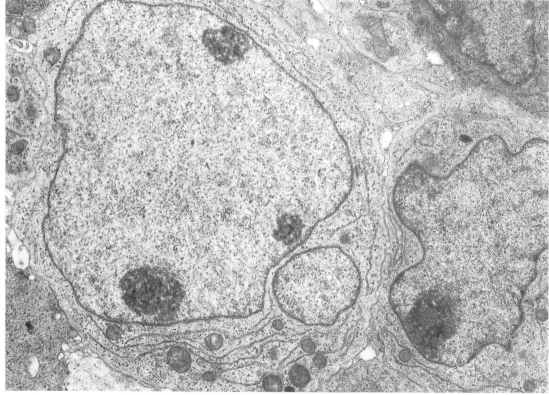

Fig. 17-20   Electron micrograph of B cell lymphoma of non-cleaved type. Note a large round nucleus with a smooth nuclear outline. Rough endoplasmic reticulum is well developed. (original magnification ×16,600)

388

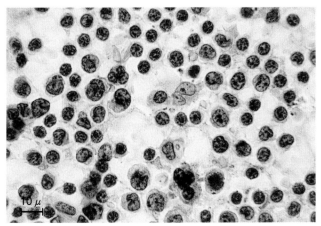

Fig. 17-26 Cytology of a variety of non-Hodgkin's lymphoma showing a mixed cellularity. Note an admixture of lymphocytic lymphoma cells with histiocytic lymphoma cells having indented nuclei and broad cytoplasm.

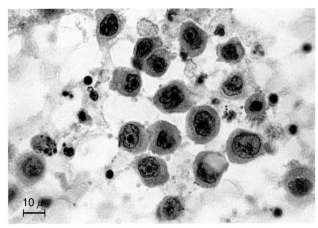

Fig. 17-27 Histiocytic lymphoma in imprints of a lymph node. The nuclei vary in shape with remarkable indentation. Note prominence of eosinophilic nucleoli. The cytoplasm is characteristically abundant.

are classified and show different proliferative activities (Ki-67 values). However, no cytomorphological distinction in needle aspirates can be made.[165)]

2) Well differentiated type (lymphocytic lymphosarcoma)

This type may include small cleaved or non-cleaved B cell lymphoma and chronic lymphocytic leukemia. Imprint or aspiration smears are characterized by a large number of fairly small, equally sized lymphocytoid cells, the nuclei of which are round or oval with delicate, smooth nuclear outline. One or two nucleoli are distinguishable (Fig. 17-23). The chromatin pattern is not so coarse as that of the poorly differentiated type, and rather compact. The cytoplasm is scant and poorly preserved. Denuded cells are frequent especially in Papanicolaou stain. Monomorphic population of immature lymphocytoid cells in imprints is favorable to lymphosarcoma (Fig. 17-24), whereas an admixture with mature lymphocytes and prolymphocytes is visible in chronic lymphocytic leukemia.

Immunophenotype: CD5 (+), CD19, 20, 79a (+), CD23 (+), CD10 (–).

3) Dysproteinemia associated with malignant lymphoma

In a broad sense of lymphoproliferative disorders including myelofibrosis, lymphocytic leukemia and non-Hodgkin's lymphoma, lymphoid cells are occasionally concerned with immunoglobulin-production as a result of either activation of immunoglobulin-producing cells or direct involvement of these cells; proliferation of lymphoid cells that is related to immunoglobulin-production will be responsible for monoclonal hyperimmunoglobulinemia. The author has experienced a case of extranodal lymphocytic lymphosarcoma affecting both ovaries measuring $20 \times 13 \times 11$ cm (left) and $10 \times 6 \times 3$ cm (right) in a 34-year-old female who showed a moderate increase in IgM of blood serum ranging from 1,500 mg to 1,800 mg/dl while IgA and IgG levels were rather depressed. Imprint smears of operated specimen showed the usual cytomorphology of monomorphous lymphocytic lymphoma except for the prominence of nucleoli (Fig. 17-25).

■ Histiocytic malignant lymphoma

This type, which has been traditionally called reticulum cell sarcoma, is characterized by predominance of atypical reticular cells showing variable maturation. The tumor cells in this type have large nuclei with polyhedral shape showing a slightly thicker nuclear rim and coarser and more vesicular chromatins than in stem cell lymphoma. Indentation of the nuclear rim or reniform configuration of the nucleus is characteristic. The presence of one or two distinct nucleoli with irregular shapes is also a distinguished finding. The cytoplasm is pale and abundant often with phagocytosis (Fig. 17-27). Small vacuoles are characteristically found in the cytoplasm.

In a recent reclassification of malignant lymphoma with use of immunochemical markers, the majority of histiocytic lymphomas fall into the B-cell category, mainly immunoblastic variety[18)](Fig. 17-26). Koh et al[81)] studied on 53 cases designated as histiocytic lymphoma with immunoperoxidase and immunofluorescence methods for histiocyte markers such as lysozyme and alpha-1-antichymotrypsin and proved only one lymphoma to be histiocytic. In order to determine histiocytic lymphoma, hereby, immunological and cytochemical approaches to the surface marker, cytoplasmic marker and functional property such as phagocytosis must be applied.

■ Hodgkin's disease (paragranuloma–granuloma–sarcoma[34,72)]) (lymphocytic predominance–nodular sclerosis–mixed cellularity–lymphocytic depletion[96)])

This type is composed of malignant histiocytic cells with specific Reed-Sternberg cells and some inflammatory cell components such as eosinophils and plasma cells. An unequivocal diagnosis of Hodgkin's disease is established by the presence of abnormal neoplastic reticulum cells, Reed-Sternberg cells, having multiple or multilobed nuclei, prominent acidophilic nucleoli with perinucleolar halo and amphophilic or slightly basophilic cytoplasm (Fig. 17-28). Out of the three or four subtypes, the mixed cellularity (granulomatous type) occupies more than 90% of Hodgkin's disease.[93)]

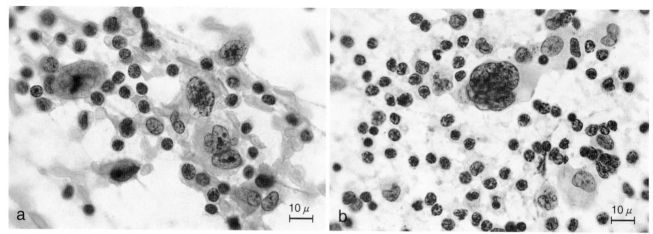

Fig. 17-28   Reed-Sternberg giant cells in imprints from Hodgkin's disease. Note multinucleated (a) or multilobulated (b) giant cells having single, centrally locating, acidophilic nucleoli. The characteristic nucleoli are occasionally basophilic in Papanicolaou stain.

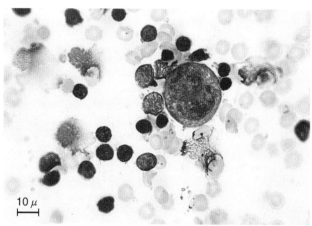

Fig. 17-29   Hodgkin's disease of lymphocytic predominance. A characteristic Reed-Sternberg cell is embedded in between many mature lymphocytes. Bimorphic cellularity containing some scattered Reed-Sternberg giant cells is a diagnostic feature of paragranuloma.

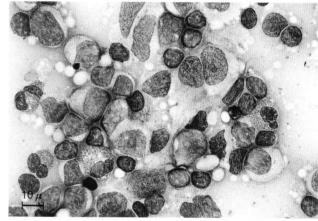

Fig. 17-30   Hodgkin's disease of mixed cellularity. Note a Reed-Sternberg giant cell with mirror image, eosinophils, histiocytic cells and lymphoid cells. This mixed cellularity reflects the histology of granuloma. Giemsa stain. (By courtesy of Dr. T. Suchi, Aichi Cancer Center, Nagoya, Japan.)

In the classification of malignant lymphoma by Rappaport (1966),[137] the directions of their possible transition and the concept of histological pattern either nodular or diffuse have been adopted. This histologic pattern and the extent of involvement of lymph nodes and organs have a prognostic significance. The cytological approach to the histology by differential cell counts of imprint smears (lymphcytogram) is available especially for the diffuse type of lymphoma. According to the report of the Nomenclature Committee on Hodgkin's disease, it is classified into four groups, (1) lymphocytic predominance (Fig. 17-29), (2) nodular sclerosis, (3) mixed cellularity (Fig. 17-30) and (4) lymphocytic depletion. Although the classification by Jackson and Parker is still used, newly added is a concept of nodular sclerosis which is histologically characterized by nodular segregation by collagenous bands and lacunar type of Reed-Sternberg cells.[96] From the aspect of clinical course, mixed cellularity is intermediate between lymphocytic predominance and lymphocytic depletion. Nodular sclerosis is a variety with favorable prognosis that tends to be arrested in the stage 2.[148]

■Histiocytic medullary reticulosis

This rare disease having several synonyms (malignant histiocytosis, malignant reticulosis, aleukemic reticulosis) and undergoing acute, progressive course with skin eruption (cutaneous form)(Fig. 17-31) and hepatosplenomegaly and lymphadenopathy (visceral form) is characterized by multifocal proliferation of reticuloendothelial cells involving the skin, lymph nodes, liver, spleen and bone marrow. A distinctive growth pattern to be distinguished from malignant lymphoma is preferentially proliferative in sites where histiocytes are physiologically present; peripheral and medullary sinuses of lymph nodes and sinusoids of the liver and spleen are the anatomical sites as such histiocytic proliferation. In spite of acute clinical manifestation cellular atypism is often mild in degree. The proliferated histiocytic cells phagocytose erythrocytes, pyknotic nuclear debris, hemosiderin, etc. (Fig. 17-31). Erythrophagocytosis and positive nonspecific esterase activity are especially of diagnostic value in cytology.[101] The nuclei are oval to elliptic with often lobulated or indented contours. The epithelioid cell-like feature and the admixture with various other types of cells are distinguishable from reticulum

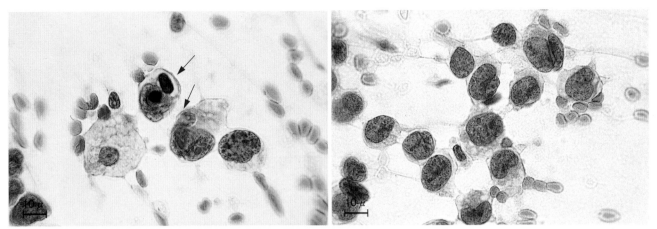

Fig. 17-31   Malignant histiocytosis (cutaneous form) in imprints of a cutaneous nodule. From a patient with acute downhill course by manifestation of lymphadenopathy, skin papules, splenomegaly and anemia. Highly cellular tumor consists of lobulated, histiocytic cells. The nuclei are hyperchromatic but delicately reticulated. The cytoplasm is of a moderate amount and ill-defined. Note phagocytotic ingestion of necrotic debris (arrows).

cell sarcoma or histiocytic lymphoma; it is therefore difficult to differentiate this entity from mixed cellularity of Hodgkin's disease when multinucleated giant cells resembling Reed-Sternberg cells are encountered. Hemorrhage and necrosis are also characteristic of histiocytic medullary reticulosis.

7 Extranodal Lymphoma

Lymphoma shows a wide-spread extension involving not only lymph nodes or lymphoid tissues but also various organs without lymphoid apparatus. When we encounter non-disseminated lymphomas, much attention should be paid to determine whether it is of primarily extranodal genesis or secondary involvement. As a site of primary extranodal lymphomas, the stomach, tonsils, small and large intestines, having lymphoid tissues, are more frequently involved than other extralymphatic organs (Fig. 17-32). There are often occurrences of malignant lymphoma in extranodal organs which were regarded as being primary simply because these were of the largest mass, while other lymphoreticular tissues were minimally involved.

Freeman et al[43] reported distribution of site-specific non-Hodgkin's lymphoma in the survey of a series of 1,467 Caucasian patients with non-disseminated extranodal lymphomas and mentioned different biological behavior being related to the site of origin; the survival rates of lymphomas arising from the lung, stomach and tonsils were better than those of all cancers in the same organs (Table 17-10).

Extranodal primary lymphoma of the liver, brain and bone marrow are associated with poor outcome.[55]

■MALT lymphoma

In recent years lymphoma originating from mucosa-associated lymphoid tissue (MALT) has been identified as distinct from other non-Hodgkin's lymphoma[68,70] (Fig. 17-32). As one of the most common extranodal lympho-proliferative disorders, gastric lymphoreticular hyperplasia most-likely related to Helicobacter pylori infection is referred to as MALT lymphom.[124,167] Chronic

antigenic stimulus by microorganism infection is the trigger of growth of lymphoma.[69]

Immunohistochemical and molecular biological properties of MALT lymphoma sustains the disease entity as a monoclonal B cell lymphoma of a low grade malignancy. In a multivariate analysis of gastrointestinal lymphomas, MALT lymphomas has been shown to be associated with a favorable outcome.[109]

MALT lymphoma of other organs such as the salivary gland,[66] thyroid,[65] thymus[171] and skin[144] has been reported.

Cytologically cell constituents vary from place to place of the tumor. Medium-sized centrocytic cells are mixed with clear cells, lympho-plasmacytoid cells, etc (Fig. 17-32).

8 Key points in the Cytodiagnosis of Lymph Node Smears

■Detection of abnormal cells specific to the disease

Even a small number of abnormal cells in lymphcytograms (lymphadenograms) or differential counts of lymph node smears may be often diagnostic.

1) Hodgkin's disease

The Reed-Sternberg cell is a giant cell measuring from 15 to 40 μm in diameter in wet preparations and from 30 to 100 μm in air-dried films. The nuclei are multinucleated, usually binucleated or trinucleated, and/or multilobulated. The characteristic feature of the Reed-Sternberg cell having binuclei face to face is called a *mirror image*. These Reed-Sternberg cells contain huge, acidophilic or amphophilic nucleoli with an "owl-eyed" appearance. The distinctive nucleoli are pyroninophilic because of rich RNA content. The chromatin is coarsely reticulated and provided with clear interchromatin spaces. The term Hodgkin's cell has been used in German literatures for the mononuclear type of giant cell with essentially the same nuclear features as those of Reed-Sternberg cells.[86,108,133] Although there are several reports that Reed-Sternberg-like cells are found in other conditions besides Hodgkin's disease such as infectious

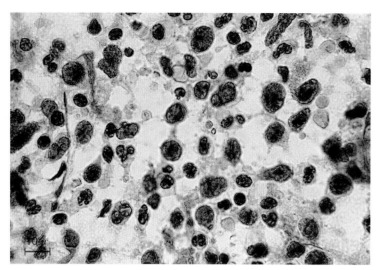

Fig. 17-32 Extranodal MALT lymphoma in imprints of gastric punch biopsy. The stomach is one of organs frequently affected by lymphoma. Primary lymphoma is rare but constitutes about 4.5% of all gastric neoplasms.[154] Most of tumor cells are denuded by degeneration but retain the feature of lymphoma; delicate nuclear rims and nuclear polymorphy with occasional indentation are seen. Note some large immunoblast-like cells with prominent nucleolus.

Table 17-10 Distributions of site-specific lymphomas with histologic types

Site	No. of cases	Lympho-sarcoma	Reticulum cell sarcoma	Other types	Male	Percent by age		
						<20	20-49	>49
Stomach	346	41%	48%	11%	57%	0%	17%	83%
Small intestine	110	45	40	15	62	11	16	73
Large intestine	82	37	46	17	50	9	16	75
Tonsil	162	41	42	17	52	2	12	86
Adenoid	37	41	35	24	54	5	35	60
Salivary gland	69	41	23	36	45	1	23	76
Skin	110	37	13	50	57	5	36	59
Connective tissue	122	32	46	22	54	5	28	67
Bone	69	4	93	3	52	16	23	61
Lung	53	64	15	21	55	0	19	81
Thyroid	36	19	56	25	6	0	8	92
Breast	33	52	24	24	6	3	24	73
Testis	23	22	61	17	100	4	4	91
Esophagus	3							
Liver, bile duct	6							
Pancreas	9							
Larynx	8							
Uterine cervix	3							
Uterine corpus	3							
Ovary	2							
Prostate	3							
Kidney	10							
Brain, spinal cord	23							
Ill-defined and other sites	145							
All cases	1,467	38%	42%	20%	53%	4%	21%	75%

(From Freeman C, et al: Cancer 29: 252, 1972.)

mononucleosis, mycosis fungoides, lymphomas and cancers,[98,149] detection of Reed-Sternberg cells is highly suggestive of Hodgkin's disease. In order to distinguish the type of Hodgkin's disease one must observe other cellular components in differential cell counts; either lymphocytic predominance or mixed cellularity can be easily suggested. As phenotypic markers the Reed-Sternberg cell and Hodgkin's cell are positive for CD30 and CD15, and often for CD25 and HLA-DR.

The other cellular components in Hodgkin's disease are different according to the subtypes of the Jackson and Parker classification. More than 90% are of mixed cellularity (granuloma type) which may disclose various kinds of cells (see Fig. 17-30); Reed-Sternberg cells are found mixed with atypical reticulum cells, eosinophils, neutrophils, plasma cells, etc. The sarcoma type shows a number of pleomorphic cells, Hodgkin's cells and Reed-Sternberg cells. The paragranuloma type is characterized by a few Reed-Sternberg giant cells isolated in a pool of lymphocytes (see Fig. 17-29).

2) Metastasis of cancer to lymph nodes
Metastatic lesions in lymph nodes are variable; early metastatic foci only histologically provable are present in the peripheral sinus, however, in advanced cases,

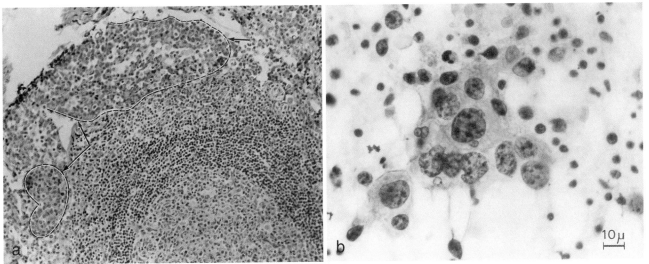

Fig. 17-33 Imprint smears from lymph node with metastatic carcinoma.
a. The subcapsular or peripheral sinus is filled with solid groups of cancer cells (arrows), which can be diagnosed by examining imprint smears.
b. Pleomorphic malignant cells with epithelial cluster surrounded by mature lymphocytes.

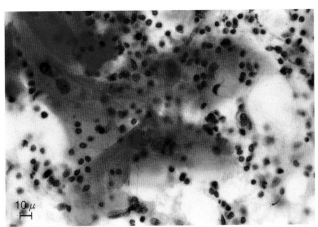

Fig. 17-34 Metastatic epidermoid carcinoma in imprints of a lymph node. Odd-shaped cells with intense orangeophilia by keratinization are found to be surrounded by lymphocytes.

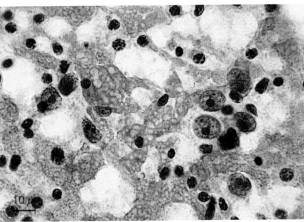

Fig. 17-35 Metastatic adenocarcinoma in imprints of a lymph node. Loosely clustered cancer cells with eosinophilic nucleolus are embedded being surrounded by mature lymphocytes and red cells.

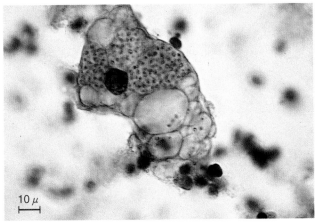

Fig. 17-36 Histoplasmosis in imprints of an enlarged lymph node showing engulfment of numerous yeast forms of *Histoplasma capsulatum* by macrophage. This progressive disseminated form showed generalized lymphadenopathy, hepatosplenomegaly in addition to pulmonary involvement.

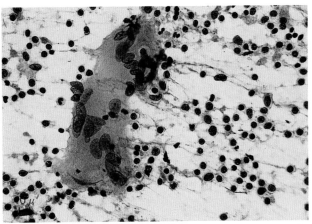

Fig. 17-37 Langhans giant cells in a needle aspirate of a lymph node. The Langhans giant cells having aggregated, elliptic nuclei with delicate nuclear membrane in the presence of amorphous debris, though invisible in this smear, are suggestive of tuberculosis. Tularemia, brucellosis should be excluded.

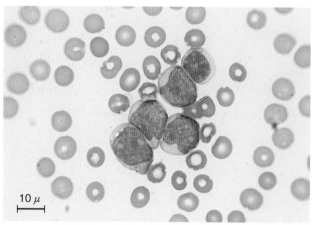

Fig. 17-38 Blood film of acute lymphatic leukemia. The nuclei are enlarged and evidently indented. Chromatin threads are more condensed than those of myelogenous leukemic cells.

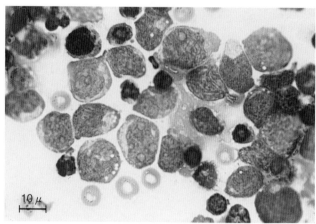

Fig. 17-39 Blood film of acute myelogenous leukemia that is occupied by myeloblasts. The nuclei show finely stranded chromatins and multiple pale nucleoli. Cytoplasmic vacuolation is seen.

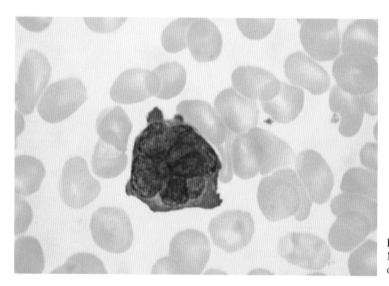

Fig. 17-40 Blood film of adult T cell leukemia. Note a characteristic convolution of nucleus like chicken foot-print.

Table 17-11 Immunological classification of malignant lymphoma and allied diseases

T cell line	Follicle center lymphoma (centroblastic/centrocytic)
ALL with mediastinal involvement	Mantle cell lymphoma (centrocytic)
Diffuse lymphoma (Peripheral T cell lymphoma unspecified)*	Diffuse large cell lymphoma (Immunoblastic lymphoma)
Sézary syndrome/Mycosis fungoides	MALT lymphoma
Angioimmunoblastic lymphoma (AILD)**	Burkitt's lymphoma
Adult T cell leukemia (ATL)	Hairy cell leukemia
Anaplastic large cell lymphoma (Ki+)	
	Null cell line
B cell line	ALL (a part of)
Small lymphocytic leukemia/Lymphoma/	Hodgkin's disease
Immunocytoma/Lymphoplasmacytic lymphoma***	Histiocytic lymphoma

  * These encompass pleomorphic, medium-sized, and large cell
  ** Acute generalized lymphadenopathy often associated with hypergammaglobulinemia and skin rash
  ***Waldenstrom's macroglobulinemia is accompanied in a part of cases

lymph nodes are almost entirely replaced by cancer cells. Even a small number of cancer cells intermingled with lymphocytes are diagnosable of metastasis, because cancer cells which are frequently encountered in clusters are distinguishable from lymphoid cells by their characteristic features such as larger nuclear size, distinctive nuclear membrane and glandular or epidermoid differentiation (Figs. 17-33 to 17-35).

3) Infectious lymphadenopathy

Although there is no difficulty in determining acute lymphadenitis from clinical manifestations, persistent enlargement of localized lymph nodes must be examined whether neoplastic or infectious. For example, melanin pigments in macrophages should be verified as to their origin; whereas melanin pigments are only found in fixed macrophages (sinus histiocytes) in dermatopathic lymphadenitis,[150] they are recognizable not only in melanoma cells but in macrophages in dissemination of melanoma. Specific organisms such as histoplasmosis (Fig. 17-36), cryptococcosis paracoccidioidomycosis and toxoplasmosis should be carefully searched in case of lymphadenopathy of unknown causes.

4) Epithelioid cell granuloma

Epithelioid cells having elliptical, pale, vesicular nuclei with finely reticular chromatin are suggestive of tuberculosis, sarcoidosis, tularemia, cat scratch fever and so on. Langhans giant cells possess exactly the same nuclear features as those of epithelioid cells (Fig. 17-37). Detection of homogeneous necrotic debris in addition to epithelioid cells may confirm the impression of tuberculosis. In the absence of caseous necrotic debris sarcoidosis and sarcoid reaction should be considered first of all.

■ Detection of abnormal patterns in differential cell counts (Lymphcytograms)

In the inflammatory process, the normal cellular pattern of lymph node smears will be displaced by a predominance of some series of cellular components, i.e., plasma cells, monocytes, polymorphonuclear leukocytes and eosinophils. However, it is not in excess of 80%. If the smear displays s monotonous pattern of abnormal lymphoid cells which are not present in normal smears, it is highly suggestive of the diffuse type of malignant lymphoma.

Distinction either of diffuse type or of nodular type can be postulated by low power observation only when careful impression of the cut surface preserves the structure of pathologic changes. Rubbing smear should be avoided.

9 Leukemia as an Allied Disease of Lymphoma

■ Acute lymphatic leukemia (ALL)

Acute lymphatic or lymphoblastic leukemia which is more common in children than adults belongs either to T cells or null cell variety,[147,156] while leukemic cells in chronic lymphatic leukemia may express monoclonal surface immunoglobulins, particularly IgM protein.[1,80] T cell ALL is encountered frequently in elder children and more commonly in boys than in girls.[118] Circulating leukemic cells are markedly increased in number (Fig. 17-38) and organ infiltration by leukemic cells in association with lymphadenopathy and hepatosplenomegaly is pronounced in T cell ALL; involvement of the mediastinum is also remarkable finding.[51,147]

Childhood lymphoma of T cell origin that is often associated with leukemic phase is arbitrarly separated from ALL according to the site of involvement,[163] i.e., whether bone marrow and peripheral blood involvement or lymphadenopathy and mediastinal tumor are of main manifestation. ALL of B cell origin is rare and comprises about 2% of all ALL and distinguished by the presence of surface immunoglobulin.[63]

■ Adult T cell leukemia-lymphoma (ATL)[59,129]

Adult T cell leukemia is of pathogenic importance because of the causative role of human T cell lymphotropic retrovirus (HTLV 1) infection. Geographical distribution of individuals positive for antibodies to HTLV 1 is congested in the southwest of Japan, Caribbean basin area, central Africa, etc. It is called endemic form of T cell leukemia/lymphoma. However, manifestation of ATL is a small portion less than 1% of sero-positive individuals. In addition to lymphadenopathy and hepatosplenomegaly, other organs such as the lung, skin and bone marrow are involved. The phenotypic property of ATL shows positive reactivity to CD3, CD5, and CD4. HLA-DR is also expressed. Hypercalcemia and hypogammaglobulinemia together with defective cell-mediated immunity are common. The patient with ATL may take a grave down-hill course.

Cytologically the tumor cells are polymorphic varying from moderate to large in size. The nuclei are markedly convoluted like gyri or chicken foot-print (Fig. 17-40).

■ Chronic lymphatic leukemia (CLL)

Chronic lymphatic or lymphocytic leukemia is more frequent in the aged than ALL and often accompanies impairment of immune reaction because of monoclonal B cell proliferation. There are several reports that IgM-possessing lymphocytes are prominent[1,31,119] in CLL and that CLL is regarded as B cell variety,[2,134] while Whiteside et al[161] has emphasized that the cell surface and functional markers do not belong to stable criteria to categorize neoplastic lymphocytes because of the presence of cross-reactivity between CLL cells and T cells. A small part of CLL is considered to be of T cell origin and often associated with erythroderma like Sézary syndrome.

General cytology of lymphatic leukemic cells (Figs. 17-43 and 17-44): Difficulties in identification of lymphatic leukemic cells do not consist in differentiation between lymphoblastic leukemic cells and lymphocytic leukemic cells but in distinction of leukemic cells from lymphosarcoma cells: (i) Lymphoblastic leukemic cells are characterized by enlarged, hyperchromatic nuclei with indentation, one to three nucleoli and scanty, agranular basophilic cytoplasm. (ii) Lymphocytic leukemic cells are medium-sized and their nuclei possess dense, somewhat stippled chromatins. The chromatin pattern is not as condensed as that of mature lymphocytes. From the immunological classification, nuclear deformity or

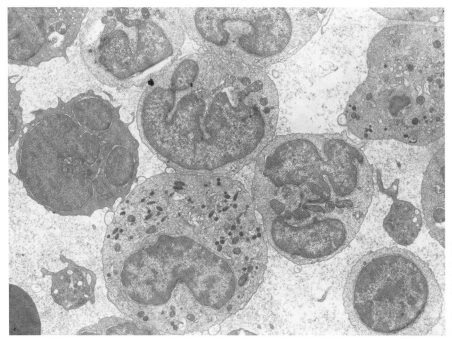

Fig. 17-41 Lymphatic leukemia of T cell line showing deep convolutions of nuclear contour. Densely increased heterochromatin is diffusely distributed. The cytoplasmic border is smooth. (original magnification × 6,600)

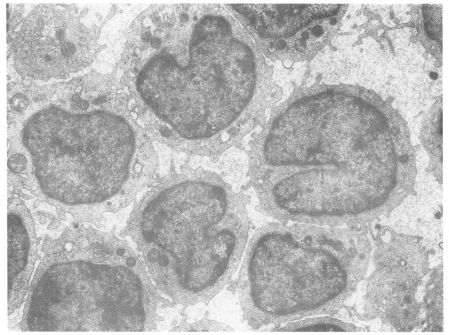

Fig. 17-42 Lymphatic leukemia of B cell line showing enlarged nuclei with occasional indentation. Indentation of nuclear contour in B cells is less marked than in T cells. In scant cytoplasm the organella are poorly developed, while polyribosomes are increased in amount. Cytoplasmic processes are present. (original magnification ×6,600)

indentation is less prominent in B cells than in T cells (Figs. 17-41 and 17-42).

Differentiation between lymphatic leukemic cells and lymphosarcoma cells is essentially difficult. The chromatin pattern of leukemic cells is more evenly and equally distributed than that of lymphosarcoma cells. A nuclear size variation, i.e., intermingling of small and large cells of the same type, is rather favorable to lymphosarcoma. The most important observation for imprints or needle aspirates is an outlook of cellularity; in case of lymphatic leukemia one sees the admixture of large immature cells (leukemic cells) with benign cells of lymphocytic series (Fig. 17-44), while the smear in lymphosarcoma is monomorphic and occupied predominantly by lymphosarcoma cells (see Figs. 17-22 to 17-24).

■Sézary syndrome (Cutaneous T cell lymphoma)

This is characterized by pruritic, exfoliative erythroderma, leonine or mask-like face, superficial lymphadenopathy, hepatomegaly and circulating atypical mononuclear cells that belong to the T cell series. The Sézary cells are slightly larger than leukocytes and possess cerebral gyrus-like nuclear convolutions[99,173](Fig. 17-45). The entity of Sézary's disease may be an erythrodermic variant with circulating atypical cells, i.e., with leukemic manifestation of mycosis fungoides. [162] The Sézary cells represent phenotypic properties for CD2, CD3 and CD5.

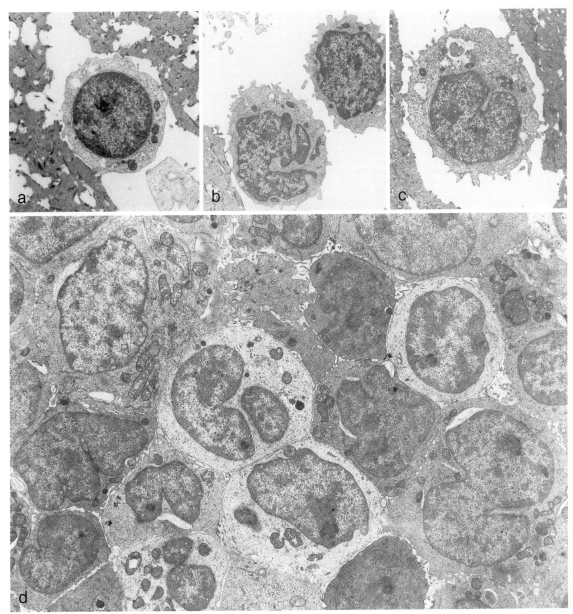

Fig. 17-43 Activated lymphocytes and chronic lymphocytic leukemic cells.
(1) A small lymphocyte (a) and activated lymphocytes (b, c) in tuberculous pleurisy. Activated lymphocytes reveal enlarged and convoluted nuclei with granular, distributed heterochromatins, whereas a small lymphocyte shows densely packed chromatins, particularly at the periphery. (2) Leukemic lymphocytes (d) have more enlarged, round to invaginated nuclei. Heterochromatin is increased in amount, but not as condensed as that of benign lymphocyte. Note excessive mitochondria. (original magnification × 4,000)

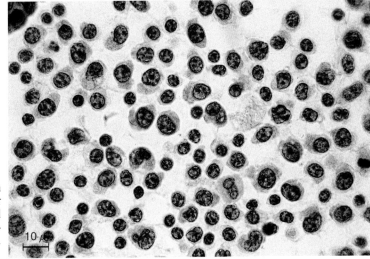

Fig. 17-44 Chronic lymphocytic leukemia in imprints of a lymph node. Note a variable number of leukemic cells similar to prolymphocytes mixed with lymphocytes. These are slightly larger than lymphocytes but smaller than lymphosarcoma cells. Note one or two nucleoli and stippled chromatins.

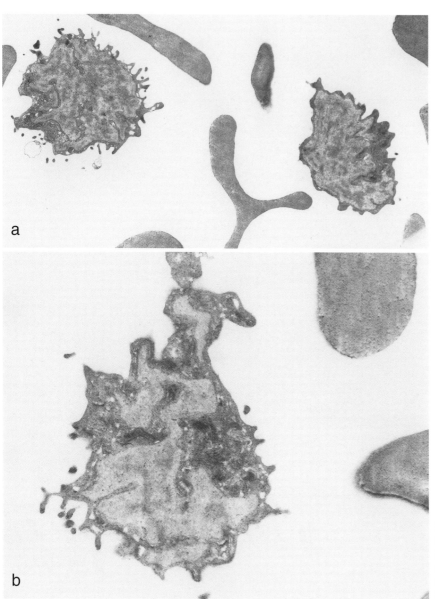

a

b

Fig. 17-45   Sézary cells which were ever called cellules monstreuses. The nuclei show characteristic cerebriform convolutions and marginated hetero-chromatins. (original magnification: a ×6,000; b ×12,000)

■ Hairy cell leukemia

Hairy cell leukemia (Schrek, 1966[146]) has been also called chronic reticulolymphocytic leukemia; this rare disease should not be confused with malignant reticulo-sis or Schilling type monocytic leukemia that takes an acute process (Fig. 17-48). The disease entity is charac-terized by (1) chronic clinical process with marked splenomegaly and mild lymphadenopathy, (2) progres-sive anemia, (3) specific cell feature with cytoplasmic processes as called hairy (Fig. 17-48) and (4) L-tartrate-resistant acid phosphatase activity in the leukemic cell.[75,76] The origin of the tumor whether of B lympho-cyte[24,50] or of monocytic/histiocytic variety[79,98] is the subject of great dispute. Their precise origin is still unknown, but recent studies have proved leukemic B lymphocytes to be facultative phagocytes[26,27,45,141](Fig. 17-49).

While surface-bound ATPase and surface immuno-globulin are favorable to B cell variety, moderate non-specific esterase activity and occasional phagocytosis are suggestive of monocytic origin. Hairy cells having hybrid features as such are moderately enlarged and possess eccentric reniform nuclei. The chromatin is delicate and nucleoli are small (Figs. 17-46 and 17-49).

■ Immunoblastic lymphadenopathy

(Fig. 17-50)

Immunoblastic lymphadenopathy has been defined by Lukes and Tindle[97] as a disease entity of hyperimmune disorder that is characterized by prominent immuno-blastic proliferation, increase in number of arborizing small vessels and deposition of PAS-positive interstitial material. Frizzera et al[44] described the similar syndrome almost at the same time as angioimmunoblastic lym-phadenopathy. Dysproteinemia, skin rash, fever and hepatosplenomegaly are frequently associated. The course of the disease is progressive[77] and often devel-ops to malignant lymphoma (immunoblastic sarco-ma).[97,110] Although the cases described by Frizzera[44] showed no development to malignant lymphoma, the entity of the disease is considered to be malignant rather than benign non-neoplastic disorder. As to the origin of

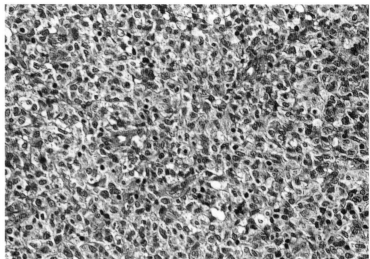

Fig. 17-46 Biopsy of a cervical lymph node from a patient with hairy cell leukemia. Note vast replacement of normal structure by atypical reticulo-histiocytic cells having phagocytosing function and L-tartrate-resistant acid phosphatase activity. Some mature lymphocytes remain.

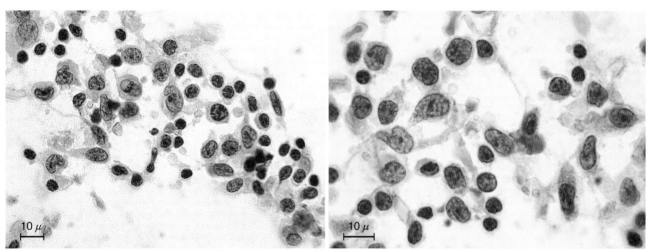

Fig. 17-47 Imprint cytology from mild lymphadenopathy in hairy cell leukemia. From the same case as Fig. 17-46. In the presence of benign lymphocytes, unfamiliar with monomorphous lymphoma, atypical endothelioid cells with elongated or reniform nuclei are seen. Nuclear membrane is delicate but distinct.

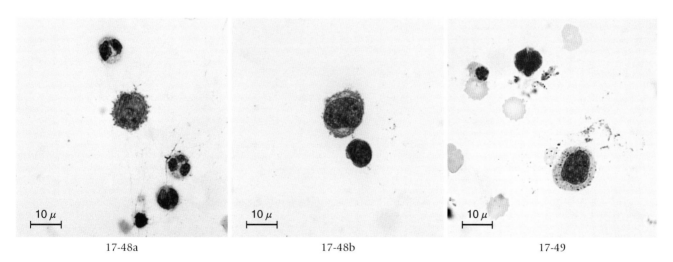

17-48a   17-48b   17-49

Fig. 17-48 Hairy cell leukemia showing fine cytoplasmic processes. This 54-year-old male who has had slight, cervical lymphadenopathy and hepatosplenomegaly developed leukemic manifestation of atypical mononuclear cells with high N/C ratio.

Fig. 17-49 Phagocytosis property of a large lymphocyte in hairy cell leukemia. India ink method for a blood film. Filamentous projections are not seen in incubation test with India ink, but atypical mononucleated cell shows phagocytosis of fine particles.

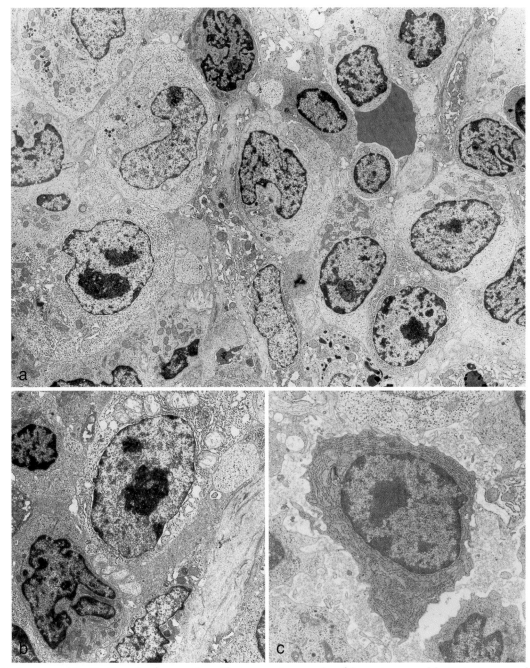

Fig. 17-50 Immunoblastic lymphadenopathy. Electron micrograph of large immunoblasts in immunoblastic lymphadenopathy. Note prominent nucleoli in large vesicular nuclei and abundant polyribosomes in the cytoplasm of a fair amount. (original magnification: a × 4,450; b × 5,390; c × 4,530)

markedly proliferating lymphoid cells, both clones, either B cell or T cell system, are concerned with this disease. The B cell type is responsive to prednisone and chemotherapy, and reveals a predominance of plasmacytoid cells, whereas the T cell type shows short survival time and is characterized by marked increase in number of arborizing vessels, eosinophils and blastoid cells.[77]

Cytology: The main component of the cellular pathology in immunoblastic lymphadenopathy is the blastoid cells having large, enrounded, hyperchromatic nuclei, in which one or two occasionally three nucleoli are present. These nucleoli often locate at the nuclear rim. The cytoplasm is of moderate amount and variable in degrees of basophilia in Giemsa stain. Cytoplasmic vac-

uolation is occasionally seen, but not specific. An admixture with plasmoblast-like cells and plasma cells is particularly remarkable in the B cell type (Fig. 17-51).

***Addendum:* Nonspecific lymphoreticular hyperplasia.** Reactive hyperplasia of lymphoreticular cells of a lymph node is cytologically similar to immunoblastic lymphadenopathy because enlarged lymphogonia are admixed with lymphocytes (Figs. 17-52 and 17-53). Differentiation should be established histologically, but the cell constituents are different and distinguishable cytologically from one to another. Firstly varying degrees of immaturity are found in lymphocytic and plasma cell series in immunoblastic lymphadenopathy. Secondly histiocytes or macrophages phagocytosing cell debris are rather conspicuous in reactive hyperplasia.

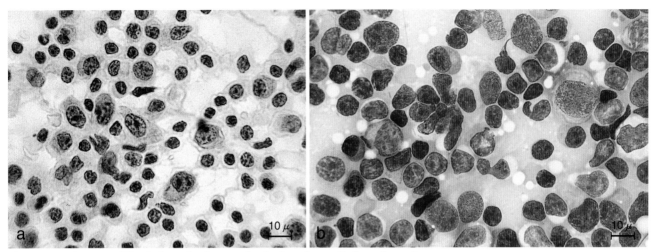

Fig. 17-51 Immunoblastic lymphadenopathy in imprints of a lymph node.
a. Note an admixture of varying number of blastoid cells with plasmacytoid cells and lymphocytes. Some of multiple nucleoli often locate at the nuclear rim.
b. Note the largest immunoblasts, immature plasmacytoid cells with perinuclear halo and cytoplasmic basophilia, and lymphocytes. Giemsa preparation is inevitably important for precise interpretation of cytology of hematopoietic disorders.

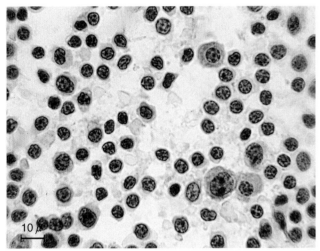

Fig. 17-52 Reactive lymphoreticular hyperplasia in imprints of a lymph node. Although a mixed cellularity consisting of lymphogonia (immunoblasts), activated lymphocytes with dotted chromatins, plasma cells and other leukocytes is of diagnostic value, it should be carefully distinguished from immunoblastic lymphadenopathy.

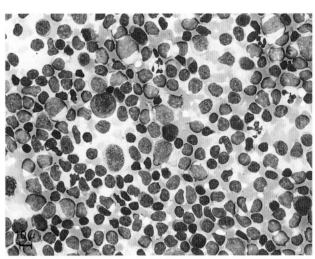

Fig. 17-53 Giemsa stain of reactive lymphoreticular hyperplasia. Giemsa stain is recognizable of cellular components more precisely than Papanicolaou stain. Immunoblasts with intense basophilia, plasma cells and a histiocyte can be identified. However, few number of immunoblasts, absence of immature plasma cells and frequent histiocytes phagocytosing cell debris are distinguishable from immunoblastic lymphadenopathy.

■Myeloma (Plasmacytoma)

Taking the plasma cells into account as a form of extreme differentiation, one may see marked variation in feature of myeloma cells according to the grade of differentiation. Morphological variations of myeloma cells are as follows (Figs. 17-54 to 17-57):

1) Differentiated or mature form has an ovoid or round nucleus with eccentric position. The chromatin is condensed at periphery. The cytoplasm is abundant and intensely stained peripherally showing a perinuclear halo. Electron microscopically, the cytoplasm is rich in mitochondria and lamellated endoplasmic reticulum which is compactly parallel at periphery. The Golgi apparatus is also well developed. Increase in size and number of nucleoli are also seen.

2) Undifferentiated or immature form has a large, centrally located nucleus, which is often indented or lobulated. The situation of nucleus and highly increased N/C ratio indicate the state of immaturity. The endoplasmic reticulum is abundant, lamellar in arrangement and often cystically dilated. The mitochondria is distorted in shape and size. Intramitochondrial granules are increased in amount and cristas are frequently disruptive. Dense bodies located inside the cisternae are considered to be responsible for formation of cytoplasmic spherules (*Russel bodies*).

3) Flame cell or thesaurocyte is another special type of myeloma cell. Much retention of M protein is shown in distended endoplasmic reticulum; it is probably the result of formation of high molecular weight polymers of carbohydrate-rich IgA.[117]

403

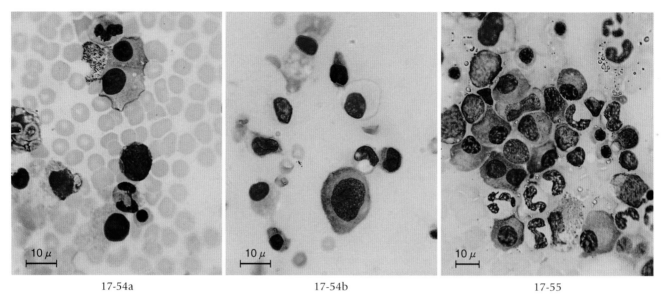

17-54a   17-54b   17-55

Fig. 17-54 Myeloma cells in needle aspirates of sternal bone marrow. A predominance of immature plasma cells (myeloma cells), aggregation in a group, is indicative of myeloma. Production of monoclonal immunoglobulin can be verified by the immunofluorescence antibody technique. **a.** mature form, **b.** immature form.

Fig. 17-55 Bone marrow aspirates showing plasmacytosis from a 48-year-old male with lymphoid interstitial pneumonia. Note difference in maturation of plasma cells from myeloma cells. This man who has had progressive dyspnea, skin eruptions, micropolyadenia of lymph nodes was found to have polyclonal gammopathy (IgG 5,540; IgA 1,380; IgM 960 mg/dl) associated with LIP.

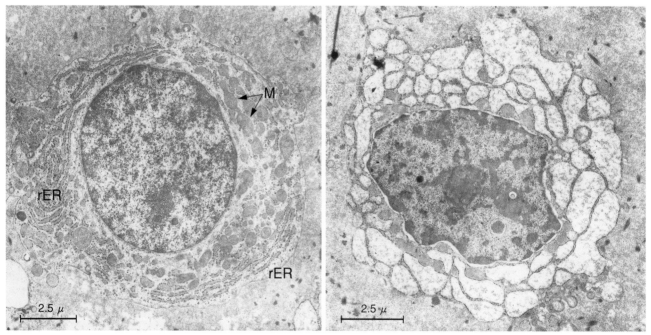

Fig. 17-56 Myeloma. Immature form of myeloma cell showing centrally locating large nucleus and lamellar rough endoplasmic reticulum at periphery. This is a common type of myeloma cell.(original magnification × 8,000) M: mitochondria; rER: rough endoplasmic reticulum.

Fig. 17-57 Myeloma. Immature form of myeloma cell showing cystic rough endoplasmic reticulum. Although deposition of granular substance is present within the dilated endoplasmic reticulum, it is not so much as that of flame cell.(original magnification ×8,000)

10 Histiocytosis X

Although eosinophilic granuloma either of solitary or multiple form should be discussed in the chapter of bone, the conventional concept of histiocytosis that includes Letterer-Siwe disease (nonlipid reticuloendotheliosis), Hand-Schüller-Christian disease and

eosinophilic granuloma have been described in the chapter of hematopoietic organs. These are fundamentally characterized by infiltrates of mononuclear or multinucleated cells with convoluted or indented nuclei and a moderate amount of acidophilic cytoplasm. Letterer-Siwe disease is the most rapid, fetal disorder of young children that involves lymph nodes, liver, spleen, and bone. On the other hand, Hand-Schüller-Christian disease is more chronic disorder in young adults or children, in which cholesterin and its esters are deposited in histiocytes. Bones especially the skull and facial bone are extensively destroyed, but the lung, thyroid, lymph nodes, spleen and liver are also affected[162](see Fig. 19-16).

References

1. Aisenberg AC, Bloch KJ: Immunoglobulins on the surface of neoplastic lymphocytes. N Engl J Med 287: 272, 1972.
2. Aisenberg AC, Bloch KJ, Long JC: Cell-surface immunoglobulins in chronic lymphocytic leukemia and allied disorders. Am J Med 55: 184, 1973.
3. Al Saati T, Caveriviere P, Gorguet B, et al: Epithelial membrane antigen in hematopoietic neoplasms. Hum Pathol 17: 533, 1986.
4. Amano S: Antibody producing cell series: A review of molecular biology of immunological phenomena. Ann Rep Inst Virus Res (Kyoto Univ) 5: 194, 1962.
5. Astaldi G, Costa G, Airo R: Phytohaemagglutinin in leukaemia. Lancet 1: 1394, 1965.
6. Bach FH, Hirschhorn K: The "in vitro" response of peripheral blood leukocytes. Semin Hematol 2: 68, 1965.
7. Bach JF, Dormont J, Dardenne M, et al: In vitro rosette inhibition by antihuman antilymphocyte serum. Transplantation 8: 265, 1969.
8. Banks PM, Chan J, Cleary ML, et al: Mantle cell lymphoma: A proposal for unification of morphologic, and molecular data. Am J Surg Pathol 16: 637, 1992.
9. Begemann H, Rastetter J: Atlas of Clinical Haematology. Berlin: Springer-Verlag, 1972.
10. Bernerd C, Geraldes A, Boiron M: Effects of phytohemagglutinin on blood-cultures of chronic lymphocytic leukaemias. Lancet 1: 667, 1964.
11. Bianco C, Patrick R, Nussenzweig V: A population of lymphocytes bearing a membrane receptor for antigen-antibody-complement complexes. I. Separation and characterization. J Exp Med 132: 702, 1970.
12. Bollum FJ: Terminal deoxynucleotidyl transferase as a haematopoietic cell marker. Blood 54: 1203, 1979.
13. Borrela L, Sen L: E rosettes on blasts from untreated acute lymphocytic leukemia (ALL). Comparison of temperature dependence of E rosettes formed by normal and leukemic lymphoid cells. J Immunol 114: 187, 1975.
14. Bøyum A: Separation of leukocytes from blood and bone marrow. Scand J Clin Lab Invest 21: Suppl 97, 1968.
15. Bøyum A: Separation of blood leukocytes, granulocytes and lymphocytes. Tissue Antigens 4: 269, 1974.
16. Brain P, Gordon J, Willetts WA: Rosette formation by peripheral lymphocytes. Clin Exp Immunol 6: 681, 1970.
17. Briellmann A: Über Plasmazellenbefunde in Lymphknoten bei Lebercirrhosen und die Bedeutung der Russellschen Körperchen. Schweitz Z Pathol 18: 335, 1955.
18. Brouchet JC, Preud'Homme JL, Flandrin G, et al: Membrane markers in "histiocytic" lymphomas (reticulum cell sarcomas). J Natl Cancer Inst 56: 631, 1976.
19. Brubakar DB, Whiteside TL: Differentiation between benign and malignant human lymph nodes by means of immunologic markers. Cancer 43: 1165, 1979.
20. Burkitt DP: A sarcoma involving the jaws of African children. Br J Surg 46: 218, 1958.
21. Carr I: The fine structure of the mammalian lymphoreticular system. Int Rev Cytol 27: 283, 1970.
22. Carstairs K: The human small lymphocyte: Its possible pleuripotential quality. Lancet 1: 829, 1962.
23. Carter TR, Feldman PS, Innes DT Jr, et al: The role of fine needle aspiration cytology in the diagnosis of lymphoma. Acta Cytol 32: 848, 1988.
24. Catovsky D: Hairy cell leukaemia and prolymphocytic leukaemia. Clin Haematol 6: 245, 1977.
25. Catovsky D, Cherchi M, Greaves MF, et al: Acid-phosphatase reaction in acute lymphoblastic leukaemia. Lancet 1: 749, 1978.
26. Catovsky D, Sperandio P, O'brien M: Facultative phagocytosis by leukemic B-lymphocytes: Further proof of the B-cell nature of hairy cells. In: Mathé G, Seligmann M, Tubiana M (eds): Recent Results in Cancer Research, Vol 64: Lymphoid Neoplasias I: Classification Categorization, Natural History. pp208-212. Berlin-Heidelberg-New York: Springer-Verlag, 1978.
27. Cawley JC, Burns GF, Hayhoe FGJ (eds): Recent Results in Cancer Research, Vol 72: Hairy Cell Leukaemia. Berlin-Heidelberg-New York: Springer-Verlag, 1980.
28. Chapman JA, Elves MW, Gough J: An electron-microscope study of the in vitro transformation of human leukocytes. J Cell Sci 2 : 359, 1967.
29. Chase MW: Delayed-type hypersensitivity and the immunology of Hodgkin's disease, with a parallel examination of sarcoidosis. Cancer Res 26: 1097, 1966.
30. Claman HN: Human thymus cell cultures: Evidence for two functional populations. Proc Soc Exp Biol Med 121: 236, 1966.
31. Clark C, Rydell RE, Kaplan ME: Frequent association of IgM with crystalline inclusions in chronic lymphatic leukemic lymphocytes. N Engl J Med 289: 113, 1973.
32. Coukle DM, Adkins B Jr: Primary-malignant tumors of the mediastinum. Ann Thorac Surg 14: 553, 1972.
33. Cowling DC, Quaglino D, Davidson E: Changes induced by tuberculin in leucocyte cultures. Lancet 2: 1091, 1963.
34. Custer RP, Bernhard WG: Interrelationship of Hodgkin's disease and other lymphatic tumors. Am J Med Sci 216: 625, 1948.
35. Davis S: The variable pattern of circulating lymphocyte subpopulation in chronic lymphocytic leukemia. N Engl J Med 294: 1150, 1976.
36. Dent PM, Gabrielsen AE, Cooper MD, et al: The secondary immunologic deficiency diseases associated with lymphoproliferative disorders. In: Miescher PA, Müller-Eberhard HJ (eds): Textbook of Immunology. New York: Grune & Stratton, 1969.
37. Dewar AE, Krajewski AS, Murray J: T-cell lymphoma in children and young adults: Clinical, immunological and pathological features. Br J Cancer 42: 659, 1980.
38. Dickler HB, Siegel EP, Bentwich ZH, et al: Lymphocyte binding of aggregated IgG and surface Ig staining in chronic lymphocytic leukemia. Clin Exp Immunol 14: 97, 1973.
39. Epstein LB, Stohlman F Jr: RNA synthesis in cultures of normal peripheral blood. Blood 24: 69, 1964.
40. Epstein MA, Barr YM: Cultivation in vitro of human lymphoblasts from Burkitt's malignant lymphoma. Lancet 1: 252, 1964.
41. Ferrata A, Storti E: Le Malattie del Sangue: Mannuale per Medici e Studenti. Milano: Soc Edit Libraria, 1946.

42. Franzén S, Giertz G, Zajicek J: Cytological diagnosis of prostatic tumours by transrectal aspiration biopsy: A preliminary report. Br J Urol 32: 193, 1960.

43. Freeman C, Berg JW, Cutler SJ: Occurrence and prognosis of extranodal lymphomas. Cancer 29: 252, 1972.

44. Frizzera G, Moran EM, Rappaport H: Angioimmunoblastic lymphadenopathy with dysproteinaemia. Lancet 1: 1070, 1974.

45. Fu SM, Winchester RJ, Rai KR, et al: Hairy cell leukemia. Proliferation of a cell with phagocytic and B-lymphocyte properties. Scand J Immunol 3: 847, 1974.

46. Gall EA: The cytological identity and interrelation of mesenchymal cells of lymphoid tissue. Ann NY Acad Sci 73: 120, 1958.

47. Gall EA, Mallory TB: Malignant lymphoma: A clinicopathological survey of 618 cases. Am J Pathol 18: 381, 1942.

48. Gall EA, Rappaport H: Seminar on diseases of lymph nodes and spleen. In: McDonald JR (ed): Proceedings of the 23rd Seminar of American Society of Clinical Pathologists, 1958.

49. Gerard-Marchant R, Hamlin I, Lennert K, et al: Classification of non-Hodgkin's lymphomas. Lancet 2: 406, 1974.

50. Golde DW, Stevens RH, Quan SG, et al: Immunoglobulin synthesis in hairy cell leukaemia. Br J Haematol 35: 359, 1977.

51. Greenberg BR, Peter CR, Glassy F, et al: A case of T-cell lymphoma with convoluted lymphocytes. Cancer 38: 1602, 1976.

52. Grundmann E: Die Bildung der Lymphozyten und Plasmazellen im lymphatischen Gewebe der Ratte: Ein zytologischer Beitrag zur Blutzellreifung. Beitr Pathol Anat 119: 217, 1958.

53. Gupta AK, Nayar M, Chandra M: Reliability and limitations of fine needle aspiration cytology of lymphadenopathies. An analysis of 1, 261 cases. Acta Cytol 35: 777, 1991.

54. Harris NL, Bhan AK: B-cell neoplasms of the lymphocytic, lymphoplasmacytoid, and plasma cell types: Immunohistologic analysis and clinical correlation. Hum Pathol 16: 829, 1985.

55. Hayward RL, Leonard RC, Prescott RJ: A critical analysis of prognostic factors for survival in intermediate and high grade non-Hodgkin's lymphoma. Scotland and Newcastle Lymphoma Group Therapy Working Party. Br J Cancer 63: 945, 1991.

56. Hellman T: Studien über das lymphoide Gewebe. Beitr Pathol Anat 68: 333, 1921.

57. Henle W, Henle G, Ho H-C, et al: Antibodies to Epstein-Barr virus in nasopharyngeal carcinoma: Other head and neck neoplasm, and control groups. J Nat Cancer Inst 44: 225, 1970.

58. Hersh EH, Oppenheim JJ: Impaired in vitro lymphocyte transformation in Hodgkin's disease. N Engl J Med 273: 1006, 1965.

59. Hinuma Y, Nagata K, Hanaoka M, et al: Adult T-cell leukemia: Antigen in an ATL cell line and detection of antibodies to the antigen in human sera. Proc Natl Acad Sci USA 78: 6476, 1981.

60. Hirschhorn K, Bach F, Kolondy RL, et al: Immune response and mitosis of human peripheral blood lymphocytes in vitro. Science 142: 1185, 1963.

61. Hirschhorn K, Schreibman RR, Bach FH, et al: In vitro studies of lymphocytes from patients with sarcoidosis and lymphoproliferative disorders. Lancet 2: 842, 1964.

62. Hirschowitz SL, Mandell D, Nieberg RK, et al: The alcohol-fixed Diff-Quik stain. A novel rapid stain for the immediate interpretation of fine needle aspiration specimens. Acta Cytol 38: 499, 1994.

63. Hoffbrand AV, Ganeshaguru K, Llewelin P, et al: Biochemical markers in leukemia and lymphoma. In: Gross R, Hellriegel KP (eds): Recent Results in Cancer Research, Vol 69: Strategies in Clinical Hematology. pp25-39. Berlin-Heidelberg-New York: Springer-Verlag, 1979.

64. Hovmark A: Acid α-naphthyl acetate esterase staining of T lymphocytes in human skin. Acta Dermatovener 57: 497, 1977.

65. Hyjek E, Isaacson PG: Primary B-cell lymphoma of the thyroid and relationship to Hashimoto's thyroiditis. Hum Pathol 19: 1315, 1988.

66. Hyjek E, Smith WJ, Issacson PG: Primary B-cell lymphoma of salivary glands and its relationship to myoepithelial sialadenitis. Hum Pathol 19: 766, 1988.

67. Imai Y, Terashima K, Matsuda M, et al: Reticulum cell and dendritic reticulum cell origin and function. Recent Adv RES Res 21: 22: 51-81, 1982.

68. Isaacson PG: Lymphomas of mucosa-associated lymphoid tissue (MALT). Histopathology 16: 617, 1990.

69. Isaacson PG, Spencer J: The biology of lowgrade MALT lymphoma. J Clin Pathol 48: 395, 1995.

70. Isaacson P, Wright D: Malignant lymphoma of mucosa-associated lymphoid tissue: A distinctive B cell lymphoma. Cancer 52: 1410, 1983.

71. Ishii Y, Koshiba H, Beno H, et al: Characterization of human B lymphocyte specific antigens. J Immunol 114: 466, 1975.

72. Jackson H Jr, Parker F Jr: Hodgkin's Disease and Allied Disorders. New York: Oxford University Press, 1947.

73. Jacobs JC, Katz RL, Shaff N, et al: Fine needle aspiration of lymphoblastic lymphoma. A multiparameter diagnostic approach. Acta Cytol 36: 887, 1992.

74. Jayaram G, Rahman NA: Cytology of Ki-1-positive anaplastic large cell lymphoma: A report of two cases. Acta Cytol 41: 1253, 1997.

75. Jancklia AJ, Yam LT: Cytochemistry of tartrate-resistant acid phosphatase. Am J Clin Pathol 71: 356, 1979.

76. Jancklia AJ, Li CY, Lam KW, et al: The cytochemistry of tartrate-resistant acid phosphatase. Am J Clin Pathol 70: 45, 1978.

77. Jones DB, Castleden M, Smith JL, et al: Immunopathology of angioimmunoblastic lymphadenopathy. Br J Cancer 37: 1053, 1978.

78. Kaiserling E: Non-Hodgkin Lymphome: Ultrastruktur und Cytogenase. Stuttgart: Gustav Fischer, 1977.

79. King GW, Hurtubise PE, Sagome AL, et al: Leukemic reticuloendotheliosis. A study of the origin of the malignant cell. Am J Med 59: 411, 1975.

80. Knapp W, Schuit HRE, Bolhuis RLH, et al: Surface immunoglobulins in chronic lymphatic leukemia, macroglobulinemia and myelomatosis. Clin Exp Immunol 16: 541, 1974.

81. Koh S-J, Vargas GF, Caces JN, et al: Malignant "histiocytic" lymphoma in childhood. Am J Clin Pathol 74: 417, 1980.

82. Kulenkampff J, Janossy G, Greaves MF: Acid esterase in human lymphoid cells and leukaemic blasts: A marker for T lymphocytes. Br J Haematol 36: 231, 1977.

83. Lamb D, Pilney F, Kelly WD, et al: A comparative study of the incidence of anergy in patients with carcinoma, leukemia, Hodgkin's disease and other lymphomas. J

Immunol 89: 555, 1962.

84. Lay WH, Mendes NF, Blanco C, et al: Binding of sheep red blood cells to a large population of human lymphocytes. Nature 230: 531, 1971.

85. Lay WH, Nussenzweig V: Receptors for complement on leukocytes. J Exp Med 128: 991, 1968.

86. Lennert K: Die Morphologie der Urethanwirkung bei Leukämien, malignen Tumoren des Lymphatischen Systems und der Lymphograunlomatose. Frankf Z Pathol 61: 339, 1950.

87. Lennert K: Lymphknoten-Diagnostik in Schnitt und Ausstrich: Zytologie und Lymphadenitis. In: Lubarsch O, Henke F (eds): Handbuch der Speziellen Pathologischen Anatomie und Histologie, I/3/A. Berlin-Göttingen-Heidelberg: Springer-Verlag, 1961.

88. Lennert K, Feller AC: Histopathology of Non-Hodgkin's Lymphomas (based on Updated Kiel Classification), 2nd ed. Berlin: Springer-Verlag, 1992.

89. Lennert K, Remmele W: Karyometrische Untersuchungen an Lymphknotenzellen des Menschen: I. Mitt. Germinoblasten, Lymphoblasten und Lymphocyten. Acta Haematol 19: 99, 1958.

90. Lennert K, Stein H, Kaisering E: Cytological and functional criteria for the classification of malignant lymphomata. Br J Cancer 31 (Suppl 2): 29, 1975.

91. Levin PH, O'conor GT, Berard CW: Antibodies to Epstein-Barr virus (EBV) in African patients with Burkitt's lymphoma. Cancer 30: 610, 1972.

92. Lukes PF: Lymph node smears in the diagnosis of lymphadenopathy: A review. Blood 10: 1030, 1955.

93. Lukes RJ: Relationship of histologic features to clinical stages in Hodgkin's disease. Am J Roentgenol 90: 944, 1963.

94. Lukes RJ: The immunological approach to the pathology of malignant lymphomas. Am J Clin Pathol 72: 657, 1979.

95. Lukes RJ, Collins RD: New approaches to the classification of the lymphoma. Br J Cancer 31 (Suppl 2): 1, 1975.

96. Lukes RJ, Craver LF, Hall TC, et al: Report of the Nomenclature Committee. Cancer Res 26: 1311, 1966.

97. Lukes RJ, Tindle BH: Immunoblastic lymphadenopathy. A hyperimmune entity resembling Hodgkin's disease. N Engl J Med 292: 1, 1975.

98. Lukes RJ, Tindle BH, Parker JW: Reed-Sternberg-like cells in infectious mononucleosis. Lancet 2: 1003, 1969.

99. Lutzner MA, Gordon HW: The ultrastructure of an abnormal cell in Sezary's syndrome. Blood 31: 719, 1968.

100. MacKinney AA Jr, Stohlman F Jr, Brecher G: The kinetics of cell proliferation in cultures of human peripheral blood. Blood 19: 349, 1962.

101. Mann RB, Jaffe ES, Berard CW: Malignant lymphomas: A conceptual understanding of morphologic diversity. Am J Pathol 94: 105, 1979.

102. Marchalonis JJ, Cone RE, Atwell JL: Isolation and partial characterization of lymphocyte surface immunoglobulins. J Exp Med 153: 956, 1972.

103. Mathé G, Misset JL, Gil-Delgado M, et al: Leukemic (or stage V) lymphosarcoma. In: Mathé G, Seligann M, Tubiana M (eds): Recent Results in Cancer Research, Vol 65: Lymphoid Neoplasias II: Clinical and Therapeutic. pp89-107. Berlin-Heidelberg-New York: Springer-Verlag, 1978.

104. McDonald GA, Dodds TC, Cruickshank B: Altas of Haematology, 4th ed. p101. Edinburgh: Churchill Livingstone, 1978.

105. McNeely TB: Diagnosis of follicular lymphoma by fine needle aspiration biopsy. Acta Cytol 36: 866, 1992.

106. Miale JB: Laboratory Medicine-Hematology. St Louis: CV Mosby, 1967.

107. Moeschlin S: Beitrag zur Morphologie der retikuloendothelialen Zellen des intravitalen Lymphknotenpunktats. Folia Haemat Leipzig 65: 181, 1941.

108. Moeschlin S, Schwarz E, Wang H: Die Hodgkinzellen als Tumorzellen. Schweitz Med Wschr 80: 1103, 1950.

109. Morton JE, Leyland MJ, Vaughan Hudson G, et al: Primary gastrointestinal non-Hodgkin's lymphoma: A review of 175 British National Lymphoma Investigation cases. Br J Cancer 67: 776, 1993.

110. Nathwani BN, Rappaport H, Moran EM, et al: Malignant lymphoma arising in angioimmunoblastic lymphadenopathy. Cancer 41: 578, 1978.

111. The Non-Hodgkin's Lymphoma Pathologic Classification Project: National Cancer Institute sponsored study of classifications of non-Hodgkin's lymphoma. Summary and description of a Working Formulation for Clinical Usage. Cancer 49: 2112, 1982.

112. Nossal GJV: Genetic control of lymphopoiesis, plasma cell formation, and antibody production. Int Rev Exp Pathol 1: 1, 1962.

113. Nowell PC: Phytohemagglutinin: An initiator of mitosis in cultures of normal human leukocytes. Cancer Res 20: 462, 1960.

114. Oertel J, Oertel B, Lobeck H, et al: Cytologic and immunocytologic studies of peripheral T-cell lymphomas. Acta Cytol 35: 285, 1991.

115. Oppenheim JJ, Whang J, Frei E III: Immunologic and cytogenetic studies of chronic lymphocytic leukemia cells. Blood 26: 121, 1965.

116. Ortega LG, Mellors RC: Cellular sites of formation of gammaglobulin. J Exp Med 106: 627, 1957.

117. Osserman EF: Multiple myeloma. In: Samter M (ed): Immunological Diseases. Boston: Little Brown, 1965.

118. Palutke M, Patt DJ, Weise R, et al: T cell leukemia-lymphoma in young adults. Am J Clin Pathol 68: 429, 1977.

119. Pangalis GA, Nathwani BN, Rappaport H: Malignant lymphoma, well differentiated, lymphocytic: Its relationship with chronic lymphatic leukemia and macroglobulinemia of Waldenstrom. Cancer 39: 999, 1977.

120. Pangalis GA, Yataganas X, Fessas PH: β-glucuronidase activity of lymph node imprints from malignant lymphomas and chronic lymphocytic leukemia. J Clin Pathol 30: 812, 1977.

121. Papac RJ: Lymphocyte transformation in malignant lymphoma. Cancer 26: 279, 1970.

122. Papadimitriou JC, Abruzzo LV, Bourquin PM, et al: Correlation of light microscopic, immunocytochemical and ultrastructural cytomorphology of anaplastic large cell Ki-1 lymphoma, an activated lymphocyte phenotype. Acta Cytol 40: 1288, 1996.

123. Parker JW, Wakasa H, Lukes RJ: The morphologic and cytochemical demonstration of lysosomes in lymphocytes incubated with phytohemagglutinin by electron microscopy. Lab Invest 14: 1736, 1965.

124. Parsonnet J, Hansen S, Rodriguez L, et al: Helicobacter pylori infection and gastric lymphoma. N Engl J Med 330: 1267, 1994.

125. Pavlowsky A: La punción ganglionar. Su contribución al diagnóstico clinico-quirúrgico de las affeciones gan-

glionares. Buenos Aires: Aniceto Lopez Impr, 1934.

126. Peapmain G, Lycette RR, Fitzgerald PH: Tuberculin-induced mitosis in peripheral blood leucocytes. Lancet 1: 637, 1963.

127. Pickren JW, Burke EM: Adjuvant cytology to frozen section. Acta Cytol 7: 164, 1963.

128. Pilotti S, Palma SDi, Alasio L: Diagnostic assessment of enlarged superficial lymph nodes by fine needle aspiration. Acta Cytol 37: 853, 1993.

129. Poiesz BJ, Ruscetti FW, Gazdar AF, et al: Detection and isolation of type C retrovirus particles from fresh and cultured lymphocytes of a patient with cutaneous T-cell lymphoma. Proc Natl Acad Sci USA 77: 7415, 1980.

130. Polliack A: Surface morphology of lymphoreticular cells: Review of data obtained from scanning electron microscopy. In: Mothé G, Seligmann M, Tubiana M (eds): Recent Results in Cancer Research, Vol 64: Lymphoid Neoplasias I: Classification Categorization, Natural History. pp66-93. Berlin-Heidelberg-New York: Springer-Verlag, 1978.

131. Polliack A, Froimovici M, Frankenburg S, et al: Altered surface morphology of Concanavalin A transformed thymic lymphocytes as seen by scanning electron microscopy. Biomedicine 24: 389, 1976.

132. Polliack A, Lampen N, Clarkson BD, et al: Identification of human B and T lymphocytes by scanning electron microscopy. J Exp Med 138: 607, 1973.

133. Potter EL: Hodgkin's disease, with special reference to its differentiation from other diseases of lymph nodes. Arch Pathol 19: 139, 1935.

134. Preud' Homme JL, Seligmann M: Surface-bound immunoglobulins as a cell marker in human lymphoproliferative diseases. Blood 40: 777, 1972.

135. Quaglino D, Cowling DC: Cytochemical studies on cells from chronic lymphocytic leukaemia and lymphosarcoma cultured with phytohemagglutinin. Br J Haematol 10: 358, 1964.

136. Ranki A: T cell acid α-naphthyl acetate esterase. Scand J Immunol 5: 1129, 1976.

137. Rappaport H: Tumors of the Hematopoietic System. Atlas of Tumor Pathology. Washington, DC: Armed Forces Institute of Pathology, 1966.

138. Rappaport H, Winter WJ, Hick EB: Follicular lymphoma, based on a survey of 253 cases. Cancer 9: 792, 1956.

139. Reed RJ, Dhurandhar AN: Stem cell (immunoblastic) lymphoma. Am J Clin Pathol 68: 8, 1977.

140. Reif AE, Allen JM: Mouse thymic isoantigens. Nature 209: 521, 1966.

141. Rieber EP, Hadam M, Linke RP, et al: Hairy cell leukemia; C-lymphocyte and monocytic properties displayed by one cell. In: Mathé G, Seligmann M, and Tubiana M (eds): Recent Results in Cancer Research, Vol 64: Lymphoid Neoplasias I: Classification Categorization, Natural History. pp204-207. Berlin-Heidelberg-New York: Springer-Verlag, 1978.

142. Roath S, Newell D, Polliack A, et al: Scanning electron microscopy and the surface morphology of human lymphocytes. Nature 273: 15, 1978.

143. Rosenberg SA: Report of the Committee on the staging of Hodgkin's disease. Cancer Res 26: 1310, 1966.

144. Satucci M, Pimpinelli N, Arganini L: Primary cutaneous B-cell lymphoma, a unique type low-grade lymphoma. Cancer 67: 2311, 1991.

145. Schrek R, Batra KV: Thymic, splenic and appendical lymphocytes. Lancet 2: 444, 1966.

146. Schrek R, Donnel WJ: "Hairy" cells in blood in lymphoreticular neoplastic disease and "flagellated" cells of normal lymph node. Blood 27: 199, 1966.

147. Sen L, Borella L: Clinical importance of lymphoblasts with T markers in childhood acute leukemia. N Engl J Med 292: 828, 1975.

148. Smithers DW: Hodgkin's disease: One entity or two ? Lancet 2: 1285, 1970.

149. Strum SB, Park JK, Rappaport H: Observation of cells resembling Sternberg-Reed cells in conditions other than Hodgkin's disease. Cancer 26: 176, 1970.

150. Sudilovsky D, Cha I: Fine needle aspiration cytology of dermatopathic lymphadenitis. Acta Cytol 42: 1341, 1998.

151. Tamaoki N, Essener E: Distribution of acid phosphatase and N-acetyl-β-glucosaminidase activities in lymphocytes of lymphatic tissues of man and rodents. J Histochem Cytochem 17: 238, 1969.

152. Tanaka H: Mesenchymal and epithelial reticulum in lymph nodes and thymus of mice as revealed in the electron microscope. Ann Rep Inst Virus Res (Kyoto Univ) 5: 146, 1962.

153. Tani E, Löwhagen T, Nasiell K, et al: Fine needle aspiration cytology and immunochemistry of large cell lymphomas expressing Ki-1 antigen. Acta Cytol 33: 359, 1989.

154. Third National Cancer Survey: Incidence data. National Cancer Institute Monograph 41, March, 1975.

155. Trowell OA: Re-utilization of lymphocytes in lymphopoiesis. J Biophys Biochem Cytol 3: 317, 1957.

156. Tsukimoto I, Wong KY, Lampkin BC: Surface markers and prognostic factors in acute lymphoblastic leukemia. N Engl J Med 294: 245, 1976.

157. Utsinger PD, Yount WJ, Fuller CR, et al: Hairy cell leukemia: B-lymphocyte and phagocytic properties. Blood 49: 19, 1977.

158. Weinstein HJ, Vance ZB, Jaffe N, et al: Improved prognosis for patients with mediastinal lymphoblastic lymphoma. Blood 53: 687, 1979.

159. Weisenburger DD, Nathwani BN, Diamond LW, et al: Malignant lymphoma, intermediate lymphocytic type: A clinicopathologic study of 42 cases. Cancer 48: 1415, 1981.

160. White RG: Functional recognition of immunologically competent cells by means of the fluorescent antibody technique. In: Wolstenholme GEW, Knight J (eds): The Immunologically Competent Cell. London: J & A Churchill, 1963.

161. White TL, Winkelstein A, Rabin BS: Immunologic characterization of chronic lymphocytic leukemia cells. Cancer 39: 1109, 1977.

162. WHO: Histological and Cytological Typing of Neoplastic Diseases of Haematopoietic and Lymphoid Tissues. International Histological Classification of Tumors, No 14. Geneva, 1976.

163. Williams AH, Taylor CR, Higgins GR, et al: Children lymphoma-leukemia. Cancer 42: 171, 1978.

164. Winkelstein A, Graddock CG: Comparative response of normal human thymus and lymph node cells to phytohemagglutinin. Br J Haematol 11: 488, 1965.

165. Wojcik EM, Katz RL, Fanning TV, et al: Diagnosis of mantle cell lymphoma on tissue acquired by fine needle aspiration in conjunction with immunochemistry and cytokinetic studies. Acta Cytol 39: 909, 1995.

166. Wood GS, Turner RR, Shiurba RA, et al: Human dendritic cells and macrophages. In situ immunophenotyp-

ic definition of subsets that exhibit specific morphologic and microenvironmental characteristics. Am J Pathol 119: 73, 1985

167. Wotherspoon AC, Oritz-Hidalgo C, Falzon MR, et al: Helicobacter pylori-associated gastritis and primary B-cell gastric lymphomas. Lancet 338: 1175, 1991.

168. Yam LT, Li CY, Crosby WH: Cytochemical identification of monocytes and granulocytes. Am J Clin Pathol 55: 283, 1971.

169. Yam LT, Li CY, Finkel HE: Leukemic reticuloendotheliosis. The role of tartrate-resistant acid phosphatase in diagnosis and splenectomy in treatment. Arch Intern Med 130: 248, 1972.

170. Yazdi HM, Burns BF: Fine needle aspiration biopsy of Ki-1-positive large-cell "anaplastic" lymphoma. Acta Cytol 35: 306, 1991.

171. Yokose T, Kodama T, Matsuno Y, et al: Low-grade B cell lymphoma of mucosa-associated lymphoid tissue in the thymus of a patient with rheumatoid arthritis. Pathol Intern 48: 74, 1998.

172. Zukerberg LR, Medeiros LJ, Ferry JA, et al: Diffuse low-grade B-cell lymphomas. Four clinically distinct subtypes defined by a combination of morphologic and immnophenotypic features. Am J Clin Pathol 100: 373, 1993.

173. Zucker-Franklin D, Melton JW III, Juagliata F: Ultrastructural, immunologic, and functional studies on Sezary cells: A neoplastic variant of thymus-derived (T) lymphocytes. Proc Natl Acad Sci USA 71: 1879, 1974.

18 Prognostic Markers in Malignant Lymphoma

Thomas M. Grogan

1 Introduction

Both immunophenotyping and genotyping have proved relevant to the biologic subclassification of the lymphomas most recently codified in the Revised European American Lymphoma (REAL) Classification.[19] As manifest in the REAL classification, new antibodies and new molecular probes directed at cellular antigens, RNA and DNA, have provided a greatly refined and expanded definition of the lymphomas.[3,4,16-18] These new assays have delineated important biologic features relevant to the diagnosis, prognosis and therapy of lymphoma. This review explores the clinical utility of immunohistochemistry assays in particular. It emphasized the use of key markers relevant to predicting patient outcome. It suggests that better predictive survival models are possible by combining laboratory and clinical parameters. In some cases, as described, new markers may become the object of alternative therapy.

2 Refined Diagnosis

Over the past 25 years batteries of antibodies have generated immunophenotypic profiles or pattern of antigen expression that distinguish lymphoma from normal variations in lymphoid tissue. In many instances a characteristic immunophenotypic "fingerprint" is associated with a specific lymphoma entity.[3,4,16-19] A case in point is found in Hodgkin's disease.[9,42] In this lymphoma, the large polylobated diagnostic Reed-Sternberg cells have a unique combination of surface antigen expression including: (1) the lymphoid activation antigen (CD30); (2) a myelomonocytic antigen (CD15); and (3) an absent universal lymphoid antigen (CD45). This unique phenotype CD15+30+45– is decidedly pathologic with no physiologic counterpart.[9,42] The neoplastic cells uniquely and inexplicably express lymphoid activation antigens (CD30) yet lack universal lymphoid (CD45) antigen. It expresses lymphoid activation antigen (CD30) yet co-expresses a myeloid antigen (CD15). The lack of any physiologic counterpart, the uniqueness of the phenotype and finally the occurrence of this unique phenotype (CD15+30+45–) in large polylobated lymphoid cells provides key adjunctive biologic data, allowing more certain and refined diagnosis of Hodgkin's disease.[9,42]

Among non-Hodgkin's lymphomas, immunophenotypic markers are also a great aid to definitive classification. A case in point is mantle cell lymphoma (MCL), a small cell lymphocytic lymphoma which is difficult to distinguish from more indolent CLL-like lymphomas by morphology alone.[2,37] The uncertain morphologic distinction from other small cell lymphoid tumors was greatly aided by detection of unique, pathologic nuclear protein, cyclin D1. The pathologic overexpression of cyclin D1 within the nuclei of mantle cell lymphoma cells is the consequence of an aberrant chromosomal translocation t (11;14) which involves the bcl-1 locus on chromosome 11 and the IgH locus of chromosome 14.[2,37] This translocation results in the overexpression of the PRAD-1 proto-oncogene which encodes the cell cycle specific cyclin D1 which is not normally expressed in physiologic lymphoid cells. The occurrence of cyclin D1-positive lymphoid cells in a mantle zone configuration is a characteristic tissue section immunohistochemical pattern of MCL. The cyclin D1 nucleo-regulatory molecule drives cells out of GI into S phase, resulting in loss of proliferative control and a less indolent, more virulent lymphoma.[2,12,37] Thus in spite of the appearance as small resting lymphoma cells, many (30%) MCL cells are in cycle, as measured by the Ki67 antibody. This increased proliferative rate (30%) contrasts with the usual indolent lymphocytic lymphoma with a proliferative rate of 1%-3% measured by Ki67. The higher proliferative rate of MCL explains the high lethality described in MCL as described below.[12]

Not all small cell lymphocytic lymphomas are defined by a single salient marker. Sometimes it is an amalgamation of markers with a matrix of negative findings which is characteristic. Mucosa-associated lymphoma (MALToma) is a case in point.[6,23,43] This heterogeneous lymphoma usually shows a predominance of small lymphocytic lymphoma cells–an entity historically known as small cell lymphocytic lymphoma (SLL). However, by the more recent definition, it is characterized by Pan B (CD20+79d+) antigen expression without CD5 (as in CLL vs MCL) without cyclin D1 (as in MCL) and without CD10 (as in follicular lymphoma) so by exclusion it is phenotypically a mucosa-associated lymphoma.[6,23,43] In most instances the adjacent mucosa shows clustered invasion by lymphoid cells (lymphoepithelial lesions) giving further morphologic definition. Importantly, this combined phenotypic/morphologic definition identifies, in the stomach or gut, a lymphoma of known etiology: an association is found to *Helicobacter pylori* bacterium in 92% of cases.[6,43] This Helicobacter driven lymphoma is initially subject to curative therapy via triple antibiotics. These antibiotics are said to be curative in 70% of cases.[6,43] Thus ultimately phenotypically and morpho-

Table 18-1 Molecular lesions in lymphomas

Lymphoma	Oncogene	Translocations	Protein
Mantle cell	BCL-1	t (11; 14)	Cyclin D1
Follicular	BCL-2	t (14; 18)	Anti-apoptosis
Small lymphocytic (SLL)	BCL-3	t (14; 19)	NF-kb inhibitor
Immunocytoma (LPL)	PAX 5	t (9;14)	Transcription factor
Large cell	BCL-6	t (32; 14)	Zin finger Transcription factor
Anaplastic	NPM-ALK	t (2; 5)	Tyrosine kinase
Burkitt's	C-MYC	t (8; 14)	Cell cycle progression

logically defined MALToma leads to an etiology which in turn leads to a specific antibiotic therapy. Markers in this circumstance lead not only to refined diagnosis but also to entity-specific curative therapy.

Since the mucosa-associated lymphoma (MALToma) is largely comprised of small lymphocytic lymphoma cells identical to SLL cells, described as category A within the Working Formulation (WF) (see Table 17-5), it would appear that it would suffice to consider a MALToma simply by the old designation of SLL. However, the need for the newly designated term MALToma is necessary for the following reasons. Firstly, a diagnosis of SLL strongly implies widespread blood borne disease frequently requiring chemotherapy, whereas a MALToma in the gut is typically entirely restricted to the gut without blood borne spread. Secondly, since MALToma may be cured by triple antibiotic therapy, chemotherapy initially would be inappropriate. In this instance, the old designation of Working Formula category A would lead to inappropriate chemotherapy and less than optimal patient care. It is this prospect of curative, etiology-based, phenotype-adapted therapy that motivates the designation of these new entities with their own separate status. In the end it may be said that the new immunophenotypic and genotypic entities have inherent therapeutic implications, suggesting the need for treatment strategies targeted toward the underlying biologic phenotypic aberrancies of the specific lymphoma types.

Among the small cell lymphomas another newly designated, phenotypically defined entity, known as the immunocytoma has also achieved status as a distinct entity due to both the specificity of its etiologic agent and its associated molecular lesion. In particular there is a newly found association between Hepatitis C virus infection and immunocytoma or lymphoplasmacytoid lymphoma (typically CIgM+CD5-10-Cyclin D1-).[11,14,36,39] There is also a newly described molecular lesion in the form of a translocation t (9;14) of the PAX5 gene.[22,34] Regarding the association of Hepatitis C virus (HCV), a strong prevalence of HCV was found in 30% of immunocytomas in a recent additional study; while HCV was a rarity (2.9%) among other lymphomas.[36,39] The majority of these HCV-positive lymphomas secrete a cryoprecipitable IgM, K component with rheumatoid factor activity and frequently present with liver or kidney involvement.[14,36,39] There are many similarities with essential mixed cytoglobulinemia (EMC) which is of interest given the therapeutic efficacy of alpha interferon

in EMC.[11,14] This implies alpha-interferon as a new novel therapeutic agent to treat immunocytoma. Regarding the molecular lesion, 50% of immunocytomas have a translocation of t(9;14). Analysis of the 9q 13 break-point reveals overexpression (11x) of the PAX 5 gene encoding B-cell specific transcription factors relevant to B-cell proliferation.[22,34] The PAX 5 gene is juxtaposed and falls under the transcriptional context of the IgH locus.[22,34] As listed in Table 18-1, this unique molecular lesion associated with immunocytoma is part of a growing list of molecular lesions specific for lymphoma entities.

3 Phenotypic Markers of Prognosis

Immunohistochemistry (IHC) has been used to detect cell surface, cytoplasmic and nuclear antigens that correlate with prognosis and this body of literature has recently been reviewed in detail.[16] The immunohistochemical study of prognostic significance to date has been largely developmental as most of the published studies are retrospective analyses of small patient groups with heterogenous clinical features, variable treatments, variable assay conditions, and variable statistical cut-points. More recently, larger prospective IHC trials among uniformly staged and treated unselected populations have been published and suggest some IHC markers may have independent clinical significance.[7,31] While more time and studies are needed to give ultimate clinical proof of utility, even the initial studies have value, as they reveal the biologic features and principles which underlay lymphoma therapy failure.[16] The specific antigens identified as factors in lymphoma tumorigenicity present new targets for therapy and present new testable treatment hypothesis.

Two major disease types are selected as examples to illustrate the utility of markers in predicting lymphoma outcome: (1) small cell lymphoma previously known in the WF (see Table 17-5) as diffuse small cleaved cell lymphoma (DSCL) also known as diffuse poorly differentiated lymphocytic lymphoma (DPDLL) (Rappaport) ; (2) Diffuse large cell lymphoma.

The first example, DPDDL or DSCL is chosen to illustrate from Southwest Oncology Group (SWOG) data how a single lymphocytic lymphoma entity has evolved into three separate clinically relevant entities over the past 20 years.[17] This evolution into separate entities was initially driven by phenotyping which revealed discrete disease specific proteins. Fig.18-1 illustrates this evolu-

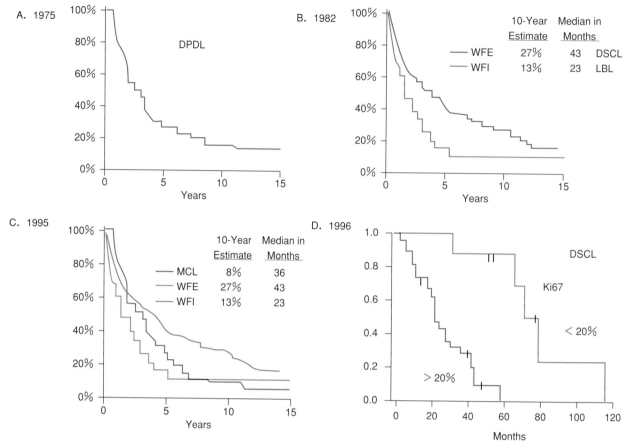

Fig. 18-1 Evolution of diffuse poorly differentiated lymphocytic lymphoma.

tionary process, beginning with a single entity of DPDLL in 1975 to the WF category of DSCL in 1982. In 1982 a DSCL entity was identified by the nuclear protein terminal deoxynucleotidyl transferase (Tdt). As illustrated, this disease proved to have a statistically significantly less favorable survival. In 1995 another entity from within DSCL was delineated: mantle cell lymphoma (cyclin D1+) which also proved to have a poor prognosis. Finally in 1996, proliferation, as measured by Ki67 index, revealed that among the remaining seemingly homogenous DSCL that yet another poor prognosis category existed.[17,26] As illustrated, protein markers continue to delineate discreet biologic entities relevant to prognosis and alternative therapies. These entity-specific markers include: cyclin D1 in MCL and Tdt in lymphoblastic lymphoma and Ki67 in poor prognosis DSCL[26]. All are nuclear proteins relevant to DSCL cell proliferation and immortalization. It may seem that these specific markers. Cyclin D1, Tdt and Ki67, only serve to delineate poor prognosis group which are incurable. In short, the markers may be seen as only delineating futility. However, as evidenced by the recent high cure rate of lymphoblastic lymphoma via marrow transplantation and by recent specific anti-cyclin D1 therapies, these high risk groups benefit from alternative phenotype-adapted therapies and yesterday's futilities becomes tomorrow's new therapeutic opportunity.

The second major lymphoma category where considerable effort has been expended regarding predicting

outcome has been with the intermediate and high grade lymphomas, in particular, large cell lymphoma. These include studies done in a retrospective manner in single institutions and those done prospectively in multi-institutional studies.

■ Single institution, retrospective study

A case in point regarding these prognostic factors is illustrated in Fig.18-2. This figure illustrates several immunophenotypic features which are predictive of poor outcome in a consecutive series of 105 diffuse large cell lymphoma patients at the University of Arizona. This figure, a composite of the 6 published papers[15,27-29,32,41] from the single study, reveals that poor outcome relates to: (1) loss or absent HLA-DR antigen; (2) a high proliferative rate with Ki67 >60%; (3) presence of T-cell lineage; (4) loss of a Pan B antigen (CD22); and (5) a deficiency of tumor infiltrating T-cells (T-TIL, CD8). Finally as shown in the illustrated model, these immunologic parameters were important independent predictors of outcome among these large cell lymphoma patients and have value in identifying patients who will not respond to currently available therapy.[40] These findings suggest at least 4 factors relevant to prognosis: proliferative status, B versus T-cell lineage cell adhesion molecules or recognition molecule status and host cell response (immunosurveillance).

Beyond this University of Arizona experience, the literature, focussed on these same four areas, indicates

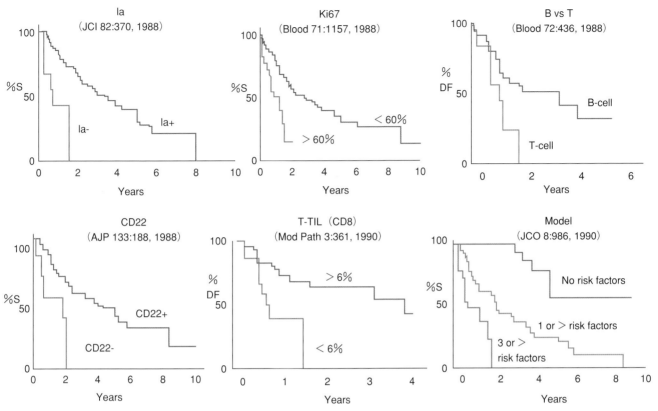

Fig. 18-2 Predictors of survival in large cell lymphoma.

mixed results. Studies from several groups have found that loss of Class I or II antigens predict a poor prognosis[20,25,32,33,38]; other do not.[30,35] Regarding cell lineage, the prognostic significance of determining B-cell versus T-cell lineage is well established among certain subtypes of lymphoma, specifically lymphoblastic T-cell lymphoma and mature T-cell subtypes of DSCL.[26] However, the utility of determining B-cell versus T-cell phenotype among other diffuse intermediate and high-grade lymphomas is more controversial with some studies showing no difference[8,21] and others showing a clear survival difference with T-cell always inferior to B-cell phenotype.[1,20,24,27] Regarding proliferation two studies of large cell lymphoma have found that poor survival correlates with elevated Ki67 of > 60%.[5] In contrast, other studies have shown the opposite with poor outcome related to a low Ki67 value.[13] However, the latter study in particular did not account for, or balance, all known prognostic or treatment factors as in the above Arizona study. Therefore, a comparison among studies can not be made and the issue of Ki67 efficacy is considered moot with a call for a prospective test.

■Multi-institutional, prospective studies

More recent studies seeking to answer this call have focussed on multi-institutional prospective design with predetermined cutpoints looking at uniformly staged and treated patients.[7,31] These studies balance known clinical prognostic factors (e.g., age, sex, stage, LDH status). One of these studies, a prospective test of Ki67 in the Southwest Oncology Group, is illustrated in Fig.18-3.[31] In this study, a high Ki67 was found to be useful in identifying a group of intermediate and high grade

patients with rapidly progressive and fatal disease. The high Ki67 patients experienced one year survival of 18% versus 82% for those with a low proliferative rate. A multivariant regression comparing the effect of Ki67 to performance status, age, stage, extranodal sites of disease and serum LDH confirmed the independent effect of proliferation and survival.[31]

In this study incorporating the Ki67 proliferative index, stage (tumor burden) and LHD, the relative risk assigned to each factor (relative risk 5.9, 1.3, and 0.8 respectively) suggests that a direct IHC measure of the growth rate with Ki67 is a more powerful prognostic tool than serum LDH levels and largely displaces LDH as a significant variable. Further evidence of the utility of the Ki67 index is the finding that 5 of 11 patients (45%) with high values were identified among the two lowest risk groups as defined by the new International Prognostic Index.[31] The utility of Ki67 has also been tested among patients with lower grades of NHL and may prove valuable in selecting patients for more aggressive therapy.[26]

Regarding lineage, a recent large multi-institutional test by GELA (Groupe d'Estude des Lymphomes Agressives) involving a consecutive series of 361 French and Belgian patients with aggressive lymphoma showed that peripheral T-cell phenotype was associated with a poor prognosis and that the T-cell phenotype is independent of other adverse clinical prognostic factors.[7] Multivariate analysis established 4 adverse factors: T phenotype, LDH level, serum albumin level and number of extranodal sites. The French, Arizona and Nebraska studies of lineage reveal a high level of relapse suggesting peripheral T-cell lymphomas may be an incurable subset

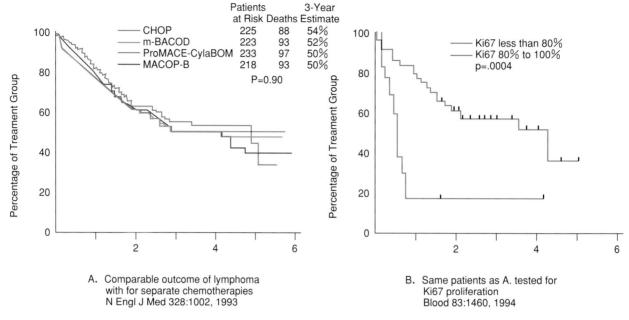

Fig. 18-3　Prognostic significance of Ki67 antigen in SWOG lymphoma.

Table 18-2　Malignant properties of lymphoma

Malignant Phenotypes	Markers	Biologic Principles
1. High proliferative rate	Ki67, P53, C-MYC	Loss of growth control
2. Low proliferative rate	BCL-2	Resting status conferred
3. Loss of CAM	LFA-1, CD54, CD22	Loss of cell cohesion
4. Gain of CAM	CD56	Extranodal localization
5. Loss of lineage Ag (PanB-T)	CD2, 5, 22, 23	Loss of cell cohesion
6. Loss of HLA	HLA I, II	Loss of immunosurveillance
7. Loss of T-TIL	CD3, 4, 8	Loss of immunosurveillance
8. Drug resistance protein	P-glycoprotein	Gain of efflux pump
9. Homing receptor exression	CD44	Gain of mobile phenotype
10. Collagenase expression	Collagenase IV	Gain in invasiveness

of intermediate and large cell lymphoma.[1,7,27]

As evidenced by the SWOG[31] and GELA[7] studies, immunophenotypic markers are showing promise for clinical utility in multigroup lymphoma trials. In particular, IHC as cited might aid patients who are highly likely to fail conventional therapy and require alternative therapy. Markers like Ki67 may serve in judging the comparability of lymphoma groups studied.

However, more important than any of these uses, IHC markers might give specific insight into why certain patients fail therapy.[16]

■Malignant phenotypes suggest biologic principles

The "malignant" phenotypes and the related biologic principles are in Table 18-2. As shown, each altered phenotypic property (e.g. absent LFA-1) suggests a principle (loss of cell adhesion or loss of immunosurveillance). It is noteworthy that many of the listed "malignant phenotypes" entail a loss or deficiency of marker expression. This suggests that loss of pivotal physiologic surface molecules is commonly adversely related to prognosis. Relevant to this hypothesis, loss of multiple recognition

molecules has been shown to result in progressively worse prognosis, whereby a patient with three lost antigens has a worse survival than one with one loss (median survival of 4 months vs. 12 months).[41] This suggests a series or progression of deletional events in lymphoma similar to the allotypic deletional events described in colon carcinoma progression.[10]

■Conclusion

The lymphoma markers mentioned in this chapter including cyclin D1, Ki67, Tdt, B and T-cell antigens and others, were utilized to aid refined diagnosis, improved prediction of patient outcome and to suggest new targets for therapy. As the lexicon of markers steadily increases we can except further refinements in diagnosis, prognosis and therapy.

References

1. Armitage JO, Vose JM, Linder J, et al: Clinical significance of immunophenotype in diffuse aggressive non-Hodgkin's lymphoma. J Clin Oncol 17: 1783, 1989.
2. Banks P, Chan J, Cleary M, et al: Mantle cell lymphoma: A proposal for unification of morphologic, immuno-

logic and molecular data. Am J Surg Pathol 16: 637, 1992.

3. Chan JKC: A new classification of lymphomas: The revised European-American lymphoma classification. Adv Anat Pathol 1: 166, 1994.

4. Chan JKC, Banks PM, Cleary ML, et al: A revised European-American classification of lymphoid neoplasms proposed by the International Lymphoma Study Group. 103: 543, 1995.

5. Chott A, Augustin I, Wrba F, et al: Peripheral T-cell lymphomas: A clinicopathologic study of 75 cases. Hum Pathol 21: 1117, 1990.

6. Cogliatti S, Sehmid U, Schumaeher U, et al: Primary B-cell gastric lymphoma: A clinicopathological study of 145 patients. Gastroenterology 101: 1159, 1991.

7. Coiffier B, Brousse N, Peuchmaur M, et al: Peripheral T-cell lymphomas have a worse prognosis than B-cell lymphomas: A prospective study of 361 immunophenotyped patients treated with the LNH-84 regimen. Ann Oncol 1: 45, 1990.

8. Cossman J, Jaffe ES, Fisher RI: Immunologic phenotypes of diffuse, aggressive non-Hodgkin's lymphomas. Correlation with clinical features. Cancer 54: 1310, 1984.

9. Falini B, Stein H, Pileri S, et al: Expression of lymphoid-associated antigens on Hodgkin's and Reed-Sternberg cells of Hodgkin's disease: An immunocytochemical study on lymph node cytospins using monoclonal antibodies. Histopathology 11: 1229, 1987.

10. Fearon ER, Cho KR, Nigro JM, et al: Identification of a chromosome 18q gene that is altered in colorectal cancers. Science 247: 49, 1990.

11. Ferri C, Marzo E, Longombardo G, et al: Interferon-alpha in mixed cryoglobulinemia patients: A randomized, crossover-controlled trial. Blood 81: 1132, 1993.

12. Fisher R, Dahlberg S, Banks P, et al: A clinical analysis of two indolent lymphoma entities: Mantle cell lymphoma and marginal zone lymphoma including MALT and monocytoid B subcategories. Blood 85: 1075, 1995.

13. Gerdes J, Stein H, Pileri S, et al: Prognostic relevance of tumor cell growth fraction in malignant non-Hodgkin's lymphomas. Lancet 2: 448, 1987.

14. Gorevic PD, Kassab HJ, Levo Y, et al: Mixed cryoglobulinemia: Clinical aspects and long-term follow-up of 40 patients. Am J Med 69: 287, 1980.

15. Grogan TM, Lippman SM, Spier CM, et al: Independent prognostic significance of a monoclonal antibody Ki67. Blood 71: 1157, 1989.

16. Grogan TM, Miller TM: Immunobiologic correlates of prognosis in lymphoma. Semin Oncol 20: 58, 1993.

17. Grogan TM, Rimsza L, Leith C: A new classification of lymphomas: An update of the Revised European American Lymphoma Classification. Adv Pathol 9: 161, 1996.

18. Grogan TM, Spier CM, Richter LC, et al: Chapter 2. Immunologic approaches to the classification of non-Hodgkin's lymphomas. In: Bennett JM, Foonk A (eds): Immunologic Approaches to the Classification and Management of Lymphomas and Leukemias. pp31-48. Norwell, MA: Kluwer Academic Publishers, 1988.

19. Harris NL, Jaffe ES, Stein H, et al: A revised European-American classification of lymphoid neoplasms: A proposal from the International Lymphoma Study Group. Blood 84: 1361, 1994.

20. Hart S, Toghill P, Vaughan-Hudson G, et al: Phenotypic analysis of diffuse large cell lymphoma in paraffin sections: The relationship to prognosis and natural history (abstract). Prog Abstr 3rd Int Conf Malignant Lymphoma, Lugano, Switzerland, 1987.

21. Horning SJ, Doggett RS, Warnke RA, et al: Clinical relevance of immunologic phenotype in diffuse large cell lymphoma. Blood 63: 1209, 1986.

22. Iida S, Rao PH, Nallasivam P, et al: The t(9; 14)(p13; q32) chromosomal translocation associated with lymphoplasmacytoid lymphoma involves the PAX-5 gene. Blood 88: 4110, 1996.

23. Isaacson P, Spencer J: Malignant lymphoma of mucosa-associated lymphoid tissue. Histopathology 11: 445, 1987.

24. Jalkanen S, Joensuu H, Klemi P: Prognostic value of lymphocyte homing receptor and S phase fraction in non-Hodgkin's lymphoma. Blood 75: 1549, 1990.

25. Kluin PK, Gromingen KV, Sandt MVD, et al: Histoimmunopathology related to survival in a regional registry of non-Hodgkin's lymphomas (abstract). Progr Abstr 3rd Int Conf Malignant Lymphomas, Lugano, Switzerland, 1987.

26. Leith C, Spier C, Grogan T, et al: Diffuse small cleaved cell lymphoma: A heterogeneous disease with distinct biologic subsets. J Clin Oncol 10: 1259, 1992.

27. Lippman SM, Miller TP, Spier CM, et al: The prognostic significance of the T-cell phenotype in diffuse large cell lymphoma: A comparative study of the T-cell and B-cell phenotype. Blood 72:436, 1988.

28. Lippman SM, Spier CM, Miller TP, et al: Tumor-infiltrating T-lymphocytes in B-cell diffuse large cell lymphoma related to disease course. Mod Pathol 3: 361, 1990.

29. List AF, Spier CM, Miller TP, et al: Deficient tumor-infiltrating T-lymphocyte response in malignant lymphoma: Relationship to HLA expression and host immunocompetence. Leukemia 7: 398 1993.

30. Medeiros LJ, Gelb AB, Wolfson K, et al: Major histocompatibility complex Class I and Class II antigen expression in diffuse large cell and large cell immunoblastic lymphomas: Absence of a correlation between antigen expression and clinical outcome. Am J Pathol 143: 1086, 1993.

31. Miller T, Grogan T, Dahlberg S, et al: Prognostic significance of the Ki67 associated proliferation antigen in aggressive non-Hodgkin's lymphomas: A prospective Southwest Oncology Group Trial. Blood 83: 1460, 1994.

32. Miller TP, Lippman SM, Spier CM, et al: HLA-DR (Ia) immune phenotype predicts outcome for patients with diffuse large cell lymphomas. J Clin Invest 82:370, 1988.

33. Momburg F, Herrmann B, Moldenhauser G, et al: B-cell lymphomas of high grade malignancy frequently lack HLA-DR, -DP and -DQ antigens and associated invariant chain. Int J Cancer 40: 598, 1987.

34. Ohno H, Furukawa T, Fukuhara S, et al: Molecular analysis of a chromosomal translocation, t(9; 14) (p13;q32), in a diffuse large-cell lymphoma cell line expressing the Ki-1 antigen. Proc Natl Acad Sci USA 87: 628, 1990.

35. O'Keane JC, Mack C, Lynch E, et al: Prognostic correlation of HLA-DR expressions in large cell lymphoma as determined by LN3 and antibody staining. Cancer 66: 1147, 1990.

36. Pozzato G, Mazzaro C, Crovatto M, et al: Low-grade

malignant lymphoma hepatitis C virus infection, and mixed cryoglobulinemia. Blood 84: 3047, 1994.

37. Rosenberg C, Wong E, Petty E, et al: Overexpression of PRAD1, a candidate BCL1 breakpoint region oncogene in centrocytic lymphomas. Proc Natl Acad Sci USA 88: 9638, 1991.

38. Rybski JA, Spier CM, Miller TP, Lippman SM, McGee D, Grogan TM: Prediction of outcome in diffuse large cell lymphoma by the major histocompatibility complex Class I (HLA-A,B,C) and Class II (HLA-DR, -DP, DQ) phenotype. Leukemia Lymphoma 6: 31, 1991.

39. Silvestri F, Pipan C, Barillari G, et al: Prevalence of Hepatitis C virus infection in patients with lymphoproliferative disorders. Blood 87: 4296, 1996.

40. Slymen DJ, Miller TM, Lippman SM, et al: Immunobiologic factors predictive of clinical outcome in diffuse large-cell lymphoma. J Clin Oncol 8: 986, 1990.

41. Spier CM, Grogan TM, Lippman SM, et al: The aberrancy of immunophenotype and immunoglobulin status as indicators of prognosis in B-cell diffuse large cell lymphoma. Am J Pathol 133:118, 1988.

42. Stein H, Uchanska-Ziegler B, Gerdes J, et al: Hodgkin and Sternberg-Reed cells contain antigens specific to late cells of granulopoiesis. Int J Cancer 29: 283, 1982.

43. Wotherspoon A, Doglioni C, Diss T, et al: Regression of primary low-grade B-cell gastric lymphoma of mucosa-associated lymphoid tissue type after irradiation of Helicobacter pylori. Lancet 342: 575, 1993.

19 Bone and Bone Marrow

Biopsy is an important tool for diagnosis of malignant tumors of bone and cartilage. Appropriateness of incisional procedure from the marrow with rich blood supply has been debated; the surgical procedure may promote distant metastases of malignant tumors *via* blood vessels. It is needless to say that needle aspiration with a special device on the tip of the needle provides less probable risk of metastasis than surgical biopsy. However, an application of needle aspiration is limited to the cases when soft tissue extension of neoplasia and osteolytic lesions have occurred; Sanerkin and Jeffree[16] mention the first choice of needle aspiration to (1) destructive lesions such as cancer metastasis, multiple myeloma and eosinophilic granuloma, (2) clinically atypical phase of chronic osteomyelitis, and (3) extraosseous extension of neoplasia as in osteosarcoma. Usually the needle for sternal marrow puncture is usable. Open needle aspiration and imprint preparation during the surgery have different significance as an adjunct to biopsy histodiagnosis; rapidity of the diagnosis and precise interpretation of cellular abnormalities are advantageous compared with tissue section prepared after decalcification.[11]

1 Osteosarcoma, Osteogenic Sarcoma

Long bones of the lower extremities are commonly affected by osteogenic sarcoma and the femur is the most common site of involvement. Osteosarcoma occurs in adolescents under 20 years of age without significant sex difference.

Fine needle aspiration cytology is preferable to biopsy incision that may result in tumor cell spillage. Cytology is an adjunct to X-ray and CT or MRI scan prior to open biopsy.

Histology of osteosarcoma is variable from case to case and from place to place; fibroblastic (Fig. 19-1), chondroblastic (Fig. 19-4), myxomatous, and classical osteoblastic patterns may be predominant. In addition to main fibrosarcomatous component with formation of osteoid matrix pleomorphic giant cells are often admixed. Multiple nuclei with atypia differ from those of giant cells in giant cell tumor. Cytologically osteolytic variety comes to the target of needle aspiration cytology. In this bulky tumor with a fleshy consistency fibroblast-like spindle cells are of the main cell constituent. Their nuclei are oval to elliptic in shape, variable in size and show delicate nuclear membrane. The cytoplasm is of moderate amount and poorly defined (Fig. 19-2). Some giant cells have atypical nuclei that vary in size and shape, thus they are easily distinguished from giant cells having uniform multiple nuclei in giant cell tumor (Fig. 19-3); giant celled osteosarcoma cells which are intensely positive for alkaline phosphatase can be distinguished from osteoclasts with strong acid phosphatase activity. Attention should be paid for the evidence that the specimen obtained from the peripheral part of the tumor is most poorly differentiated and lacks osteoid tissue.[7]

Since no significant difference among histologic subtypes to predict the prognosis has been established,[13] interpretation of cell typing is not contributory. An attempt[9] to examine the coexistence of lymphocytes with tumor as the host's immune response is the subject for further investigation. Practically important is the frequent hematogenous metastasis to the lung, especially with bronchial involvement, that gives a positive sputum cytology.

2 Chondrosarcoma

Chondrosarcoma occurring from cartilage *de novo* is more frequent than that secondary to chondroma or osteochondroma. Since the centrally arising chondrosarcoma develops to infiltrative outgrowth destroying the cortical bone, it becomes the target of needle aspiration. Chondrosarcoma cells are characterized by plump cells with large round nuclei. In spite of increased cellularity their nuclei are often fairly uniform (Fig. 19-7). Double nuclei are occasionally encountered. The distinctive feature of chondrosarcoma consists in the cytoplasm enwrapping the nuclei. The cytoplasm is palely stained, glassy in appearance and positive for PAS reaction (Figs. 19-5 and 19-6).

While in osteosarcoma cytological grading of malignancy or subclassification of the histological types do not take a part in prognosis, grading of cellular atypism such as nuclear hyperchromatism, pleomorphism, cellularity and mitotic activity is worth reading in chondrosarcoma.[4, 8]

Chondrosarcoma accounts for about 20% of malignant bone tumors and occurs more frequently in men than women. Middle age groups over 40 years are affected. Although any bone is involved, the pelvis, femur, ribs and scapular girdle are more common sites.[17]

3 Chondroblastoma

Chondroblastoma is a benign tumor of rare occurrence arising from the epiphysis of long bones. This well demarcated tumor is globular and lobulated. Histo-

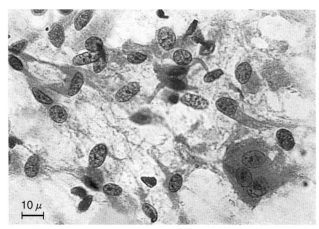

Fig. 19-1 Fibroblastic proliferation in chronic osteomyelitis. Many fibroblasts possess enlarged but uniform nuclei with bland chromatins. Note a benign osteoblast with abundant cytoplasm and bland nuclear figure.

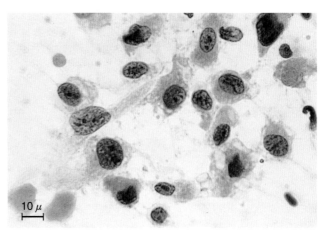

Fig. 19-2 Giant cell component of osteogenic sarcoma. From the same case as Fig. 19-2. It differs from a benign multinucleated osteoclast. The cytoplasm is scant and ill-defined. The multiple nuclei are hyperchromatic and variable in size.

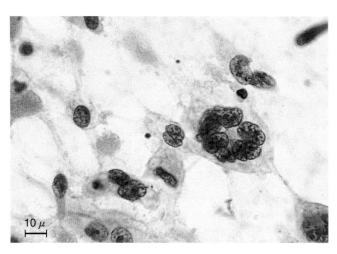

Fig. 19-3 Osteogenic sarcoma cells of poorly differentiated fibroblastic type in imprints. A case of 10-year-old girl. Spindle-shaped, anaplastic tumor cells are characterized by round to elliptic nuclei with coarse chromatins. The nuclear rim is distinct but delicate.

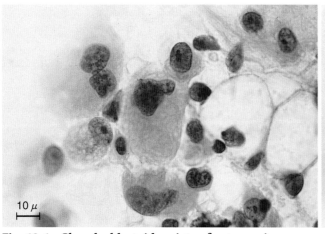
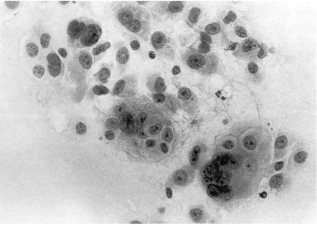

Fig. 19-4 Chondroblastoid variety of osteogenic sarcoma. Fine needle aspiration from the pelvis of an 11-year-old girl. The nuclei are enlarged, hyperchromatic and binucleated. The abundant, glassy cytoplasm mimics chondrocytoid matrix.

logically the tumor is highly cellular and made up of polygonal or rounded cells and some multinucleated giant cells of osteoclast type. Cytologic feature that characterizes chondroblastoma is not found because of resemblance to giant cell tumor. The clinical evidence that the tumor occurs in young people under 20 years of age and the common sites are the epiphysis of tibia, femur and upper humerus should be referred for differential diagnosis. Single plasma cell-like polygonal or round tumor cells are closely packed and possess fairly

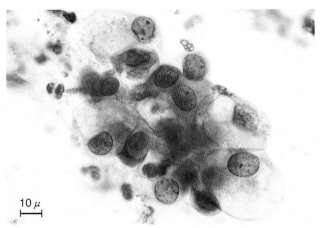

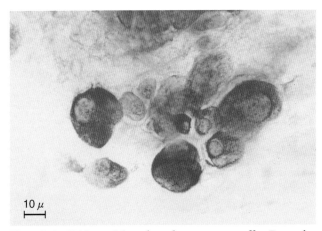

Fig. 19-5 Chondrosarcoma cells shed in cluster by needle aspiration. Enrounded tumor cells are aggregated and characterized by oval to round nuclei with bland chromatin and clear cytoplasm.

Fig. 19-6 PAS-positive chondrosarcoma cells. From the same specimen as Fig. 19-5. Note plump cytoplasm that is intensely positive for PAS reaction. Their surrounding matrix shows also positive stain.

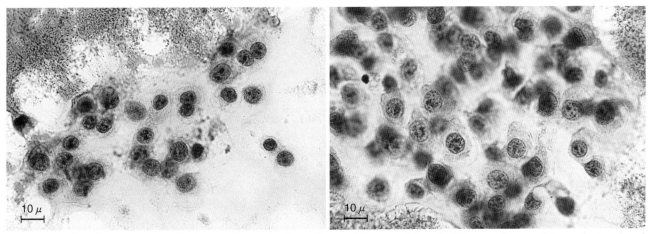

Fig. 19-7 Chondrosarcoma cells in needle aspirations from a 30-year-old male. Three or four nuclei are clustered and surrounded by glassy matrix that is indicative of chondromatous origin. Although the nuclei are uniformly sized, hyperchromatism and coarsely granular chromatin are of malignant criteria.

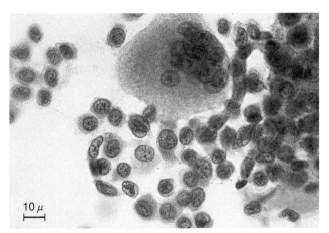

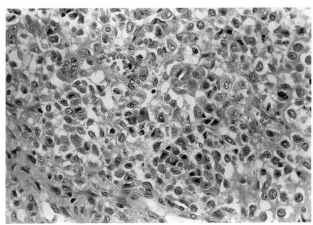

Fig. 19-8 Cytology of chondroblastoma showing multinucleated giant cells and closely gathered single polygonal cells. From the distal end of the femur of a 16-year-old boy. Their nuclei are round to oval, and clearly outlined. Nucleoli are discernible. The chromatin is bland.

Fig. 19-9 Histology of chondroblastoma. From a radiolucent tumor at the distal epiphysis of the femur of a boy aged 16. The tumor is highly cellular and consists of polygonal cells occasionally interspersed by multinucleated giant cells.

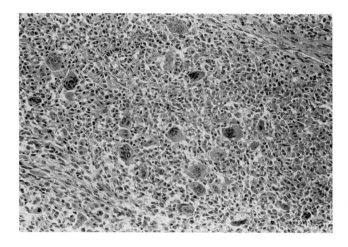

Fig. 19-10 Histology of giant cell tumor reflecting the cytologic pattern. The multinucleated giant cells having many round nuclei are scattered in densely packed polygonal stromal cells. There is neither cartilaginous nor steroid matrix.

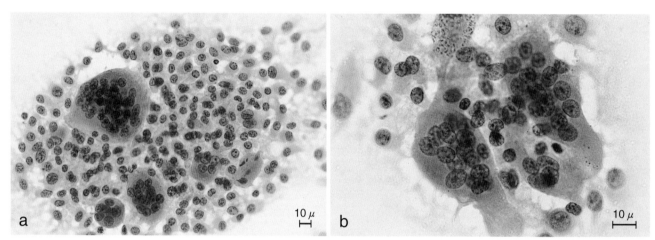

a 10 μ

b 10 μ

Fig. 19-11 Giant cell tumor in imprint smear. The conventional form consisting of fairly uniform stromal cells.
a. Note multinucleated giant cells and single polygonal cells having bland oval or round nuclei. Multiple round nuclei in giant cells are uniform in size and shape.
b. Higher magnification. Multiple nuclei are centrally locating and uniform in size and shape. The specimen is taken from a reddish yellow, granular mass. From an 18-year-old girl.

Table 19-1 Characteristic cytopathology of giant celled lesions

Disease	Common age group	Location in bone	Cytopathology of lesion	
			Giant cells	Stromal cells
Giant cell tumor; benign-malignant	3rd and 4th decades; after the age	Epi-metaphysis of long bones	Osteoclast-like giant cells	Atypical fibroblast-like spindle cells ; malignancy reflected in atypism
Aneurysmal bone cyst; benign	1st and 2nd decades	Metaphysis of long bones and any bone	Small number of giant cells	Hemosiderin-laden stromal cells
Osteogenic sarcoma; malignant	2nd and 3rd decades	Metaphysis of long bones	Apaplastic esteoblasts	Fibrosarcoma cells, occasionally chondrosarcoma cells admixed
Chondroblastoma; benign	2nd decade	Epiphysis of long bones	Small number of giant cells	Plump, round or polygonal stromal cells
Giant cell epulis; benign	2nd and 3rd decades	Mandible and maxilla	Giant cells with many small nuclei	Mature, uniform fibrocytes
Osteoblastoma; benign	2nd and 3rd decades	Any bone	Small number of obsteoclasts	Small number of fibrocytes

uniform nuclei with distinct nuclear rims. Single nucleoli are centrally present. The chromatin pattern is bland. The cytoplasm is from small to fair in amount and ill-defined (Fig. 19-8) Multi-nucleated giant cells with multiple round nuclei simulate those of giant cell tumor but not as many as in the giant cell tumor (Fig. 19-9). Cartilaginous matrix is rarely found in cytology.

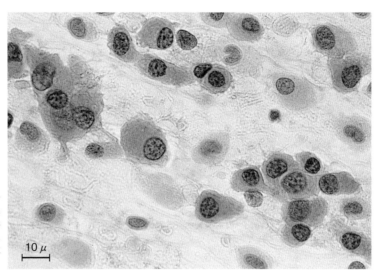

Fig. 19-12 Embryonal rhabdomyoblasts in aspiration during operation. This tumor growing with bone destruction in the pelvis of a 5-year-old boy is highly cellular and composed of small cells with characteristically peripheral nuclei and light greenophilic cytoplasm.

10 μ

4 Giant Cell Tumor

The middle age group of the third and fourth decades are commonly affected by giant cell tumor or osteoclastoma. It occurs primarily from the epiphysis of long bones and extends to the metaphysis; the lower end of the femur and radius, the upper end of the tibia and fibula, and sacrum are the common sites of occurrence.[2] Cystic rarefaction of the tumor discloses a characteristic X-ray picture called "soap bubble appearance." The cellular constituents that characterize giant cell tumor are ovoid cells with plump cytoplasm and multinucleated giant cells having a large number of uniform nuclei (Figs. 19-10 and 19-11). Conventional giant cell tumor or grade I giant cell tumor is abundant in giant cells, but these giant cells decrease in number as the grade of malignancy progresses; in the malignant grade III tumor anaplastic spindle cells overwhelm the giant cells.[7] Therefore, histological grading of malignancy that is dependent on atypism of stromal single cells[10] is only possible in cytodiagnosis if representative areas are smeared. A cytochemical approach to identify the giant cells of osteoclast type is based on increased activity of acid phosphatase, nonspecific esterase, and beta glucuronidase. The fact that increased activity of acid phosphatase is present in the giant cells, but absent in stromal cells is verified by abundance of lysosomes.[1, 3] Doty and Schofield utilize the cytochemical technique to distinguish giant cells that show high lysosomal reactions from the similar-looking osteoclasts that do not display neutral phosphatase activity.

Addendum: Aneurysmal bone cyst

Primary aneurysmal bone cyst is caused by development of arteriovenous fistula within bone in any site. The cavity contains honeycombed vascular spaces surrounded by tissues consisting of fibroblasts and multinucleated giant cells.

5 Rhabdomyosarcoma

Rhabdomyosarcoma is a malignant tumor of soft tissue in the pediatric age period, but becomes the subject of differential diagnosis in bone tumors when it arises from extremities. The typical polymorphic rhabdomyoblasts in adult rhabdomyosarcoma can be distinguished from giant cell tumor and osteogenic sarcoma by their peculiar shape and cross striations. Embryonal rhabdomyoblasts are characterized by small hyperchromatic cells with acidophilic, scanty cytoplasm which are often admixed with strap-shaped myoblasts (Fig 19-12).

6 Bone Marrow Tumor

Malignant tumor in the bone marrow is classified into the primary and secondary neoplasms (Figs. 19-13 and 19-14). The former are those arising from the hematopoietic tissue such as plasma cell myeloma, Ewing's sarcoma and lymphoma.

1) Multiple myeloma

In multiple myeloma or plasma cell myeloma, multiple osteolytic lesions as called "punched out lesions" are recognized in various bones, vertebrae, ribs, skull, pelvis, femur, etc. Visceral organs are also involved and immature plasma cells (myeloma cells) producing monoclonal immunoglobulins are identified morphologically as well as immunocytochemically.

2) Ewing's sarcoma

Ewing's sarcoma is also called endothelial myeloma or endothelial sarcoma. This rare tumor arising from the marrow is found in childhood before puberty with frequent association with intermittent fever, anemia and leukocytosis. The shaft and metaphysis of long bones such as humerus, femur, tibia and fibula are the common sites of Ewing's tumor. It arises from the marrow and extends to the periosteum and soft tissues. Fine needle aspiration becomes an appropriate diagnostic procedure after cortical breakthrough. Cytology is characterized by packed, almost denuded, round cells. The hyperchromatic nuclei are fairly uniform in size and

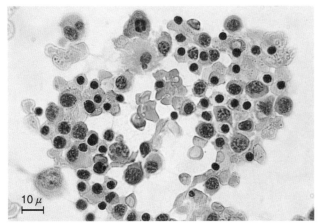

Fig. 19-13 Papanicolaou-stained needle aspirate of bone marrow showing many normoblasts with small round nuclei and immature myeloid and erythroid cells. A few plasma cells with perinuclear halo are also seen.

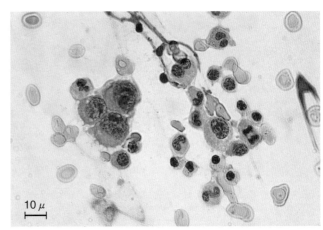

Fig. 19-14 Metastatic adenocarcinoma cells in a bone marrow aspirate. From the same smear as Fig. 19-13. Note clustered cancer cells with enormously large nuclei in eccentric position.

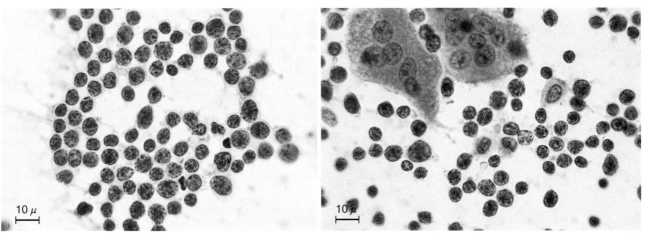

Fig. 19-15 Ewing's sarcoma in imprint smears from a 9-year-old girl. Sheets of enrounded, densely stained tumor cells and two osteoblasts with multiple nuclei are found. These are hardly distinguished from lymphosarcoma cells. But uniformity in shape of the nuclei and sheet or cord-like arrangement of the tumor cells are favorable to Ewing's sarcoma. Osteoblasts are considered to be derived from subperiosteal bone formation.

shape, enrounded like lymphocytes. A few nucleoli and scattered karyosomes are recognizable. Nuclear outlines are delicate but distinct. The cytoplasm is scant, pale and poorly defined (Fig. 19-15). In addition to these cells another cell population comprised of lightly stained larger cells with ill-defined pale cytoplasm are recognizable.[14] The PAS-positive granules which are digested by diastase are present in the cytoplasm,[6] although the PAS stain is not an absolute criterion to classify the "round celled sarcoma" of the bone.[15]

3) Differential diagnosis of look-alike

A difficulty in establishing Ewing's sarcoma consists in the lack of specific features. The monomorphous population of round tumor cells looks like that of malignant lymphoma. The content of diastase-sensitive glycogen by PAS stain is helpful to distinguish from lymphoma.

Occasionally Ewing's cells that aggregate around foci of cell debris may display pseudorosettes. Metastatic neuroblastoma often to the skull is the subject of differential diagnosis. The clinical history of the age younger than 5 years, the light microscopic feature forming Homer Wright rosettes and urinalysis positive for vanillylmandelic acid and/or catecholamines are valuable

findings in neuroblastoma.[5]

Lastly atypical myeloma should be put in mind for differential diagnosis.

7 Histiocytosis X

Histiocytosis involving bones such as Letterer-Siwe disease, Hand-Schüller-Christian disease and eosinophilic granuloma will be the target of needle aspiration cytology. Multiple osseous lesions in early childhood with minor visceral involvement (Hand-Schüller-Christian disease) and localized osteolytic lesions in adolescence or adults (eosinophilic granuloma) are diagnosable cytologically.

1) Hand-Schüller-Christian disease is clinically manifested by diabetes insipidus, exophthalmos and multiple defects of the bone. The skin is occasionally involved and manifested by disseminated yellowish xanthomata.

Cytology: Histological characteristics composed of massive infiltration of histiocytes with occasional positive fat stain and certain eosinophils will be simply reflected in smears; a large number of single histiocytes and multi-

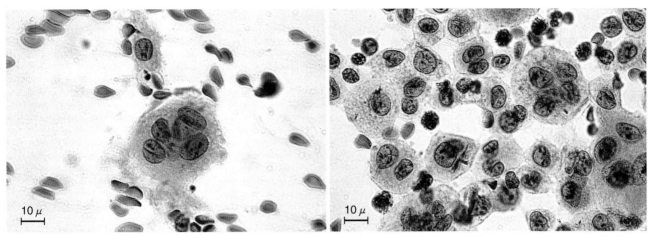

Fig. 19-16 Hand-Schüller-Christian disease showing a characteristic multinucleated histiocyte having centrally locating ovoid nuclei. Needle aspirates from a fronto-temporal lytic region of the skull. A two year and five months old boy disclosed generalized lymphadenopathy and tumorous defects in the skull. Note eosinophilia and intracellular Charcot-Leyden crystals.

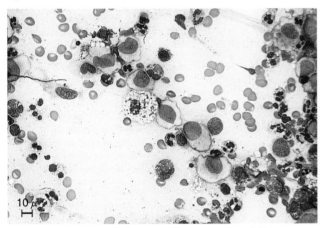

Fig. 19-17 Eosinophilic granuloma aspirated from granulomatous lesion of the skull. There is an admixture of mononucleated histiocytic cells with many eosinophils, neutrophils, and lymphocytes.

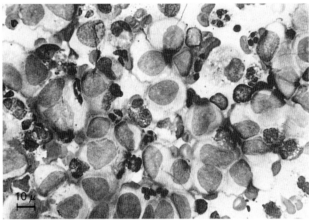

Fig. 19-18 Eosinophilic granuloma of a skull in needle aspiration. A number of histiocytes with bland nuclei are mixed with eosinophils. (By courtesy of Dr. K. Hayashi and Mr. M. Okada, CT, IAC, Prefectural Tajimi Hospital, Tajimi, Gifu, Japan.)

nucleated histiocytes either of the foreign body type or of the Touton type are characteristically seen in mixture with eosinophils. The nuclei are oval or kidney-shaped, peripherally located in the cytoplasm of a moderate amount, and show characteristic cleftlike or coffee bean wrinkle-like foldings of the nuclear membrane (Fig 19-16). Electron microscopically, intracytoplasmic granules were considered to be identical with those of Langerhans cells.[12] In author's experience of a 2-year-old boy with diabetes insipidus and multiple osteolytic lesions in Medical School of Botucatu, Brazil, bacterioid or fine granular substances in the cytoplasm could be demonstrated by differential interference microscopy. The evidence may support the hypothesis on histogenesis of histiocytosis X by Zemel et al[18] who referred to it as impaired lysosomal degradation.

2) Eosinophilic granuloma is of the least severity in this group. Involvement of bones is less marked than in Hand-Schüller-Christian disease, although various sites such as the skull, facial bones, femur, pelvis and spines are affected.

Cytology is characterized by admixture of mononuclear or multinucleated histiocytes with many eosino-

phils (Figs. 19-17 and 19-18).

3) Letterer-Siwe disease is of the worst prognosis and involves rapidly visceral organs; hepatosplenomegaly, lymphadenopathy, pulmonary infiltration and anemia are associated. Osseous infiltration and cutaneous eruption can be cytologically accessible.

References

1. Boouist L, Larsson SE, Lorentzon R: Genuine giant-cell tumor of bone: A combined cytological, histopathological and ultrastructural study. Pathol Eur 11: 117, 1976.
2. Dahlin DC, Cupps RE, Johnson EW Jr: Giant cell tumor. A study of 195 cases. Cancer 25: 1061, 1970.
3. Doty SB, Schofield BH: Enzyme histochemistry of bone and cartilage cells. Prog Histochem Cytochem 8: 1, 1976.
4. Evans HL, Ayala AG, Romsdahl MM: Prognostic factors in chondrosarcoma of bone. A clinicopathologic analysis with emphasis on histologic grading. Cancer 40: 818, 1977.
5. Fechner RE, Mills SE: Small cell sarcomas. In: Tumors of the Bones and Joints. Atlas of Tumor Pathology. pp187-201. Washington, DC: Armed Forces Institute of Pathology, 1993.

6. Gompel C: Atlas of Diagnostic Cytology. New York: John Wiley and Sons, 1978.

7. Huvos AG: Bone Tumors: Diagnosis, Treatment and Prognosis. Philadelphia: WB Saunders, 1979.

8. Marcove RC, Lewis MM, Huvos AG: Cartilaginous tumors of the ribs. Cancer 27: 794, 1971.

9. Marsh B, Flynn L, Enneking W: Immunologic aspects of osteosarcoma and their applications to therapy. J Bone Joint Surg 54: 1367, 1972.

10. Miller FN: Peery and Miller's Pathology. Boston: Little Brown, 1978.

11. Naib ZM: Exfoliative Cytopathology. Boston: Little Brown, 1976.

12. Niebauer G, Krawczyk W, Wilgram GF: Über die Langerhans-Zellorganelle bei Morbus Letterer-Siwe. Arch Klin Exp Derm 239: 125, 1970.

13. O'hara JM, Hutter RVP, Foote FW Jr: An analysis of thirty patients surviving longer than ten years after treatment for osteogenic sarcoma. J. Bone Joint Surg 50: 335, 1968.

14. Salzer-Kuntschik M: Cytologic and cytochemical behavior of primary malignant bone tumors. In: Grundmann E (ed): Malignant Bone Tumors. pp145-156. New York: Springer-Verlag, 1976.

15. Salzer-Kuntschik M, Wunderlich M: Das Ewing-Sarkom in der Literatur: Kritische Studien zur Prognose. Arch Orthop Unfallchir 71: 297, 1971.

15. Sanerkin NG, Jeffree GM: Cytology of Bone Tumours. Bristol: John Wright, 1980.

17. Toogood I: Bone sarcomas. In: Love RR, et al (eds): Manual of Clinical Oncology (UICC). pp464-471. Berlin: Springer-Verlag, 1994.

18. Zemel H, Deeken J, Asel N, et al.: The ultrastructural features of normolipemic plane xanthoma. Arch Pathol 89: 111, 1970.

20 Central Nervous System

1 Collection of Specimens and Preparation of Smears

Collection of cerebrospinal fluid

The cerebrospinal fluid is usually obtained by lumbar puncture at the 3rd or 4th intervertebral space and occasionally by ventricular and occipital punctures with strictly aseptic technique. The normal pressure is less than 180 mmH$_2$O in the lateral decubitus position with the spine strictly horizontal. In cases of increased intracranial pressure as caused by brain tumor, cerebral edema, hemorrhage and meningeal inflammation, withdrawal of the fluid may cause a fatal pressure cone by cerebral herniation. Therefore, there may be occasions when only 2 or 3 ml are available for cytodiagnosis. In suspected brain tumor, abscess or hemorrhage, the fluid should be removed slowly under manometric control.

Centrifugation method

The fluid must be immediately centrifuged at 1,500 rpm for five minutes. If the fluid is insufficient in amount for centrifugation, balanced saline solution is added before centrifugation. When the fluid cannot be processed to centrifugation within the first hour, pretreatment of the fluid with an equal amount of 2% formalin-40% alcohol-balanced saline solution is preferable to prevent autolysis; the solution composed of 2 ml of formalin, 40 ml of ethyl alcohol and 58 ml of balanced saline solution can be preserved as a stock solution for the pretreatment of the cerebrospinal fluid.

The sediment is smeared on thinly albuminized clean glass slides; this procedure is useful to retain the cells on glass slides during fixation. Unless Millipore filtration is performed, one air-dried smear should be prepared for Giemsa stain to avoid the loss of cells during wet film fixation.

The use of polylysine-coated glass slides is able to collect more cells than conventional glass slides. According to the study by van Oostenbrugge et al,[36] only minor improvement in diagnostic sensitivity was described, their comparative study has been carried out using Cyto-Tek Centrifuge (Sakura Finetechnical Co., Tokyo, Japan) that gains superior cell recovery to ordinary centrifugation.

The Cyto-Tek Centrifuge designed for minute fluid

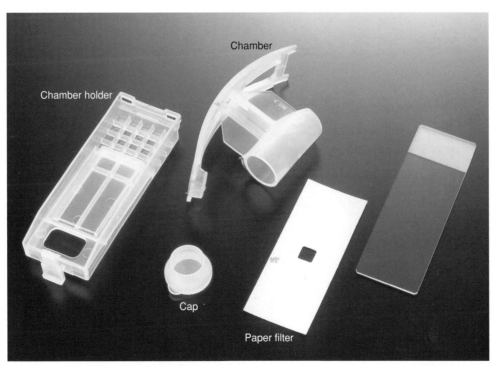

Fig. 20-1 The Cyto-Tek Centrifuge. For preparation of minute amount of fluid specimen. (By courtesy of Sakura Finetechnical Co., Tokyo, Japan.)

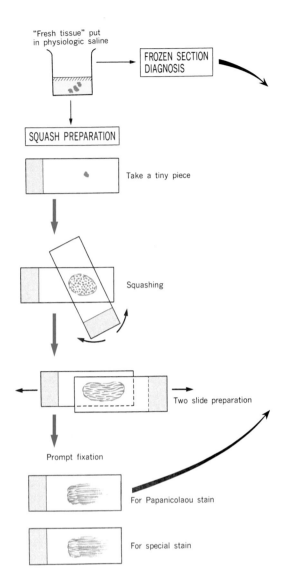

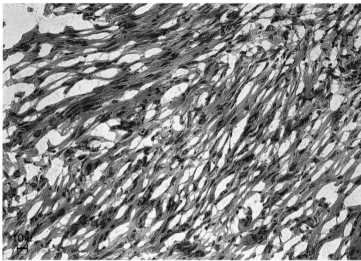

Fig. 20-3 Frozen section showing fascicular streams of spindly cells. Although nuclear palisading and Verocay bodies were not found, acoustic neurinoma was suggested by HE stain.

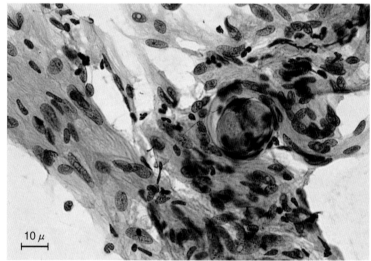

Fig. 20-2 Method of squash (crush) preparation for the central nervous system.

Fig. 20-4 Rapid imprint cytology showing a characteristic whorled pattern of meningothelial cells. Note elongated nuclei with blunt poles. The chromatins are lacy and bland. The presence of whorled nests is diagnostic of meningotheliomatous meningioma.

specimen like cerebrospinal fluid is a unique device consisting of three different size chambers, chamber holder and disposable paper filters. The device is able to make automatically even cellular smears using adjustable rotating speeds from 500 to 2,500 rpm (Fig. 20-1).

■Membrane filtration method

Millipore filter or Nuclepore filter having a pore size of 5 μm is used in conjunction with filtration apparatus set: It consists of a glass funnel, a holding clamp, stainless steel clips, a filtering flask and a suction tube. Cerebrospinal fluids, which are usually small in quantity and scant in cellular components, are the most appropriate specimens for cell concentration by membrane filtration.[18,26] Unfixed specimens must be immediately processed to filtration, but pretreatment of such a mucus-free specimen as cerebrospinal fluid with 50% ethyl alcohol is advisable.[8] Although Millipore filter is dissolved by certain fixatives such as methyl alcohol, formalin and ether-ethanol, Nuclepore filter has advantages

as follows; (a) the membrane is unaffected by most fixatives, (b) various stains are available, (c) mounting is easily performed, and (d) microscopy is not interrupted by the background because of the transparency of the membrane.

■Sedimentation technique

Bots and associates[4] made good use of the sedimentation apparatus designed by Sayk[32] for cytology of cerebrospinal fluid. The principle of sedimentation is absorption of fluid by a hard filter paper put below a glass tube with open ends. The tube 15 mm in diameter and 20 mm long is placed upright on a clean glass slide. A hard filter paper (Green 602 or Schleicher & Scüll 602 hard) with a hole that fits exactly the size of inner wall of the tube is inserted between the tube and the glass slide. A thick rubber block with a hole accommodating the glass tube is used to hold the above arrangement. Fixing is accomplished by pressure upon the rubber block. The fluid is slowly absorbed through the bottom of the glass tube and the cells settle on the glass slide.

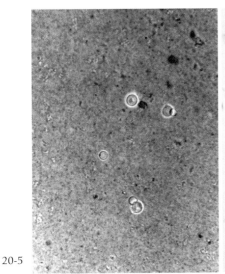 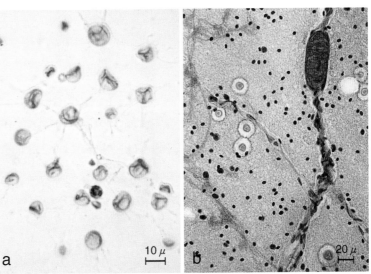

20-5

a

10 μ

b

20 μ

20-6

Fig. 20-5 *Cryptococcus neoformans* in cerebrospinal fluid demonstrated by India ink preparation.

Fig. 20-6 *Cryptococcus neoformans*. **Smear from sediments of cerebral fluid (a), and squash smear (b).** The organisms are demonstrated as light brownish spheres measuring about 8 to 20 μm in diameter by Papanicolaou stain. Although the capsules are not visible (a), these are clearly shown in a squash smear (b).

Bots et al state that 20 to 30% cellular components in the fluid are placed on the glass slide.

■Aspiration cytology

Aspiration cytology under vision directly from a tumor or cystic lesion using a 21 gauge hypodermic needle is valuable when a biopsy specimen cannot be taken. Blind needle aspiration should be avoided.

■Imprint smear

Imprint smears are made from a fresh excised tissue specimen in the same way as it is stamped on clean glass slides; if the specimen is large enough to cut, the fresh cut-surface should be provided for good preparation. This technique is very important as an adjuvant to frozen section diagnosis, because frozen sections of a tiny specimen are occasionally technically impossible. Thus, imprint smear is made before processing to frozen sections. If enough tissue is submitted, squash smears of a minute piece of tissue can be done by the two slide pull-apart method.

■Squash method

A whole-mount technique by squashing a tiny piece of cerebral tissue is superior to the imprint preparation because of precise cellular features and preservation of tissue structures such as blood capillaries. For diagnosis of glioma, cellularity of abnormal cells and behavioral relationship of abnormal cells to glial cells and to capillaries are important. The technique is easy and performed as follows, although its disadvantages are present such as deformity of fragile cells and difficulty of squash for fibrous tumors like meningioma (Figs. 20-2 to 20-4).

Clinical significance: The squash preparation is worthy for intraoperative rapid diagnosis. When a minute piece of specimen is submitted for frozen section diagnosis, cytology using squash samples has definitely priority in identification of tumor cells and abnormally proliferat-

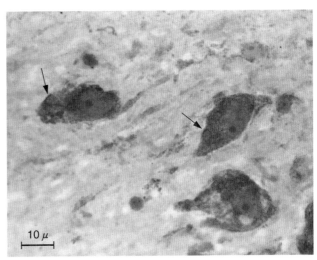

10 μ

Fig. 20-7 Negri body in imprints of dog brain with rabies. The Negri bodies are found in the cytoplasm of neurons as round or oval, eosinophilic inclusions (arrows). They are often multiple. The most common site of localization of the Negri bodies is the Ammo's horn of the temporal lobe.

ing vascular endothelial cells. Torres and Collaco have described 307 cases of intraoperative cytodiagnosis and correctly diagnosed in 92.2% after review of paraffin sections. Difficult interpretation was mentioned to be grading of astrocytoma, confirmation of mixed glioma, identification of oligodendroglioma, etc.[35] Determination of different components such as oligoastrocytoma, ependymoastrocytoma and oligoependymoma can be proved if enough material is submitted to squash preparation. Regarding diagnostic rates as adjunct to frozen section diagnosis, nearly the same results are described.[6, 41]

A marked decrease in misinterpretation for intraoperative rapid diagnosis by combination of squash preparation and frozen section is reported by Kobayashi[15]; the author emphasizes the applicability of immunoperoxidase staining to demounted glass slides.

427

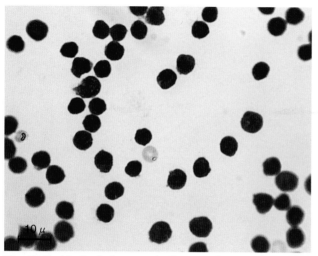

Fig. 20-8 Marked pleocytosis due to tuberculous meningitis in cerebrospinal fluid. From an 18-year-old boy. The cell count in tuberculous meningitis is usually 100 to 500 per cubic millimeter with lymphocytes predominating in association with increase in protein content and decrease in glucose level.

Technique:

1) A tip of match-sized brain tissue is placed between two clean glass slides as early as possible after excision.

2) The tissue is squashed with finger press squeezing so strongly as to make a paste. Any cellular constituent is never destroyed because a matrix of cerebral tissue plays a role to preserve cellular features.

3) Thin films are made by a "pull-apart method" and promptly fixed in absolute ethanol (Fig. 20-2).

Addendum 1: Cryptococcosis and India ink preparation

Cryptococcosis: Cryptococcus neoformans spreads widely from the primary lesion in the lung and shows an affinity to the central nervous system. Wide-spread dissemination may be accelerated in immune-suppressive state, and cryptococcus meningitis may occur. In Papanicolaou stain the organism appears as a pale, brownish, refractile sphere which is surrounded by clear halo. Gomori's methenamine silver and PAS or Gridleys stains are able to visualize the structure.[30]

India ink preparation: This technique is used to examine *Cryptococcus neoformans (Torula histolytica):* A drop of India ink and a drop of the sediment of cerebrospinal fluid are mixed on a slide. A cover slip is placed before microscopic examination. Cryptococcus, with a wide gelatinous capsule, measures from 5 to 10 μm in diameter, and is recognized by a clear halo around each cell (Figs. 20-5 and 20-6).

Addendum 2: Rabies

Rabies rarely becomes the subject of cytology. Pathologic changes in the brain are perivascular lymphocytosis and occurrence of acidophilic Negri bodies in ganglion cells in the hippocampus, medulla oblongata and cerebellar Purkinje cells (Fig. 20-7).

2 Gross Appearance of Cerebrospinal Fluid

Cerebrospinal fluid is normally clear and colorless like water. It has an alkaline reaction with a pH of 7.4 to 7.5 and a specific gravity of 1.003 to 1.008. In purulent inflammation it becomes cloudy and greyish white. It is pink or red with fresh erythrocytes, and xanthochromic from bilirubin with old cerebral or subarachnoid hemorrhage and severe jaundice, and from elevation of total protein; in the presence of prolonged subarachnoid obstruction and spinal block, the protein content is markedly elevated. In acute suppurative or tuberculous meningitis, a pellicle or web-like fibrin net is formed.

3 Benign Cellular Components in Cerebrospinal Fluid

Normal cerebrospinal fluid has essentially no cells or at most five cells per cubic millimeter. They are mostly lymphocytes. Ventricular puncture, however, may contain tissue fragments of ependymal origin. Cytological examination of the cerebrospinal fluid is not limited to diagnosis of tumor cells, but applicable to the detection of inflammation.

■ Non-tumorous pleocytosis

"Sonnenstäubchen" is a phenomenon in which minute dust-like particles appear to be floating if the tube with fluid is shaken and then observed through the sunlight against a dark background. This condition is responsible for slight pleocytosis. Slight haziness of fluid is caused by moderate pleocytosis ranging from 100 to 500/mm³. Slight to moderate lymphocytic pleocytosis is encountered in cases such as tuberculous meningitis, syphilitic meningitis, viral infection, lymphocytic choriomeningitis (Fig. 20-8), etc. A visible turbidity of the fluid occurs with polymorphonuclear leukocytic pleocytosis over 500/mm³ and is indicative of bacterial infection.

Rarely, superficial squamous cells, ependymal cells and erythrocytes can be observed. Squamous cells may be derived from the skin during lumbar puncture. Ependymal cells originate from the ventricular epithelium. They are low columnar or cuboidal with eccentric small nuclei.

■ Cell counting in cerebrospinal fluid

Pappenheim solution for cell counting is prepared by dilution of 4 ml acetic acid with 40 ml distilled water and dissolution of 0.2 g methyl violet. The staining solution is filtered after preparation. Pappenheim solution is taken to mark 1 and the cerebrospinal fluid is drawn up to mark 11 in a white blood cell counting pipette. All the cells in the entire areas of a Fuchs-Rosenthal chamber are counted. One-third of the total count represents the number of cells per cubic milliliter of undiluted fluid.

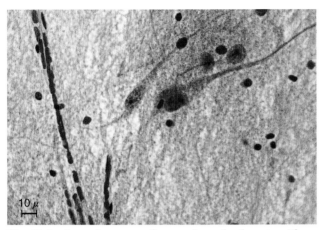

Fig. 20-9   Squash smear of normal cerebral cortex. There are nerve cells, glial cells and capillaries with thin endothelial cell lining in amorphous fibrillar matrix.

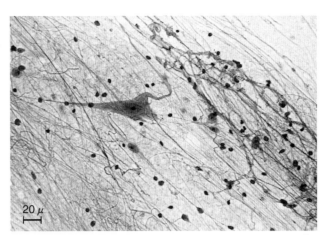

Fig. 20-10   Squash smear of normal cerebral cortex. The Bodian stain demonstrates a pyramidal nerve cell, axis cylinders, and glial cells so well as shown in a tissue section.

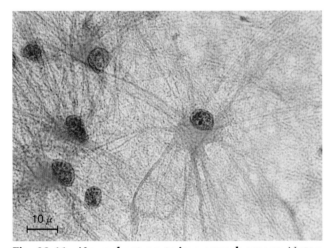

Fig. 20-11   Normal astrocyte in a squash smear. Note a round nucleus and a moderate amount of stellate cytoplasm, from which cytoplasmic processes and fine fibrils are protruded.

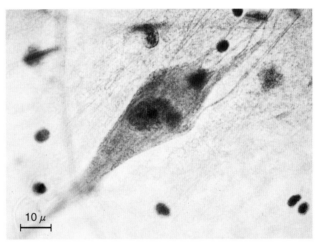

Fig. 20-12   Normal ganglion cell with ample pyramidal cytoplasm in a squash smear. An elliptic nucleus having a large nucleolus locates centrally. Note elongated cytoplasmic processes.

4 Cellular Features of Benign Cells

Neurons and glial cells are not cellular constituents of cerebrospinal fluid, but they are representative cells in a fresh squash or imprint smear (Figs. 20-9 to 20-13).

Ependymal cells and choroidal cells

The lining cells of the ventricular system are of neuroectodermal origin and possess cilia and blepharoplasts along the free edge side. The latter is only detectable histologically by phosphotungstic-acid hematoxylin stain (PTAH). Practically, they rarely participate into the cell constituents of cerebrospinal fluid. The presence of ependymal cells in cerebrospinal fluid means some degenerative disorder of cerebral parenchyma involving the ventricles. The cytoplasm is low columnar and occasionally ciliated. Cilia are not always visible because of degeneration in the fluid. The nuclei are oval, paracentrally located and bland in chromatin pattern. The choroidal cells have no cilia in adults and possess small, pyknotic nuclei.

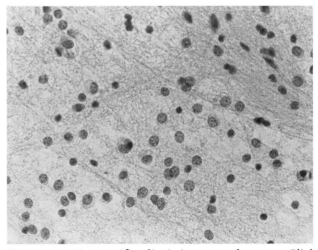

Fig. 20-13   Nonspecific gliosis in a squash smear. Glial cells with uniform, round or oval nuclei are scattered regularly, though slightly increased in number.

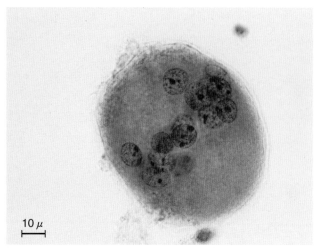

Fig. 20-14 **Multinucleated giant cell of foreign body type in cerebrospinal fluid.** From a case with metastatic carcinoma. Multiple nuclei are equal in size and shape. In an activated state one may find prominent nucleoli.

■Pia-arachnoid cells

The pia-arachnoid system has varied properties such as the function of reticuloendothelial system and the tissue-repairing fibroblastic activity; histiocytes, foreign body giant cells and squamous pia-arachnoid mesothelial cells to be transformed to fibroblasts belong to the pia-arachnoid system.

■Astrocytes

Astrocytes do not appear in cerebrospinal fluid as free cells unless ventricular walls are destroyed by neoplasm, infarction and surgical maneuver. However, these are of main cell constituents in the squash and imprint smear (Figs. 20-11 and 20-13). Round or oval nuclei are characterized by fine dotted chromatin particles. Nuclear rims are not thickened at all. The cytoplasm is broad, polyhedral, and shows cyanophilic staining reaction. Cytoplasmic processes of well preserved astrocytes are visible in the squash smear.

■Oligodendrocytes

The nuclei are small, round or oval, and densely stained. The cytoplasm is transparent and liable to undergo degeneration. In the gray matter the location of oligodendrocytes that adjoin the nerve cells is an informative finding for identification. The naked cells are similar to lymphocytes.[17]

■Microglia

Microglia which are thought to belong to the reticuloendothelial system in the central nervous system are practically impossible to be identified. Reticulum cell-like cells with polygonal or indented, dusky nuclei are called microglia, when they are in the vicinity of the blood capillaries. They may appear as free cells but only discernible in the squash smear.

■Nerve cells

The cytoplasm is pyramidal with multipolar processes. A distinctive neurite can be found in the squash smear. The nucleus is round or oval and centrally located. The

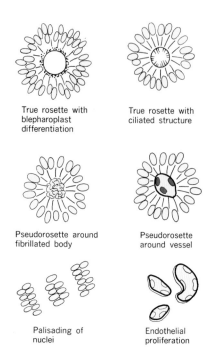

Fig. 20-15 **Variations of malignant cell formation.**

nuclear rim is poorly preserved and indistinctive. There is a prominent, single nucleolus (Figs. 20-10 and 20-12). The neuron varies in size within a physiological condition.

■Macrophages

Mononuclear and multinucleated macrophages in cerebrospinal fluid are considered to be derived from fibrovascular tissue with a tissue reparatory purpose. Phagocytosis of hemosiderin, melanin, lipid and other foreign particles is often recognizable (Fig. 20-14).

5 Abnormal Cytology

Since the cerebrospinal fluid is acellular, the occurrence of abnormal cells alien to the cerebrospinal fluid (CSF) may be informative, regardless of malignant criteria. For example, the cells shed from astrocytoma grade 1 or grade 2 do not meet the criteria of malignancy. The relatively low diagnostic rate of glioma depends on the infrequent involvement of the leptomeninges and the ventricular walls by tumor invasion. According to the study by Wagner et al[37] extension of primary glioma to the leptomeninges is proven histologically in about only 10% of biopsy specimens and only in 35% of autopsies. These facts are in accordance with the low accuracy of cytodiagnosis; 42.9% by Kline[13] and 29.4% by Naylor[21](Table 20-1).

Bigner and Johnston have studied 318 malignant specimens from 232 patients out of 12,026 CSF. It is noteworthy that only 25 patients of 220 patients could be initially diagnosed. Distribution of types of positive CSF cytology in their study shows markedly high frequencies in generalized lymphoma or leukemia (49.1%) and

Table 20-1 Types of tumors and accuracy of cytodiagnosis of cerebrospinal fluid

	Naylor[21]		Kline[13]	
Types	No. of cases	Positive cases	No. of cases	Positive cases
Astrocytoma (1, 2)	17	2	13	4
Astrocytoma (3, 4)	7	3	26	11
Ependymoma	15	5	2	0
Medulloblastoma	7	3	2	2
Oligodendroglioma	1	1	1	0
Pinealoma	4	1	0	0
Choroid plexus carcinoma	0	0	1	1
Chromophobe adenoma	12	3	0	0
Craniopharyngioma	6	0	2	1
Hemangioblastoma	4	0	0	0
Hemangiosarcoma	1	0	1*	0
Meningioma	6	0	6	3
Neurilemmoma	4	0	5	0
Olfactory neuroblastoma	1	1	0	0
Teratoma	0	0	1	1
Metastatic carcinoma	14	5	33	16
Metastatic melanoblastoma	2	1	0	0
Reticulosarcoma	1	1	0	0
Hodgkin's disease	0	0	3	0

* Perithelial sarcoma in the original description.

Table 20-2 Distribution of types of positive CSF cytology

Neoplasms	Number of patients		Initial diagnosis
Lymphoma, leukemia	108		2
Metastatic tumors	98		16
Breast ca.		18	0
Lung, small cell ca.		16	2
Lung, adenoca.		15	3
Lung, undiff. ca.		13	3
Melanoma		12	1
Primary unknown		7	5
Stomach ca.		3	2
Retinoblastoma		2	0
Orbit squamous ca.		2	0
Bile duct ca.		2	0
Others		8	0
Brain tumors	14		7
Medulloblastoma		4	2
Glioblastoma		2	2
Ependymoblastoma		2	0
Pineal germinoma		1	0
Choroid papilloma		1	0
Primary lymphoma		1	0
Others		3	3
Total	220		25

(From Bigner SH, Johnston WW: Acta Cytol 28: 29, 1984.)

Table 20-3 Frequency of brain metastases of primary extracranial neoplasms

Authors	Tom[34]	Knights[14]	Störtebecker[33]	Lesse and Netsky[16]	Aronson[2]
No. of cases	82	94	158	207	397
Lung	22%	22%	16%	24%	46%
Breast	16	18	7	34	13
Gastrointestinal tract	13	7	9	5	9
Melanoma	9	6	8	1	3
Reproductive organs, F.	5				2
Reproductive organs, M.	1	3			2
Urinary tract	5	3	20	5	3
Endocrine organs	4	3		3	4
Lymphoma and sarcoma	6	13		12	10
Others or unknown	19	25	40	16	8

metastatic tumors (44.5%) compared with primary brain tumors (6.3%).[3]

■Structure profile diagnostic of brain tumors

Although malignancy is diagnosable based on single cells in CSF, little information will be given for histological classification. Cell clusters observed in aspirated and/or squashed smears provide not infrequently distinctive patterns that characterize the histological classification (Fig. 20-15).

1) True rosette: Tumor cells are placed around the lumen that resembles the central canal of the spinal canal. The supranuclear cytoplasm closed to the lumen shows often granular blepharoplasts and cilia. Formation of true rosette is characteristically seen in ependymoma and retinoblastoma.

2) Pseudorosette: Homer Wright rosette is simply formed by tumor cells which are radially arranged around lightly eosinophilic areas. Fibrillated material by silver impregnation in the central luminal area is regarded as the site suggestive of neuroblastic differentiation. Formation of Homer Wright rosette is found in medulloblastoma.

3) Perivascular pseudorosette: Radial arrangement of tumor cells around the blood vessels is characteristic of astroblastoma. Other gliomas such as ependymoma, gemistocytic astrocytoma, glioblastoma multiforme and schwannoma occasionally reveal pseudorosette.

4) Nuclear palisading: Parallel fashion with nuclear palisading side by side is characteristic of spongioblastoma. A similar nuclear pattern arranged side by side in rows towards necrosis (pseudopalisading) is distinctively seen in glioblastoma multiforme.

5) Endothelial proliferation: Endothelial proliferation of blood capillaries is a distinctive finding of glioblastoma multiforme, although similar vascular hyperplasia is a phenomenon of oligodendroglioma.

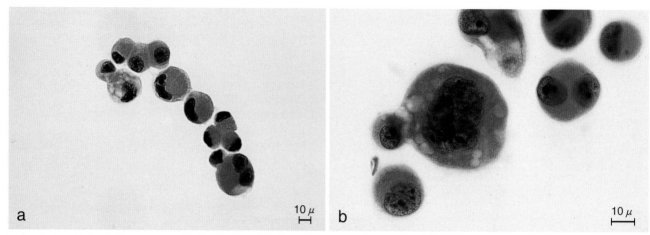

Fig. 20-16　Metastatic cancer cells from gastric adenocarcinoma in cerebral fluid.
a. Low power view showing hyperchromatic nuclei eccentric in position. The cytoplasm is well preserved and sharply bordered.
b. Higher magnification. The cellular features such as hyperchromatism with coarse chromatins and intense stain with distinct cytoplasmic borders differ from those of glioma cells.

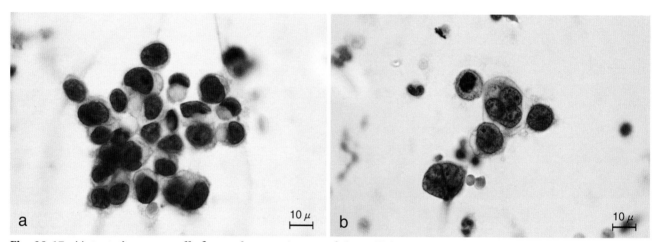

Fig. 20-17　Metastatic cancer cells from adenocarcinoma of the gallbladder in cerebral fluid.
a. Clustered cancer cells. The nuclei are of medium size, but hyperchromatic and possess prominent nucleoli. Note translucent columnar cytoplasm.
b. Well preserved, isolated cancer cells. Note hyperchromatism with coarse chromatins and distinct nuclear rims.

6　Metastatic Carcinoma

Involvement of the pia-arachnoid by leukemia and metastasis of neoplasms (carcinomatous meningitis) take the majority of positive CSF cytology (Fig. 20-18, Table 20-2).

Metastatic carcinoma comprises about 30% of intracranial neoplasms. The common primary extracranial neoplasms that produce frequent metastases to the brain are bronchogenic carcinoma (Figs. 20-20 and 20-22), breast carcinoma, melanoma (Fig. 20-23), carcinoma of the gastrointestinal tract and hypernephroma of the kidney (Fig. 20-19, Table 20-3). The diagnostic rate of neoplastic cells in the CSF from metastatic carcinoma is higher than primary glioma. The identification of cancer cells is based on the following features of cells: (1) Cancer cells are larger than glioma cells and tend to form clusters. (2) The cell types of carcinoma causing frequent metastases are adenocarcinoma and undifferentiated carcinoma.[5, 21)] Differentiated glandular type of

cancer cells may reveal characteristic features distinguishable from glioma cells; in addition to the tendency to form cell clumps, a columnar, abundant cytoplasm with occasional vacuolation and distinct boundaries and eccentric position of nuclei can be found. (3) Cancer cells of any histologic type possess distinctive nuclear membranes, irrespective of chromatin amount, whereas glioma cells disclose delicate nuclear rims. Metastasis from gastric carcinoma is not frequent, but it may cause diffuse meningeal carcinomatosis (Figs. 20-16 and 20-24). Exfoliation of cancer cells from such a condition is more frequent than from solid tumors in the brain.

7　Glioma

Of the primary neoplasms of the brain, 40% to 45% belong to a glioma group (Table 20-4). As general features of glioma cells in CSF, they are smaller than cancer cells and appear singly or in loose clusters. The cytoplasm is scant and indistinctly bordered. The size and shape variations of the nuclei depend on grading of

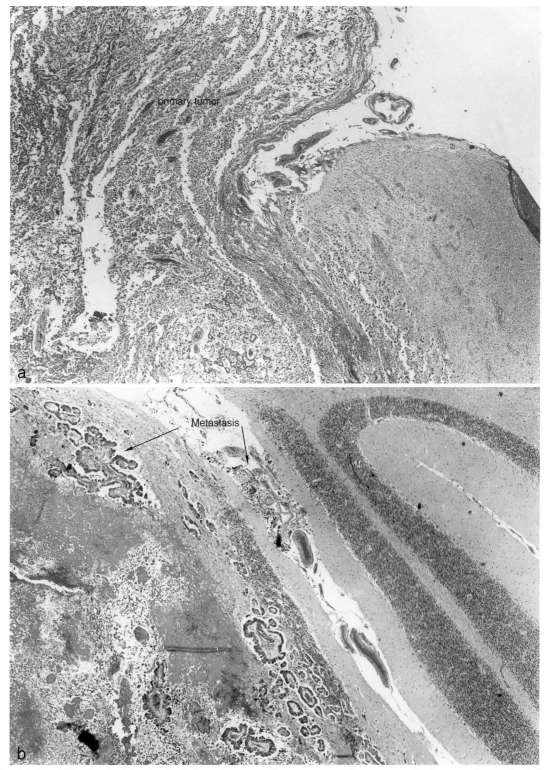

Fig. 20-18 Histological basis shedding glioma cells or metastatic cancer cells into cerebrospinal fluid.
a. Direct extension of primary glioma to the ventricle.
b. Metastatic meningeal carcinomatosis from adenocarcinoma of the stomach.

malignancy and histological types. Glioblastoma multiforme may shed pleomorphic cells and occasionally multinucleated giant cells with indistinct cytoplasm. The cells from medulloblastoma are likely to be, without cytoplasm, nearly the same size as large lymphocytes and oat cells of lung cancer. They occur in loose clusters suggestive of a rosette pattern. A high accuracy of cytodiagnosis can be expected for this tumor, because this

malignant tumor in children grows rapidly and involves frequently the fourth ventricle.[24] Ependymoma is in contact with the ventricular walls and may shed the tumor cells in clusters forming true rosettes. The tumor cells are cuboidal or columnar with eccentric nuclei.

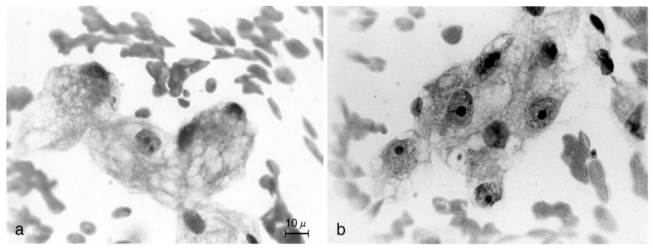

Fig. 20-19 Metastatic adenocarcinoma cells arising from renal carcinoma in cerebral fluid.
a. This hemorrhagic fluid contains clustered cancer cells which are characterized by translucent, vacuolar cytoplasm and small nuclei containing prominent acidophilic nucleoli.
b. Note abundance of vacuolar cytoplasm and prominence of single, round nucleoli locating in the center of nuclei.

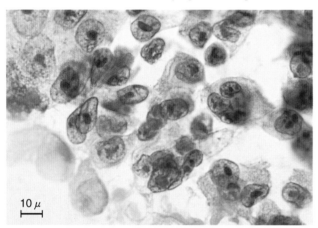

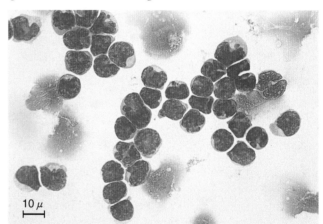

Fig. 20-20 Metastatic adenocarcinoma cells in a squash smear from a case with pulmonary adenocarcinoma. The nuclei are polyhedral and vary in size. Single but prominent, amphoteric nucleoli and thickened nuclear rims are favorable to adenocarcinoma cells.

Fig. 20-21 Leukemic cells (ALL) in cerebral fluid from a 6-year-old boy. Nuclear convolution is seen. Meningeal involvement by ALL is seen in about 25% of the cases and much more common than in acute myelogenous leukemia. Degenerative shadowy cells are present. Giemsa stain.

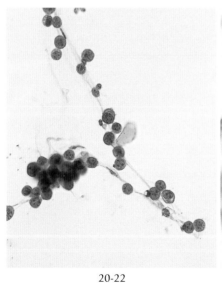

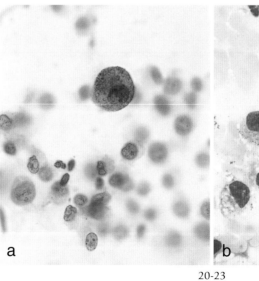

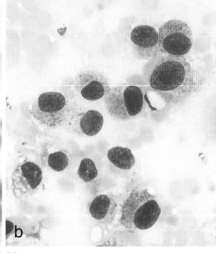

20-22

20-23

Fig. 20-22 Metastatic small cell carcinoma of the lung in cerebral fluid. Cancer cells are distinguished from lymphocytes or leukemic cells. Note aggregation of epithelial cells in cluster or in side by side arrangement.

Fig. 20-23 Metastatic melanoma cells in cerebral fluid.
a. Papanicolaou stain showing a tumor cell rich in melanin pigments.
b. Giemsa stain distinguishable fine melanin pigments in the cytoplasm.

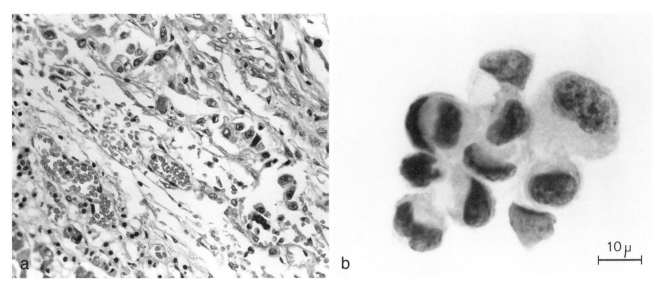

Fig. 20-24   Diffuse meningeal carcinomatosis caused by gastric adenocarcinoma.
a. Histology. Small clusters of adenocarcinoma cells are present in vascular spaces of the leptomeninges.
b. Cytology showing malignant cells in cerebrospinal fluid. These metastatic cancer cells show typical grouping with acinar or glandular pattern, the nuclei of which are eccentrically located in the columnar cytoplasm.

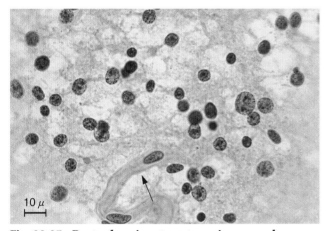

Fig. 20-25   Protoplasmic astrocytoma in a squash smear. There are only slight increase in number (cellularity) and variation in size of astrocytes. Although a capillary is included, endothelial proliferation is not found (arrow).

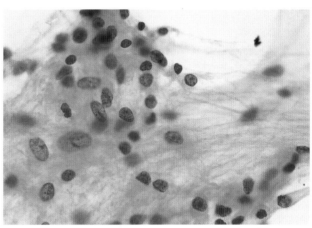

Fig. 20-26   Fibrillary astrocytoma showing fibrillated texture. Squash method. Note slight increase in cellular density and variation in nuclear size. No endothelial proliferation is seen.

▮Astrocytoma and glioblastoma multiforme

Astrocytoma grade 1 or grade 2 may shed tumor cells rarely into CSF. These are observed only by squash or imprint preparation (Figs. 20-25 and 20-26). The nuclei are uniform in size and shape, round to oval, and show delicate nuclear membranes. In another word astrocytoma can be characterized by increased cellular density of astrocyte without endothelial proliferation, necrosis and hemorrhage. Fibrillary and protoplasmic astrocytomas are hardly differentiated in cytology: fibrillary astrocytoma reveals fibrillated processes which can be demonstrated by special stains. Special varieties of astrocytoma display a distinguishable cellular pattern.

1) Gemistocytic astrocytoma

A pure form of gemistocytic astrocytoma is uncommon but found as mixture with astrocytoma. This tumor consists of characteristic glioma cells with plump cytoplasm that stains bluish green with homogeneous, thickened, ground glass appearance. The nuclei take an eccentric position (Fig. 20-27).

Table 20-4   Incidence of primary tumor of the brain

Tumors	Percent of total
Astrocytoma	8-10
Glioblastoma and spongioblastoma	18-25
Ependymoma	5-7
Oligodendroglioma	4-6
Medulloblastoma (and neuroblastoma)	4
Meningioma	12-16
Choroid plexus papilloma	0.5-0.7
Pinealoma	0.4
Craniopharyngioma	2.5-4
Pituitary adenoma	8
Neurinoma	7-8
Fibroma, sarcoma, chordoma, lipoma, osteoma, chondroma	4

(From Adams RD, Sidman RL: Introduction to Neuropathology. New York: McGraw-Hill, 1968.)

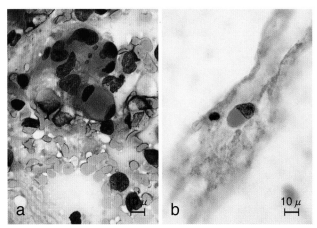

Fig. 20-27　Gemistocytic astrocytoma in imprints. The nuclei are placed peripherally in the plump cytoplasm. Note an opaque and thickened cytoplasm that is diagnostic of gemistocyte origin. **a.** Giemsa stain; **b.** Papanicolaou stain.

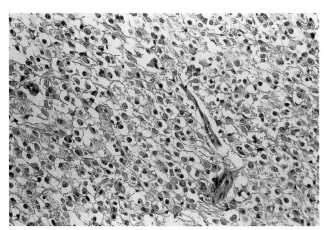

Fig. 20-28　Histology of gemistocytic astrocytoma. This pure form of gemistocytic astrocytoma is composed of diffusely scattered gemistocytic cells with polyhedral or globoid figures. Note characteristic plump, homogeneous cytoplasm. Perivascular pseudorosettes are not noticeable.

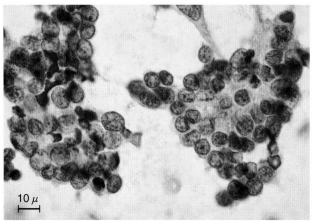

Fig. 20-29　Astroblastoma of a male aged 28 in a squash smear. This highly cellular tumor is characterized by an arrangement forming perivascular pseudorosettes. Crowding of uniform oval cells surrounding red cells is notable.

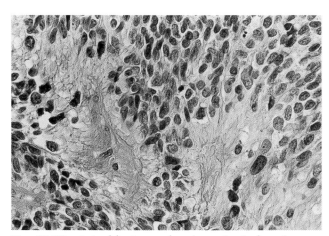

Fig. 20-30　Astroblastoma showing a typical perivascular pseudorosette. These highly cellular tumor cells are arranged towards blood vessels with fibrillar structures.

2) Astroblastoma

Astroblastoma, a variety of astrocytoma, was defined by resemblance to developing astroblasts in the fetus. This tumor is of rare occurrence in the third or fourth decade and comprises about 0.45% of all gliomas.[11] It is localized in cerebral hemisphere and does not invade the ventricular wall. The diagnosis of astroblastoma is allowable to use when the whole neoplasm can be examined. However, a distinctive feature is formation of pseudorosette around blood vessels (Figs. 20-29 and 20-30). Foot plates with tangled fibrils can be demonstrated in a squash smear. The nuclei are uniform in shape and equally sized.

3) Pilocytic astrocytoma, polar spongioblastoma

The term of spongioblastoma was defined by Bailey and Cushing 1926, and Penfield 1931 because of resemblance to spongioblasts having long, slender cytoplasmic processes. The tumor is called pilocytic astrocytoma and belongs to grade 1 astrocytoma. Most optic nerve gliomas are low-grade, juvenile pilocytic astrocytomas.[23] Cytologically pilocytic astrocytoma has characteristic features such as fibrillar meshwork associated with

eosinophilic Rosenthal fibers and bipolar cytoplasmic processes. This histologically distinctive variety of astrocytoma involves the third and fourth ventricles in childhood or adolescence. Cytology displays fibrillated ovoid cells arranged in a palisading pattern.

4) Anaplastic astrocytoma

The pattern is that of astrocytoma in association with anaplasia (grade 3) such as increased cellularity, nuclear polymorphy and occasional mitoses.[29]

The occurrence is frequent in frontal and temporal lobes of the cerebrum. Absence of perinecrotic pseudopalisading and pronounced endothelial proliferation is the target of observation to be distinguished from glioblastoma.

5) Glioblastoma multiforme

As this name illustrates, bizarre, multinucleated or multilobed cells are found in addition to smaller round glioma cells (Figs. 20-31 to 20-33).[37] Looking at glioma cell clumps, there is a pattern of nuclear palisading. Although the nuclei are arranged side by side in rows, the interval between the cellular rows is composed of necrotic tissue; thus the arrangement is called

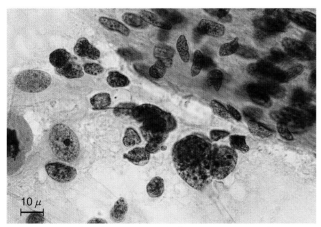

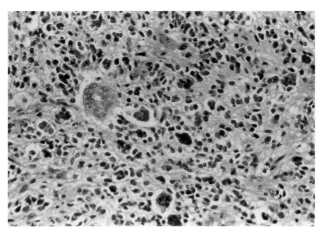

Fig. 20-31　Glioblastoma multiforme in a squash smear. This highly cellular tumor is composed of polymorphic glioma cells, including giant cells, and proliferated endothelial cells.

Fig. 20-32　Glioblastoma multiforme showing diffuse infiltration of undifferentiated tumor cells with marked hyperchromatism and pleomorphism. The main components of the tumor cells are spindly spongioblasts. Multinucleated giant cells are scattered and prominent proliferation of endothelial cells is seen.

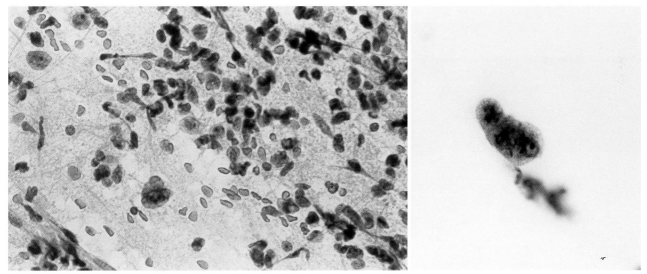

Fig. 20-33　Squash smear from glioblastoma multiforme. High cellularity and polymorphism of the nuclei are the same as histology (Fig. 20-32). Note a characteristic feature of capillary proliferation.

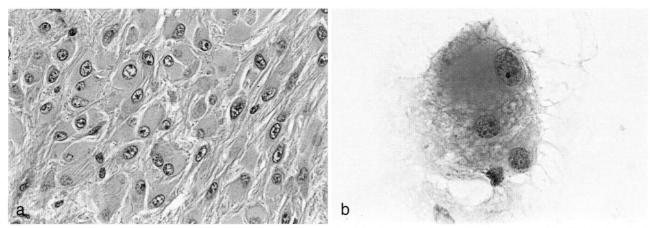

Fig. 20-34　Subependymal giant cell astrocytoma arising beside the lateral ventricle. A 9-year-old girl who has had epilepsy and retardation.
a. Histology consisting of characteristic large, gemistocyte-like tumor cells.
b. Imprint smear showing large plump cells. (By courtesy of Dr. K. Mizuguchi, Teikyo University, Kawasaki, Japan.)

437

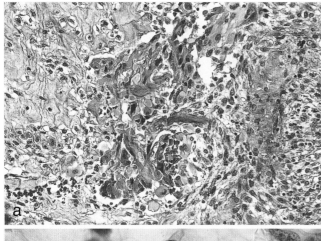

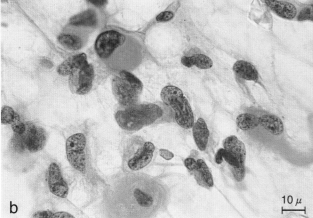

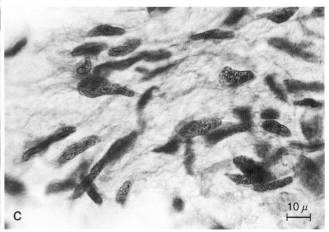

Fig. 20-35 Mixed glioblastoma and fibrosarcoma in 53-year-old female.
a. Histology by Azan-Mallory stain. The central zone of glioblastoma in association with vessel proliferation is demarcated from the fibrosarcoma zone in the right side.
b. Squash smear. There is an admixture of plump neoplastic astrocytes with fusiform mesenchymal cells; the typical texture is represented by clumped astrocytic cells as though they were intersected by bundles of spindly sarcoma cells.
c. Squash smear showing fibrosarcoma cells lying in a bundle. Elongated, fibrillar cells are from the zone of sarcoma. Very fine chromatins spread evenly without thickening of the membrane.

pseudopalisading. The elongated tumor cells with tapering processes of the cytoplasm are likely polar spongioblasts. Increase in blood vessels with endothelial proliferation is also an important finding if a squash or imprint smear is examined (Fig. 20-33).

6) Subependymal giant cell astrocytoma

This special variety of astrocytoma is almost always associated with tuberous sclerosis and arises from the wall of lateral ventricles. Cytology of subependymal giant cell astrocytoma is characterized by large plump cells with eccentric nuclei (Fig. 20-34).

7) Mixed glioblastoma and sarcoma (glioblastoma with sarcomatous component, gliosarcoma)

This variety is not so rare as ever considered. In the series of glioblastomas by Moranz et al[19] about 8% of their cases revealed this special type of histology. It is noticed as a circumscribed tumor at the temporal lobe involving the leptomeninges. An admixture of glioblastoma with fibrosarcoma is of the most common histologic feature.

The sarcomatous component is considered to originate from hyperplastic vascular elements. Cytologically the sarcoma cells are identifiable in squash preparation, although they seldom shed in imprinting (Fig. 20-35).

■Medulloblastoma

The tumor affects frequently the cerebellum of children especially in midline. It invades the roof of the fourth ventricle and occasionally sheds tumor cells into the CSF (Fig. 20-39). The nuclei are about the same as activated lymphocytes in size and densely stained but often angu-

lated. Mostly, the tumor cells are denuded (Figs. 20-37 to 20-39). They show a characteristic arrangement in a pseudorosette (Homer-Wright rosette)(Fig. 20-36). No tubular structures exist in the middle of pseudorosettes.

■Ependymoma

The fluid aspirated from a ventricle contains ependymoma cells having uniform, oval nuclei (Fig. 20-41). True rosette formation with "blepharoplast" structures is a diagnostic feature but is not always present even if a squash smear was prepared (Fig. 20-40). The best stain to demonstrate blepharoplasts is not Papanicolaou stain but PTAH. It must be remembered that perivascular rosettes are more frequently encountered than true ependymal rosettes.

Plexus papilloma sheds cell clusters arranged in a papillary fashion. The nuclei are small, round or oval, and uniform in size (Figs. 20-42 and 20-43). No blepharoplastic differentiation can be found. These develop, in order of decreasing order, in the fourth ventricle, the lateral ventricles and in the third ventricle.

Anaplastic ependymoma or malignant ependymoma has invasive properties in association with cellular atypia such as polymorphy, giant cell formation and increased cellularity. Clinically the tumor shows invasive growth and diffuse spread (Figs. 20-44 and 20-45).

■Oligodendroglioma

Oligodendroglioma occurs usually in the frontal lobe with a long-standing clinical history in adults of the

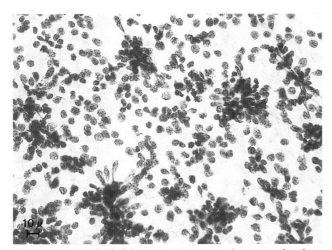

Fig. 20-36 Medulloblastoma in a squash smear displaying characteristic Homer Wright rosettes. Darkly staining nuclei tend to aggregate radially.

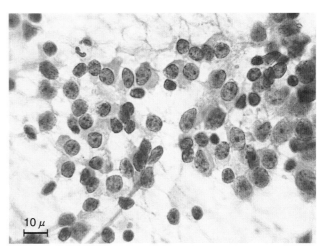

Fig. 20-37 Medulloblastoma in a squash smear. The nuclei are stained darkly and shaped irregularly as called "carrot-shaped cells." There is a small amount of poorly stained cytoplasm.

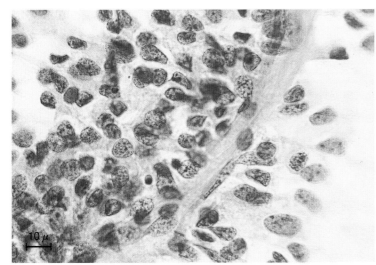

Fig. 20-38 Medulloblastoma in a squash smear showing a perivascular pseudorosette. Diagnostic points: (1) absence of endothelial proliferation in thin-walled vessels, (2) monomorphic tumor cells consisting of darkly stained, denuded nuclei.

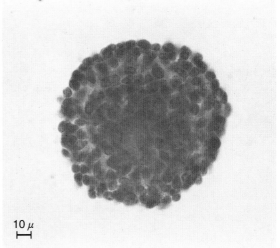

Fig. 20-39 Medulloblastoma clustered in a cell ball in cerebral fluid. Lymphocyte-sized hyperchromatic nuclei tend to aggregate radially. They appear to be denuded.

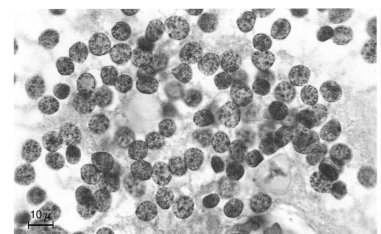

Fig. 20-40 Common type of ependymoma in a squash smear. This highly cellular tumor is characterized by ependymal rosettes around central canals. Blepharoplastic differentiation is suggestive. The nuclei are equal in size and shape. Their chromatin pattern, though increased in amount, is finely granular and bland.

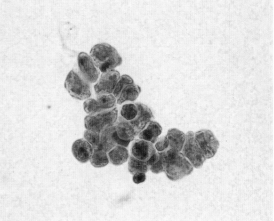

Fig. 20-41 Ependymoma cells in cerebral fluid. Note closely grouped round to polygonal cells suggestive of "ependymal tubule" arrangement.

439

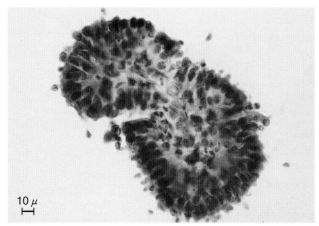

Fig. 20-42　Choroid plexus papilloma in a squash smear. Compactly piled, low columnar cells are arranged characteristically in a radial fashion. Supporting stromal tissue is also associated.

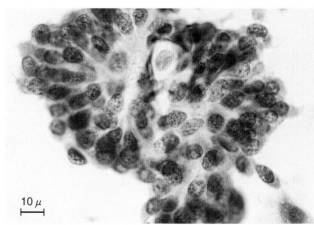

Fig. 20-43　Higher magnification of choroid plexus papilloma. The nuclei of the principal tumor component are uniformly oval in shape with a bland chromatin pattern and arranged radially.

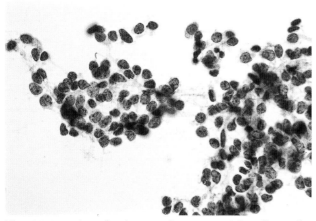

Fig. 20-44　Anaplastic ependymoma arising from the fourth ventricle. Imprint smear of moderate magnification shows highly cellular, hyperchromatic cells arranged in pseudorosette pattern. True ependymal rosette is not found. (By courtesy of Dr. K. Mizuguchi, Teikyo University, Kawasaki, Japan.)

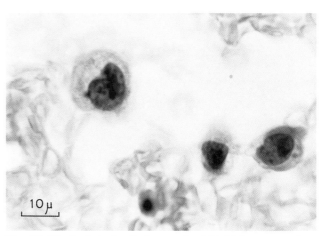

Fig. 20-45　Isolated cells in cerebral fluid from anaplastic ependymoma. These cells reveal cuboidal or low columnar cytoplasm with indistinct borders and vacuolation. The nuclei vary in size and shape. Note hyperchromatism and coarse clumping of chromatins.

fourth or fifth decade. It accounts for about 5% of intracranial gliomas.

Oligodendroglioma is liable to undergo degeneration. Well preserved oligodendroglioma cells in the squash method appear in sheets or as isolated cells with finely vesicular cytoplasm. The nuclei are generally round with delicate nuclear rim. Since the cytoplasm is often poorly preserved, the nuclei appear to be almost denuded and look like chromophobe adenoma. However fibrillary, wispy cytoplasm can be recognized.[22] There is marked nuclear size variation. Calcification is occasionally associated (Fig. 20-46).

Addendum: Rosettes in glioma. Rosette formation is seen in various types of glioma such as ependymoma, retinoblastoma and medulloblastoma. True rosettes as designated originally by Flexner[9] are those arranged around the central canal and are found in retinoblastoma as well as in ependymoma. On the other hand, the rosette that is made as though it surrounds faintly stained, fibrillated material (Homer Wright[38]) is shown in medulloblastoma.

8 Pinealoma

Many pinealomas are circumscribed tumors having thin capsules and grow expansively; therefore, cytological diagnosis from the cerebrospinal fluid is usually poor. The term pinealoma is not strictly used for tumors originating from the pineal parenchyma but includes many varieties developed in the pineal region. They were classified by Friedman (1947)[10] into germinoma, undifferentiated teratoma (teratocarcinoma), differentiated teratoma and neural tumor. Ringertz[28] classified the tumor into the seminoma-like pattern (germinoma-type), classic type with two-cell pattern, and the adult type with fibroglial vessel-carrying stroma which resembles mature pineal tissue. About half of pinealomas fall into the classic type that is composed of large, pale, epithelial-like cells and darkly stained lymphoid cells[25](Fig. 20-48).

Teratoma affects young males within the first two decades and histologically identical with embryonal carcinoma of the testis. The tumor cells are arranged in

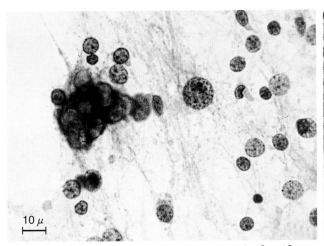

Fig. 20-46   Oligodendroglioma comprised of uniform, enrounded glioma cells in a squash smear. The spheroidal nuclei, though vary in size, are equally shaped. The clear cytoplasm that characterizes oligodendroglia is invisible in the smear. Calcification is one of the distinctive features.

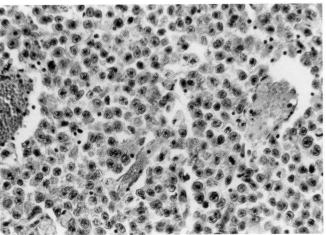

Fig. 20-47   Pineal germinoma simulating seminoma. Histology of germinoma showing sheets of large polygonal tumor cells and a few small lymphoid cells. The larger cells have clear, round to polyhedral nuclei with prominent nucleoli. The cytoplasm is indistinct.

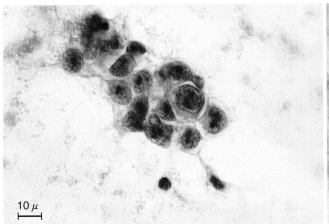

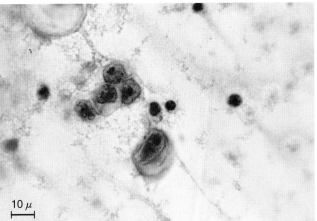

Fig. 20-48   Pineal germinoma cells in cerebral fluid. Loose clusters of polygonal or spheroidal cells with vesicular nuclei are seen. Diagnostic points: (1) Centrally locating, single prominent nucleolus with acidophilia, (2) clear, vesicular nuclei, and (3) mixture with lymphoid cells are diagnostic of germinoma.

solid sheets and tubules. Clear vacuolated cytoplasm with eccentric nuclei is a characteristic feature of the tumor.

Germinoma is the most common type of pinealoma and frequently affects young people aged 15 to 25.[30]

This seminoma-like type is commonly used as pinealoma of a broad sense; histologically the tumor is composed of two types of cells, i.e., large polygonal cells and lymphoid cells. The former is characterized by clear cytoplasm and centrally locating round nucleus, in which a prominent nucleolus is present (Figs. 20-47 and 20-48).

Pineoblastoma arising from pineal parenchymal cells is rare and hardly distinguished from medulloblastoma. The cytoplasm is scant and ill-defined. Small dark-staining nuclei are the same as those of medulloblastoma.

Note: The recent WHO classification of tumors of the central nervous system separates pineal cell tumors from germ cell tumors comprised of germinoma, embryonal carcinoma, choriocarcinoma and teratoma.[39]

Table 20-5   Brain squash cytology as adjunct to frozen diagnosis

Pathology	Case	False negative	Incorrect typing	Correct typing
Epidermal cyst (craniopharyngioma)	7	0	0	7
Neurinoma	9	1	2	6
Meningioma	18	0	3	15
Meningeal sarcoma	1	0	1	0
Melanosis	1	1	0	0
Astrocytoma	16	0	2	14
Glioblastoma	12	0	0	12
Medulloblastoma	4	0	0	4
Oligodendroglioma	2	0	1	1
Ependymoma	2	0	1	1
Pituitary adenoma	16	0	1	15
Pinealoma	1	0	0	1
Metastatic tumor	13	0	0	13
Total	102	2 (2.0%)	11 (10.8%)	89 (87.2%)

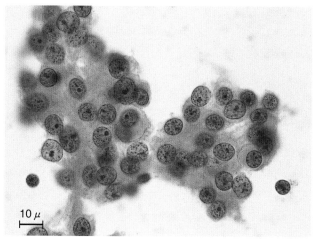

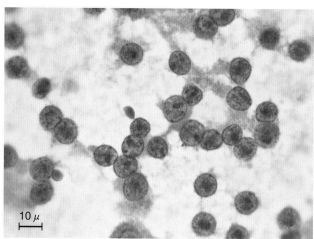

Fig. 20-49 Pituitary chromophobe adenoma in a squash smear. From a case with alveolar pattern of chromophobe adenoma that is characterized by formation of cell cluster and well preservation of the cytoplasm. Note uniformity in nuclear size and stainability.

Fig. 20-50 Pituitary chromophobe adenoma in a squash smear. From a case with diffuse cellular pattern of chromophobe adenoma. The tumor cells are almost devoid of the cytoplasm. The nuclei are round and uniformly sized.

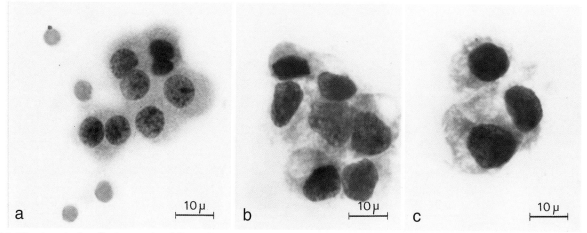

Fig. 20-51 Atypical cells in cerebral fluid from chromophobe adenoma.
a. Benign adenoma cells exfoliated in the cerebral fluid following incomplete resection of the tumor. The nuclei are small, round and uniform in size.
b, c. Giemsa-stained adenoma cells in clusters obtained by needle aspiration from a cystic area of chromophobe adenoma. This smear is stained by Giemsa air-dry method, thus the cells appear larger than those by Papanicolaou method. However, the nuclei are considerably pleomorphic and hyperchromatic. The cytoplasm is fairly abundant but poorly outlined.

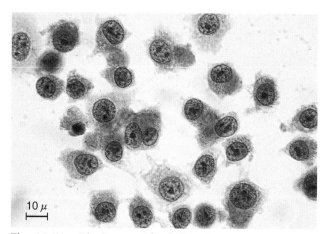

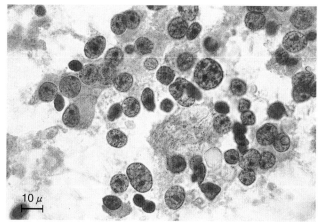

Fig. 20-52 Pituitary acidophilic adenoma in a squash smear. The nuclei are hyperchromatic and uniformly round. The cytoplasm is of a moderate amount, somewhat granular and intensely stained bluish green.

Fig. 20-53 Malignant chromophobe adenoma in a squash smear. The tumor was first noted in the sphenoidal air sinus because of marked invasive and destructive behavior. Such pronounced variation in chromatin content and nuclear size as in this tumor is seldom found in ordinary chromophobe adenoma; these features may argue for a true malignant change.[27]

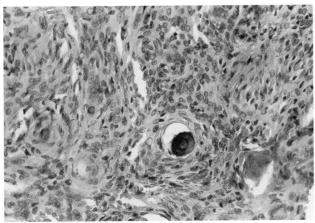

Fig. 20-54 **Meningothelial meningioma consisting of compact whorls of meningothelial cells with ovoid vesicular nuclei and eosinophilic cytoplasm.** The cell boundaries are indistinct. Note a distinct psammoma body formation in the center of the field.

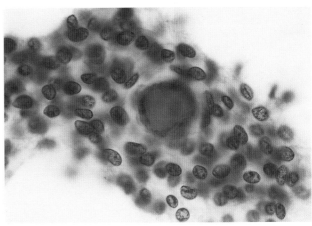

Fig. 20-55 **Meningothelial meningioma in a squash smear.** The tumor is composed of densely aggregated enrounded cells. A whorling pattern around a psammoma body is indicative of meningioma.

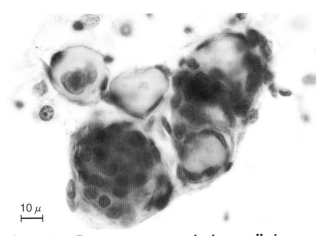

Fig. 20-56 **Psammomatous meningioma cells in cerebral fluid.** Ovoid meningothelial cells are arranged characteristically in a whorl pattern, the center of which shows an imperfect psammoma body.

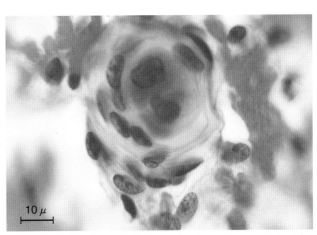

Fig. 20-57 **Meningothelial meningioma in imprints.** The nuclei of whorled meningioma cells are elliptic and compressed. The chromatin distribution is smooth or homogeneous.

9 Pituitary Adenoma

Chromophobe adenoma, functionally inactive adenoma, is the major part of pituitary adenoma. This slow-growing tumor undergoes suprasellar expansion upward to the third ventricle as well as intrasellar erosion of the floor of the sella. Since chromophobe adenoma is not always completely excised, exfoliation of adenoma cells in cerebrospinal fluid may occur in recurrence (Fig. 20-51). The nuclei are moderately enlarged, round to oval, but equal in size. The nuclear rim is delicate, and clearly outlined. Usually, single prominent nucleoli are found. The cytoplasm is small to moderate in amount, and usually poorly preserved (Figs. 20-49 and 20-50).

Invasive chromophobe adenoma is a variety of chromophobe adenoma undergoing destructive growth through the sphenoidal air sinus or the nasopharynx. Cytologically the tumor shows more pronounced variation in nuclear size and chromatin content than benign chromophobe adenoma (Fig. 20-53).

Acidophil adenoma is clinically manifested by gigantism or acromegaly. The nuclei are spherical and larger than those of chromophobe adenoma. To demonstrate the specific granules, Mallory-Heidenhain method is satisfactory. Their cytoplasm is finely granular with more intense staining reaction. Eosinophilia by hematoxylin eosin stain will be demonstrated in Papanicolaou stain as affinity to Light Green which belongs to the acidic dye group (Fig. 20-52).

Basophil adenoma is too small to recognize as a tumor and localized in the sella. There is association with Cushing's syndrome clinically. Basophil adenoma cells are variable in size and show Schiff-positive staining reaction.

10 Meningioma

Of the non-gliomatous tumors of the brain, meningiomas are the most common and comprise about 12% to 16% of all tumors of the central nervous system. Meningioma is a well circumscribed tumor of local growth, but is occasionally hazardous when operated upon because of its tendency to invade the blood vessels and cranial nerves.[7,12] It may shed tumor cells into the cerebral fluid after incomplete resection. Malignant vari-

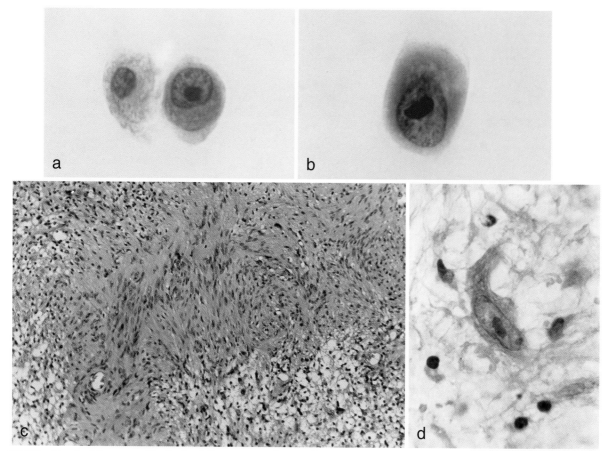

Fig. 20-58 Atypical tumor cells derived from acoustic neurinoma.

a, b. Atypical single cells in cerebral fluid from recurrent acoustic neurinoma. These atypical cells were collected by Millipore filter after incomplete surgical removal of the tumor, and malignancy was suspected because of nuclear enlargement, high nuclear-cytoplasmic ratio and prominent nucleolus. The occurrence of benign giant cells could be verified histologically and it is referred to as degenerative process.

c. Histology of surgically removed acoustic neurinoma. The tumor is composed of interlacing fasciculi of elongated, fusiform cells, in which typical nuclear palisading pattern is recognized. Giant cells are found in focal areas with cystic degeneration.

d. Mononuclear giant cell in histology of acoustic neurinoma. A high-power field of histology showing a giant cell in loose, degenerative tumor tissue. This will be correspondent to the large single cell in cerebral fluid (a, b).

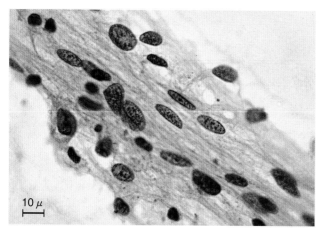

Fig. 20-59 Acoustic schwannoma in imprint smear. Schwannoma or neurinoma is characterized by compact bundles of elongated, fibrillated cells which are arranged in a parallel fashion. Presence of hyperchromatic large nuclei is of occasional occurrence and has no relation to malignancy.

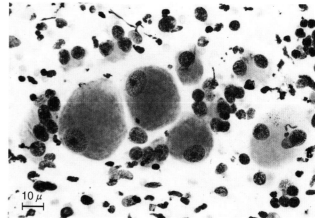

Fig. 20-60 Ganglioneuroblastoma in imprints from adrenal tumor. Note a preponderance of plump ganglion cells with round eccentric nuclei admixed with round sympathoblasts resembling glial cells.

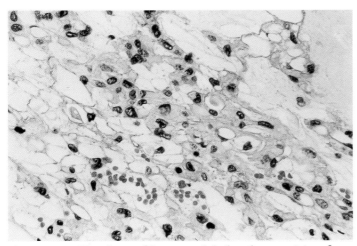

Fig. 20-61　Histology of intracranial chordoma arising from the clivus of a case of 42-year-old female. Note a loose mucinous texture consisting of vacuolated "physaliferous" cells.

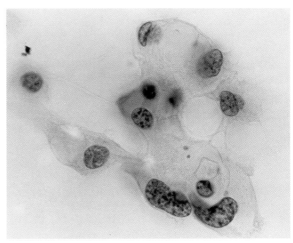

Fig. 20-62　Imprint smear from chordoma showing large vacuoles in the cytoplasm. The nuclei are enlarged, varying in size and rich in chromatin granules.

eties of meningioma should not be abandoned. In Craig's series of 57 meningiomas, 19.3% were of the malignant form.[7] Since the malignant criteria in histopathology are based on strictly histologic grounds, imprint smears or squash smears demonstrating nuclear details are useful as an adjunct to frozen section diagnosis during the surgical procedures (see Figs. 20-2 to 20-4).

■Classification

Classification of meningioma containing components of meningocytes and mesenchymal cells of the pia arachnoid is variable according to the predominant cellularity and distinctive features of the cell component.

Roussy and Bucy adopted nine types; (1) mesenchymal, (2) meningotheliomatous, (3) angioblastic, (4) psammomatous, (5) osteoblastic, (6) fibroblastic, (7) melanoblastic, (8) sarcomatous, and (9) lipomatous. Russell uses a simpler classification; (1) syncytial, (2) transitional, (3) fibrous, (4) angioblastic, and (5) sarcomatous. Some histologic features as above are recognizable in cytodiagnosis based on their cell patterns and features. As a general aspect of meningothelial cells, the nuclei are oval to elliptic with delicate outlines. The chromatin is finely granular and evenly distributed. The cytoplasm showing fibrillar structure is poorly defined. The features that characterize meningioma are formation of psammoma bodies (Figs. 20-54 and 20-55) and cell arrangement with whorls (Figs. 20-56 and 20-57).

11 Neurinoma (Schwannoma)

Neurinoma or neurilemmoma that is most likely to arise from the neuro-ectodermal sheath of Schwann is also called Schwannoma, and affects the cranial and spinal nerve roots. The entity of neurinoma differs from neurofibroma that is referred to as fibroblastic neoplasm derived from the epineurium and perineurium. Cytologically, the tumor is composed of bundles of long bipolar spindle cells. Histological characteristics of Antoni A and B will not be demonstrated in smears. However, a palisading pattern of elliptic nuclei resem-

bling a school of small fish is one of the characteristic textures of neurinoma. Lipoid-laden foamy cells may be intermingled as a result of degeneration. It should be kept in mind that giant nuclei are included in clusters of elliptic nuclei to avoid misinterpretation as malignant. A case in the author's hospital revealed a false positive cytology in the cerebral fluid after incomplete removal of a large acoustic neurinoma (Fig. 20-58a, b). The appearance of giant cells in neurinoma has been described by Kernohan and Sayre.[12] It is not related to malignancy and is referred to as degenerating process (Fig. 20-58c, d).

12 Ganglioglioma and Ganglioneuroblastoma

Ganglioglioma (ganglioneuroma) is a rare tumor occurring in the floor of the third ventricle in young age groups under 30 of age. Cytologically the tumor consists of large ganglion cells and stromal glial cells.

Ganglioneuroblastoma arising from the peripheral nervous system and adrenal medulla reveals cytologically a transitional form between ganglioglioma and more primitive neuroblastoma; it is characterized by admixture of ganglion cells and sympathoblasts (Fig. 20-60).

13 Chordoma

Chordoma means a benign tumor originating from the notochordal remnants. Histologically, the tumor is composed of pleomorphic cells embedded in a loose mucinous matrix. The term of physaliferous (bubble-bearing) cells is applied to those cells having vacuolated cytoplasm because of content of glycogen and mucus (Figs. 20-61 and 20-62).

14 Dermoid Cyst and Craniopharyngioma

Dermoid cyst is a congenital malformation that has a predilection at parapituitary region. The cyst is walled by fibrous tissue, the inner surface of which is covered

445

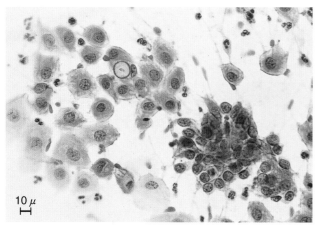

Fig. 20-63 **Craniopharyngioma in aspiration smear from a cystic suprasellar nodule.** There are two types of cells: (1) clustered columnar cells with uniform oval nuclei, and (2) squamous cells of parabasal type.

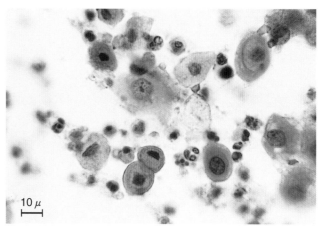

Fig. 30-64 **Craniopharyngioma showing squamous cell components in aspirated smear.** Note well differentiated squamous cells either with or without keratinization.

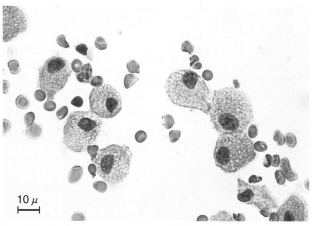

Fig. 20-65 **Imprint cytology of infundibuloma.** The term of infundibuloma based on the localization is under the discussion whether it derives from pituicytes or neurohypophysis; thus it is also called choristoma and myoblastoma. The tumor is composed of round cells with granular dusky cytoplasm.

by stratified squamous epithelium. Aspirates of the cyst contain mature squamous cells.

Craniopharyngioma is a common benign tumor either originating in remnants of Rathke's pouch or relating to metaplasia of adenohypophyseal cells.

The tumor takes a suprasellar location, but intrasellar growth is occasionally noted. The cyst is surrounded by solid areas which are composed of anastomosing bands of stratified squamous epithelium with columnar-shaped peripheral cell layers; the histology is referred to as adamantinoma of the pituitary gland. Aspirates from the cyst contain mature squamous cells singly or as cell clusters, lipid-laden foamy cells and cholesterol crystals (Figs. 20-63 and 20-64).

Choristoma that is found in the pars nervosa is also called pituicytoma or granular cell myoblastoma because of abundant, eosinophilic cytoplasm with granular texture (Fig. 20-65). The term "infundibuloma" of the pituitary is applied to the tumor essentially of neurohypophyseal origin.

15 Neurocutaneous Melanosis and Meningeal Melanoma

Neurocutaneous melanosis is characterized by giant pigmented nevi of the skin and diffuse or occasionally nodular proliferation of melanocytes that are normally recognizable in the pia mater. Cerebrospinal fluid contains a number of melanin-containing cells (Fig. 20-66).

Primary meningeal melanoma without associated cutaneous nevi can be diagnosed from sediments of CSF.

16 Cerebral Paragonimiasis Westermani

Paragonimiasis is widely distributed in the Far East and affects the respiratory tract. Yellowish brown operculated ova are found in sputum smears. Although the lung is the target organ, the brain is occasionally the site of granuloma formation (Fig. 20-67). It is clinically manifested as a space-occupying lesion. Cerebrospinal fluid may show eosinophilia and increase in protein content.

17 Retinoblastoma

Retinoblastoma is found before the age of three and varies in malignancy with medulloblastoma. It undergoes the direct spread from one eye to the other through the optic nerve. If the subarachnoid space is involved, dissemination through the cerebrospinal fluid may occur (Figs. 20-68 and 20-69).

The tumor is composed of masses of lymphocyte-like or oat-celled hyperchromatic cells which are characteristically arranged in rosettes of Flexner. Aspiration cytology from the chamber of the eye ball may disclose closely packed, hyperchromatic cells which are almost devoid of the cytoplasm. The arrangement suggestive of rosettes is of diagnostic value cytologically.

Since the extension along the optic nerve into the subarachnoid space occurs occasionally, positive cytology with rosettes in the spinal fluid of infants should need differentiation among retinoblastoma, medulloblastoma and metastatic neuroblastoma.

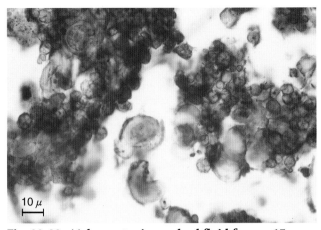

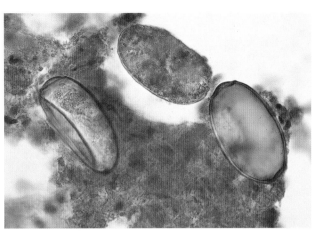

Fig. 20-66 Melanocytes in cerebral fluid from a 15-year-old boy with "mélanoses neurocutanées"(Touraine). This rare case, probably of congenital maldevelopmental origin, is characterized by diffuse melanoblastic proliferation in the skin and meninges. Note a large number of melanin-containing cells varying in size.

Fig. 20-67 Oval operculated ova of the lung fluke in a squash smear taken from granulomatous lesion of the brain. From a 37-year-old woman. Note miracidium in a thick-walled ovum.

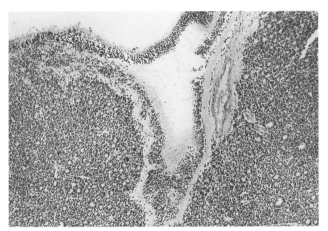

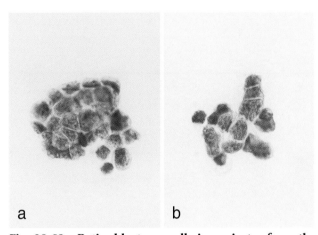

Fig. 20-68 Histology of retinoblastoma from a 2-year-old infant. There is massive infiltration of lymphocyte-like, darkly staining tumor cells. Note formation of true rosettes of Flexner.(By courtesy of Dr. J. Kawachi, Miyazaki Prefectural Hospital, Miyazaki, Japan.)

Fig. 20-69 Retinoblastoma cells in aspirates from the anterior chamber of the eye ball.
a. Closely packed cells, almost denuded, are arranged circularly with rows of cells in tandem.
b. Densely stained nuclei are polyhedral because of close facing from one to another.

References

1. Adams RD, Sidman RL: Introduction to Neuropathology. New York: McGraw-Hill, 1968.
2. Aronson SM, Garcia JH, Aronson BE: Metastatic neoplasms of the brain: Their frequency in relation to age. Cancer 17: 558, 1964.
3. Bigner SH, Johnston WW: The diagnostic challenge of tumors manifested initially by the shedding of cells into cerebrospinal fluid. Acta Cytol 28: 29, 1984.
4. Bots G Th. AM, Went LN, Schaberg A: Results of a sedimentation technique for cytology of cerebrospinal fluid. Acta Cytol 8: 234, 1964.
5. Burton JF: Tumor cells in cerebrospinal fluid. Med Radiography Photography 37: 22, 1961.
6. Cahill EM, Hidvegi DF: Crush preparations of lesions of the central nervous system: A useful adjunct to the frozen section. Acta Cytol 29: 279, 1985.
7. Craig W McK: Malignant intracranial endotheliomata. Surg Gynecol Obstet 45: 760, 1927.
8. Durfee GR: Cytologic techniques. In: Koss LG (ed): Diagnostic Cytology and Its Histopathologic Bases, 2nd ed. Philadelphia: JB Lippincott, 1968.
9. Flexmer S: A peculiar glioma (neuroepithelioma?) of the retina. Johns Hopkins Hosp Bull 2: 115, 1891.
10. Friedman NB: Germinoma of the pineal: Its identity with germinoma ("seminoma") of testis. Cancer Res 7: 363, 1947.
11. Hoag G, Sima AA, Rozdilsky B: Astroblastoma revisited: A report of three cases. Acta Neuropathol 70: 10, 1986.
12. Kernohan JW, Salre GP: Tumors of the Central Nervous System. Atlas of Tumor Pathology. Washington, DC: Armed Forces Institute of Pathology, 1952.
13. Kline TS: Cytological examination of the cerebrospinal fluid. Cancer 15: 591, 1962.
14. Knights EM Jr: Metastatic tumors of brain and their relation to primary and secondary pulmonary cancer. Cancer 7: 259, 1954.
15. Kobayashi S: Meningioma, neurilemmoma and astrocytoma specimens obtained with the squash method for cytodiagnosis: A cytologic and immunochemical study. Acta Cytol 37: 913, 1993.
16. Lesse S, Netsky MG: Metastasis of neoplasms to the cen-

tral nervous system and meninges. AMA Arch Neurol Psychiat 72: 133, 1954.

17. Mathios AJ, Nielsen SL, Barret D, et al: Cerebrospinal fluid cytomorphology: Identification of benign cells originating in the central nervous system. Acta Cytol 21: 403, 1977.

18. McCormick WF, Coleman SA: A membrane filter technic for cytology of spinal fluid. Am J Clin Pathol 38: 191, 1962.

19. Moranz RA, Feigin I, Ransohoff J: Gliosarcoma: A clinical and pathological survey of 24 cases. J Neurosurg 45: 398, 1976.

20. National Health Education Committee: Survival experience of patients with malignant neoplasms; The end-results group for the 4th National Cancer Conference, Minneapolis 1960, in facts on the major killing and crippling diseases in the United States today. New York: National Health Education Committee Inc, 1961.

21. Naylor B: The cytologic diagnosis of cerebrospinal fluid. Acta Cytol 8: 141, 1964.

22. Nguyen G-K, Johnson ES, Mielke BW: Comparative cytomorphology of pituitary adenomas and oligodendrogliomas in intraoperative crush preparations. Acta Cytol 36: 661, 1992.

23. Pennelli N, Montaguti A, Carteri A, et al: Juvenile pilocytic astrocytoma of the optic nerve diagnosed by fine needle aspiration biopsy. Acta Cytol 32: 395, 1988.

24. Polmetter FE, Kernohan JW: Meningeal gliomatosis: A study of 42 cases. Arch Neurol Psychiat 57: 593, 1947.

25. Ramsey HJ: Ultrastructure of a pineal tumor. Cancer 18: 1014, 1965.

26. Reynaud AJ, King EB: A new filter for diagnostic cytology. Acta Cytol 11: 289, 1967.

27. Ricoy J, Carrillo R, Garcia J, et al: Dissemination of pituitary adenomas. Acta Neurochir 31: 123, 1974.

28. Ringertz N, Nordenstam H, Flyger G: Tumors of the pineal region. J Neuropathol Exp Neurol 13: 540, 1954.

29. Rorke LB, Gilles FH, Davis RL, et al: Revision of the World Health Organization classification for childhood brain tumors. Cancer 56 (Suppl): 1869, 1985.

30. Russell DS, Rubinstein LJ: Pathology of Tumours of the Nervous System, 4th ed. London: Edward Arnold, 1977.

31. Saigo P, Rosen PP, Kaplan MH, et al: Identification of cryptococcus neoformans in cytologic preparations of cerebrospinal fluid. Am J Clin Pathol 67: 141, 1977.

32. Sayk J: Zytologie der Zerebrospinalflüssigkeit. Jena: Fischer Verlag, 1962.

33. Störtebecker TP: Metastatic tumors of the brain from a neurosurgical point of view: A follow-up study of 158 cases. J Neurosurg 11: 84, 1954.

34. Tom MI: Metastatic tumours of brain. Canad Med Assoc J 54: 265, 1946.

35. Torres LFB, Collaco LM: Smear technique for the intraoperative examination of nervous system lesions. Acta Cytol 37: 34, 1993.

36. van Oostenbrugge RJ, Arends JW, Buchholtz R, et al: Cytology of cerebrospinal fluid. Are polylysine-coated slides useful? Acta Cytol 41: 1510, 1997.

37. Wagner JA, Frost JK, Wisotzskey H Jr: Subarachnoid neoplasia incidence and problems of diagnosis. Southern Med J 53: 1503, 1960.

38. Watson CW, Hajdu SI: Cytology of primary neoplasms of the central nervous system. Acta Cytol 24: 40, 1977.

39. WHO International Histological Classification of Tumours. In: Zulch KJ, et al (eds): Histological Typing of Tumours of the Central Nervous System. Geneva: 1979.

40. Wright JH: Neurocytoma or neuroblastoma: A kind of tumor not generally recognized. J Exp Med 12: 556, 1910.

41. Zhang YX, Luo KS, Liv JC, et al: Cytological diagnosis of 500 cases of intracranial tumors during craniotomy. Chin J Clin Cytol 3: 19, 1987.

21 Skin and Its Related Mucous Membrane

1 Collection of Specimens and Preparation of Smears

Since the skin has a thick cornified layer, scraping the lesion with a sharp curette is necessary. Otherwise only anucleated squamous cells and keratin material are submitted. An attempt to conquer the problem became successful to some extent by Scotch tape strip method as described below. Bullous lesions should be scraped from the base after removal of the roof. Buccal smears for demonstration of sex chromatins or X bodies are made by wiping off the saliva with gauze and scraping with a wooden tongue blade. Since the scraped specimen is deficient in muco-proteinaceous material which may aid adherence of the specimen to glass slides during fixation, the use of a thinly albuminized slide is recommended for wet preparation.

In order to prepare good wet films, one must take care to put the slide promptly into the fixative as soon as the touch smear is made, because it is apt to dry up through delay. Commercially made coating fixatives are suitable for such a rapid procedure. Air-dried films followed by methanol fixation should be prepared for the Giemsa method.

1) Scotch tape strip method
As an attempt to strip off the cornified layer for obser-

vation of deeper epithelial cells or to examine cornified cells, a strip method with use of commercial Scotch tape has been carried out (Fig. 21-1).[5,36] The procedure as Wolf[56] and other researchers[5,36] described consists of (1) application of the transparent Scotch tape on the cleaned skin rubbing to assure adhesion, (2) stripping off with a quick movement and repeating this procedure as many times as required, (3) adhering the sticky side of the tape down to a clean glass slide, on which a few drops of ether-xylol were previously placed, and (4) taking off the tape to be placed in the fixative. The last procedure will be successfully done if the slides are put in a sealed pot containing ether-ethanol.* After this strip method, a good smear can be obtained if a clean glass slide is attached against the oozing skin with finger press.

*Note: Flammable ether has to be replaced by alcohol. The above description is left as an original method.

2) Skin surface biopsy cytology[28-30]
After cleaning the skin with alcohol, a drop of clear surgery adhesive is placed on the skin to be examined. A clean glass slide that is promptly placed and pressed on the lesion is quickly removed after 30 seconds.

Most dermatologists prefer biopsy giving sufficient information to cytology. However the effort to understand the advantage of cytodiagnosis[55] that several

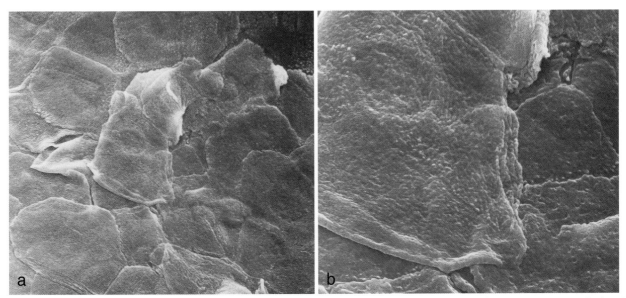

a   b

Fig. 21-1 Scanning electron micrograph showing good detachment of anucleated superficial squamous cells by the "Scotch tape method" in a case of psoriasis. The Scotch tape method is a good mechanical scraping technique to prove the "Auspitz sign" in psoriasis.

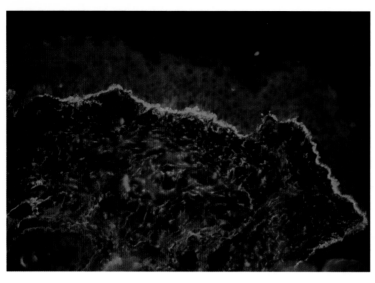

Fig. 21-2 Fluorescence lupus band at the basement membrane. The FITC-labeled anti-IgG antiserum emits specifically fluorescent band along the basement membrane. From sun-exposed, clinically normal skin in a patient with lupus erythematosus. (By courtesy of Prof. Marcello F. de Franco, Faculdade de Medicina, Universidade de Botucatu, São Paulo, Brazil.)

Table 21-1 Application of cytology to dermatology

Method	Diseases	Diagnostic point
	A. Bullous lesions	
	1. Intraepidermal bulla	
	Pemphigus vulgaris and pemphigus vegetans	Acantholysis and deposition of IgG and C_3
Incisional touch (imprint) method	2. Intraepidermal viral bulla	
	Herpes simplex, Herpes zoster and Varicella	Intranuclear inclusion cells
	3. Subepidermal bulla	
	Bullous pemphigoid	Clustered basal cells
	Dermatitis herpetiformis	Neutrophils and eosinophils
	B. Nonbullous epidermal lesions	
	1. Disorder of granular and horny layers	
	Psoriasis vulgaris	Parakeratotic cells, exocytosis and IgG and C_3 deposition
	Darier's disease	Dyskeratosis, corps ronds and acantholytic cells
Scotch tape strip method	Dominant congenital ichthyosis	Irregular keratohyaline and loss of distinct cell borders
	2. Disorder of prickle cells	
	Bowen's disease and erythroplasia (Queyrat)	Dyskaryosis and dyskeratosis
	Paget's disease	Mixture with Paget's cells
	Basal cell epithelioma	Clustered atypical basal cells (basalioma cells)
	C. Dermal and subcutaneous lesions	
	1. Noninfectious granuloma	
	Sarcoidosis	Epithelioid cells with noncaseous background
	Gout tophi	Urate crystals and foreign body giant cells
	2. Bacterial granuloma	
	Tuberculosis	Proof of tuberculous bacilli and Langhans cells
	Leprosy	Lepra cells
	3. Lymphoma and pseudolymphoma	
	Non-Hodgkin's lymphoma	Lymphosarcoma cells
Needle aspiration and biopsy cytology	Hodgkin's disease	Hodgkin's and Reed-Sternberg cells
	Mycosis fungoides	Mycosis cells and exocytosis referable to as Pautrier's abscess
	Sézary syndrome	Sézary cells
	Lymphocytoma cutis (Spiegler-Fendt)	Admixture with lymphocytes and histiocytes
	4. Deep fungal infection	
	North American blastomycosis, Paracoccidioidomycosis, etc.	Proof of fungus
	5. Protozoa infection	
	Leishmaniasis	Proof of organism
	6. Metastatic carcinoma	Identification of cancer cells

lesions can be examined without incision, whereas biopsy examination is limited in a site of resection, may widen the application.

3) Fine needle aspiration

This procedure is limited in practical application[25,27] because wide varieties of skin adnexal tumors can be histologically studied with ease by excisional procedures. Dey et al[6] emphasize its worthy application for diagnosis of metastatic tumor and determination of benign or malignant nature.

2 Immunofluorescence Dermatologic Cytology

The immunofluorescence technique is recently utilized in dermatopathology to diagnose or to evaluate the prognosis in lupus erythematosus,[33] dermatitis herpetiformis,[49] lichen planus,[49] necrotizing vasculitis,[10,26,57] drug rashes,[44] etc. For instance, IgA deposition forming irregular aggregates is seen below the dermal-epidermal junction at the tip of papillae in dermatitis herpetiformis,[38,46] and IgG and IgM depositions can be recognized as a homogeneous band, threads or stipples at the basement membrane in lupus erythematosus.[20,35,41,49,54] (Fig. 21-2). This lupus band test for nonsunexposed, clinically normal skin is not only available for distinction from discoid lupus erythematosus,[8,12,35] but for the account of its prognosis.[49] The correlation between clinical disease activity of lupus erythematosus and properdin deposition has been also confirmed by immunofluorescence technique. However, the practice of immunofluorescence method in dermatocytology is limited to the pemphigus group and occasionally psoriasis. While in bullous pemphigoid IgG and C_3 are shown at the basement membrane,[12] in pemphigus vulgaris as well as in pemphigus vegetans the IgG deposition of all its subclasses[40] is characteristically proved in the intercellular space. Although the nature of the antigen is still unknown, it is suggested by Hashimoto[13] that the pemphigus antibody binding site is either carbohydrate or glycoprotein moiety containing alpha-D-glucopyranosides or alpha-N-acetyl-D-gulcosides. The C_3 complement is specifically present in areas of acantholysis of pemphigus vulgaris.[19] According to the report by Harrist and Mihm, the positive rate of direct immunofluorescence for IgG and C_3 is as high as 80 to 95%.[12] The observation of deposition of immunoglobulins in the stratum corneum of psoriatic epidermis by Krogh and Tönder[24] was confirmed by other researchers.[1, 17, 23]

3 General Aspect of Cutaneous Cytology

The lesion of the skin is visible and accessible with ease, but thick cornified layer prohibits exfoliation of prickle cells of the epidermis. Thus dermatologists prefer biopsy to cytology. The strip method is valuable to observe the behavior of cornification in several epidermal diseases such as ichthyosis, psoriasis, seborrheic keratosis, etc; parakeratosis, i.e., the appearance of nucleated superficial cells in the uppermost layer, and exocytosis are in favor of psoriasis. Marks and co-workers improved the original tape method by application of transparent adhesives of cyanoacrylate. Direct observation and availability of scanning electron microscopy widened the value of cutaneous cytology.[28-30] For the lesion of dermis and skin appendages cytodiagnosis cannot be successful unless erosion or ulceration accompanies. The advantage of cytology consists in limited conditions when biopsy should be avoided or refused by patients; if malignant melanoma is suspected, a touch smear or a lightly scraped smear is worthy of consideration. Viral and fungus infections are more accurately diagnosed if cytology is added to histology. Cutaneous blastomycosis, leishmaniasis and amebiasis are the target diseases of cutaneous cytology. Various blister-forming diseases such as herpes infection and pemphigus will be diagnosed by scraping the bottom of the blister. Imprint smears from a resected specimen become the adjunctive to reach the correct histodiagnosis. For example, classification of Hodgkin's disease and non-Hodgkin's lymphoma is done with great assistance of cytology (Table 21-1).

4 Histology of the Skin and Normal Cellular Components

The epidermis and its adjacent mucous membrane reveal histologically a stratified squamous epithelium with various degrees of keratinization: The stratified squamous epithelium consists of stratum corneum lacking nuclei and full of keratin, stratum lucidum with eleidin, stratum granulosum with keratohyaline granules, stratum spinosum with a polyhedral, abundant cytoplasm showing intercellular bridges and the stratum germinativum. The prickle cells of the stratum spinosum are several layers thick and encountered predominantly; they are glycogen-containing and similar to those in vaginal smears except for the tendency to keratinization with intense orangeophilia and keratohyaline granules. The superficial keratinized cells which have lost their nuclei and cytoplasmic organelles desquamate physiologically and are replaced by cells arising from the basal layer and differentiating towards the keratinized cells. The thickness of the stratum corneum differs markedly from place to place according to the site of the body surface.

Although prickle cells in which the nuclei are retained are shed by scraping from the normal mucous membrane covering the vaginal wall, oral mucosa and esophageal mucosa, these cells do not exfoliate from the normal skin even by vigorous scraping.

5 Abnormal Cytology

Since Tzanck's report in 1947,[50] the cytologic technique has been applied to the diagnosis of skin disease by viral infection such as herpes simplex, herpes zoster, varicella,[3,4,22,45,48,50,52] and molluscum contagiosum.[41] Generally speaking, however, the field of application of cytology is limited to the diseases of the epidermis forming bullae or erosion. Koss states that the method has not found acceptance in his institution, probably because of the ease with which biopsies are obtained.[21]

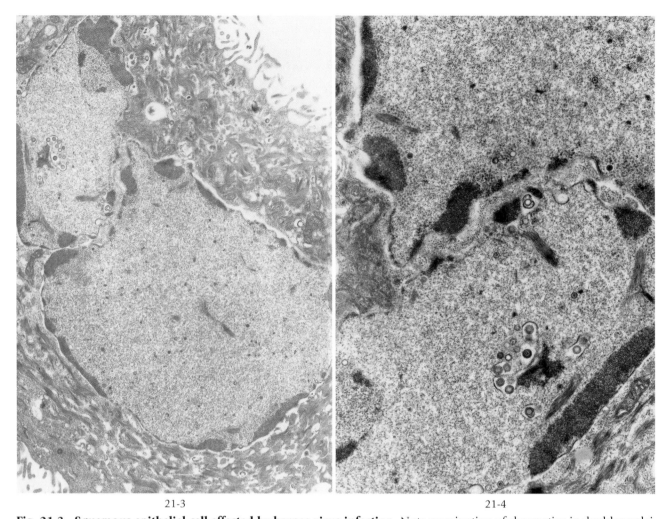

21-3   21-4

Fig. 21-3   Squamous epithelial cell affected by herpes virus infection. Note margination of chromatins in double nuclei. Virus particles are scattered in the nuclei as well as in the cytoplasm. (original magnification × 10,000)
Fig. 21-4   Higher magnification showing virions that possess electron dense cores surrounded by capsid. (original magnification × 16,000)

■Viral infection

Viruses contain either DNA or RNA and possess no mitochondria or ribosomes. Although bacteria containing DNA, RNA and many chemicals can multiply by simple division, viruses are able to multiply within the affected cells; viral nucleic acid enters the nucleic acid pool of the host cell and are incorporated into the genetic material of the host cell. Thus the virus infection is latent maintaining the symbiotic relationship with the host cell. Herpes simplex infection is a good example. It manifests when this symbiosis is disturbed by intercurrent infections such as common cold, pneumonia and in physically depressed conditions; exposure to sunlight, menstruation and emotional strain are susceptible to manifestation.

1) Herpes simplex

Cytologic examination is of value if the smear is taken from the floor of freshly opened vesicle. There are two types of herpes virus immunologically different from each other. Type 1 herpes simplex affects the skin and oral mucosa, while type 2 affects the genital mucosa. This viral infection demonstrates characteristic changes in the squamous epithelium (Figs. 21-5 and 21-6): (a) multinucleation and crowding of infected cells, (b) ground glass appearance of the nuclei with distinct nuclear margination by chromatin condensation; basophilic homogeneous material filling the nucleus can be proved to be positive for Feulgen reaction, (c) appearance of eosinophilic inclusion body with a clearing between a central eosinophilic body and chromatin material at the nuclear rim.

Electron microscopically the affected cells are rich in virus particles (virions) showing electron dense, round to oval cores which are surrounded by the capsid (Figs. 21-3 and 21-4).

2) Herpes zoster

It shows the same cytological changes as those in herpes simplex infection. The same type of intranuclear eosinophilic inclusion bodies will be found.

3) Molluscum contagiosum

The lesion shows small, waxy, umbilicated papules. Scraping the lesion, one obtains large, brownish or eosinophilic intracytoplasmic inclusion bodies (Fig. 21-8). The displaced nuclei are only visible as thin crescents. The molluscum bodies increase in size and turn basophilic as the infected cells move toward the granular layer (Fig. 21-7). These large bodies enmeshed in the thickened horny layer measure about 25-35 μm in diam-

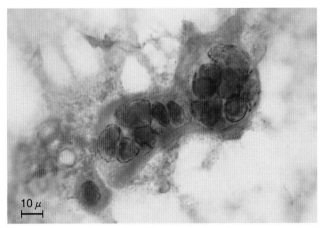

Fig. 21-5 Epidermoid cells affected by herpes simplex. Note syncytial multinucleation of epidermoid cells with homogeneous stain and margination of chromatin. Molding of the nuclei is of diagnostic feature.

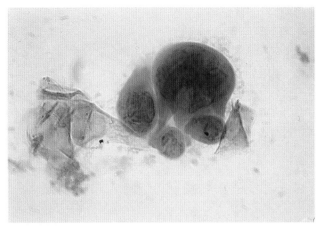

Fig. 21-6 Mononuclear and multinucleated giant cells from vesicular lesions in herpes zoster infection. Giant cell formation is considered to be the result of adhesion of affected cells rather than amitotic multiplication of cells.

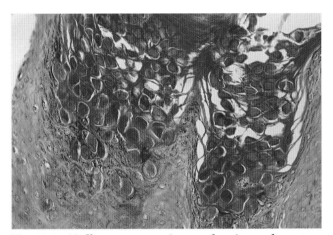

Fig. 21-7 Molluscum contagiosum showing a characteristic degeneration at the granular and horny layers. Note the mass of homogeneous rounded molluscum bodies. The infection produces lobulated epidermal growths compressing the papillae as thin fibrous septa.

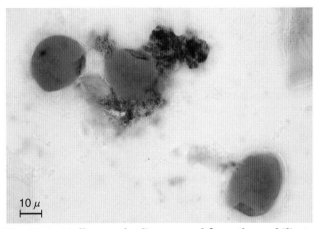

Fig. 21-8 Molluscum bodies scraped form the umbilicated, waxy center of molluscum contagiosum. The molluscum bodies in a smear are characterized by large enrounded concretions with homogeneous, eosinophilic hue. The pyknotic small nuclei are displaced at periphery.

eter. The DNA virus that belongs to the poxvirus group is brick-shaped and measures about 300×200 nm.

■Pemphigus vulgaris

Pemphigus vulgaris forms acantholytic suprabasal bulla intraepithelially as a result of degeneration of epidermal cells losing their intercellular bridges.[26,47] Intercellular edema and loss of intercellular bridges result in detachment of the epidermal prickle cells into the bulla as single cells or loose clusters of cells (Fig. 21-9). These acantholytic cells were advocated by Tzanck to be demonstrable by simple cytologic technique and valuable for the diagnosis,[2,51] although acantholysis could be found in Darier's disease, familial benign pemphigus (Hailey and Hailey), bullous herpes simplex, squamous cell carcinoma, etc. Buccal mucosa is also affected not only by the extension but as the first manifestation of pemphigus vulgaris. A bullous lesion is easily deroofed, but acantholytic cells are diagnosable by scraping the lesion (Figs. 21-10 and 21-11).

Cytology: The typical acantholytic cell from the prickle cell layer contains a centrally located large nucleus with prominent nucleolus and scanty, thick cytoplasm with a perinuclear clear zone (bull's-eye cells)(Fig. 21-13). Multinucleation of epithelial cells is not rare.[31] However, there may be variations from the typical type in size, shape and stainability of the cells. Pemphigus vegetans is characterized by the admixture of the acantholytic cells with many leukocytes, particularly eosinophils; these cell constituents are reflective of intraepidermal abscess. The Tzanck test is useful for a prompt diagnosis at the outpatient clinic, but it should be reemphasized that the simulating cells can be observed in other non-acantholytic bullous or pustular diseases.[11] Recently immunofluorescence studies[16,18,34] have revealed autoimmune antibodies mainly IgG to be deposited on the cellular membrane of acantholytic cells in all forms of pemphigus such as pemphigus vulgaris, pemphigus foliaceus and pemphigus vegetans. The presence of complement (C_3)[19] only in areas of acantholysis proves it to be the result of true immune reaction (Fig. 21-12). The exact role of the antibody in relation to dissolution of desmosomes is, as yet, not clarified.[12]

453

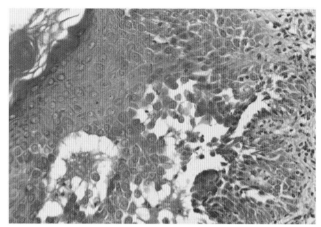

Fig. 21-9　Pemphigus vulgaris showing suprabasal bulla.
The prickle cells separate from one to another through the disappearance of intercellular bridges. Thus the bulla contains detached, so-called acantholytic cells. No inflammatory reaction is seen.

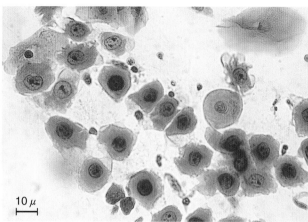

Fig. 21-10　Acantholytic cells from the bulla of pemphigus vulgaris. Note disintegration of cellular borders caused by acantholysis (Fig. 21-9). Nuclear enlargement, high N/C ratio, prominence of nucleoli should not be misinterpreted as malignant. Uniform configuration of the nuclei and bland chromatin pattern are excludable of malignancy. Shorr stain. (By courtesy of Dr. C.R. Vieira e Silva, Faculdade de Medicina, Universidade de Botucatu, São Paulo, Brazil.)

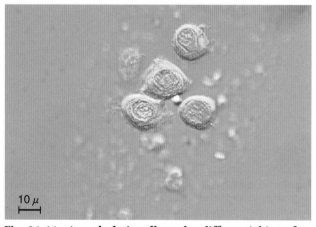

Fig. 21-11　Acantholytic cells under differential interference microscopy from pemphigus vegetans. The acantholytic cells are spherically enrounded and sharply outlined. Nuclear predominance is well demonstrated.

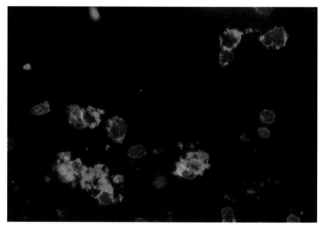

Fig. 21-12　Acantholytic cells in pemphigus vulgaris showing specific immunofluorescence. Direct stain with anti-C_3 fluorescent antiserum. IgG was also positive at the cellular membrane.

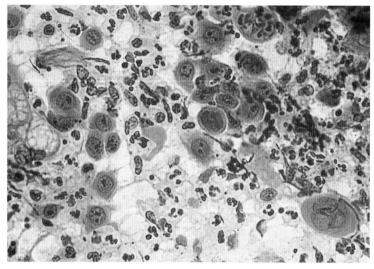

Fig. 21-13　Pemphigus vulgaris associated with Kaposi's varicelliform eruption. This 15-year-old boy who had been treated for pemphigus vulgaris with immunosuppressive drugs noticed extensive eruption composed of pustules. Note the mixture of multinucleated giant cells by herpes inoculation with characteristic acantholytic cells. (By courtesy of Dr. C.R. Vieira e Silva, Faculdade de Medicina, Universidade de Botucatu, São Paulo, Brazil.)

■Disorders of horny and granular layers

The Scotch tape strip method is a good tool to examine the disorders of horny and granular layers of the epidermis, because each cell layer can be examined by stepwise strippings.

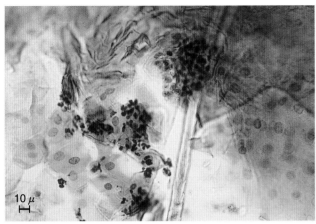

Fig. 21-14 *Psoriasis vulgaris* **showing exocytosis in the superficial layer by the strip method.** From a 27-year-old female with chronically progressive scaly papules. Note parakeratosis and leukocytic exocytosis that is referred to as Munro microabscess.

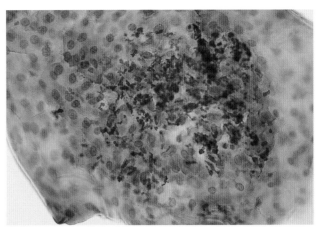

Fig. 21-15 *Psoriasis vulgaris* **by the strip method.** Accumulation of leukocytes is illustrated within the upper prickle cell layer. The Scotch tape strip preparation demonstrates figures coincident with tangential tissue sections.

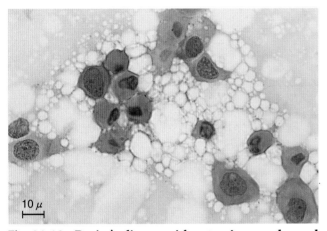

Fig. 21-16 Darier's disease with extensive papular and nodular eruptions. The strip method reveals many parabasal cells with acantholysis in deep layers. Premature pyknosis and pronounced dyskeratosis are distinguishable from acantholytic cells in pemphigus.

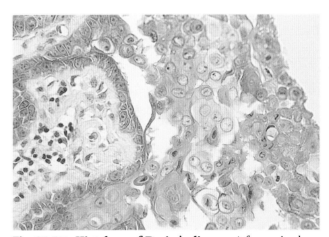

Fig. 21-17 Histology of Darier's disease. A few grains having a grain-like nucleus and eosinophilic cytoplasm and acantholytic prickle cells are recognizable. Note suprabasal acantholysis and lacunae formation.

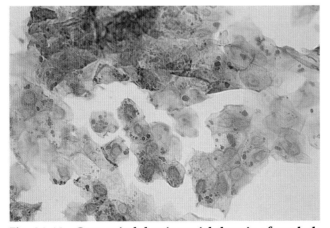

Fig. 21-18 Congenital dominant ichthyosis of nonbullous form from a 5-year-old boy. The cytoplasm is poorly stained, ill-defined and shows irregular keratohyaline granules that is referred to as granular degeneration.

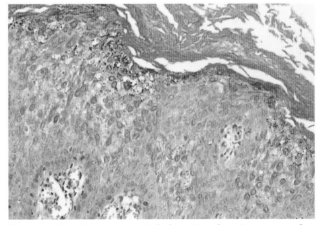

Fig. 21-19 Dominant ichthyosis showing granular degeneration. The upper two thirds of the epidermis shows indistinctive cellular boundaries.

1) Psoriasis

This chronic epidermal disorder starts with localized, scaly papules in the scalp, trunk and extensor surfaces of the extremities. Psoriasis vulgaris is characterized by elongation of rete ridges, thinning of suprapapillary portions, parakeratosis (absence of granular cells) and exocytosis that is called Munro microabscess. The strip method often demonstrates parakeratosis and exocytosis

455

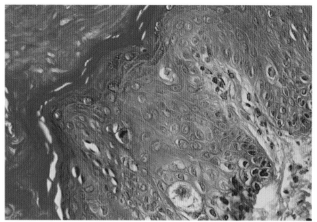

Fig. 21-20 Bowenoid type solar keratosis arising from the dorsa of the hand of a 64-year-old female. The scaly lesion is characterized by marked hyperkeratosis, parakeratosis and acanthosis with atypia; the lesion is also called squamous cell carcinoma grade 1/2. Note atypical cells which are disorderly arranged and varying in size and shape. Abnormal mitoses are frequently seen.

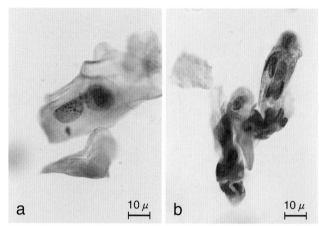

Fig. 21-21 Dyskaryotic squamous cells in solar keratosis from varying layers after detachment of keratinized layer. An intermediate cell has a large nucleus with even, finely granular chromatins. The N/C ratio is within dysplastic limits. Individual cell keratinization is suggestive (a).

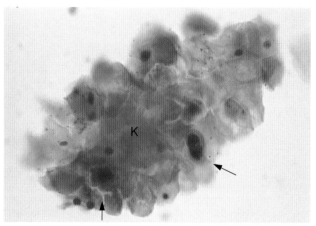

Fig. 21-22 Hyperkeratotic atypical cells embedded in the dense keratinous layer from Bowenoid solar keratosis. Single atypical cells retaining large densely stained nuclei (arrows) within thick orangeophilic keratin material (K) are diagnostic of Bowenoid change.

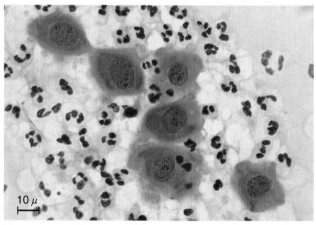

Fig. 21-23 Hypertrophic squamous cells simulating malignancy in pseudoepitheliomatous hyperplasia. These cells having worm-eaten appearance are from the edge of nonspecific granulation. The enlarged nuclei are bland and uniform in size and shape. Note a mixture with many leukocytes.

in preparations from initial stripping (Figs. 21-14 and 21-15).

2) Darier's disease

This slowly progressive, occasionally dominantly transmitted disease shows keratotic papules in the scalp and inguinal or axillary regions. The peculiar dyskeratosis forming corps ronds and grains is distinctively figured out in cytology.[48] Superimposition of dyskeratosis and premature pyknosis of the nuclei on acantholysis differs from the acantholysis in pemphigus (Figs. 21-16 and 21-17).

3) Congenital dominant ichthyosis

The nonbullous form dominantly inherited ichthyosis is characterized by marked vacuolation of the cytoplasm with indistinct borders, large irregular keratohyaline granules and decrease in stainability (Figs. 21-18 and 21-19). So-called granular degeneration can be found by skin surface biopsy cytology.

■Carcinoma

1) Squamous cell carcinoma

Selbach and Heisel[43] described high accuracy in detection of basal cell epithelioma (88.9% of 27 cases), whereas the diagnosis for squamous cell carcinoma was disappointing. This conflicting matter depends on inappropriate specimen obtained by ordinary scraping because of hyperkeratosis or crust formation. Early squamous cell carcinoma, precancerous lesions and its simulating dermatosis cannot be distinguished merely by cytological examination: Senile keratosis of hypertrophic type (carcinoma grade 1/2), keratoacanthoma and seborrheic keratosis are all associated with pronounced keratosis, which may prevent shedding of epidermal cells from the deeper layers (Figs. 21-20 to 21-23). For the distinction of early squamous cell carcinoma and Bowen's disease from hypertrophic senile keratosis and pseudocarcinomatous hyperplasia (pseudoepitheliomatous hyperplasia), further precise studies on morpholog-

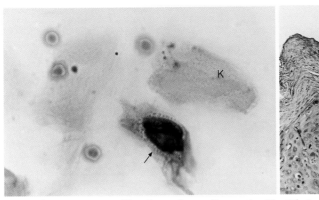
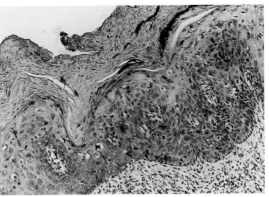

21-24   21-25

Fig. 21-24 High power magnification of a malignant cell with India ink droplet-like nucleus from Bowen's disease.
An atypical cell in the center of the field showing increased nuclear-cytoplasmic ratio, intense orangeophilia of thick cytoplasm and pyknotic nucleus with irregular contour is consistent with a keratinizing cancer cell (arrow).
Fig. 21-25 Bowen's disease with marked hyperkeratosis. The epidermis is thickened by atypical proliferation of epidermal cells in association with disorderly arrangement, hyperchromatism and variation in size and shape of the nuclei. Mitoses are frequent. The basement membrane is intact. Histology suggests that sufficient cytology specimen from the deep prickle cell layer is hardly obtained unless marked hyperkeratosis (K) is excluded beforehand.

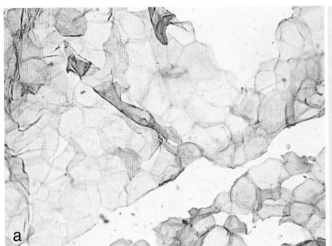
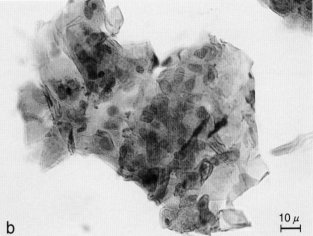

Fig. 21-26 Dyskeratotic atypical cells of parabasal or intermediate type. Scotch tape method for Bowen's disease.
These cells from a deeper layer were successfully submitted by the strip method. The nuclei, occasionally binucleated, are hyperchromatic and deformed in shape. The cytoplasm is thickened and intensely orangeophilic because of dyskeratosis. However, nuclear atypism is not so pronounced as in squamous cell carcinoma.
a. From superficial zone consisting of anucleated superficial squamous cells.
b. From a deeper intermediate zone following several strippings.

ical cellular atypism and on biochemical or biological behaviors are needed. In conclusion, cytodiagnosis cannot be a substitute for biopsy. The strip method provides some information to become the clue distinguishing Bowen's disease and Paget's disease from frank squamous cell carcinoma (Figs. 21-28 and 21-29).

a) Bowen's disease: This is the state of intraepidermal squamous cell carcinoma that is also referred to as squamous cell carcinoma in situ (Fig. 21-25). The appearance of dyskeratotic atypical cells including individual cell keratinization without tumor diathesis is in favor of Bowen's disease (Figs. 21-24 to 21-26).

b) Leukoplakia: The term of leukoplakia is used for the white patchy lesion of the oral mucosa or the vulva that is correspondent to hypertrophic solar keratosis of the skin. Histologically the lesion shows keratosis and acanthosis with cellular atypia and disorderly arrangement (Fig. 21-27).

c) Basal cell epithelioma: Some neoplasms, although

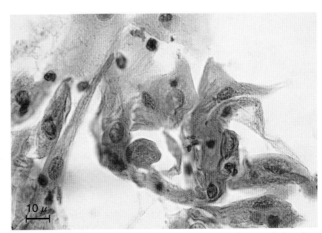

Fig. 21-27 Leukoplakia of the tongue. Dyskeratotic benign atypical cells scraped from the tongue resemble squamous cancer cells. They are distinguishable from the cancer cells by low N/C ratio, and fairly small nuclei despite the marked dyskeratosis.

457

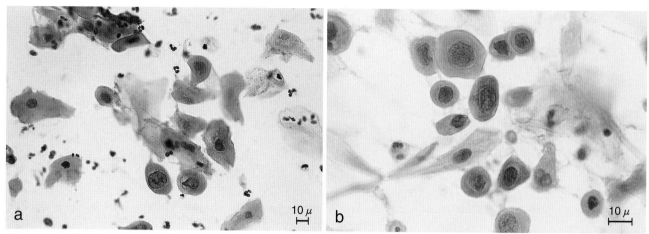

Fig. 21-28 Well demonstrated squamous cancer cells (in a scraping smear) from the site with unpronounced hyper-keratosis.
a. Scraping from cancer developed in the skin graft. Keratinizing and nonkeratinizing cancer cells of parabasal type are exhibited with anucleated degenerate cells.
b. Scraping from cancer in the penis. Keratinizing and nonkeratinizing cancer cells.
　Diagnostic points: (1) Isolation as single cells, (2) pronounced variation of the nuclei in size and shape, and (3) high N/C ratio are of differential diagnostic points from precancerous lesion such as solar keratosis, Bowen's disease, etc.

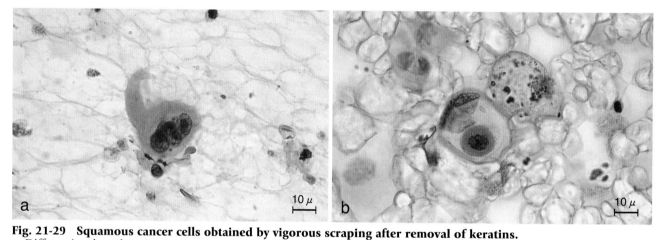

Fig. 21-29 Squamous cancer cells obtained by vigorous scraping after removal of keratins.
a. Differentiated nonkeratinizing cancer cell. Note thickened bluish cytoplasm and multiple pyknotic nuclei.
b. Squamous cancer cells showing cell inclusion. An ingested cell shows a typical feature of parabasal type squamous cancer cell.

quite limited in number, may develop to ulceration by their outgrowth and shed tumor cells. A type of basal cell epithelioma, i.e., rodent ulcer or nodulo-ulcerative type, is a slowly growing benign tumor, but tends to develop central ulceration, by which high accuracy of cytodiagnosis is obtained (Selbach and Heisel)[43](Fig. 21-30).

2) Extramammary Paget's disease
The genital, axillary and perianal regions are the common sites of extramammary Paget's disease. Histology and cytology are the same as those of breast Paget's disease. The Paget cells in the epidermis appear to be derived from apocrine glands in the dermis. These are found as tight cell clusters consisting of round cells with often diastase-resistant PAS positivity. Their nuclei are enrounded and fairly uniform in shape (Fig. 21-31).

■Melanoma
Malignant melanoma originating in the epidermo-dermal junction shows invasive growth into the epidermis as well as downward growth. Thus, epidermal ulceration

often ensues and melanoma cells are obtainable by scraping the lesion (Fig. 21-34). As Miescher[32] states, excision of the lesion for biopsy may release melanoma cells into lymphatic and blood vessels. Exfoliative cytology, a more conservative method than biopsy, is applicable to the diagnosis of malignant melanoma, but still limited to some portion.

1) Lentigo
The lentigo is characterized by concentration of melanocytes at the basal layer. In senile lentigo, the rete ridges with basal melanocytes are elongated. Malignant lentigo reveals more atypical melanocytes of a spindle shape; histologically these atypical melanocytes are condensed along the epidermal-dermal junction.

2) Pagetoid melanoma
Pagetoid melanoma in situ is characterized by scattered cell groups of rounded, large malanocytes with an epithelioid appearance (epithelioid melanocytes)(Fig. 21-35).

3) Malignant melanoma
Epithelioid melanoma cells are predominant compared

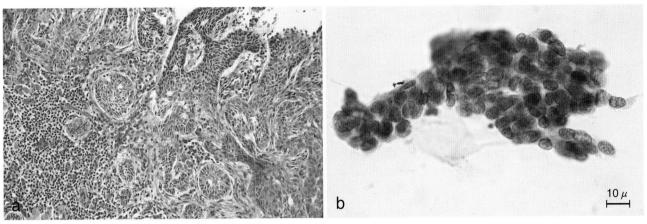

Fig. 21-30 Basal cell epithelioma form the lid.
a. Histology of the primordial type showing solid bands or cords of densely stained, basal type cells. The peripherally locating cells in the nests are arranged in a palisade fashion.
b. Cytology by vigorous scraping from rodent ulcer. Note a trabecular cell cluster having oval to elliptic nuclei and very scanty cytoplasm. The nuclei are hyperchromatic, but bland in chromatin pattern.

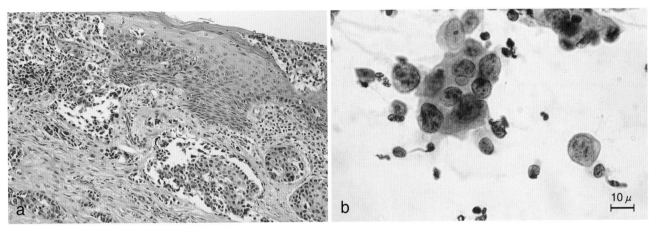

Fig. 21-31 Paget disease in genital region.
a. Histology. This tumor not only involves epidermis of the scrotum but invades lymphatic channels.
b. Cytology by strip method. An appearance of tumor cells displaying overlapping cluster from the deeper layer in strippings differs from common squamous cell carcinoma.

with spindle melanoma cells. As in other malignant tumors, the melanoma cells reveal high N/C ratio, thickening of nuclear rim and nuclear size and shape variations. A central macronucleolus is of characteristic feature (Fig. 21-34). Malignant melanoma cells are polymorphic and vary in size and shape; these is a variety that is composed of fusiform melanoma cells that simulate sarcoma cells particularly if the tumor is amelanotic.

The juvenile melanoma is hardly distinguishable from malignant melanoma. In our experience in imprint smears, juvenile melanoma cells were more uniform in size and shape and more bland in chromatin pattern than malignant melanoma cells. The age of the patient is important for decision.

◼Cutaneous lymphoma

Cutaneous manifestation of various types of lymphomas (see chapter on "Lymph Node") and leukemia occurs as a part of the disease. On the other hand, mycosis fungoides and Sézary syndrome affect the skin predominantly.

Cytology by imprint smears is helpful to distinguish lymphocytoma cutis (pseudolymphoma of Spiegler-

Fendt) from malignant lymphoma (Figs. 21-36 and 21-37). While lymphocytoma cutis is composed of mature lymphocytes massively infiltrated and larger follicle center cells, malignant lymphoma is mainly monomorphous in cell constituents consisting of either lymphoblast-like or immature histiocytic cells. Mycosis fungoides that begins from erythematous stage and advance to the plaque stage and finally to the tumor stage may reveal distinctive atypical mononuclear (mycosis) cells with large, hyperchromatic, convoluted nuclei. This long lasting disease is considered to be of T lymphocyte origin. Sézary erythroderma is likely a variant of mycosis fungoides with leukemic manifestation. The Sézary cells are slightly larger than leukocytes but possess large nucleus with irregular shapes. These are indistinguishable from the mycosis cells light microscopically.

◼Dermal lesions

Dermal lesions are inaccessible even by vigorous scraping. Imprint smears from resected lesions or fine needle aspiration smears provide useful information for granulomatous or tumorous lesions of the dermis (Figs. 21-38 to 21-42). Cutaneous sarcoid, tuberculosis, lepromatous

459

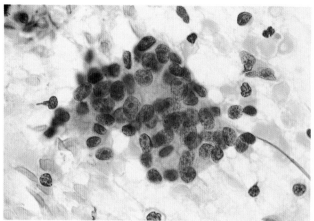

Fig. 21-32　Cylindroma in imprints of a tumor arising from the lacrimal gland. This tumor of the skin appendages is not the target of routine skin cytology, but the tumor cells occurring from the same site resemble basal cell epithelioma. Acinar pattern of cylindroma differs from the trabecular column of basal cell epithelioma. The presence of nucleoli is also distinctive.

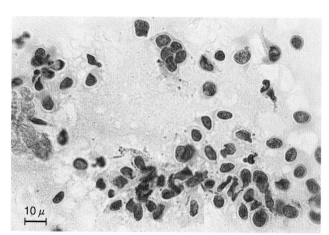

Fig. 21-33　Nevus cells in imprint smear from intradermal nevus. Benign nevus cells are characterized by uniform oval nuclei. Their chromatins are smoothly and homogeneously distributed. A small amount of melanin pigment is present.

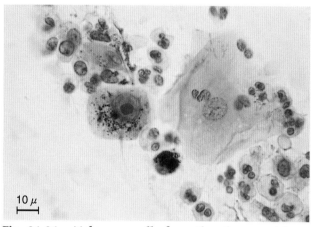

Fig. 21-34　Melanoma cells from the pharynx. A large amount of melanin is contained. Note a single, centrally locating basophilic nucleolus that is characteristic of malignant melanoma.

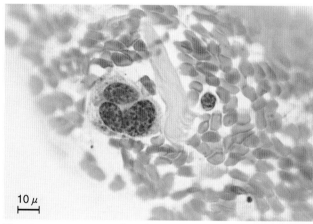

Fig. 21-35　Pagetoid melanoma cells lying in a small group. The large enrounded nuclei are equally sized and possess single eosinophilic macronucleoli. Uniformity in size and shape of the nuclei and their arrangement in nest can be distinguished from the classical malignant melanoma cells.

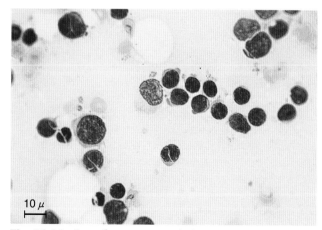

Fig. 21-36　Lymphocytoma cutis in imprints. Main components of the cells are small mature lymphocytes and a small number of large lymphogonia from the follicle center. Some lymphocytes show a distinctive cleavage that is suggestive of follicle center origin.

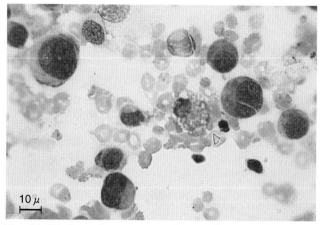

Fig. 21-37　Lymphocytoma cutis. Another case of lymphocytoma cutis of a 36-year-old female. Variable cell components such as large lymphogonia with nucleolus, plasma cells and histiocyte. Such a compound cellularity differs from monotonous lymphosarcoma cells.

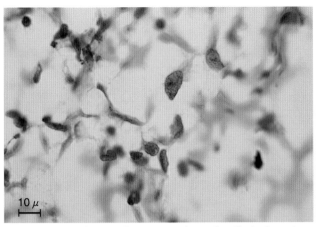

Fig. 21-38   Polymorphic mesenchymal cells in imprints from dermatofibrosarcoma protuberans. The nuclei vary in size and shape; these are spindly or stellate. The cytoplasm is stained feebly and poorly defined.

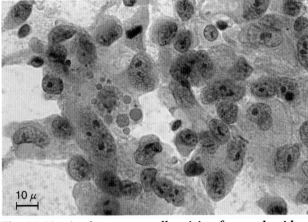

Fig. 21-39   Angiosarcoma cells arising from subepidermal region in imprints. The nuclei are markedly enlarged, oval to elliptic, and occasionally curved. Prominence of multiple nucleoli, thin nuclear membrane, indistinct cellular borders are characteristic of sarcoma. Note phagocytosis.

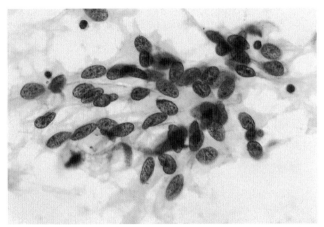

Fig. 21-40   Kaposi's sarcoma in imprints. From a dark bluish brown nodule of a male aged 51. The clinical history in the lower extremity of AIDS is unknown. Highly cellular components consisting of hyperchromatic spindly cells arranged in a fascicular pattern. These were positive for factor VIII in immunochemistry.

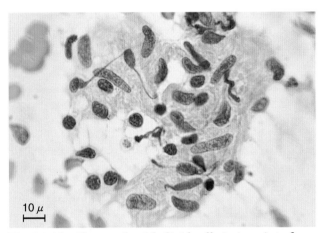

Fig. 21-41   Clustered epithelioid cells in imprints from lepromatous leprosy. In the absence of caseous necrosis, the lesion is hardly distinguished from sarcoidosis. Bacterial investigation is important.

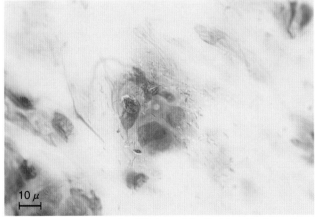

Fig. 21-42   Leprosy bacilli in epithelioid cells from leproma. This is the same case as Fig. 21-41. Acid-fast property of the bacilli is not so intense as that of tuberculous bacilli. Ziehl-Neelsen stain.

leprosy, granulomatous tertiary syphilis and mixed cellular (lymphocytes, eosinophils, neutrophils, plasma cells and histiocytes) granuloma caused by fungal infec-

tion will be illustrated by cytology: (1) Determination of micro-organism by specific stains such as PAS, Gomori's methenamin silver nitrate methods, (2) account for cellularity, (3) differential count of the cell components, (4) identification of histiocytes and their phagocytosed substance and (5) observation on whether or not necrosis is associated are important for consideration of the dermal cytopathology (Figs. 21-43 and 21-44). It should be noted that mesenchymal cells are hardly detached even from malignant neoplasm. Mesenchymal tumor cells are isolated or loosely textured. The cytoplasmic borders are poorly defined and the nuclear rims are not thickened. There are marked nuclear variation in size and shape and prominence of one or two nucleoli irregular shapes.

Kaposi's sarcoma

Nodular Kaposi's sarcoma irrespective of acquired immune deficiency syndrome (AIDS) develops commonly in extremities as a dark brownish plaque that may undergo ulceration. Kaposi's sarcoma is frequent clinical manifestation of AIDS and involves anywhere in the body as a visceral disease. [7,15] The tumor is com-

461

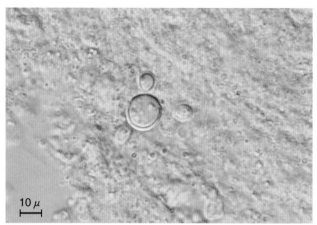

Fig. 21-43 *Paracoccidioides brasiliensis* **from ulcerated lesion of the lip.** Paracoccidioidomycosis or South American blastomycosis involving the nasopharynx, skin, lymph nodes and respiratory tract can be cytologically diagnosed with ease. Differential interference microscopy was available for immediate detection of yeast cells with buds varying in numbers. Note a "steering wheel appearance".

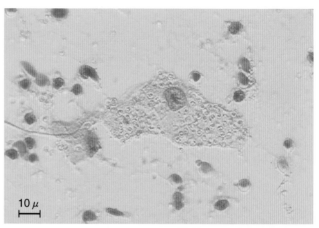

Fig. 21-44 **Macrophage engorged with numerous spores of *Histoplasma capsulatum* under differential interference microscopy.** From a cutaneous lesion with systemic histoplasmosis. Leishmania organisms with the similar size can be discriminated by the presence of clear halos. This 40-year-old male who had been treated with immunosuppressive therapy for CLL has had persistent dyspnea, generalized lymphadenopathy and hepatomegaly.

posed of two types of sarcoma; hemangiomatous with vascular meshes or fibroblastic predominance resembling fibrosarcoma. The latter form can be identified by immunocytochemistry for factor VIII (Fig. 21-40).

■Tumors of the epidermal appendages

Epidermal disorders become the target of scraping and/or stripping smears. However, primary tumors arising from epidermal appendages are commonly diagnosed by excisional histology. There have been only a few reports by fine needle aspirations such as hidradenoma,[37] basal cell epithelioma,[6] sebaceous carcinoma,[14] eccrine spiradenoma,[53] etc. Rash imprint cytodiagnosis is available as an adjunct to histodiagnosis especially when rash interpretations are in demands not only for neoplasia but for granulomatous lesions (see Fig. 21-32).

References

1. Beutner EH, Jablonska S, Jarzabek-Chorzelska M, et al: Studies in immunodermatology. VI. IF studies of autoantibodies to the stratum corneum and of *in vivo* fixed IgG in stratum corneum of psoriatic lesions. Int Arch Allergy Appl Immunol 48: 301, 1975.
2. Blank H, Burgoon CF: Abnormal cytology of epithelial cells in pemphigus vulgaris: A diagnostic aid. J Invest Dermatol 18: 213, 1952.
3. Blank H, Burgoon CF, Baldridge GD, et al: Cytologic smears in diagnosis of herpes simplex, herpes zoster and varicella. JAMA 146: 1410, 1951.
4. Blank H, Rake G: Viral and Rickettsial Diseases of the Skin, Eye and Mucous Membranes of Man. Boston: Little Brown, 1955.
5. Brehmer-Andersson E, Brunk U: Tapestripping method for cytological diagnosis of mycosis fungoides. Acta Dermatovener 47: 177, 1967.
6. Dey P, Das A, Radhika S, et al: Cytology of primary skin tumors. Acta Cytol 40: 708, 1996.
7. Francis ND, Parkin JM, Weder J, et al: Kaposi's sarcoma in acquired immune deficiency syndrome (AIDS). J Clin Pathol 39: 469, 1986.
8. Gilliam JN: The significance of cutaneous immunoglobulin deposits in lupus erythematosus and NZB/NZW F_1 hybrid mice. J Invest Dermatol 65: 154, 1975.
9. Goldman L, McCabe RM, Sawyer F: The importance of cytology technic for the dermatologist in office practice. AMA Arch Dermatol 81: 359, 1960.
10. Gower RG, Sams WM Jr, Thorne EG, et al: Leukocytoclastic vasculitis: Sequential appearance of immunoreactants and cellular changes in serial biopsies. J Invest Dermatol 69: 477, 1977.
11. Graham JB, Bingul O, Burgoon CB: Cytodiagnosis of inflammatory dermatoses. Arch Dermatol 87: 118, 1963.
12. Harrist TJ, Mihm MC Jr: Cutaneous immunopathology: The diagnostic use of direct and indirect immunofluorescence techniques in dermatologic disease. Hum Pathol 10: 625, 1979.
13. Hashimoto K, King LE Jr, Yamanishi Y, et al: Identification of the substance binding pemphigus antibody and concanavalin A in the skin. J Invest Dermatol 62: 423, 1974.
14. Hood IC, Qizilbash AH, Salama SS, et al: Needle aspiration cytology of sebaceous carcinoma. Acta Cytol 28: 305, 1984.
15. Hutt MSR: Kaposi's sarcoma. Br Med Bull 40: 355, 1984.
16. Jablonska S, Chorzelski TP, Beutner EH, et al: Herpetiform pemphigus; A variable pattern of pemphigus. Int J Dermatol 14: 353, 1975.
17. Jablonska S, Chorzelski TP, Jarzaber-Chorzelska M, et al: Studies in immunodermatology. VII. Four-compartment system studies of IgG in stratum corneum and of stratum corneum antigen in biopsies of psoriasis and control dermatoses. Int Arch Allergy Appl Immunol 48: 324, 1975.
18. Jordon RE: Complement activation in bullous skin diseases. J Invest Dermatol 65: 162, 1975.
19. Jordon RE: Complement activation in pemphigus and bullous pemphigoid. J Invest Dermatol 67: 366, 1976.
20. Jordon RE, Schroeter AL, Winkelmann RK: Dermal-epidermal deposition of complement components and

properdin in systemic lupus erythematosus. Br J Dermatol 92: 263, 1975.

21. Koss LG: Diagnostic Cytology and Its Histopathologic Bases. Philadelphia: JB Lippincott, 1968.

22. Kotcher E, Gray LA, James QC, et al: Cervical cell inclusion bodies and viral infections of the cervix. Ann NY Acad Sci 97: 571, 1962.

23. Krogh HK: The occurrence of antigens to stratum corneum in man. Int Arch Allergy 37: 649, 1970.

24. Krogh HK, Tönder O: Immunoglobulins and anti-immunoglobulin factors in psoriatic lesions. Clin Exp Immunol 10: 623, 1972.

25. Layfield LJ, Glasgow BJ: Aspiration biopsy cytology of primary cutaneous tumors. Acta Cytol 37: 679, 1993.

26. Lever WF: Pemphigus and Pemphigoid. Springfield, Ill: Charles C Thomas, 1965.

27. Malberger E, Tillinger R, Lichtig C: Diagnoais of basal cell carcinoma with aspiration cytology. Acta Cytol 28: 301, 1984.

28. Marks R, Dawber RPR: Skin surface biopsy: An improved technique for the examination of the horny layer. Br J Dermatol 84: 117, 1971.

29. Marks R, Dawber RPR: In situ microbiology of the stratum corneum. Arch Dermatol 105: 216, 1972.

30. Marks R, Saylan T: The surface structure of the stratum corneum. Acta Dermatovener 52: 119, 1972.

31. Medak H, Burlakow P, McGrew EA, et al: The cytology of vesicular conditions affecting the oral mucosa: Pemphigus vulgaris. Acta Cytol 14: 11, 1970.

32. Miescher G: Über Klinik und Therapie der Melanome. Arch f Derm u Syph 200: 215, 1955.

33. Monroe EW: Lupus band test. Arch Dermatol 113: 830, 1977.

34. Nishikawa T, Kurihara S, Harada T, et al: Capability of complement fixation by in vivo bound antibodies in pemphigus skin lesions. Clin Exp Dermatol 3: 57, 1978.

35. O'Loughlin S, Schroeter AL, Jordon RE: A study of lupus erythematosus with particular reference to generalized discoid lupus. Br J Dermatol 99: 1, 1978.

36. Pinkus H: Examination of the epidermis by the strip method of removing horny layers. J Invest Dermatol 16: 383, 1951.

37. Ray R, Day P: Fine needle aspiration cytology of malignant hidradenoma. Acta Cytol 37: 842,1993.

38. Reunala T: Gluten-free diet in dermatitis herpetiformis. Br J Dermatol 98: 69, 1978.

39. Sams WM Jr, Claman HN, Kohler PF, et al: Human necrotizing vasculitis: Immunoglobulins and complement in vessel walls of cutaneous lesions and normal skin. J Invest Dermatol 64: 441, 1975.

40. Sams WM Jr, Schur PH: Studies of the antibodies in pemphigoid and pemphigus. J Lab Clin Med 82: 249, 1973.

41. Schrager MA, Rothfield NF: Clinical significance of serum properdin levels and properdin deposition in the dermal-epidermal junction in systemic lupus erythematosus. J Clin Invest 57: 212, 1976.

42. Schroeter AL, Winkelmann RK: Livedo vasculitis (the vasculitis of atrophie blanche): Immunohistopathologic study. Arch Dermatol 111: 188, 1975.

43. Selbach G, Heisel E: The cytological approach to skin disease. Acta Cytol 6: 439, 1962.

44. Stein KM, Schlappner QLA, Heaton CL, et al: Demonstration of basal cell immunofluorescence in drug-induced toxic epidermal necrolysis. Br J Dermatol 86: 246, 1972.

45. Stern E, Longo LD: Identification of herpes simplex virus in a case showing cytological features of viral vaginitis. Acta Cytol 7: 259, 1963.

46. Stingel G, Hönigsmann H, Holubar K, et al: Ultrastructural localization of immunoglobulins in skin of patients with dermatitis herpetiformis. J Invest Dermatol 67: 507, 1976.

47. Tappeiner J, Pfleger L: Pemphigus vulgaris: Dermatitis herpetiformis. Arch Klin Exp Derm 214: 415, 1962.

48. Tomb JA, Matta MT: The cytology of vesiculo-bullous conditions affecting the oral mucosa. Lab Med J 31: 125, 1980.

49. Tuffanelli DL: Cutaneous immunopathology: Recent observations. J Invest Dermatol 65: 143, 1975.

50. Tzanck A: Le cytodiagnostic immédiat en dermatologie. Bull Soc Franc Derm Syph 7: 68, 1947.

51. Tzanck A: Le cytodiagnostic immédiat en dermatologie. Ann de Dermat et Syph 8: 205, 1948.

52. Tzanck A, Aron-Brunetiere Le "cytodiagnostic immédiat" des dermatoses bulleuses. Gas Med Port 3: 667, 1949.

53. Varsa EW, Jordan SW: Fine needle aspiration cytology of malignant spiradenoma arising in congenital eccrine spiradenoma. Acta Cytol 34: 275, 1990.

54. Wierzchowiecki MO, Quismorio FP, Friou GJ: Immunoglobulin deposits in skin in systemic lupus erythematosus. Arthritis Rheumatol 18: 77, 1975.

55. Wilson JE: Cytodiagnosis. In: Rook A, et al (eds): Textbook of Dermatology. pp72-75. Oxford: Blackwell, 1979.

56. Wolf J: Das Oberflächenrelief der menschlichen Haut. Z Mikr-Anat Forsch 47: 351, 1940.

57. Wolff HH, Maciejewski W, Scherer R, et al: Immuno-electron-microscopic examination of early lesions in histamine induced immune complex vasculitis in man. Br J Dermatol 99: 13, 1978.

22 Telepathology and Telecytology

In recent years long distance teleconsultations transmitting digital images of pathology, cytology, endoscopy, CT and MRI have come to be of practical usage in a broad sense of telemedicine since Weinstein and his associates introduced an advanced information technology[11-13]: technological innovations in information engineering, medical electronics and computer science have solved problems in transmission of digital compressed microscopic images such as rapid multimedia communication, accurate recovery of compressed images and remote-controlled microscope operated at a diagnostic workstation. The speed of image transmission depends on the communication modality, i.e., ordinary telephone line (33.6 kbps), ISDN (64 kbps-1.5 Mbps), LAN (10 Mbps, 100 Mbps), etc. The compression technology, magnification (x4, x10, x20, x40), memory size also affect the speed of transmission.

1 Telepathology/Telecytology Systems

There are three basic system models for telecommunication, namely static, dynamic and hybrid. The pure dynamic system is more or less replaced by the hybrid system which has a bias either to static or dynamic.

A static system is common and available to real time and off-line transmissions[6]; after very rapid transmission of low power images in series under JPEG-based compression, a whole image of tissue section or smear can be retrieved at the remote site of diagnostic workstation (OLMICOS/WX: Olympus Optical Co. Ltd., Tokyo, Japan) (T2000/T2000R: Nikon Corporation, Tokyo, Japan). Diagnostically valuable fields can be selected and required for transfer. Internet connection may become very popular for teleconsultation, education and conference because of low cost and wide accessibility.[6] The hybrid system is preferred to the static system that may need longer time, because dynamic and static combined mechanics are able to transmit live images either by interactive manual or remote-controlled robotic operation and subsequently static images with higher resolutions.[4,5] As a representative hybrid system, AutoCyte LINK (TriPath, Burlington, North Carolina, U.S.A.) is designed to have the function transmitting live and subsequently still images. Priority of orientation of fields and selection of diagnosable images consists in the site of diagnostic workstation. The Hi-SPEED (Sakura Finetechnical Co. Ltd., Tokyo, Japan) in which the author has participated for development is equipped with unique transmission modalities; digital images captured by 3CCD camera (HC 2000 Fujix, 1300000 × 3

pixels, Tokyo, Japan) are transferred in any mode of live scan with low resolution (320 × 240, 1-3 frame/sec by H261 Codec) and still image transmissions with moderate and high resolutions (640 × 480, 7 sec/frame by JPEG and 1280 × 960, 40 sec/frame by JPEG).[9] The highest resolution mode is effective for cytologic interpretation. Now the telepathology/telecytology systems go on improving the function. Fast advancement of information technology may change the modality of multimedia communication.

2 Telepathology

Although telepathology and telecytology appear to be essentially the same in transmitting microscopic images, there is much difference in the purpose and results of interpretation. A telepathology system is applicable to both primary diagnosis and second opinion consultation. Intraoperative frozen section diagnosis that gives an important guide of operation such as indication of radical resection or simple lumpectomy and absence of malignancies at resection ends was available only at large referral hospitals having a well equipped pathology laboratory with full-time pathologists. However, real time telepathology has enabled a surgeon in a small or rural hospital to perform appropriate surgeries.

Second opinion surgical pathology is comprised of single organ system review by expert pathologists and mandatory second opinion diagnosis for patients who are referred to a large referral hospital for further therapies. The former category is what is requested by pathologists who are unsure of the diagnosis. Disagreements may occur in special organ pathology, i.e., in 8.8% for the nervous system[1] and 12.7% for the ovary.[3] Mandatory second opinion pathology for patients referred to the Johns Hopkins Medical Institutions found major modifications from the outside previous diagnosis in 86 of 6,178 cases (1.4%).[2] There is no doubt that second opinion consultation used to be carried out by conveying glass slides via mail or courier can be largely replaced by telepathology without delay.

3 Telecytology

Cytology is managed by cytotechnologists for screening and cytopathologists for diagnosis. Such a cooperative process is not always universally available except for large laboratories. Telecytology is suitable for consultation on significant cells picked up in screening which can be transferred as images to the pathologist/cytologist

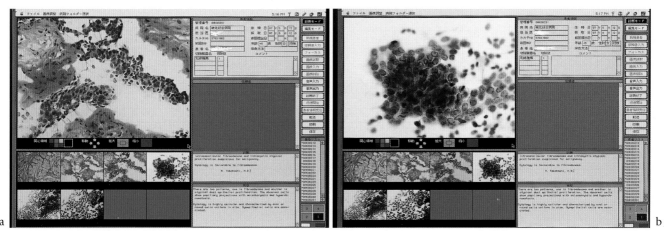

a
b

Fig. 22-1 Real time intraoperative telepathology/telecytology. From a case of 45-year-old woman who was operated because of a fairly defined tumor in the left breast.
a. Histological image of frozen section showing typical intracanalicular fibroadenoma (lower left). Another image displaying intracystic trabecular growths of hyperchromatic epithelial cells was hardly distinguished whether benign or malignant.
b. Imprint smear images simultaneously transmitted illustrate clustered cells uniform in size and shape. The association with small pyknotic myoepithelial cells within or around the cell cluster is in favor of benignancy.

Fig. 22-2 Workstation of telepathology/telecytology (Hi-SPEED). With my associates Dr. Maria Mernyei (right) and Dr. Tang Xlaoyan (left). Standardization of videomicroscopy among the telepathology/telecytology team is important.

by means of real time or off-line telecommunication. When specimens are smeared widely, transmission of general view as live images needs much time. Therefore, priority of cell selection to be transferred consists in the sender's side.

In other words, sampling adequate and appropriate fields in smears remains as a problem of technical restriction in telecytology unless a remote controllable system is available.[10]

Although telecytology of intraoperative imprint smears has limitation in use, cytologic images which are simultaneously transmitted with histologic images are often valuable as an adjunct to frozen section diagnosis (Fig. 22-1).

In order to receive precise cell images at workstation, higher magnification over × 20 and higher resolution at image capturing station are required. The difficulty of focus adjustment for three-dimensional cell clusters must be kept in mind. The accuracy of cervico-vaginal

screening on video has been studied by Raab et al who found reduced accuracy compared with glass slide review (85.6% vs 95.6%).[7] The necessity of intralaboratory standardization of monitor observation (Fig. 22-2) is reported by Takahashi et al who found disagreements among pathologist/cytotechnologists in AutoCyte SCREEN interactive monitor review.[8]

Telecytology and telepathology consultations involve electronic transfer of patient information. All protocols must be stored in the state of patient confidentiality that restricts strictly access.[6]

References

1. Brunner JM, Inouye L, Fuller GN, et al: Diagnostic discrepancies and their clinical impact in a neuropathology referral practice. Cancer 79: 796, 1997.
2. Kronz JD, Westra WH, Epstein JI: Mandatory second opinion surgical pathology at a large referral hospital. Cancer 86: 2426, 1999.
3. McGowan L, Norris HJ: The mistaken diagnosis of carcinoma of the ovary. Surg Gynecol Obstet 173: 211, 1991.
4. Nordrum I, Eide TJ: Remote frozen section service in Norway. Arch Anat Cytol Pathol 43: 253, 1995.
5. Oberholzer M, Fisher HR, Christen H, et al: Telepathology: Frozen section diagnosis at a distance. Virchows Arch 426: 3, 1995.
6. O'Brien MJ, Brugal G, Takahashi M: Digital imagery/telecytology. IAC Task Force Summary. Acta Cytol 42: 148, 1998.
7. Raab SS, Zaleski MS, Thomas PA, Niemann TH, et al: Telecytology: Diagnostic accuracy in cervical-vaginal smears. Am J Clin Pathol 105: 599, 1996.
8. Takahashi M, Kimura M, Akagi A, et al: AutoCyte SCREEN interactive automated primary cytology screening system: A preliminary evaluation. Acta Cytol 42: 185, 1998.
9. Takahashi M, Mernyei M, Shibuya C, et al: Present and prospects of telepathology. Jpn J Clin Pathol 47: 27, 1999.

10. Vacher-Lavenu MC, Marsan C: Telecytoconsultation: Application du systeme TRANSPATH a la pathologie cervico-vaginale. Arch Anat Cytol Pathol 43: 262, 1995.

11. Weinstein RS: Prospects for telepathology. Hum Pathol 17: 433, 1986.

12. Weinstein RS, Bloom KJ, Krupinski EA, et al: Human performance studies of the videomicroscopy component of a dynamic telepathology system. Zentralbl Pathol 138: 399, 1992.

13. Weinstein RS, Bloom KJ, Rozek LS: Telepathology and the networking of pathology diagnostic services. Arch Pathol Lab Med 111: 646, 1987.

Index

A

ABC (aspiration biopsy cytology)
  of brain, 427
  of breast, 136, 137
  of liver, 264
  of lung, 169
  of lymph nodes, 374
  of prostate, 369
  of salivary glands, 243
  of skin, 449-451
  of thymus, 238
  of thyroid, 229, 230
  tumor transplantation, 137
ABC (avidin-biotin enzyme complex method), 4
Abnormal mitosis
  multipolar, 91, 93
  in carcinoma in situ, 93
  three group, 93
  two group, 93
Abortion, threatened, 63
Abscess, amebic, of liver, 265
Acantholytic cells, of skin, 453
Acinar cell carcinoma, of pancreas, 274
Acinic cell carcinoma, of salivary glands, 246
Acquired immune deficiency syndrome (AIDS), 461
Acridine orange, 330
Actin, 7
Actinomycosis, of lung, 191
Acute inflammatory carcinoma, of breast, 144
Acute lymphatic leukemia (ALL), 397
Acute lymphocytic leukemia (ALL), 43
Acute myelogenous leukemia (AML), 43
Acute rejection, of renal transplantation, 363
Adenoacanthoma
  of cervix, 106
  of endometrium, 115
  of esophagus, 255
Adenocarcinoma
  in effusions, 295
    globular type, 295
    mucus-secreting type, 295
    single columnar type, 295
    tubular type, 295
  of bile ducts, 272
  of breast, 154
  of cervix, 101
    adenoid cystic type, 102, 103
    clear cell, in young girls, 108
    endocervical type, 102
    endometrioid type, 102, 103
    invasive, 102
  of endometrium, 114
  of gallbladder, 272
  of liver, 268
  of lung, 194
    clear cell, 206

of pancreas
  papillary, 272
  tubular, 272
of prostate, 371, 372
of rectum
  mucinous, 261
  tubular, 261
of urinary bladder, 353
Adenocarcinoma in situ, of cervix, 101
Adenoid cystic carcinoma
  of esophagus, 255
  of lung, 216, 217
  of salivary glands, 247
Adenoid cystic type, of adenocarcinoma, of cervix, 102, 103
Adenolymphoma, of salivary glands, 243
Adenoma
  basal cell, of salivary glands, 247
  follicular, of thyroid, 231
  liver cell, 266
  of breast, 150
  of pituitary gland, 443
  oxyphilic, of thyroid, 232
  pleomorphic, of salivary glands, 243
Adenoma malignum, of endometrium, 114
Adenomatosis, of lung, 199
Adenomatous goiter, 230
Adenomatous polyp, of intestine, 260
Adenosis
  of breast, 149
  vaginal, in young girls, 108
Adenosquamous carcinoma, *see also* Adenoacanthoma
  of cervix, 105
  of pancreas, 274
Adenovirus infection, of lung, 186
Adult T cell leukemia (ATL), 397
Adult T cell lymphoma (ATL), 397
Aerosol induced sputum 166
Ag-As stain, 297
AGUS (atypical glandular cells of undetermined significance), 101
ALL (acute lymphatic leukemia), 397
ALL (acute lymphocytic leukemia), 43
Allograft rejection, of kidney, 362
Alpha-fetoprotein (AFP), 10
  and yolk sac tumor, 110
Alternaria
  in urine, 342
  of cervix, 75
Alveolar cells, *see* Alveolar surface epithelium
Alveolar microlithiasis, of lung, 187
Alveolar surface epithelium, 171
  type I (A) cells, 171
  type II (B) cells, 171
Amebiasis
  of cervix, 78
  of vagina, 78
AML (acute myelogenous leukemia), 43

AML1, 43
Anaplastic astrocytoma, of brain, 436
Anaplastic carcinoma, of thyroid, 236
Anaplastic large cell Ki-1-positive lymphoma, 390
Androgen hyperactivity, 62
Aneurysmal bone cyst, 421
Angioimmunoblastic lymphadenopathy, 400
Angiomyolipoma
  of kidney, 342
  of liver, 266
Anti-chromogranins, 8
Anti-cytokeratin antibodies, 19
Anti-neuroendocrine antibodies, 8
Antibody sensitivity, 3
Antigen, 3, 5
Apocrine carcinoma, of breast, 161
  effusions with, 317
Apocrine metaplasia, of breast, 143, 145
Apudoma, of cervix, 101
Arias-Stella phenomenon, of endometrium, 64
Asbestos body, in lung, 188, 189
ASCUS (atypical squamous cells of undetermined significance), 82
Askanazy cells, of thyroid, 231
Aspergillosis, of lung, 191
Aspergillus, of cervical smear, 75
Aspiration biopsy cytology (ABC)
  of brain, 427
  of breast, 136, 137
  of liver, 264
  of lung, 169
  of lymph nodes, 374
  of prostate, 369
  of salivary glands, 243
  of skin, 449-451
  of thymus, 238
  of thyroid, 229, 230
  tumor transplantation, 137
Asteroid body, in sputum, 187
Astroblastoma, of brain, 436
Astrocytes, of brain, 430
Astrocytoma, of brain, 435
  anaplastic, 436
  gemistocytic, 435
  pilocytic, 436
  subependymal giant cell, 438
ATL, *see* Adult T cell leukemia/lymphoma
Atypical glandular cells of undetermined significance (AGUS), 101
Atypical squamous cells of undetermined significance (ASCUS), 82
AutoCyte PREP, 328
Avidin-biotin enzyme complex method (ABC), 4
  fixatives, coagulant, 4
    formaldehyde-based, 4
    glutaraldehyde-bond, 4
    mercuric-based, 4
Ayre's wooden spatula, for cervix, 47

B

B cells
  identification of, 378
  lymphoma, 386
  population of, 378
BAL (bronchoalveolar lavage), 169
Barr body, 24
Barrett's esophagus, 251, 253, 255
Basal cell adenoma, of salivary
  glands, 247
Basal cell epithelioma, of skin, 457,
  462
Basal cell hyperplasia, of lung, 178
Basal cells
  of lung, 178
  of vagina, 52
Baseball catcher's mit appearance, of
  lymphoma, 383
Basedow's disease, of thyroid, 231
bcr/abl chimera, 43
Bence Jones proteins, 332
Ber-Ep4, for effusions, 288
Berylliosis, of lung, 187
Bethesda System, 125
Bile ducts (biliary tract)
  adenocarcinoma, 272
  epithelial cells, 257
  extrahepatic, 271
Bile pigment-laden macrophages, 186
Blastic transformation, of lympho-
  cytes, 381
Blastoma, of lung, 210, 211
Blastomyces dermatitidis, 192
Blastomycosis
  North American, 192
  South American, 192
Blepharoplasts, in rosette, 438
Blue blods, 72
Body cavities, serous, 279
Bone
  aneurysmal bone cyst, 421
  bone marrow tumor, 421
  chondroblastoma, 417
  chondrosarcoma, 417
  eosinophilic granuloma, 423
  giant cell tumor, 421
  Hand-Schüller-Christian disease,
    422
  histiocytosis X, 422
  Letterer-Siwe disease, 423
  osteogenic sarcoma, 417
  osteosarcoma, 417
    giant celled, 417
  rhabdomyosarcoma, 421
Bone cyst, aneurysmal, 421
Bone marrow tumor, 421
Bowen's disease, of skin, 457
Brain (central nervous system)
  aspiration biopsy cytology (ABC),
    427
  astrocytes, 430
  carcinoma, metastatic, 432
  cerebrospinal fluid
    benign cellular components, 428
    cell counting, 428
    centrifugation method, 425
    collection of, 425
    Cyto-Tek Centrifuge, 426
    gross appearance, 428
    membrane filtration method,
      426
    sedimentation technique, 426
  chordoma, 445
  choristoma, 446
  choroid plexus papilloma, 438
  choroidal cells, 429
  craniopharyngioma, 441, 446
  cryptococcosis, 428

dermoid cyst, 446
endothelial proliferation, 431
ependymal cells, 429
ganglioglioma, 445
ganglioneuroblastoma, 445
ganglioneuroma, 445
germinoma, of epiphysis, 441
glioma, 432
  anaplastic astrocytoma, 436
  astroblastoma, 436
  astrocytoma, 435
  ependymoma, 438
  gemistocytic astrocytoma, 435
  general features, 432
  glioblastoma multiforme, 436
  medulloblastoma, 438
  mixed glioblastoma, 438
  oligodendroglioma, 438
  pilocytic astrocytoma, 436
  polar spongioblastoma, 436
  subependymal giant cell astrocy-
    toma, 438
Homer Wright rosette, 438, 439
imprint smear, of cerebral tissue,
  427
India ink preparation for, 428
macrophages, 430
meningeal melanoma, 446
meningioma, 443
metastases, frequency of, 432
microglia, 430
nerve cells, 430
neurinoma, 445
neurocutaneous melanosis, 446
nuclear palisading, 431
oligodendrocytes, 430
paragonimiasis westermani, 446
perivascular pseudorosette, 431
pia-arachnoid cells, 430
pinealoma, 440
pineoblastoma, 441
pituitary adenoma, 443
plexus papilloma, 438
pseudorosette, 431
rabies, 428
retinoblastoma, 446
schwannoma, 445
"Sonnenstäubchen," 428
squash method, 426, 427
teratoma, of epiphysis, 441
true rosette, 431
tumors, structure profile, 431
BrDU (bromodeoxyuridine), 20
Breast
  adenoma, 150
  adenosis, 149
    sclerosing, 149
  apocrine metaplasia, 143, 145
  aspiration biopsy cytology, 136,
    137
  benign cellular components
    duct epithelial cells, 139
    foamy cells, 138
    myoepithelial cells, 143
  carcinoma
    acute inflammatory carcinoma,
      144
    adenocarcinoma, 154
    apocrine carcinoma, 161
    colloid carcinoma, 156
    comedocarcinoma, 157
    duct carcinoma, 154
    effusions with, 308
    intracystic papillary carcinoma,
      156
    lobular carcinoma, 154
    malignant lymphoma, 162
    medullary carcinoma, 157
    metastasis in lung, 220

    mucinous carcinoma, 156
    Paget's disease, 157
    prognostic view, 162
    proliferation markers, 163
    signet ring cell carcinoma, 156
    squamous cell carcinoma, 161
    tubular carcinoma, 157
  colostrum, 139
  cystic hyperplasia, 148
  duct papillomatosis, 142
  fibroadenoma, 150
  fibrocystic disease, 144
  fibrosclerosis, 144
  foreign body granuloma, 153
  galactocele, 139
  gasoline pump cells, 145, 149
  gynecomastia, 153
  histology, of mammary gland, 137
  imprint smear cytology, 137
  intracystic papilloma, 149
  intraduct papilloma, 149
  lactating breast, 139
  mammary duct ectasia, 144
  mastitis, acute lactation, 141, 143
    plasma cell, 144
  mastopathia, 144
  needle aspiration, 136, 137
  nipple discharge, 136
  phyllodes tumor, 151
  resting mammary gland, 139
Broder's classification, of genito-urinary carcinoma, 346
Bromodeoxyuridine (BrDU), 20
Bronchial brushing, 168, 214
Bronchial carcinoid, 214
Bronchial curetting, 168
Bronchial gland carcinoma, 214
Bronchial hyperplastic cells, atypical,
  180
Bronchial scraping cytology, selective,
  168
Bronchial washings, 169
  cell block of, 169
Bronchiectasis, infective follicular,
  207
Bronchiolo-alveolar carcinoma, 197
  Clara cell bronchiolar carcinoma,
    199
  type II cell carcinoma, 197, 199
Bronchoalveolar lavage (BAL), 169
Bronchoscopic cytology, 168
Brushing cytology, bronchial, 214
Brushing method
  for esophagus, 249
  for lung, 168, 214
Bull's-eye cells, 453
Bullous lesions, of skin, 453
Burkitt's lymphoma, 389

C

c-myc amplification, 41
Calcific concretions, of lung, 187
Calcospherites, of thyroid, 234
Cancer screening, *see* Screening
Candidiasis (*Candida albicans*)
  of cervix, 74
  of esophagus, 252
  of lung, 191
Cannonball, of breast cancer cells, 98,
  308
Capsulated sponge method, for
  esophagus, 249
Carcinoembryonic antigen (CEA), 7,
  10
  for effusions, 288
Carcinoid
  of lung, 215, 216

of mediastinum, 242
Carcinoma in situ
  of cervix, 85, 88
  of lung, 213
  of urinary tract, 349
Carcinoma
  adenosquamous, of cervix, 105
  clear cell
    of cervix, 108
    of endometrium, 117
    of lung, 203
  combined
    of cervix, 105
    of lung, 210
  dual, of lung, 210
  metastatic, of lung, 218, 219
  mixed, 105
  mucoepidermoid, 106
  small cell
    of cervix, 101
    of lung, 199
Carcinosarcoma
  of liver, 269
  of lung, 211, 212
  of uterus, 123
CD22, 412
CD45, 8
CEA (carcinoembryonic antigen), 7,
  10
Cell block technique, for effusions,
  277
Cell concentration method, of sputum, 166
Cell cycle, 13
Central nervous system, see Brain
Centroblastic malignant lymphoma,
  390
Cerebral paragonimiasis westermani,
  446
Cerebrospinal fluid, see Brain, cerebrospinal fluid
Cervical intraepithelial neoplasia
  (CIN), 82, 90, 91
Cervical scraping smear, 47
Cervicitis
  chronic lymphocytic, 120, 121
  follicular, 122
Cervix uteri (uterine cervix)
  adenoacanthoma, 106
  adenocarcinoma, 101
    clear cell, in young girls, 108
    endocervical-type, 102
    endometrioid-type, 102, 103
    invasive, 102
  adenocarcinoma in situ, 101
  AGUS, 101
  Alternaria, 75
  amebiasis, 78
  apudoma, 101
  ASCUS, 82
  Aspergillus, 75
  benign cellular components
    basal cells, 52
    endocervical cells, 52
    endometrial cells, 53
    histiocytes, 54
    navicular cells, 52
    parabasal cells, 52
    squamous cells, 50
  borderline lesions, 80
    dysplasia, 73, 82
    dysplasia in pregnancy, 86
    HSIL, 85, 91
    koilocytotic atypia, 82
    LSIL, 84, 91
    postirradiation dysplasia, 85
    squamous intraepithelial lesion
      (SIL), 80
  candidiasis, 74

carcinoma, 96
  adenoacanthoma, 106
  adenocarcinoma in situ, 101
  adenosquamous carcinoma, 105
  clear cell carcinoma, 108
  combined carcinoma, 105
  endodermal sinus tumor, 110
  extrauterine carcinoma, 108
  giant cell carcinoma, 100
  glassy cell carcinoma, 65, 99
  invasive adenocarcinoma, 102
  large cell carcinoma, 100
  mixed carcinoma, 65, 105
  mucoepidermoid carcinoma,
    106
  small cell carcinoma, 101
  squamous cell carcinoma, 96
  verrucous carcinoma, 96, 99
carcinoma in situ, 85, 88
  abnormal mitoses, 93
  colpomicroscopy, 91
  colposcopy, 91
  concept of, 89
  cytology, 91
  histology, 91
carcinosarcoma, 123
Chlamydia, 76
clue cells, of Haemophilus vaginitis,
  74
condyloma, 88
contaminants
  pollen, 78
  starch powder, 78
cytologic indices, for cytohormonal
  patterns, 58
cytolysis, 58
Döderlein bacilli, 58
Enterobius vermicularis, 78
folic acid deficiency, 73
Haemophilus vaginitis, 74
  clue cells, 74
herpes virus infection, 76
histology
  of uterus, 49
  of vagina, 50
hormonal change
  cyclic, 56
  in childhood, 60
  in infant, 59
  in menarche, 60
  in menopause, 58
  in postpartum, 63
  in pregnancy, 60
  in puberty, 58
hormonal dysfunction, 60
  androgen hyperactivity, 62
  estrogen hyperactivity, 60
  estrogen hypoactivity, 61
  progesterone hyperactivity, 61
human papillomavirus (HPV)
  infection, 88
Indian file appearance, 85, 93
inflammatory cell changes, 71
intrauterine device (IUD), 72
Leptothrix infestation, 74
mesodermal mixed tumor, 124
moniliasis, 74
Müllerian mixed tumor, 122
precancerous glandular lesions, 101
preparation of smears
  Ayre's wooden spatula, 47
  Cytobrush, 47, 48
  scraping smear, 47
  self sampling device, 48
  VCE smear, 48
psammoma bodies, 79
sarcoma, 122
sexually transmitted disease (STD),
  75

squamous metaplasia, 67
tissue repair, 66
trichomoniasis, 76
tuberculosis, 78
tumor diathesis, 96, 97
vaginal flora, 73
vernix caseosa cells, 64
Charcot-Leiden crystals, in lung, 189
Chimera, 39
Chlamydia infection, of cervix, 76
Cholangiocellular carcinoma, 268
  adenocarcinoma, 268
  carcinosarcoma, 269
  mucus-secreting carcinoma, 268
  squamous cell carcinoma, 268
Chondroblastoma, 417
Chondrosarcoma, 417
Chordoma, 445
Choriocarcinoma, 119
  metastatic to lung, 223
Chorionic tumors, of endometrium,
  118
Choristoma, 446
Choroid plexus papilloma, of brain,
  438
Choroidal cells, of brain, 429
Chromogranin A, and pulmonary car-
  cinoid, 218
Chromatin-positive gonadal dysgene-
  sis, 26
Chromosomal abnormality (aberra-
  tion), in neoplasia, 39
Chronic lymphatic leukemia (CLL),
  386, 397
Chronic lymphocytic cervicitis, 120,
  121
Chronic myelogenous leukemia
  (CML), 43
Chronic rejection, of renal transplan-
  tation, 363
Chronic thyroiditis, 230
Chyliform effusion, 293
Chylous fluid, 279
Ciliocytophthoria, of lung, 184
CIN (cervical intraepithelial neopla-
  sia), 82, 90, 91
Clara cell bronchiolar carcinoma, 199
Clara cells, 171
Class I/II antigen, loss of, 413
Clear cell adenocarcinoma
  in young girls, 108
  of lung, 206
Clear cell carcinoma
  of cervix, 108
  of endometrium, 117
  of lung, 203
Clear cell tumor
  of lung, 207
  of ovary, 305
CLL (chronic lymphatic leukemia),
  386, 397
Clue cells, of cervix, 74
CML (chronic myelogenous
  leukemia), 43
Coagulant fixatives, 4
Coating adhesives, 328
Cold nodules, of thyroid, 230
Colloid carcinoma, of breast, 156
Colon
  carcinoma, effusion with, 297
  epithelial cells, 257
  histology, 257
Colostrum, 139
Colpomicroscopy, 67, 91
Colposcopy, 91
Combined carcinoma
  of cervix, 105
  of lung, 210
Comedocarcinoma, of breast, 157

Comet cells, in urine, 361
Condyloma, of cervix, 88
Congestive heart failure, effusion with, 293
Contraceptives, oral, 64, 72
Convoluted lymphoblastic malignant lymphoma, 390
Corpora amylacea
  in prostatic fluid, 369
  of lung, 188
Craniopharyngioma, 441, 446
Crowded cell index, 59
Crowded menopause, 58
Cryptococcosis
  India ink preparation for, 428
  of brain, 428
  of lung, 191
CSF, see Brain, cerebrospinal fluid
Curity-Issacs sampler, for cervix, 48
Curschmann's spirals, in sputum, 187
Cutaneous cytology
  dermal lesions, 459
  general aspects, 451
Cutaneous lymphoma, 459
Cutaneous sarcoid, 459
Cutaneous T cell lymphoma, 398
Cyclic hormonal secretion, 55
Cyclin D1, 410, 412
Cylindroma
  of esophagus, 255
  of lung, 215, 216
Cystadenocarcinoma, of pancreas, 274
Cystadenoma, of pancreas, 274
Cystic hyperplasia, of breast, 148
Cyto-Tek Centrifuge, for cerebrospinal fluid, 426
Cytobrush, for cervix, 47, 48
Cytocentrifugation, 328
Cytohormonal patterns, in pregnancy
  cytolytic type, 63
  estrogenic type 63
  inflammatory type, 63
  navicular type, 62
Cytokeratins, 6
  high molecular weight, 6
  low molecular weight, 6
  profile, 6
Cytologic indices, 58
  crowded cell, 59
  eosinophilic, 59
  folded cell, 59
  karyopyknotic, 59
  maturation cell, 59
Cytomegalovirus infection
  of esophagus, 252
  of kidney, 358
  of lung, 186
Cytoplasmic inclusion body, of urinary tract, 360
Cytosedimentation, 328
Cytoskeletal proteins, 7
Cytospin Mark 1, 328
Cytotrophoblasts, see Trophoblasts

D

Darier's disease, of skin, 453, 456
Decidual cells, of cervix, 64
Decoy cells, in urine, 361
Dermatitis herpetiformis, 451
Dermatologic cytology, immunofluorescence method, 451
Dermoid cyst, of brain, 446
Desmin, 6, 7
Diethylstilbestrol, 108
Diff-Quik stain
  of lymph nodes, 374

of urinary tract, 326
Differential cell count, of lymph nodes, 397
Differential interference microscopy, 330
Diffuse poorly differentiated lymphocytic lymphoma (DPDLL), 411, 412
DNA-specific fluorochrome, 13
  acridine orange, 17
  propidium iodide, 15, 17
Döderlein bacilli, 58, 74
Dominant ichthyosis, congenital, 456
DPDLL, see Diffuse poorly differentiated lymphocytic lymphoma
Drumsticks, 27, 31
Dual carcinoma, of lung, 210
Duct carcinoma, of breast, 154
Duct cell carcinoma, of pancreas, 272
Duct papillomatosis, of breast, 142
Duodenum
  endoscopic aspiration, 256
  epithelial cells, 257
  pancreozymin-secretin method, 256
Dynamic system, in telemedicine, 464
Dysgerminoma, of ovary, 304
Dyskaryosis, of cervix, 85
Dyskaryotic index, 84
Dyskeratosis, of cervix, 85
Dysplasia
  mammary, 144
  of cervix, 82
    mild, 84
    classification, 82
    in pregnancy, 86
    keratinizing, 85
    metaplastic, 85
    mimic change, 73
    moderate, 85
    postirradiation, 85
    with koilocytosis, 82
  of esophagus, 252
Dysproteinemia, malignant lymphoma with, 391

E

E rosette
  in effusions, 289
  of lymph nodes, 379
Eccrine spiradenoma, 462
Echinococcus granulosus
  of lung, 194
  of liver, 265
Effusions, 277
  benign cellular components
    histiocytes, 280
    inflammatory cells, 289
    lymphocytes, 289
    macrophages, 280
    mesothelial cells, 282
  body cavities, 279
  chyliform, 293
  chylous fluid, 279
  E rosette, 289
  eosinophilia, 292
  exudate, 279
  immunocytochemistry, 288
  LDH isoenzyme stain, 286
  malignant cells, 295
    adenocarcinoma, 295
    epidermoid carcinoma, 295
  mesothelial cells, 282
  mesothelium, 279
  mucicarmine stain, 286
  neutral red supravital stain, 286

non-suppurative inflammatory, 292
periodic acid-Schiff (PAS) reaction, 283, 286
peroxidase reaction, 286
phagocytosis, 286
preparation
  cell block method, 277
  centrifugation method, 277
  membrane filter method, 277
  thin layer preparation method, 278
  ThinPrep Processor, 278
signet ring cells, 290
special stains for, 283, 286
sudan black B stain, 286
suppurative inflammatory, 292
T cell population, 290
transudate, 279
tuberculous, 292
with breast carcinoma, 308
with colon carcinoma, 297
with congestive heart failure, 293
with gastric carcinoma, 297
with granulosa cell tumor of ovary, 298
with hepatoma, 308
with leukemia, 310
with liver carcinoma, 308
with liver cirrhosis, 293
with lung carcinoma, 305
with malignant lymphoma, 310
with malignant mesothelioma, 314
with myeloma, multiple, 314
with ovarian carcinoma, 297
EGF-R (epidermal growth factor receptor), and breast carcinoma, 163
Eggs of liver fluke, 257
EMA (epithelial membrane antigen), 10
Embryonal carcinoma, of ovary, 304
Endobrush, for endometrium, 49
Endocervical cells, 52
Endocervical glandular atypia, 101
Endocervical-type adenocarcinoma, of cervix, 102
Endocrine intersex, 25
Endocyte, for endometrium, 49
Endodermal sinus tumor, of cervix, 110
Endometrial aspiration, 48
Endometrial carcinoma, 111, 114
  high risks, 47
Endometrial cells, 53
  epithelial cells, 53
  stromal cells, 54
Endometrial glands, 111
  of proliferative phase, 111
  of secretory phase, 111
Endometrial hyperplasia, 113
  complex, 113
  simple, 113
Endometrial scraping smear 49
Endometrial stromal cells, 54
Endometrial stromal sarcoma, 124
Endometrioid adenocarcinoma, of endometrium, 114
Endometrioid-type adenocarcinoma, of cervix, 102, 103
Endometrium
  adenoacanthoma, 115
  benign cellular components
    endometrial stromal cells, 54
    endometrial epithelial cells, 53
    exodus, 54
  carcinoma, 111, 114
    adenoacanthoma, 115
    choriocarcinoma, 119

chorionic tumors, 118
clear cell carcinoma, 117
endometrioid adenocarcinoma, 114
hydatidiform mole, 118
risk factors, 111
secretory carcinoma, 115
serous carcinoma, 115
squamous cell carcinoma, 117
histology, 49
hyperplasia, 113
placental site trophoblastic tumor, 118
preparation of smears
aspiration, 48
Curity-Issacs sampler, 48
Endobrush, 49
Endocyte 49
scraping smear, 49
stromal sarcoma, 124
Endoscopic aspiration, of duodenum, 256
Endoscopic retrograde cholangiopancreatography (ERCP), 256
Endoscopic washing method, 256
Endothelial proliferation, in brain, 431
Entamoeba histolica, in colon, 259
Enterobius vermicularis, of cervix, 78
Enzymatic digestion, of sputum, 166
Eosinophilia, in effusions, 292
Eosinophilic granuloma, of bone, 423
Eosinophilic index, 59
Ependymal cells, of brain, 429
Ependymoma, of brain, 438
Epidermal appendages, tumors of, 462
Epidermal growth factor receptor (EGF-R), and breast carcinoma, 163
Epidermoid carcinoma
in effusions, 295
of lung, 194
Epithelial membrane antigen (EMA), 10
for effusions, 288
Epithelial thymoma, 239
Epithelioid cell granuloma
of lung, 187
of lymph node, 397
Epithelioid cells, of lung, 186
Epithelioma, basal cell, of skin, 457, 462
Epitope, 3
erb-B2, 42
and breast carcinoma, 163
Erythrocyte cast, in urine, 335
Erythrocyte-amboceptor-complement (EAC) technique, of rosette formation, 379
Esophageal brushings, 249
Esophageal washings, 249
Esophagitis, 251
herpetic, 252
reflux, 251
Esophagus
adenoacanthoma, 255
adenoid cystic carcinoma, 255
Barrett's esophagus, 251, 253, 255
benign cellular components, 250
brushing method, 249
candidiasis, 252
capsulated sponge method, 249
cylindroma, 255
cytomegalovirus infection , 252
dysplasia, 252
herpetic esophagitis, 252
histology, 250

malignant melanoma, 256
precancerous dysplasia, 253
reflux esophagitis, 251
small cell carcinoma, 256
washing technique, 249
Estrogen
hyperactivity, 60
hypoactivity, 61
receptors, 9
Estrogenic type, of cytohormonal pattern, 63
Euthyroidism, 229
Ewing's sarcoma, 421
Exodus, of endometrium, 54
Extrahepatic bile ducts, 271
carcinoma, 271, 272
Extranodal lymphoma, 393
Extrauterine carcinoma, 108
Exudate, of effusions, 279

F

FCM (flow cytometry), 14
Ferruginous bodies, in lung, 188
Feulgen stain, 16
Fiber cells, 98
Fiber-like substance, in sputum, 187
Fibroadenoma, of breast, 150
Fibrocystic disease, of breast, 144
Fibronectin, for effusions, 288
Fibrosarcoma, metastatic to lung, 221
Fibrosclerosis, of breast, 144
Fine needle aspiration, see Aspiration biopsy cytology (ABC)
FISH (fluorescence in situ hybridization), 39
Flame cell, in myeloma, 403
Flow cytometry (FCM), 14
Fluorescence in situ hybridization (FISH), 39
multicolor, 41
multitarget, 41
technique, 40, 42
FNH, see Focal nodular hyperplasia
Foamy cells, of breast, 138
Focal nodular hyperplasia (FNH), of liver, 266
Folded cell index, 59
Folic acid deficiency, of cervix, 73
Follicle center cell-derived follicular lymphoma, 387
Follicle-stimulating hormone (FSH), 55
Follicular adenoma, of thyroid, 231
Follicular carcinoma, of thyroid, 233
Follicular cervicitis, 122
Follicular dendritic cells, of lymph nodes, 378
Foreign body granuloma, of breast, 153
Formaldehyde-based fixatives, 4
Franzen-Stormby's apparatus, 374
FSH (follicle-stimulating hormone), 55

G

Galactocele, of breast, 139
Gallbladder, 271
adenocarcinoma, 272
carcinoma, 271, 272
Ganglioglioma (ganglioneuroma), 445
Ganglioneuroblastoma, 20-181
Ganglioneuroma (ganglioglioma), 445
Gartner's duct cyst, of vagina, 55

Gasoline pump cells, of breast, 145, 149
Gastric carcinoma, effusion with, 297
Gemistocytic astrocytoma, of brain, 435
Germ cell tumors, 10
of mediastinum, 239
Germinoblasts, of lymph nodes, 377
Germinoma
metastatic to lung, 223
of epiphysis, 441
of mediastinum, 239, 241
GFAP (glial fibrillary acidic protein), 5, 6
Giant cells
multinucleated
foreign body type, 186
Langhans type, 186
Reed-Sternberg, 209
Giant cell carcinoma
of cervix, 100
of lung, 203
phagocytic activity, 203
of thyroid, 235
Giant cell tumor, of bone, 421
Giant celled osteosarcoma, 417
Giardia lamblia, of intestine, 257, 259
Giemsa stain, instant, 330
Glassy cell carcinoma, of cervix, 65, 99
Glial fibrillary acidic protein (GFAP), 5, 6
Glioblastoma, mixed, 438
Glioblastoma multiforme, 436
Glioma, of brain, 432
Gliomatosis
peritoneal, 305
pleural, 305
Glitter cells, in urine, 361
Glutaraldehyde-bond fixatives, 4
Goblet cells, of intestine, 257
Goiter
adenomatous, 230
nodular, 230
Gonadal dysgenesis
chromatin-positive, 26
pure, 26
Gonorrhea, in urine, 342
Granular cast, in urine, 337
Granuloma
eosinophilic, of bone, 423
epithelioid cell, 397
Granulomatous tertiary syphilis, 461
Granulomatous thyroiditis, 231
Granulosa cell tumor, of ovary, 298
Graves' disease, of thyroid, 231
Grawitz's tumor, of kidney, 354, 356
Gynecomastia, of breast, 153

H

Haemophilus vaginalis, 74
Hairy cell leukemia, 400
Hamartoma, of lung, 212, 203
Hand-Schüller-Christian disease, of bone, 422
Hashimoto's disease, of thyroid, 230
HCG (human chorionic gonadotropin), 9, 10, 120
Heart failure cells, 175, 186
Heavy smokers, 212
Hemangioma, of liver, 266
Hemangiosarcoma, of liver, 270
Hematuria, 339, 341
Hemosiderin-laden macrophages, 175, 186
Hepatitis C virus (HCV)-positive lymphoma, 411

Hepatitis virus infection, 309
  hepatitis B virus, 309
  hepatitis C virus, 309
Hepatoblastoma, of liver, 269
Hepatocellular carcinoma
    (hepatoma), 266
  alpha-fetoprotein (AFP), 310
  effusions with, 308
  malignant, 266
  metastatic to lung, 220
HER2/neu, 42
Hermaphroditism, 24
Herpes genitalis, 76
Herpes virus infection
  of cervix, 76
  of lung, 186
  of skin, 452
Herpetic esophagitis, 252
HHF-35, 7
Hidradenoma, 462
High-grade squamous intraepithelial
    lesion (HSIL), of cervix, 85
High risks
  for endometrial carcinoma, 111
  for lung cancer, 165, 188
Histiocytes, of uterus, 54
  mononuclear, 54
  multinucleated, 55
Histiocytic malignant lymphoma,
    391
Histiocytic medullary reticulosis, 392
Histiocytosis X
  of bone, 422
  of lymph nodes, 404
HLA-DR, 397
HLA-DR antigen, 412
HMB 45, 9
Hodgkin's cell, mononuclear, 209
Hodgkin's disease, of lymph nodes,
    314, 391, 393
Homer Wright rosette, 438, 439
Hormonal change, cyclic
  in cervix, 56
  in childhood, 60
  in infant, 59
  in menarche, 60
  in menopause, 58
  in postpartum, 63
  in pregnancy, 60
  in puberty, 59
Hormonal cytology, 55
  contraceptives, 64, 72, 73
  estrogen effect, 57, 60
  hormonal dysfunction, 60
  indices, 58
  pathologic changes, 60
  postpartum, 63
  pregnancy, 62
  progesterone effect, 57, 61
Hormonal dysfunction, 60
Hot nodules, of thyroid, 230
HPV (human papillomavirus), 47
HTLV, see Human T cell lymphotropic
    virus
Human chorionic gonadotropin
    (HCG), 9, 10, 120
Human papillomavirus (HPV)
  infection, 88
  subtypes, 47
Human T cell lymphotropic virus
    type 1 (HTLV-1), 397
Hürthle cells, of thyroid, 231
Hyaline cast, in urine, 332
Hybrid system, in telemedicine, 464
Hydatidiform mole, of endometrium,
    118
Hydatidosis, of lung, 194
Hyperacute rejection, in renal trans-
    plantation, 362

Hyperplasia
  cystic, of breast, 148
  endometrial, 113
  focal nodular, of liver, 266
  lymphoreticular, nonspecific, 402
  nodular, of prostate, 370
  prickle cell, in esophagus, 251
Hyperthyroidism, 229
  primary, 231
Hypothyroidism, 229

I

Iatrogenic effect, 118
Ichthyosis, congenital dominant, 456
IHC, see Immunohistochemistry
Image analysis, 15
Image transmission, in telemedicine,
    464
Immunoblastic lymphadenopathy,
    400
Immunoblastic lymphosarcoma, 390
Immunoblastic sarcoma, 389
Immunoblasts, of lymph nodes, 377
Immunochemistry, principles, 3
Immunocytochemistry
  and metastasis in lung, 221
  for effusions, 288
    Ber-Ep4, 288
    carcinoembryonic antigen
      (CEA), 288
    epithelial membrane antigen
      (EMA), 288
    fibronectin, 288
    keratin, 288
    MOC-31, 288
    vimentin, 288
  reporting, results of, 5
Immunocytoma, 386, 411
Immunofluorescence dermatologic
    cytology, 451
Immunohistochemistry, diagnostic
    interpretation, 5
Immunosuppressive agent, 88
Imprint smear cytology
  of breast, 137
  of cerebral tissue, 427
Incisional touch (imprint) method,
    for skin, 450
Index, cytologic, see Cytologic indices
India ink phagocytic test, 286
India ink preparation, for brain, 428
Indian file appearance, 85, 93
Infectious lymphadenopathy, 397
Infective follicular bronchiectasis, 207
Inflammatory cell changes, of cervix,
    71
Inflammatory pseudotumor, of liver,
    266
Inflammatory type, of cytohormonal
    pattern, 63
Interdigitating cells, of lymph nodes,
    378
Intermediate cells, of vagina, 50
Intermediate filaments, 5
  cytokeratins, 6
  desmin, 7
  glial fibrillary acidic protein
    (GFAP), 5, 6
  neurofilaments, 6, 7
  vimentin, 5
Internet, in telemedicine, 464
Intestine, 256
  adenocarcinoma, 261
    mucinous, 261
    tubular, 261
  adenoma, tubular, 260
  adenomatous polyp, 260

benign cellular components
    bile duct epithelial cells, 257
    colonic epithelial cells, 257
    duodenal epithelial cells, 257
    goblet cells, 257
  carcinoma, 262
  endoscopic washing, 256
  Entamoeba histolica, 259
  Giardia lamblia, 257, 259
  histology, 257
  liver fluke, eggs of, 257
  ulcerative colitis, chronic, 260
Intracystic papillary carcinoma, of
    breast, 156
Intracystic papilloma, of breast, 149
Intraduct papilloma, of breast, 149
Intrauterine device (IUD), 72
Invasive adenocarcinoma, of cervix,
    102
Invasive mole, of endometrium, 118
Iodine stain, 330
Islet cell carcinoma, of pancreas, 274
IUD (intrauterine device), 72
IUD users
  with calcific bodies, 72
  with infection, 73

K

Kaposi's sarcoma, 461
Karyopyknotic index, 59
Keratin, for effusions, 14-172
Keratinizing dysplasia, of cervix, 85
Ki67, 412
Kidney
  allograft rejection of, 362
  angiomyolipoma, 342
  cytomegalovirus infection, 358
  Grawitz's tumor, 354, 356
  malignant tumors, 345
    staging and grading, 345
  nephroblastoma, 356
  nephron, 331
  renal cell carcinoma, 354
  tuberculosis, 343
  tubular epithelial cells, 331
  urinary casts, 332
  Wilms' tumor, 356
Kiel classification, updated, of lym-
    phoma, 383, 386
Klinefelter's syndrome, 26, 34
Koilocytotic atypia, of cervix, 82
Kulchitsky cells, 202, 216

L

Lactate dehydrogenase (LDH), serum,
    413
Lactating breast, 139
Langhans giant cells
  in uterus, 78
  of lung, 186
Langhans type trophoblastic cancer
    cells, 119
Large cell carcinoma
  of cervix, 100
  of lung, 203
Large lymphoid cells, 377
Laser scanning cytometry, 17
LCA (leukocyte common antigen), 8
LDH, see Lactate dehydrogenase
LDH isoenzyme stain, for effusions,
    286
Leiomyosarcoma, metastatic to lung,
    221
Lentigo, 458
Leprosy, lepromatous, 459

Leptothrix, of cervix, 74
Letterer-Siwe disease, of bone, 423
LEU M1, 8
Leukemia
  acute lymphatic (ALL), 397
  adult T cell (ATL), 397
  chronic lymphatic (CLL), 397
  hairy cell, 400
Leukemic cells, in effusions, 310
Leukocyte cast, in urine, 335
Leukocyte common antigen (LCA), 8
Leukoplakia, of skin, 457
LH (luteinizing hormone), 55
Lipofuscin, 184
Lipophagocytes, 175, 186
Liver, 264
  abscess, 265
  angiomyolipoma, 266
  aspiration biopsy cytology, 264
  carcinoma
    cholangiocellular carcinoma, 268
    effusions with, 308
    hepatocellular carcinoma, 266
    liver cell carcinoma, 266
  cirrhosis, effusion with, 293
  cyst, 265
  Echinococcus granulosus, 265
  fluke, eggs of, 257
  focal nodular hyperplasia, 266
  hemangioma, 266
  hemangiosarcoma, 270
  hepatoblastoma, 269
  hydatid cyst, 265
  inflammatory pseudotumor, 266
  liver cell adenoma, 266
  malignant lymphoma, 270
  malignant tumors, 266
  normal cell components, 264
  preparation of specimens, 264
  regenerative nodule, 266
  WHO classification of tumors, 267
Liver fluke, eggs of, 257
Lobular carcinoma, of breast, 154
Low-grade squamous intraepithelial lesion (LSIL), of cervix, 84
LSIL (low-grade squamous intraepithelial lesion), of cervix, 84
Lukes-Collins classification, of lymphoma, 383
Lung
  actinomycosis, 191
  adenocarcinoma, 194
  adenoid cystic carcinoma, 216, 217
  adenomatosis, 199
  adenovirus infection, 186
  asbestos body, 188, 189
  aspergillosis, 191
  aspiration biopsy cytology (ABC), 169
  atypical hyperplasia, bronchial and bronchiolar, 180
  basal cell hyperplasia, 178
  benign clear cell tumor, 207
  berylliosis, 187
  blastoma, 210, 211
  blastomycosis, 192
  bronchial brushing, 168, 214
  bronchial carcinoid, 215, 216
  bronchial curetting, 168
  bronchial scraping cytology, 168
  bronchial washings, 169
  bronchoalveolar lavage (BAL), 169
  calcific concretions, 187
  candidiasis, 191
  carcinoid, 215, 216
  carcinoma
    bronchial gland carcinoma, 215
    bronchiolo-alveolar carcinoma, 197

cell typing of, 194
    clear cell carcinoma, 203
    combined carcinoma, 210
    dual carcinoma, 210
    effusions with, 305
    epidermoid carcinoma, 194
    giant cell carcinoma, 203
    high risk for, 165, 188
    large cell carcinoma, 203
    metastatic to, 218, 219
    screening, 212, 213
    small cell carcinoma, 199
    spindle cell squamous carcinoma, 194
    squamous cell carcinoma, 194
  carcinoma in situ, 213
  carcinosarcoma, 211, 212
  Charcot-Leiden crystals, 189
  ciliocytophthoria, 184
  clear cell adenocarcinoma, 206
  corpora amylacea, 188
  cryptococcosis, 191
  Curschmann's spirals, in sputum, 187
  cylindromas, 215, 216
  cytomegalovirus infection, 186
  early carcinoma, 213
  Echinococcus granulosus, 194
  epithelioid cell granuloma, 187
  epithelioid cells, 186
  ferruginous bodies, 188
  fiber-like substance, 187
  fine needle aspiration, 169
  hamartoma, 212, 213
  heavy smokers and, 212
  herpes virus infection, 186
  histology, 171
  hydatidosis, 194
  infective follicular bronchiectasis, 207
  Langhans giant cells, 186
  lymphocytic interstitial pneumonia, 207
  lymphoid interstitial pneumonitis, 208
  malignant lymphoma, 207, 208
  MALT lymphoma, 207
  micropapillomatosis, 182
  mucoepidermoid tumor, 218
  multinucleation, 184
  normal cell components
    basal cells, 175
    ciliated columnar cells, 172
    deep cells, 175
    goblet cells, 175
    macrophages, 175
    squamous cells, 175
  Pancoast tumor, 195
  paragonimiasis, 192
  Pneumocystis carinii, 194
  pollen, 190
  proteinaceous materials, 189
  psammoma bodies, 187-189
  radiation change, 223
  Saccomanno's method, 166
  sarcoidosis, 187
  sarcoma, 218
  schistosomiasis, 193
  sputum
    asteroid body, 187
    cell concentration method, 166
    macroscopic observation, 165
    Saccomanno's method, 166
    Schaumann body, 187
  squamous metaplasia, 178
  Strongyloides stercoralis, 193
  sugar tumor, 207
  surfactant, 171
  tuberculosis, 187

tumorlet, 180, 181
Lupus band test, 451
Lupus erythematosus, 451
Luteinizing hormone (LH), 55
Lymph nodes
  aspiration biopsy cytology (ABC), 374
  B cell identification, 378
  collection of specimens, 374
  Diff-Quik stain, 374
  epithelioid cell granuloma, 397
  Franzen-Stormby's apparatus, 374
  histiocytosis X, 404
  histology of, 374
  Hodgkin's disease, 393
  infectious lymphadenopathy, 397
  leukemia
    acute lymphatic (ALL), 397
    adult T cell (ATL), 397
    chronic lymphatic (CLL), 397
    hairy cell, 400
  lymphadenopathy
    angioimmunoblastic, 400
    immunoblastic, 400
    infectious, 397
  lymphoma, extranodal, 393
  MALT lymphoma (MALToma), 393
  metastasis to, 394, 397
  myeloma, 403
  normal cellular components, 375
    follicular dendritic cells, 378
    interdigitating cells, 378
    large lymphoid cells, 377
    lymphocytes, 376
    plasma cells, 378
    prolymphocytes, 377
    reticulum cells, 378
  plasmacytoma, 403
  preparation of smears, 374
  reticulosis, histiocytic medullary, 392
  Sézary syndrome, 398
  T cell identification, 379
Lymphadenopathy
  angioimmunoblastic, 400
  immunoblastic, 400
  infectious, 397
Lymphatic leukemia
  acute (ALL), 397
  chronic (CLL), 397
Lymphcytogram (lymphadenogram), 393, 397
Lymphoblastic lymphosarcoma
  poorly differentiated, 390
  well differentiated, 391
Lymphoblasts (immunoblasts, germinoblasts), of lymph nodes, 377
Lymphocyte phenotyping, 289
Lymphocytes, 376
  blastic transformation, by mitogens, 381
  in effusion, 289
  medium-sized, 377
Lymphocytic interstitial pneumonia, 207
Lymphocytic malignant lymphoma, 390
Lymphocytic thymoma, 239
Lymphocytoma cutis, 459
Lymphoepithelial thymoma, 239
Lymphoid interstitial pneumonitis, 208
Lymphoma
  anaplastic large cell Ki-1-positive, 390
  B cell, 386
  baseball catcher's mit appearance, 383
  Burkitt's, 389

Lymphoma (*cont.*)
  CD22, 412
  classification
    Lukes-Collins, 383
    Kiel classification, updated, 383, 386
    NCI International Working Formulation, 383, 385
    REAL, 385, 410
  cutaneous, 459
  cutaneous T cell, 398
  cyclin D1, 412
  extranodal, 393
  HCV-positive, 411
  histiocytic, 391
  histiocytoid, 312
  HLA-DR antigen, 412
  Hodgkin's lymphoma (disease), 391, 393
  immunocytoma, 411
  Ki67, 412
  lymphocytic, diffuse poorly differentiated, 411, 412
  malignant, 122, 383, 410
    centroblastic, 390
    convoluted lymphoblastic, 390
    cyclin D1, 410
    effusions with, 310
    histiocytic, 391
    immunocytoma, 411
    lymphocytic, 390
    of breast, 162
    of liver, 270
    of lung, 207, 208
    of urinary tract, 356
    prognostic markers, 410
    refined diagnosis, 410
    undifferentiated, 389
  MALT (MALToma), 393, 410
  mantle cell (MCL), 390, 410, 412
  mucosa-associated lymphoid tissue (MALT) lymphoma, 393, 410
  multivariate analysis, 413
  Pan B antigen, 412
  poorly differentiated, 390
  prognostic factors, 412
  small cell lymphocytic, 410, 411
  stem cell, 389
  T cell, 383
  T-cell lineage, 412
  Tdt, 412
  well differentiated, 391
  with dysproteinemia, 391
Lymphoreticular hyperplasia, nonspecific, 402
Lymphosarcoma
  immunoblastic, 390
  lymphoblastic
    poorly differentiated, 390
    well differentiated, 391

M

Macrophages
  in cerebrospinal fluid, 430
  in sputum
    bile pigment-laden, 186
    hemosiderin-laden, 175, 186
    melanin-laden, 186
  lipophagocytes, 175, 186
Malakoplakia, of urinary tract, 360
Malignancy-associated changes, 31
Malignant effusions, 293
Malignant lymphoma, *see* Lymphoma
Malignant melanoma, *see* Melanoma
Malignant mesothelioma, *see* Mesothelioma

Malignant pearl, 98
Malignant phenotypes, of lymphoma, 414
MALT lymphoma (MALToma), 393
  of lung, 207
Maltese cross appearance, 326
Mammary duct ectasia, 144
Mammary dysplasia, 144
Mantle cell lymphoma (MCL), 390, 410, 412
Mastitis
  acute, 144
  acute lactation, 141
  plasma cell, 144
Mastopathia, 144
Maturation cell index, 59
Mayo lung project, 165
MCL (mantle cell lymphoma), 390, 410, 412
Mediastinum, 238
  anterior, 238
  carcinoid tumor, 242
  germ cell tumors, 239
  germinoma, 239, 249
  middle, 238
  posterior, 238
  squamous cell carcinoma, 239
Medroxyprogesterone, 118
Medullary carcinoma
  of breast, 157
  of thyroid, 236
Medulloblastoma, of brain, 438
Melanin-laden macrophages, 186
Melanoma
  lentigo, 458
  metastatic, 322
    to lung, 220
  malignant
    of esophagus, 256
    of skin, 458
    of vulva, 120
  meningeal, 446
  pagetoid, 458
Melanophages, 186
Melanosis, neurocutaneous, 446
Membrane filtration method
  for cerebrospinal fluid, 426
  for effusions, 277
Menarche, 60
Meningeal melanoma, 446
Meningioma, 443
  classification, 445
  metastatic to lung, 221, 223
Menopause, 58
  advanced, 58
  crowded, 58
Menstrual cycles, 113
  ovulation phase, 113
  proliferative phase, 113
  secretory phase, 113
Mercuric-based fixatives, 4
Mesodermal mixed tumor, of uterus, 124
Mesonephroma, 108
Mesothelial cells, in effusions, 282
Mesothelioma
  epithelial, 315
  fibrous, 315
  malignant, effusions with, 314
Mesothelium, histology of, 279
Metaplasia, squamous, of cervix, 67
Metaplastic dysplasia, of cervix, 85
Metastasis
  to brain, 432
  to liver, 270
  to lung, 218, 219
  to lymph nodes, 394, 397
Method sensitivity, 3

Michaelis-Guttmann bodies, in urine, 360
Microglia, of brain, 430
Micropapillomatosis, of lung, 182
Migratory peritonitis, 292
Mild dysplasia, of cervix, 84
Millipore filter, *see* Membrane filtration
Mitosis, abnormal, 85
Mixed carcinoma, of cervix, 65, 105
Mixed glioblastoma, of brain, 438
MOC-31, for effusions, 288
Moderate dysplasia, of cervix, 85
Molluscum contagiosum, of skin, 452
Moniliasis, of cervix, 74
Monoclonal antibody, 3
  HHF-35, 7
Mononuclear histiocytes, of uterus, 54
Monosomy, 39
Monospecific antibody, 3
Morula formation, 283
Mucicarmine stain, for effusions, 286
Mucinous adenocarcinoma
  of pancreas, 274
  of rectum, 261
Mucinous carcinoma, of breast, 156
Mucoepidermoid carcinoma (tumor)
  of cervix, 106
  of lung, 218
  of salivary glands, 247
Mucosa-associated lymphoid tissue (MALT) lymphoma, 393
  of lung, 207
Mucus-secreting carcinoma, of liver, 268
Müllerian mixed tumor, 122
Multinucleated cells
  epithelial
    of breast, 141, 144
    of cervix, 71
    of lung, 184
    of urothelium, 339
  histiocytes, of uterus, 55
Multinucleated giant cells
  foreign body type, 186
  Langhans type, 79, 186
Multiparameter DNA content, 19
Multiple myeloma, 421
Multivariate analysis, of lymphoma, 413
Munro microabscess, of skin, 455
Myeloepithelioma, of salivary glands, 246
Myeloma
  cast, in urine, 332
  multiple, 421
    in effusions, 314
  of lymph nodes, 403
Myeloma cells, in effusions, 314
Myoepithelial cells, of breast, 143

N

Navicular cells, of vagina, 52, 62
NCI International Working Formulation, for lymphoma classification, 383, 385
Near-diploid tumors, 18
Needle aspiration, *see* Aspiration biopsy cytology (ABC)
Nephroblastoma, 356
Nephron, 331
Nerve cells, of brain, 430
Neurinoma (schwannoma), 445
Neuroblastoma, 322
Neurocutaneous melanosis, 446
Neurofilaments, 6, 7

Neuron-specific enolase (NSE), 8
  and pulmonary carcinoid, 218
Neutral red supravital stain, for effusions, 286
Newly acquired antibodies, 4
Nipple discharge, 136
Nodular goiter, 230
Nodular hyperplasia, of prostate, 370
Nodule, regenerative, of liver, 266
Non-Hodgkin's lymphoma, 207
Non-suppurative inflammatory effusions, acute, 292
Non-tumorous pleocytosis, of brain, 428
Nonspecific lymphoreticular hyperplasia, 402
North American blastomycosis, of lung, 192
NSE (neurone-specific enolase), 8, 218
Nuclear palisading, in cerebrospinal fluid cells, 431
Nullisomy, 39

O

Oligodendrocytes, of brain, 430
Oligodendroglioma, of brain, 438
Osteogenic sarcoma, 417
  metastatic to lung, 220
Osteosarcoma, 417
  giant celled, 417
Oval fat bodies, in urine, 332
Ovarian carcinoma, effusion with, 297
Ovary, specific tumors of, effusion with, 298
Oxyphilic adenoma, of thyroid, 232
Oxyphilic cells, of thyroid, 231
Oyster shell cell, 63

P

Paget's disease
  extramammary, of skin, 458
  of breast, 157
Pagetoid melanoma, of skin, 458
Pan B antigen, 412
Pancoast tumor, of lung, 195
Pancreas, 272
  acinar cell carcinoma, 274
  adenocarcinoma
    papillary, 272
    tubular, 272
  adenosquamous carcinoma, 274
  cystadenocarcinoma, 274
  cystadenoma, 274
  duct cell carcinoma, 272
  islet cell carcinoma, 274
  mucinous adenocarcinoma, 274
  squamous cell carcinoma, 274
  tumors, 272
  undifferentiated carcinoma, 274
Pancreozymin secretin method, 256
PAP (prostate acid phosphatase), 9
Papillary carcinoma
  of breast, intracystic, 156
  of thyroid, 234
Papilloma, choroid plexus
  of brain, 438
  of breast, intracystic, 149
Parabasal cells, of vagina, 52
Paracoccidioides brasiliensis, 192
Parafollicular C cell, of thyroid, 236
Paragonimiasis
  of brain, 446
  of lung, 192

Paramesonephric clear cell carcinoma, 108
Pemphigus
  foliaceus, 453
  vegetans, 451, 453
  vulgaris, 451, 453
Periodic acid-Schiff (PAS) reaction, for effusions, 283, 286
Peritonitis
  migratory, 292
  transdiaphragmatic, 292
Perivascular pseudorosette, of brain tumor, 431
Peroxidase reaction, for effusions, 286
Ph^1 (Philadelphia chromosome), 43
Phagocytosis, for effusions, 286
Phenotypes, malignant, of lymphoma, 414
Phenotypic markers, of prognosis, of lymphoma, 411
Philadelphia chromosome (Ph^1), 43
Phyllodes tumor, of breast, 151
Pia-arachnoid cells, of brain, 430
Pigment cast, in urine, 337
Pill, 65, 73
Pilocytic astrocytoma, of brain, 436
Pinealoma, of brain, 440
Pineoblastoma, 441
Pituitary adenoma, 443
  acidophil, 443
  basophil, 443
  chromophobe, 443
  invasive chromophobe, 443
Placental site trophoblastic tumor, 118
Plasma cell mastitis, 144
Plasma cells, of lymph nodes, 378
Plasmacytoid lymphocytic lymphoma, 386
Plasmacytoma, of lymph nodes, 403
Pleocytosis, non-tumorous, of brain, 428
Pleomorphic adenoma, of salivary glands, 243
Plexus papilloma, of brain, 438
PML/RARA, 43
Pneumaturia, 327
Pneumocystis carinii, 169, 186, 194
Pneumonia, lymphocytic (lymphoid) interstitial, 207, 208
Polar spongioblastoma, of brain, 436
Pollen, 78
Polyclonal antibody, 3
Polyomavirus infection, in urinary tract, 358
Polyp, adenomatous, of intestine, 260
Polysomy, 39
Pool smears, vaginal, 56
Postirradiation dysplasia, of cervix, 85
Postmenopausal atrophy, 113
Postpartum cells, 63
Postpartum period, cytology in, 63
Precancerous dysplasia, of esophagus, 253
Precancerous glandular lesions, of cervix, 101
Pregnancy, cytology of, 62
  at term, 63
Prickle cell hyperplasia, of esophagus, 251
Progesterone hyperactivity, 61
Progesterone receptors, 9
Prognostic factors (markers), of lymphoma, 410, 412
Prolymphocytes (medium-sized lymphocytes, small cleaved cells), 377
Prostate
  adenocarcinoma, 371, 372

aspiration biopsy cytology (ABC), 369
  carcinoma, 371
  collection of specimens, 368
  cytology of, 369
  "monstrous" cells, 373
  nodular hyperplasia, 370
  prostatic fluid, 369
  prostatic massage, 368
  prostatitis, 369
  stilbestrol therapy, 373
  structure of, 368
  therapeutic effect of cancer, 373
Prostate acid phosphatase (PAP), 9
Prostatic-specific antigen (PSA), 9
Proteinaceous cast, in urine, 332
Proteinaceous material, in lung, 189
PSA (prostatic-specific antigen), 9
Psammoma bodies
  in effusions with ovarian carcinoma, 297
  of cervix, 78
  of lung, 187-189
  of thyroid, 234
Pseudolymphoma, of lung, 207, 209
Pseudomyxoma peritonei, effusions with, 297
Pseudorosette
  of brain tumor, 431
  of Ewing's cells, 422
Pseudotumor, inflammatory, of liver, 266
Psoriasis vulgaris, 455
Puberty, 59
Pulmonary adenomatosis, 199
Pulmonary blastoma, 210, 211
Pulmonary cytology, steps for, 165

Q, R

Q-staining, 35
Quinacrine mustard, 35
Rabies, 428
Radiation cell changes
  malignancy-mimic changes, 223
  of cervix, 110
  of lung, 223
Radiation cystitis, 364
REAL (Revised European American Lymphoma) classification, 385, 410
Real-time transmissions, 464
Reed-Sternberg (giant) cells, 209, 314, 393
Reflux esophagitis, 251
Rejection
  acute, 363
  chronic, 363
  hyperacute, 362
  in renal graft, 363
Renal cell carcinoma, 354
  effusions with, 317
  metastatic to lung, 219
Reserve cell hyperplasia, of cervix, 66
Respiratory tract, see Lung
Resting mammary gland, 139
Reticulosis, histiocytic medullary, of lymph nodes, 392
Reticulum cell sarcoma, 314
Reticulum cells, of lymph nodes, 377
Retinoblastoma, 440, 446
Revised European American Lymphoma (REAL) classification, 385, 410
Rhabdomyosarcoma, 322, 421
Rosette formation, 379
  blepharoplast structure, 438
  E rosette formation, 379

Rosetle formation (*cont.*)
  EAC (erythrocyte-amboceptor-complement) rosette, 379
  Homer Wright rosette, 438, 439
  in glioma, 440
  pseudorosette, of brain, 431
  true rosette, of brain, 431
Russel bodies, 403

S

S100 protein, 8
Saccomanno's method, for sputum, 166
Salivary glands
  aspiration biopsy cytology (ABC), 243
  adenolymphoma, 243
  basal cell adenoma, 247
  carcinoma
    acinic cell carcinoma, 246
    adenoid cystic carcinoma, 247
    mucoepidermoid carcinoma, 247
  myeloepithelioma, 246
  pleomorphic adenoma, 243
  Warthin's tumor, 243
Sarcoidosis
  cutaneous, 459
  of lung, 187
Sarcoma
  endometrial stromal, 124
  Ewing's, 421
  immunoblastic, 389
  Kaposi's sarcoma, 461
  of lung, 218
  of urinary bladder, 358
  of uterus, 122
  osteogenic, 417
  reticulum cell, 314
Schaumann body, in sputum, 187
Schistosoma haematobium, in urine, 341, 358
Schistosomiasis
  of lung, 193
  of urinary bladder, 341, 358
Schwannoma (neurinoma), 445
Sclerosing adenosis, of breast, 149
Scotch tape strip method, for skin, 449
Screening
  for cervical cancer, 89
  for lung cancer, 212, 213
Sebaceous carcinoma, 462
Second opinion consultation, 464
Secretory carcinoma, of endometrium, 115
Sedimentation technique, for cerebrospinal fluid, 426
Self sampling device, for cervix, 48
Senile colpitis, 72
Serous carcinoma, of endometrium, 115
Serum LDH (lactate dehydrogenase), 413
Sex chromatin, 24
Sex reversals, 26
Sexually transmitted disease (STD), 75
Sézary erythroderma, 459
Sézary syndrome, 398
Signet ring cell carcinoma
  in effusions, 290
  of breast, 156
  of stomach, 297
SIL (squamous intraepithelial lesion), of cervix, 80
Single parameter DNA content, 17

Skin
  acantholytic cells, 453
  application of cytology, 450
  aspiration biopsy cytology (ABC), 449, 450, 451
  bull's-eye cells, 453
  bullous lesions of, 453
  carcinoma, 456
    basal cell epithelioma, 457
    Bowen's disease, 457
    Paget's disease, extramammary, 458
    squamous cell carcinoma, 456
  collection of specimens, 449
  cutaneous sarcoid, 459
  Darier's disease, 453, 456
  granulomatous tertiary syphilis, 461
  histology, 451
  ichthyosis, congenital dominant, 456
  immunofluorescence technique, 451
  incisional touch (imprint) method, 450
  lepromatous leprosy, 459
  leukoplakia, 457
  lymphocytoma cutis, 459
  melanoma, 458
    lentigo, 458
    malignant melanoma, 458
    pagetoid melanoma, 458
  Munro microabscess, 455
  normal cellular components, 451
  pemphigus foliaceus, 453
  pemphigus vegetans, 451, 453
  pemphigus vulgaris, 451, 453
  preparation of smears, 449
    Scotch tape strip method, 449
    skin surface biopsy cytology, 449
  psoriasis vulgaris, 455
  Sézary erythroderma, 459
  tuberculosis of, 459
  Tzanck test, 453
  viral infection
    herpes simplex, 452
    herpes zoster, 452
    molluscum contagiosum, of skin, 452
Skin surface biopsy cytology, 449
SLL, *see* Small cell lymphocytic lymphoma
Small cell carcinoma
  of cervix, 101
  of esophagus, 256
  of lung, 199
    intermediate type, 203
    lymphocyte-like, 203
Small cell lymphocytic lymphoma (SLL), 386, 410, 411
Small cleaved cells, of lymph nodes, 377
Smudge cell, in urine, 358
Snake cells, 96
"Sonnenstäubchen," of cerebrospinal fluid, 428
South American blastomycosis, of lung, 192
Southwest Oncology Group (SWOG), 411
Spindle cell squamous carcinoma, of lung, 194
Spindle-celled thymoma, 241
Spiradenoma, eccrine 462
Spongioblastoma, polar, of brain, 436
Sputum, 165
  aerosol induction of, 166
  cell concentration method, 166

  enzymatic digestion, 166
  gross finding, 165
  Saccomanno's method, 166
Squamo-columnar junction (SCJ), 67
Squamous cell carcinoma
  of breast, 161
  of cervix, 96
  of endometrium, 117
  of liver, 268
  of lung, 194
  of mediastinum, 239
  of pancreas, 274
  of skin, 456
  of urinary tract, 353
Squamous cells, of vagina, 50
Squamous intraepithelial lesion (SIL), of cervix, 80
Squamous metaplasia
  of cervix, 67
  of lung, 178
    atypical, 178
    regular, 178
Squash method, for cerebral tissue, 426, 427
Staining
  acridine orange, 330
  Diff-Quik stain, 326
  Giemsa stain, instant, 330
  iodine stain, 330
  Sternheimer (Sternheimer-Malbin) stain, 329
  sudan black B stain, 286
Starch powder, 78
Static system, in telemedicine, 464
STD (sexually transmitted disease), 75
Stein-Leventhal syndrome, 61
Stem cell lymphoma, 389
Sternheimer (Sternheimer-Malbin) stain, 329
Stilbestrol therapy, of prostatic carcinoma, 373
Strongyloides stercoralis, of lung, 193
Subacute thyroiditis, 231
Subependymal giant cell astrocytoma, of brain, 438
Sudan black B stain, for effusions, 286
Sugar tumor, of lung, 207
Superfemale, 27
Superficial cells, of vagina, 50
Suppurative inflammatory effusions, acute, 292
Surfactant, 171
SWOG (Southwest Oncology Group), 411
Sympathicoblastoma, in effusions, 318
Syncytiotrophoblasts, *see* Trophoblasts
Syphilis, granulomatous tertiary, 461

T

T cell
  identification, 379
  lymphoma, 373
    cutaneous, 398
  marker, cytochemical property as, 380
  population, in effusions, 290
  of lymph nodes, 379
T-cell lineage, 412
T200 antigen, 8
T3, 229
T4, 229
Tadpole cells, 96, 98
Tamm-Horsfall protein, 329
Tamoxifen, 117
Tdt, 412

Teleconsultations, 464
Telecytology/telepathology (telemedicine), 464
  systems of, 464
Teratoma
  of epiphysis, 441
  of ovary, cystic, 305
  of ovary, solid malignant, 304
Testicular dysgenesis, 26
Testicular feminization, 26
Thesaurocyte, in myeloma, 403
Thin layer preparation method, for effusions, 278
ThinPrep Processor, for effusions, 278
Threatened abortion, 63
Three-glass urine, 327
Thymic carcinoma, 239
Thymocyte maturation, 238
Thymoma, *see* Thymus
Thymus
  aspiration biopsy cytology (ABC), 238
  carcinoma, 239
  thymocyte maturation, 238
  thymoma, 239
    epithelial, 239
    lymphocytic, 239
    lymphoepithelial, 239
    spindle-celled, 241
Thyroglobulin, 229
Thyroid
  adenoma
    follicular, 231
    oxyphilic, 232
  Askanazy cells in, 231
  aspiration biopsy cytology (ABC), 230
  Basedow's disease, 237
  calcospherites of, 234
  carcinoma, 233
    anaplastic carcinoma, 236
    follicular carcinoma, 233
    giant cell carcinoma, 235
    medullary carcinoma, 236
    metastatic to lung, 220
    papillary carcinoma, 234
  chronic thyroiditis, 230
  cold nodules, 230
  euthyroidism, 229
  goiter, adenomatous (nodular), 230
  Graves' disease, 231
  Hashimoto's disease, 230
  hot nodules, 230
  Hürthle cells, 231
  hyperthyroidism, 229
    primary, 231
  hypothyroidism, 229
  oxyphilic cells, 231
  parafollicular C cells, 236
  psammoma bodies, 234
  T3, 229
  T4, 229
  thyroglobulin, 229
  TSH, 229
Thyroiditis
  chronic, 230
  subacute, 231
Tissue repair, of cervix, 66
TPT, *see* Triphenyltetrazolium
Transdiaphragmatic peritonitis, 292
Transitional cell carcinoma
  in effusions, 317
  of urinary tract, 346
Transitional cells, of urinary tract, 339
Transudate, of effusions, 279
Trichomoniasis, of cervix, 76
Trichomonas, in urine, 341
Triphenyltetrazolium (TPT) reaction, 296

Trophoblasts
  cytotrophoblasts, 64
  syncytiotrophoblasts, 64
True rosette, of brain tumor, 431
TSH (thyroid-stimulating hormone), 229
Tuberculosis
  in effusions, 292
  of cervix, 78
  of kidney, 343
  of lung, 187
  of skin, 459
Tuberculous cervicitis, 78
Tubular adenocarcinoma, of rectum, 261
Tubular adenoma, of intestine, 260
Tubular carcinoma, of breast, 157
Tubular epithelial cast, in urine, 335
Tubular epithelial cells, of urinary tract, 338
Tumor diathesis, of cervix, 96, 97
Tumorlet, of lung, 180
Tumors of the female genital tract (WHO 1994), 125
Turner's syndrome, 26, 61
Tzanck test, of skin, 453

U

Ulcerative colitis, chronic, 260
Urinalysis
  acridine orange, 330
  AutoCyte PREP, 328
  centrifugation method, 328
  coating adhesives, 328
  cytocentrifugation, 328
  cytosedimentation, 328
  Cytospin Mark 1, 328
  differential interference microscopy, 330
  Giemsa stain, instant, 330
  iodine stain, 330
  irrigation method, 329
  Maltese cross appearance, 332
  NCCLS recommendation on, 331
  oval fat bodies, 332
  polarized light illumination for, 330
  pretreatment solution, 329
  routine procedure, 326
  standardization of, 330
  Sternheimer (Sternheimer-Malbin) stain, 329
  Tamm-Horsfall protein, 329
  three-glass urine, 327
Urinary casts
  erythrocyte cast, 335
  granular cast, 337
  hyaline cast, 332
  leukocyte cast, 335
  myeloma cast, 332
  pigment cast, 337
  proteinaceous cast, 332
  tubular epithelial cast, 335
  waxy cast, 337
Urinary cytology, serial, 327
Urinary tract
  adenocarcinoma, 353
  allograft rejection cytology, 362
  benign atypical cells, 342
  benign cellular components, 338
    multinucleated cells, 339
    neutrophil leukocytes, 339
    red blood cells, 339
    squamous cells, 339
    transitional cells, 339
    tubular epithelial cells, 338
  carcinoma
    Broder's classification, 346

    carcinoma in situ, 349
    squamous cell carcinoma, 353
    transitional cell carcinoma, 346
  cytomegalic inclusion, 358
  cytoplasmic inclusion, 360
  gonorrhea, 342
  hematuria, 339, 341
  histology, 337
  lymphoma, malignant, 356
  malakoplakia, 360
  polyomavirus infection, 358
  radiation cystitis, 364
  sarcoma, 358
  Schistosoma haematobium, 341
  Trichomonas, 341
  urolithiasis, 342
Urine
  collection of specimen, 327
  coloration, abnormal, 327
  comet cells, 361
  decoy cells, 361
  glitter cells, 361
  gross appearance, 327
  Michaelis-Guttmann bodies, 360
  smudge cell, 358
  three-glass, 327
  turbidity of, 327
Urolithiasis, 342
Uterine sarcoma, 122
Uterus, *see* Cervix uteri

V

Vagina
  Gartner's duct cyst, 55
  histology
    basal cells, 52
    intermediate cells, 50
    navicular cells, 52
    parabasal cells, 52
    squamous cells, 50
    superficial cells, 50
  pool smears, 56
Vaginal adenosis, 108
Vaginal flora, 73
VCE smear, for cervix, 48
Vernix caseosa cells, of cervix, 64
Verrucous carcinoma, of cervix, 96, 99
Vimentin, 5
  for effusions, 288
Vitamin A deficiency, 60
Vitamin B$_{12}$ deficiency, 73

W, X, Y

Warthin's tumor, of salivary glands, 243
Waxy cast, in urine, 337
Wilms' tumor, of kidney, 356
Working Formulation (WF), for classification of lymphomas, 383, 385, 411
X body, 24, *see also* X chromatin
  and breast carcinoma, 163
X chromatin, 24
  cancer cells, 29
  configuration, 24
  technique of determination, 27
Y body, 34, *see also* Y chromatin
Y chromatin, 34
  cancer cells, 35
  configuration, 34
  evaluation, 34
  technique of determination, 35
YY syndrome, 34